The Physical Geography of South America

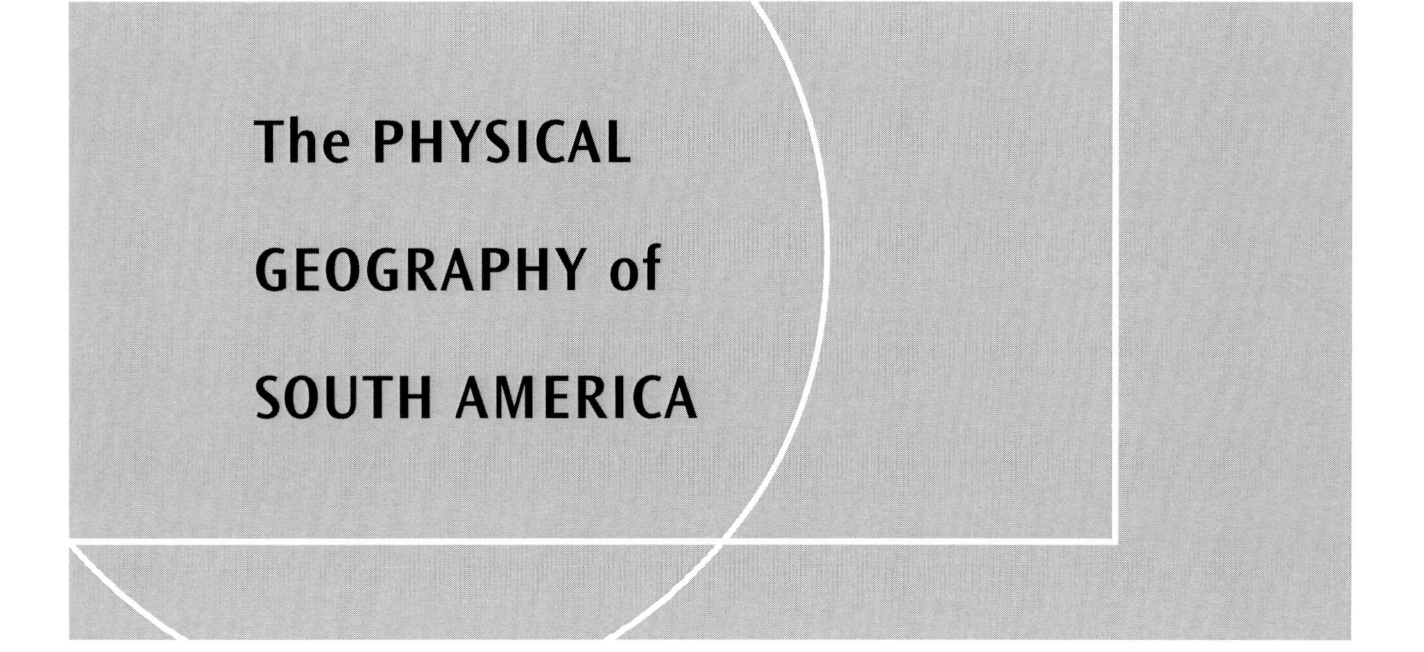

Edited by

Thomas T. Veblen

Kenneth R. Young

Antony R. Orme

2007

OXFORD
UNIVERSITY PRESS

Oxford University Press, Inc., publishes works that further
Oxford University's objective of excellence
in research, scholarship, and education.

Oxford New York
Auckland Cape Town Dar es Salaam Hong Kong Karachi
Kuala Lumpur Madrid Melbourne Mexico City Nairobi
New Delhi Shanghai Taipei Toronto

With offices in
Argentina Austria Brazil Chile Czech Republic France Greece
Guatemala Hungary Italy Japan Poland Portugal Singapore
South Korea Switzerland Thailand Turkey Ukraine Vietnam

Published by Oxford University Press, Inc.
198 Madison Avenue, New York, New York 10016

www.oup.com

Library of Congress Cataloging-in-Publication Data
The physical geography of South America / edited by
Thomas T. Veblen, Kenneth R. Young, and Antony R. Orme.
p. cm.
Includes bibliographical references.
ISBN 978-0-19-531341-3
1. South America—Geography. I. Veblen, Thomas T., 1947–
II. Young, Kenneth R. III. Orme, A. R.
F2211.5.P49 2007
918.02—dc22 2006016266

To
Leal Anne Kerry Mertes
Geoffrey O. Seltzer

Distinguished Scientists and Friends

Foreword

The Physical Geography of South America is the seventh in a series of advanced books that is being published by Oxford University Press under the rubric of Oxford Regional Environments.

The aim of the series is to provide a durable statement of physical conditions on each of the continents. Each volume includes a discussion of the systematic framework (for instance, tectonism, climate, and biogeography), followed by an evaluation of dominant environments (such as mountains, forests, and deserts) and their linkages, and concludes with a consideration of the main environmental issues related to the human use and misuse of the land (for example, interactions of nature and people through time, modern agricultural and urban impacts, resource issues, and conservation).

While books in the series are framed within an agreed context, individual volumes seek to emphasize the distinctive qualities of each continent. We hope that this approach will provide a coherent and informative basis for physical geography and related sciences, and that each volume will be an important and useful reference source for those concerned with understanding the varied environments of the continents.

Andrew Goudie, University of Oxford
Antony Orme, University of California, Los Angeles

Preface

This book, *The Physical Geography of South America*, is a contribution to the Oxford Regional Environments series being published by Oxford University Press. It presents a detailed and current statement of knowledge written by specialists in the many research fields of physical geography and related fields. With this book we aspire to fill a void in recent scientific literature, namely, the lack of high-quality interpretative and correlative work that seeks to integrate knowledge of South America across the environmental spectrum.

South America is an unusually fascinating continent. Although now tenuously linked physically with other lands to the north, and increasingly enmeshed in human globalization, its character owes much to prolonged isolation from neighboring continents, diluted only quite recently by infusions of plants, animals, and peoples from elsewhere. The continent thus reflects a strong element of endemism—of plants and animals that developed more or less in isolation over long intervals of geologic time, and later of peoples, particularly in the High Andes, who had achieved much long before they were impacted by colonists from beyond the seas. It is also a continent of dramatic gradients and elevational patterns. For example, vertical relief of over 13,000 m, in a distance of little more than 200 km, separates peaks in the High Andes from the depths of the Perú-Chile Trench offshore, while on land alone the highest Andes descend equally abruptly from alpine tundra into rain forest just above sea level in the Amazon lowlands.

Accordingly, this book seeks to emphasize and explain South America's distinctive qualities, while also examining those features that can only be explained by reference to factors, from tectonic forces to colonization, triggered beyond its shores. It contains 21 chapters that are broadly divided into three groups: systematic framework, regional environments, and human impacts. The first eight chapters focus systematically on the broad physical and biogeographic character of South America. Of these, chapters 1 to 5 examine the shaping of the continent's physical framework, whereas chapters 6 to 8 provide an overview of South America's distinctive biogeography. Chapters 9 to 15 examine some of the continent's more distinctive regional environments, based on recognition of differing forest, shrubland, grassland, steppe, and desert biomes, and of ocean coasts. Last, chapters 16 to 21 examine nature in a human context, showing how nature has influenced people and has in turn been reshaped by them, producing landscapes that are often quite different from those inherited from nature. This involves examining the roles of native and immigrant peoples, agriculture and urbanism, El Niño oscillations, and finally natural and human impacts and resource issues of relevance to future environments.

A book of this nature inevitably involves much synthesis and some subjectivity. Whereas the chapters are framed within an agreed context, individual authors have been encouraged to be original, flexible, and selective in their approach. To aid the reader in pursuing various themes, each chapter offers a substantial bibliography, which includes both selected classic works and much that has been published within the past two decades.

Thomas T. Veblen
Kenneth R. Young
Antony R. Orme

Acknowledgments

The authors, editors, and publisher thank the following who have kindly given permission for the use of copyrighted material:

Chapter 1: Oxford University Press for figure 1.3, from Orme, 2002, based on Ziegler et al., 1997. William A. Bowen, California Geographical Survey, for figures 1.5, 1.6, and 1.7.

Chapter 2: Oxford University Press for figure 2.1, from Orme, 2002; Department of Geography, UCLA, for figures 2.5 and 2.8, from the Fairchild Collection; Diana Blackburn for figure 2.9.

Chapter 4: The American Association for the Advancement of Science for figures 4.4 and 4.7; Elsevier Ltd. for figures 4.6 and 4.9; The Geological Society of America for figures 4.10 and 4.11; The International Glaciological Society and B.G. Mark for figure 4.12.

Chapter 5: The University of Chicago Press for figure 5.3; V.M. Ponce for figure 5.4B; The Geological Society of America for figures 5.7C–F; Elsevier Ltd. for figure 5.8.

Chapter 6: Figure 6.1 is modified from Josse et al., 2003, by permission of NatureServe.

Chapter 8: Blackwell Science for figure 8.1 from G.S. Helfman, B.B. Collette, and D.E. Facey, 1997, *The Diversity of Fishes*; The Johns Hopkins University Press for figure 8.2, from W.E. Duellman, 1999, *Patterns of Distributions of Amphibians*; The University of Chicago Press for figure 8.6, from D.F. Stotz, J.F. Fitzpatrick, T.A. Parker III, and D.K. Moskovits, 1996, *Neotropical Birds: Ecology and Conservation.*

Chapter 9: Belhaven Press for figure 9.4; Elsevier Ltd. for figure 9.5; Oxford University Press (Clarendon Press) for figure 9.7; C. Bohrer for figure 9.9; Blackwell Publishing for figure 9.10.

Chapter 13: Bellwether Publishing, V.H. Winston & Son, Inc., and Antony R. Orme, Editor-in-Chief for figure 13.3, from L.D. Daniels and T.T. Veblen, 2000, *Physical Geography*, 21, 223–243.

Chapter 16: Universiteit van Amsterdam for figure 16.1, from A.J. Ranere, in: O.R. Ortiz-Troncoso and T. van der Hammen (Editors), 1992, *Archaeology and Environment in Latin America*; University of California Press for figure 16.2, from B.L. Gordon, 1957, *Human Geography and Ecology in the Sinú Country of Colombia*, Ibero-Americana, 39; University of Pennsylvania Museum for figure 18.3, from C. L. Erickson, 2001, *Pre-Columbian Roads of the Amazon*, Expedition, 43(2): 28; and The American

Geographical Society for figure 8.4, from J.J. Parsons and W.A. Bowen, 1966, *Geographical Review*, 56.

Chapter 18: S. Wood, K. Sebastian, and S. J. Scherr for figure 18.1, from S. Wood et al., 2000, *Pilot Analysis of Global Ecosystems: Agrosystems*, International Food Policy Research Institute and World Resources Institute, Washington, D.C.

Chapter 19: César Caviedes for figures 19.2 and 19.3 from his book *El Niño in History*, 2001, University Presses of Florida

Chapter 20: *Environment and Urbanization* for figure 20.1, from J. Morello et al., 2000, *Environment and Urbanization*, 12 (2), 119–131.

We are also grateful to individual authors and friends and to various government and international public agencies for freely providing many other figures and data used in this book, as acknowledged therein. Certain figures are reproduced, with full acknowledgment, under the "fair use" clause accorded authors by journals in which the material was originally published. Although every effort has been made to trace and contact copyright holders, we apologize for any apparent negligence. Finally, we thank the following reviewers and cartographers for their generous advice and assistance, and various universities and agencies for support of the research projects that contributed to these chapters:

Chapters 1 and 2: Thomas Dunne, Leal Ann Kerry Mertes, P.E. Rundel, and Chase Langford.

Chapter 3: J. Rutllant and R. Rondanelli; NOAA Climate Diagnostics Center for the NCEP–NCAR reanalysis; JISAO, University of Washington, for precipitation data; and research support from FONDECYT (Chile) Grant 1000913 to the Department of Geophysics, Universidad de Chile, and from the Department of Research and Development, Universidad de Chile, Grant I002–99/2.

Chapter 4: The Earth System History Program of the U.S. National Science Foundation for continued support of paleoclimate research in the tropical Andes.

Chapter 5: James Wells for his careful review of the manuscript, and the University of Washington CAMREX Project, directed by Jeffrey E. Richey.

Chapter 8: John Bates, John Cadle, William Duellman, Michael Parrish, Bruce Patterson, Alan Resetar, Robert Ricklefs, Mary Ann Rogers, Richard Vari, and Harlan Walley; and support from the U.S. National Science Foundation (DEB98–73708).

Chapter 17: Karl Butzer and William M. Denevan

Chapter 18: Alan Moore and Edgar Amezquita

Chapter 20: Graham Haughton and Ian Douglas; and Jorge Hardoy, who inspired our work.

The Editors conducted independent reviews of all chapters.

Contents

Foreword, vii

Contributors, xv

I. Systematic Framework

1. The Tectonic Framework of South America, 3
   Antony R. Orme

2. Tectonism, Climate, and Landscape Change, 23
   Antony R. Orme

3. Atmospheric Circulation and Climatic Variability, 45
   René D. Garreaud and Patricio Aceituno

4. Late Quaternary Glaciation of the Tropical Andes, 60
   Geoffrey O. Seltzer

5. Rivers, 76
   Thomas Dunne and Leal Anne Kerry Mertes

6. Flora and Vegetation, 91
   Kenneth R. Young, Paul E. Berry, and Thomas T. Veblen

7. Soils, 101
   Stanley W. Buol

8. Zoogeography, 112
   Peter L. Meserve

II. Regional Environments

9. Tropical Forests of the Lowlands, 135
   Peter A. Furley

10. Arid and Semi-Arid Ecosystems, 158
   P.W. Rundel, P.E. Villagra, M.O. Dillon, S. Roig-Juñent, and G. Debandi

11. The Mediterranean Environment of Central Chile, 184
   Juan J. Armesto, Mary T.K. Arroyo, and Luis F. Hinojosa

12. Tropical and Subtropical Landscapes of the Andes, 200
   Kenneth R. Young, Blanca León, Peter M. Jørgensen, and Carmen Ulloa Ulloa

13. Temperate Forests of the Southern Andean Region, 217
   Thomas T. Veblen

14. The Grasslands and Steppes of Patagonia and the Río de la Plata Plains, 232
   José María Paruelo, Estebán G. Jobbágy, Martín Oesterheld, Rodolfo A. Golluscio, and Martín R. Aguiar

15. Ocean Coasts and Continental Shelves, 249
José Araya-Vergara

III. Nature in the Human Context

16. Pre-European Human Impacts on Tropical Lowland Environments, 265
William M. Denevan

17. The Legacy of European Colonialism, 279
Gregory Knapp

18. Agriculture and Soil Erosion, 289
Carol P. Harden and Glenn G. Hyman

19. Impacts of El Niño-Southern Oscillation on Natural and Human Systems, 305
César Caviedes

20. Environmental Impacts of Urbanism, 322
Jorgelina Hardoy and David Satterthwaite

21. Future Environments of South America, 340
Thomas T. Veblen, Kenneth R. Young, and Antony R. Orme

Index, 353

Contributors

Patricio Aceituno (Ph.D., 1987, University of Wisconsin, Madison) is Professor of Geophysics in the Universidad de Chile, Santiago, with research interests in atmospheric dynamics and precipitation over South America and adjacent oceans, and in seismology and volcanology.

Martín R. Aguiar is a research scientist in the Laboratorio de Análisis Regional y Teledetección, IFEVA-Facultad de Ciencias, in the Universidad de Buenos Aires/CONICET, Argentina.

José Araya-Vergara (Ph.D., 1967, University of Chile) is Emeritus Professor of Geography in the Universidad de Chile, Santiago. His research interests lie in coastal geomorphology, with particular reference to the variety of coastal processes and landforms of Chile.

Juan J. Armesto (Ph.D., 1984, Rutgers University, New Jersey) is Professor of Forest Ecology and Ecosystems at the Universidad Catolica de Chile and at the Facultad de Ciencias, Universidad de Chile, Santiago. His principal research interests are in ecology and biodiversity.

Mary T.K. Arroyo (Ph.D., 1971, University of California, Berkeley) is Professor of Botany and Director of the Millennium Center for Advanced Studies in Ecology and Research on Biodiversity at the Facultad de Ciencias, Universidad de Chile, Santiago.

Paul E. Berry (Ph.D., 1980, Washington University, St. Louis) is Professor of Ecology and Evolutionary Biology, and Herbarium Director, in the University of Michigan, Ann Arbor. His main research interests concern plant systematics, evolution, and biogeography.

Stanley W. Buol (Ph.D., 1960, University of Wisconsin, Madison) is Emeritus Professor of Soil Science in North Carolina State University, Raleigh. His research interests concern soil genesis and classification.

César Caviedes (D.Sc., 1969, University of Freiburg, Germany) is Professor of Geography in the University of Florida, Gainesville. His research interests involve environmental systems, human implications of environmental change, El Niño events, and Latin America.

G. Debandi (Ph.D., Universidad Nacional de La Plata) is with the Laboratory of Entomology, IADIZA-CRICYT, Mendoza, Argentina. He is currently working on the design of nature reserves and biodiversity protection using predictive models of taxa distributions.

William M. Denevan (Ph.D., 1963, University of California, Berkeley) is Professor Emeritus of Geography at the University of Wisconsin, Madison. His research interests are historical ecology, prehistoric and traditional cultivation, human impacts on environment, and indigenous demography, with a focus on Amazonia and the Andes.

M.O. Dillon (Ph.D., 1976, University of Texas, Austin) is Chair and Curator of Flowering Plants in the Department of Botany, The Field Museum, Chicago. His research focuses on the Andes, especially the floras of Perú and Chile. In 1995, he launched the Andean Botanical Information System website [www.sacha.org].

Thomas Dunne (Ph.D., 1970, University of Washington) is Professor in the Bren School of Environmental Science and Management in the University of California, Santa Barbara. His research focuses of hydrology and fluvial geomorphology.

Peter A. Furley (D. Phil., University of Oxford) is Emeritus Professor of Geography in the School of GeoSciences, University of Edinburgh. His research focuses on tropical biogeography, soils, and land development, and on forest and savanna ecology of the New World tropics.

René D. Garreaud (Ph.D., 1996, University of Washington, Seattle) is Associate Professor of Geophysics in the Universidad de Chile, Santiago, with interests in atmospheric dynamics and precipitation over the south Pacific Ocean, the Pacific coast of South America, and the Andes.

Carol P. Harden (Ph.D., 1987, University of Colorado, Boulder) is Professor of Geography in the University of Tennessee, Knoxville. Her research interests involve geomorphology, soils, and mountain environments, with a particular focus on the Andes.

Jorgelina Hardoy (University of Buenos Aires and Rutgers University, New Jersey) is a research scholar at the Instituto Internacional de Medio Ambiente y Desarrollo (IIED-América Latina) in Buenos Aires, Argentina.

Luis F. Hinojosa (Ph.D., 2003, Universidad de Chile) is Assistant Professor in the Facultad de Ciencias, Universidad de Chile, Santiago, with research interests in paleobotany and phytogeography.

Glenn G. Hyman (Ph.D., 1996, University of Tennessee) is a Research Scientist with the Centro Internacional de Agricultura Tropical (CIAT) in Cali, Colombia. His research interests involve geomorphology and environmental issues related to tropical agriculture in Latin America.

Estebán G. Jobbágy, Martín Oesterheld, and Rodolfo A. Golluscio are scientists with the Grupo de Estudios Ambientales—IMASL, Universidad Nacional de San Luis/CONICET, Argentina.

Peter M. Jørgensen (Ph.D., 1993, University of Aarhus, Denmark) is Associate Curator of the Missouri Botanical Garden, St. Louis, with research interests in Neotropical floristics and conservation.

Gregory Knapp (Ph.D., 1984, University of Wisconsin, Madison) is Associate Professor of Geography and the Environment at the University of Texas, Austin. His research applies an adaptive dynamics approach to cultural ecology, especially prehistoric, colonial, and traditional resource management practices in the Americas.

Blanca León (Ph.D., 1993, University of Aarhus, Denmark) is a Research Fellow in Geography and the Environment at the University of Texas, Austin. Her research interests involve plant systematics and conservation with a focus on the flora of Perú and the Neotropics.

Leal Mertes (Ph.D., 1990, University of Washington, Seattle) was Professor of Geography in the University of California, Santa Barbara. Her principal research interests focused on geomorphic and hydrologic processes responsible for the development of wetlands and floodplains in large river systems. Sadly, she died in 2005 while this book was in production. Her zest for life, and her passion for the environmental sciences and for people will long be remembered.

Peter L. Meserve (Ph.D., 1972, University of California, Irvine) is Distinguished Professor of Biology at Northern Illinois University, DeKalb. His research interests include biogeography, and population and community ecology of birds and mammals, especially in South America.

Antony R. Orme (Ph.D., 1961, University of Birmingham, England) is Professor of Geography at the University of California, Los Angeles. His research involves geomorphology, Quaternary studies, and environmental management, with a focus on landscape change and interactions between natural systems and human activity.

José María Paruelo (Ph.D., Colorado State University) is a Professor in the Laboratorio de Análisis Regional y Teledetección, in IFEVA-Facultad de Agronomía at the Universidad de Buenos Aires/CONICET, Argentina, with particular interests in biogeography and remote sensing.

S. Roig-Juñent (Ph.D., Universidad de La Plata), is with the Laboratory of Entomology, IADIZA-CRICYT, Mendoza, Argentina. His research interests in entomology focus on Coleoptera systematics, biogeography, and conservation.

P.W. Rundel (Ph.D., 1969, Duke University, North Carolina) is Professor of Biology at the University of California, Los Angeles. His principal research interests are in ecology and ecophysiology, with particular reference to South America and the Atacama Desert.

David Satterthwaite (Ph.D., 1998, London School of Economics) is a Senior Fellow at the International Institute for Environment and development (IIED) in London, and Editor of the journal *Environment and Urbanization.* He was awarded the Volvo Environment Prize in 2004.

Geoffrey O. Seltzer (Ph.D., 1991, University of Minnesota, Minneapolis) was Professor of Earth Sciences in Syracuse University, New York, with research interests in ice-cap dynamics, glaciation, lake systems, and paleoclimates of the tropical Andes. Sadly, he died in 2005 while this book was in production. His scientific talents, generosity, and humor are missed, but his scientific legacy lives on.

Carmen Ulloa Ulloa (Ph.D., 1993, University of Aarhus, Denmark) is an Ecuadorean botanist and Associate Curator of the Missouri Botanical Garden, St. Louis. Her research interests focus on the floristics and phytogeography of Andean flora.

Thomas T. Veblen (Ph.D., 1975, University of California, Berkeley) is Professor of Geography at the University of Colorado, Boulder. His research interests in biogeography and ecology focus on the forests of southern South America, Guatemala, New Zealand, and the Rocky Mountains.

P.E. Villagra (Ph.D., Universidad Nacional de Cuyo) is in the Department of Dendrochronology and Environmental History, IANIGLA-CONICET, Mendoza, Argentina. His current research interests focus on the ecology and management of *Prosopis* woodlands in Argentina's arid lands.

Kenneth R. Young (Ph.D., 1990, University of Colorado, Boulder) is Associate Professor of Geography and the Environment at the University of Texas, Austin. His research interests involve ecology, biogeography, and conservation of tropical environments and organisms.

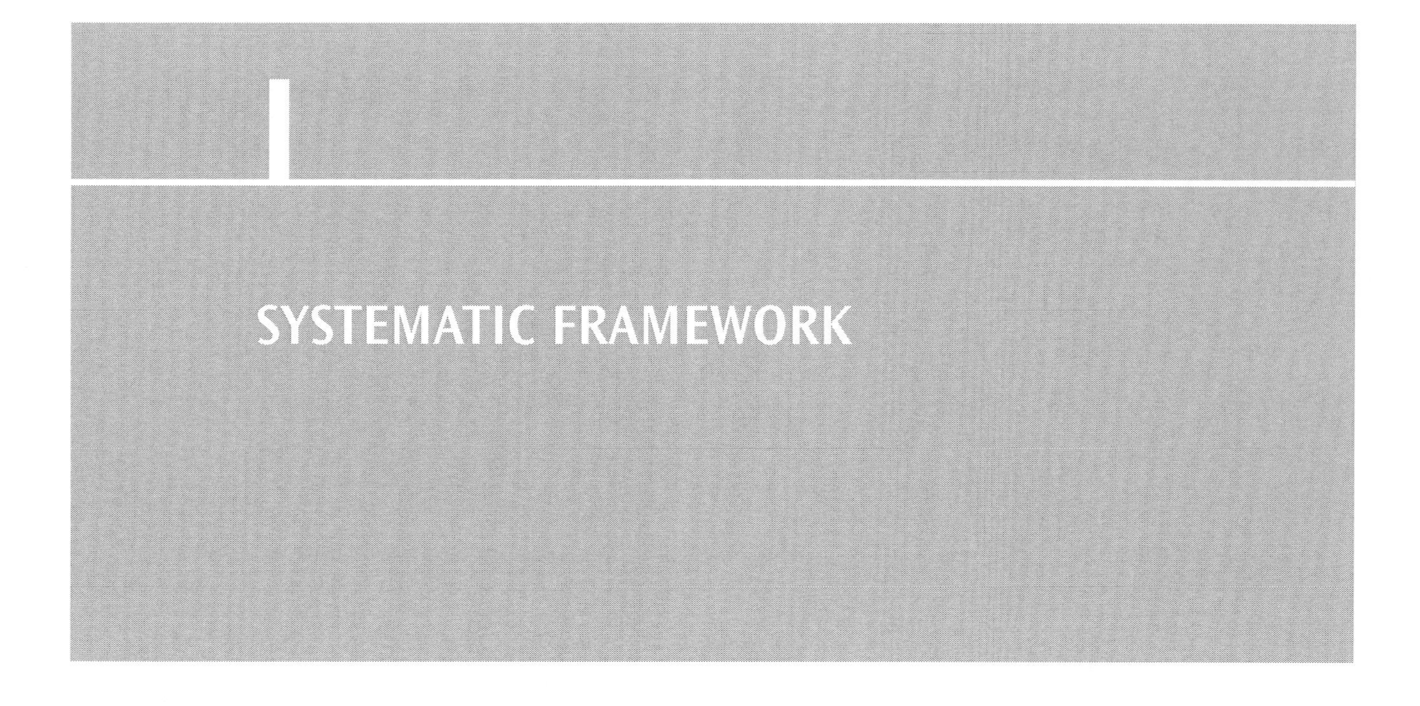

I

SYSTEMATIC FRAMEWORK

1

The Tectonic Framework of South America

Antony R. Orme

Tectonism is the science of Earth movements and the rocks and structures involved therein. These movements build the structural framework that supports the stage on which surface processes, plants, animals and, most recently, people pursue their various roles under an atmospheric canopy. An appreciation of this tectonic framework is thus a desirable starting point for understanding the physical geography of South America, from its roots in the distant past through the many and varied changes that have shaped the landscapes visible today.

Tectonic science recognizes that Earth's lithosphere comprises rocks of varying density that mobilize as relatively rigid plates, some continental in origin, some oceanic, and some, like the South American plate, amalgams of both continental and oceanic rocks. These plates shift in response to deep-seated forces, such as convection in the upper mantle, and crustal forces involving push and pull mechanics between plates. Crustal motions, augmented by magmatism, erosion, and deposition, in turn generate complex three-dimensional patterns. Although plate architecture has changed over geologic time, Earth's lithosphere is presently organized into seven major plates, including the South American plate, and numerous smaller plates and slivers.

The crustal mobility implicit in plate tectonics often focuses more attention on plate margins than on plate interiors. In this respect, it is usual to distinguish between passive margins, where plates are rifting and diverging, and active margins, where plates are either converging or shearing laterally alongside one another. At passive or divergent margins, such as the present eastern margin of the South American plate, severe crustal deformation is rare but crustal flexuring (epeirogeny), faulting, and volcanism occur as plates shift away from spreading centers, such as the Mid-Atlantic Ridge, where new crust is forming. Despite this lack of severe postrift deformation, however, passive margins commonly involve the separation of highly deformed rocks and structures that were involved in the earlier assembly of continental plates, as shown by similar structural legacies in the facing continental margins of eastern South America and western Africa. At active convergent margins, mountain building (orogeny) commonly results from subduction of oceanic plates, collision of continental plates, or accretion of displaced terranes. Along the present western margin of the South American plate, oceanic crust forming the Nazca plate is being consumed as it subducts toward the mantle, often accompanied by intense magmatic and seismic activity, even as new crust is being generated at spreading centers beneath the Pacific Ocean farther west. At active shearing or transform margins, such as those along South America's northern and southern coasts, orogenic activity is less conspicuous but seismicity is high and earthquakes are frequent. In contrast, the nucleus of the South American plate, the continental cratons exposed in the Guiana and Brazilian shields, is more stable but by no means quiescent.

1.1 The South American Plate and Its Neighbors

The South American continent covers nearly 18 × 10^6 km^2 or 12% of Earth's present land area (table 1.1). It is smaller than North America or Africa, its neighbors to the north and east, but larger than Antarctica to the south. Addition of the continental shelf and slope forms a continental block of more than 22 × 10^6 km^2 in area. Thus, about 80% of the continental block is presently freeboard, although this value neared 90% when the continental shelf, widest off Argentina, was more exposed during low Pleistocene sea levels. The land mass is roughly triangular in shape, broadest toward the north where it extends over 5,000 km from east to west just south of the Equator, but tapering southward, rapidly so beyond the Tropic of Capricorn. Even so, the continent extends over nearly 70° of latitude and almost 8,000 km from north to south. Unlike North America, a similarly shaped but mainly temperate continent, South America has a predominantly tropical and subtropical character, tempered mostly by elevation and in the far south by polar influences.

In tectonic terms, the continental crustal block is but part of the larger South American plate whose passive eastern margin includes trailing oceanic crust generated at the Mid-Atlantic Ridge since the rupture of Gondwana and South America's separation from Africa (fig. 1.1). Elsewhere, the South American plate and continental limits closely coincide in active margins. The plate's western margin with the Nazca and Antarctic plates occurs where the Andes abut the subduction zone represented on the ocean floor by the Perú-Chile Trench, while the Caribbean and Scotia plates shear eastward and impinge on the far northern and southern coasts, respectively (Table 1.2).

These relationships between tectonic plates and continental block are, together with the internal architecture of the South American continent, the product of prolonged tectonic development (Coffin et al., 2000; Cordani et al., 2000). As a guide to understanding the continent's present physique, this record is divided into five major phases: (1) a lengthy period of Precambrian time, before the assembly of Gondwana, during which the continent's cratonic nucleus was formed; (2) the assembly of Gondwana in late Neoproterozoic and Cambrian time, the subsequent provision of cover rocks and orogenic imprints, the suturing of Patagonia, and the inclusion of Gondwana in the vastness of Pangea from late Paleozoic to early Mesozoic time; (3) Mesozoic rupture, first of Pangea and then of West Gondwana, which interposed the Atlantic Ocean between Africa and South America, and moved the latter westward; (4) the episodic formation of the Andes; and (5) Cenozoic tectonism in the far north and south involving the Caribbean and Scotia plates, respectively.

1.2 The Precambrian Nucleus

The ancestral nucleus of the continent, the South American Platform, is an amalgam of Precambrian cratons and orogenic belts that underlie about 15 × 10^6 km^2 or 83% of the continental freeboard (Cordani et al., 2000). The remaining 3 × 10^6 km^2 of freeboard lie within the Andean cordillera, which formed along the continent's western margin during Phanerozoic time, dramatically so during the Cenozoic Era.

The Precambrian cratons comprise the large Amazonian craton (4.3 × 10^6 km^2) exposed in the Guiana and Guaporé (central Brazil) shields, and the São Francisco, São Luis,

Table 1.1 South America's dimensions compared with other continents

Continent	Land Area (10^6 km^2)	Continental Shelf (10^6 km^2)	Continental Slope (10^6 km^2)	Continental Block[1] (10^6 km^2)	Freeboard[2] (%)	Mean Altitude (m)	Coastline (10^3 km)
South America	17.79	2.43	2.14	22.36	79.56	590	29
North America	24.24	6.74	6.68	37.66	64.36	720	76
Africa	30.30	1.28	2.25	33.83	89.56	750	30
Antarctica	14.10	0.36	5.41	19.87	70.96	2200	
Asia	44.50	9.38	7.01	60.89	72.57	960	71
Australia	8.56	2.70	1.64	12.90	66.36	340	20
Europe	9.96	3.11	3.13	16.20	61.48	340	37

[1]The Continental Block is defined as Land Area plus Continental Shelf plus Continental Slope. It is not coterminous with the continental plate (see text).

[2]Freeboard refers to the land area above present sea level as a percentage of the Continental Block. Antarctica is a special case because most of its land area is covered by ice sheets that extend onto the continental shelf. All values are approximations.

Modified from Kossinna (1933) and Orme (2002).

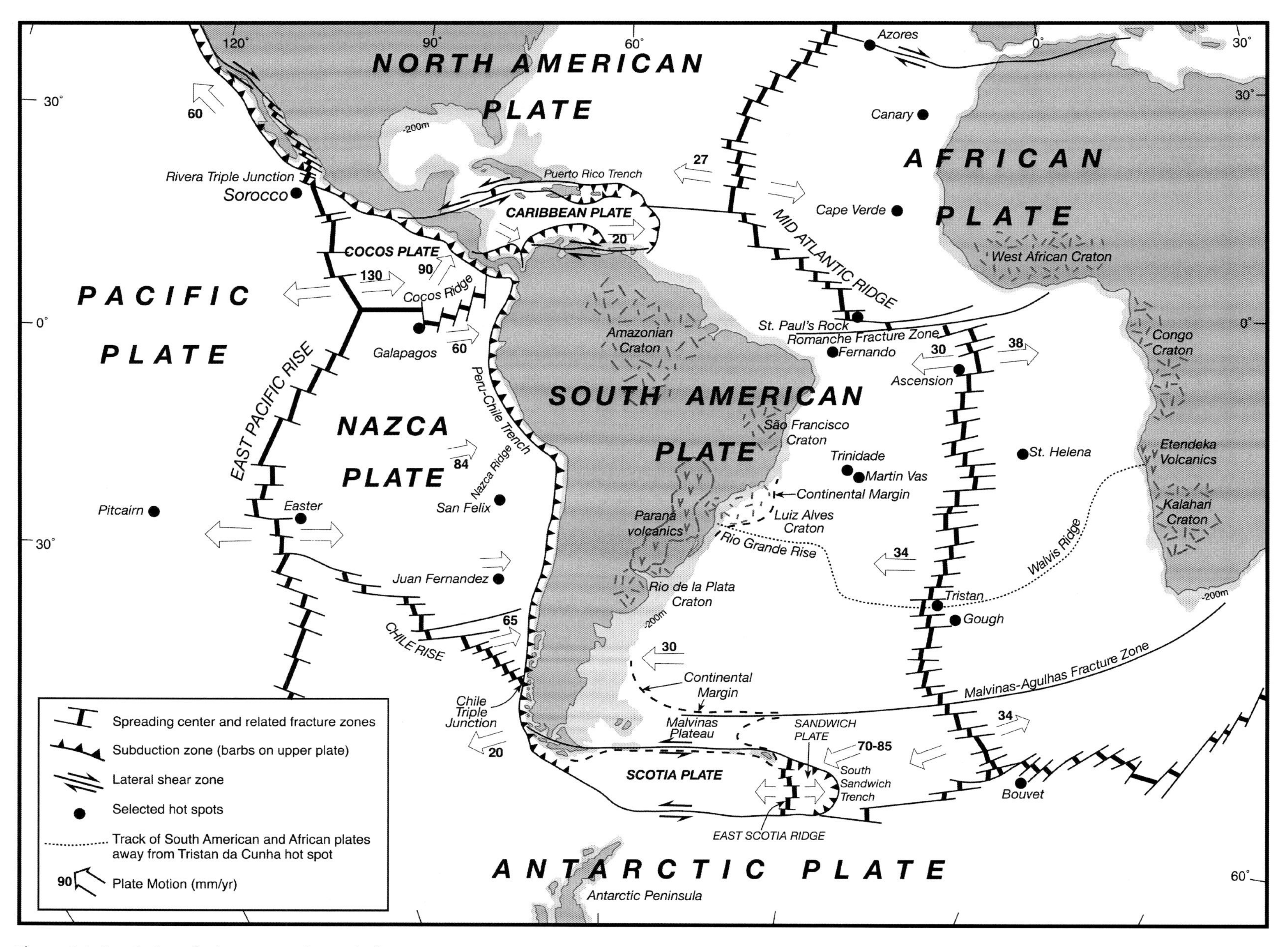

Figure 1.1 South America's present plate relations

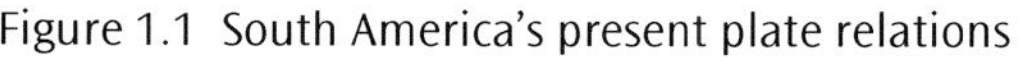

Table 1.2 Principal tectonic, climatic, and biotic events affecting South America over geologic time. The logarithmic timescale condenses the distant past, thereby enhancing Mesozoic and Cenozoic events relevant to the present landscape.

Age (Ma)	Eon	Era	Period [Epoch]	Plate Motion	Orogenic Events	Climate-Forcing Events	Biotic Events
1			(Quaternary) [Pleistocene]		Andean continuing	Patagonian glaciation	Megafaunal extinctions
					Panamanian	Central American isthmus closes	Great American biotic interchange
			[Pliocene]	Gulf of Guayaquil opens		Northern Andes nearing present height	
				Nazca plate moving east	Andean maximizes		raccoons swim to South America
		Cenozoic	Neogene			Tethys closing	Beringia pathway fully open
10			[Miocene]	Scotia plate moving east	severe Andean compression	Antarctic Circumpolar Current stronger	hominids in Africa
	PHANEROZOIC			Caribbean plate moving east	Andean main uplift begins	Drake Passage widens	
			[Oligocene]	Farallón plate ruptures		Global cooling Antarctic ice	rodents appear
			Paleogene [Eocene]	Australia leaves Antarctica		Southern Ocean opening India hits Asia Altiplano rising	early horses appear grasses appear mammals diffuse
			[Paleocene]	Caribbean and Scotia plates form	Altiplano near sea level	Atlantic fully open but narrow	primates diffuse dinosaur extinctions angiosperms spread
100		Mesozoic	Cretaceous			Brazil separates from Guinea	placental mammals
				Paraná volcanics			Gondwana biota separate
			Jurassic	West Gondwana ruptures	Andean begins	Atlantic begins unzipping	early birds early mammals
			Triassic	Pangea ruptures		magmatism	'modern' vertebrates
			Permian		Gondwanan	continentality	major extinctions
			Carboniferous	Pangea forms	Alleghanian	glaciation	conifers appear
		Paleozoic	Devonian				gymnosperms appear
			Silurian		Precordilleran		land plants appear
			Ordovician		Famatinian	glaciation	fish appear
			Cambrian		Pampean		
				Gondwana forms	Brasiliano-Panafrican	continentality	multicelled organisms
				Rodinia ruptures		oceanicity	
1000 (1 Ga)	PROTEROZOIC	Precambrian		Rodinia forms	Sunsás (~Grenvillian)	continentality	
				Roraima Supergroup	several older orogenies		
	ARCHEAN			South American protocratons form			bacteria & algae oldest life forms
4.6 Ga				Origin of Earth			

Luis Alves, and Río de la Plata cratons, which collectively form Brazil's Atlantic Shield (fig. 1.2). These cratons are composed of complex igneous and metamorphic rocks, flanked by Neoproterozoic orogenic belts, and partly concealed beneath later cover rocks (Cordani et al., 2000). The Amazonian craton involves six major tectonic provinces, including Archean crystalline basement and Proterozoic igneous and metamorphic units that were welded together in several orogenies some time before 1 Ga (billion years before present). The São Francisco craton, which contains 3.7 Ga gneiss, and the other cratons, including fragments in the Andes, also had complex histories in different locations.

These cratons have various origins. The Amazonian craton was part of Rodinia, a Mesoproterozoic supercontinent that formed with the suturing of Laurentia, the ancestral nucleus of North America, and other plates during the Grenvillian orogeny, 1.2–1.0 Ga. The Grenvillian event, widely recognized in eastern North America, may be reflected in the Sunsás orogeny of similar age in the Amazonian Shield and in certain basement rocks of the Northern and Central Andes. The subsequent breakup of Rodinia after 900 Ma (million years before present) saw its cratons disperse and Laurentia separate from Amazonia, although the Arequipa craton that later docked against the proto-Andes may have Laurentian affinities. Changing plate arrangements then paved the way for the assembly of a new supercontinent, Gondwana, during later Neoproterozoic time.

1.3 The Assembly of Gondwana

The assembly of Gondwana involved the closure of existing seaways, subduction of oceanic crust, and multiple collisions between continental cratons, culminating in the Brasiliano-Panafrican orogeny between 650 and 550 Ma. The orogeny generated fold belts, among which the Borborema, Tocantins, and Mantiquiere provinces of Brazil are noteworthy. Suturing of continental plates also formed enduring cratonic continua, such as the São Luis–West African and São Francisco-Congo cratons, that only ruptured with the breakup of West Gondwana and the subsequent opening of the Atlantic Ocean in later Mesozoic time (Alkmim et al., 2001).

The Brasiliano-Panafrican orogeny involved in the assembly of Gondwana provided not only high relief to South America's ancestral framework but also intracratonic basins that have been the locus of prolonged if intermittent sedimentation throughout Phanerozoic time. Thus, while the exposed cratonic shields are important features of the present landscape, they are buried elsewhere by great thicknesses of sedimentary and volcanic rock, which impart a quite different texture to the landscape. These basins may have been zones of crustal weakness in the Brasiliano-Panafrican framework, but their continued if sporadic subsidence during Phanerozoic time is still difficult to explain.

There are five major intracratonic basins—the Solimões, Amazonas, and Parnaíba basins in northern Brazil, and the Paraná and Chaco-Paraná basins farther south (fig. 1.2; Milani and Filho, 2000). These basins, which together occupy about 3.5×10^6 km^2, or almost 20% of the continent, are separated from one another by structural arches whose influence on drainage and sedimentation has persisted to the present, notably in the Amazon River basin (see chapter 5).

The intracratonic basins contain 3,000 to 7,000-m thick sequences of mostly siliciclastic rocks related to several Paleozoic transgressive-regressive depositional cycles and accompanying subsidence, interrupted by episodes of uplift and erosion, and capped by later continental and volcanic rocks. In general, these sequences begin with an Ordovician marine transgression, continue from Silurian to Carboniferous time with alternating shallow marine, deltaic, fluvial, and aeolian sedimentation, with carbonates and evaporites in the northern basins, and are topped off by fluvial, lacustrine, and aeolian "red beds" of Permian and Triassic age. These red beds were deposited following the final withdrawal of Paleozoic seas from Gondwana's western interior and reflect the desiccation of the intracratonic basins under truly continental conditions. Aridity continued through the Jurassic and early Cretaceous with the widespread accumulation of aeolian sands under desert conditions. As the Mesozoic progressed, however, these deserts were invaded by large volumes of magma, mostly basalt lavas related to the impending breakup of Gondwana.

Tillites and other glaciogenic deposits are noteworthy features of the Paleozoic rock record, indicating that the South American Platform and its Gondwana neighbors then lay near enough to the South Pole to be affected by repeated continental glaciation. As Gondwana shifted location during the era, so the South Pole appeared to wander from northwest Africa in the Ordovician to off southeast Africa in the Permian (Crowley and North, 1991). Earlier in the twentieth century, observed similarities between Carboniferous and Permian stratigraphic sequences of southern Brazil and southern Africa, now separated by 6,000 km of ocean, provided support for the struggling theory of continental drift (du Toit, 1927, 1937). In both areas, late Carboniferous to early Permian glaciogenic deposits, up to 1,500 m thick in the Paraná Basin, underlie postglacial Permian shales containing the distinctive *Glossopteris* flora and the reptile *Mesosaurus*. These rocks are now far apart, a circumstance more readily explained by mobilist concepts involving plate separation than by former beliefs in continental stability and land bridges.

Despite their thickness, the Paleozoic and Mesozoic rocks of these intracratonic basins make a variable contribution to the present landscape. In the Solimões Basin (0.6×10^6 km^2), they are wholly concealed by sandy late Cretaceous and Cenozoic continental sediment. In the Chaco-Paraná Basin (0.5×10^6 km^2), they are concealed beneath a

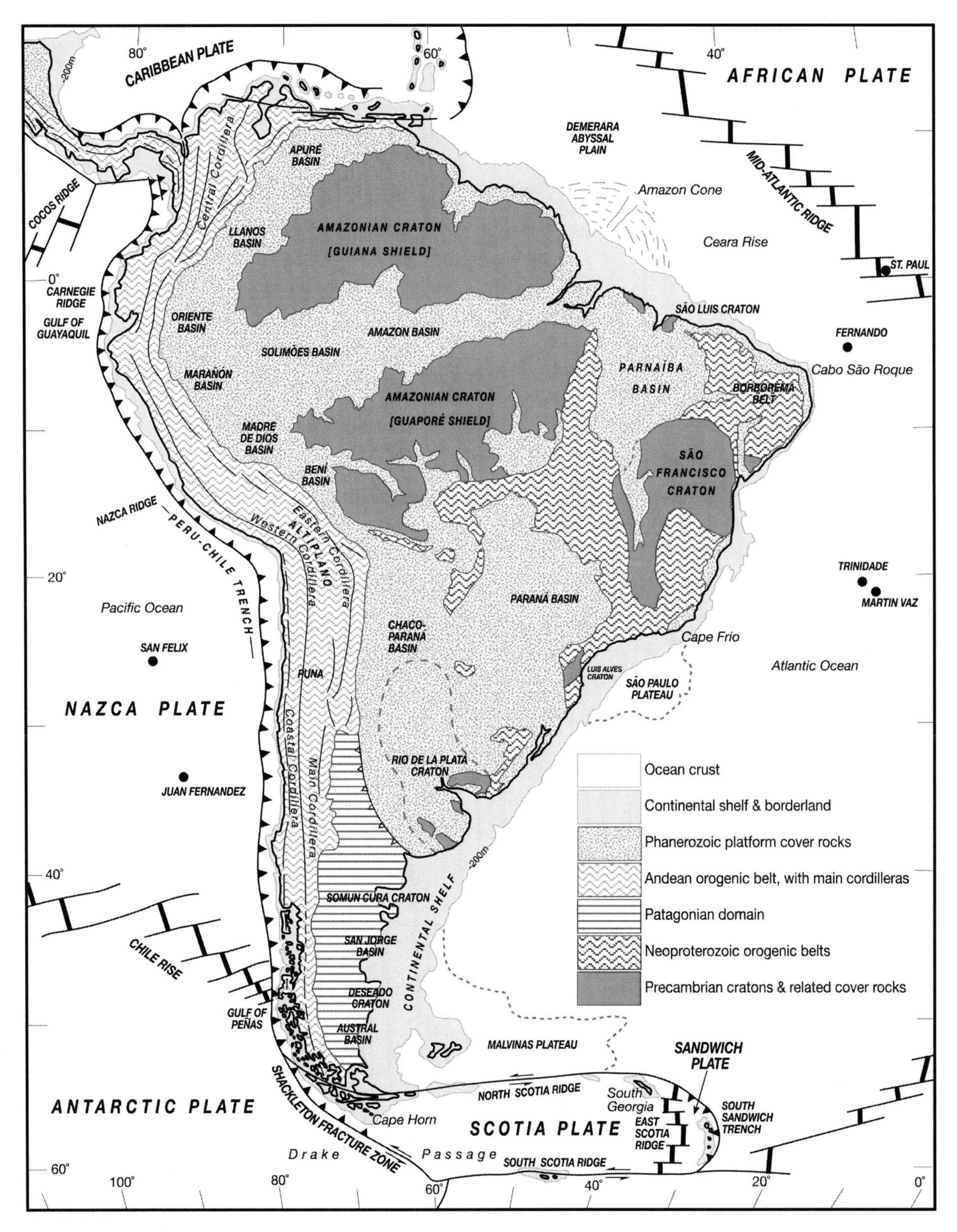

Figure 1.2 Tectonic provinces of South America

featureless plain formed of Cenozoic fluvial debris shed from the rising Andes to the west. However, in the Amazonas (0.5 × 10^6 km^2) and Parnaíba basins (0.5 × 10^6 km^2), Paleozoic rocks crop out against the margins of adjacent shields, while late Jurassic and early Cretaceous aeolian sandstones and basaltic lavas are widely exposed across the Paraná Basin (1.4 × 10^6 km^2).

During Gondwana's existence, the South American Platform was augmented to the south, in Patagonia, by the suture of the Gondwanides, a discrete Paleozoic deformational belt of uncertain origin formed from several episodes of terrane accretion and orogenic activity from the late Neoproterozoic onward. The culminating Gondwanan orogeny is well expressed in the Sierra de la Ventana of southern Buenos Aires province from whence the Permian suture describes an arc extending northwest toward the Andes (fig. 1.2). The Gondwanides comprise two distinct cratons, the Somun Cura and Deseado massifs that fused in early Paleozoic time to form the Austral continental plate, together with old mountain chains such as the Patagonian Precordillera farther west (Ramos and Aguirre-Urreta, 2000). The Proterozoic and early Paleozoic crystalline basement of the Gondwanides is generally overlain by Paleozoic marine, continental, and glaciogenic sediment, intruded by Paleozoic and early Mesozoic plutons, and capped by Mesozoic volcanics portending the rupture of Gondwana. This Patagonian domain, which extends eastward beneath the broad continental shelf, was later modified by post-Paleozoic subsidence, notably in the San Jorge and Austral basins. In Cenozoic time, Patagonia has been mantled by basaltic lavas and ash-fall tuffs, derived from volcanic activity linked to transient hot-spots or to plate subduction farther west, and by continental sediments derived from the rising Andes and containing a rich mammalian fauna. Subsequent dissection of these volcanic and sedimentary formations has provided the Patagonian landscape with its characteristic *mesetas*.

Meanwhile, the tectonic forces that had assembled Gondwana also transformed its western edge from a passive margin, following the rupture of Rodinia, to an active convergent margin, subject to subduction, peripheral subsidence, and episodic orogenic and magmatic activity as the supercontinent converged from time to time with various plates farther west (Dalziel, 1997; Bahlburg and Hervé, 1997). In this way, the stage was set for the initiation of the Andes, long before the events that culminated in massive Cenozoic uplift. Thus, in Paleozoic time, a series of pre-Andean foreland basins developed along the western margin of the South American Platform. Although their stratigraphies were similar to those of intracratonic basins farther east, rocks in these foreland basins experienced episodic compression, usually in a back-arc setting, and eventually became involved in the development of the Andes.

Of the orogenic events affecting the southwestern margin of Gondwana, three are particularly noteworthy (Pankhurst and Rapela, 1998; Rapela, 2000). First, the Pampean orogeny (the last phase of the Brasiliano-Panafrican orogeny) and related magmatism of Cambrian time (~530–515 Ma) saw Proterozoic terrane from the west collide with Gondwana, thus providing the crystalline roots of the Sierra de Cordoba and eastern Sierras Pampeanas in west-central Argentina. Second, the Famatinian orogeny of Ordovician time (~490–450 Ma) was associated with magmatic arc subduction, which provided granite batholiths to several later Andean ranges in western Argentina. Third, by Silurian time (~430–410 Ma), the distinctive Precordilleran terrane of western Argentina, comprising Proterozoic crystalline basement overlain by Cambrian and Ordovician passive margin shelf carbonates, had docked against Gondwana, to be followed by intrusion of post-orogenic Devonian batholiths.

The Precordilleran terrane of the Andean foothills is noteworthy for its similarity to Neoproterozoic-Cambrian terranes and faunas along the Appalachian-Ouachita margin of Laurentia. This has suggested a Laurentian origin for the Precordilleran terrane, implying that North America had collided with West Gondwana by Silurian time (e.g., Ramos et al., 1984; Dalziel, 1997; Keller, 1999; Thomas and Astini, 2003). Indeed, collision and rifting may have occurred earlier, perhaps several times, as Laurentia bounced clockwise along South America's western margin. Thus, the earlier Famatinian orogeny has been linked with the Taconian orogeny (480–440 Ma) of the Appalachians, reflecting an earlier collision between Laurentia and West Gondwana that stranded the Arequipa craton in southern Perú. Similarly, the later Appalachian Acadian orogeny (430–360 Ma) may be mirrored in structures in northern Perú (Dalla Salda et al., 1992).

Finally, during the late Paleozoic, Gondwana as a whole became sutured to the more northerly supercontinent of Laurussia to form a vast single landmass, Pangea (fig. 1.3). The Iapetus Ocean between the two land masses closed. This suturing process again involved collision with Laurentia, this time along the northwest margin of the Amazonian craton in the Alleghanian orogeny (330–260 Ma). This collision, which probably involved the Yucatán terrane and the Ouachita fold-and-thrust belt of North America, is reflected in the northern Andes. Pangea was not to last, however, because, after crustal extension in Permian and Triassic time, Laurentia began to separate from Gondwana in the mid-Jurassic (~180 Ma), thereby returning oceanic conditions to South America's north coast. Farther south, crustal extension created subparallel rifts along the proto-Andean margin of Gondwana. At this time, however, the northwest continental margin lay 200 km inland of the present coast of Colombia and Ecuador, where it was later masked by Cenozoic terrane accretion.

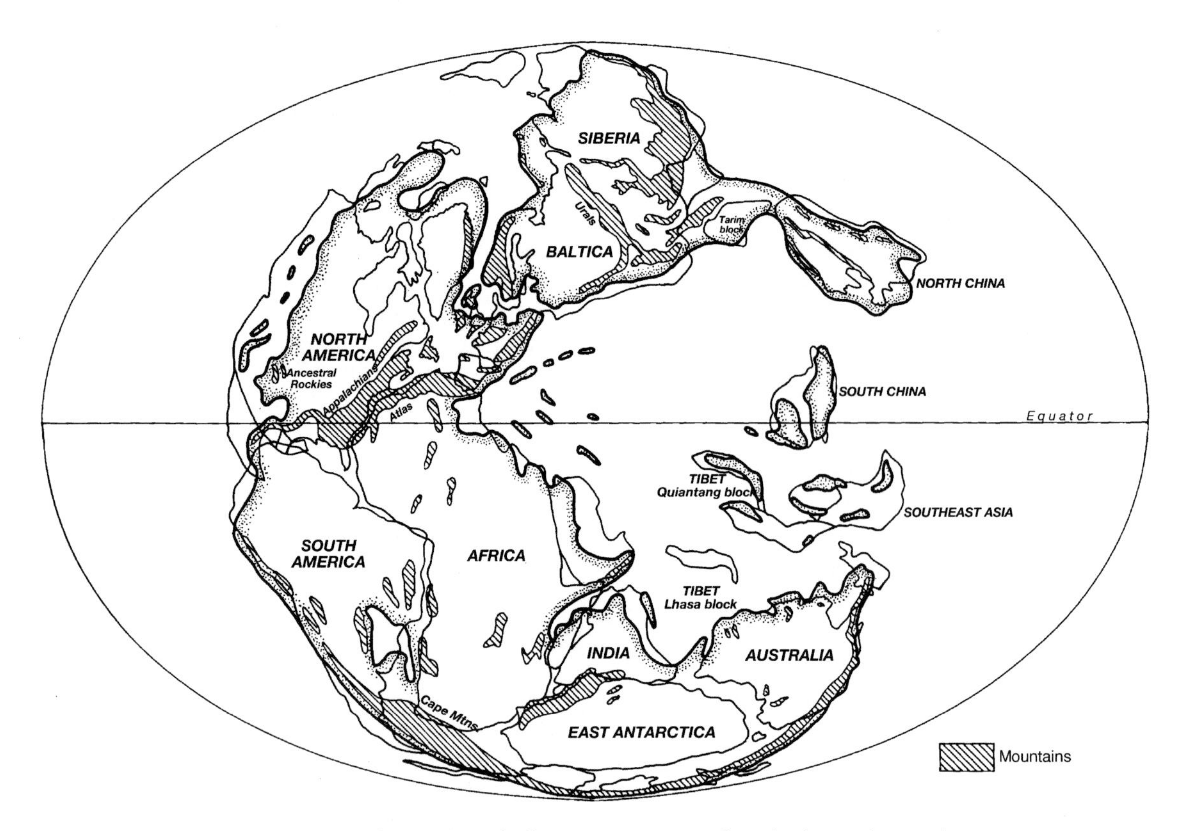

Figure 1.3 Pangea at the close of the Paleozoic (from Orme, 2002, after Ziegler et al., 1997)

1.4 The Rupture of Gondwana and Opening of the Atlantic Ocean

The breakup of Pangea was the precursor of a global plate reorganization that later involved the rupture of Gondwana and the isolation of South America as a discrete continent (fig. 1.4). Rifting of South America from Africa and the opening of the Atlantic Ocean were diachronous events that ranged over 100 Ma, from early Jurassic to middle Cretaceous time, and occurred along a 12,000 km divergent margin from northern Venezuela to the Malvinas Plateau. Farther south, South America may have separated from Antarctica as early as the late Jurassic (Smith et al., 1994) or as late as the Paleocene (Hay et al., 1999) or Eocene (Brundin, 1988).

Opening of the South Atlantic Ocean involved three distinct tectonic styles and locations (fig. 1.4; Milani and Filho, 2000). In the north, in the central Atlantic domain, crustal extension began in the late Triassic (~210 Ma), and sea-floor spreading in early Jurassic time (190–180 Ma) introduced the western Tethys Ocean to the Venezuela-Guiana margin (Steiner et al., 1998). In the south, in the southern Atlantic domain, the Malvinas Plateau began rifting from southern Africa in the early Jurassic (200–180 Ma). Then, as fracturing propagated northward, so the South Atlantic Ocean unzipped and the Paraná flood basalts spewed to the surface around 135–130 Ma, probably from the Tristan da Cunha hotspot (Peate et al., 1990; Glen et al., 1997). In the intervening equatorial Atlantic domain, and in response to these stresses, right-lateral wrenching along a massive transform system began moving northern Brazil westward along the Guinea coast of Africa in middle Cretaceous time (120–100 Ma).

All three domains show certain sequential similarities: initial crustal extension and enhanced magmatism, subsequent rifting of Precambrian cratons and Paleozoic cover rocks, basin formation along the rifted margin, followed by transitional then oceanic tectonic and sedimentologic regimes. In the north, opening of the central Atlantic Ocean was heralded by widespread late Triassic-early Jurassic magmatism in the Solimões, Amazonas, and Parnaíba basins, followed by rifting, the onset of sea-floor spreading, and the opening of marginal basins to repeated transgressive-regressive sedimentary cycles. Late Mesozoic and Cenozoic sandstones of these cycles are found in the Guiana Highlands and beneath the Demerera Plateau offshore. Farther east, basin development across Brazil's north coast was strongly influenced by transtensional rifting along a complex east-west transform plate boundary. Thick carbonate sequences in these basins indicate frequent shallow-water

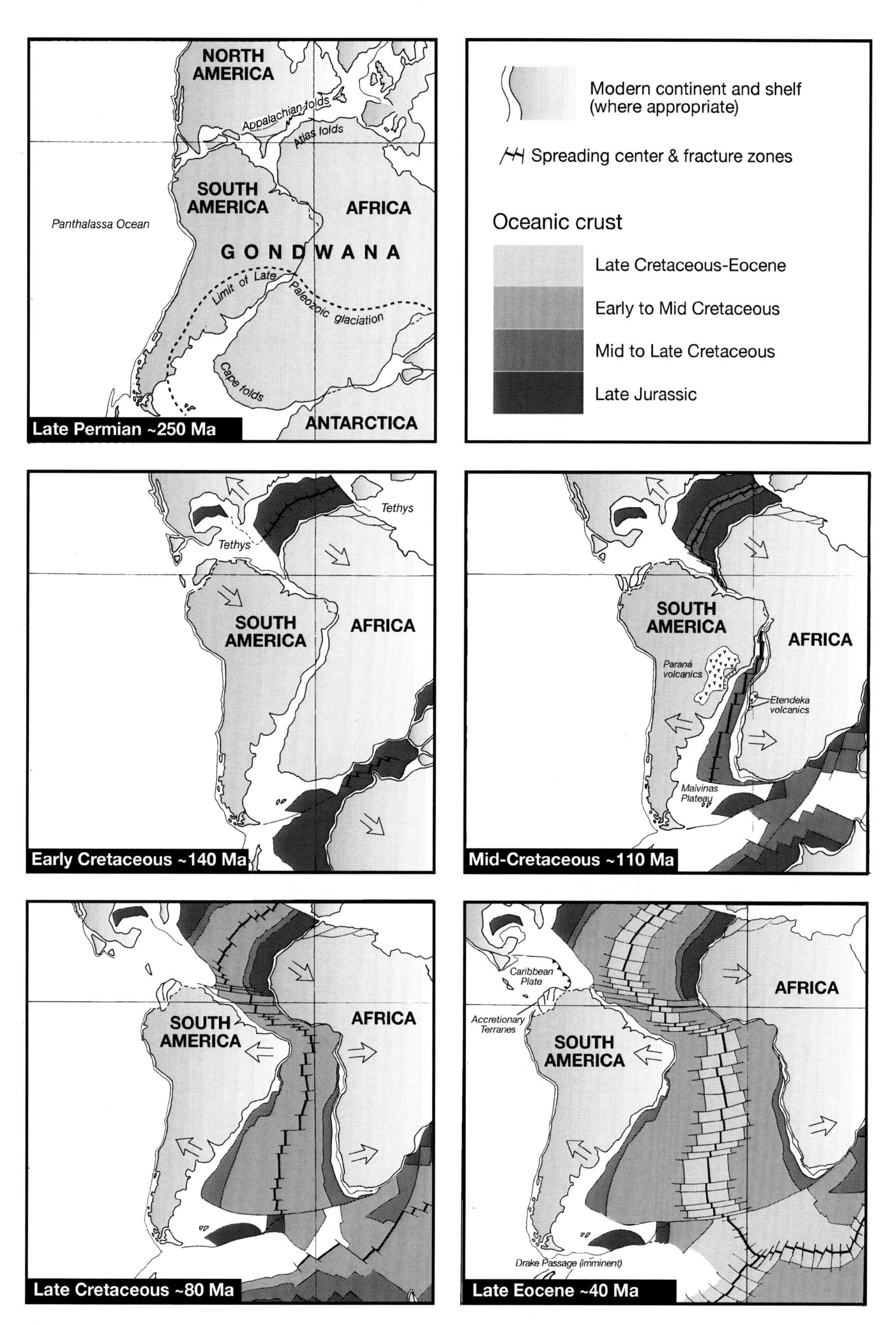

Figure 1.4 Breakup of Pangea and opening of the South Atlantic Ocean

conditions along the continental margin from the late Cretaceous onward.

The southern Atlantic domain is a classic divergent continental margin. Its impending rupture was marked by crustal uplift, extension, and volcanism reflected today in the 2,000-m thick basaltic lavas of the Paraná Basin. Later, massive normal faulting occurred as the unsupported eastern margin of the South American Platform slipped into the widening rift zone. The coastal escarpment of southeast Brazil, up to 2,200 m high and 1,000 km long in the Serra do Mar, reflects this rifted margin, extended westward by erosional scarp retreat of the Santos fault zone, now offshore (Almeida and Carneiro, 1998). As these normal faults were offset by transverse faulting related to east-west fracture zones in the spreading sea floor farther east, so a series of marginal basins formed, earlier in the south than in the north. Thus late Jurassic and early Cretaceous rifting off Argentina and Uruguay produced grabens and half-grabens with axes perpendicular or oblique to the continental margin which, below later sediments, still influence the shape of the Río de la Plata and Bahía Blanca embayments (Ramos and Aguirre-Urreta, 2000). The Pernambuco Basin at Brazil's northeast tip became the youngest such feature when it was initiated in mid-Cretaceous time (120–110 Ma). Most of the basins offshore, whether parallel or perpendicular to the continental margin, filled with pre-rift continental sediment and volcanic rocks, syn-rift deltaic, evaporite, and shallow-water carbonate rocks, and post-rift open marine sediment. The location and extent of each unit reflect the timing of basin formation and changing relationships among basin tectonics, sea level, and continental freeboard. In general, continental facies thicken and coarsen to the west and onshore, while marine facies thicken and fine seaward and offshore. Owing to continued faulting, seaward tilting, and thermal subsidence, these basins now lie mostly offshore, beneath Cenozoic sediment mantling the continental shelf and slope. There are also several failed-rift basins along this margin, systems initiated by early rifting but later abandoned. Among these are the San Jorge Basin between the Somun Cura and Deseado massifs in Patagonia, and the Recôncavo-Tucano Basin in northeast Brazil, a north-south graben initiated in early Cretaceous time that was later abandoned when rifting shifted to the present coastal margin.

Noteworthy features of this divergent margin include the São Paulo Plateau, a submerged fragment of continental crust now 2,000–3,000 m below sea level off east-central Brazil, and the Malvinas Plateau, which forms a 2,000-km wide continental shelf off Patagonia. The latter plateau supports the Malvinas (Falkland) Islands and is flanked eastward by a rifted margin descending to oceanic crust, northward by an east-west fracture zone defining the 3,000-m high Malvinas Escarpment, and southward by the active transform boundary with the Scotia plate. Its Precambrian-Paleozoic basement, which has close affinities with southernmost Africa, supports 4,000 m of post-rift sediment.

As the Mesozoic and Cenozoic progressed, a broad distinction thus developed within the eastern part of the South American plate between the rifted continental margin and its subsiding basins, and the widening mass of oceanic crust offset by transform faults farther east. As the continent moved away from the Mid-Atlantic Ridge, so the mantle plumes partly responsible for earlier continental magmatism emerged as hot spots in the oceanic crust. They occur today on or near the central ridge (e.g., St. Paul, Ascension, Tristan da Cunha) and farther west (e.g., Fernando, Trinidade) (fig. 1.1). These hot spots, fixed with respect to the mantle, indicate that the South American plate is presently moving westward away from the Mid-Atlantic Ridge at a rate of about 25 to 35 mm yr^{-1}.

1.5 The Formation of the Andes

The Andes are a superb example of a mountain chain raised by plate subduction, magmatism, crustal shortening, terrain accretion, and isostatic adjustments. The modern Andes extend over 8,500 km, from latitude 12°N to 56°S, and vary in width from 250 km to 750 km (fig. 1.2). Structurally, they comprise three segments: (1) the 2,000-km long Northern Andes trending NNE-SSW from 12°N to 5°S; (2) the 5,200-km long Central Andes comprising the 2,000-km long North-Central Andes, trending NW-SE from the Huancabamba Bend at 5°S to the Arica Bend at 18°S, and the 3,200-km long South-Central Andes trending N-S from there to the Chile triple junction near the Gulf of Peñas at 46°S; and (3) the 1,300-km long Southern Andes from 46°S to 56°S, forming a broad arc through 90° from the Gulf of Peñas to the transform boundary with the Scotia plate in Tierra del Fuego.

Temporally, the formation of the Andes involves a complex sequence of events that may be divided into four major stages (Ramos and Aleman, 2000): (1) terrane accretion and orogenesis along the proto-Andean margin of Gondwana during Proterozoic and early Paleozoic time; (2) late Paleozoic terrane accretion, episodic subduction, and orogenesis during the assembly of Pangea, involving collision of North and South America in the Alleghanian orogeny and suturing of the Gondwanides in the south; (3) crustal extension along the proto-Andean margin during the breakup of Pangea, punctuated by collision of island arcs and batholith emplacement; and (4) the Andean orogeny of later Mesozoic and Cenozoic time that saw subduction of oceanic terranes beneath the westward-moving South American plate and closure of remaining trans-Andean seaways in Venezuela, Ecuador, and Patagonia.

The first three stages, discussed earlier, are only summarized here. Following the assembly of Gondwana, several plates foundered against its active western margin during the Paleozoic. The resulting Pampean, Famatinian, Precordilleran, and Alleghanian orogenies may have involved independent peri-Gondwanan terranes, or repeated collisions and terrane transfers with Laurentia as it bounced northward along Gondwana's western flanks. Precambrian cratonic fragments and Paleozoic igneous and sedimentary rocks involved in these orogenies were to be further deformed, metamorphosed, and mineralized during the fourth stage, the Andean orogeny.

The present Andean orogenic system was initiated in late Mesozoic time and has peaked with massive uplift over the past 30 Ma, the magnitude and timing of which have varied from north to south, owing to the nature and rate of subduction, and from west to east as orogenic forcing has migrated eastward. During this uplift phase, the South American continental plate, moving west at 25 to 35 mm yr^{-1}, has overridden the Nazca oceanic plate as it subducts eastward into the Perú-Chile Trench at 50 to 85 mm yr^{-1}. Mountain building and magmatism are thus responding to net plate convergence of between 75 and 120 mm yr^{-1} (Jaillard et al., 2000). Over time, however, orogenic activity has been influenced by differing geometric and kinematic relations between convergent plates and by variable crustal shortening, which help to explain anomalies in the elevation and trend of the modern Andes. In the Andes of northern and central Perú between 3°S and 15°S, for example, late Miocene orogenic and igneous activity was initially driven by steep subduction of the Nazca plate beneath the continent. Over the past 5 Ma, however, subduction of the Nazca Ridge has led to plate flattening beneath the Andes, the so-called Peruvian flat slab, which in turn has driven further tectonic uplift, detachment faulting, and erosional unloading of the Cordillera Blanca batholith, although volcanic activity has ceased north of the ridge (McNulty and Farber, 2002).

Volcanism is also linked to the angle at which discrete oceanic slabs dip beneath the continent. Extensive volcanism occurs where these slabs subduct at 25–30°, notably in the Northern, Central, Southern, and Austral Volcanic Zones (see chapter 2). In contrast, volcanic gaps occur where slabs and ocean ridges subduct more gently and flatten at shallow depth (Gansser, 1973). Flat-slab subduction inhibits arc-magma generation by preventing wedges of mobile asthenosphere from penetrating between subducting and overriding plates. Thus, volcanic gaps coincide with the Bucaramanga, Peruvian, and Pampean flat slabs, and farther south with subduction of the Chile Rise (see fig. 2.4).

The Northern Andes form a distinctive tectonic realm combining deformed continental crust, accreted oceanic terranes, magmatism, and strike-slip faulting (fig. 1.5). Initially, the Jurassic crustal extension that opened the western Tethys Ocean also separated continental crust in Colombia's Central Cordillera from that in the Eastern Cordillera. Following an interlude of passive margin sedimentation, oceanic terranes related to the Farallón plate began docking obliquely against the continent from late Cretaceous time onward, notably the mafic rocks and deep-sea sediment of the Piñón-Dagua terrane, which now forms part of the Western Cordillera in Colombia and Ecuador. Following the breakup of the Farallón plate into separate Nazca and Cocos plates in the late Oligocene (~28 Ma), the Nazca plate converged orthogonally on the west coast, generating massive crustal shortening, which peaked in Miocene time but continues today. These events also coincided with the formation of the Caribbean plate, whose eastward thrust from Eocene time onward transformed the Venezuelan or Mérida Andes into a transpressional orogenic belt characterized by extensive strike-slip displacement. Meanwhile, Miocene accretion of the Panamá-Baudó oceanic arc, now the Serranía de Baudó in northwest Colombia, reflected processes that would lead to closure of the Central American isthmus in late Pliocene time.

While the collision of these oceanic volcanic arcs with the continent formed the Western Cordillera and the Serranía de Baudó, it also caused compressional uplift of the Central and Eastern Cordilleras beyond the Romeral fault zone farther east (Cooper et al., 1995). The latter are amalgams of ancient and modern fold-thrust structures, including Proterozoic crystalline basement and Paleozoic metasedimentary rocks of continental affinity and Cretaceous marine sediments (Gómez et al., 2003), which have been raised by continuing uplift and isostatic adjustments to great heights in a very short time. Paleobotanical evidence from the Eastern Cordillera suggests that between 15 Ma and 5 Ma this range had reached no more than 40% of its modern height, but that from 5 Ma to 2 Ma it rose rapidly, at rates of 0.5–3 mm yr^{-1} to reach modern elevations by ~2.7 Ma (Gregory-Wodzicki, 2000). Meanwhile, subsidence formed linear basins along bounding faults, notably along the present Magdalena and Cauca valleys. The middle Magdalena valley contains 7,000 m of clastic debris derived from erosion of the Central Cordillera (Gómez et al., 2003). The Cauca basin continues south into the Interandean Graben of Ecuador, where the bounding Peltetec and Pujilí dextral faults are the focus of active volcanism. West of the graben, the Western Cordillera is built of deformed oceanic rocks. To the east, Ecuador's Eastern Cordillera is composed of Paleozoic and Mesozoic metamorphic rocks topped and flanked, like Colombia's Central Cordillera, by active volcanoes (see chapter 2). These three linear features disappear south of 3°S.

The Central Andes are the product of deformation that began in late Cretaceous time, forming a fold-thrust belt that subsequently migrated eastward (Horton and DeCelles, 1997). In general terms, as orogenic activity propagated eastward, it generated a forearc zone, a magmatic arc, and a backarc region, with actively deforming foreland basins farther east (Isacks, 1988; Jordan et al., 1997). The present

Figure 1.5 Panorama looking north over the Northern Andes of Colombia and western Venezuela. The main features, from left to right, are accreted oceanic terranes of the Pacific coastlands and Western Cordillera, the Cauca Valley, the Central Cordillera, the Magdalena Valley, and the Eastern Cordillera (splitting northward into the Sierra de Perija and Mérida Andes on opposite sides of the Maracaibo basin). To the lower right lies the Serranía de la Macarena, an arch in the South American Platform separating the Llanos and Apuré foreland basins to the north from the Putumayo and Oriente foreland basins to the south. This image is a mathematical simulation of satellite data generated by combining USGS digital elevation models, NASA Shuttle Radar Topography Mission (SRTM) data, and georeferenced satellite images from NASA's Moderate Resolution Imaging Spectroradiometer (MODIS) (courtesy: William A. Bowen, California Geographical Survey).

relief of this region is largely a result of continuing late Cenozoic uplift, igneous activity, and erosional exhumation. The thick crust and high altitudes of the Central Andes reflect 250–500 km of crustal shortening that began around 25 Ma when reorganized plate motions accelerated Nazca plate convergence normal to the coast (McQuarrie, 2002). In the north, oblique plate convergence has raised Perú's Cordillera Blanca to 6,768 m in Huascarán, exhuming a granodiorite batholith that was emplaced a mere 5 to 8 Ma ago (McNulty et al., 1998).

Farther southeast, in the widest part of the Andes where continental crust is 75–80 km thick (Zandt et al., 1994), the region comprises two major cordilleras reaching over 6,000 m, separated by an elevated plateau averaging 4,000 m above sea level whose internal drainage flows to lakes Titicaca (8,135 km^2) and Poopo (fig. 1.6). The Western Cordillera is a late Cenozoic magmatic arc with many active volcanic piles such as Sajama (6,520 m). The intervening Altiplano-Puna plateau occupies a 120 to 250-km wide structural basin containing approximately 10,000 m of Cenozoic continental deposits whose lowest units were derived from Brazilian cratons when the Altiplano was still lowland (Isacks, 1988; Horton et al., 2001). This plateau was at sea level until about 60 Ma, and paleobotanical evidence suggests that it had attained no more than a third of its present elevation by 20 Ma, and no more than half its elevation by 10 Ma (Gregory-Wodzicki, 2000). The Eastern Cordillera, whose late uplift echoes the Altiplano, is a fold-thrust belt of mostly Ordovician sediment intruded by granite batholiths, which supports extinct volcanoes such as Illampu (6,550 m) and Illimani (6,457 m) near La Paz (DeCelles and Horton, 2003). Folding and thrusting have

Figure 1.6 Panorama looking northwest across the Central Andes. Features shown include the Main Cordillera, Argentinean Precordillera, and Sierra de Famatina (left center); the Eastern and Western Cordilleras enclosing the Altiplano (center); Lake Titicaca, Lake Poopo, and the Salar de Uyuni on the Altiplano; and the narrower ranges of northwest Perú. The tectonic inflection of the Pacific coastline, the Arica Bend, also separates the Atacama Desert from the Peruvian Desert. Active fold systems of the Subandean Ranges of eastern Perú and Bolivia are prominent below the Eastern Cordillera, transected by antecedent rivers such as the Pilcomayo, Parapeti, and Grande (courtesy: William A. Bowen, see fig. 1.5 for explanation of image).

since propagated into foreland belts east of the High Andes, notably into the Subandean Ranges of Perú, Bolivia, and northern Argentina where antecedent rivers such as the Pilcomayo, Parapeti, and Grande flow into the Bolivian lowlands through actively rising folds (fig. 1.6). Similar movements have also reactivated basement uplift in the Sierras Pampeanas. Farther east, basins in the Amazon and Chaco lowlands have been deformed by forebulge tectonism (Horton and DeCelles, 1997).

South of the Arica Bend, the High Andes are flanked to the west by an active forearc of discontinuous north-south lineaments comprising, from east to west, the 2,300-m high Salar de Atacama depression containing 10,000 m of Permian-Holocene continental debris; the 4,300-m high, copper-rich Precordillera; a 1,500-m high Central Depression that runs southward through Chile; and, beyond the 1,000-km long Atacama strike-slip fault zone, the 1,000 to 3,000-m high Coastal Cordillera, a Mesozoic magmatic arc that is again deforming and dislocating, and where Pleistocene marine terraces reach 600 m above sea level on the Mejillones Peninsula (Hartley et al., 2000). Space-geodetic data, which define Nazca plate convergence relative to stable cratons in Brazil and Argentina, indicate that the Central Andes continue to rise in response to crustal shortening of 10–25 mm yr^{-1} (Kendrick et al., 1999).

The Southern Andes reflect uplift along that portion of the South American plate that is now converging with the subducting Antarctic plate (fig. 1.7). This occurs south of the Chile (Aysén) triple junction where the Chile Rise, the spreading center between the Nazca and Antarctic plates, dips beneath the South American plate. While tectonism here has long been dictated by oblique plate convergence, the present phase began around 14 Ma when the triple junction first collided with South America (Hervé et al., 2000). Since then, this junction has migrated northward, subducting crab-like beneath the continent to bring the Antarctic plate progressively into contact with the South American plate. Structurally, the Southern

Andes comprise the south-trending southern Patagonian Andes and, south of the fault-guided Strait of Magellan, the east-trending Fuegan Andes at the northern margin of the Scotia plate. Lithologically, the Patagonian Andes are similar both north and south of the triple junction—a core batholith of Mesozoic and Cenozoic plutons that extends to Cape Horn; a western metamorphic belt of late Paleozoic through Cenozoic oceanic subduction complexes; an eastern, weakly metamorphic, fold-and-thrust belt of Mesozoic and Cenozoic oceanic and terrigenous sediment; and a variable mantle of Cenozoic volcanic and sedimentary rocks that thin eastward across Patagonia.

Finally, why have the Central Andes become massively broader than the Northern and Southern Andes during late Cenozoic time, despite the fact that the entire western margin of the South American plate has been overriding the Nazca plate since the breakup of Gondwana? This anomaly has been attributed to various events over the past 30 Ma, for example: (1) steepening of the subduction angle of the Nazca plate (e.g., Isacks, 1988); (2) more intense crustal shortening, plate reorganization, and increased plate convergence (e.g., Somoza, 1998); (3) accelerated westward movement of the South American plate (e.g., Silver et al., 1998); and (4) increased shear stress between the two plates caused by increasing aridity and thus sediment starvation in the Perú-Chile Trench (e.g., Lamb and Davis, 2003). Using a coupled thermodynamic model and assuming a 40–45-km thick crust in a 100–130-km thick continental lithosphere, Sobolev and Babeyko (2005) have found that accelerated westward drift of the South American plate, from 20 to 30 mm yr^{-1}, coupled with more intense friction with the subducting Nazca plate, can produce > 300 km of tectonic shortening for the Central Andes for the past 30 to 35 Ma, and massive uplift of the Altiplano over the past 10 Ma. They also found that mechanically strong but thinner continental crust and less friction produce < 40 km of shortening over the same period, as seen in the narrower and lower Southern Andes. These model results agree well with observed data.

1.6 Tectonism at the Extremities

The South American plate is flanked in the far north and south by active transform margins marking the relative eastward movement of the Caribbean and Scotia-Sandwich

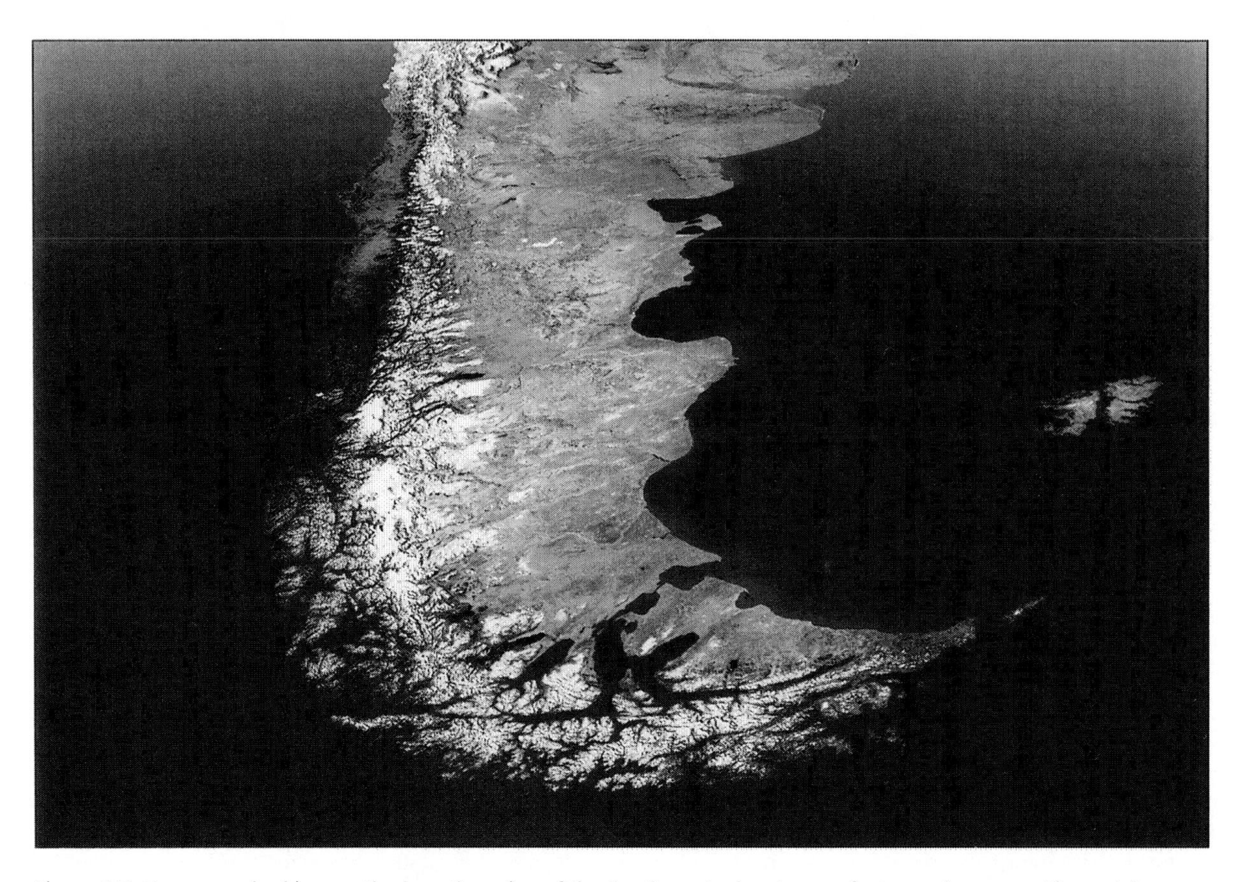

Figure 1.7 Panorama looking north along the spine of the Southern Andes. Among features shown are Tierra del Fuego, the Strait of Magellan, the Patagonian Cordillera and Ice Field, Chilean archipelagoes, Patagonian plateaus, and the Río de la Plata plains (courtesy: William A. Bowen; see fig. 1.5 for explanation of image).

plates, respectively, during Cenozoic time. These plates may not have much presence within South America but, as the Cenozoic progressed, their motion has exerted a major influence on oceanic, climatic, and biogeographic processes affecting the continent.

In the far north, the Caribbean plate and the accreted volcanic terranes of western Colombian and Ecuador are remnants of one or more oceanic plateaus that formed in late Cretaceous time (~90 Ma) in the vicinity of the present Galapagos Islands, or beyond. The Caribbean plate then migrated northeast, colliding with the South American plate in early Cenozoic time before moving east (Hoernle et al., 2002; Kerr and Tarney, 2005). Prolonged interaction between the Caribbean plate and pursuing volcanic arcs led eventually, around 3 Ma, to final uplift of the Central American Isthmus and with it closure of the post-Pangean Tethys Seaway that had linked the Atlantic and Pacific Oceans (Coates and Obando, 1996; Coates et al., 2003). The southern boundary of the Caribbean plate is presently marked by the Oca-Sebastian-El Pilar fault system that runs from near Santa Marta eastward to beyond Trinidad (fig. 1.2). These are right-lateral, strike-slip faults, dipping vertically into the crust, that have displaced the northern coastlands from the Guajira Peninsula to the Paria Peninsula eastward relative to the South American plate. However, this is also a transpressional margin in that, as the Caribbean plate elbowed its way east, portions of its oceanic crust were thrust southward as nappes onto the Mérida Andes. The Caribbean Andes thus formed from a combination of terrane accretion, crustal shortening, flexural subsidence, and transcurrent faulting involving both Caribbean and Andean rocks (Bellizzia and Dengo, 1990; Ramos and Aleman, 2000). The oil-rich Maracaibo Basin, which originated along the Mesozoic passive margin with the Tethys Ocean, has been progressively modified during Cenozoic time by the uplift of the Mérida and Perija Andes, to the east and west respectively, and by transpressional tectonics as the Caribbean plate passed by to the north. This almost-closed basin has filled with 7,000 m of sediment during the Cenozoic. Global positioning data show that the Caribbean plate is presently moving due east at ~20 mm yr^{-1} relative to the South American plate (Weber et al., 2001).

In the far south, eastward movement of the Scotia-Sandwich plate complex relative to the South American plate during later Cenozoic time has deformed and dismembered the Fuegan Andes and enlarged the Drake Passage with major implications for the oceanic and atmospheric circulation of the region (fig. 1.1, 1.7). This plate complex involves the East Scotia Ridge, an oceanic spreading center from which the Scotia and Sandwich plates are diverging at a rate of 65–70 mm yr^{-1} (Barker and Lawver, 1988; Cunningham et al., 1995; Livermore et al., 1997). The oceanic part of the South American plate is thus now converging at 70–85 mm yr^{-1} on the Sandwich plate, subducting into the South Sandwich Trench, and generating the volcanic arc of the South Sandwich Islands (Pelayo and Wiens, 1989; Vanneste et al., 2002). The margins of the Scotia-Sandwich plate complex are marked by transform faults paralleling the North Scotia and South Scotia ridges, respectively.

1.7 Rock Distribution and Relief Hierarchy

Spatial links between past tectonism and present relief features are reflected in the distribution of different rock types within the continent. Of the major rock types exposed, South America comprises 25% metamorphic, 2% intrusive, 11% extrusive, and 62% sedimentary (Blatt and Jones, 1975). The higher proportion of metamorphic rocks relative to the global average (17%) reflects mainly their outcrop in the Precambrian shields. The high value of extrusive rocks (global average 8%) reflects widespread volcanism during the rupture of West Gondwana, notably in the Paraná Basin, and the intimate association of volcanism with subduction and accretion in the Andes. Although less extensive than the global average (8%), intrusive rocks are subsumed in the Precambrian shields and exhumed in the Andes. Sedimentary rocks (global average 66%) are widespread, ranging from undeformed Proterozoic platform covers to resistant Paleozoic sediments in Andean fold belts and poorly consolidated Cenozoic debris subject to reworking in lowland basins and Andean intermontane valleys. Inland basins such as those along the Andean foreland also retain sediment deposited in early Miocene seas and later Miocene lakes (Ramos and Alonso, 1995).

The impact of tectonism is reflected in the highest three orders of relief (fig. 1.2). At the primary level, tectonic controls are well expressed along the eastern edge of the continental platform where the relatively late separation of South America from Africa is reflected in the mirror image presented by opposing passive continental margins, and indeed by much of the present coastline. It was this mirror image that attracted early attention to the possibility of continental drift, initially by circumstantial matching of structures, glaciogenic deposits, and fossil remains across the South Atlantic, and later by the panoply of evidence invoked to support plate tectonic concepts during the later twentieth century. Farther south, off Patagonia where the continental platform is shallowly submerged, the mirror image is reflected far offshore, between the outer margin of the Malvinas Plateau and southernmost Africa. In the far north, any prior geometric fit with the North American margin has been destroyed by the Cenozoic intrusion of the Caribbean plate. The western edge of the continent does of course coincide closely with the boundary between the South American and Nazca/Antarctic plates.

At the secondary level, tectonism provides the basis for the continent's principal relief triad of shields, lowlands, and western cordilleras. The Guiana and Brazilian shields are the exposed portions of Precambrian cratons and orogenic belts that were reactivated during the rupture of Gondwana. The vast lowlands of the Amazon and Paraguay-Paraná basins overlie ancient intracratonic sags and arches, still active and variably mantled with Phanerozoic sedimentary and volcanic rocks. The Andes as a whole are a Phanerozoic amalgam of orogens and accreted terranes whose present elevation reflects massive late Cenozoic uplift supported by continental crust up to 80 km thick. Their bold western face provides the Earth's largest relative relief within such a short distance. In southern Perú, for example, Coropuna volcano (6,425 m) is separated from the Perú-Chile Trench (−6,865 m) by a mere 240 km, a vertical change of 13,290 m.

At the tertiary level are the distinctive morphotectonic provinces within the main relief triad. By definition, these provinces were initiated by tectonism and have been reshaped by subsequent erosion and sedimentation. Morphotectonic relations are vividly expressed in the Northern Andes where, depending on relative plate motions, individual cordilleras are the products of episodic compression or transpression, enhanced by terrane accretion and igneous underplating (fig. 1.5). Similarly, major valleys, such as those of the Magdalena and Cauca rivers, often originated from structural subsidence along extensional or transtensional fault systems between cordilleras. In the Central Andes, the Altiplano owes its mean elevation of 4000 m to massive crustal shortening and regional uplift during the late Cenozoic (fig. 1.6; Isacks, 1988; Vandervoort et al., 1995; Sobolev and Babeyko, 2005). Tectonism is also reflected in linear relief farther south where the Coastal Cordillera, Central Valley, and Main Cordillera of central Chile pass into the Chonos Archipelago, Moraleda Canal, and Patagonian Cordillera of southern Chile, respectively (fig. 1.7).

Beyond the Andes, morphotectonic relations are more subtle, expressed, for example, in the differing orogenic provenance of Argentina's Sierra de Cordoba and Sierra de la Ventana, in Patagonia's volcanic *mesetas*, and in differential weathering and drainage patterns dictated by ancient rock types and fracture patterns within the Precambrian cratons and fold belts of the Guiana and Brazilian shields. Immediately east of the Andes, tectonic compression has caused structural bow waves to raise the Subandean ranges across the paths of outflowing rivers in Bolivia (fig. 1.6). Beyond these ranges, orogenic loading has formed a series of foreland basins now choked by late Cenozoic alluvial waste which supports wetlands and lakes from the swampy Apuré Basin in western Venezuela to the Bení Basin in northeast Bolivia and beyond. During early Miocene time, shallow seas may have reached these foreland basins from one or more directions (Ramos and Alonso, 1995). By late Miocene time, however, such seaways as may have existed had been closed by uplift and their legacy of marine sediments was being progressively buried by terrigenous deposits.

The rifted continental margin of the Brazilian Shield, much modified by fault-scarp retreat, is reflected in the 1,000-km long, 2,200-m high Serra do Mar. Buried rift systems initiated during the breakup of Gondwana also underlie and influence the estuaries of the Amazon and Río de la Plata (see chapter 5). Farther west, in the Paraná basin, the shield's crystalline basement descends beneath Phanerozoic cover rocks, which in turn provide some prominent relief features. In Mato Grosso state, for example, the Serra de São Lourenço is a bold escarpment formed in Devonian Chapada sandstone across which rivers such as the Itiquire and Taquari fall swiftly into the marshy alluvium of the Pantanal Basin, where late Cenozoic climate forcing is reflected in relict and modern alluvial fans and stream patterns.

Jurassic rifting, uplift, and subsequent erosion of the Guiana Shield have also produced 1,000-m high escarpments in the Proterozoic platform cover overlying the basement complex, notably along the erosional margins of *tepuis* (table mountains) such as the Pacaraima Plateau (2,870 m), a very old surface underlain by much fractured, mostly continental, sedimentary rocks of the ~1.8 Ga Roraima Supergroup (Dalton, 1912; Santos et al., 2003). Spectacular waterfalls cascade over resistant formations of both the crystalline shield and cover rocks. These include Venezuela's Angel Falls (979 m) on a Caroni tributary, the world's highest single drop, and Cuquenán Falls (610 m), Guyana's Kaieteur Falls (226 m), where the Potaro drops off the Pacaraima Plateau, and numerous falls on the Essequibo and Courantyne rivers farther east. Water also flows from the plateau through intricate networks of subsurface fractures, solution pipes, and cave systems (Urbani, 1986).

1.8 Tectonism and Mineral Resources

South America's considerable mineral wealth reflects past tectonic and geomorphic events, and also identifies with present morphotectonic patterns. Thus, Precambrian rocks exposed in the Guiana and Brazilian shields contain both ancient minerals and ore-rich residues from prolonged weathering; the Andes form one of the world's richest orogenic belts in terms of metallic ores; and lowland sedimentary basins contain about 9% of Earth's known petroleum resources.

The amalgam of Precambrian cratons and orogenic belts forming the South American Platform is rich in minerals, while Paleozoic intracratonic sags, Mesozoic rifted margins, and Mesozoic and Cenozoic sedimentary and volcanic cover rocks are each reflected in the region's mineral resources (Dardenne and Schobbenhaus, 2000). Precambrian igneous and metamorphic rocks of the Guiana and

Brazilian shields contain gold, diamonds, iron, copper, lead, zinc, and other minerals emplaced in the distant past. Paleozoic cover rocks contain iron and evaporite minerals. Mesozoic rocks contain vein fluorite and diamond-bearing kimberlite pipes related to syn-rift volcanism, and evaporites and phosphates deposited in post-rift coastal basins. Cenozoic sediments include placer gold, diamonds, and tin released by erosion to rivers, while Cenozoic weathering has produced extensive deposits of lateritic iron, bauxite, kaolin, and nickel.

The Andes owe their metallic mineral wealth to tectonic and magmatic events repeated throughout several orogenic episodes and to the heterogeneity of the rocks involved. Some nickel-chromium ores in Perú may be of Precambrian age, while the porphyry copper ores of northern Chile have been forming since at least the Carboniferous (Oyazún, 2000). Nevertheless, most minerals relate to the latest orogenic events of Cretaceous and Cenozoic age. In essence, magmatic activity has provided the direct source and mechanisms for ore generation, while tectonic forces have controlled the location and movement of intrusive and extrusive magmas and ore-bearing fluids. Four major metallic belts run more or less parallel with the strike of the Andes. Nearest the Pacific is the iron belt, comprising ores ranging in age from Proterozoic to Neogene, and richest in southern Perú and northern Chile. Farther inland is the copper belt, dominated by the rich porphyry copper ores of the same regions, which formed as metal-rich magmas ascended through thick continental crust. This belt accounts for 25–30% of the world's reserves of copper and associated molybdenum. Farther east, and throughout the length of the Andes, lies a polymetallic belt rich in gold, silver, lead, zinc, and copper. Finally, in the eastern Andes of Bolivia, the 500-km long tin belt is associated with Mesozoic batholiths and Cenozoic subvolcanic activity. Many of these minerals have long attracted interest—from prehistoric workings for gold and silver, which lured later colonial adventurers seeking El Dorado, to mining of tin and copper that began in earnest during the later nineteenth century and continues today, together with mining of evaporite minerals such as nitrate from the desert regions.

Petroleum and natural gas resources occur in five megasystems involving similar formative conditions, mostly in marine gulfs and deep lakes, in which hydrocarbons were generated from algae and other organisms, and then selectively preserved in lowland basins by later tectonic events (Figueiredo and Milani, 2000). Foremost among these is the Andean foreland megasystem extending from the Caribbean coast to the foreland basins of southern Venezuela, eastern Colombia, Ecuador, and northern Perú. These basins contain 90% of South America's known reserves, in Cretaceous and Cenozoic reservoir rocks ranging from the century-old Maracaibo and Maturín oil fields of Venezuela to developing fields within the Llanos, Oriente, Marañon, and other basins east of the Andes (fig. 1.2). The second largest megasystem, with ~5% of the continent's reserves, comprises passive-margin basins of post-Jurassic age associated with the South Atlantic Rift system, from the Recôncavo Basin in coastal Bahía state and the Campos Basin offshore from Rio de Janeiro to the Magallanes Basin in the far south. The three other megasystems comprise the Subandean basins of southern Perú, eastern Bolivia, and northern Argentina, where petroleum derived from Paleozoic sources is trapped; the Austral Rift system of Cretaceous and Cenozoic back-arc basins in Argentina, notably the Neuquén Basin; and the Paleozoic intracratonic sags overlain by later rocks beneath the central Amazon, Parnaíba, and Paraná river basins. Coal deposits formed in late Paleozoic wetlands prior to rifting of the South Atlantic are preserved beneath later rocks in the intracratonic Paraná Basin, while coal formed in Mesozoic and Cenozoic lakes is preserved within and flanking the Andes (Lopes and Ferreira, 2000).

Exploitation of these mineral resources to date has generated environmental impacts locally from surface mines such as Chile's vast Chuquicamata copper pit, and more widely from air and water pollution related to mineral processing, as in the Atacama Desert, from subsidence related to fluid withdrawal, for example in the Venezuelan oil fields, and from toxic spills from trans-Andean and other pipelines. Future discoveries and exploitation, especially of hydrocarbons in the Andean foreland and off the Atlantic coast from the Guianas to the Malvinas Plateau, will generate further environmental impacts from transportation and industrial activities. As recent events have shown, mineral exploitation may also trigger political changes and even conflict within and between nations.

1.9 Conclusion

The fundamental character of the South American landscape is driven by tectonic forcing, which has given the continent its location and general architecture and provided it with its major relief features and mineral resources. While tectonism is a mostly slow process on human timescales, it involves pulses that are frequent reminders of underlying forces. Earthquakes along the active Pacific Rim are a primary expression of these forces, and great earthquakes (Magnitude [M] > 8) occur somewhere along the continent's western margin every few years. For example, a M8.8 earthquake off northern Ecuador in January 1906 caused widespread destruction. The M9.5 earthquake along the Chilean coast in May 1960 caused over 2,000 deaths, rendered two million persons homeless, and triggered massive landslides and surface ruptures between Concepción and Puerto Montt. It also generated a tsunami that swept across low-lying areas of Hawaii and Japan. The M8.0 Antofagasta earthquake of July 1995 was related similarly to thrusting of the South American plate over the subducting Nazca plate, with rupture propagating upward to the Perú-Chile Trench from a depth of 46 km (Delouis et al., 1997). Though less common, the more passive eastern portions of the South American plate also suffer occasional modest earthquakes related to the compression of Brazil between the continent's westward push of 25 to 35 mm yr^{-1} from the Mid-Atlantic Ridge and the eastward subduction of the Nazca plate of 50 to 85 mm yr^{-1} into the Perú-Chile Trench (DeMets et al., 1990; Bezerra and Vita-Finzi, 2000). Such events are but recent expressions of continuing tectonic adjustments within the South American plate and of the latter's changing relationships with its neighbors.

References

Alkmim, F.F., S. Marshak, and M.A. Fonseca, 2001. Assembling West Gondwana in the Neoproterozoic: Clues from the São Francisco craton region, Brazil. *Geology*, 29, 319–322.

Almeida, F.F.M., and C.D.R. Carneiro, 1998. Origem e evolução da Serra do Mar. *Revista Brasiliera de Geosciências*, 28, 135–150.

Bahlburg, H., and F. Hervé, 1997. Geodynamic evolution and tectonostratigraphic terranes of northwestern Argentina and northern Chile. *Geological Society of America Bulletin*, 109, 869–884.

Barker, P.F., and L.A. Lawver, 1988. South American-Antarctic plate motion over the past 50 Myr, and the evolution of the South American-Antarctic Ridge. *Royal Astronomical Society Geophysical Journal*, 94, 377–386.

Bellizzia, A., and G. Dengo, 1990. The Caribbean mountain system, northern South America: A summary. In: G. Dengo and J.E. Case (Editors), *The Caribbean Region*. The geology of North America, volume H, Geological Society of America, Boulder, 167–175.

Bezerra, F.H.R., and C. Vita-Finzi, 2000. How active is a passive margin? Paleoseismicity in northeastern Brazil. *Geology*, 28, 591–594.

Blatt, H., and R.L. Jones, 1975. Proportions of exposed igneous, metamorphic, and sedimentary rocks. *Geological Society of America Bulletin*, 86, 1085–1088.

Brundin, L., 1988. Phylogenetic biogeography. In: A.A. Myers and P.S. Gillers (Editors), *Analytic Biogeography: An Integrated Approach to Plant and Animal Distributions*. Chapman Hall. London.

Coates, A.G., and J.A. Obando, 1996. The geological evolution of the Central American isthmus. In: J.B.C. Jackson et al. (Editors), *Evolution and Environment in Tropical America*. University of Chicago Press, Chicago.

Coates, A.G., M-P. Aubry, W.A. Berggren, L.S. Collins, and M. Kunk, 2003. Early Neogene history of the Central American arc from Bocas del Toro, western Panama. *Geological Society of America Bulletin*, 115, 271–287.

Coffin, M.F., L.A. Lawver, L.M. Gahagan, and D.A. Campbell, 2000. *The Plates Project 2000; Atlas of Plate Reconstructions (750 Ma to present day)*. Plates Project Progress Report, Institute for Geophysics, University of Texas, Austin.

Cooper, M.A., and 11 others, (1995). Basin development and tectonic history of the Llano basin, Eastern Cordillera, and middle Magdalena Valley, Colombia. *American Association of Petroleum Geologists Bulletin*, 79, 1421–1432.

Cordani, U.G., K. Sato, W. Teixeira, C.C.G. Tassinari, and M.A.S. Basei, 2000. Crustal evolution of the South American Platform. In: U.G. Cordani, E.J. Milani, A.T. Filho, and D.A. Campos (Editors), *Tectonic Evolution of South America*. 31st International Geological Congress, Rio de Janeiro, 19–40.

Crowley, T.J., and G.R. North, 1991. *Paleoclimatology*. Oxford University Press, New York.

Cunningham, W.D., J.W.DS. Dalzeil, T. Lee, and L.A. Lawver, 1995. Southernmost South America–Antarctic Peninsula relative plate motions since 84 Ma: Implications for the tectonic evolution of the Scotia Arc region. *Journal of Geophysical Research*, 100 (B5), 8257–8266.

Dalla Salda, L., I.W.D. Dalziel, C.A. Cingulani, and R. Varela, 1992. Did the Taconic Appalachians continue into South America? *Geology*, 20, 1059–1062.

Dalton, L.V., 1912. On the geology of Venezuela. *Geological Magazine*, 9, 203–210.

Dalziel, I.W.D., 1997. Neoproterozoic-Paleozoic geography and tectonics: Review, hypothesis, environmental speculation. *Geological Society of America Bulletin*, 109, 16–42.

Dardenne, M.A., and C. Schobbenhaus, 2000. The metallogenesis of the South American platform. In: U.G. Cordani, E.J. Milani, A.T. Filho, and D.A. Campos (Editors), *Tectonic Evolution of South America*. 31st International Geological Congress, Rio de Janeiro, 755–850.

DeCelles, P.G., and B.K. Horton, 2003. Early to middle Tertiary foreland basin development and the history of Andean crustal shortening in Bolivia. *Geological Society of America Bulletin*, 115, 58–77.

Delouis, B., T. Monfret, L. Dorbath, M. Pardo, L. Rivera, D. Comte, H. Haessler, J.P. Caminade, L. Ponce, E. Kausel, and A. Cisternos, 1997. The M_w8.0 Antofagasta (northern Chile) earthquake of 30 July 1995: A precursor to the end of the large 1877 gap. *Seismological Society of America Bulletin*, 87, 427–445.

DeMets, C., R.G. Gordon, D.F. Argus, and S. Stein, 1990. Current plate motions. *Geophysical Journal International*, 101, 427–478.

Du Toit, A.L., 1927. A geological comparison of South America with South Africa. *Carnegie Institution of Washington*, Publication 381, 1–157.

Du Toit, A.L., 1937. *Our Wandering Continents: An Hypothesis of Continental Drifting*. Oliver and Boyd, Edinburgh.

Figueiredo, A.M.F. de, and E.J. Milani, 2000. Petroleum systems of South American basins. In: U.G. Cordani, E.J. Milani, A.T. Filho, and D.A. Campos (Editors), *Tectonic Evolution of South America*. 31st International Geological Congress, Rio de Janeiro, 689–718.

Gansser, A., 1973. Facts and theories on the Andes. *Journal of the Geological Society*, London, 129, 93–131.

Glen, J.M.G., P.R. Renne, S.C. Milner, and R.S. Coe, 1997. Magma flow inferred from anistrophy of magnetic susceptibility in the coastal Paraná-Etendeka igneous province: Evidence of rifting before flood volcanism. *Geology*, 25, 1131–1134.

Gómez, E., T.E. Jordan, R.W, Allmendinger, K. Hegarty, S. Kelley, and M. Heizler, 2003. Controls on architecture of the late Cretaceous to Cenozoic Middle Magdalena Valley Basin, Columbia. *Geological Society of America Bulletin*, 115, 131–147.

Gregory-Wodzicki, K.M., 2000. Uplift history of the Central and Northern Andes: A review. *Geological Society of America Bulletin*, 112, 1091–1105.

Hartley, A.J., G. May, G. Chong, P. Turner, S.J. Kape, and E.J. Jolley, 2000. Development of a continental forearc: A Cenozoic example from the Central Andes, northern Chile. *Geology*, 28, 331–334.

Hay, W.W., R.M. DeConto, C.N. Wold, K.M. Willson, S. Voigt, M. Schulz, A. Rossby-Wold, W-C. Dullo, A.B. Ronov, A.N. Balukhovsky, and E. Soeding, 1999. An alternative global Cretaceous paleogeography. In: E. Barrera and C. Johnson (Editors), *Evolution of the Cretaceous Ocean-Climate System*. Geological Society of America Special Paper 332, 12–48.

Hervé, F., A. Demant, V.A. Ramos, R.J. Pankhurst, and M. Suarez, 2000. The southern Andes. In: U.G. Cordani, E.J. Milani, A.T. Filho, and D.A. Campos (Editors), *Tectonic Evolution of South America*. 31st International Geological Congress, Rio de Janeiro, 605–634.

Hoernle, K., P. van den Bogaard, R. Werner, B. Lissinna, F. Hauff, G. Alvarado, and D. Garber-Schönberg, 2002. Missing history (16–71 Ma) of the Galápagos hotspot: Implications for the tectonic and biological evolution of the Americas. *Geology*, 30, 795–798

Horton, B.K., and P.G. DeCelles, 1997. The modern foreland basin system adjacent to the Central Andes. *Geology*, 25, 895–898.

Horton, B.K., B.A. Hampton, and G.L. Waanders, 2001. Paleogene synorogenic sedimentation in the Altiplano plateau and implications for initial mountain building in the central Andes. *Geological Society of America Bulletin*, 113, 1387–1400.

Isacks, B.L., 1988. Uplift of the central Andean plateau and bending of the Bolivian orocline. *Journal of Geophysical Research-Solid Earth and Planets*, 93 (B4), 3211–3231.

Jaillard, E., G. Hérail, T. Monfret, E. Diaz-Martinez, P. Baby, A. Lavenu, and J.F. Dumon, 2000. Tectonic evolution of the Andes of Ecuador, Perú, Bolivia, and northernmost Chile. In: U.G. Cordani, E.J. Milani, A.T. Filho, and D.A. Campos (Editors), *Tectonic Evolution of South America*. 31st International Geological Congress, Rio de Janeiro, 481–559.

Jordan, T.E., J.H. Reynolds, and J.P. Erickson, 1997. Variability in age of initial shortening and uplift in the Central Andes, 16–33°30'S. In: W.F. Ruddiman (Editor), *Tectonic Uplift and Climate Change*, Plenum Press, New York, 41–61.

Keller, M. (Editor), 1999. *Argentine Precordillera: Sedimentary and Plate Tectonic History of a Laurentian Crustal Fragment in South America*. Geological Society of America, Special Paper 341.

Kendrick, E.C., M. Bevis, R.F. Smalley, O. Diluents, and F. Galban, 1999. Current rates of convergence across the central Andes: Estimates from continuous GPS observations. *Geophysical Research Letters*, 26, 541–544.

Kerr, A.C., and J. Tarney, 2005. Tectonic evolution of the Caribbean and northwestern South America: The case for accretion of two late Cretaceous oceanic plateaus. *Geology*, 33, 269–272.

Kossinna, E., 1933. Die Erdoberflache. In: B. Gutenberg (Editor), *Handbuch der Geophysik*, 2, Berlin, Borntraeger, 869–954.

Lamb, S., and P. Davis, 2003. Cenozoic climate change as a possible cause for the rise of the Andes. *Nature*, 425, 792–797.

Livermore, R., A. Cunningham, L. Vanneste, and R. Larter, 1997. Subduction influence in magma supply at the East Scotia Ridge. *Earth and Planetary Science Letters*, 150, 261–275.

Lopes, R. da C., and J.A. Ferreira, 2000. An overview of the coal deposits of South America. In: U.G. Cordani, E.J. Milani, A.T. Filho, and D.A. Campos (Editors), *Tectonic Evolution of South America*. 31st International Geological Congress, Rio de Janeiro, 719–723.

McNulty, B.A., and D. Farber, 2002. Active detachment faulting above the Peruvian flat slab. *Geology*, 30, 567–570.

McNulty, B.A., D.L. Farber, G.S. Wallace, R. Lopez, and O. Palacios, 1998. Role of plate kinematics and plate-slip-vector partitioning in continental magmatic arcs: Evidence from the Cordillera Blanca, Peru. *Geology*, 26, 827–830.

McQuarrie, N., 2002. Initial plate geometry, shortening variations and evolution of the Bolivian orocline. *Geology*, 30, 867–870.

Milani, E.J., and A.T. Filho, 2000. Sedimentary basins of South America. In: U.G. Cordani, E.J. Milani, A.T. Filho, and D.A. Campos (Editors), *Tectonic Evolution of South America*. 31st International Geological Congress, Rio de Janeiro, 389–449.

Orme, A.R., 2002. Tectonism, climate, and landscape. In: A.R. Orme (Editor), *The Physical Geography of North America*, Oxford University Press, New York, 3–35.

Oyazún, J., 2000. Andean metallogenesis: A synoptical review and interpretation. In: U.G. Cordani, E.J. Milani, A.T. Filho, and D.A. Campos (Editors), *Tectonic Evolution of South America*. 31st International Geological Congress, Rio de Janeiro, 725–753.

Pankhurst, R., and C.W. Rapela (Editors), 1998. *The Proto-Andean Margin of Gondwana*. Geological Society of London, Special Publication 142.

Peate, D.W., C.J. Hawkesworth, M.S.M. Mantovani, and W. Shukowsky, 1990. Mantle plumes and flood basalt stratigraphy in the Paraná, South America. *Geology*, 18, 1223–1226.

Pelayo, A.M., and D.A. Wiens, 1989. Seismotectonics and relative plate motion in the Scotia Sea region. *Journal of Geophysical Research*, 94, 7293–7320.

Ramos, V.A., and R.N. Alonso, 1995. El Mar Paranense en la provincia de Jujuy. *Revista Geológica de Jujuy*, 10, 73–80.

Ramos, V.A., and M.B. Aguirre-Urreta, 2000. Patagonia. In: U.G. Cordani, E.J. Milani, A.T. Filho, and D.A. Campos (Editors), *Tectonic Evolution of South America*. 31st International Geological Congress, Rio de Janeiro, 369–380.

Ramos, V.A., and A. Aleman, 2000. Tectonic evolution of the Andes. In: U.G. Cordani, E.J. Milani, A.T. Filho, and D.A. Campos (Editors), *Tectonic Evolution of South America*. 31st International Geological Congress, Rio de Janeiro, 635–685.

Ramos, V.A., T.E. Jordan, R.W. Allmendinger, S.M. Kay, J.M. Cortés, and M.A. Palma, 1984. Chilenia: Un terrano election en la evolución Paleozoica de los Andes Centrales. *9° Congreso Geologic Argentines, San Carlos de Bariloche*, 2, 84–106.

Rapela, C.W., 2000. The Sierras Pampeanas of Argentina: Paleozoic building of the southern proto-Andes. In:

U.G. Cordani, E.J. Milani, A.T. Filho, and D.A. Campos (Editors), *Tectonic Evolution of South America.* 31st International Geological Congress, Rio de Janeiro, 381–387.

Santos, J.O.S., P.E. Potter, N.J. Reis, L.A. Hartmann, I.R. Fletcher, and N.J. McNaughton, 2003. Age, source, and regional stratigraphy of the Roraima Supergroup and Roraima-like outliers in northern South America based on U-Pb geochronology. *Geological Society of America Bulletin*, 115, 331–348.

Silver, P.G., R.M. Russo, and C. Lithgow-Bertelloni, 1998. Coupling of South America and African plate motion and plate deformation. *Science*, 279, 60–63.

Smith, A.G., D.G. Smith, and B.M. Funnell, 1994. *Atlas of Mesozoic and Cenozoic Coastlines.* Cambridge University Press, Cambridge.

Sobolev, S.V., and A.Y. Babeyko, 2005. What drives orogeny in the Andes? *Geology*, 33, 617–620.

Somoza, R., 1998. Updated Nazca (Farallon)-South America relative plate motions during the last 40 m.y.: Implications for mountain building in the central Andean region. *Journal of South American Earth Sciences*, 11, 211–215.

Steiner, C., A. Hobson, P. Five, G.M. Stampfli, and J. Hernandez, 1998. Mesozoic sequence of Fuerteventura (Canary Islands): Witness of early Jurassic seafloor spreading in the central Atlantic. *Geological Society of America Bulletin*, 110, 1304–1317.

Thomas, W.A., and R.A. Astini, 2003. Ordovician accretion of the Argentine Precordillera terrane to Gondwana: A review. *Journal of South American Earth Sciences*, 16, 67–79.

Urbani, F., 1986. Notas sobre el origen de las cavidades en rocas cuarciferas precámbricas del Grupo Roraima, Venezuela. *Interciencia*, 11, 298–300.

Vandervoort, D.S., T.E Jordan, P.K. Zeitler, and R.N. Alonso, 1995. Chronology of internal drainage development and uplift, southern Puna plateau, Argentine central Andes. *Geology*, 23, 145–148.

Vanneste, L.E., R.D. Larter, and D.K. Smythe, 2002. Slice of intraoceanic arc: Insights from the first multichannel reflection profile across the South Sandwich island arc. *Geology*, 3, 819–822.

Weber, J.C., T.H. Dixon, C. DeMets, W.B. Ambeh, P. Jansma, G. Mattioli, J. Saleh, G. Sella. R. Bilham, and O. Perez, 2001. GPS estimates of relative motion between the Caribbean and South American plates, and geological implications for Trinidad and Venezuela. *Geology*, 29, 75–78.

Zandt, G., A.A. Velasco, and S.L. Beck, 1994. Composition and thickness of the southern Altiplano crust, Bolivia. *Geology*, 22, 1003–1006.

Ziegler, A.M., M.L. Hulver, and D.B. Rowley, 1997. Permian world topography and climate. In: I.P. Martini (Editor), Late Glacial and Postglacial Environmental Changes: Quaternary, Carboniferous-Permian, and Proterozoic. Oxford University Press, New York, 111–146.

2

Tectonism, Climate, and Landscape Change

Antony R. Orme

Earth's physical landscapes are framed initially by tectonism, reshaped by climate, garnished by plants and animals, and modified by human activity. Tectonism constructs the physical framework of the continents and ocean floors. Climate, the synthesis of weather, generates the surface processes that reshape this framework through erosion and sedimentation, and also provides the conditions necessary to support life. In various guises, tectonism and climate have played these roles from early in Earth history, although 90% of Earth time had passed before the continents began to acquire vascular plants and land animals. However, because Earth's crust is ponderously mobile and climate depends ultimately on the variable receipt of solar radiation, tectonic and climatic forcing vary across time and space. Consequently, continents come to acquire distinctive suites of landscapes that reflect changing locational, tectonic, climatic, and biotic influences over time. South America exemplifies this concept—a continent whose distinctive qualities owe much to the roles played by tectonism and climate over time, including their impacts on landforms and biota.

Tectonism and climate are interactive forces. By determining the distribution and shape of land masses and ocean basins, tectonism influences the relative importance of continentality and oceanicity to climate. Over time, tectonism also influences climate change by promoting uplift favorable to prolonged cooling and perhaps glaciation, by opening and closing seaways to ocean circulation, and by influencing atmospheric composition by the generation and consumption of crustal rocks. Though more subtle, climate may in turn affect tectonism by redistributing continental mass through erosion and deposition, thereby generating isostatic adjustments to crustal loading and unloading. Tectonism also influences plant and animal distributions directly, for example by providing linkages or barriers to migration, while climate and biota are intimately linked in the biome concept and the feedback effect of biomes on climatic processes.

This chapter examines the interactive roles of tectonism and climate in changing the South American landscape over the 200 million years that have passed since the initial breakup of Pangea. It then discusses the implications of these changes for geomorphology and biogeography, and concludes with a brief evaluation of the pace of landscape change.

2.1 Tectonism and Environment: Premises and Linkages

The basic premise of this chapter is that the many components of the physical environment are interdependent, linked by couplings and feedbacks at various spatial and temporal scales (fig. 2.1). Under the pervading effect of gravity, the tectonic architecture of the continents both shapes climatic responses and is in turn reshaped by climate-induced erosion and sediment transfers. However, as high tectonic relief is lowered by erosion, so Earth's crust responds to unloading by rising isostatically, even as tectonic basins subside to accommodate eroded waste. Meanwhile, from habitats provided initially by tectonism and climate, plants and animals exert feedback effects on climate and relief through their impact on atmospheric composition, surface albedo, water budgets, mass wasting, and soil formation.

Simple notions of dependence, of cause and effect, have long been part of environmental science, for example in the assumed dependence of relief on rock type and of vegetation distributions on climate. In the earlier twentieth century, despite some awareness of isostasy, the prevailing model of landform development assumed structural uplift as its starting point and ignored it thereafter. Notions of interdependence emerged more slowly but, during the later twentieth century, the concept of plate tectonics and a re-awakened interest in climate change paved the way for improved understanding of Earth's surface as a shifting stage that both influences and responds to physical and biological processes.

Acceptance of plate tectonic concepts has had significant implications for the understanding of changes in climate and biota. Earth's climate is driven by solar radiation whose receipt is influenced by Earth's orbital relations with the Sun which change in cycles over timescales of 10^4–10^5 years. While such changes are affecting Earth's radiation environment, however, tectonic forces are also slowly changing the location and shape of continents and ocean basins, the dimensions of seaways so important for ocean heat transfers, and the nature and composition of the atmosphere. Such changes are more or less continuous but become more obvious at timescales exceeding 10^6 years. Thus climate change is intimately linked both to the shifting tectonic geometry of Earth's surface and to Earth's orbital relations with the Sun. In addition, biota are influenced directly by Earth's crustal mobility, for example by the provision of structural links between continents that favor exchanges of terrestrial biota. These influences on biota augment the indirect effects of tectonism expressed through changing climates and may complicate the biological roles of evolution, competition, extinction, and chance.

2.2 Tectonism and Climate

Driven by solar radiation, Earth's climates change through time in response to solar output, asteroid impacts, and Earth's variable relations with the Sun. These relations involve the eccentricity of Earth's orbit, the tilt of its rotational axis, and the wobble on that axis that change on cycles of ~95 ka, ~41 ka, and ~22 ka (thousands of years), respectively. Cyclic change is thus a basic property of Earth's climate systems, more or less predictable on timescales ranging from seasonal to 10^5 years.

But climates also respond to changes in Earth's surface features. Because a significant amount of solar radiation penetrates the atmosphere, how it is treated depends largely on the nature of this surface. This implies differences between land and sea, mountain and plain, bare soil and vegetation cover, and between surfaces that are covered by snow and ice and those that are not. It is in this context that tectonism exerts a profound influence, directly or indirectly, on climate. Tectonism and climate interact at three scales: the global scale related to the formation, location, and shape of continents and ocean basins; the continental scale related to major relief features such as the Andes; and regional scales involving individual mountain ranges and volcanic eruptions. The effect of tectonism through time is to change the playing field on which climate performs and to influence the composition and mobility of the atmosphere overlying that playing field. Thus the ostensibly independent effects of tectonic and solar forcing become closely interwoven.

2.2.1 Changing Mesozoic and Cenozoic Climates

South America's climatic patterns have developed over the past 200 Ma in response to four major tectonic events: (1) the breakup of Pangea; (2) the related opening of the Atlantic and Southern oceans; (3) formation of the Central American Isthmus; and (4) uplift of the Andes (see chapter 1 and table 1.2). These events, intimately linked with global climatic and eustatic changes, saw the greenhouse world of mid-Cretaceous (110–90 Ma [million years before present]) and late Paleocene-early Eocene times (61–49 Ma), when frost-intolerant palms lived in Tierra del Fuego (Romero, 1993), replaced by more variable climates. Late Eocene-early Oligocene cooling (37–29 Ma) then allowed Antarctic ice to expand; late Miocene (8–6 Ma) cooling and drying favored expansion of tropical and temperate grasslands and mammal radiation (Pascual et al., 1985); and warm-cold oscillations typified the Quaternary (2.5–0 Ma). Owing in part to orogenic forcing, Cenozoic oceans reached high levels in mid-Miocene time (15–12 Ma) when seas entered the Orinoco and Paraná basins (Ramos and Alonso, 1995), and may have invaded the persistent freshwater lakes and wetlands of the western Amazon lowlands, perhaps as late as 11–8 Ma through the Paraná Seaway from the south (Webb, 1995; Vonhof et al., 2003; Hovikoski et al., 2005).

The formation of Pangea caused South America to become embedded in a supercontinent, much of it far removed from deep oceans. Except along its western margin, the climate was profoundly continental. Most reconstructions place South America wholly within the Southern

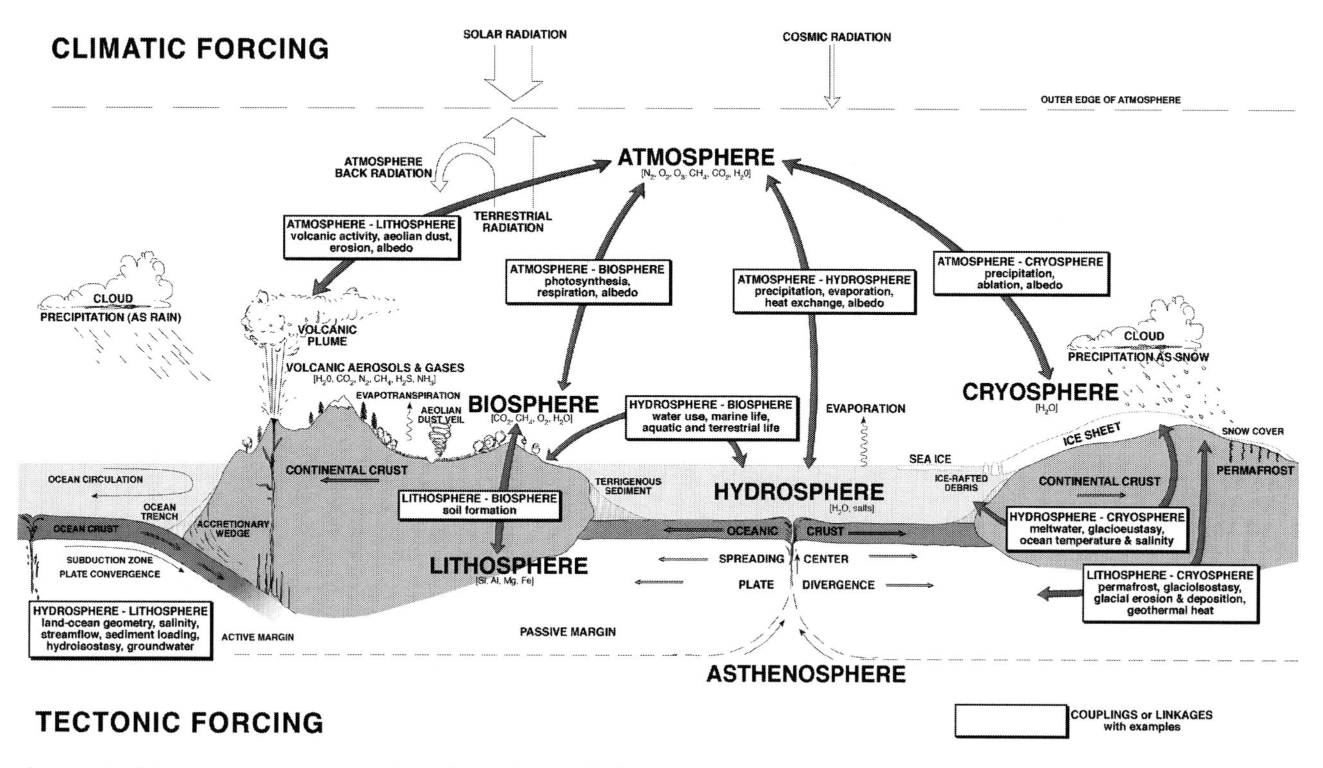

Figure 2.1 Linkages between Earth systems (from Orme, 2002)

Hemisphere during late Paleozoic time, from the Equator to just within the Antarctic Circle (fig. 1.3; Ziegler et al., 1997; Coffin et al., 2000). More southerly areas were thus well placed for the repeated continental glaciations found in the late Paleozoic stratigraphic record while continentality elsewhere was reflected in red beds and evaporites.

The subsequent rupture of Pangea began isolating South America as Atlantic rifting introduced ocean waters to the north and east coasts between 180 Ma and 120 Ma. Initially, narrow seas limited marine influences but, as the Tethys Ocean to the north and the Atlantic Ocean to the east widened, oceanicity strengthened and ocean-atmosphere coupling moved surface winds and ocean currents westward toward South America's ruptured margins. While the continent still lay south of the Equator, however, equatorial winds and currents must have passed mostly to the north. Then, as the continent has shifted northward by 10–12° of latitude over the past 120 Ma, the South Atlantic gyre has strengthened, tropicality linked to the Intertropical Convergence Zone has increased, summer monsoons have become common over the Amazon basin, while polar influences farther south have diminished.

In the south, even as the South Atlantic Ocean widened, South America remained tenuously linked to Antarctica for most of the Cretaceous (Hay et al., 1999) and beyond. This, together with Antarctica's bonds with Australia and the Kergeulen Plateau, favored meridional ocean flows and postponed the development of a circumpolar ocean. Fed by the Brazil Current, surface waters of the South Atlantic remained warm, as shown by oxygen isotope evidence from the Malvinas Plateau (Huber et al., 1995).

The subsequent separation of South America and Australia from Antarctica between 80 Ma and 40 Ma, and the widening of the Drake Passage from 30 Ma to 20 Ma, led to progressive opening of the Southern Ocean (Kennett, 1977; Cunningham et al., 1995; Lawver and Gahagan, 2003). Although the timing and rate of opening are still uncertain, the resulting increase in oceanicity around South America's southern margins strengthened westerly winds, reduced meridional flows and heat transport, and initiated the modern ocean circulation.

The Antarctic Circumpolar Current could now encircle the globe from west to east, uninterrupted by any land barrier or, after 20 Ma, by any major submarine sill. Today, while wind-driven surface flows in the 800-km wide Drake Passage (56°–62°S) rarely exceed 0.4 m s^{-1}, the Antarctic Circumpolar Current extends 4000 m down to the sea floor and yields a total mean discharge of around 130 × 10^6 m^3 s^{-1}, by far the largest of Earth's ocean currents. Also today, this current sends nutrient-rich flows northward off South America, allowing the Humboldt Current off the west coast to move cool water north to the Equator, and the cool Malvinas (Falkland) Current to invade the South Atlantic and reach northward across the Malvinas Plateau to the Río de la Plata and Cabo Frio. The Humboldt and Malvinas currents would certainly have been affected by the opening of the Drake Passage and the strengthening of ocean-atmosphere interactions directed eastward across the Southern Ocean. Although these impacts are as yet poorly understood, maritime polar air masses and cool windy conditions probably penetrated farther north into Patagonia.

Ocean-atmosphere interaction in the Southern Ocean has also been affected in later Cenozoic time by cold water gyres and dense brines in the Ross and Weddell seas, by ocean-atmosphere-cryosphere coupling reflected in the Antarctic Circumpolar Wave, and by pulses of Antarctic glaciation and related eustatic and sea-ice changes, which may have begun as early as 40 Ma and reached present conditions by 14 Ma (Hannah, 1994; Diester-Haass, 1996).

In contrast to the expanding Southern Ocean, the Cenozoic shrinkage of the Tethys Ocean and formation of the Central American Isthmus progressively restricted ocean-atmosphere interaction across northern South America. By late Miocene time, Central America existed as a complex island-arc archipelago. Three principal corridors linked Pacific and Caribbean waters: the San Carlos Strait across Nicaragua, the Panama Strait, and the deeper Atrato Strait across northwest Colombia (Coates and Obando, 1996; Coates et al., 2003). These straits varied in size from ~25 Ma to 3 Ma, as seen in marine sediment in Panama, where benthic foraminifera in the late Miocene Gatun Formation (~12–8 Ma) show strong Caribbean affinity, whereas those in the overlying Chagres Formation (8–5 Ma) have Pacific affinity. By 8 Ma, the waters between these islands were <150 m deep and by 4 Ma only 50 m deep (Coates, 1997). Even before closure, shallow waters and islands must have caused east-west ocean currents to falter (Mikolajewitz and Crowley, 1997).

With closure of the Central American Isthmus around 3 Ma, decoupling of the Atlantic and Pacific systems was finalized. The present circulation of the Atlantic Ocean was confirmed and reinforced meridional transfers, most notably a stronger Gulf Stream and enhanced production of North Atlantic Deep Water, may have been major triggers for high latitude cooling and ensuing Northern Hemisphere glaciation. Ocean cores indicate higher temperatures and increased salinity for Caribbean waters after 3 Ma (Haug et al., 2001). Conversely, deprived of a warm Atlantic influx, tropical Pacific coastal waters became less saline and may have seen equatorward extensions of the cool Humboldt and California currents. With Pacific Ocean circulation now barred to the east, the El Niño–La Niña cycle, dependent on thermally induced gravity forcing of equatorial ocean waters, may have intensified when the isthmus closed, while the Intertropical Convergence Zone over both oceans may have relocated southward (Hovan, 1995; Billups et al., 1999).

Lastly, Cenozoic uplift of the Andes interposed a massive north-south barrier to zonal atmospheric circulation across South America. Some 8,500 km long, this "climatic wall" is now the most continuous mountain barrier on Earth and influences weather systems and climate patterns across the continent and beyond. Within the Andes, paleosols and fossil floras suggest that Cenozoic uplift generated distinctive local climates for individual ranges and intermontane valleys. The rain-shadow effects of the rising Eastern Cordillera had helped impose an orographic desert on the Puna by 14 Ma (Vandervoort et al., 1995). Farther south, uplift of the Sierra Pampeanas to 4,000–5,500 m similarly restricted inflow of Atlantic-derived moisture, transforming the Santa María valley from a hot, seasonally wet foreland around 12 Ma to a semiarid intermontane basin by 3 Ma (Kleinert and Strecker, 2001). Since at least Miocene time, Andean uplift has imposed rain shadows westward on the Pacific coastlands from the Sechura Desert around 5–7°S in northern Perú through the Atacama Desert of northern Chile to as far as 30°S. Farther south, Andean uplift has formed rain shadows eastward on the Monte and Patagonia from 28° to 50°S and has seen the warm humid Miocene climate of Patagonia, reflected in coal swamp floras, replaced by colder drier conditions (Volkheimer, 1983).

The present hyper-aridity of the Atacama Desert reflects the combination of this rain shadow with the stabilizing effects of cool offshore waters and tropospheric subsidence, both of which further inhibit precipitation. These latter effects were linked to later Cenozoic cooling trends related to closure of warm oceanic seaways through the rising Andes, poleward shifts in the westerly wind belt, opening of the Drake Passage, growth of Antarctic ice, and to their collective impacts on global climate. The onset of hyper-aridity in the Atacama Desert is more controversial. It has been placed as early as ~25 Ma, based on the lack of erosion since then (Dunai et al., 2005) and the cessation of supergene enrichment of ore bodies after 14 Ma (Alpers and Brimhall, 1988), and as late as 3 Ma based on curtailed fluvial sediment inputs and increased evaporite precipitation in northern Chile's Central Depression (Hartley and Chong, 2002). Sediment starvation in the Perú-Chile Trench, induced by increasing aridity, may even have influenced the uplift of the High Andes (Lamb and Davis, 2003).

Meanwhile, climate patterns induced by tectonism were also responding to changing Earth-Sun relations. During quasi-periodic changes in the Pleistocene, colder, wetter phases combined with uplift to favor glaciation at higher elevations and in the far south, rainier conditions on the eastern slopes of the Andes farther north, and episodic expansions of lakes Titicaca and Poopo on the Altiplano (see chapter 4). Conversely, during drier phases, savanna-type climates characterized eastern Andean slopes, as well as large portions of the Amazon basin, while Altiplano lakes contracted to expose broad evaporite basins or *salars*. During the Holocene, glacier advances in the Patagonian Andes may correlate with millennial-scale changes in the path and intensity of the snow-bearing westerly winds (Douglass et al., 2005).

2.2.2 Present Climate Patterns

Three-quarters of South America now lie within tropical latitudes and within the influence of the Intertropical Convergence Zone and related circulation systems. With ocean-atmosphere coupling in the lower troposphere directed mostly westward, the bulk of this area is exposed to Atlantic influences whereas, owing to the marginal location of the Andes, only a narrow western coastal strip is directly influenced by the Pacific Ocean (see chapter 3). Tropical Atlantic influences are favored by onshore surface winds and poleward ocean heat transport along the continent's eastern margin (fig. 2.2). The upward tilting of ancient cratons along Brazil's southeast coast enhances seasonal tradewind rainfall against their seaward slopes, but hinders penetration farther inland. Farther north, there is no such barrier and, attracted by deep seasonal convection over Amazonia, moist surface winds off the equatorial Atlantic penetrate far inland toward the Andes, bringing much cloud and rain to the foothills of the Eastern Cordillera. During the summer monsoon, low-level northwesterly jets in turn deflect this moisture southward into the Gran Chaco or carry it aloft to irrigate the Altiplano, which in turn may recharge aquifers in the central Atacama Desert (Nogués-Paegle and Mo, 1997; Rech et al., 2002). Tectonic relief favors highly variable local climates within the Andes, including distinct altitudinal zonation on major ranges and volcanic piles. Along the mountain-girt west coast, tropical Pacific influences are more restricted, and aided by tropospheric subsidence, the cool Humboldt Current, and local upwelling of coastal waters, arid conditions prevail from near the Equator to beyond the Tropic of Capricorn. The limited moisture available to this arid coast is mostly squeezed from stratus and fog moving onshore over cool ocean waters (fig. 2.3; see chapter 10).

Despite plate motion to the north and west over the past 120 Ma, however, the continent still protrudes farthest into the Atlantic south of the equator. Cabo São Roque, at 5°S, forces much of the west-flowing South Equatorial Current along the north coast, as the Guyana Current, to feed the formidable North Atlantic gyre. Denied this flow, the Brazil Current and the South Atlantic gyre have long been weaker than the Gulf Stream and the North Atlantic gyre, with major implications for regional climates. A further change in coastal tectonic alignment near 23°S forces the Brazil Current offshore toward the Southern Convergence Zone, thereby allowing the cool waters of the Malvinas Current to penetrate weakly northward to the aptly named Cabo Frio.

The continent's temperate southern cone is also influenced by tectonic relief asymmetry, with a narrow windward belt of wet or foggy coastlands and mountains, the latter sufficiently cold to support glaciers, and a leeward semidesert in the Andean rain shadow of the Monte and Patagonia. Near the ragged southern tip, annual precipitation ranges from 2,600 mm on islands off the Pacific coast to 4,000 mm at sea level on the western mainland coast to around 10,000 mm atop the Andes, but declines rapidly eastward to 430 mm at Punta Arenas on the Strait of Magellan (Schneider et al., 2003). Owing to South America's southward taper, continental polar influences are muted but still sufficient to support glaciers near sea level in the far south and for cold winds to sweep northward across Patagonia. Subtle changes in Earth-Sun relations during Pleistocene time favored expanded glaciation in these southern mountains and foothills. However, unlike North America, which widens poleward, temperate South America has never been large enough in post-Pangean time to support continental ice sheets or the widespread permafrost, tundra, and boreal forest found in interglacial and postglacial North America.

2.3 Tectonism, Climate, and Geomorphology

Apart from tectonism's impact on rock distributions and morphotectonic provinces, discussed in chapter 1, interactions between tectonism and climate are also reflected in the tectonic forcing of volcanism and its impact on climate; the continent's asymmetric drainage patterns; the tectonic and climatic forcing of geomorphic processes, past and present; and complex relationships between denudation and isostasy.

2.3.1 Volcanism

Volcanism is a recurrent feature of South America's tectonic record, having peaked during the crustal extension that prefaced plate separation in Mesozoic time, notably in the Paraná flood basalts that heralded the opening of the South Atlantic, and again during orogenies involving plate convergence and subduction, most recently during the late Cenozoic phase of the Andean orogeny. The focus here is on recent volcanism in the Andes, where four distinct volcanic zones are linked to relatively steep subduction of oceanic crust beneath the continental plate (fig. 2.4). These are mainly stratovolcanoes composed of andesitic lavas and pyroclastic deposits which, with the Andes as their base, are high enough to sustain glaciers and thus the potential for devastating lahars (volcanic debris flows) when volcanic eruptions melt snow and ice.

The Northern Volcanic Zone extends NNE-SSW from 5°N to 3°S, where the Nazca plate is subducting at 30° beneath the continent and active faults and old terrane sutures provide vents for escaping magma, mostly basaltic andesite. Thus, in Ecuador, the Pujilí and Peltetec dex-

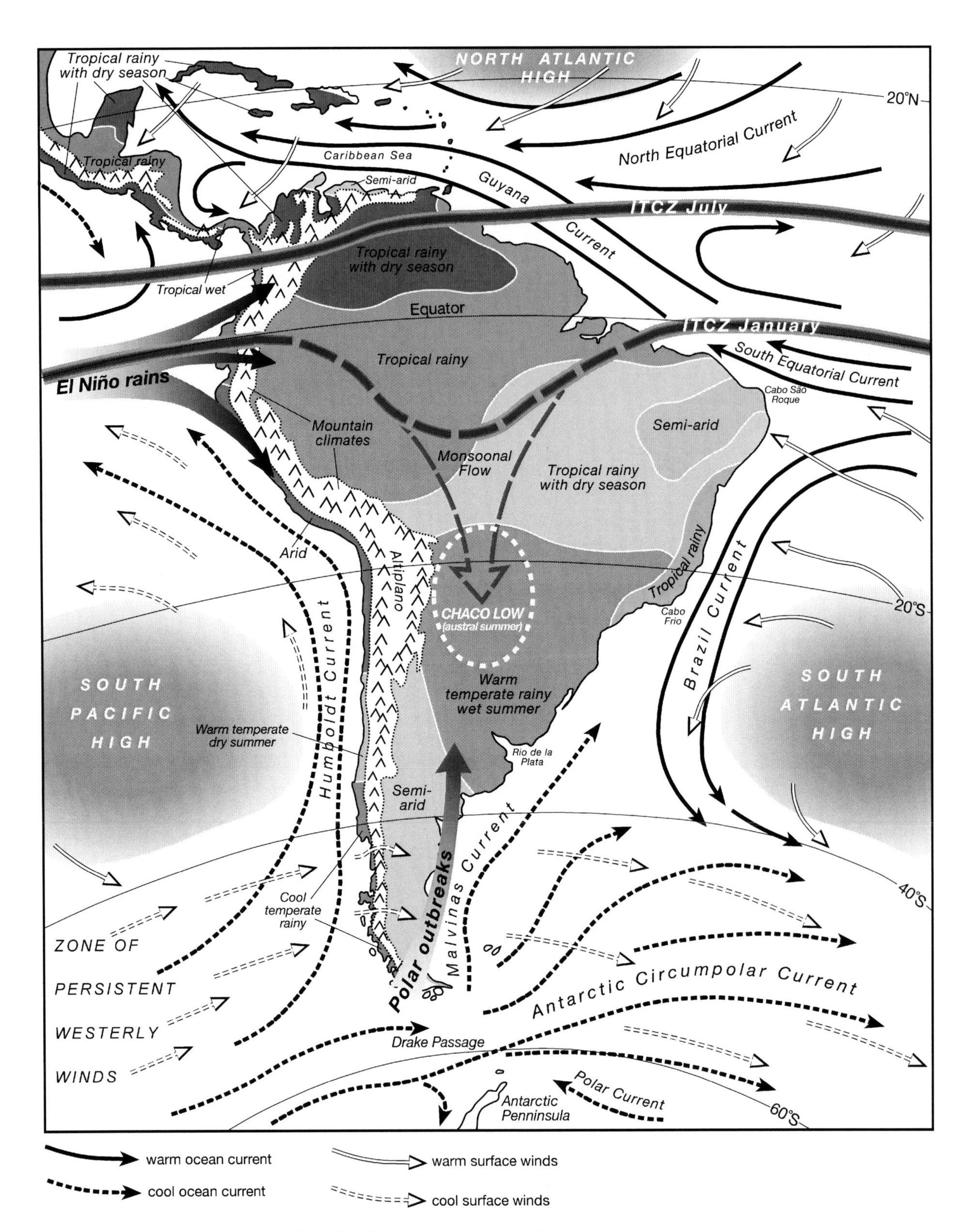

Figure 2.2 Climate regions and climatic influences in South America

Figure 2.3 Characteristic weather patterns over South America. This, the first image from NOAA's GOES satellite on 25 October 1975, shows (1) the ITCZ cloud belt extending across Colombia into the central Atlantic; (2) extensive stratus and fog off the Peruvian and Atacama deserts; (3) scattered clouds of the austral spring over Brazil; and (4) midlatitude cyclonic systems off the Pacific and Atlantic coasts (photo: GOES image, NOAA, 1975).

tral fault systems bounding the Interandean Graben are the loci of widespread volcanism, culminating in the great piles of dormant Chimborazo (6,310 m), and active Cotopaxi (5,911 m) and Sangay (5,230 m, fig. 2.5). This is also a zone where devastating lahars punctuate the Holocene landscape. Nevado del Ruiz (5321 m) in Colombia's Central Cordillera has seen many historic lahars, for example in 1595 when they killed 600 people, in 1845 when 1,000 people died, and in 1985 when pyroclastic flows mixed with melting snow and ice to trigger pulses of lahars that swept down Río Lagunillas to entomb 23,000 people in Armero (Pierson et al., 1990). This last tragedy, Earth's fourth deadliest volcanic event in historic time, could have been mitigated with more effective disaster preparedness. In 1994, earthquakes and heavy rains combined to mobilize lahars from volcanic ash on Nevado del Huila (5,700 m). Cotopaxi has seen recurrent lahars, notably around 4,700 B.P. (radiocarbon years before 1950) and historically in 1744 and 1768, and in 1877 when one lahar traveled more than 240 km and buried 1,000 people (USGS, 2005). Nevado Cayambe has also seen many lahars over the past 4,000 years (Samaniego et al., 1998).

On the Equator 1,000 km offshore, the Galapagos volcanic archipelago rises above a hot spot fed by a mantle plume that now lies near the spreading center between the Cocos and Nazca plates. Submarine basalts spewing from this hot spot have been involved with plate motions northwest of the continent over the past 95 Ma and probably formed the oceanic plateau that later became the Caribbean plate (Hoernle et al., 2002). The present islands, which emerged above sea level around 3 Ma, comprise mostly basaltic lavas and pyroclastic deposits. Because plate movement across this hot spot is eastward, active volcanism propagates westward. The western islands are thus more active than those farther east and include six impressive calderas on Isabela and the solitary active caldera of Fernandina.

The Central Volcanic Zone extends along the Western Cordillera from 16°S around the Arica Bend to 27°S. The most recent volcanic episode began here around 26 Ma and has continued episodically ever since, although few of the 600 volcanoes that rise above 5,000 m are now active. Nevertheless, majestic stratovolcanoes provide some of the region's highest peaks, notably Coropuna (6,425 m) and Ampato (6,310 m) in southern Perú, Sajama (6,522 m) in western Bolivia, Llullaillaco (6,750 m) at 25°S, and the highest active volcano on Earth—Ojos del Salado (6,887 m) at 27°S. The andesitic and dacitic lavas and ignimbrites (welded tuffs) of this zone owe their differentiation to lengthy ascent through 70 km of continental crust. The 100 km^3 of ignimbrites that erupted from the Cerro Galán caldera around 2.5 Ma are noteworthy. The largest explosive eruption in historic time in the Andes occurred in A.D. 1600 at the Huaynaputina volcano in southern Perú (Thouret et al., 1999). Some 13 km^3 of tephra, pyroclastic flows, and surge deposits devastated an area of 2,800 km^2

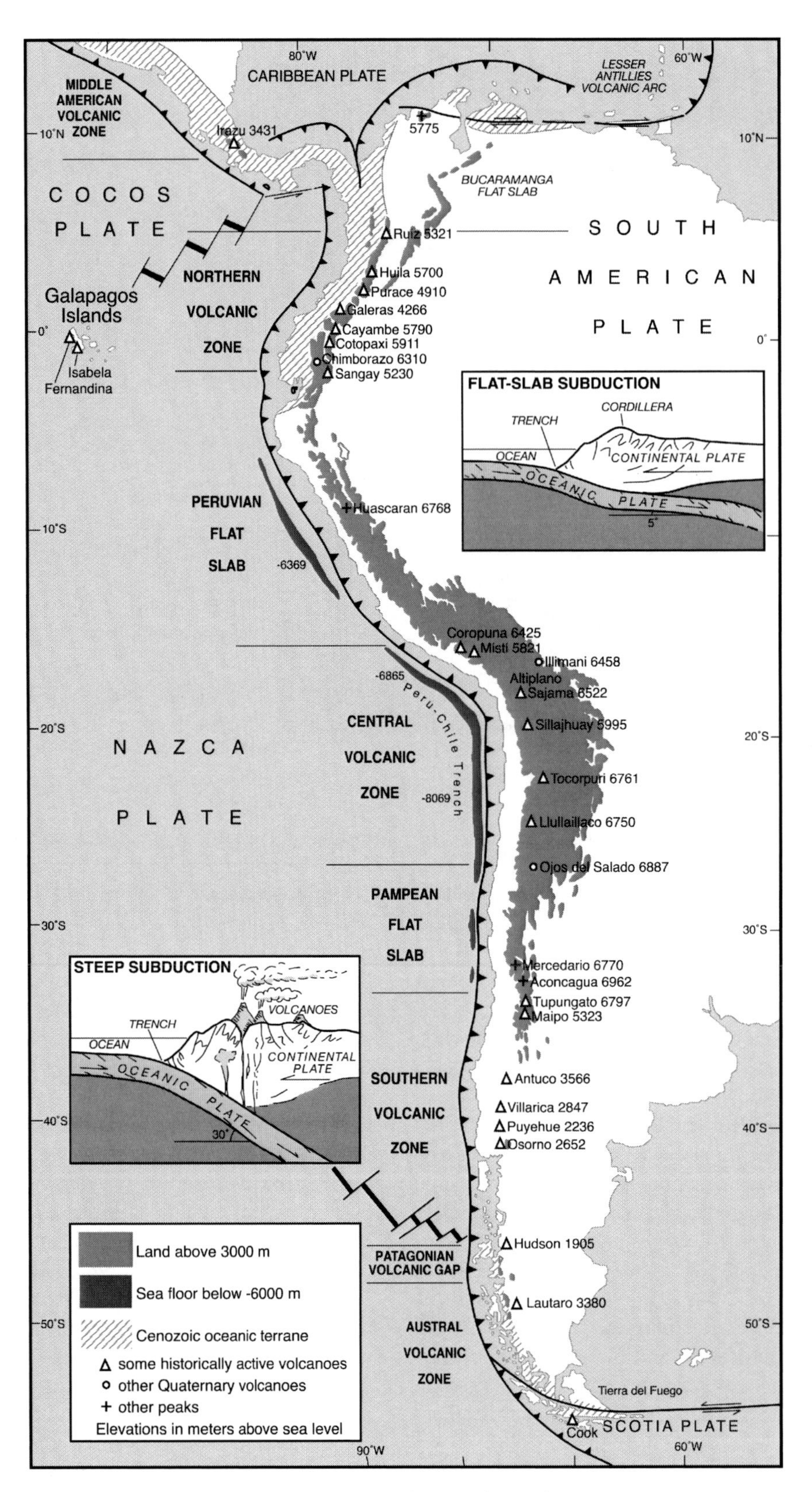

Figure 2.4 Volcanoes of the Andes and their tectonic relations

Figure 2.5 Sangay (5,230 m), a frequently active, dominantly andesitic stratovolcano in the Eastern Cordillera, Ecuador (photo: Fairchild Collection, UCLA, October 1939)

and killed 1,500 people. Such an eruption, if repeated today, could affect 750,000 people in and around the city of Arequipa.

The Southern Volcanic Zone, which extends 1,000 km from 33°S to 46°S, is linked to subduction of the Nazca plate at a steep angle of 30–40°. At these latitudes, repeated Quaternary volcanism has been intimately involved with glacial erosion and deposition, notably in the Tatara–San Pedro eruptive center (Singer et al., 1997). Volcanic ejecta are mostly andesite and dacite in the north, rhyolite and basalt farther south. The Pacific slope of the Main Cordillera contains several active stratovolcanoes, notably Tupungato (6,797 m), San José, Longuimay, and Hudson, the last of which shed huge volumes of airborne tephra across Patagonia in 1991.

By comparison, active volcanoes in the Austral Volcanic Zone south of 47°S are fewer and smaller. Their andesite cones reflect subduction of the Antarctic plate, while their lavas and pyroclastic flows disrupt drainage and glacier movement, dam lakes, and interfinger with glacial and marine sediment. Mount Cook in the Fuegan Andes is the southernmost active volcano in South America.

Volcanism plays a significant role in the present landscape, presenting both hazards and opportunities to human life and livelihood. During eruptive phases, volcanoes pose major threats to nearby life as a result of lava and pyroclastic flows, airfall tephra, and surges of solid, liquid, and gaseous matter. Lava flows and coarse airfall tephra tend to remain near their vents, although ignimbrite flows erupted from large calderas cover more than 500,000 km^2 of the Puna. Lahars generated when volcanic waste is mobilized by water, including glacier meltwater, may pose hazards far beyond their source areas. Over the longer term, volcanic dust veils and CO_2, SO_2, and other gasses generated by large eruptions may diffuse widely and linger long enough to reduce insolation and disrupt radiation transfers through the atmosphere, and thus impact regional climate. Although debate continues regarding the impact of volcanic emissions on global climate change, the high elevation of most active Andean volcanoes, many 6,000 m above sea level, means that their eruptive columns easily reach the stratosphere, where dust and gas may be transported far downwind. On the positive side, volcanism also provides South America with some of its most fertile soils and for this reason, despite the risks, the slopes of volcanoes have long been favored by farmers.

2.3.2 Drainage Asymmetry

Tectonic uplift of the Andes has imposed a remarkable asymmetry to South America's drainage patterns. Rivers draining the Pacific and Caribbean slopes are mostly short and steep; those draining to the Atlantic have, after steep initial descents from the Andes, lengthy runouts across broad sedimentary basins. This asymmetry is a response to the late Cenozoic completion of a continuous intercontinental divide along the Andes and thereby tectonic closure of earlier outlets to the Pacific. Although the evidence is fragmentary, and allowing for areas of internal drainage, it is likely that West Gondwana's major rivers probably flowed seaward toward a western ocean. Shallow marine and floodplain deposits of late Cretaceous and Paleocene age beneath the Altiplano, for example, indicate a provenance in Brazilian cratons, whereas overlying Eocene and Oligocene fluvial deposits derive from rising land to the west (Horton et al., 2001). Farther north, until the middle Miocene, the drainage of the northwest Amazon and upper Orinoco basins flowed northward to a delta beyond modern Lake Maracaibo (Hoorn et al., 1995). Then, around 12–10 Ma, uplift of Colombia's Eastern Cordillera closed this outlet, separated the Magdalena River from the Orinoco, and diverted the latter's drainage eastward. Uplift of the Vaupes arch between the Llanos and Oriente basins then separated the Orinoco and Amazon systems. Meanwhile, with final closure of pathways through the Ecuadorian Andes, the Amazon adopted its present route to the Atlantic. Epeirogenic flexuring of structural arches continues to influence drainage within the Orinoco and Amazon basins (see chapter 5). Drainage asymmetry also characterizes temperate South America, although Atlantic rivers become shorter as the continent tapers southward and Andean basins have been modified by glaciation.

Drainage asymmetry and variations in erosion intensity are also evident along and across the Andes. Between 2°S and 42°S, most of the Andes lie east of the intercontinental drainage divide, whereas farther north and south most of the cordillera lies west of the divide. This may reflect the dominance of precipitation-driven erosional processes in the humid north and south, while tectonic processes remain dominant in the drier central sector (Montgomery et al., 2001). Late Cenozoic alluvial fans along the Pacific slope of the Andes, though impressive, are modest compared with the vast fans that spill debris eastward into the Andean foreland.

2.3.3 Geomorphic Processes

In South America, as elsewhere, a general relationship exists between tectonism, climate, and surface processes. Tectonism provides the physical stage for erosion and deposition, while average climate conditions explain the particular mix of geomorphic processes normally found in different regions, for example in the persistent weathering and stream flows of the perennially humid tropics and the effect of winds in arid areas largely devoid of vegetation. Over millennial or longer timescales, however, tectonism and climatic forcing may change locales and emphasis, thereby clouding their relative roles in shaping the landscape. Nevertheless, despite massive Andean uplift to the west, terrigenous oxide minerals trapped on the Ceara Rise off the Atlantic coast appear to favor climatic forcing of erosion in the Amazon basin under relatively dry conditions over the past 13 Ma (Harris and Mix, 2002).

During historic time, however, it is easier to assign cause and effect to tectonism or climate because major events such as earthquakes and prolonged droughts may be documented. Even so, from the instrumental record of the past century, it is clear that departures from average conditions often do most of the work of erosion and deposition, and, because they are less predictable, cause most of the resulting human problems. Such variability includes synoptic-scale weather systems lasting a few days that generate storm seas, single floods, and strong winds; intraseasonal changes that, over several weeks, may change conditions from frequent rain to searing drought; and fluctuations measured over annual, decadal, and longer time scales. Of these, the episodic return of El Niño rains, on time scales from 3 to 8 years, may cause lasting changes to hillslopes, mountain valleys, and coastal floodplains (see chapters 3 and 19).

The ancient rocks of the Guiana and Brazilian shields have long been subdued by recurrent episodes of weathering and erosion, particularly by the persistently warm humid climates of Cretaceous and early Cenozoic time. A large volume of rock has been affected by these processes, generating deep weathering residues and plinthites rich in quartz, iron, manganese, and clay minerals such as kaolinite and gibbsite, but poor in many nutrients and trace elements such as boron, copper, sulfur, and zinc. With prolonged weathering, these iron-rich but nutrient-poor residues endow cratonic landscapes with resistant plinthite crusts, such as laterites cemented by iron and silica, that limit infiltration and erosion. The weathering residues often provide low grade but extensive and economically viable minerals—iron and manganese ores from laterites; bauxite ores from residues rich in gibbsite and other aluminum oxides but leached of iron, manganese, and silica. However, their naturally infertile soils (Oxisols, Latosols)

Figure 2.6 Geomorphic provinces of South America

favor scrub and open savanna rather than forest, and restrict agricultural opportunities (see chapters 7 and 9). With modest slopes and stream gradients compared to the Andes, the removal of these weathering products by mass movement and streams is relatively slow—although accelerating where forest clearance has occurred. Waste removal may also exhume relatively resistant crystalline rocks, such as the granitic Sugar Loaf near Rio de Janeiro.

In wetter areas of the Andes, heavy rains and snowmelt have long triggered rockfalls, landslides, and debris flows. The vast avalanches from Huascarán (6,768 m) in Perú probably owe their lengthy travel paths to reduced friction arising from rock waste mixed with melting ice (Plafker and Erickson, 1978). Intense rains related to El Niño events generate debris flows *(huaicos)* that cause considerable damage along the coasts of Ecuador and Perú (see chapter 19). In recent times, mass movement has been often exacerbated by mining and road construction. In March 1993, for example, 30×10^6 m^3 of rock waste slipped into the Río Paute at La Josefina in Ecuador, forming a temporary dam whose later breaching caused devastating floods and erosion downstream (Harden, 2001). This rockslide, caused by a combination of heavy rain and mining activity, killed 35 people and displaced a further 7,000. Landslides induced by heavy rains are now hazardous to many Andean cities where settlement has spread onto unstable hillsides (see chapter 20). Along the mountainous coast of Vargas state, Venezuela, heavy rains in December 1999 (1,100 mm from 2 to 16 December) led to massive flooding and debris flows on piedmont alluvial fans, killing some 19,000 persons in Caraballeda and other urbanized communities (fig. 2.7; Larsen et al., 2001). Similarly, heavy rains triggered devastating landslides along the Brazilian coast near Rio de Janeiro in 2002. Mass movement is also triggered by earthquakes and the collapse of volcanic cones. For example, a mid-Holocene earthquake may have caused the Alemania rockfall in northwest Argentina, which for several hundred years dammed the Río de las Conchas before dam breaching led to massive downstream flooding (Wayne, 1999).

Cenozoic uplift of the Andes has generated steep unstable slopes and massive transfers of rock waste downslope onto foreland wedges and into stream channels, aggravated in recent times by overgrazing and poor tillage practices (see chapter 18). Deeply incised river valleys, notably those descending from the Central Andes, reflect continuing adjustments between tectonic deformation and fluvial processes (Le Roux et al., 2000). Many valleys have become choked with fluvial debris shed from the mountains, more so during wetter Pleistocene intervals than today.

Just how much of this debris ultimately reaches the ocean depends on stream competence, and thus on hydroclimate and gradient, on stream capacity as a reflection of discharge, and on the trap efficiency of downstream basins. Steep Andean rivers draining onto Pacific coastal deserts often have insufficient discharge to deliver much sediment to the coast, depositing most of their load instead onto piedmont alluvial fans and *bajadas*. Nevertheless, flash floods may be unpredictably effective, especially during peak El Niño events. Conversely, tropical rivers draining toward the Atlantic, supplied by frequent precipitation, are more efficient conveyors of rock waste. However, long runout distances and low gradients cause much of the sediment delivered by Andean headwaters to be deposited and stored, at least temporarily, in foreland basins and floodplains farther downstream, for example along the Napo River in the Oriente Basin of Ecuador (fig. 2.8). Nevertheless, much sediment eventually reaches the Atlantic Ocean, where suspended clay and fine silt are entrained by density currents, coarse silt and sand move as littoral drift, and the continental shelf becomes plastered with a mix of sediment textures and types. Thus, the broad shelf off the mouths of the Amazon sees suspended mud

Figure 2.7 Caraballeda alluvial fan at the mouth of the Rio San Julián basin below the 2,000-m high Coastal Cordillera, Venezuela: (a) the highly urbanized fan before December 1999, looking north to the Caribbean Sea; (b) destruction of the urbanized area involving 1.8 million tons of fresh sediment delivered by floods and debris flows in December 1999, looking south toward the Coastal Cordillera (photos: USGS, Larsen et al., 2001).

and littoral sand moving northwest toward the Orinoco delta, while waste reaching the Río de la Plata is reworked along the coast to feed barrier beaches in southern Brazil (see chapter 15). The vast Amazon fan, a terrigenous pile up to 8000-m thick beyond the continental shelf, testifies to the long-term effectiveness of Cenozoic denudation (Flood and Piper, 1997). Sediment cores from this fan suggest an increased influx of terrigenous sediment after ~10 Ma for which either Andean uplift or climate change, or both, may be invoked (Curry et al., 1995).

Because South American rivers are discussed more fully in chapter 5, only a few salient points are introduced here. Rivers reaching the mouths of the Amazon, primarily the 6,400-km long main stem of the Amazon, drain around one-third of the continent and, despite some storage en route, are by far the largest suppliers of rock waste to the ocean, with a total sediment load of around 1200×10^6 t yr^{-1} and a sediment yield (load divided by basin area) of 195 t km^{-2} yr^{-1} (Milliman and Syvitski, 1992; see tables 5.1, 5.2). The 2,600-km long Orinoco River drains a smaller area (1.1×10^6 km^2) and carries less total sediment but, draining in part from the wet unstable Northern Andes, its sediment yield of 140 t km^{-2} yr^{-1} approaches that of the Amazon. Smaller Andean rivers generate much higher sediment yields—from Colombia's sizeable Magdalena basin (716 t km^{-2} yr^{-1} from 270,000 km^2) to the spectacular yields of many small rivers such as Venezuela's Tuy (1,800 t km^{-2} yr^{-1} from a 6,600 km^2 basin) and Maticora (2,200 t km^{-2} yr^{-1} from a 2,500 km^2 basin)(Milliman and Syvitski, 1992).

In contrast, rivers draining long denuded shields and resistant platform covers transport much less waste. Thus the 3,200-km long São Francisco River, draining South America's fourth largest but seasonally dry basin, carries relatively little sediment from the Brazilian Shield, while the 1,650-km long Uruguay River yields little more. The Paraguay-Paraná system is more complex because the 2,500-km long Paraguay River, whose headwaters drain

Figure 2.8 Confluence of Río Napo and Río Yasuní in the Amazon rain forest, Oriente Basin, eastern Ecuador, 400 km from the Napo's confluence with the Amazon and 3,400 km by river from the Atlantic Ocean (photo: Fairchild Collection, UCLA, October 1939)

the seasonally dry Mato Grosso, loses water and sediment downstream to the alluvial fans and wetlands of the Pantanal, while the 4,000-km long Paraná carries relatively little sediment from the Brazilian Shield but downstream receives waste from the Paraguay and Andean tributaries. The Iguaçu Falls discharge an impressive volume of water from the Paraná volcanic terrain of southern Brazil (fig. 2.9).

Wind is most effective as a geomorphic process where the vegetation cover is thin or absent, notably along the Pacific coastlands from the barchan fields of the Sechura Desert of northern Perú to the hyper-arid Atacama Desert of northern Chile. On the Paracas peninsula of central Perú, star dunes, yardangs, and corridors of lag gravels bear witness to the strength of prevailing southerly to southeasterly winds and northwesterly onshore breezes (McCauley et al., 1977). The streamlined yardangs, up to 2 km long between Pisco and Icá, have been sculpted mostly from erodible Miocene siltstones. On the windswept Altiplano, vast ignimbrite sheets have been similarly sculpted into yardangs by northwesterly winds and, aided by flowing water, into *quebradas* (gullies). Across the "arid diagonal," wind is similarly effective in the semi-arid rain shadow of the more southerly Andes (see chapters 10 and 14). Thus did Pleistocene rivers descending the eastern Andes introduce vast quantities of heterogeneous fluvial debris to the Monte and Patagonia, much of it since winnowed and reworked by Holocene winds into aeolian dunes and loess sheets, leaving a legacy of lag gravels, desert pavement, and discrete dune systems to the west but also providing for the fertile loess plains and sand seas of the Pampas to the east. In the latter area, paleosols interbedded with aeolian deposits testify to alternating humid and dry phases over the past 100,000 years. During winter, the Monte along the eastern foothills of the Andes in Argentina sees much aeolian activity related to the *zonda*, the strong downslope wind triggered by lee troughs that form as low pressure systems pass eastward across the cordillera (see chapters 3 and 10). In recent decades, wind erosion and dust storms have been aggravated by farming activities spreading into sensitive semi-arid regions such as the Chaco of northwest Argentina (see chapter 18). The semi-arid *caatinga* of northeast Brazil is also vulnerable to wind erosion.

Glacial processes are generally restricted by South America's predominantly tropical climates, but many Andean peaks and volcanoes above 5,000 m are sufficiently high to support modern perennial snowfields and glaciers near the Equator, notably on Chimborazo (6,272 m) and Cotopaxi (5,896 m) in Ecuador. The highest peaks of Perú, western Bolivia, and northern Chile also support glaciers amid perennial snowfields but, as precipitation lessens from east to west across the Central Andes, so modern snowlines rise from about 4,400 m on the eastern slopes of the Eastern Cordillera to near 6,000 m in the Western Cordillera (see chapter 4; Klein et al., 1999). From ~27°S in northern Chile, modern snowlines decline southward to below 1,000 m in Tierra del Fuego.

In the far south, where uplift and climate change spawned glaciation in the Patagonian Andes as early as late Miocene time, a major linear ice cap covered the Andes and adjacent foothills during the Pleistocene (Mercer, 1976). At its maximum around 1.2 Ma, this ice cap extended 3,000 km from Tierra del Fuego northward to 32°S, spilled westward as far as tidewaters on the narrow Chilean continental shelf would allow, and eastward onto the broad shallow shelf off the Strait of Magellan. Subsequent glacial denudation and tectonic subsidence may have re-

Figure 2.9 Iguaçu Falls occur as a tributary of the Río Paraná spills off the Jurassic basalt plateau of southern Brazil. Although only 70 m high, these falls are among the world's largest in terms of annual discharge (photo: Diane Blackburn, 2001).

duced the relief available for ice accumulation and may explain why Pliocene and early Pleistocene glacial deposits lie far beyond the limits of later Pleistocene glaciations (Clapperton, 1990). Even so, a large ice cap, 2,000-km long and 100–300-km wide, still existed during the Last Glacial Maximum around 20 ka, and the present spectacular relief of the Patagonian Andes owes much to glacial processes past and present. Similarly, much of the Patagonian landscape to the east, notably its widespread plateau gravels, reflects fluvial reworking of glacial and glaciofluvial debris derived from the Andes (see chapter 14). Today, fragmented Patagonian glaciers still form Earth's third largest ice field. Ice covers 30,000 km^2, mostly in two large masses, one extending 200 km southward from Hudson volcano and the other 300 km south past Lautaro volcano. A small ice field occurs in Tierra del Fuego.

2.3.4 Denudation and Isostasy

The rapid uplift and impressive denudation of the Andes pose interesting questions regarding South America's crustal response to massive transfers of rock waste from the mountains to the lowlands and oceans. These questions involve complex issues relating denudation to isostasy.

Denudation is the lowering of the landscape by the combined processes of mass wasting and erosion. It is conditioned partly by the resistance of the underlying rocks, reflecting tectonism and crustal processes, and partly by the efficacy of climate-induced geomorphic processes. Denudation is often measured indirectly by reference to the volume and density of the solid load (mechanical denudation) and dissolved load (chemical denudation) removed from a drainage basin by rivers. The resulting data are translated into a millennial lowering of the basin surface (mm ka^{-1}) or as annual volume lost per unit area (m^3 km^{-2} yr^{-1}). This approach is subject to errors of measurement and extrapolation, essentially because denudation is spatially variable and episodic rather than continuous. Even so, useful data often emerge, as shown in table 2.1. Clearly, denudation rates for the three South American basins lie below the high rates from Asian basins draining erodible loess (Huang He) and the rugged Himalayas (Ganges), but above the low rates recorded from arid basins (Nile, Rio Grande), glaciated cratonic oldlands (St. Lawrence), and forested lowlands (Congo, Yenisei). Given their Andean links, higher denudation rates might be anticipated for the Amazon and Orinoco basins but, apart from storage upstream, sediment supplied from the Andes becomes diluted in the denudation equation when joined by lower yields from large subbasins, such as those of the Negro and Xingu draining to the Amazon from the drier, more resistant Guiana and Brazilian shields, respectively. Nevertheless, the Andes are unraveling remarkably fast, as shown by the high sediment yields by Venezuelan rivers and for the Beni and Pilcomayo rivers in the foothills of the Bolivian Andes (see chapter 5). This raises the issue of isostatic response to denudation.

Relations between surface relief and Earth's lithosphere have long intrigued scientists. Thus, when geodesist Pierre Bouguer sought to define Earth's equatorial shape in Ecuador in the eighteenth century, he was puzzled by the lack of deflection of survey plumb lines predicted by the apparent mass of the Andes (Bouguer, 1749). His observation triggered much debate regarding the density of Earth's crust and its gravity anomalies. Then, from the 1830s, questions began to be raised about

Table 2.1 Estimated denudation rates for selected drainage basins in South America and elsewhere. The two groups of basins are ranked by total denudation.

River	Basin Area (10^6 km^2)	Total Denudation (mm ka^{-1})	Mechanical Denudation (mm ka^{-1})	Chemical Denudation (mm ka^{-1})	Mechanical Denudation of Total (%)
Orinoco	1.10	91	78	13	86
Amazon	6.15	70	57	13	81
Paraná	2.60	19	14	5	74
Huang He	0.77	529	518	11	98
Ganges	0.98	271	249	22	92
Colorado (USA)	0.63	84	78	6	93
Mississippi	3.27	44	35	9	80
Yukon	0.84	37	27	10	73
Mackenzie	1.81	30	20	10	67
Nile	2.96	15	13	2	87
St. Lawrence	1.03	13	1	12	8
Rio Grande	0.67	9	6	3	67
Yenisei	2.58	9	2	7	22
Congo	3.82	7	4	3	57

Sources: Meybeck (1976); Summerfield (1991); Milliman and Syvitski (1992).

the effects of denudation and sedimentation on the behavior of Earth's lithosphere, leading Dutton (1882) to coin the term *isostasy*. Isostasy describes the potential hydrostatic equilibrium that Earth's crust seeks in response to changes in mass at and below the surface. Just as wood floats in water, so Earth's relatively light and rigid outer crust usually floats in a denser, more mobile lower crust and upper mantle. Isostasy was soon invoked to explain regional crustal uplift following deglaciation (as in southern Patagonia) and crustal subsidence caused by massive sediment loading (as in the Amazon and Orinoco deltas), but the concept was long ignored at larger scales, until England and Molnar (1990) distinguished between surface uplift, which reflects tectonic forcing, and rock uplift, which combines tectonic forcing with isostatic rebound driven by denudation (Orme, 2007). Invoking evidence from survey data and crustal properties, Gregory-Wodzicki (2000) has suggested that the rapid uplift of Columbia's dissected Northern Andes, which rose at a remarkable rate of up to 3 mm yr^{-1} during the Pliocene, incorporates both tectonic forcing and isostatic rebound resulting from denudation. Isostatic adjustments of this magnitude imply that, while denudation may offset tectonic uplift, the Andes are likely to persist for a very long time. In contrast, rapid tectonic uplift has raised the Altiplano into an arid zone where, apart from wind action, dissection is limited and support for denudation-driven isostatic rebound is less strong.

2.4 Tectonism, Climate, and Biogeography

Because biomes are closely linked to climate, much that has been said about tectonism and climate also applies to biogeography. The tectonic forces that brought South America to its present location, shaped its initial relief, and influenced its climate, also affected the nature and distribution of its plants and animals. There are thus broad correlations among terrain, climate, and biomes. But there are also significant differences. Plants and animals exhibit preferences for particular habitats that may become inaccessible because of tectonic or climatic change. Further, through time, direct correlation of biota with climate may become clouded by evolution, opportunism, radiation, competition, selective survival, extinction, and delayed response, or simply by chance. Past and present relationships are further complicated by seed dispersal by winds and birds, and by floating rafts of vegetation and animal stowaways moving with ocean currents. And although land plants tend to be more responsive to environmental change, land animals, being mobile, are more adaptable. Some land animals also swim, at least for short distances between island archipelagoes. Thus tectonism may not generate a predictable sequence and distribution of biota in quite the same way that climate responds to tectonism.

2.4.1 Developing Mesozoic and Cenozoic Biogeographies

The four principal forcing factors identified in post-Pangean relations between tectonism and climate also influenced developing biogeographies although, for the reasons outlined above, the effect on plants and animals was less predictable (see table 1.2).

First, the breakup of Pangea initiated a prolonged episode of biogeographic fragmentation set against a backcloth of continuing evolution, diversification, and extinction. Initially, the narrow seas, continental fragments, and volcanic islands between diverging plates probably allowed episodic exchanges of terrestrial biota to continue for a time. Later, however, the progressive separation of the northern continents from Gondwana favored evolutionary divergence as many elements in the flora and fauna were rafted away from their common heritage.

By late Mesozoic time, distinct dinosaur faunas had evolved on each landmass, with titanosaurs and ceratosaurs in Gondwana while tyrannosaurs, hadrosaurs, and ceratopsids dominated Laurasia (Sereno, 1991). One titanosaur, *Argentinosaurus huinculensis*, in Patagonia was 40 m long and weighed 90 tonnes. When dinosaurs met their end at the close of the Mesozoic, probably from the environmental effects of massive asteroid impact (Alvarez et al., 1980), perhaps for other reasons such as global warming initiated by prolonged volcanic emissions of dust and greenhouse gasses (Officer and Page, 1996), South America was well isolated from North America and mainland Africa. In contrast, similarities between certain dinosaurs, crocodiles, gondwanathere mammals, and insects from late Cretaceous deposits in South America, India, and Madagascar suggest that these areas remained linked via Antarctica and the Kerguelen Plateau to the close of the Cretaceous (Krause et al., 1997, 1999), perhaps later (Brundin, 1988). By early Cenozoic time, however, South America's land mammals were developing qualities distinct from those elsewhere.

Among the unusual South American land mammals that evolved during the later Cenozoic were giant anteaters (Myrmecophagidae), almost modern forms of which appeared around 15 Ma; tree sloths (Bradypodidae) and ground sloths (Megalonychidae, Mylondontidae, Megatheriidae); armadillos (Dasypodidae, Chlamytheriidae) and glyptodonts (Glyptodontidae); opossums (Didelphidae); porcupines (Erethizontidae); and rhinoceros-like ungulates (Toxidontidae). A tusked gomphothere, *Amahuacatherium peruvium*, found below 9 Ma tephra in southwest Amazonia, poses intriguing questions regarding the arrival or survival of gomphotheres in South America (Campbell et al., 2001).

Sites of Miocene age at La Venta (Colombia) and Urumaco (Venezuela) also contain noteworthy fossils. The latter, in the vast Caribbean delta of the proto-Orinoco, contains the world's largest fossil rodent (the 700 kg *Phoberomys pattersoni*), which was prey for the largest crocodiles (*Purussaurus* spp.) (Sanchez-Villagra et al., 2003). There were also flightless, predatory birds of the Phorusrhachidae family in Brazil and Argentina, one 4–m tall descendant of which, *Titanis walleri*, migrated to Florida. Several genera of crocodiles, gavialids, and turtles evolved from Mesozoic stock.

Second, the Cenozoic opening of the Southern Ocean severed land links between South America, Antarctica, Australia, and New Zealand. That these land masses once shared a common biotic heritage is suggested by Cretaceous monotremes in South America and Eocene marsupials in Antarctica, mammals now most strongly associated with Australia (Woodburne and Case, 1996). Former links between these southern continents are also suggested by floristic similarities, for example between *Nothofagus* trees (beech family), whose antecedents lived in Antarctica between 80 and 30 Ma, and whose successors still occupy parts of southern South America, Australia, and New Zealand (see chapter 13). Nevertheless, the loss of these land links favored the exchange of marine organisms across the Southern Ocean. Upwelling in the divergence zone between the east-flowing Antarctic Circumpolar Current and the west-flowing Polar Current farther south now yields one the world's biologically most productive regions. High primary productivity supports the large populations of zooplankton and larger organisms, ranging from krill to seals, whales, and penguins, that characterize coastal seas off South America.

Third, no tectonic forcing of biogeography has been more dramatic than the closure of the Central American Isthmus and, with it, initiation of the Great American Biotic Interchange—the mass migration of plants and animals between North and South America and the ensuing, at least temporary, enrichment of both continents (Stehli and Webb, 1985). Although final closure of the isthmus came around 3 Ma, biotic exchange between these continents began earlier, perhaps much earlier, depending on how the Caribbean plate and the Aves Ridge are reconstructed (Hoernle et al., 2002). More certainly, fossil evidence shows that by 8 Ma swimming mammals, such as a now extinct North American raccoon (Procyonidae) and two kinds of South American ground sloth, were island-hopping across warm shallow seas between the continents (Webb, 1997). After 3 Ma, biotic exchange suddenly became commonplace. North American land mammals, waiting in the Central American highlands, flooded southward onto the South American scene, where they began displacing the long resident fauna. Diffusion between 3 Ma and 2 Ma was rapid. Late Pliocene and early Pleistocene deposits in seacliffs southeast of Buenos Aires and in the Patagonian steppes contain abundant remains of North American mammals—llamas, horses, tapirs, bears, cats, peccaries, deer, and mice—whose early Pliocene records are confined to North America. Most of these immigrants were grazers or mixed-feeding herbivores suited to grassy savannas and open woodlands. After 2 Ma, migration seems to have slowed and some late Eurasian immigrants to North America, such as mammoth and bison, never reached South America. This was perhaps because the grassy savannas began developing into dense tropical rainforest, especially after 0.8 Ma (Colinvaux, 1997). However, several families of carnivores—bears, raccoons, otters, dogs, and cats, including the saber-tooth cat (*Smilodon*)—did cross, with devastating impact on native South American animals who had no previous experience of efficient mammalian carnivores (Webb, 1997).

In contrast, the northward migration of South American land mammals, accustomed to mostly tropical habitats, was less successful, and many that crossed between islands or over land bridges, such as the now submerged Aves Ridge off Venezuela, did not survive (Webb, 1997; Orme, 2002). Toxidonts died out; ground sloths diffused north to Alaska but did not survive the Pleistocene; anteaters are now found no farther north than Guatemala, while only one armadillo species survives farther north.

Conversely, closure of the isthmus prevented the exchange of marine animals between the tropical Atlantic and Pacific Oceans. Marine invertebrates of the Caribbean and eastern Pacific are similar until around 5 Ma but then diverge rapidly as the isthmus shallowed and closed (Jackson and d'Croz, 1997). A remarkable pulse of origination, evolution, and extinction in marine mollusks and reef-building corals occurred between 3 Ma and 2 Ma, probably due to the changing ecology and opportunities presented by closure of the isthmus. Thereafter, shallow-water branching corals in the genus *Acropora* came to dominate Caribbean reefs, while finger corals of the genus *Pocillopora* typify less-developed Pacific reefs across the isthmus. In general, Pacific waters have rich pelagic fisheries while Caribbean waters are dominated by carbonate-associated benthic foraminifera (Coates et al., 2003).

Fourth, uplift of the Andes had a dramatic impact on biotic distributions by imposing physical barriers and elevational constraints on plant distributions and animal migration, and by generating greater habitat diversity and resulting biotic specialization (see chapters 12 and 13). For example, fossil evidence suggests that Andean uplift fragmented what was once a rich earlier Cenozoic subtropical flora that straddled the continent from southern Brazil through Bolivia and Argentina to Chile (Hinojosa and Villagrán, 1997). This helps to explain why the distinctive semi-arid vegetation of central Chile exhibits close affinities with similar formations east of the Andes, and why the relict Olivilla forest of the Chilean coast, now so dependent on sea fog, may be descended from subtropical rainforests that survived the summer drought brought on by Neogene uplift of the Andes (see chapter 11). Furthermore, pollen and macrofossil evidence from Columbia's Eastern Cordillera indicate that Miocene floras were mostly lowland types (van der Hammen et al., 1973; Wijninga, 1996). Miocene floras from the Altiplano formerly lived 1500–2000 m lower than where they are now found (Gregory-Wodzicki et al., 1998). Such evidence illustrates the disruptive ecological impacts of tectonism.

Andean uplift has also been invoked to explain the divergent lineages of many birds and iguanas in western Amazonia (Blumfield and Capparella, 1996; Schulte et al., 2000). Fish in the Magdalena basin of Colombia once shared common drainages with fish in the Guianas and Amazon basin (see chapter 7). In addition, uplift combined with Earth's changing orbital relations to produce episodic changes in biome patterns. Thus the eastern slopes of the tropical Andes may have seen the rainforest of wetter, warmer stages, such as today, alternate with open woodland and savanna during drier, cooler stages. Following closure of the Central American Isthmus, North American animals may have used these more open corridors to spread rapidly southward (Marshall and Sempere, 1993).

In the Amazon lowlands farther east, Neogene flexuring has been invoked to explain diversification among small mammals and amphibians on opposing sides of low structural arches straddling the basin (Lougheed et al., 1999; Lara and Patton, 2000). Superimposed on these slow tectonic flexures, however, are the faster environmental changes triggered by late Cenozoic climate changes, most notably those involving the changing patterns of tropical rain forest and savanna in the Amazon Basin. A variety of geological and biological evidence suggests that during drier, cooler intervals, such as the Last Glacial Maximum (~20 ka), the savannas expanded at the expense of the tropical rain forest (e.g., Haffer, 1969; van der Hammen, 1974; Whitmore and Prance, 1992; Behling and Hooghiemstra, 2001; see chapters 8 and 9). The forest retreated into discrete refugia, preserved its diversity, and then reemerged to supplant the savannas during wetter, warmer intervals, such as the present Holocene. Further, the complex interplay of late Cenozoic tectonic, hydroclimatic, and sedimentary processes in Amazonia may have stressed the biota, resulting in faster speciation and thus greater modern biodiversity (Rossetti et al., 2005). Understandably, the nature and details of these interesting scenarios are contentious issues (e.g., Colinvaux et al., 2000; Kastner and Goñi, 2003).

Finally, the Cocos Ridge between the 3 Ma-old Galapagos Islands and Costa Rica may have been an emergent volcanic archipelago for some time after 15 Ma (Werner et al., 1999), thus increasing the time over which evolution of Galapagos biota could occur. The presence of a long emergent ridge would help solve the riddle of speciation and divergence of Galapagos marine and land iguanas that so puzzled Charles Darwin.

2.4.2 Present Biogeographic Patterns

The present distribution of biomes reflects the broad interplay of fluctuating climatic factors and biotic processes on a slowly changing tectonic stage. Most components of the present vegetation of the forest, savanna, steppe, and desert biomes are readily identified with specific ranges of thermal and moisture conditions, modified by local relief and edaphic constraints. Nevertheless, there are anomalies, components of the vegetation that challenge simple interpretation. In central Chile, for example, fragments of the fog-dependent Olivilla forest may have survived from an earlier range disrupted by Andean uplift, while the *Nothofagus* forest here seems to have survived from colder, wetter stages of the Pleistocene (see chapters 11 and 13). The *caatinga* woodlands of northeast Brazil may have been isolated by climate changes combining Earth's orbital cycles with regional crustal uplift (see chapter 9). A further imponderable is the extent to which certain modern plants may have evolved from lineages in the deep past, for example from organisms that flourished in the Pangean landscapes of the Brazilian and Guiana Shields (see chapter 6).

Over much of the continent, primeval vegetation patterns and composition have been changed directly or indirectly by a wide range of human impacts involving forest clearance, livestock grazing, changing fire frequencies, drainage and irrigation works, demise of native herbivores such as the guanaco and the spread of exotic deer, goats, wild boar, and rabbits, and competition between native plants and exotic trees, shrubs, crops, weeds, and above all annual grasses (see chapters 13, 14, 16, and 17).

The modern fauna of South America has evolved in part from isolated Gondwana roots and in part from North American immigrants. Diffusion and speciation of the latter have been so effective over the past 3 Ma that around half of the South American animals now have North American origins. The emergence of the Panama isthmus changed South America from a continent with no deer, cats, or mice to one where they are now common, and where cats large and small became major predators. The large cats include jaguar (*Panthera onca*), puma (*Puma concolor*), and a rare Andean cat (*Felis jacobita*). Nevertheless, ecological changes during and since the Pleistocene have taken their toll on both immigrant and native stock (see chapter 8, table 8.1). Horses and saber-tooth cats became extinct throughout the Americas, while in South America jaguars retreated to the shelter of the forest, and bears were reduced to one species, *Tremarctos*, the spectacled bear. Among the native stock, ground sloths did not survive the Pleistocene, while tree sloths retreated to forest habitats.

The effect on animals of prehistoric human predation has been the subject of much debate. Many Pleistocene genera, such as the saber-tooth cat, became extinct before people arrived, while the latter's impact on other animals was probably highly selective. Thus, some large game animals, such as the gomphothere (mastodon), *Cuvieronius tropicus*, may have been hunted to extinction by early immigrants in the more open habitats of the savannas and Andean foothills, but there is no evidence for animal extinctions by pre-European peoples in Amazonia (see chapter 16). Many herbivores of the deer and llama families were also hunted to extinction, but related genera survived and, in the absence of larger competitors, became the dominant animals of the savannas and grasslands (see chapter 14). Some rheas, large flightless birds, also lingered on, notably in the Argentinean grasslands and steppes. Even so, all those that survived suffered increasing competition for food from introduced herbivores such as cattle and sheep, some as early as the sixteenth century, but others in Patagonia only as late as the twentieth century. The net result today is a depleted yet still distinct Neotropical fauna, embracing both ancient and immigrant elements, that extends across South America and has penetrated as far north as southern Mexico.

2.5 Conclusion

The fundamental character of the South American landscape is driven by tectonic and climatic forcing. Tectonism has given the continent its location and general architecture. Climate has generated the contexts whereby this crude tectonic framework has been refashioned by surface processes. Abetted by evolution, migration, and extinction, biological responses to these physical scenarios have in turn clothed this landscape with vegetation and colonized it with animals and, ultimately, with people.

These forcing factors operate over vastly different timescales and rates of change. Tectonic changes are normally very slow and, except for earthquakes and volcanic eruptions, become more apparent over periods exceeding 10^6 years. For example, the South American plate is presently moving westward at barely perceptible rates of 25 to 35 mm yr^{-1}, but over geologic timescales this is a respectable 25 to 35 km Ma^{-1}. Similarly extrapolated, net convergence of South America with the eastward-moving Nazca plate is occurring at rates of between 75 and 120 km Ma^{-1}.

In contrast, major climate changes related to Earth's orbital relations with the Sun operate at intermediate timescales of 10^4–10^5 years. These are expressed, for example, in asymmetric rates of change—from full interglacial to full glacial conditions over 100,000 years, but returning more quickly to full interglacial conditions in less than 20,000 years. When two or more orbital cycles coincide, feedback within the ocean-atmosphere system and resulting climate changes may occur over millennial or even centennial timescales. Further, as the synthesis of weather, climate also contains short-term departures from mean conditions that function at decadal, annual, and seasonal timescales—the timescales best understood by, and most likely to affect people during their lifetimes. These pulses are expressed in exceptional rainfalls, high winds, and desiccating droughts, which in turn impact the landscape through landslides, floods, storm damage, and desertification. The coincidence of tectonic instability and climatic variability is particularly troubling to Andean landscapes and the people that inhabit them.

While plants and animals may well respond within these longer timescales, for example in changes from forest to savanna and back over an interglacial-glacial cycle, many biological changes are more rapid, while those involving extinction are irreversible. Rates of change are now fastest among plant and animal communities because, regardless of natural forcing factors, their habitats are most vulnerable to human interference. The impacts generated by prehistoric peoples were mostly subtle and reversible, although in the tropical and subtropical Andes centuries of lengthy human occupance, vegetation change, and domestication of selected species were more enduring. Nevertheless, over the past 500 years, these early impacts have been augmented and accelerated by those of colonizers from Europe and elsewhere whose cultural ethos, practical needs, and technical abilities are now firmly imprinted across much of the South American landscape. The native plants and animals of the savannas and grasslands have been diluted or displaced by introduced species; native grasslands have been converted to cropland; steppes have been irrigated or afforested; the vast forests of Amazonia, once thought almost impenetrable beyond their waterways, are being cleared; and, despite some resilience, many landscapes may be lost forever. Only in recent decades have the feedback effects of these changes, for example on Earth's carbon cycle, begun to be measured and understood.

To a greater or lesser extent everywhere, those environmental changes wrought by nature on a slowly shifting tectonic stage beneath a pulsing climatic canopy are now being augmented, often swiftly, by the direct and indirect effects of human activity. These processes generate complex responses, understanding of which must be incorporated into the proper management of the South American landscape.

References

Alpers, C., and G. Brimhall, 1988. Middle Miocene climatic change in the Atacama Desert, northern Chile: Evidence from supergene mineralization at La Escondida. *Geological Society of America Bulletin*, 100, 1640–1656.

Alvarez, L.W., W. Alvarez, F. Asaro, and H.V. Michel, 1980. Extraterrestrial cause for the Cretaceous-Tertiary extinction. *Science*, 208, 1095–1108.

Behling, H., and H. Hooghiemstra, 2001. Neotropical savanna environments in space and time: Late Quaternary interhemispheric comparisons. In: V. Markgraf (Editor), *Interhemispheric Climate Linkages*. Academic Press, London, 307–323.

Billups, K., A.C. Ravelo, J.C. Zachos, and R.D. Norris, 1999. Link between oceanic heat transport, thermohaline circulation, and the Intertropical Convergence Zone in the early Pliocene Atlantic. *Geology*, 27, 319–322.

Blumfield, R.T., and A.P. Capparella, 1996. Historical diversification of birds in northwestern South America: A molecular perspective on the role of vicariant events. *Evolution*, 50, 1607–1624.

Bouguer, P., 1749. *La Figure de la Terre*. Quay des Augustins, Paris.

Brundin, L., 1988. Phylogenetic biogeography. In: A.A. Myers and P.S. Gillers (Editors), *Analytic Biogeography: An Integrated Approach to Plant and Animal Distributions*. Chapman Hall, London.

Campbell, K.E., M. Heizler, C.D. Frailey, L. Romero-Pitman, and D.R. Prothero, 2001. Upper Cenozoic chronostratigraphy of the southwestern Amazon basin. *Geology*, 29, 595–598.

Clapperton, C., 1990. Quaternary glaciations in the Southern Hemisphere: An overview. *Quaternary Science Reviews*, 9, 299–305.

Coates, A.G., 1997. The forging of Central America. In: A.G. Coates (Editor). *Central America: A Natural and Cultural History*. Yale University Press, New Haven, 1–37.

Coates, A.G., and J.A. Obando, 1996. The geological evolution of the Central American isthmus. In: J.B.C. Jackson et al. (Editors), *Evolution and Environment in Tropical America*. University of Chicago Press, Chicago.

Coates, A.G., M-P. Aubry, W.A. Berggren, L.S. Collins, and M. Kunk, 2003. Early Neogene history of the Central American arc from Bocas del Toro, western Panama. *Geological Society of America Bulletin*, 115, 271–287.

Coffin, M.F., L.A. Lawver, L.M. Gahagan, and D.A. Campbell, 2000. *The Plates Project 2000; Atlas of Plate Reconstructions (750 Ma to Present Day)*. Plates Project Progress Report, Institute for Geophysics, University of Texas, Austin.

Colinvaux, P., 1997. The history of forests on the isthmus from the ice age to the present. In: A.G. Coates (Editor), *Central America: A Natural and Cultural History*. Yale University Press, New Haven, 123–136.

Colinvaux, P., P.E. de Oliveira, and M.B. Bush, 2000. Amazonian and tropical plant communities on glacial time-scales: The failure of the aridity and refuge hypotheses. *Quaternary Science Reviews*, 19, 141–169.

Cunningham, W.D., I.W.D. Dalziel, T. Lee, and L.A. Lawver, 1995. Southernmost South America-Antarctic Peninsula relative plate motions since 84 Ma: Implications for the tectonic evolution of the Scotia Arc region. *Journal of Geophysical Research*, 100 (B5), 8257–8266.

Curry, W.B., N.J. Shackleton, C. Richter, and others, 1995. Leg 154 synthesis. In: *Proceedings of the Ocean Drilling Program, Initial Results, Leg 154*. Ocean Drilling Program, 154, College Station, Texas, 421–442.

Diester-Haass, L., 1996. The Eocene-Oligocene preglacial-glacial transition in the Atlantic sector of the Southern Ocean (ODP Site 690). *Marine Geology*, 131, 123–149.

Douglass, D.C., B.S. Singer, M.R. Kaplan, R.P. Ackert, D.M. Mickleson, and M.W. Caffee, 2005. Evidence of early Holocene glacial advances in southern South America from cosmogenic surface-exposure dating. *Geology*, 33, 237–240.

Dunai, T.J., G.A. González Lopez, and J. Juez-Larré, 2005. Oligocene-Miocene age of aridity in the Atacama Desert revealed by exposure dating of erosion-sensitive landforms. *Geology*, 33, 321–324.

Dutton, C.E., 1882. Review of *Physics of the Earth's Crust*, by the Rev. Osmond Fisher. *American Journal of Science*, 3rd Series, 23 (136), 283–290.

England, P., and P. Molnar, 1990. Surface uplift, uplift of rocks, and exhumation of rocks. *Geology*, 18, 1173–1177.

Flood, R.D., and D.J.W. Piper, 1997. Amazon fan sedimentation: The relationship to equatorial climate change, continental denudation, and sea-level fluctuations. In: R.D. Flood, D.J.W. Piper, A. Klaus, and L.C. Peterson (Editors). *Proceedings of the Ocean Drilling Program, Scientific Results*, 155, College Station, Texas, 653–675.

Gregory-Wodzicki, K.M., 2000. Uplift history of the Central and Northern Andes: A review. *Geological Society of America Bulletin*, 112, 1091–1105.

Gregory-Wodzicki, K.M., W.C. McIntosh, and K. Velazquez, 1998. Paleoclimate and paleoelevation of the late Miocene Jakokkota flora, Bolivian Altiplano. *Journal of South American Earth Sciences*, 11, 533–560.

Haffer, J., 1969. Speciation in Amazon forest birds. *Science*, 165, 131–137.

Hannah, M.J., 1994. Eocene dinoflagellates from CIROS-1 Drill Hole, McMurdo Sound, Antarctica. *Terra Antarctica*, 1, 371.

Harden, C., 2001. Sediment movement and catastrophic events: The 1993 rockslide at La Josefina, Ecuador. *Physical Geography*, 22, 305–320.

Harris, S.E., and A.C. Mix, (2002). Climate and tectonic influences on continental erosion of tropical South America, 0–13 Ma. *Geology*, 30, 447–450.

Hartley, A.J., and G. Chong, 2002. Late Pliocene age for the Atacama Desert: Implications of the desertification of western South America. *Geology*, 30, 43–46.

Haug, G.H., R. Tiedmann, R. Zahn, and A.C. Ravelo, 2001. Role of Panama uplift on oceanic freshwater balance. *Geology*, 29, 207–210.

Hay, W.W., R.M. DeConto, C.N. Wold, K.M. Wilson, S. Voigt, M. Schulz, A. Rossby Wold, W.C. Dullo, A.B. Ronov, A.N. Balukhovsky, and E. Söding, 1999. An alternative global Cretaceous palaeogeography. In: E Barrera and C. Johnson (Editors), *Evolution of the Cretaceous Ocean/Climate System*, Geological Society of America Special Paper 332, 1–47.

Hinojosa, L.F., and C. Villagrán, 1997. Historia de los bosques del sur de Sudamérica: I. Antecedentes paleobotanicos, geológicos y climáticos del Terciario del cono sur de América. *Revista Chilena de Historia Natural*, 70, 225–239.

Hoernle, K., P. van den Bogaard, R. Werner, B. Lissinna, F. Hauff, G. Alvarado, and D. Garbe-Schönberg, 2002. Missing history (16–71 Ma) of the Galapagos hotspot: Implications for the tectonic and biological evolution of the Americas. *Geology*, 30, 795–798.

Hoorn, C., J. Guerrero, G.A. Sarmiento, and M.A. Lorente, 1995. Andean tectonics as a cause for changing drain-

age patterns in Miocene northern South America. *Geology*, 23, 237–240.
Horton, B.K., B.A. Hampton, and G.L. Waanders, 2001. Paleogene synorogenic sedimentation in the Altiplano plateau and implications for initial mountain building in the central Andes. *Geological Society of America Bulletin*, 113, 1387–1400.
Hovan, S., 1995. Late Cenozoic atmospheric circulation intensity and climate history recorded in eolian deposits in the eastern equatorial Pacific, Leg 138. In: *Proceedings of the Ocean Drilling Program, 138*, College Station, Texas, 615–625.
Hovikoski, J, M. Räsänen, M. Gringas, M. Roddaz, S. Brusset, W. Hermoza, L.R. Pittman, and K. Lertola, 2005. Miocene semidiurnal tidal rhythmites in Madre de Dios, Peru. *Geology*, 33, 177–180.
Huber, B.T., D.A. Hodell, and C.P. Hamilton, 1995. Middle-late Cretaceous climate of southern high latitudes: Stable isotopic evidence for minimal equator-to-pole thermal gradients. *Geological Society of America Bulletin*, 107, 1164–1191.
Jackson, J.B.C., and L. D'Croz, 1997. The ocean divided. In: A.G. Coates (Editor), *Central America: A Natural and Cultural History*. Yale University Press, New Haven, 38–71.
Kastner, T.P., and M.A Goñi, 2003. Constancy in the vegetation of the Amazon Basin during the late Pleistocene: Evidence from the organic matter composition of deep sea fan sediments. *Geology*, 31, 291–294.
Kennett, J.P., 1977. Cenozoic evolution of Antarctic glaciation, the circum-Antarctic ocean, and their impact on global palaeoceanography. *Journal of Geophysical Research*, 82 (27), 3843–3860.
Klein, A.G., G.O. Seltzer, and B.L. Isacks, 1999. Modern and Late Glacial Maximum snowlines in the central Andes of Peru, Bolivia, and northern Chile. *Quaternary Science Reviews*, 18, 63–84.
Kleinert, K., and M.R. Strecker, 2001. Climate change in response to orographic barrier uplift: Paleosol and stable isotope evidence from the late Neogene Santa María basin, northwestern Argentina. *Geological Society of America Bulletin*, 113, 728–742.
Krause, D.W., G.V.R. Prasad, W. von Koenigswald, A. Sahni, and F.E. Grine, 1997. Cosmopolitanism among late Cretaceous mammals. *Nature*, 390, 504–507.
Krause, G.W., R.R. Rogers, C.A. Forsetr, J.H. Hartman, G.A. Buckley, and S.D. Sampson, 1999. The late Cretaceous vertebrate fauna of Madagascar: Implications for Gondwanan paleobiogeography. *GSA Today*, 9, 1–7.
Lamb, S., and P. Davis, 2003. Cenozoic climate change as a possible cause for the rise of the Andes. *Nature*, 425, 792–797.
Lara, M.C., and J.L. Patton, 2000. Evolutionary diversification of spiny rats (genus Trinomys, Rodentia: Echimyidae) in the Atlantic Forest of Brazil. *Zoological Journal of the Linnaean Society*, 130, 661–686.
Larsen, M.C., G.F. Wieczorek, L.S. Eaton, B.A. Morgan, and H. Torres-Sierra, 2001. Natural hazards on alluvial fans: The Venezuelan debris flow and flash flood disaster [1999]. *U.S. Geological Survey Fact Sheet*, FSS-103–01, 4.
Lawver, L.A., and L.M. Gahagan, 2003. Evolution of Cenozoic seaways in the circum-Antarctic region. *Palaeogeography, Palaeoclimatology, and Palaeoecology*, 198, 11–37.
Le Roux, J.P., C. Tavares, and F. Alayza, 2000. Sedimentology of the Rimac-Chillon alluvial fan at Lima, Peru, as related to Plio-Pleistocene sea-level changes, glacial cycles, and tectonics. *Journal of South American Earth Sciences*, 13, 499–510.
Lougheed, S.C., C. Gascon, D.A. Jones, J.P. Bogart, and P.T. Boag, 1999. Ridges and rivers: A test of competing hypotheses of Amazonian diversification using the dart-poison frog *(Epipedobates femoralis)*. *Proceedings of the Royal Society*, London, 266, 1829–1835.
Marshall, L.G., and T. Sempere, 1993. Evolution of the Neotropical Cenozoic land mammal fauna in its geochronologic, stratigraphic, and tectonic context. In: P. Goldblatt (Editor), *Biological Relationships between Africa and South America*. Yale University Press, New Haven, 329–392.
McCauley, J.F., M.J. Grolier, and C.S. Breed, 1977. Yardangs. In: D.O. Doehring (Editor), *Geomorphology in Arid Regions*, Proceedings of the 8th Annual Geomorphology Symposium, Binghamton, New York, 233–269.
Mercer, J.H., 1976. Glacial history of southernmost South America. *Quaternary Research*, 6, 125–166.
Meybeck, M., 1976. Total annual dissolved transport by world major rivers. *Hydrological Science Bulletin*, 21, 265–289.
Mikolajewitz, U., and T.J. Crowley, 1997. Response of a coupled ocean/energy balance model to restricted flow through the Central American isthmus. *Paleoceanography*, 12, 429–441.
Milliman, J.D., and J.P.M. Syvitski, 1992. Geomorphic/tectonic control of sediment discharge to the ocean: The importance of small mountain rivers. *Journal of Geology*, 100, 525–544.
Montgomery, D.R., G. Balco, and S.D. Willett, 2001. Climate, tectonics, and the morphology of the Andes. *Geology*, 29, 579–582.
Nogués-Paegle, J., and K.C. Mo, 1997. Alternating wet and dry conditions over South America during summer. *Monthly Weather Review*, 125, 279–291.
Officer, C.B., and J. Page, 1996. *The Great Dinosaur Extinction Controversy*. Addison-Wesley-Longman, Reading, 209 pp.
Orme, A.R., 2002. Tectonism, climate, and landscape. In: A.R. Orme (Editor), *The Physical Geography of North America*, Oxford University Press, New York, 3–35.
Orme, A.R., 2007. Clarence Edward Dutton (1841–1912): Soldier, polymath, and aesthete. In: P. Wyse Jackson (Editor), *Geological Travelers*, International Union for the History of the Geological Sciences, Geological Society, London, Special Publications, in press.
Pascual, R., M.G. Vucetich, G.J. Scillato-Yané, and M. Bond, 1985. Main pathways and mammalian diversification in South America. In: F.G. Stehli and S.D. Webb (Editors), *The Great American Biotic Interchange*. Plenum Press, New York, 219–247.
Pierson, T.C., R.J. Janda, J-C. Thouret, and C.A. Borrero, 1990. Perturbation and melting of snow and ice by the 13 November 1985 eruption of Nevado del Ruiz, Colombia, and subsequent mobilization, flow, and deposition of lahars. *Journal of Volcanology and Geothermal Research*, 41, 17–66.
Plafker, G., and G.E. Erickson, 1978. Nevado Huascarán avalanches, Peru. *Developments in Geotechnical Engineering*, 14A, 277–314.
Ramos, V.A., and R.N. Alonso, 1995. El Mar Paranense en la provincia de Jujuy. *Revista Geológica de Jujuy*, 10, 73–80.

Rech, J.A., J. Quade, and J.L. Betancourt, 2002. Late Quaternary paleohydrology of the central Atacama Desert (lat. 22°–24°S), Chile. *Geological Society of America Bulletin*, 114, 334–348.

Romero, E.J., 1993. South American paleofloras. In: P. Goldblatt (Editor), *Biological Relationships between Africa and South America*. Yale University Press, New Haven, 62–85.

Rossetti, D.F., P.M. Toledo, and A.M. Góes, 2005. New geological framework for western Amazonia (Brazil) and implications for biogeography and evolution. *Quaternary Research*, 63, 78–89.

Samaniego, P., M. Monzier, C. Robin, and M.L. Hall, 1998. Late Holocene eruptive activity at Nevado Cayambe volcano, Ecuador. *Bulletin of Volcanology*, 59, 451–459.

Sanchez-Villagra, M.R., O. Aguilera, and I. Horovitz, 2003. The anatomy of the world's largest extinct rodent. *Science*, 30 (5640), 1708–1710.

Schneider, C., M. Glaser, R. Kilian, A. Santana, N. Butorovic, and G. Casassa, 2003. Weather observations across the Southern Andes at 53°S. *Physical Geography*, 24, 97–119.

Schulte, J.A., J.R. Macey, R.E. Espinoza, and A. Larson, 2000. Phylogenetic relationships in the iguanid lizard genus *Liolaemus*: Multiple origins of viviparous reproduction and evidence for recurring Andean vicariance and dispersal. *Biological Journal of the Linnaean Society*, 69, 75–102.

Sereno, P.C., 1991. Ruling reptiles and wandering continents: A global look at dinosaur evolution. *GSA Today*, 1 (7), 141–145.

Singer, B.S., R.A. Thompson, M.A. Dungan, T.C. Feeley, S.T. Nelson, J.C. Pickens, L.L. Brown, A.W. Wulff, J.P. Davidson, and J. Metzger, 1997. Volcanism and erosion during the past 930 k.y. at the Tatara-San Pedro complex, Chilean Andes. *Geological Society of America Bulletin*, 109, 127–142.

Stehli, F.G., and S.D. Webb (Editors), 1985. *The Great American Biotic Interchange*. Plenum Press, New York.

Summerfield, M.A., 1991. *Global Geomorphology*. Longman, London.

Thouret, J-C., J. Davila, and J-P. Eissen, 1999. Largest explosive eruption in historical times in the Andes at Huaynaputina volcano, A.D. 1600, southern Perú. *Geology*, 27, 435–438.

USGS, 2005. USGS/Cascades Volcanic Observatory: http://vulcan.wr.usgs.gov/volcanoes/.

van der Hammen, T., 1974. The Pleistocene changes in vegetation and climate in tropical South America. *Journal of Biogeography*, 1, 3–26.

van der Hammen, T., J.H. Werner, and H. van Dommelen, 1973. Palynological record of the upheaval of the Northern Andes: A study of the Pliocene and lower Quaternary of the Colombian Eastern Cordillera and the early evolution of its High-Andean biota. *Review of Palaeobotany and Palynology*, 16, 1–122.

Vandervoort, D.S., T.E. Jordan, P.K. Zeitler, and R.N. Alonso, 1995. Chronology of internal drainage development and uplift, southern Puna plateau, Argentine central Andes. *Geology*, 23, 145–148.

Volkheimer, W., 1983. Geology of extra-andean Patagonia. In: N. West (Editor), *Deserts and Semideserts of Patagonia*, Elsevier, Amsterdam, 425–429.

Vonhof, H.B., F.P. Wesselingh, R.J.G. Kaandorp, G.R. Davies, J.E. van Hinte, J. Guerrero, M. Räsänen, L. Romero-Pittman, and A. Ranzi, 2003. Paleogeography of Miocene western Amazonia: Isotopic composition of molluscan shells constrains the influence of marine incursions. *Geological Society of America Bulletin*, 115, 983–993.

Wayne, W.J., 1999. The Alemania rockfall dam: A record of a mid-Holocene earthquake and catastrophic flood in northwestern Argentina. *Geomorphology*, 27, 295–306.

Webb, S.D., 1995. Biological implications of the middle Miocene Amazon seaway. *Science*, 269, 361–362.

Webb, S.D., 1997. The great American faunal interchange. In: A.G. Coates (Editor), *Central America: A Natural and Cultural History*. Yale University Press, New Haven, 97–122.

Werner, R., K. Hoernle, P. van den Bogaard, C. Ranero, R. von Huene, and D. Korich, 1999. Drowned 14–m.y.-old Galapagos archipelago off the coast of Costa Rica: Implications for tectonic and evolutionary models. *Geology*, 27, 499–502.

Whitmore, T.C., and G.T. Prance (Editors), 1992. *Biogeography and Quaternary History in Tropical America*. Clarendon Press, Oxford.

Wijninga, V.M., 1996. *Palaeobotany and Palynology of Neogene Sediments from the High Plains of Bogotá (Colombia)*. Ponsen and Looingen, Wageningen, Netherlands.

Woodburne, M.O., and J.A. Case, 1996. Dispersal, vicariance, and the late Cretaceous to early Tertiary land mammal biogeography from South America to Australia. *Journal of Mammalian Evolution*, 3, 121–161.

Ziegler, A.M., M.L. Hulver, and D.B. Rowley, 1997. Permian world topography and climate. In: I.P. Martini (Editor), *Late Glacial and Postglacial Environmental Changes: Quaternary, Carboniferous-Permian, and Proterozoic*. Oxford University Press, New York, 111–146.

3

Atmospheric Circulation and Climatic Variability

René D. Garreaud
Patricio Aceituno

Regional variations in South America's weather and climate reflect the atmospheric circulation over the continent and adjacent oceans, involving mean climatic conditions and regular cycles, as well as their variability on timescales ranging from less than a few months to longer than a year. Rather than surveying mean climatic conditions and variability over different parts of South America, as provided by Schwerdtfeger and Landsberg (1976) and Hobbs et al. (1998), this chapter presents a physical understanding of the atmospheric phenomena and precipitation patterns that explain the continent's weather and climate.

These atmospheric phenomena are strongly affected by the topographic features and vegetation patterns over the continent, as well as by the slowly varying boundary conditions provided by the adjacent oceans. The diverse patterns of weather, climate, and climatic variability over South America, including tropical, subtropical, and midlatitude features, arise from the long meridional span of the continent, from north of the equator south to 55°S. The Andes cordillera, running continuously along the west coast of the continent, reaches elevations in excess of 4 km from the equator to about 40°S and, therefore, represents a formidable obstacle for tropospheric flow. As shown later, the Andes not only acts as a "climatic wall" with dry conditions to the west and moist conditions to the east in the subtropics (the pattern is reversed in midlatitudes), but it also fosters tropical-extratropical interactions, especially along its eastern side. The Brazilian plateau also tends to block the low-level circulation over subtropical South America. Another important feature is the large area of continental landmass at low latitudes (10°N–20°S), conducive to the development of intense convective activity that supports the world's largest rain forest in the Amazon basin. The El Niño–Southern Oscillation phenomenon, rooted in the ocean-atmosphere system of the tropical Pacific, has a direct strong influence over most of tropical and subtropical South America. Similarly, sea surface temperature anomalies over the Atlantic Ocean have a profound impact on the climate and weather along the eastern coast of the continent.

3.1 Mean Climatic Features and Regular Cycles

In this section we describe the long-term annual and monthly mean fields of several meteorological variables. These climatological fields are obtained by averaging many daily fields, each of them constructed on the basis of surface, upper-air, and satellite meteorological observations. The atmospheric circulation (winds, pressure, etc.) is characterized by using the NCEP-NCAR reanalysis from 1979 to 1995 (Kalnay et al., 1996). The precipitation fields are a blend of station data (Legates and Willmott, 1990) and precipitation estimates from the Microwave Sounding Unit (MSU) covering the period 1979 to 1992 (Spencer, 1993). While the averaging procedure does not have any physical a priori significance, the regional climate is defined by the relevant features in the mean fields, which in turn are forced by the fixed (or very slowly varying) boundary conditions of the atmosphere: land-sea distribution, continental topography, and the time/space changes of the solar radiation reaching the surface. Figure 3.1 shows the annual mean precipitation, sea-level pressure, and low-level winds [1,000–850 hecto-Pascal (hPa) average]. The maximum precipitation occurs over the tropical oceans along a band at approximately 8°N that coincides with a belt of low pressure and low-level wind convergence. This band, the so-called Intertropical Convergence Zone (ITCZ), is a major feature of the global circulation, and its year-round position to the north of the equator is ultimately related to the land-sea distribution and orientation of the coastlines (Mitchell and Wallace, 1992). The rainfall in the ITCZ decreases slightly as it straddles northern South America, in part due to the decrease in surface evaporation, but still produces the highest continental precipitation over the equatorial Andes, the western Amazon basin, and near the mouth of the Amazon River. The rainfall in this part of the continent is produced by deep, moist convection—the very energetic ascent of buoyant air from near the surface to the tropopause that sustains the largest rainforest of the world (see chapter 9).

Two other bands of high annual mean rainfall are evident in figure 3.1. The western band has its origin in the western equatorial Pacific (central Pacific during El Niño years), reaching the continent between its tip and 40°S. Precipitation in the South Pacific Convergence Zone (SPCZ; Vincent, 1998) is largely produced by extratropical frontal systems. The annual mean rainfall is high in southern Chile due to the enhanced uplift over the western slopes of the Andes (Lenters and Cook, 1995), but it decreases sharply to the east, producing rather dry conditions in Argentina's Patagonia (see also chapters 13 and 14). The eastern band has its root over the central part of the continent, and extends southeastward forming the South Atlantic Convergence Zone (SACZ; Kodoma, 1992; Figueroa et al., 1995). Rainfall over the central part of the continent is largely produced by deep convection, but as one moves into the subtropics, Southern Hemisphere (SH) frontal systems become more important in promoting deep convection and eventually produce most of the precipitation (Lenters and Cook, 1995). For instance, the uniformly large amounts of rainfall observed year round on the coast of southern Brazil and Uruguay are produced by deep con-

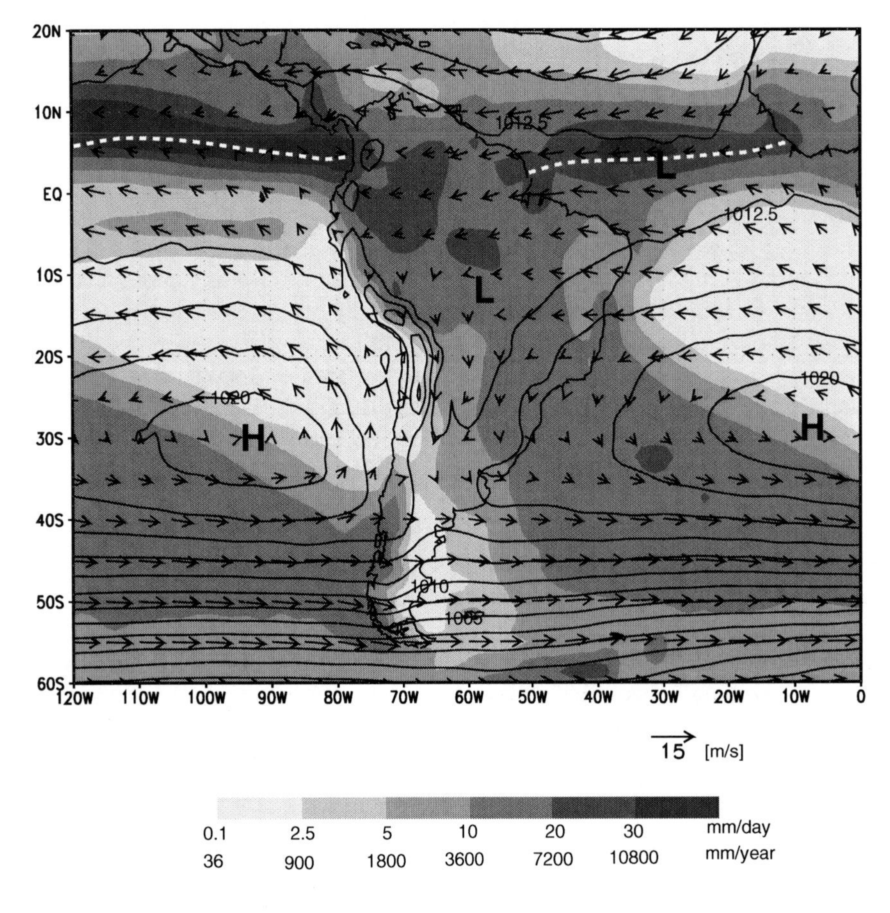

Figure 3.1 Annual mean rainfall (shading scale at bottom), sea level pressure (contoured every 2.5 hPa), and low-level winds (arrow scale at bottom). Dashed white line indicates mean position of the ITCZ. Letters H and L indicate approximate center of the subtropical anticyclones and continental low, respectively.

vection during the warm season (November to March) and by frontal systems during the cold season (Montecinos et al., 2000).

Rainfall is nearly absent over broad areas of the subtropical oceans owing to large-scale midtropospheric subsidence (fig. 3.1). The rate of subsidence is mainly dictated by the radiative cooling of the air parcels that have reached the upper troposphere over the ITCZ (Rodwell and Hoskins, 1996). The subtropical subsidence together with the ascent over the ITCZ, the low-level trade winds, and the upper-level poleward flow form the Hadley cell, a major feature of the atmosphere's general circulation. The subsiding air also maintains the very persistent cells of surface high pressure and low-level anticyclonic circulation over the southeast Pacific and Atlantic oceans, with only minor seasonal variations. Over the continent, deep convection (either moist or dry) offsets the large-scale subsidence, so that mean upward motion prevails over the central part of South America, enhancing the subsidence over the southeast Pacific Ocean (Gandu and Silva Dias, 1998).

Consequently, sea-level pressure over the continent is lower than the corresponding value over the adjacent oceans throughout the year, thus forcing the trade winds over the tropical Atlantic to penetrate into the continent in a nearly east-west direction, until they become convergent near the Andes cordillera. On the western side of the tropical-subtropical Andes (which act as an effective barrier to the flow), the equatorward low-level winds promote coastal upwelling of cold waters, maintaining a coastal tongue of low sea surface temperature (SST) that extends westward at the equator, where it is enhanced by equatorial upwelling. The adiabatically warmed air aloft and the cold SST result in a cool, moist marine boundary layer of 500–1,000 m thick capped by a strong temperature inversion over the subtropical southeast Pacific (e.g. Garreaud et al., 2002). A very extensive deck of shallow, nonprecipitating stratocumulus clouds is typically observed at the top of the marine boundary layer (e.g., Klein and Hartmann, 1993). This cloud layer plays an important role in the regional and global climate by substantially reducing the amount of solar radiation reaching the ocean surface (the so-called albedo effect; e.g., Hartmann et al., 1992) and by cooling the lower troposphere due to the strong upward emission of infrared radiation at the top of the clouds (Nigam, 1997).

Two conspicuous dry regions occur over tropical-subtropical South America (fig. 3.1; see also chapter 11). The first region encompasses the western side of the continent, a 100–300 km strip of land between the coastline and the Andes cordillera from ~30°S as far north as 5°S. The coastal desert of northern Chile and Perú is primarily explained by the strong, large-scale subsidence over the subtropical southeast Pacific Ocean, but its extreme aridity (places with no precipitation for several years) seems related to regional factors (Abreu and Bannon, 1993; Rutllant et al., 2000). At these latitudes, the central Andes (including the Altiplano) become a truly climatic border between the extremely arid conditions to the west and the wet conditions to the east (e.g., Garreaud 2000a; see also chapters 10 and 12). The second region corresponds to the eastern tip of the continent, in northeast Brazil, where the annual mean precipitation is less than half of the inland values at the same latitude and the rainy season is restricted to March to May, when the ITCZ reaches its southernmost position (e.g., Kousky and Ferreira, 1981; Hastenrath, 1982). The dryness of this region appears to result from a local intensification of the Hadley cell in connection with the strong convection over the equatorial Atlantic (Moura and Shukla, 1981; Mitchell and Wallace, 1992).

The area affected by deep convection experiences significant changes during the year, leading to a pronounced mean annual cycle in rainfall over tropical and subtropical South America, as shown by seasonal maps of mean precipitation (fig. 3.2). Such changes are controlled by the annual north-south march of insolation, resulting in changes in land surface temperatures, and also by complex interactions with changes in low-level moisture transport (Fu et al., 1999). During the austral winter (June–July–August, JJA), the heaviest precipitation and thunderstorms are found over northern South America and southern Central America, connected with the oceanic ITCZ (fig. 3.2a). At this time of year, the central part of the continent (including southern Amazonia) experiences its dry season, interrupted occasionally by the passage of modified cold fronts from the southern midlatitudes. By the end of October, there is a rapid shift of the area of intense convection from the northern extreme of the continent into the central Amazon basin (e.g., Horel et al., 1989), marking the onset of the so-called South American summer monsoon (fig. 3.2b, Zhou and Lau, 1998; Vera et al., 2006).

The area of convective precipitation reaches its southernmost position during the austral summer (December–January–February, DJF), encompassing the southern Amazon basin, the Altiplano, and the subtropical plains of the continent (southern Bolivia and Brazil, Uruguay, Paraguay, and central Argentina), and extending over the South Atlantic Convergence Zone (fig. 3.2c). In this season, a low pressure cell forms over the extremely hot and dry Chaco region (Seluchi and Marengo, 2000), forcing the southward flow of the trade winds and their subsequent convergence over the subtropical plains. Observational and modeling evidence has shown that the northerly flow between the Andes and the Brazilian highlands is often organized in a low-level jet with its core (wind speed often in excess of 12 m/s) at about 1 km above the ground and less than 100 km from the eastern slopes of the Andes (e.g. Saulo et al., 2000). The poleward moisture transport by this low-level jet feeds the convective rainfall over the subtropical plains, a major agricultural region and heavily populated area (Berbery and Collini, 2000; Saulo et al., 2000).

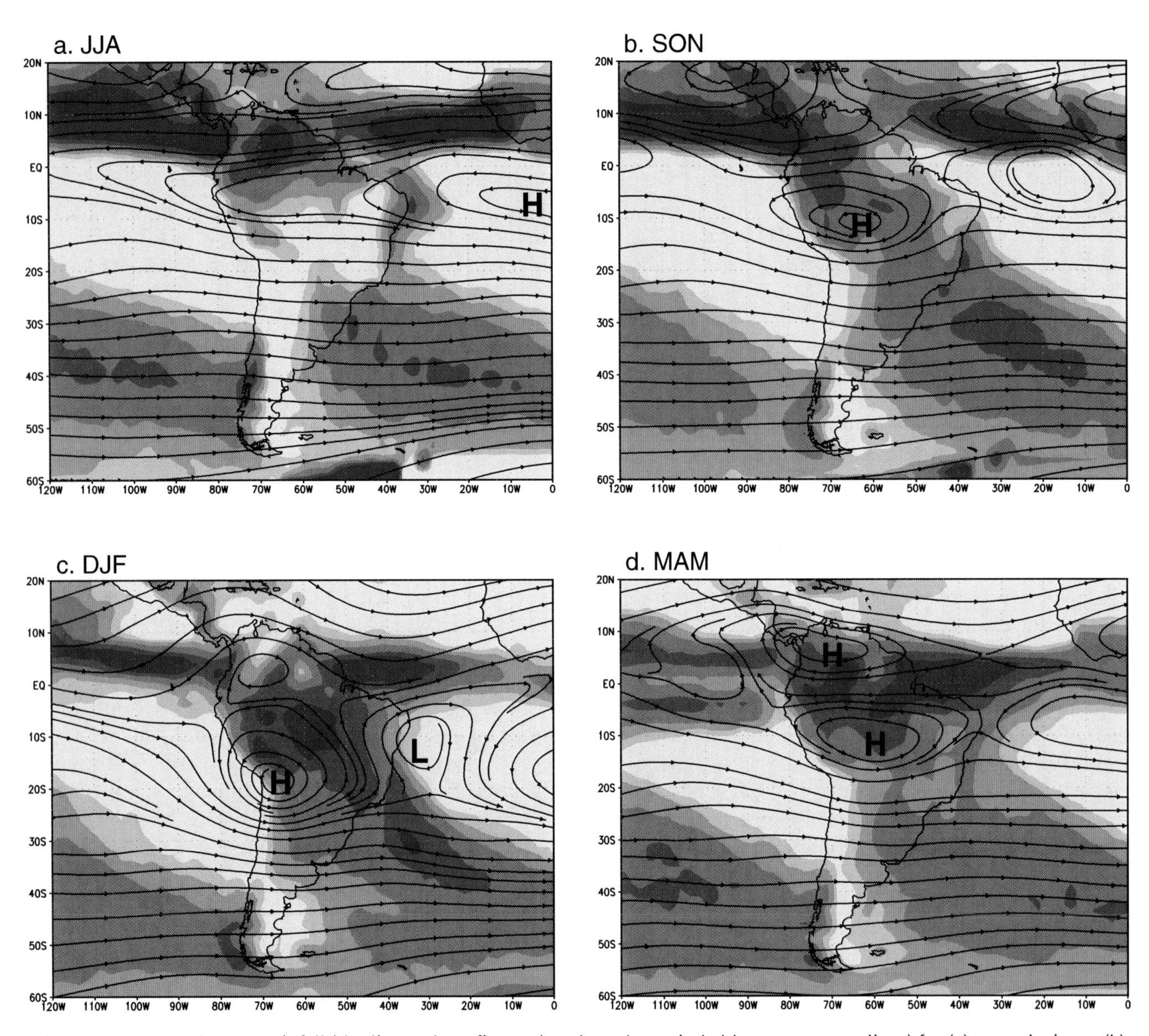

Figure 3.2 Seasonal mean rainfall (shading scale as figure 1) and 200 hPa winds (shown as streamlines) for (a) austral winter, (b) spring, (c) summer, and (d) fall. Maximum wind speed in midlatitudes is 60 m/s. Letter H indicates the approximate center of upper level anticyclones. Letter L in (c) indicates the center of the upper level trough over northeast Brazil.

During early fall, deep convection gradually diminishes over the subtropics and begins to shift northward, thus leading to the demise of the South American summer monsoon by the end of April (fig. 3.2d).

The seasonal maps in figure 3.2 show the upper-level atmospheric circulation in terms of the mean 200–hPa winds superimposed on the precipitation field. During the austral winter (JJA), strong Southern Hemisphere (SH) westerly flows prevail over South America as far north as 5°S, consistent with the equatorward displacement of the maximum tropical-extratropical thermal gradient (fig. 3.2a). During the course of the austral spring (SON), the subtropical jet stream moves southward, while a weak anticyclonic circulation develops over the tropical part of the continent (fig. 3.2b). During the austral summer (DJF), the SH westerlies over the continent are restricted to the south of 22°S, while the Northern Hemisphere (NH) westerlies reach the northern extreme of the continent. In this season, an upper-level anticyclonic circulation, referred as the Bolivian High, becomes firmly established over the central part of the continent (with its center located at 15°S and 65°W). This High is accompanied by a cyclonic circulation downstream over the northeast coast of Brazil, and a convergence region along the coast of Perú and Ecuador (fig. 3.2c; Virji 1981; Lenters and Cook, 1997).

It was originally proposed that the Bolivian High had a thermal origin, maintained by the strong sensible heating over the central Andes and the release of latent heat in the summertime convection over the Altiplano (Schwerdtfeger, 1961; Gutman and Schwerdtfeger, 1965). Modeling studies (reviewed in Lenters and Cook, 1997), however, indicate that the Bolivian High is instead a dynamical response to the warming of the upper troposphere generated by the cumulus convection over the Amazon basin. Diabatic heating over the Altiplano does not appear essential for the existence of the Bolivian High, although the Andes play an indirect role by organizing the low-level flow and convection over the central part of the continent. On the other hand, the presence of the Bolivian High is instrumental for the occurrence of summertime precipitation over the Altiplano, since the easterly flow aloft favors the transport of moist air from the lowlands toward the central Andes (Garreaud, 1999b). Furthermore, the Bolivian High contributes to the intensification of the SACZ during summer, owing to vorticity advection aloft (Figueroa et al., 1995). During the austral fall (MAM), the SH westerlies return to the subtropics, and a pair of anticyclones is found over tropical South America, consistent with the convection centered on the equator (fig. 3.2d).

During the rainy season, there is a well-known preference for an afternoon / evening maximum of deep convection over continental areas, since land-surface heating tends to destabilize the lower troposphere (e.g. Meisner and Arkin, 1987). Nevertheless, detailed analyses of the mean diurnal cycle of rainfall (Negri et al., 1994) and convective clouds (Garreaud and Wallace, 1997) demonstrate that the timing of the maximum convection is location-dependent, and closely linked to regional topographic features such as mountain ranges and concave coastlines. To illustrate this point, figure 3.3 shows the evening and early morning frequencies of cold cloudiness, a proxy for convective rainfall, during the austral summer (DJF). Maximum convection during the evening is very pronounced over the central Andes, along the northeast coast of the continent, and in two parallel bands over central Amazonia (fig. 3.3a). The first two bands arise from the concurrent timing of the thermodynamic destabilization and the maximum strength of dynamical forcing: plain-to-mountain wind convergence over the Andes, and land-sea breeze convergence along the coast. The intervening parallel bands are interpreted as the afternoon reactivation of recurrent Amazon coastal squall lines (ACSL; e.g., Cohen et al., 1995). ACSLs are forced at the coast by sea breezes and move inland, maintaining their identity for 12–36 hours, most notably in early autumn. On the other hand, convection tends to peak during night-time and early morning along the eastern slopes of the central Andes, over the subtropical plains, and off the northeast coast of the continent, highlighting the dynamical forcing in these areas (fig. 3.3b; Berbery and Collini, 2000).

The mean annual and diurnal cycles of the precipitation to the south of 40°S are generally less pronounced than those at lower latitudes, because of the dynamical forcing (in contrast with thermodynamic forcing) of frontal rainfall. A notable feature of the SH midlatitudes is the circumpolar ring of strong westerly flow throughout the depth of the troposphere; this intersects the southern tip of South America between 50°S and 40°S. As described later, the zonal symmetry observed in figure 3.2 is the result of averaging many daily maps, each of them containing significant departures from zonal symmetry and strong meridional flow. The

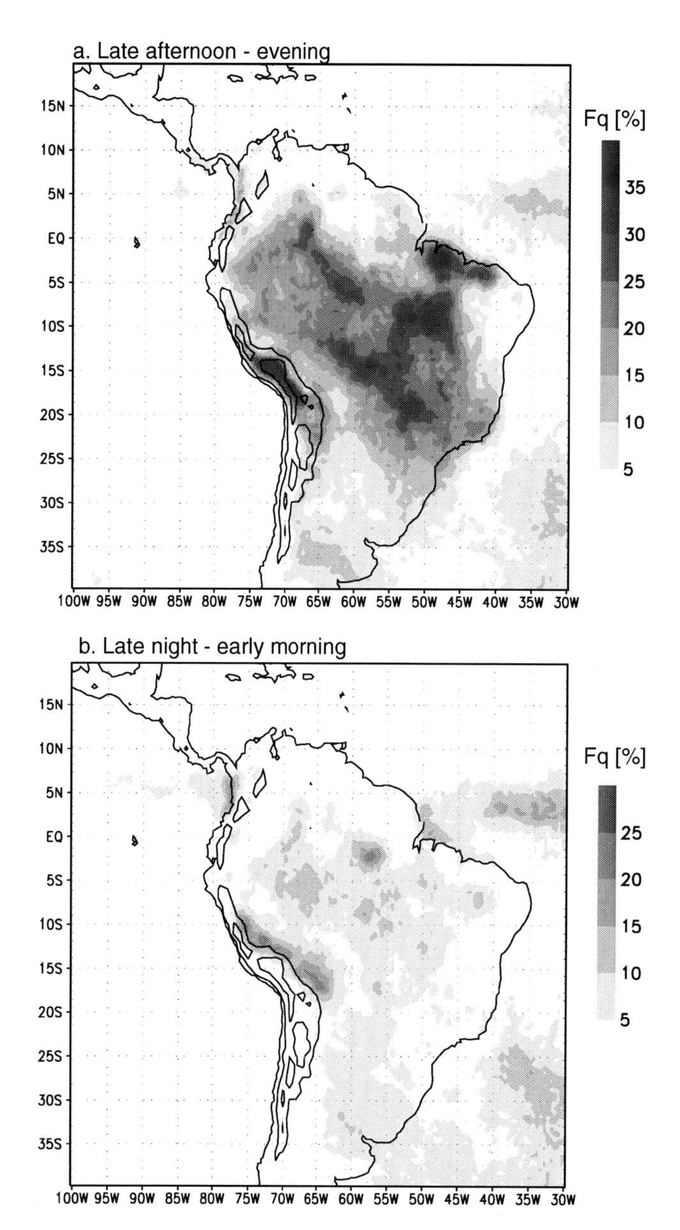

Figure 3.3 (a) Late afternoon and evening (1,800–2,100 UTC average; UTC: Universal Time Coordinated = Greenwich Time) frequency of cold clouds (cloud-top temperature less than 235°K) during the austral summer (DJF). Solid lines within the continent indicate terrain elevation at 2,000 and 4,000 m above mean sea level. (b) As (a) but for late night and early morning (0900–1,200 UTC average; modified from Garreaud and Wallace, 1997).

position of the jet axis coincides with a zone of maximum meridional temperature gradient, and therefore is a proxy for the preferred path, or storm track, of synoptic-scale disturbances (Trenberth, 1991). Although subtle, there is an intensification and equatorward shift of the storm track and similar changes in the strength and position of the subtropical anticyclones during the austral winter (JJA; Physick, 1981), thus leading to the rainy season in central Chile and Argentina (30–40°S) (fig. 3.2; see also chapter 11).

3.2 High-Frequency Climatic Variability

Temporal changes in atmospheric conditions, such as temperature, rainfall, and wind, over a given region exhibit nonregular fluctuations across a broad range of scales superimposed on the mean diurnal and annual cycles. These fluctuations include synoptic-scale variability, broadly associated with weather, as well as intraseasonal, interannual, interdecadal, and longer-scale variations, that arise from the internal variability of the atmosphere and its coupling with other components of the Earth system such as the oceans, land vegetation, and sea-ice. In this section, we describe fluctuations in the atmospheric circulation and rainfall that occur on timescales ranging from a few days to a few months, usually termed high-frequency variability. These fluctuations are important in themselves because they influence weather patterns, while changes in their frequency and intensity constitute the building blocks of longer-scale variability.

3.2.1 Synoptic-Scale Variability

Synoptic variability refers to changes in atmospheric conditions over periods ranging from 3 to 14 days. These changes are caused by synoptic-scale disturbances, that is, troposphere-deep phenomena with horizontal dimensions of thousands of kilometers in at least one direction that maintain their identity for several days. The prime source of synoptic disturbances are the baroclinic waves that grow, reach maturity, and decay while embedded in the midlatitude westerly flows. Details on the dynamics, structure, and evolution of baroclinic waves can be found in most textbooks of meteorology (e.g. Wallace and Hobbs, 1977; Holton, 1992). In addition to midlatitude waves, synoptic variability can also be produced by phenomena of pure tropical origin, such as tropical depressions. Except for the influence of easterly waves along the northeast coast, these kinds of synoptic phenomena are seldom observed in South America, a fact attributed to the relatively cold sea surface temperature in the adjoining oceans (Satyamurti et al., 1998).

As mentioned earlier, the circumpolar SH storm track intersects the continent to the south of 40°S, with only minor seasonal changes. More detailed analysis by Berbery and Vera (1996) indicates the existence of two preferred west-to-east paths of baroclinic waves in the South American region: one along the subpolar jet, about 60°S, and the other along the subtropical jet that crosses the Andes between 30°S and 45°S. The baroclinic waves in the former path are little affected by continental topography, and they evolve smoothly as they move to east-southeast over the southern oceans (Vigliarolo et al., 2000). In contrast, waves evolving along the subtropical path are blocked by the Andes cordillera and experience substantial distortion in their structure. The effect of the Andes on baroclinic waves has been addressed in statistical analysis by Gan and Rao (1994b), Berbery and Vera (1996) and Seluchi et al. (1998), among others. At upper levels, the waves tend to move to the northeast, slightly departing from the general zonality. The equatorward dispersion of the wave activity is consistent with the conservation of potential vorticity of the air columns crossing the Andes cordillera (Seluchi et al., 1998). Upper level mid-latitude waves can also spawn closed cyclonic vortices (cutoff lows) that subsequently move irregularly in the subtropics, as discussed later. At lower levels, midlatitude disturbances experience major distortion due to mechanical blocking. To the west of the subtropical Andes, pressure anomalies tend to be out of phase with respect to the incoming disturbance farther south, especially in the anticyclonic case (Garreaud et al., 2001). To the east of the Andes, both low- and high-pressure cells are markedly deflected to lower latitudes. Near the east coast of the continent, cyclones usually deepen dramatically (Taljaard, 1972; Sinclair, 1994), in part because of the diabatic effect of the warm waters of the subtropical Atlantic (Seluchi and Saulo, 1998).

Low-Level Effects To describe further the effects of topography on weather systems, the life cycle of low-level low- and high-pressure cells (steered by a trough-ridge wave in the middle troposphere) moving across South America is schematized in figure 3.4. Over the southeast Pacific Ocean, the persistent subtropical anticyclone limits the equatorward extent of the low cell to about 35°S (fig. 3.4a), while the blocking effect of the Andes upon the northwesterly flow (ahead of the surface low) tends to delay the arrival of the cold fronts to the west coast of the continent (Rutllant and Garreaud, 1995). Eventually, the low-pressure cell moves into the southern tip of the continent and the cold front is able to reach the coast, producing rainfall in southern and central Chile and snowfall over the high subtropical Andes (fig. 3.4b).

At the same time, strong westerly winds over the Andes produce lee-side subsidence in western-central Argentina, leading to the formation of a thermal-orographic depression. This later effect, acting in concert with cyclonic vorticity advection aloft, produces the advance of the surface low toward lower latitudes (fig. 3.4b). During the warm

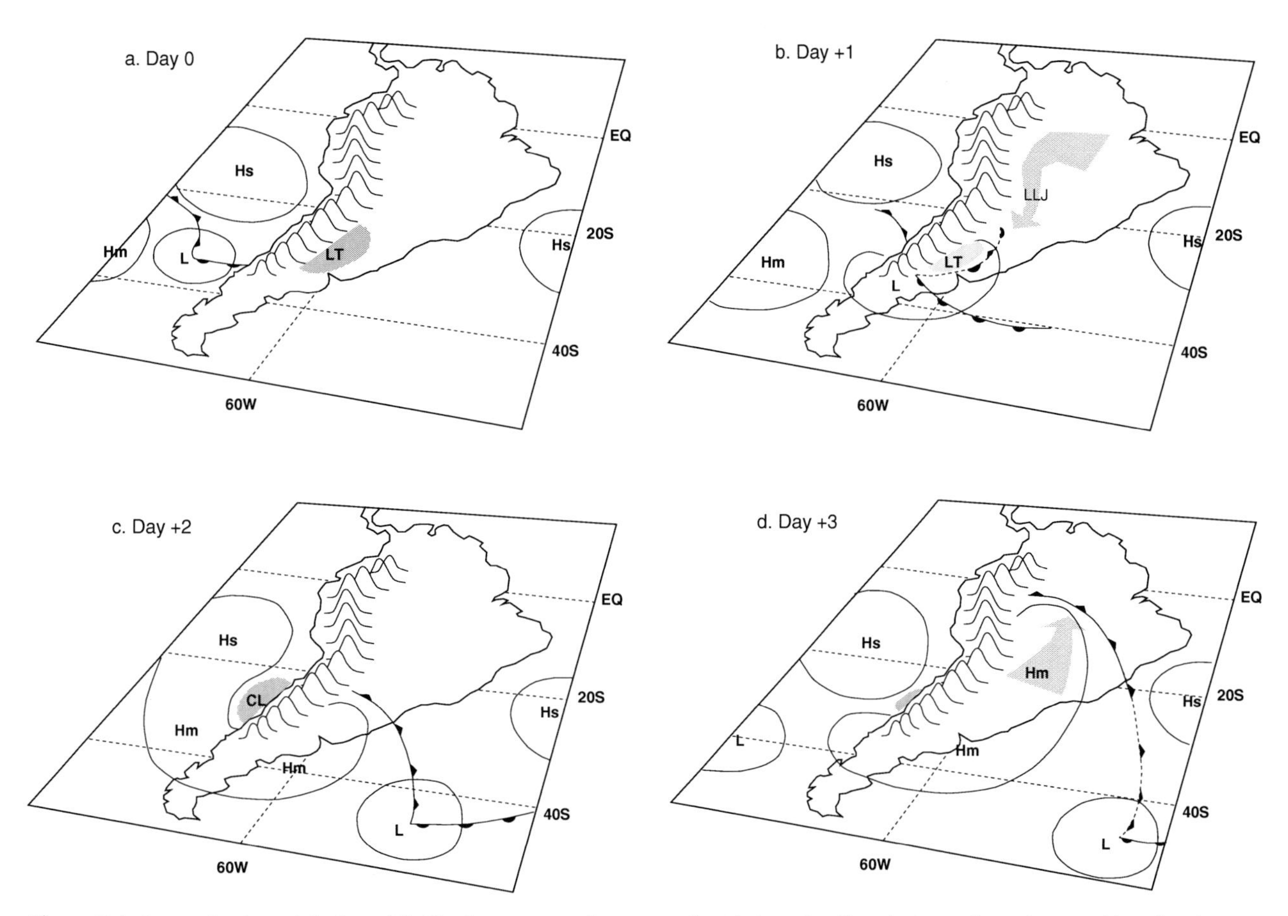

Figure 3.4 Conceptual model of a midlatitude wave moving across South America (for clarity, only surface and low-level features are shown). Solid lines represent isobars at the surface. Recall the clockwise (counter-clockwise) circulation of geostrophic winds around a low (high) pressure cell in the Southern Hemisphere. Hs = subtropical anticyclone; Hm = migratory (cold) anticyclone; L = migratory low pressure cell; LT = lee trough; CL = coastal low. Also shown are surface cold fronts (solid lines with filled triangles), warm fronts (solid lines with filled semicircles), the approximate position of the low-level jet (LLJ, shaded arrow), and the path of the cold air surge (shaded arrow pointing northward). The whole sequence takes about 4 days.

season, the lee-side subsidence deepens the climatological Chaco low, which in turn intensifies the transport of moist, warm air from the Amazon basin toward the subtropical plains, fueling severe storms in the form of prefrontal squall lines or other mesoscale convective complexes (Velasco and Fritsch, 1987; Garreaud and Wallace, 1998; Seluchi and Marengo, 2000). The latent heat release in these storms further deepens the surface low as it moves eastward, thus triggering a positive feedback mechanism (Seluchi and Saulo, 1998). The increasing availability of moist air toward the east coast also contributes to storm intensification. Consistently, explosive cyclogenesis (or cyclone redevelopment) is observed near the east coast of the continent between 25–35°S, especially in the spring and fall, when the synoptic forcing at these latitudes is still strong and the troposphere is unstable (Gan and Rao 1994a;Sinclair, 1994). During the austral winter, thermodynamic conditions over the subtropical plains are less conducive to moist convection but the formation of the lee-trough can, under some conditions, forces severe damaging downslope wind storms along the eastern foothills of the Andes, locally known as *Zonda* events (Norte and Silva, 1995).

To return to the synoptic sequence of figure 3.4, as the surface low moves eastward into the Atlantic, a migratory cold-core anticyclone approaches the west coast of the continent and merges with the subtropical anticyclone over the southeast Pacific (fig. 3.4c). Off the coast of central Chile, the easterly geostrophic winds in the lower troposphere force downslope flow over the Andean slope, so that cool marine air is replaced by warm, continental air. Consequently, the surface pressure drops along the subtropical west coast of the continent, leading to the formation of a coastal low and broad clearing of the stratocumulus clouds over the subtropical Pacific (Garreaud et al., 2001), with significant impacts on regional weather. To the east of the subtropical Andes, the easterly flow along the northern flank of the migratory high is dammed by the mountains, breaking down the geostrophic balance and leading to southerly wind over a band of country about 1,000 km wide beyond the Andean foothills. The cold, dry air surges equatorward between the Andes and the Brazilian plateau, displacing relatively warm, moist air (fig.3.4d). This latter effect results in a hydrostatic rise in surface pressure, explaining the expansion of the surface anticyclone into subtropical and tropical latitudes (Garreaud, 1999a).

Cold surges are a year-round feature of the synoptic climatology of the South American continent (e.g., Kousky and Cavalcanti, 1997; Garreaud, 2000b; Vera and Vigliarolo, 1999; Seluchi and Marengo, 2000), with a near-weekly periodicity but a large range in their intensity and meridional extent. Extreme wintertime episodes (one every few years) produce freezing conditions and severe agricultural damage from central Argentina to southern Bolivia and Brazil (locally known as *friagems* or *geadas*), and have motivated case studies by Hamilton and Tarifa (1978), Fortune and Kousky (1983), and Marengo et al. (1997), among others. Summertime episodes produce less dramatic fluctuations in temperature and pressure, owing to the smaller seasonal temperature gradient between mid- and low latitudes, but they are accompanied by synoptic-scale bands of deep convection at the leading edge of the cool air (Ratisbona, 1976; Kousky, 1979; Garreaud and Wallace, 1998; Liebmann et al., 1999). The banded cloud pattern extends from the eastern slopes of the Andes southeastward into the South Atlantic Ocean, where it intensifies the SACZ. These bands account for up to 40% of the summertime precipitation over subtropical South America (Garreaud and Wallace, 1998) and can reach as far as the northeast coast of Brazil (Kousky, 1979) and across the equator (Parmenter, 1976; Kiladis and Weickmann, 1997).

Cutoff Lows Cutoff lows are isolated cyclonic vortices in the upper troposphere at subtropical and higher latitudes, which develop from preexisting cold troughs (Carlson, 1998). Over South America, cutoff lows are observed over the northeast coast of Brazil and to the west of the subtropical Andes. In the former location, they are a frequent summertime phenomenon (Kousky and Gan, 1981) that dramatically intensifies the climatological trough in this region (fig. 3.2c). Kousky and Gan documented that cutoff lows tend to form near the axis of a SH trough crossing the eastern side of the continent, and they exhibit a direct thermal circulation, with colder air subsiding in their core (inhibiting convection) and warmer air ascending on their periphery (fostering deep convection). Once formed, the vortex may persist for several days, with an irregular horizontal movement, and is associated with excessively dry or wet conditions over northeast Brazil (Kousky and Gan, 1981, Satyamurty et al., 1998).

Cutoff lows to the west of the subtropical Andes also form from a SH trough approaching this region. The segregation of the cold vortex occurs when the zonal flow farther south is very strong (Pizarro and Montecinos, 2000). Once formed, the cutoff low moves rather irregularly along the west side of the Andes, causing strong windstorms and heavy precipitation on the Andean slopes. In contrast with cold fronts that rarely affect latitudes to the north of 30°S, wintertime cutoff lows can produce significant snowfall at higher elevations in northern Chile and Argentina (Pizarro and Montecinos, 2000) and the southern Altiplano (Vuille and Ammann, 1997) a few times a year.

3.2.2 Intraseasonal Variability

Atmospheric fluctuations with periods ranging from 10 to 90 days are generally termed intraseasonal (IS) variability. IS variability in the tropics has been the subject of substantial research since a planetary-scale tropical oscillation with a 40- to 50-day period was discovered in the early 1970s (Madden and Julian, 1971, 1972). It is believed that the Madden-Julian Oscillation (MJO) is primarily forced by anomalies in tropical SST and their feedback in circulation and convection (Madden and Julian, 1994). The MJO-related convection does not directly affect South America, since the associated tropical region of active convection normally moves from the eastern Indian Ocean to the western Pacific and then decays over the eastern Pacific (Salby and Hendon, 1994). In contrast, the MJO-related circulation anomalies at low-latitudes are circumglobal, causing a reversal in the upper level zonal wind over tropical (10°S to 10°N) South America over an ~30-day period. Whether these circulation fluctuations could produce IS variability in the convection over the equatorial Andes and western Amazonia needs to be addressed. IS variability in the subtropics and higher latitudes, although generally less pronounced than the fluctuations associated with individual disturbances, can modulate the regional weather over extended periods of time. IS variability in the extratropics can result from internal nonlinear atmospheric dynamics, or be remotely forced from the tropics by so-called atmospheric teleconnections (see details in Kiladis and Mo, 1998).

Over tropical and subtropical South America, the most notable IS fluctuation is a seesaw of dry and wet conditions with periods ranging between 2 and 3 weeks during the austral summer (Nogués-Paegle and Mo, 1997; Liebmann et al., 1999, Aceituno and Montecinos, 1997). Convection and precipitation over the Altiplano is also organized in rainy episodes of about 1 to 3 weeks, interrupted by similar dry episodes (Aceituno and Montecinos, 1997, Garreaud, 1999b, Lenters and Cook, 1999), and tend to be out-of-phase with convection over the eastern side of the continent (Garreaud, 1999b). The IS variability of the convective rainfall is associated with continental-scale anomalies of the tropospheric circulation, illustrated in figure 3.5 by the 200-hPa height difference and 850-hPa wind difference between positive and negative events (each of them lasting a week or longer) of the rainfall seesaw identified by Nogués-Paegle and Mo (1997). This difference is associated with rainy conditions over the subtropical plains and dry conditions over the SACZ (wet-continent phase of the seesaw). The anticyclonic anomalies at lower-levels enhance the northerly flow and moisture transport to the east of the Andes that in turn feeds convection over the subtropical plains. At upper level, the anticyclonic anomalies intensify and extend southward the region of easterly flow over the Andes associated with the Bolivian High. Easterly flow aloft is typically connected with rainy conditions over the Altiplano, since it fosters the transport of moist air from the Bolivian lowlands to higher elevations (Garreaud, 1999b). Finally, the dry conditions over the SACZ are explained by subsidence downstream of the upper-level anticyclone. Roughly opposite mechanisms act during the dry-continent phase of the seesaw.

It has been proposed that the summertime circulation anomalies over South America (fig. 3.5) are part of large-scale wave train emerging from the South Pacific, referred to as the Pacific South American (PSA) modes. During wintertime, the PSA modes dominate the intraseasonal variability in the SH (Ghil and Mo, 1991; Mo and Higgins, 1997). Mo and Higgins (1997) suggest that tropical convection over the western Pacific serves as a catalyst in the development of standing PSA modes, therefore connecting intraseasonal variability in the remote tropics with intraseasonal variability in subtropical South America.

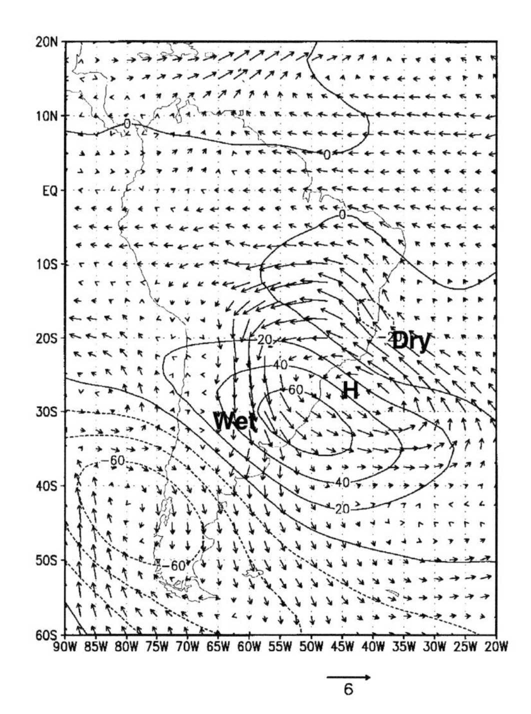

Figure 3.5 200 hPa height difference (contoured every 20 m) and 850 hPa winds (scale in m/s at bottom) difference between the wet-continent and dry-continent phases of the subtropical rainfall seesaw during summer (from Nogués-Paegle and Mo, 1997).

The PSA modes have also been associated with the onset of blocking anticyclones in the south Pacific to the west of the Antarctic Peninsula (Renwick and Revell, 1999), and whose subsequent maintenance arises from a complex interaction between the mean flow and the transient disturbances (Marques and Rao, 1999). The tropospheric-deep anticyclonic anomalies (barotropic structure) tend to split the mid-latitude zonal flow into equatorward and poleward branches. As the blocks in the southeast Pacific remain stationary for 5 to 15 days (Sinclair, 1996; Renwick, 1998), the mid-latitude storm track shifts equatorward, producing stormy conditions in central Chile and dry, cold conditions at the southern tip of the continent (Rutllant and Fuenzalida, 1991).

3.3 Low-Frequency Climatic Variability

Low-frequency climate variability includes changes occurring between consecutive years, usually identified as interannual variability (IA), and slower changes at timescales of decades (interdecadal variability) or longer periods (trends).

Many factors determine interannual climate variability in South America. As elsewhere in the world, those associated with the occurrences of El Niño (EN) and La Niña phenomena in the tropical Pacific, during the extreme phases of the Southern Oscillation (SO), have received much more attention, particularly concerning their impacts on rainfall. The worldwide signatures of ENSO in surface pressure and rainfall were first recognized by Walker and Bliss (1932), and the link between the atmospheric anomalies and tropical Pacific SST anomalies was discovered in the late 1950s and 1960s (e.g. Bjerknes, 1966). Because of its impact and complex dynamics, research literature on ENSO is extensive, and the reader is referred to the volumes of Philander (1990), Diaz and Markgraf (2000), and Glantz et al. (1991) for in-depth treatments of this phenomenon (see also chapter 19).

The sign and strength of the ENSO-related anomalies are geographically and seasonally dependent, rendering a complex picture of the functioning of this phenomenon in the South American region (Aceituno, 1988; Kiladis and Diaz, 1989; Ropelewski and Halpert, 1987, 1989). The negative phase of the SO is typically accompanied by positive SST anomalies in the tropical Pacific, while relatively cold conditions prevail during the positive phase of the SO. When these SST anomalies are relatively intense and last for several months, an episode of El Niño (warm conditions) or La Niña (cold conditions) is defined (Trenberth, 1997).

Studies of ENSO-related rainfall anomalies at a global scale indicate that El Niño episodes are typically associated with below-normal rainfall in the northern part of

South America and anomalously wet conditions in the southeastern portion of the continent, including southern Brazil, Uruguay, southern Paraguay, and northeast Argentina (fig. 3.6; Ropelewski and Halpert, 1987). Opposite rainfall anomalies are typically observed in both regions during La Niña events (Ropelewski and Halpert, 1989). This large picture of the ENSO impacts on rainfall in South America exhibits considerable variation when analyzed at a regional scale.

Hydrological records for the two largest rivers in the northern Andes, the Magdalena and Cauca, show a distinct ENSO signal, indicating a tendency for drought during El Niño episodes and flooding conditions during La Niña events (Aceituno and Garreaud, 1995). The strength of this signal reaches a maximum during the boreal winter (January–February) that coincides with one minimum in the rainfall semi-annual cycle that characterizes this region (Aceituno, 1988; Hastenrath, 1990). Mechanisms involved in this ENSO-related anomaly in the rainfall regime are not well understood. The weakening of the temperature contrast between the continent and adjacent oceans (Pacific and the Caribbean) and the enhanced convection over the eastern tropical Pacific during an El Niño episode are two factors favoring a weakening of convection over the Colombian Andes.

Flood conditions along the semi-arid coastal areas in southern Ecuador and northern Peru during El Niño episodes are one of the trademarks of the climatic impacts of this phenomenon in South America (see Chapter 19). In fact, many decades before coherent global-scale climate anomalies were linked to ENSO, the damaging floods during episodes of anomalously warm waters along the coast of northern Peru had been well documented (e.g., Murphy, 1926). As in most regions having a significant ENSO signal on climate variability, rainfall anomalies here are phase-locked with the annual rainfall cycle. Thus, flood conditions during El Niño events develop from around December to May, and there is evidence that strong convection-producing rainfall is triggered by anomalously high values of SST in the adjacent ocean (Horel and Cornejo-Garrido, 1986). Although enhanced convection and rainfall are confined to lowland areas near the coast, the impact on river discharges is extremely large. As an example, at the peak of the major 1982–1983 El Niño episode, in May 1983 when SST surpassed 28.0°C along the coast of northern Perú, the discharge of the Piura River neared 1,200 m^3/s, well above its average flow of about 40 m^3/s for that month!

Over the central Andes, especially along the western side of the Altiplano, a relatively weak tendency for drier-than-average conditions during the wet season (December–March), when positive SST anomalies prevail in the tropical Pacific, has been documented in several studies (Francou and Pizarro, 1985; Ronchail, 1995; Vuille, 1999). This anomalous behavior in the rainfall regime produces a significant mark on hydrological and glaciological records. Specifically, El Niño episodes are associated with a relatively low seasonal increment in the level of Lake Titicaca (Kessler, 1974; Aceituno and Garreaud, 1995)) and below-average snow accumulation on glaciers (Thompson et al., 1984; Wagnon et al., 2001). Negative rainfall anomalies during El Niño are consistent with stronger-than-average subtropical westerlies over the Altiplano, conditions that inhibit the advection of moisture from the warm and humid environment of the Bolivian lowlands. Opposite upper-level circulation anomalies tend to occur during wet seasons coinciding with La Niña episodes, thus enhancing the moisture transport toward the Altiplano and favoring wetter-than-average conditions.

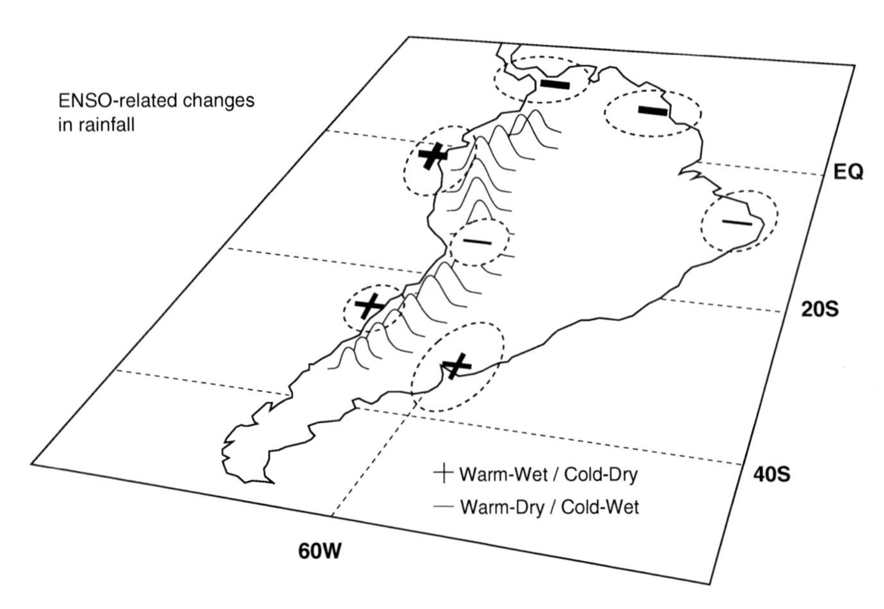

Figure 3.6 Main areas with ENSO-related changes in rainfall. The affected regions are approximately outlined by dashed lines. The correlation between rainfall anomalies and tropical Pacific SST (an index of ENSO) is indicated by plus and minus signs, the thickness of which indicates the relative strength of the correlation. The strongest rainfall anomalies tend to occur during the corresponding climatological wet season.

Long recognized is the tendency for drought to occur in semiarid northeast Brazil (the Brazilian Nordeste) when El Niño conditions prevail in the tropical Pacific (e.g., Caviedes, 1973). Nevertheless, the interannual rainfall variability in this part of the continent has stronger links to anomalous SST, surface wind, and convection patterns in the tropical Atlantic, conditions that are partially modulated by ENSO (Hastenrath and Heller, 1977; Moura and Shukla, 1981). A dipole of SST anomalies over the Atlantic on both sides of the equator sometime persists for several months. Warm SST anomalies to the north of the equator lead to enhanced convection in the ITCZ and stronger-than-normal subsidence and drought over the Nordeste (Moura and Shukla, 1981). When SST warm anomalies occur to the south of the equator, the ITCZ encompasses the Nordeste, causing wetter than normal conditions there.

Changes in the global atmospheric circulation induced by anomalies in the ocean-atmosphere system in the tropical Pacific are at the origin of many ENSO-related climate anomalies in subtropical and extratropical areas around the globe (that is, teleconnection patterns). In South America, these signals are significant along the subtropical western border (central Chile) and over the southeast portion of the continent. Regarding the links between SST anomalies in the tropical Pacific and rainfall, a warm-wet / cold-dry relationship over central Chile and the subtropical Andes during the wet season (May–September) has been documented in several studies (Quinn and Neal, 1983; Kiladis and Diaz, 1989; Aceituno and Garreaud, 1995). A detailed analysis of this ENSO signal revealed that the tendency for positive (negative) rainfall anomalies during El Niño (La Niña) is significant in the 33°S to 36°S latitudinal band during the austral winter, while from 36°S to 39°S the same signal is best defined during the spring (Montecinos et al., 2000). The tendency to above average wintertime rainfall during El Niño years is consistent with a relatively higher frequency of blocking anticyclones to the west of the Antarctic Peninsula, and the subsequent northward displacement of the South Pacific storm track (Rutllant and Fuenzalida, 1991).

Over southeastern South America, anomalously wet conditions typically occur during El Niño that generate extensive flooding in the lowlands of southern Brazil, southern Paraguay, Uruguay, and northeast Argentina. Several studies have described this ENSO-related signal based on the analysis or rainfall and streamflow records (Kousky et al., 1984; Aceituno, 1988; Pisciottano et al., 1994; Diaz et al., 1998; Grimm et al., 1998). This relationship between ENSO and rainfall has been linked to the intensification of the subtropical jet and its associated baroclinic activity (Kousky et al., 1984; Lenters and Cook, 1999). Its strength is greatest during the austral spring and early summer (Montecinos et al., 2000).

It has been recognized that central Pacific SST anomalies during warm and cold episodes are accompanied by temperature anomalies of the same sign in the tropical troposphere throughout the globe. Over South America, there is a clear tendency for above (below) normal air temperature during warm (cold) ENSO years, from the northern margins of the continent south to about 30°S (Aceituno, 1988, 1989). The opposite relationship (that is, cold air anomalies during El Niño years) is observed in midlatitudes year-round and along the subtropics during the early summer, probably because ENSO-related increased precipitation over these regions is also associated with a reduction of solar radiation (Aceituno, 1988).

One must keep in mind that the relationships between ENSO and regional rainfall and temperature, as described above, are typically obtained from data records during the last 30 to 40 years. The strength of these relationships might vary significantly when viewed on longer timescales (Aceituno and Montecinos, 1993). It is also important to remember that ENSO is only one of the significant factors modulating interannual climatic variability in South America. In fact, many studies have demonstrated the significant influence that changes in the tropical Atlantic ocean-atmospheric system have on rainfall variability over many areas of tropical South America (Hastenrath and Heller, 1977; Marengo, 1992; Nobre and Shukla, 1996; Diaz et al., 1998)

Recent work has focused on interdecadal variability of the ocean-atmosphere system that may explain slow or abrupt changes in climate. The broadly known "regime shift" of 1976 appears to be an outstanding example (e.g., Graham, 1994). Yet our capacity to identify, describe, and explain this interdecadal climate variability is hampered by the relatively short length of the instrumental records relative to the timescale to be addressed (Garreaud and Battisti, 1999). At even longer timescales, slow changes in the climate are present as trends in almost all climatic records. Slowly evolving boundary conditions of the atmosphere (mostly SST) seem to be the most important factors influencing this type of variability. In subtropical South America, a marked contrast is apparent in the long-term evolution of rainfall on both sides of the continent. On the western margin (central Chile), a negative trend characterized most of the rainfall records during the twentieth century (Aceituno et al., 1992), while on the eastern side a remarkable upward trend has been observed on rainfall and streamflow records, especially after 1950 (Krepper et al., 1989; Castañeda and Barros, 1994; Minetti and Vargas, 1997; Robertson and Mechoso, 1998).

References

Abreu, M., and P. Bannon, 1993. Dynamics of the South American Coastal Desert. *Journal of Atmospheric Sciences*, 50, 2952–2964.

Aceituno, P., 1988. On the functioning of the Southern Oscillation in the South American sector. Part I: Surface climate. *Monthly Weather Review*, 116, 505–524.

Aceituno, P., 1989. On the functioning of the Southern Oscillation in the South American sector. Part II: Upper-air circulation. *Journal of Climate*, 21, 341–355.

Aceituno, P., H. Fuenzalida, and B. Rosenbluth, 1992. Climate along the extratropical west coast of South America. In: H.A. Mooney, B. Kronberg, and E.R. Fuentes (Editors), *Earth system responses to global change*. Academic Press, New York, 61–96.

Aceituno, P., and R.D. Garreaud, 1995. The impact of the ENSO phenomenon in the pluviometric regime over the Andes. *Revista Chilena de Ingeniería Hidraulica*, 2, 33–43.

Aceituno, P., and A. Montecinos, 1993. Estabilidad de la relación entre la Oscilación del Sur y la precipitación en América del Sur. *Bulletin Institut Français Etudes Andines*, 22, 53–64.

Aceituno, P., and A. Montecinos, 1997. Patterns of convective cloudiness in South America during austral summer from OLR pentads. In: Preprints, Fifth International Conference on Southern Hemisphere Meteorology and Oceanography, Pretoria, South Africa. *American Meteorological Society*, Boston, 328–329.

Berbery, E.H., and E. Collini, 2000. Springtime precipitation and water-vapor flux convergence over southeastern South America. *Monthly Weather Review*, 128, 1328–1346.

Berbery, E.H., and C.S. Vera, 1996. Characteristics of the Southern Hemisphere winter storm track with filtered and unfiltered data. *Journal of Atmospheric Sciences*, 53, 468–481.

Bjerknes, J., 1966. A possible response of the atmospheric Hadley circulation to equatorial anomalies of ocean temperature. *Tellus*, 18, 820–829.

Carlson, T.N., 1998. *Mid-latitude weather systems*. American Meteorological Society. Boston.

Castañeda, M.E., and V.R. Barros, 1994. Long-term trends in rainfall along southern South America eastward from the Andes. *Meteorológica*, 19, 23–32.

Caviedes, C.N., 1973. Secas and El Niño: Two simultaneous climatological hazards in South America. *Proceedings of the Association of American Geographers*, 5, 44–49.

Cohen J., M. Silva Dias, and C. Nobre, 1995. Environmental conditions associated with Amazonian squall lines: A case study. *Monthly Weather Review*, 123, 3163–3174.

Diaz, H.F., and V. Markgraf, 1992. *El Niño*. Cambridge University Press, Cambridge.

Diaz, A.F., C.D. Studzinski, and C.R. Mechoso, 1998. Relationships between precipitation anomalies in Uruguay and southern Brazil and sea surface temperature in the Pacific and Atlantic Oceans. *Journal of Climate*, 11, 251–271.

Figueroa, S.N., P. Satyamurty, and P.L. Silva Dias, 1995. Simulation of the summer circulation over the South American region with an Eta coordinate model. *Journal of Atmospheric Sciences*, 52, 1573–1584.

Fortune, M.A., and V.E. Kousky, 1983. Two severe freezes in Brazil: Precursors and synoptic evolution. *Monthly Weather Review*, 111, 181–196.

Francou, B., and L. Pizarro, 1985. El fenómeno de El Niño y las sequias en los Andes centrales (Perú y Bolivia). *Bulletin Institut Français Etudes Andines*, 14, 1–18.

Fu, R., B. Zhu, and R.E. Dickinson, 1999. How do atmosphere and land surface influence seasonal changes of convection in the tropical Amazon? *Journal of Climate*, 12, 1306–1321.

Gan, M.A., and V. Rao, 1994a. Surface cyclogenesis over South America. *Monthly Weather Review*, 119, 1923–1302.

Gan, M.A., and V. Rao, 1994b. The influence of the Andes cordillera on transient disturbances. *Monthly Weather Review*, 122, 1141–1157.

Gandu, A.W., and P.L. Silva Dias, 1998. Impact of tropical heat sources on South American tropospheric circulation and subsidence. *Journal of Geophysical Research*, 103, 6001–6015.

Garreaud, R.D., 1999a. Cold air incursions over subtropical and tropical South America. A numerical case study. *Monthly Weather Review*, 127, 2823–2853.

Garreaud, R.D., 1999b. A multi-scale analysis of the summertime precipitation over the central Andes. *Monthly Weather Review*, 127, 901–921.

Garreaud, R.D., 2000a. Intraseasonal variability of moisture and rainfall over the South American Altiplano. *Monthly Weather Review*, 128, 3379–3346.

Garreaud, R.D., 2000b. Cold air incursions over Subtropical South America: Mean structure and dynamics. *Monthly Weather Review*, 128, 2544–2559.

Garreaud, R.D., and D.S. Battisti, 1999. Interannual (ENSO) and interdecadal (ENSO-like) variability in the Southern Hemisphere tropospheric circulation. *Journal of Climate*, 12, 2113–2123.

Garreaud, R.D, and J.M. Wallace, 1997. The diurnal march of convective cloudiness over the Americas. *Monthly Weather Review*, 125, 3157–3171.

Garreaud, R.D., and J.M. Wallace, 1998. Summertime incursions of midlatitude air into tropical and subtropical South America. *Monthly Weather Review*, 126, 2713–2733.

Garreaud, R.D., J. Rutllant, and H. Fuenzalida, 2002. Coastal lows in the subtropical western coast of South America: Mean structure and evolution. *Monthly Weather Review*, 130, 75–88.

Garreaud, R.D., J. Rutllant, J. Quintana, J. Carrasco, and P. Minnis, 2001. CIMAR-5: A snapshot of the lower troposphere over the subtropical Southeastern Pacific. *Bulletin of the American Meteorological Society*, 82, 2193–2207.

Ghil, M., and K.C. Mo, 1991. Intraseasonal oscillations in the global atmosphere. Part II: Southern Hemisphere. *Journal of Atmospheric Sciences*, 48, 780–790.

Glantz, M., R. Katz, and N. Nicholls (Editors), 1991. *Teleconnections linking worldwide climate anomalies*. Cambridge University Press. Cambridge.

Graham, N.E., 1994. Decadal-scale climate variability in the 1970s and 1980s: Observations and model results. *Climate Dynamics*, 10, 135–162.

Grimm, A.M., S.E. Ferraz, and J. Gomes, 1998. Precipitation anomalies in southern Brasil associated with El Niño and La Niña events. *Journal of Climate*, 11, 2863–2880.

Gutmann, G., and W.S., Schwerdtfeger, 1965. The role of the latent and sensible heat for the development of a high pressure system over the subtropical Andes in the summer. *Meteorologische Rundschau*, 18, 1–17.

Hamilton, M.G., and J.R. Tarifa, 1978. Synoptic aspects of a polar outbreak leading to frost in tropical Brazil, July 1972. *Monthly Weather Review*, 106, 1545–1556.

Hartmann, D.L., M.E. Ockert-Bell, and M.L. Michelsen, 1992. The effect of cloud type on Earth´s energy balance: Global analysis. *Journal of Climate*, 5, 1281–1304.

Hastenrath, S., 1982. *Climate dynamics of the tropics*. 2nd ed., Kluwer, New York.
Hastenrath, S., 1990. Diagnostics and prediction of anomalous river discharge in northern South America. *Journal of Climate*, 3, 1080–1096.
Hastenrath, S., and L. Heller, 1977. Dynamic of climate hazards in Northeast Brazil. *Quarterly Journal of the Royal Meteorological Society*, 103, 77–92.
Hobbs, J.E., J.A., Lindeasay, and H.A. Bridgman (Editors), 1998. *Climate of the southern continents: Present, past and future*. John Wiley. Chichester.
Holton, J.R., 1992. *An introduction to dynamic meteorology*. 3rd ed., Academic Press. New York.
Horel, J., and A.G. Cornejo-Garrido, 1986. Convection along the coast of Northern Peru during 1983: Spatial and temporal variations in cloud and rainfall. *Monthly Weather Review*, 114, 2091–2105.
Horel, J., A. Hahmann, and J. Geisler, 1989. An investigation of the annual cycle of the convective activity over the tropical Americas. *Journal of Climate*, 2, 1388–1403.
Kalnay, E., and others, 1996. The NCEP/NCAR 40–years reanalysis project. *Bulletin of the American Meteorological Society*, 77, 437–472.
Kessler, A., 1974. Atmospheric circulation anomalies and level fluctuations of Lake Titicaca. *Bonner Meteorologisch Abh*. 17, 361–372.
Kiladis, G.N., and H. Diaz, 1989. Global climatic anomalies associated with extremes in the Southern Oscillation. *Journal of Climate*, 2, 1069–1090.
Kiladis, G.N., and K.M. Weickmann, 1997. Horizontal structure and seasonality of large-scale circulations associated with submonthly tropical convection. *Monthly Weather Review*, 125, 1997–2013.
Kiladis, G.N., and K.C. Mo, 1998. Interannual and intraseasonal variability in the Southern Hemisphere. In: D. Karoly and D. Vincent (Editors), *Meteorology of the Southern Hemisphere. Meteorological Monographs,* No. 49. American Meteorological Society. Boston, 307–336.
Klein, S.A., and D.L. Hartmann, 1993. The seasonal cycle of low stratiform clouds. *Journal of Climate*, 6, 1587–1606.
Kodoma, Y., 1992. Large-scale common features of subtropical precipitation zones (the Baiu frontal zone, the SPCZ and the SACZ). Part I: Characteristic of subtropical frontal zones. *Journal of the Meteorological Society of Japan*, 70, 813–836.
Kousky, V.E., 1979. Frontal influences on northeast Brazil. *Monthly Weather Review*, 107, 1140–1153.
Kousky, V.E., and I. Cavalcanti, 1997. The principal modes of high-frequency variability over the South American region. In: Preprints Fifth International Conference on Southern Hemisphere Meteorology & Oceanography, Pretoria, South Africa. American Meteorological Society, Boston, 7B2–7B3.
Kousky, V.E., and J. Ferreira, 1981. Frontal influences on northeast Brazil: their spatial distributions, origins and effects. *Monthly Weather Review*, 109, 1999–2008.
Kousky, V.E., and M.A. Gan, 1981. Upper tropospheric cyclonic vortices in the tropical South Atlantic. *Tellus*, 33, 538–551.
Kousky, V.E., M.T. Kagano, and I.F. Cavalcanti, 1984. A review of the Southern Oscillation: Oceanic-atmospheric circulation changes and related rainfall anomalies. *Tellus*, 36a, 490–504.
Krepper, C.M., B. Scian, and J. Pierini, 1989. Time and space variability of rainfall in central-east Argentina. *Journal of Climate*, 2, 39–47.
Legates, D.R., and C. Willmott, 1990. Mean seasonal and spatial variability in gauge-corrected global precipitation. *International Journal of Climatology*, 10, 111–127.
Lenters, J.D., and K.H. Cook, 1995. Simulation and diagnosis of the regional summertime precipitation climatology of South America. *Journal of Climate*, 8, 2298–3005.
Lenters, J.D., and K.H. Cook, 1997. On the origin of the Bolivian High and related circulation features of the South American climate. *Journal of Atmospheric Sciences*, 54, 656–677.
Lenters, J.D., and K.H. Cook, 1999. Summertime precipitation variability over South America: Role of the large-scale circulation. *Monthly Weather Review*, 127, 409–431.
Liebmann, B., G. Kiladis, J. Marengo, T. Ambrizzi, and J.D. Glick, 1999. Submonthly convective variability over South America and the South Atlantic Convergence Zone. *Journal of Climate*, 11, 2898–2909.
Madden, R., and P. Julian, 1971. Detection of a 40–50 day oscillation in the zonal wind. Journal of Atmospheric Sciences, 28, 702–708.
Madden, R., and P. Julian, 1972. Description of global-scale circulation cells in the Tropics with a 40–50–day period. *Journal of Atmospheric Sciences*, 29, 1109–1123.
Madden, R., and P. Julian, 1994. A review of the intraseasonal oscillation in the Tropics. *Monthly Weather Review*, 122, 814–837.
Marengo, J., 1992. Interannual variability of surface climate in the Amazon basin. *International Journal of Climatology*, 12, 853–863.
Marengo, J., A. Cornejo, P. Satymurty, C. Nobre, and W. Sea, 1997. Cold surges in tropical and extratropical South America: The strong event in June 1994. *Monthly Weather Review*, 125, 2759–2786.
Marques, R., and V.B. Rao, 1999. A diagnosis of a long-lasting blocking event over the southeast Pacific Ocean. *Monthly Weather Review*, 127, 1761–1776.
Meisner, B., and P. Arkin, 1987. Spatial and annual variations in the diurnal cycle of the large-scale tropical convective cloudiness and precipitation. *Monthly Weather Review*, 115, 2009–2032.
Minetti, J.L., and W.M. Vargas, 1997. Trends and jumps in the annual precipitation in South America south of the 15°S. *Atmósfera*, 11, 205–221.
Mitchell, T.P., and J.M. Wallace, 1992. The annual cycle in the equatorial convection and sea surface temperature. *Journal of Climate*, 5, 1140–1156.
Mo, K.C., and R.W. Higgins, 1997. The Pacific-South American mode and tropical convection during the Southern Hemisphere winter. *Monthly Weather Review*, 126, 1581–1596.
Montecinos, A., A. Diaz, and P. Aceituno, 2000. Seasonal diagnostic and predictability of rainfall in subtropical South America based on Tropical Pacific SST. *Journal of Climate*, 13, 746–758.
Moura, A.D., and J. Shukla, 1981. On the dynamics of droughts in northeast Brazil: Observations, theory and numerical experiments with a general circulation model. *Journal of Atmospheric Sciences*, 38, 2653–2675.
Murphy, R.C., 1926. Oceanic and climatic phenomena along the west coast of South America during 1925. *Geographical Review*, 16, 26–64.

Negri, A., R. Adler, E. Nelkin, and G. Huffman, 1994. Regional rainfall climatologies derived from special sensor microwave imager (SSM/I) data. *Bulletin of the American Meteorological Society*, 75, 1165–1182.

Nigam, S., 1997. The annual warm to cold phase transition in the eastern equatorial Pacific: Diagnosis of the role of stratus cloud-top cooling. *Journal of Climate,* 10, 2447–2467.

Nobre, P., and J. Shukla, 1996. Variations of sea-surface temperature, wind stress and rainfall over the tropical Atlantic and South America. *Journal of Climate*, 9, 2464–2479.

Nogués-Paegle, J., and K.C. Mo, 1997. Alternating wet and dry conditions over South America during summer. *Monthly Weather Review*, 125, 279–291.

Norte, F., and M. Silva, 1995. Predicting severe versus moderate Zonda wind in Argentina. In: Preprints of the 14th Conference on Weather Analysis and Forecasting. American Meteorological Society. Boston, 128–129.

Parmenter, F.C., 1976. A Southern Hemisphere cold front passage at the equator. *Bulletin of the American Meteorological Society*, 57, 1435–1440.

Philander, S.G., 1990. *El Niño, La Niña, and the Southern Oscillation*. Academic Press. New York.

Physick, W.L., 1981. Winter depression tracks and climatological jet streams in the Southern Hemisphere during the FGGE year. *Quarterly Journal of the Royal Meteorological Society*, 107, 883–898.

Pisciottano, G., A. Diaz, G. Cazes, and C.R. Mechoso, 1994. El Niño-Southern Oscillation impact on rainfall in Uruguay. *Journal of Climate*, 7, 1286–1302.

Pizarro, J., and A. Montecinos, 2000. Cutoff cyclones off the subtropical coast of Chile. In: Proceedings of the 6th International Conference on Southern Hemisphere Meteorology and Oceanography. Santiago, Chile, 2000. American Meteorological Society. Boston, 278–279.

Quinn, W.H., and V.T. Neal, 1983. Long-term variations in the Southern Oscillation, El Nino, and the Chilean subtropical rainfall. *Fishery Bulletin*, 81, 1258–1288.

Ratisbona, C. R., 1976. The climate of Brazil. In: W. Schwerdtfeger and H.E. Landsberg (Editors), *Climate of Central and South America*. World Survey of Climatology, 12, Elsevier. New York, 219–293.

Renwick, J.A., 1998. ENSO-related variability in the frequency of South Pacific blocking. *Monthly Weather Review*, 126, 3117–3123.

Renwick, J.A., and M. Revell, 1999. Blocking over the South Pacific and Rossby wave propagation. *Monthly Weather Review*, 127, 2233–2247.

Robertson, A.W., and C.R. Mechoso, 1998. Interannual and decadal cycles in river flows of southeastern South America. *Journal of Climate*, 11, 2570–2581.

Rodwell M.J., and B.J. Hoskins, 1996. Monsoons and the dynamics of deserts. *Quarterly Journal of the Royal Meteorological Society*, 122, 1385–1404.

Ronchail, J., 1995. Variabilidad interanual de las precipitaciones en Bolivia. *Bulletin Institute Français Etudes Andines*, 24, 369–378.

Ropelewski, C.F., and M.S. Halpert, 1987. Global and regional-scale precipitation patterns associated with the El Niño/Southern Oscillation. *Monthly Weather Review*, 115, 1606–1626.

Ropelewski, C.F., and M.S. Halpert, 1989. Precipitation patterns associated with the high index phase of the Southern Oscillation. *Journal of Climate*, 2, 1606–1626.

Rutllant, J., and H. Fuenzalida, 1991. Synoptic aspects of the central Chile rainfall variability associated with the Southern Oscillation. *International Journal of Climatology*, 11, 63–76.

Rutllant, J., and R. Garreaud, 1995. Meteorological air pollution potential for Santiago, Chile: Towards an objective episode forecasting. *Environmental Monitoring and Assessment*, 34, 223–244.

Rutllant, J., H. Fuenzalida, P. Aceituno, A. Montecinos, R. Sanchez, H. Salinas, J. Inzunsa, and R. Zuleta, 2000. Coastal climate dynamics of the Antofagasta region (Chile, 23°S): The 1997–1998 DICLIMA Experiment. In: Proceedings of the 6th International Conference on Southern Hemisphere Meteorology and Oceanography. Santiago, Chile, 2000. American Meteorological Society, Boston, 268–269.

Salby, M. L., and H.H. Hendon, 1994. Intraseasonal behavior of clouds, temperature, and motion in the tropics. *Journal of Atmospheric Sciences*, 51, 2207–2224.

Satyamurty, P., C. Nobre, and P.L. Silva Dias, 1998. South America. In: D. Karoly and D. Vincent (Editors), *Meteorology of the Southern Hemisphere*. Meteorological Monographs No. 49. American Meteorological Society. Boston, 119–140.

Saulo, C., M. Nicolini, and S. Chou, 2000. Model characterization of the South American low-level flow during the 1997–1998 spring-summer season. *Climate Dynamics*, 16, 867–881.

Schwerdtfeger, W.C., 1961. Stromings und Temperatufeld der freien Atmosphare uber den Andes. *Meteorologische Rundshau*, 14, 1–6.

Schwerdtfeger, W., and L. Landsberg (Editors), 1976. *Climate of Central and South America*. Elsevier, New York.

Seluchi, M., and J. Marengo, 2000. Tropical-midlatitude exchange of air masses during summer and winter in South America: Climatic aspects and examples of intense events. *International Journal of Climatology*, 20, 1167–1190.

Seluchi, M., and A.C. Saulo, 1998. Possible mechanisms yielding an explosive coastal cyclogenesis over South America: Experiments using a limited area model. *Australian Meteorology Magazine*, 47, 309–320.

Seluchi, M., Y.V. Serafini, and H. Le Treut, 1998. The impact of the Andes on transient atmospheric systems: A comparison between observations and GCM results. *Monthly Weather Review*, 126, 895–912.

Sinclair, M. R., 1994. An objective cyclone climatology for the Southern Hemisphere. *Monthly Weather Review*, 123, 2239–2256.

Sinclair, M. R., 1996. A climatology of anticyclones and blocking for the Southern Hemisphere. *Monthly Weather Review*, 124, 254–263.

Spencer, R. W., 1993. Global oceanic precipitation from the MSU during 1979–91 and comparisons to other climatologies. *Journal of Climate*, 6, 1301–1326.

Taljaard, J.J., 1972. Synoptic meteorology of the Southern Hemisphere. In: *Meteorology of the Southern Hemisphere*. Meteorological Monographs, No. 35, American. Meteorological Society, 139–213.

Thompson, L.G., E. Mosley-Thompson, and B. Morales-Arnao, 1984. El Niño Southern Oscillation as recorded in the stratigraphy of the tropical Quelccaya ice cap, Peru. *Science*, 226, 50–52.

Trenberth, K. E., 1991. Storm tracks in the Southern Hemisphere. *Journal of Atmospheric Sciences*, 48, 2159–2178.

Trenberth, K.E., 1997. The definition of El Niño. *Bulletin of the American Meteorological Society*, 78, 2771–2777.

Velasco, I., and J.M. Fritsch, 1987. Mesoscale convective complexes in the Americas. *Journal of Geophysical Research*, 92, 9591–9613.

Vera, C., W. Higgins, J. Amador, T. Ambrizzi, R. Garreaud, D. Gochis, D. Gutzler, D. Lettenmaier, J. Marengo, C.R. Mechoso, J. Nogues-Paegle, P.L. Silva Dias, and C. Zhang, 2006. A unified view of the American monsoon systems. *Journal of Climate,* 19, 4977–5000.

Vera, C.S., and P.K. Vigliarolo, 1999. A diagnostic study of cold-air outbreaks over South America. *Monthly Weather Review*, 128, 3–24.

Vigliarolo, P.K., C. Vera, and S. Diaz., 2000. Southern Hemisphere winter ozone fluctuations. *Quarterly Journal of the Royal Meteorological Society*, 61, 1–20.

Vincent, D.G., 1998. Pacific Ocean. Meteorology of the Southern Hemisphere. In: D. Karoly and D. Vincent (Editors), *Meteorology of the Southern Hemisphere.* Meteorological Monographs, No. 49. American Meteorological Society. Boston, 101–118.

Virji, H., 1981. A preliminary study of summertime tropospheric circulation patterns over South America estimated from cloud winds. *Monthly Weather Review*, 109, 599–610.

Vuille, M., 1999. Atmospheric circulation over the Bolivian Altiplano during dry and wet periods and extreme phases of the Southern Oscillation. *International Journal of Climatology*, 19, 1579–1600.

Vuille, M., and C. Ammann, 1997. Regional snowfall patterns in the high arid Andes. *Climatic Change*, 36, 413–423.

Wagnon P., P. Ribstein, B. Francou, and J.M. Sicart, 2001. Anomalous heat and mass budget on the Zongo glacier during the 1997–98 El Niño year. *Journal of Glaciology*, 156, 21–28.

Walker, G.T., and E.W. Bliss, 1932. World Weather V. *Memoire of the Royal Meteorological Society*, 4, 53–84.

Wallace, J.M., and P. Hobbs, 1977. *Atmospheric science: An introductory survey.* Academic Press, San Diego.

Zhou, J., and K.M. Lau, 1998. Does a monsoon climate exist over South America? *Journal of Climate*, 11, 1020–1040.

4

Late Quaternary Glaciation of the Tropical Andes

Geoffrey O. Seltzer

The effects of climate change are intrinsic features of Earth's landscapes, and South America is no exception. Abundant evidence bears witness to the changes that have shaped the continent over time—from the glacial tillites inherited from late Paleozoic Gondwana to recent terrigenous sediments and life forms trapped in alluvial, lacustrine, and nearby marine deposits (see chapters 1 and 2). Preeminent among this evidence are the landforms and sediments derived from the late Cenozoic glaciations of the Andes, which have been the focus of so much recent and ongoing research. Because South America has long been a mainly tropical and subtropical continent, most of it escaped the direct effects of these glaciations. Nevertheless, portions of the continent extend sufficiently far poleward and rise high enough to attract snowfall and promote glaciers today. Glaciers were more emphatically present during Pliocene and Pleistocene cold stages, and it is their legacies that provide information about the changing environments of those times, and more especially of the past 30,000 years.

There is evidence for glaciation in the tropical and extratropical Andes as early as Pliocene time (Clapperton, 1993). In southern South America, along the eastern side of the Patagonian Andes, Mercer (1976) dated a series of basalts interbedded with glacial tills that suggest multiple glacial advances after ~3.6 Ma (million years before present). In the La Paz Valley, Bolivia, volcanic ashes dated by K/Ar (potassium/argon) methods are interbedded with glacial tills indicative of at least two phases of glaciation in the late Pliocene, at 3.27 and 2.20 Ma (Clapperton, 1979, 1993). This evidence for early glaciation in disparate parts of the Andes indicates that portions of the cordillera were high enough and climatic variations were great enough in the Pliocene for glaciers to form long before the cold episodes of the Pleistocene.

Glacial deposits in Ecuador, Perú, and Bolivia provide evidence for climate variability in tropical South America in the recent geological past. In the late Pleistocene, glacier equilibrium-line altitudes were as much as 1,200 m lower than they are today on the eastern slopes of the Andes, indicative of a significant depression in mean annual temperature in the tropics at maximum glaciation (e.g., Klein et al., 1999). After the last maximum glaciation (~21 ka, thousand years before present), glaciers appear to have expanded and retreated quickly in response to relatively rapid climate changes prior to retreating well within their modern limits by the start of the Holocene (e.g., Seltzer et al., 1995; Rodbell and Seltzer, 2000). Today, glaciers are in fast retreat, thus providing evidence for continual climate change during and since the twentieth century that, in part, may be exacerbated by global warming (e.g., Hastenrath and Ames, 1995; Brecher and Thompson, 1993). Glaciers are tuned to contemporary climatic conditions by their relationship between the accumulation and ablation (primarily through melting) of snow on an annual basis at the glacier surface. If conditions should become warmer and drier, glaciers retreat, such as is happening today. In contrast, should conditions become colder and wetter, there is a likelihood of glacier advances, such as occurred during the late Pleistocene glacial maximum. The challenge for scientists is to determine not only the timing of these variations in the extent of glaciation, but also to apply reasonable climatic interpretations to the glacial record.

Although there is evidence from deep sea sediments for multiple climate changes at a global scale, and for frequent glaciations at regional scales, over the course of the Quaternary (e.g., Shackelton and Opdyke, 1973), the stratigraphc and geomorphic record of glaciation in the Andes is discontinuous and primarily represents the late Quaternary. Subsequent erosion and later glaciations that overrode earlier glacial features have made the older glacial record difficult to determine. In the Northern Andes of Venezuela and Colombia, Helmens (1988) reported four glacial moraine stages, three of which are related to the most recent episodes of Pleistocene glaciation. High-elevation areas around Bogotá, in the Eastern Cordillera of Colombia, and in the Andes of central Venezuela all show morphologic indicators of these four stages of glaciation, supported by some radiocarbon chronologies for the three most recent stages. The focus in this chapter, however, is on late Quaternary glaciation in the tropics south of the Equator (fig. 4.1), as this is where most of the new data on the topic have emerged since the most recent reviews of South American glaciation (Clapperton, 1990, 1993, 2000; Clapperton and Seltzer, 2000).

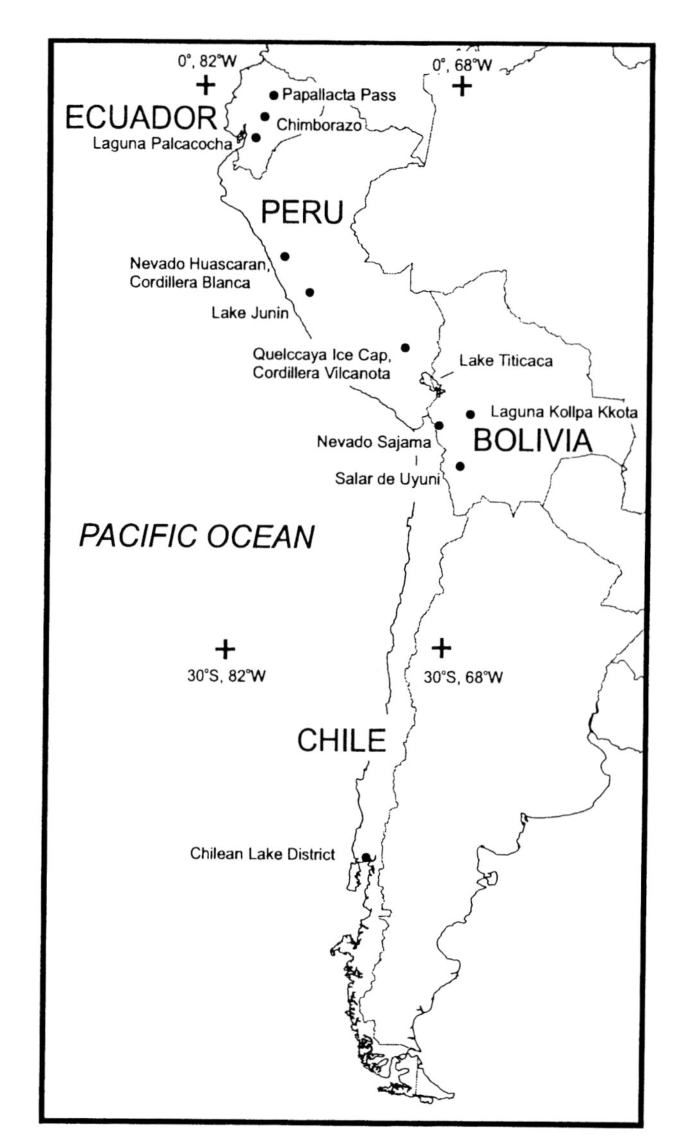

Figure 4.1 Location of sites mentioned in the text

4.1 Glacial Chronology, Evidence, and Implications

The absolute chronology of late Quaternary glaciation depends primarily on the association with glacial moraines of organic material that is suitable for radiocarbon dating. Glacial moraines in the tropical Andes are typically ridges of rock debris that were deposited by glaciers in terminal and lateral positions. Radiocarbon dating of organic material, either buried by moraines or deposited in lakes and wetland basins since deglaciation, can provide important absolute chronologies of glaciation. Other means of assembling relative or semi-quantitative glacial chronologies include lichenometry and the degree of soil development (fig. 4.2; e.g., Rodbell, 1992, 1993; Goodman et al., 2001). The dating of organic material buried beneath glacial deposits provides a maximum age for glaciation because the organics accumulated prior to burial by advancing ice (fig. 4.3). These types of deposits are very limited in the tropical Andes because the high and dry environment limits primary productivity and the accumulation of organic material on the landscape. This contrasts with the extensive data set of maximum ages for glaciation from the Lake District of Chile, where trees were buried during glacial advances in the late Pleistocene (fig. 4.4; Lowell et al., 1995). Much more common in the tropical Andes are lakes and wetlands that formed during deglaciation from glacial maxima (fig. 4.5). If the basal organic material in these basins can be retrieved by sediment coring, then radiocarbon dating of this material provides minimum ages for deglaciation. In this manner, much of the absolute chronology of glaciation in the tropical Andes has been established by bracketing glacial episodes between maximum and minimum limiting radiocarbon ages on organic material. More recently, the dating of cosmogenic nuclides on bedrock surfaces and erratic boulders exposed by deglaciation has offered a new dimension to problems of

Figure 4.2 Examination of a soil pit on a moraine that dates to ~16,000 cal yr B.P. on the north side of the Cordillera Vilcanota, Perú

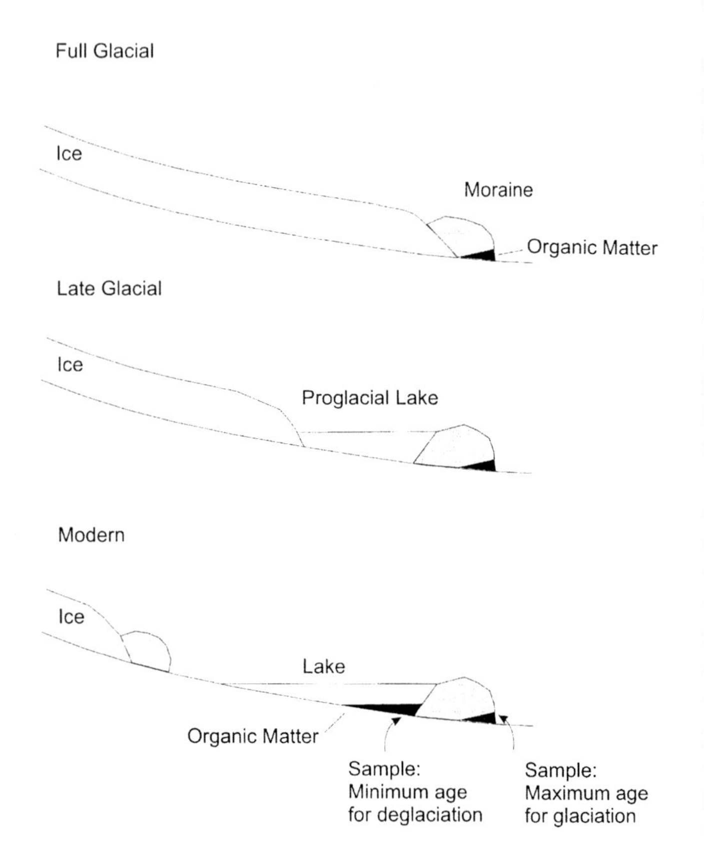

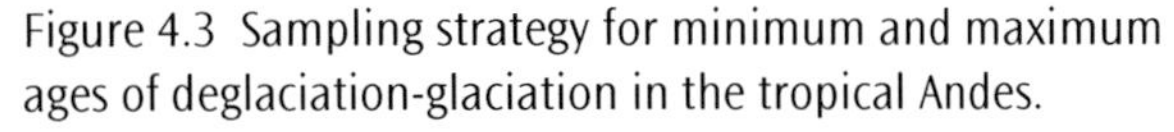

Figure 4.3 Sampling strategy for minimum and maximum ages of deglaciation-glaciation in the tropical Andes.

chronology. This method is beginning to be applied in the tropical Andes (e.g., Smith et al., 2001) and may provide an unparalleled chronology of glaciation extending back several hundred thousand years.

A further challenge to deciphering the paleoclimatic signal from glacial deposits lies in understanding the variety of climate changes that can impact glacial mass balance. The mass balance of a glacier, which ultimately controls the extent of glaciation, reflects a complex interplay between the energy balance and the delivery of snow to a glacier surface. The energy balance of a glacier surface is impacted by a number of factors, including solar radiation, the net longwave and shortwave energy balance, and the temperature and humidity of the overlying atmosphere (Seltzer, 1994a). As a result, the climatic interpretation of glaciation is often non-unique. It is therefore necessary to make a number of simplifying assumptions in order to interpret glaciation in terms of climate change.

Other proxy data derived from lake deposits and ice cores complement the glacial record and provide a more complete picture of late Quaternary climate change in the tropical Andes. Lacustrine deposits are found throughout the region and are distributed similarly to glacial deposits. The advantage of these records is that they are continuous and contain a number of potential proxies of both terrestrial and aquatic change. Glacial moraines typically provide snapshots in time of when relative glacial maxima occurred, but subsequent glaciations bury or destroy the moraine record of earlier, less extensive glaciations. In contrast, lakes accumulate sediment more or less continuously, unless they have desiccated or been destroyed by later glaciers, and this provides the opportunity to moni-

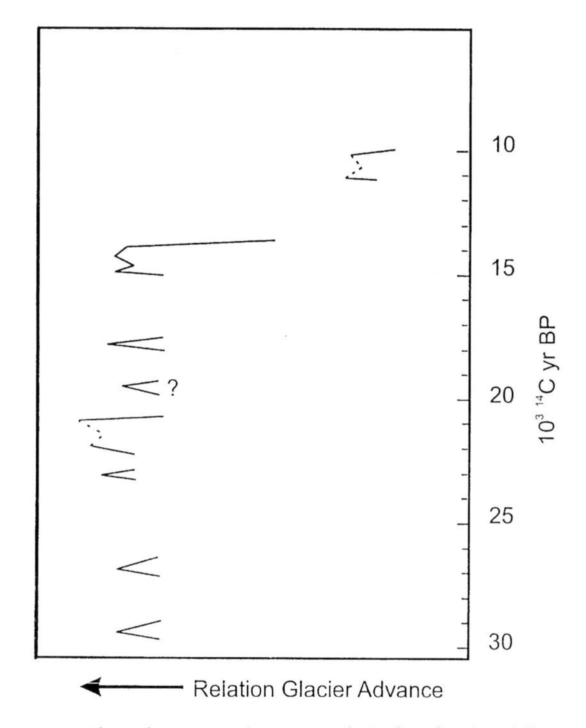

Figure 4.4 The advance and retreat of glaciers in the Chilean Lake District and the Southern Alps, New Zealand (after Lowell et al., 1995).

tor paleoenvironmental conditions throughout all stages of glaciation. Most lakes in the Andes formed during deglaciation as a result of the damming of glacial drainage by terminal and recessional moraines across valleys. However, there are a number of intermontane basins, such as Lake Titicaca, beyond the limits of glaciation where thick lacustrine sequences have accumulated. These provide unprecedented records of environmental change over long periods (e.g., Baker et al., 2001a, 2001b; D'Agostino et al., 2002; Seltzer et al., 2002).

Ice cores from high-altitude glaciers also provide important records of past climatic change (e.g., Thompson et al., 1985, 1995, 1998). At these sites little or no ablation has occurred, so the geochemical, dust, and mass accumulation records are nearly pristine. However, the ice-core records have presented some of their own challenges, such as those involving the development of firm chronologies and the interpretation of geochemical proxies (e.g., Vuille et al., 2003). Nonetheless, these records provide evidence of significant variability of the tropical Andean climate over the late Quaternary.

The establishment of the glacial record in the tropical Andes provides the opportunity to assess several important issues in paleoclimatology. For example, the magnitude of temperature change at the Last Glacial Maximum, around 21,000 cal yr B.P. (calendar years before present), when glaciers were at their maximum extent in middle to high latitudes of the Northern Hemisphere, has remained a significant issue in paleoclimatology (e.g., Rind and Peteet, 1985; Hostetler and Clark, 2000). The resolution of this issue for the tropical Andes would address the potential magnitude of temperature change in the tropics more generally. Associated with this are the questions of whether or not the tropics changed in synchrony with regions at higher latitude and, if so, what the mechanisms were that could explain these episodes of global change.

The late glacial period, between the maximum of the last glaciation and about 11,500 cal yr B.P., also provides an opportunity to test ideas about the nature and synchrony of global climate change over shorter periods. In the terminal Pleistocene of northern Europe, for example, the

Figure 4.5 Example of a moraine-dammed lake in the Cordillera Real, Bolivia. Lago Taypi Chaca Kkota lies within a glacial valley on the west side of the Cordillera Real and is dammed by a late-glacial terminal moraine.

Younger Dryas interval was a brief episode of cooling between 13,000 and 11,500 cal yr B.P., preceded and followed by warming trends (e.g., Alley et al., 1993). Was a similar climatic oscillation manifested in the tropical Andes, and if so, what are some of the mechanisms that could explain these changes? A Younger Dryas equivalent has been reported from glacial deposits in the middle latitudes of the Southern Hemisphere (Denton and Hendy, 1994), but unequivocal deposits related to the Younger Dryas have yet to be identified in the tropics (e.g., Rodbell and Seltzer, 2000).

The record of glaciation for the Holocene, the last 11,500 years, may also provide insight into the occurrence of millennial-scale changes as detected in other regions, specifically the North Atlantic (e.g., Bond and Lotti, 1995; Bond et al., 1997). These types of climate changes, similar to the Younger Dryas in duration but not necessarily in magnitude, occur on suborbital time scales, which means that they are not subject to the same interpretation of climate changes that may occur with periods of ~20,000, ~41,000, or ~100,000 years (see chapter 2). Again, the geographic range of such changes, and their phasing, may provide important clues to the mechanisms that could explain millennial-scale climate change. Related to the intriguing possibility of climate changes that occur on suborbital time scales is whether or not there has been modulation of the frequency of El Niño climatic events over the Holocene (e.g., Moy et al., 2002; Rodbell et al., 1999). If El Niño has varied in magnitude and frequency over the Holocene, this may be tied to internal oscillations in the tropical climate system or some other mechanism operating on suborbital time scales (see chapter 19).

Finally, the behavior of tropical glaciers over the twentieth century may provide an opportunity to assess the impact of global warming (Mann et al., 1998). The recent retreat of glaciers in the tropical Andes needs to be analyzed carefully to determine the rate at which glacier volume is actually decreasing. This change can then be related to potential climate forcing involving temperature, radiation balance, and precipitation. The recent retreat of tropical glaciers will also have an important impact on alpine hydrology and the downstream use of water, including domestic water supplies, irrigation, and hydroelectric power. Thus we can anticipate that glacier records, in concert with other proxy records, will provide a wealth of paleoclimatic information on a variety of timescales. The challenge is to provide plausible interpretations of these records.

4.2 Glaciers in the Tropical Andes

The first step in deciphering the glacier record for the tropical Andes is to understand the modern distribution of glaciers. An essential condition for glaciers to form in the tropics is that sufficient topography must lie at elevations high enough for the predominant form of precipitation to occur as snow. Compared with midlatitude regions, diurnal temperature variations in the tropics are greater than the change in monthly mean temperatures. This lack of temperature seasonality means that the atmospheric 0°C isotherm occurs at about the same elevation throughout year, and for the tropical Andes this is between 4,800 and 5,000 m above sea level (Klein et al., 1999). The elevation of the 0°C isotherm puts some constraints on the lower limit of the equilibrium line altitude (ELA) in the region. The ELA is the unique zone on a glacier where annual snow accumulation is balanced by the ablation of snow, primarily through melting. For example, on the eastern slope of the Bolivian Andes, modern snowlines are found at about 4,400 m (fig. 4.6; Klein et al., 1999). Snowline was defined by Klein et al. as the lower limit of perennial snow, which is probably lower than a true glaciological ELA. Below 4,400 m on the eastern slope of the Bolivian Andes, temperatures are warm enough such that melting and rainfall are common throughout the year. In contrast, to the west, snowlines rise to about 5,000 m on the western slope of the Eastern Cordillera of the Andes. This is because of the strong orographic effect of the Andes on precipitation totals. Mountains less than 4,800 to 5,000 m above sea level in the Eastern Cordillera of Bolivia do not support modern glaciers because of the predominance of melting at these elevations. Snowlines rise significantly (to altitudes >5,800 m) above the 0°C isotherm to the south and west in the Bolivian Andes. This is a result of a strong precipitation gradient from >1,000 mm yr^{-1} on the eastern slope of the Andes to <200 mm yr^{-1} in the Western Cordillera of southern Bolivia. These observations indicate that glaciers in the Eastern Cordillera would be most sensitive to temperature variation as they occur in an environment where precipitation is not the main limitation on their elevational extent. In contrast, in the Western Cordillera where glaciers are almost entirely above the elevation of the 0°C isotherm, a precipitation increase would be imperative to produce a more extensive glaciation.

4.3 The Last Glacial Maximum (~21,000 cal yr B.P.)

Considerable debate surrounds the magnitude of temperature change that occurred in the tropics during the Last Glacial Maximum (LGM). This debate emerged from the observation that tropical sea-surface temperatures, as determined by CLIMAP (1981), appeared to have changed little in comparison to significant temperature reductions interpreted from palynological and snowline studies. For example, terrestrial evidence from various tropical mountains suggested to Rind and Peteet (1985) that treelines

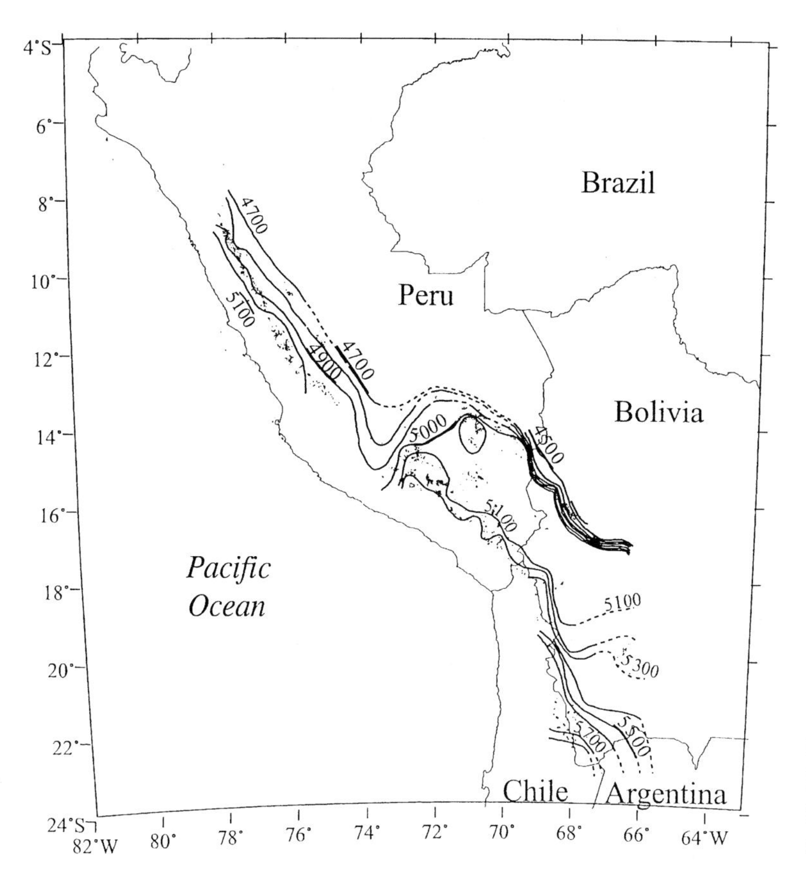

Figure 4.6 Modern snowline in the Peruvian and Bolivian Andes. Contour interval is 100 m, dashed where approximate. Dots represent individual snowline observations (after Klein et al., 1999).

and snowlines were depressed by at least 1,000 m compared to today, and when multiplied by a temperature lapse rate for a moist atmosphere, mean annual temperature appeared to have been >5°C colder than today. This is not easily reconciled with a tropical ocean where sea-surface temperatures were not much colder than today (e.g., Webster and Streten, 1978; Rind and Peteet, 1985). Thus the resolution of this debate, at least with regard to snowlines, has centered on two issues: the timing of the LGM in the tropics and the temperature reduction associated with glaciation.

An important objective of CLIMAP (1981) was to map sea-surface temperatures for the LGM based on microfossil assemblages contained in deep sea cores. The target age for the LGM was ~21,000 cal yr B.P., which represents the maximum enrichment in $\delta^{18}O$ of calcite in planktonic foraminifera (e.g., Shackleton and Opdyke, 1973). CLIMAP (1981) recognized that, given the variations in sedimentation rates and other uncertainties, the target time frame for the LGM was really ~16,800 to ~26,000 cal yr B.P. Given what we know today about the frequency of potential climate changes, this is a broad time span and may integrate a number of important events, such as the Heinrich I event, with other climatic events (Bond and Lotti, 1995). Nonetheless, the challenge in dating the glacial sequence is to see whether or not the LGM in the tropical Andes corresponds to marine isotope stage 2, or ~30,000 to ~20,000 cal yr B.P., when presumably Northern Hemisphere continental glaciers were at their maximum extent.

4.3.1 The Last Glacial Maximum in the Tropical Andes

Until recently, the best data on the timing of the LGM in the tropical Andes have been minimum ages for deglaciation. One of the few sites that have close limiting ages is Laguna Kollpa Kkota, Bolivia, located in a small cirque with a headwall elevation of 4,560 m (Seltzer, 1994b). The glacier that formed here covered <l km^2 and formed a distinct end moraine that dammed a lake in the cirque following deglaciation. That lake is now dry. This site represents the lowest elevation headwall-lake-moraine complex that has been sampled in the tropical Andes, and it is a reasonable assumption that the age of the basal lake sediments at the site should provide a good minimum age for deglaciation. Various organic fractions (i.e., humin, humic, and bulk organic carbon) from the basal lake sediments have yielded a minimum age of at least 21,000 cal yr B.P. for deglaciation here. At the time of the LGM, glacier ELA was about 600 m lower than it is today. At other

sites in the Eastern Cordillera of Bolivia, basal ages from lakes formed during deglaciation are ~14,000 cal yr B.P., which clearly postdate the onset of deglaciation from the LGM (e.g., Seltzer, 1992). This discrepancy probably arises because these other sites have high headwall elevations (>5,000 m) and continued to experience glaciation following the LGM.

Sediment cores from Lake Titicaca, Perú-Bolivia, and Lake Junin, central Perú, confirm an age for deglaciation from the LGM at ~19,500 to 22,000 cal yr B.P. (Seltzer et al., 2000; Baker et al., 2001a; Seltzer et al., 2002). Both of these lakes lie beyond the glacial limit and received glaciofluvial outwash during the last glaciation. In both records a drop in magnetic susceptibility of the sediments and a decrease in sediment accumulation rates signify when glaciers retreated within their outermost limits at the end of the LGM (fig. 4.7). At this time proglacial lakes formed in the surrounding cordillera and acted as sediment traps for the glacially derived silts. At Lake Junin there is an apparent shift to glacial sedimentation manifested by high magnetic susceptibility and low organic carbon around 30,000 cal yr B.P., which would date the onset of the LGM in this region. The sedimentology from these two sites indicates that the LGM in the tropical Andes roughly corresponded with the LGM of the Northern Hemisphere, in other words, marine isotope stage 2, from ~30,000 to 20,000 cal yr BP.

Deglaciation from the LGM appears to have been associated with an increase in mean annual temperatures as climatic conditions remained wet at this time (Seltzer et al., 2002). Both Lake Titicaca and Lake Junin appear to have been deep and fresh during and after maximum glaciation. The principal controlling factor on glacier mass balance that could lead to deglaciation would therefore be an increase in mean annual temperature. This warming in the tropics predates the significant warming of the Northern Hemisphere indicated by ice-core records from Greenland (fig. 4.7), which suggests that the tropics and the Southern Hemisphere may have played important roles in the last global deglaciation.

Cosmogenic isotope dating of erratic boulders on glacial moraines and of glacially eroded bedrock surfaces in the Peruvian Andes does, however, challenge some of our concepts concerning which moraines in the landscape may date to the LGM. For example, ^{10}Be ages of erratics in the Alcacocha Valley east of Lake Junin indicate that the LGM most likely dates to marine isotope stage 2, but the LGM moraines do not represent the most extensive Pleistocene glaciation in the valley (Smith et al., 2001). Moraines >250,000 years old are found at the mouth of this valley (fig. 4.8). Thus it should not be assumed that the outermost moraines in a valley always date to the last glaciation, because it is clear that there have been a number of earlier, more extensive glaciations during Pleistocene time in the tropical Andes.

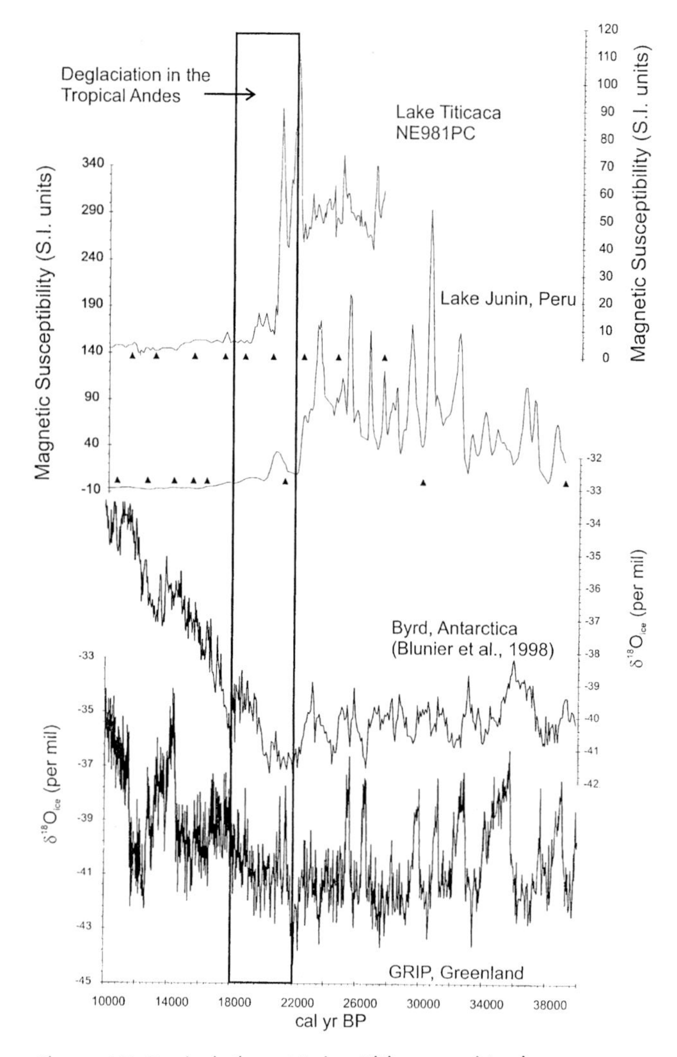

Figure 4.7 Deglaciation at Lakes Titicaca and Junin represented by a sharp drop in the magnetic susceptibility of lake sediments compared to the isotopic proxy for temperature from ice cores in Antarctica and Greenland. Solid triangles represent radiocarbon ages used to develop the chronology of the lake-core sequences (after Seltzer et al., 2002).

4.3.2 *Climatic Significance*

Determination of the climatic significance of past glaciations requires both a quantitative reconstruction of the magnitude of glaciation and modeling of the impact of possible climatic changes on glacier mass balance. Often this type of paleoclimatic reconstruction produces nonunique solutions and is best viewed as an attempt to understand the range in possible past climate scenarios. Additional proxy records from lakes, ice cores, and other archives can be used to constrain some of the potential solutions in these reconstructions. In the case of the tropical Andes, the reconstruction of temperature change at the LGM has been a major objective, and it has therefore been

Figure 4.8 Left and right (in shadow) lateral moraines in the Alcacocha Valley on the Junin Plain, Perú. These moraines are >250,000 years old and are several kilometers downvalley from the LGM (~21,000 cal yr B.P.) moraines.

important to determine the glacial environment that might produce the most reliable paleotemperature information.

The simplest way of seeking climatic interpretations for past glaciations is to assess the magnitude of glacial ELA depression from the modern values. Methods have been derived from observations of modern ELAs of midlatitude glaciers that use a ratio based on a glacier's accumulation area relative to the entire reconstructed glacier surface area. The application of these methods for past glaciations may rely on the hypsometry of reconstructed glacier surfaces or simply on a determination of the elevations of terminal moraines and associated cirque headwalls (e.g., Meierding, 1982). For midlatitude glaciers, typical ratios for the accumulation area relative to the entire surface area range from 0.4 to 0.65. Some studies of tropical glaciers, however, have suggested that this ratio may be as high as 0.8 (Kaser and Osmaston, 2002). The reason that tropical glaciers may have larger accumulation areas than midlatitude glaciers is that ablation occurs throughout the year in the tropics and therefore the ablation zone does not have to be as large to waste the requisite mass that is accumulated in the upper reaches of a glacier.

During the LGM in the tropical Andes, glacier ELAs were depressed as much as 1,200 m beneath modern values on the eastern slopes (fig. 4.9; Klein et al., 1999). To the west, especially over the Altiplano, ELA depression appears to have been significantly less, reaching about half the magnitude of ELA depression on the eastern slopes. To the south, in the more arid parts of the Central Andes, total ELA depression was also less than on the eastern slope. This pattern of ELA depression is similar to the modern pattern of lower snowlines to the north and east, and reflects the orographic impact of the Andes on precipitation totals.

The occurrence of the modern snowline at or below the elevation of the 0°C atmospheric isotherm on the eastern slope of the Andes suggests that glacial ELAs there are limited not by the amount of moisture but by temperature. Following this reasoning, Klein et al. (1999) used an approach first proposed by Kuhn (1981, 1989) to relate a change in ELA to climate change based on energy-balance parameters of a glacier surface. In their analysis, Klein et al. considered changes in temperature, insolation, precipitation, and humidity and their impact on glacier ELA along the eastern slope of the Andes. They found that even large changes in precipitation had little impact on the ELA. In contrast, changing the altitude of the 0°C isotherm, or essentially the mean annual temperature, had the most significant impact on the ELA. Thus the modeled estimates of depression of mean annual temperature varied from 5° to 9°C. The range in reconstructed values for mean annual temperature is related to uncertainties in the magnitude of ELA depression and the factors that make up the energy balance of a glacier. Nonetheless, these results confirmed that there was a substantial depression in mean annual temperature in the tropical Andes during the LGM that would not be consistent with little change in sea-surface temperature (e.g., Hostetler and Clark, 2000).

Why were glacial ELAs depressed twice as much along the eastern slopes of the Bolivian Andes compared to the Altiplano to the west? Assuming that ELA depression on the eastern slopes reflects a significant reduction in mean annual temperature, it is reasonable to expect this amount of reduction to be similar across the region. If this were the case, then it seems that glaciers in the Altiplano were either limited by the amount of precipitation change or that other factors caused significantly less reduction in ELA. One possibility is that, as glaciers expanded beyond the

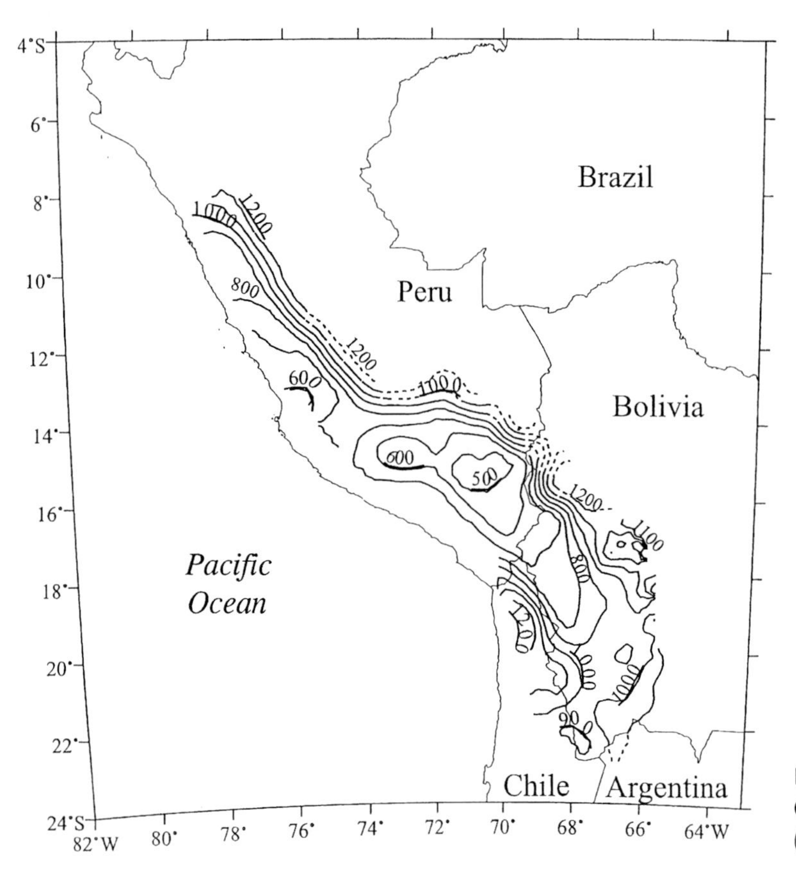

Figure 4.9 Change in snowline, LGM minus modern. Contour interval 100 m, dashed where approximate (after Klein et al., 1999).

cordillera onto the Altiplano, their hypsometry changed such that much more of their surface area was subject to ablation. Thus topography may explain the difference in the magnitude of ELA depression between these two regions.

There is support from other paleoclimatic archives for the climatic interpretation of a major LGM temperature depression associated with glaciation in the eastern tropical Andes. Piston cores from Lake Titicaca indicate that the lake was fresh and overflowing during the LGM, compared to the moderately saline conditions that developed in postglacial times (Baker et al., 2001a). Thus precipitation was most likely higher at the LGM and appears to have remained so until at least 15,000 cal yr B.P. even while deglaciation was taking place (Seltzer et al., 2002). This further supports the belief that deglaciation was responding primarily to an increase in mean annual temperature (Klein et al., 1999).

Isotopic evidence from Nevado Sajama, Bolivia, and Nevado Huascarán, Perú, also indicates much colder temperatures at the LGM (Thompson et al., 1995, 1998). The oxygen isotope ratio in ice of the last glacial stage has been interpreted to indicate a temperature reduction of 8° to 12°C. This interpretation assumes that a midlatitude linear relationship between isotopic depletion and mean annual temperature can be applied to these ice-core records. In the tropics, however, the isotopic composition of glacial ice is probably also strongly related to the amount of precipitation in a given year (e.g., Grootes et al., 1989). Studies of isotopic variability of glacial ice at high altitudes in the Andes also suggest that there may be a close relationship between synoptic climate patterns over the tropical Pacific and the ice-core record, further complicating the quantitative reconstruction of temperature change from the ice cores (Vuille et al., 2003). At this time, given the uncertain relationship between temperature and the isotopic composition of precipitation in the tropics, it is equivocal whether or not the ice cores provide an independent test of the ELA interpretations.

Archives from low elevations in the tropics do provide some verification that mean annual temperatures in tropical South America were significantly colder at the LGM. Noble gases trapped in Brazilian groundwater during the LGM indicate that the recharge areas for these waters were 4° to 5°C cooler (Stute et al., 1995). Paleoecological evidence also indicates that temperatures were ~5°C cooler in tropical South America and in Central America (Bush et al., 2001). Recent reevaluations of sea surface temperatures in the tropical Pacific further indicate that temperatures were significantly cooler at the LGM (Hostetler and

Mix, 1999; Lea et al., 2000) and that warming at the end of the LGM preceded deglaciation in high northern latitudes (Lea et al., 2000).

4.4 The Late Glacial Interval in the Tropical Andes (~21,000 to 11,500 cal yr B.P.)

The late glacial interval in the tropical Andes spans the period from the end of the LGM (~21,000 cal yr B.P.) to the beginning of the Holocene (11,500 cal yr B.P.). Elsewhere, this has been a period characterized by abrupt climate changes as glaciers retreated from maximum positions. In the North Atlantic region, the late glacial was interrupted by both the Heinrich I event (Bond and Lotti, 1995) and the Younger Dryas event (e.g., Alley et al., 1993). The Heinrich events were large discharges of icebergs into the North Atlantic Ocean associated with cooler atmospheric and sea surface temperatures. The Heinrich I event occurred ~17,000 cal yr B.P., after a period of cooling that lasted ~3,000 years. The Younger Dryas was a return to cold, often glacial conditions between 13,000 and 11,500 cal yr B.P. Both events indicate that, at least in the North Atlantic region, climate can change abruptly. Such changes may have an impact at a hemispheric or global scale. In the glacial record from the Southern Hemisphere, both the Heinrich events and the Younger Dryas have been identified in the Lake District of Chile and in the Southern Alps of New Zealand (Lowell et al., 1995). If these events were truly global in extent, then it should be possible to detect them in the tropics as well. Complicating this search for global millennial-scale climate changes in the glacial record is the difficulty in establishing a chronology sufficiently accurate for comparison with the North Atlantic records.

In the Cordillera Vilcanota, southern Perú, there is a moraine sequence dated to <16,000 cal yr B.P. (Mercer and Palacios, 1977; Mercer, 1984; Goodman et al., 2001; Mark et al., 2002). These moraines occur ~8 km downvalley from the modern glacier and overlie peat that has been independently radiocarbon dated in a number of studies (fig. 4.2). Thus the age from this peat is a maximum age for glaciation in the valley. Upvalley from this site, a modern wetland has a basal age of 12,250 cal yr B.P., and provides a minimum age for the moraines. This site has been considered a type locality for several other studies in the Peruvian and Bolivian Andes that identify a late glacial advance around 16,000 cal yr B.P. (e.g., Seltzer, 1992; Rodbell, 1993).

The recognition of a moraine stage equivalent to the Younger Dryas in the tropical Andes has been contentious. At Papallacta Pass, Ecuador, Clapperton et al. (1997) dated a glacial till to the Younger Dryas time interval. They contended that this till was deposited by an ~140 km^2 ice cap that formed as a result of a global Younger Dryas cooling event. In contrast, Heine and Heine (1996) viewed the Younger Dryas time period at Papallacta Pass as a period of soil formation and no glaciation. Both groups apparently worked in the same region but in different stratigraphic sections through the late Pleistocene deposits. Earlier, Clapperton (1993) had also identified proglacial lacustrine deposits at Chimborazo, farther south in Ecuador, which he attributed to a Younger Dryas glacial advance. Again, Heine and Heine (1995) questioned this interpretation and suggested alternatively that the Chimborazo organic deposits, considered interstadial by Clapperton, formed as a result of lake-level lowering in a proglacial environment.

In the Cordillera Blanca, Perú, a glacial advance that culminated at the beginning of the Younger Dryas may also correlate with a similar glacial advance at the Quelccaya Ice Cap (Rodbell and Seltzer, 2000). The evidence from both sites suggests that glaciers were in fast retreat during Younger Dryas time. Thus the evidence for a late glacial equivalent of the Younger Dryas in the North Atlantic region remains equivocal, given the variety of dating results on glacial moraines. At this time, the moraines in Perú identified as late glacial in age most likely do not date to the Younger Dryas interval. Throughout the Ecuadorian, Peruvian, and Bolivian Andes, glaciers had retreated to within their modern limits by 11,500 cal yr B.P., which marked the end of the late glacial interval in the region (Seltzer et al., 1995).

Additional information on late glacial climatic conditions in the tropical Andes comes from ice-core and lake records. Two ice cores recovered in 1993 from Nevado Huascarán, Perú, were drilled some 60 m through the ice to bedrock. Based on an ice-thinning model and fitting to an isotopic record of planktonic foraminifera off the coast of Portugal, Thompson et al. (1995) hypothesized that the record extended back to 25,000 cal yr B.P. and that it revealed a record of Younger Dryas cooling. In the basal meters of the ice cores, they measured an ~2 per mil depletion in $\delta^{18}O$ in the ice, which they attributed to millennial-scale cooling in the late glacial. The chronology for this event was derived by fitting the isotopic record from the ice core to the planktonic foraminifera record of $\delta^{18}O$ from Bard et al. (1987) off the coast of Portugal. Their reasoning for this correlation was that the source of moisture for Nevado Huascarán was the tropical Atlantic and that therefore the isotopic record off Portugal should reflect how the source of precipitation to the region might have changed. A problem with this reasoning is the assumption that the isotopic event revealed in the ice core is time synchronous with the event off Portugal. Once this assumption is made, there is no means to test independently whether, in fact, the Younger Dryas event occurred in the tropical Andes.

The chronology of the ice-core record from Nevado Sajama, Bolivia, is better constrained by radiocarbon dates on organic material at the base of the core and in the Holocene section of the record (Thompson et al., 1998). However, there is a lack of absolute age control between ~25,000 and 10,000 cal yr B.P., and therefore their recognition of late glacial cooling in the stable isotope record has only tentative correlation with the Younger Dryas. In fact, Thompson et al. refer to this cooling event at Nevado Sajama as the Deglacial Climate Reversal because it started one thousand years before the Younger Dryas and lasted ~2,500 years (~14,000 to 11,500 cal yr B.P.). This interval also appears to coincide with higher snow accumulation. The oxygen isotopes, if correctly interpreted, thus suggest a longer, cooler interval, and the ice accumulation rate suggests wetter conditions than either before or immediately after the Deglacial Climate Reversal.

At Lake Titicaca, overflow conditions are recognized from >25,000 to ~15,000 cal yr B.P., followed by intervals during which the lake may have dropped below its outlet and then overflowed again in the Younger Dryas, before falling below its outlet for an extended period (Baker et al., 2001a). The important proxy used for making these lake-level interpretations was the relative quantity of benthic diatoms in the sedimentary record. Essentially, when the percentage of benthic diatoms in a standard assay exceeded 20%, then the lake level was thought to be at most 30 m above the coring site (Tapia et al., 2003). An alternative hypothesis for lake-level variation in the late glacial at Lake Titicaca can be derived from the carbon isotopic composition of bulk organic matter in the sediment. The isotopic composition of bulk organic matter provides a guide to the mix of planktonic versus nearshore organic carbon that has accumulated in the sediment. Presumably distinct isotopic end members can be identified and used in this way to trace the organic carbon sources to the coring site. From the isotopic composition of bulk organic matter, Rowe et al (2002) suggested that lake levels were becoming progressively shallower throughout the Younger Dryas. Thus, whether or not the Younger Dryas coincides with a high lake level at Lake Titicaca remains equivocal.

At the Salar de Uyuni in the southern Altiplano, conditions remained wet until ~15,000 cal yr B.P., when lake levels fell and the Salar began to desiccate. Based on the natural gamma-radiation log of the drill hole in the Salar, Baker et al. (2001b) suggested that a return to lacustrine conditions may have been associated with the Younger Dryas, but there is presently no independent dating to support this hypothesis. There was a brief return to wet conditions in the early Holocene, which Sylvestre et al. (1999) referred to as the Coipasa event, followed by dry conditions throughout the remainder of the Holocene. Thus, in general terms, the Altiplano experienced wet conditions until ~15,000 cal yr B.P., followed by progressive desiccation that may or may not have been punctuated by brief wet intervals.

At Lake Junin (11°S), Perú, authigenic carbonates provide a record of the isotopic composition of lake water throughout late glacial and Holocene time. This record generally shows progressive drying from ~16,000 to 10,000 cal yr B.P., without any wet interludes (Seltzer et al., 2000). The isotopic record of ice at Nevado Huascarán (9°S) can be used as a monitor of the changing isotopic composition of precipitation in the region. By juxtaposing these two records, Seltzer et al. (2000) deconvolved the contribution of changes in temperature, source precipitation, and precipitation-minus-evaporation balance to the isotopic composition of the lacustrine carbonates. The resulting interpretation, shown in figure 4.10, allows for a unique determination of the moisture balance of the region.

In summary, the late glacial interval in the tropical Andes includes periods of glaciation, particularly at

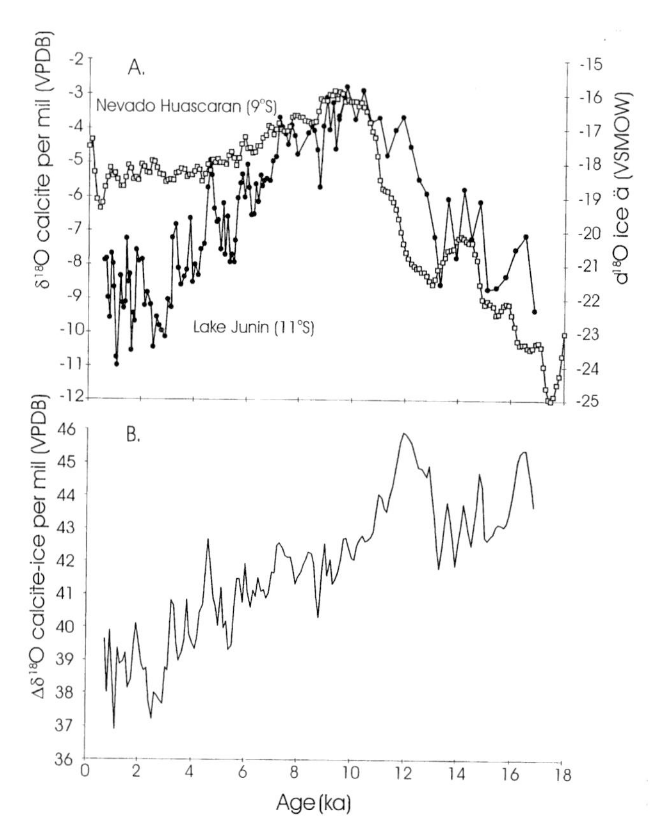

Figure 4.10 (A) Stable isotope records from Lake Junin and Nevado Huascaran, Perú. (B) The difference between the Huascaran and Junin isotope records (Junin minus Huascaran) on the VPDB scale. Larger values indicate drier conditions in the watershed and more evaporative enrichment in ^{18}O in the lake water (after Seltzer et al., 2000).

~16,000 and ~13,000 cal yr B.P., but it is uncertain how ubiquitous these glacial advances were and what constitutes a regionally coherent signal of climate change. In the lake records, there is equivocal evidence for wet conditions during the Younger Dryas at Lake Titicaca, while dry conditions apparently prevailed at the Salar de Uyuni and Lake Junin. At Nevado Sajama, the ice core record contains an apparent deglacial cold reversal from ~14,000 to 11,500 yr cal yr B.P. associated with wet conditions, but the chronology of the isotopic record at Nevado Huascarán has been fitted to a higher-latitude site in the North Atlantic, which reduces its usefulness for assessing independently the occurrence of the Younger Dryas in the tropical Andes. By 11,500 cal yr B.P., glaciers were within their modern limits and climatic conditions were generally drier and warmer.

4.5 Holocene Glaciation

4.5.1 The Early to Middle Holocene (11,500 to 3500 cal yr B.P.)

The early to middle Holocene is characterized by little evidence for glacial activity. Glaciers were at or near their modern limits throughout this period. Evidence for renewed glaciation does not become widespread until after ~4,000 cal yr B.P., when climatic conditions became cooler and wetter (Seltzer, 1992). Many of the summits <6,000 m above sea level in the region may have deglaciated entirely during the early and middle Holocene (e.g., Abbott et al., 1997). For example, the Quelccaya Ice Cap has a basal age of only ~1,500 years (Thompson et al., 1985). Apparently the glacier is frozen to its base, so it should not be losing mass over time at its base. The summit of the ice cap is ~5,400 m above sea level, which is 400 m above the current 0°C isotherm for the region. A temperature rise of ~2°C would raise the ELA to >5,400 m and would result in complete deglaciation of the plateau that supports the ice cap. It is possible that this happened in the early to middle Holocene, which would explain the truncation of the ice-core record and the observation that the Quelccaya Ice Cap was less extensive than today, between ~2,750 and 1,500 cal yr B.P. (Mercer and Palacios, 1977; Seltzer, 1990).

Additional evidence for a dry early to middle Holocene comes from lakes in the region. Lake Titicaca dropped ~85 m below its current level in the early to middle Holocene as a result of a decrease in effective moisture, or precipitation-minus-evaporation (fig. 4.11; Seltzer et al., 1998; Cross et al., 2000; Baker et al., 2001a; D'Agostino et al., 2002). The lake level began to rise again around 4,000 cal yr B.P., and the shallow basin of the lake flooded ~3,500 cal yr B.P. (Abbott et al., 1996; Baker et al., 2001a). The Lake Junin isotopic record shows a maximum in aridity at ~10,000 cal yr B.P., followed by an increase in effective moisture throughout the Holocene, with some significant steps to higher levels of effective moisture in the late Holocene (Seltzer et al., 2000). At Laguna Palcacocha, Ecuador, a record of detrital laminae in a lake core has been interpreted as an emergence of modern El Niño-Southern Oscillation (ENSO) variability about 6,000 cal yr B.P., culminating about 1,100 cal yr B.P. (Rodbell et al., 1999; Moy et al., 2002). Generally, the warm phase of ENSO (El Niño), which brings heavy rainfall to coastal Perú, is associated with drier conditions in the tropical Andes, whereas the cool phase (La Niña) is associated with wetter conditions (see chapter 19). For the Holocene as a whole, an increase in ENSO variability may be linked with increasing effective

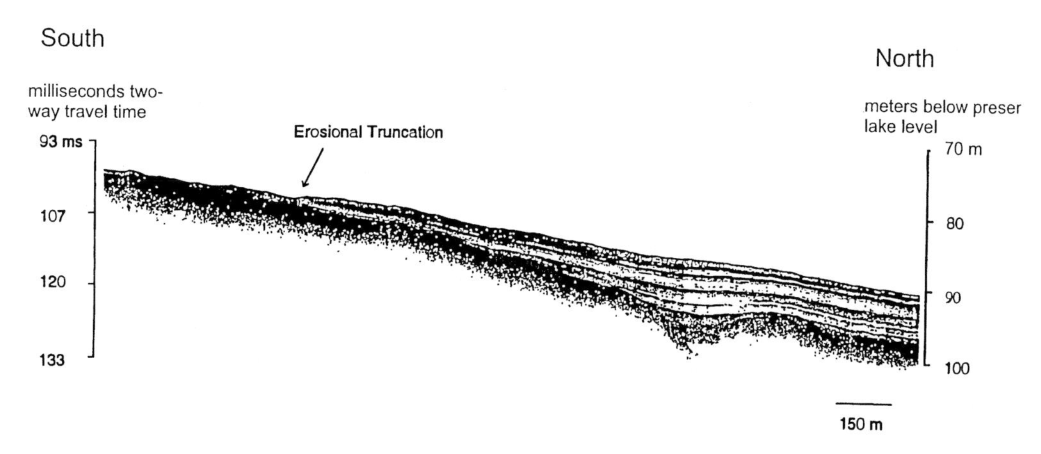

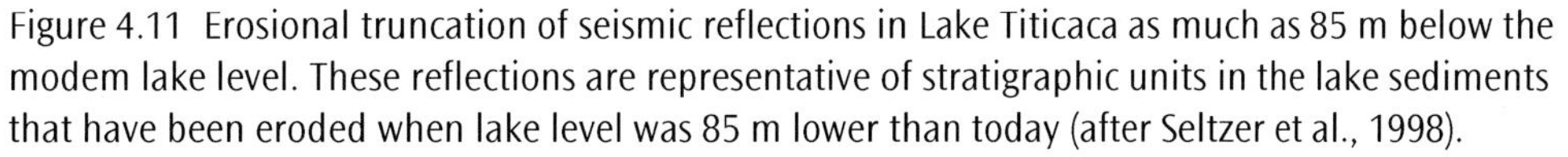
Figure 4.11 Erosional truncation of seismic reflections in Lake Titicaca as much as 85 m below the modem lake level. These reflections are representative of stratigraphic units in the lake sediments that have been eroded when lake level was 85 m lower than today (after Seltzer et al., 1998).

moisture in the tropical Andes, while increased summer insolation may also have been important for driving changes in effective moisture (Abbott et al., 1997; Seltzer et al., 2000; Wolfe et al., 2001).

4.5.2 Neoglaciation and Recent Glacier Retreat

Relatively minor glacier advances (neoglaciation) have occurred in the late Holocene throughout the tropical Andes (e.g., Seltzer, 1990, 1992; Seltzer et al., 1995; Rodbell, 1992). Accurate dating of these advances is difficult, although in some cases there are maximum limiting ages on organic material that was buried by advancing glaciers. These glacial advances may have occurred as a result of relatively subtle climate changes, certainly modest when compared to the presumed temperature reduction that resulted in the LGM in the region.

Recent glacier retreat has been widespread in the tropical Andes (Brecher and Thompson, 1993; Hastenrath and Ames, 1995; Mark et al., 2002). This retreat phase probably began with the waning of the global Little Ice Age during the eighteenth and nineteenth centuries, but has accelerated in the latter half of the twentieth century (Seltzer, 1990). Accelerated glacier retreat has been attributed by some scientists to recent climatic warming of the tropical Andes, which can be clearly detected in the instrumental records from the region (Mark, 2002). The challenge in the study of modern glacier retreat is to measure the volumetric loss of ice, not just the retreat of ice fronts, to determine quantitatively the energy transfer necessary to produce glacier recession. In this manner it is possible to evaluate more definitively the forcing factors that have produced glacier retreat in the twentieth century (e.g., Mark, 2002).

Modern deglaciation in the Andes is also important in terms of the hazards it poses and the hydrologic changes that affect alpine watersheds. For example, proglacial lakes may form as glaciers retreat and meltwaters are dammed by moraines. If these moraine dams burst, the large volumes of water and mud suddenly released pose a major hazard to people living downstream (Kaser and Osmaston, 2002). Melting of glacier ice during the extended dry season (November to April) is also important in moderating the impact of seasonal drought in the tropical Andes (fig. 4.12; Francou et al., 1995; Ribstein et al, 1995; Mark and Seltzer, 2003). The potential loss of the water reservoir trapped in glacier ice may produce stream flows that are more variable and the availability of water for hydropower, irrigation, and domestic water supplies will be diminished during the dry half of the year (Seltzer, 2001). Thus the fate of the tropical Andean glaciers may have a serious impact on the growing populations in Ecuador, Perú, and Bolivia that depend on waters discharged from high altitude glaciers for a variety of uses.

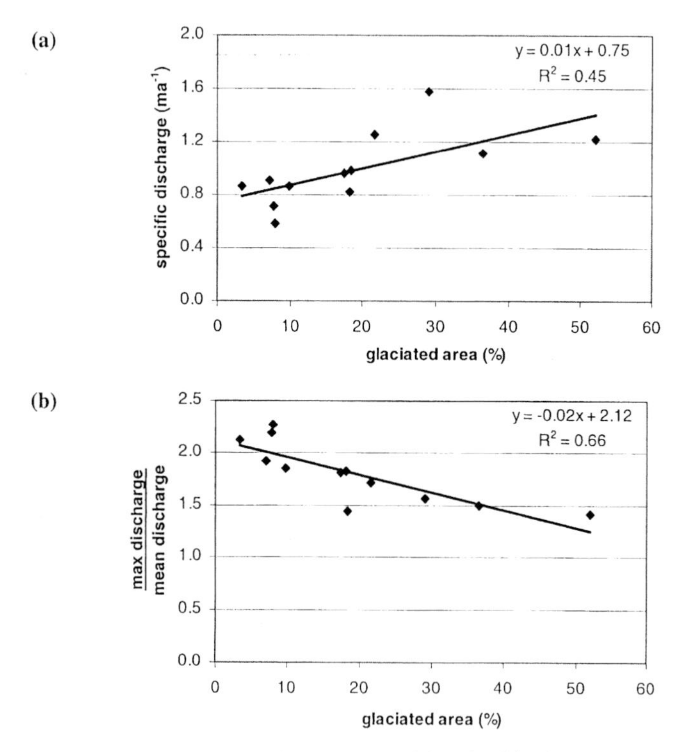

Figure 4.12 (a) Specific discharge and (b) variability in discharge in glaciated watersheds of the Cordillera Blanca, Perú (after Mark and Seltzer, 2003).

4.6 Conclusion

Tropical Andean glaciers are dynamic and indicative of significant climatic variation in tropical South America over at least the last 30,000 years. Generally, Andean glaciers were more extensive during the same period (i.e., marine isotope stage 2) as glaciers expanded in high latitudes of the Northern Hemisphere, reaching their maximum extent around 21,000 cal yr B.P. At this time climatic conditions were colder and wetter than later. This observation differs from earlier assessments of minimal temperature change in the tropics during the last glaciation (CLIMAP, 1981), and is more in line with several proxies that now indicate that the LGM was significantly colder than today in the tropical Andes. Glaciers retreated rapidly at the end of the LGM, and this may have predated similar temperature warming at higher latitudes, a possibility that suggests that the Southern Hemisphere and the tropics may have had an important role in contributing to global deglaciation (e.g., Seltzer et al., 2002).

In the late glacial there are several documented advances of glaciers, but their regional coherency in terms of timing and how they may be related to millennial-scale changes in the extent of glaciers in the North Atlantic region has still to be determined. By the early Holocene, glaciers had retreated within their modern limits, and they may have disappeared entirely from lower summits by mid-Holocene time in response to warmer and drier conditions. Neoglaciation was under way by about 4,000 cal yr B.P., yet glaciers never expanded more than a few kilometers beyond their modern limits. Glaciers have been in retreat throughout the tropical Andes since at least the end of the nineteenth century, and this recession appears to be accelerating. The loss of glaciers on many tropical mountains in the twenty-first century could cause significant problems in terms of natural hazards and water availability.

Intriguing yet unresolved problems are generated by the late Quaternary glaciation of the tropical Andes. For example, what produced the cold glacial conditions that impacted the Southern Hemisphere and the tropics at the LGM? Why might the tropics have warmed more rapidly than higher latitudes in the Northern Hemisphere at the end of the LGM? Were there millennial-scale climatic oscillations in the tropical Andes, and can glacial sequences be used to evaluate this question? What may have led to dry conditions in the early to mid Holocene in the tropical Andes—was it a decrease in summer insolation? Is there a consistent relationship between ENSO variability and the overall hydrologic status of the tropical Andes? Finally, do we fully understand what the impact on regional hydrology will be if current rapid rates of deglaciation continue? The glaciers of the tropical Andes are not large compared with the ice fields of Patagonia, but they do pose intriguing questions of regional and even global significance, and may yet yield some answers.

References

Abbott, M., M.B. Binford, M.W. Brenner, and K.R. Kelts, 1996. A 3500 ^{14}C yr high resolution record of lake level changes in Lake Titicaca, South America. *Quaternary Research*, 47, 169–180.

Abbott, M., G.O. Seltzer, K.R. Kelts, K., and J. Southton, 1997. Holocene paleohydrology of the Tropical Andes from Lake Records. *Quaternary Research*, 47, 70–80.

Alley, R., D.A. Meese, C.A. Shuman, A.J. Gow, K.C. Taylor, P.M. Grootes, J.W.C. White, M. Ram, E.D. Waddington, P.A. Mayewski, and G.A. Zielinski, 1993. Abrupt increase in Greenland snow accumulation at the end of the Younger Dryas event. *Nature*, 362, 527–529.

Baker, P.A., C.A. Rigsby, G.O. Seltzer, S.C. Fritz, T.K. Lowenstein, N.P. Bacher, and C. Veliz, 2001b. Tropical climate changes at millennial and orbital timescales revealed by deep drilling on the South American Altiplano. *Nature*, 409, 698–701.

Baker, P.A., G.O. Seltzer, S.C. Fritz, R.B. Dunbar, M.J. Grove, P.M. Tapia, S.L. Cross, H.D. Rowe, and J.P. Broda, 2001a. The history of South American tropical precipitation for the past 25,000 years. *Science*, 291, 640–643.

Bard, E., M. Arnold, P. Maurice, J. Duprat, J. Moyes, and J-C. Duplessy, 1987. Reheat velocity of the North Atlantic polar front during the last deglaciation determined by ^{14}C accelerator mass spectrometry. *Nature*, 328, 791–794.

Bond, G., W. Showers, M. Cheseby, R. Lotti, P. Almasi, P. deMenocal, P. Priore, J. Cullen, I. Hajdas, and G. Bonani, 1997. A pervasive millennial-scale cycle in the North Atlantic Holocene and glacial climates. *Science*, 278, 1257–1266.

Bond, G.C., and R. Lotti, 1995. Iceberg discharges into the North Atlantic on millennial time scales during the Last Glaciation. *Science*, 267, 1005–1009.

Brecher, H.H., and L.G. Thompson, 1993. Measurement of the retreat of Qori Kalis glacier in the tropical Andes of Perú by terrestrial photogrammetry. *Photogrammetric Engineering and Remote Sensing*, 59, 1017–1022.

Bush, M., M. Stute, M. Ledru, H. Behling, P. Colinvaux, P.E. De Oliveira, E.C. Grimm, H. Hooghiemstra, S. Haberle, B. Leyden, M.L. Salgado-Labouriau, and R.S. Webb, 2001. Paleotemperature estimates for the lowland Americas between 30°S and 30°N at the Last Glacial Maximum. In: V. Markgraf (Editor), *Interhemispheric Climate Linkages*. Academic Press, New York, 293–306.

Clapperton, C.M., 1979. Glaciation in Bolivia before 3.27 Myr. *Nature*, 277, 375–376.

Clapperton, C.M., 1990. Quaternary glaciations in the Southern Hemisphere: An overview. *Quaternary Science Reviews*, 9, 299–304.

Clapperton, C.M., 1993. *Quaternary Geology and Geomorphology of South America*. Elsevier, Amsterdam.

Clapperton, C.M., 2000. Interhemispheric synchroneity of Marine Oxygen Isotope Stage 2 glacier fluctuations along the American cordilleras transect. *Journal of Quaternary Science*, 15, 435–468.

Clapperton, C.M., and Seltzer, G.O., 2000. Glaciation during Marine Isotope Stage 2 in the American Cordillera. In V. Markgraf (Editor), *Interhemispheric Climate Linkages*, Academic Press, New York.

Clapperton, C.M., M. Hall, P. Mothes, M.J. Hole, J.W. Still, K.F. Helmens, P. Kuhry, and A.M.D. Gemmell, 1997. A Younger Dryas icecap in the equatorial Andes. *Quaternary Research*, 47, 13–28.

CLIMAP, 1981. Seasonal reconstructions of the Earth's surface at the Last Glacial Maximum: *Geological Society of America Map and Chart Series*, MC-36, Geological Society of America, Boulder.

Cross, S. L., A. Baker, G.O. Seltzer, S.C. Fritz, and R.B. Dunbar, 2000. A new estimate of the Holocene lowstand level of Lake Titicaca, and implications for tropical paleohydrology. *The Holocene*, 10, 1–32.

D'Agostino, K., G.O. Seltzer, A. Baker, S.C. Fritz, and R.B. Dunbar, 2002. Late-Quaternary seismic stratigraphy of Lake Titicaca (Perú/Bolivia). *Palaeogeography, Palaeoclimatology, Palaeoecology*, 179, 97–111.

Denton, G.H., and C.H. Hendy, 1994. Younger Dryas age advance of Franz Josef Glacier in the Southern Alps of New Zealand. *Science*, 264, 1434–1437.

Francou, B., P. Ribstein, R. Saravia, and E. Tiriau, 1995. Monthly balance and water discharge of an inter-tropical glacier: Zongo Glacier, Cordillera Real, Bolivia, 16°S. *Journal of Glaciology*, 41, 61–67.

Goodman, A.Y., G.O. Seltzer, D.T. Rodbell, and B.G. Mark, 2001. Subdivision of glacial deposits in southeastern Perú based on pedogenic development and radiometric ages. *Quaternary Research*, 56, 31–50.

Grootes, P.M., M. Shiver, L.G. Thompson, and E. Mosley-Thompson, 1989. Oxygen isotope changes in tropical ice, Quelccaya, Perú. *Journal of Geophysical Research*, 94. 1187–1194.

Hastenrath, S., and A. Ames, 1995. Recession of Yanamarey Glacier in Cordillera Blanca, Perú, during the 20th century: *Journal of Glaciology*, 41, 191–196.

Heine, K., and J.T. Heine, 1996. Late Glacial climatic fluctuations in Ecuador: Glacier retreat during Younger Dryas time. *Arctic and Alpine Research*, 28, 496–501.

Helmens, K. F., 1988. Late Pleistocene glacial sequence in the area of the high plain of Bogotá (Eastern Cordillera, Colombia). *Palaeogeography, Palaeoclimatology, Palaeoecology*, 67, 263–283.

Hostetler, S.W., and P.U. Clark, 2000. Tropical climate at the Last Glacial Maximum inferred from glacier mass-balance modeling. *Science*, 290, 1747–1750.

Hostetler, S.W., and A.C. Mix, 1999. Reassessment of ice-age cooling of the tropical ocean and atmosphere. *Nature*, 399, 673–676.

Kaser, G., and H. Osmaston, 2002. *Tropical Glaciers*. Cambridge University Press, Cambridge.

Klein, A.G., G.O. Seltzer, and B.L. Isacks, 1999. Modern and Last Glacial Maximum snowlines in the Central Andes of Perú, Bolivia, and Northern Chile. *Quaternary Science Reviews*, 18, 63–84.

Kuhn, M., 1981. Climate and glaciers. *International Association of Hydrological Sciences*, 131, 3–20.

Kuhn, N., 1989. The response of the equilibrium line altitude to climate fluctuations: Theory and observations. In: J. Oerlemans (Editor), *Glacier Fluctuations and Climatic Change*. Kluwer Academic Publishers, Dordrecht, 407–417.

Lea, D.W., D.K. Pak, and H.J. Spero, 2000. Climate impact of Late Quaternary equatorial Pacific sea surface temperature variations. *Science*, 289, 1719–1724.

Lowell, T., C.J. Heusser, B.G. Andersen, I. Moreno, A. Hausser, L.E. Heusser, C. Schluchter, D.R. Marchant, and G.H. Denton, 1995. Interhemispheric correlation of late Pleistocene glacial events. *Science*, 269, 1541–1549.

Mann, M.E., R.S. Bradley, and M.K. Hughes, 1998. Global-scale temperature patterns and climate forcing over the past six centuries. *Nature*, 392, 779–787.

Mark, B.G., 2002. Observations of modern deglaciation and hydrology in the Cordillera Blanca. *Acta Montaña*, A19 (123), 23–36.

Mark, B.G., and G.O. Seltzer, 2003. Tropical glacial meltwater contribution to stream discharge: A case study in the Cordillera Blanca, Perú: *Journal of Glaciology*, 49, 271–281.

Mark, B.G., G.O. Seltzer, D.T. Rodbell, and A.Y. Goodman, 2002. Rates of deglaciation during the Last Glaciation and Holocene in the Cordillera Vilcanota-Quelccaya Ice Cap region, southeastern Perú. *Quaternary Research*, 57, 287–298.

Meierding, T.C., 1982. Late Pleistocene glacial equilibrium-line altitudes in the Colorado Front Range: A comparison of methods. *Quaternary Research*, 18, 289–310.

Mercer, J.H., 1976. Glacial history of southernmost South America. *Quaternary Research*, 6, 125–166.

Mercer, J.H., 1984. Late Cainozoic glacial variation in South America south of the Equator In: J.C. Vogel (Editor), *Late Cainozoic Paleoclimates of the Southern Hemisphere*. A.A. Balkema, Rotterdam, 45–58.

Mercer, J.H., and M.O. Palacios, 1977. Radiocarbon dating of the last glaciation in Perú. *Geology*, 5, 600–604.

Moy, C., G.O. Seltzer, D.T. Rodbell, and D.M. Anderson, 2002. Variability of El Nino/Southern Oscillation activity at millennial time scales during the Holocene Epoch, *Nature*, 420, 162–165.

Ribstein, P., E. Tiriau, B. Francou, and R. Saravia, R., 1995. Tropical climate and glacier hydrology: A case study in Bolivia. *Journal of Hydrology*, 165, 221–234.

Rind, D., and D. Peteet, 1985. Terrestrial conditions at the Last Glacial Maximum and CLIMAP sea-surface temperature estimates: Are they consistent? *Quaternary Research*, 24, 1–22.

Rodbell, D.T., 1992. Lichenometric and radiocarbon dating of Holocene glaciation, Cordillera Blanca Perú. *The Holocene*, 2, 19–29.

Rodbell, D., 1993. Subdivision of late Pleistocene moraines in the Cordillera Blanca, Perú, based on rock-weathering features, soils, and radiocarbon dates. *Quaternary Research*, 39, 133–143.

Rodbell, D.T., and G.O. Seltzer, 2000. Rapid ice margin fluctuations during the Younger Dryas in the tropical Andes. *Quaternary Research*, 54, 328–338.

Rodbell, D.T., G.O. Seltzer, D.M. Anderson, M.B. Abbott, D.B. Enfield, and J.H. Newman, 1999. An ~15,000-year record of El Niño-driven alluviation in southwestern Ecuador. *Science*, 283, 516–520.

Rowe, H. D., R.B. Dunbar, D. Mucciarone, G.O. Seltzer, P.A. Baker, and S.C. Fritz, 2002. Insolation, moisture balance, and climate change on the South American Altiplano since the Last Glacial Maximum. *Climatic Change*, 52, 175–199.

Seltzer, G.O., 1990. Recent glacial history and paleoclimate of the Peruvian-Bolivian Andes. *Quaternary Science Reviews*, 9, 137–152.

Seltzer, G.O., 1992. Late Quaternary glaciation of the Cordillera Real, Bolivia. *Journal of Quaternary Science*, 7, 87–98.

Seltzer, G.O., 1994a. Climatic interpretation of alpine snowline variations on millennia1 time scales. *Quaternary Research*, 41, 154–159.

Seltzer, G.O., 1994b. A lacustrine record of late-Pleistocene climatic change in the subtropical Andes. *Boreas*, 23, 105–111.

Seltzer, G.O., 2001. Late-Quaternary Glaciation in the Tropics: Future Research Directions. *Quaternary Science Reviews*, 20, 1063–1066.

Seltzer, G.O., P.B. Baker, S. Cross, S.C. Fritz, and R. Dunbar, 1998. High-resolution seismic reflection profiles from Lake Titicaca, Perú/Bolivia: Evidence for Holocene aridity in the tropical Andes. *Geology*, 26, 167–170.

Seltzer, G.O., D.T. Rodbell, and M. Abbott, 1995. Andean glacial lakes and climate variability since the Last Glacial Maximum. *Bulletin de l'Institut des Etudes Andines*, 24, 539–549.

Seltzer, G.O., D.T. Rodbell, P.A. Baker, S.C. Fritz, P.M. Tapia, H.D. Rowe, and R.B. Dunbar, 2002. Early warming of tropical South America at the last glacial-interglacial transition. *Science*, 296, 1685–1686.

Seltzer, G.O., D.T. Rodbell, and S. Burns, 2000. Isotopic evidence for late Quaternary climatic change in tropical South America. *Geology*, 28, 35–38.

Shackleton, N.J., and N.D. Opdyke, 1973. Oxygen-isotope and paleomagnetic stratigraphy of equatorial Pacific core V28–238: Oxygen isotope temperatures and ice volumes on a 10^5 year and 10^6 year scale. *Quaternary Research*, 3, 39–55.

Smith, J.A., G.O. Seltzer, D. Rodbell, R.C. Finkel, and D.L. Farber, 2001. Cosmogenic dating of glaciation in the Peruvian Andes: >^{10}Be ka to Last Glacial Maximum. 2001 Abstracts Annual Meeting, Boston, Geological Society of America, Boulder, A441.

Stute, M., M. Forster, H. Frischkorn, A. Serejo, J.F. Clark, P. Schlosser, W.S. Broecker, and G. Bonani, 1995. Cooling of tropical Brazil (5 degrees C) during the Last Glacial Maximum. *Science*, 269, 379–383.

Sylvestre, F., M. Servant, S. Servant-Vildary, C. Causse, M. Fournier, and J.P. Ybert, 1999. Lake-level chronology on the southern Bolivian Altiplano (18°–23°S) during late-Glacial time and the early Holocene. *Quaternary Research*, 51, 54–66.

Tapia, P.M., S.C. Fritz, P.A. Baker, G.O. Seltzer, and R.B. Dunbar, 2003. A late Quaternary diatom record of tropical climate history from Lake Titicaca (Perú and Bolivia). *Palaeogeography, Palaeoclimatology, Palaeoecology,* 194, 139–164.

Thompson, L.G., M.E. Davis, E. Mosley-Thompson, T.A. Sowers, K.A. Henderson, V.S. Zagorodnov, P.N. Lin, V.N. Mikhalenko, R.K. Campen, J.F. Bolzan, J. Cole-Dai, and B. Francou, 1998. A 25,000-year tropical climate history from Bolivian ice core. *Science*, 282, 1858–1864.

Thompson, L.G., E. Mosley-Thompson, J.F. Bolzan, and B.R. Koci, 1985. A 1500-year record of tropical precipitation in ice cores from the Quelccaya ice cap, Perú. *Science*, 229, 971–973.

Thompson, L.G., E. Mosley-Thompson, M.E. Davis, P-N. Lin, K.A. Henderson, J. Cole-Dai, J.F. Bolzon, and K-b. Liu, 1995. Late glacial stage and Holocene tropical ice-core records from Huascarán, Perú. *Science*, 269, 46–50.

Vuille, M., R.S. Bradley, R. Healy, M. Werner, D.R. Hardy, L.G. Thompson, and F. Keiming, 2003. Modeling $\delta^{18}O$ in precipitation over the tropical Americas, Part II: Simulation of the stable isotope signal in Andean ice cores. *Journal of Geophysical Research—Atmospheres*, 108 (D6), dol 10.1029/2001JD002039.

Webster, P.N., and N. Streten, 1978. Late Quaternary ice age climates of tropical Australasia, interpretation and reconstruction. *Quaternary Research*, 10, 279–309.

Wolfe, B.B., R. Aravena, M.B. Abbott, G.O. Seltzer, and J.J. Gibson, 2001. Reconstruction of paleohydrology and paleohumidity from oxygen isotope records in the Bolivian Andes: *Palaeogeography, Palaeoclimatology, Palaeoecology*, 176, 177–192.

5

Rivers

Thomas Dunne
Leal Anne Kerry Mertes

River basins and river characteristics are controlled in part by their tectonic setting, in part by climate, and increasingly by human activity. River basins are defined by the tectonic and topographic features of a continent, which determine the general pattern of water drainage. If a major river drains to the ocean, its mouth is usually fixed by some enduring geologic structure, such as a graben, a downwarp, or a suture between two crustal blocks. The largest river basins constitute drainage areas of extensive low-lying portions of Earth's crust, often involving tectonic downwarps. The magnitude of river flow is determined by the balance between precipitation and evaporation, summed over the drainage area. Seasonality of flow and water storage within any basin are determined by the seasonality of precipitation in excess of evaporation, modified in some regions by water stored in snow packs and released by melting, and by water stored in wetlands, lakes, and reservoirs. Increasingly the flows of rivers are influenced by human land use and engineering works, including dams, but in South America these anthropogenic influences are generally less intense and widespread than in North America, Europe, and much of Asia. Thus the major rivers of South America can be viewed in the context of global and regional tectonics and climatology (see chapters 1, 2, and 3). For reference, figure 5.1 outlines South America's three largest river basins—the Orinoco, Amazon, and Paraguay-Paraná systems—while figure 5.2 shows the locations of rivers referred to in the text against a background of the continent's density of population per square kilometer.

5.1 Structural Controls over River Basins

The geologic history of South America has bequeathed to the continent a number of structural elements that are relevant to the form and behavior of its three major river systems (fig. 1.2). These structural elements are (1) the Andes; (2) a series of foreland basins, approximately 500 km wide immediately east of the Andes and extending southward from the mouth of the Orinoco to the Chaco-Paraná basin, where the crust is depressed by the weight of the Andes and the sediment derived from the mountains; (3) the Guiana and Brazilian shields reflecting Precambrian cratons and orogenic belts of mostly crystalline metamorphic rocks, partly covered with flat-lying sedimentary rocks and deeply weathered regolith; and (4) the Central Amazon Basin, a large cratonic downwarp with some graben structures dating back to early Paleozoic time, which runs generally east-west between the two shields, connecting the foreland basins to the west with a graben that localizes the Amazon estuary at the Atlantic coast.

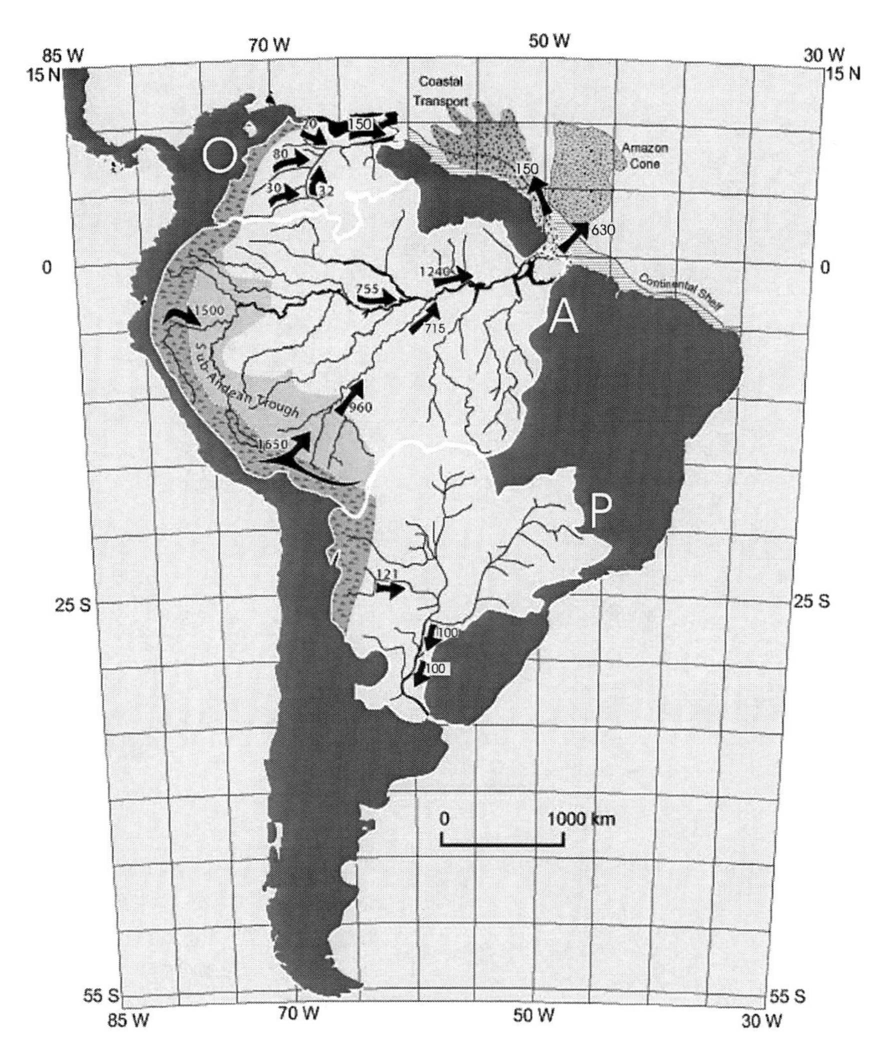

Figure 5.1 The three largest river basins of South America: Orinoco (O), Amazon (A), and Paraguay-Paraná or Rio de la Plata system (P). Data for suspended sediment transport are shown for each river (Orinoco, R.H. Meade, personal communication; Amazon—Subandean Trough and Madeira River, Aalto et al., 2006; Amazon mainstem, Dunne et al., 1998).

In the above context, the assembly and development of these river basins have been plausibly linked to several tectonic and climatic factors. The Mesozoic opening of the Atlantic Ocean fixed the mouths of the Amazon and Paraguay-Paraná systems in rifts that extended into the continent (Potter, 1978). The Oligocene-Pliocene rise of the Andes then closed off the marine embayments in the Central Amazon Basin from former outlets westward to the Gulf of Guayaquil and northward to the Gulf of Maracaibo. This closure caused the drainage of a vast area (soon to be the modern Amazon Basin) to seek a path eastward towards the topographic low provided by the still active graben that presently localizes the mouth of the Amazon (fig. 5.1). Simultaneously, orogenic uplift also created the foreland basins east of the Andes, as the mountains pushed over the downwarping cratons, thereby forming a partial trap for a large fraction of the sediment emanating from the mountains. Meanwhile, mantle-plume driven fracturing of the crust at triple junctions and aulacogens, and hotspot doming during continental separation, caused topographic highs to develop near the rifted margin of southeast Brazil, producing a drainage system that converges first inland into the Paraná basin before leaving the continent through the Río de la Plata estuary (Cox, 1989). Within the cratonic shields, transcurrent sets of faults at angles of 90 to 100° to one another appear to have resulted from crustal tension since at least Miocene time when collapse of the Nazca and Caribbean plates began. Most rivers and lakes on the central Brazilian Shield are aligned with these fracture systems and have reticulated courses with sharp bends (Sternberg and Russell, 1952). Finally, tilting and folding have occurred within the basins between the cratons, as exemplified by the presence of structural highs or arches along the Amazon River valley (Tricart, 1977; Caputo, 1984, 1991). These apparently ongoing structural processes have been shown to impact the fluvial geomorphology in their vicinity, with the arches associated with the narrowing of the valley and changes in modern water-surface gradients (Birkett et al., 2002; Dunne et al., 1998; Latrubesse and Franzinelli, 2002; Mertes et al., 1996).

Potter (1978) originally systematized the relationships between large rivers and tectonic context, illustrating that

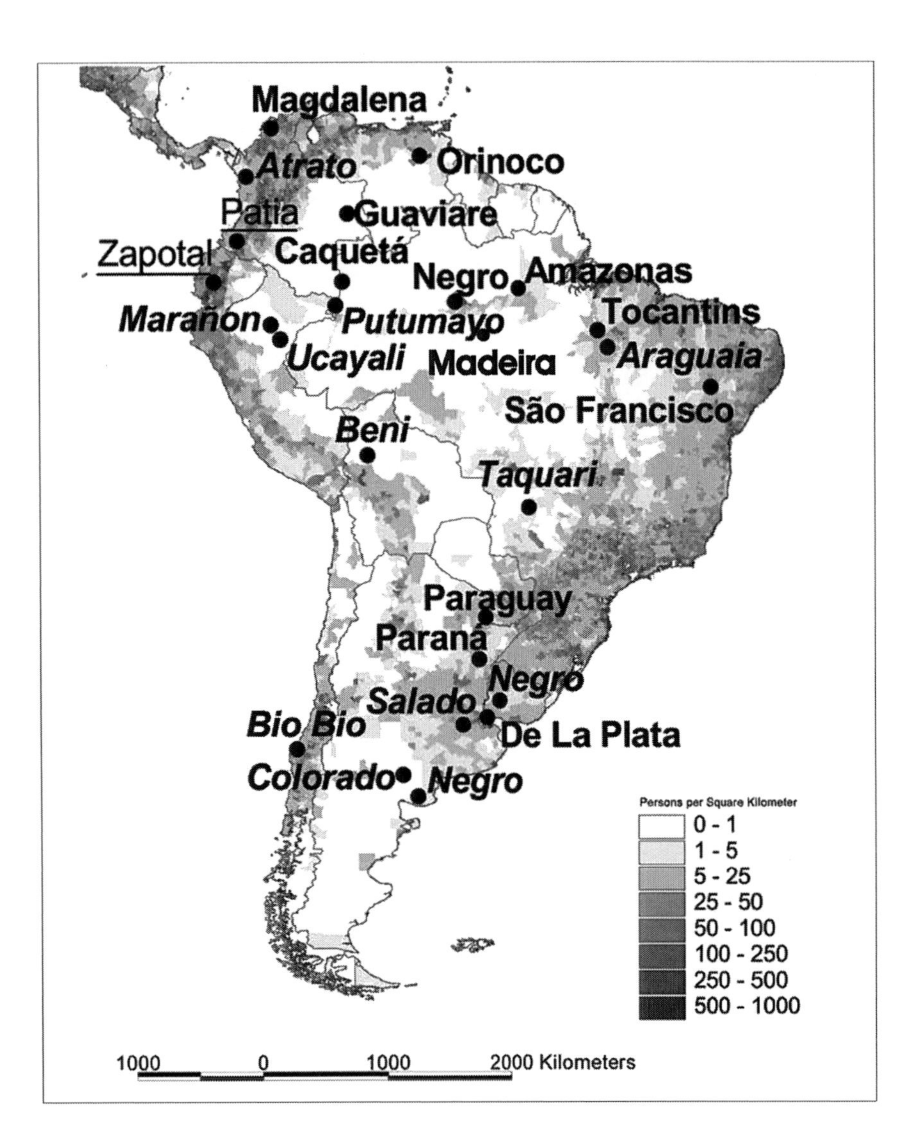

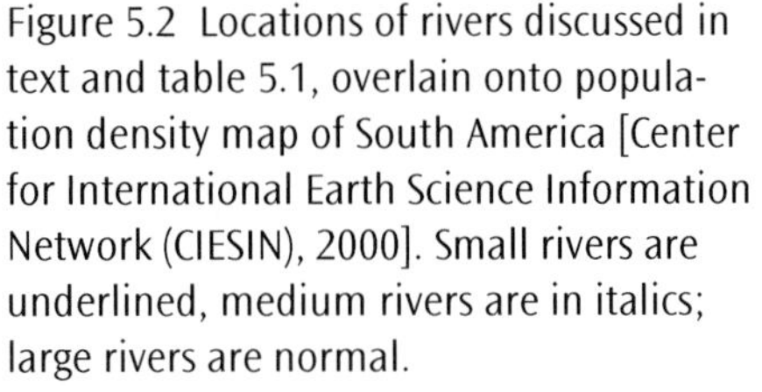

Figure 5.2 Locations of rivers discussed in text and table 5.1, overlain onto population density map of South America [Center for International Earth Science Information Network (CIESIN), 2000]. Small rivers are underlined, medium rivers are in italics; large rivers are normal.

not only their sizes but also other features such as their sediment loads and valley geomorphology could be related to their geological context. Figure 5.3 summarizes the classification scheme that Potter developed to express the tectonic contexts of certain kinds of rivers across the globe. Examples of each class in his scheme, which we have labeled with P-values in the figure, can be recognized in South America.

P1 rivers are those that flow from an active orogen that fringes a craton (Amazon) or a stable crustal platform such as Patagonia (Colorado). A large debris load is injected into the upper end of the river system, but little is supplied from the downstream parts of the basin. River systems transport a portion of their load across the lowland, depositing large fractions of it along the way, especially in foreland or intracratonic basins. Floodplains usually increase in width with distance downstream, but with significant variations in width because of structural patterns. Since the Amazon is exceptionally long, only a small fraction of the input reaches the ocean.

P2 rivers, such as the Paraguay, flow roughly parallel to an orogen and receive all their sediment loads from the tributaries on one bank. These rivers therefore often construct depositional wedges and large alluvial fans at tributary junctions, thereby diverting the main stem away from the mountain front. The Orinoco is intermediate between the P1 and P2 types.

P3 rivers flow in tectonic depressions along the strike of mountain ranges. South American examples include the Magdalena in Colombia, with an alluvial plain 30 to 100 km wide (Garcia Lozano and Dister, 1990), and the Marañon in Perú, which travels northwest through the high north-central Andes before crossing to the east to join the Amazon main flow.

P4 rivers are those that flow across mountain ranges, exemplified by the irregular courses of the Ucayali in eastern Perú and the Pilcomayo in eastern Bolivia, which flow through Subandean ranges toward the Amazon and Paraguay-Paraná systems, respectively.

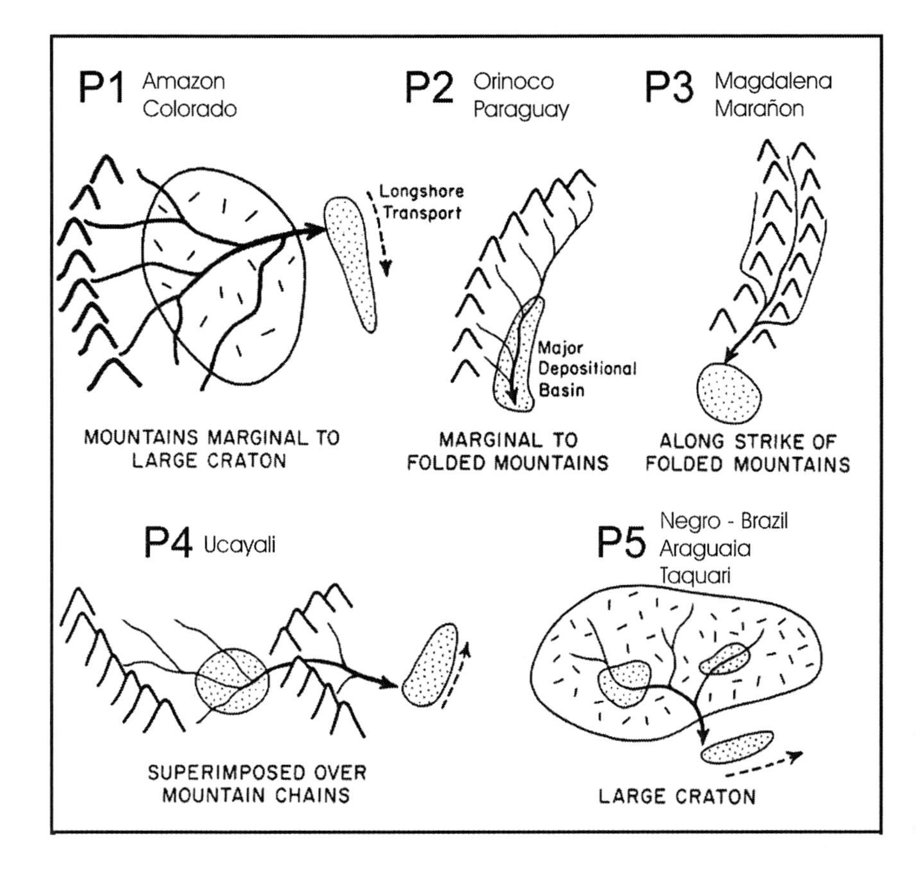

Figure 5.3 Potter classification for continental setting of large rivers (Potter, 1978, fig. 7). P1-P5 labels are assigned for reference. South American rivers cover all categories.

P5 rivers drain large cratonic areas without a significant mountain source. These rivers, such as Brazil's Negro, Araguaia, and Taquari, have relatively low sediment loads and thus narrow or no floodplains, except where they flow into large structural depressions, such as the Taquari, which flows into the Pantanal wetland before reaching the Paraguay River (Ussami et al., 1999).

5.2 Long-term Climate Controls over River Basins

Late Cenozoic climate changes have directly affected the modern form and behavior of South American rivers. Glacial-interglacial cycles of the past few million years have influenced the magnitude, frequency, and seasonality of precipitation and runoff, the nature of sedimentation on the megafans of the Andean foreland and in river floodplains, and the vegetation cover of river basins (see chapters 2 and 9). The growth and decay of glaciers in the tropical Andes have affected the Amazon system (see chapter 4), just as the waxing and waning of the Patagonian ice cap have shaped proglacial lakes and drainage systems in the southern Andes (Clapperton, 1993).

Indirectly, the late Cenozoic eustatic sea-level changes that accompanied these climate cycles have had major impacts on the mouths of most rivers, and these have been transmitted upstream for varying distances. The low sea levels that coincided with Pleistocene glacial stages lowered the base levels of erosion for rivers, causing them to scour sediment from their estuaries and floodplains. Many of the river valleys on the western slopes of the Andes in central and southern Chile were also deeply scoured by glaciers and then dammed by glacial debris, causing modern rivers to flow through glacial lakes into fjords (see chapter 15; Clapperton, 1993). In contrast, the higher sea levels of warm interglacial stages raised base levels and favored sedimentation in these scoured lower reaches. Thus, in response to high Holocene sea level, the scoured depression of the lower Orinoco floodplain and delta, which lies relatively close to its Andean sediment source, has filled with sediment. Similar infilling is occurring in the lower reaches of many smaller rivers flowing from the Andes.

By contrast, the mouth of the Amazon, though having a huge sediment source in the Andes, has not yet filled during the Holocene because the mountains are far from the coast and separated from it by large foreland basins that trap sediment. During episodes of low sea level in the lower Amazon valley, knickpoint recession appears to have extended upstream at least to the vicinity of the Purús arch, a structural high several hundred kilometers above Manaus (Mertes and Dunne, in press; Mertes et al., 1996), and up each of the tributaries farther downstream (Tricart, 1977), and to have removed a wedge of sediment that exceeded 100 m thick at the coast. Postglacial sea-level rise has caused sedimentation in the lower river valley, but this reach remains underfilled with sediment, with large irregular lakes and a large delta plain several hundred kilometers from the mouth of the estuary, in which 300–400 × 10^6 t yr^{-1} of sediment are being trapped (Dunne et al., 1998). Rapid aggradation of the lower Amazon floodplain has also raised the base level of major tributaries draining the forested cratons. However, because these rivers supply only small amounts of sediment, they are unable to fill in behind the natural dams produced by main-stem sedimentation, thereby forming ria lakes with small deltas at their heads (Siolli, 1968; Tricart, 1977).

During episodes of low sea level, the Amazon flushed sediment across the continental shelf to feed the massive Amazon Cone beyond the continental slope (fig. 5.1; Damuth and Kumar, 1975; Manley and Flood, 1988; Milliman et al., 1975; Nittrouer et al., 1986b). In contrast, under present high sea level at the mouth of the Amazon, deltaic sedimentation is minimal owing to the strength of tidal currents and littoral drift on the shelf, with currents transporting the majority of the sediment plume to the north (see chapter 15; Augustinus, 1982; Kuehl et al., 1982, 1986; Nittrouer et al., 1986a, 1986b; Pujos et al., 1997). Similarly, in the lower Paraguay-Paraná valley, modern sedimentation is focused on the delta at the head of the Río de la Plata estuary, hundreds of kilometers from the open ocean.

5.3 Patterns of River Flow

A convenient separation of the rivers of South America occurs at about 15°S latitude, dividing rivers flowing northward from those flowing to the south. The northern rivers drain a very wet region which, together with the sizes of the tectonic basins, creates rivers with exceptionally large discharge, width, depth, and extensive floodplain inundation lasting for months, particularly for the Amazon, Orinoco, Tocantins, Magdalena, and São Francisco rivers. Lewis et al. (1989) calculated that the northern region of 12 × 10^6 km^2, with a mean annual precipitation of 1,700 mm, generates more river runoff (averaging 850 mm) than any other comparably sized region in the world. The southern portion of the continent is smaller and drier, and therefore its rivers are generally smaller, and the largest by far, the Paraguay-Paraná system, drains the subtropical and tectonically most active southern basin. For present purposes, we will compare rivers according to their annual volume of flow, expressed as depth of runoff (annual runoff as volume per unit area of drainage basin in table 5.1); their mean annual discharge; and their seasonality, expressed in terms of the range and timing of monthly flows. Table 5.1 presents these statistics for a sample of rivers, selected to illustrate differences within the continent.

5.3.1 Inter-Annual Variability

Even in the largest rivers of the continent, there is a significant and persistent deviation from average annual flows (fig. 5.4). Richey et al. (1989) demonstrated that low annual flows for the Amazon have occurred during El Niño years when ocean circulation anomalies associated with El Niño–Southern Oscillation modify both the intensity and location of the energy source for atmospheric heating and circulation (see chapter 19; Molion, 1990). The ascending east-west branch of the Walker Circulation that usually covers the Amazon basin is displaced westward over the abnormally warm east Pacific Ocean, and intensifies. The descending branch of the circulation then covers the Amazon, reducing rainfall. Molion (1990) quotes reduc-

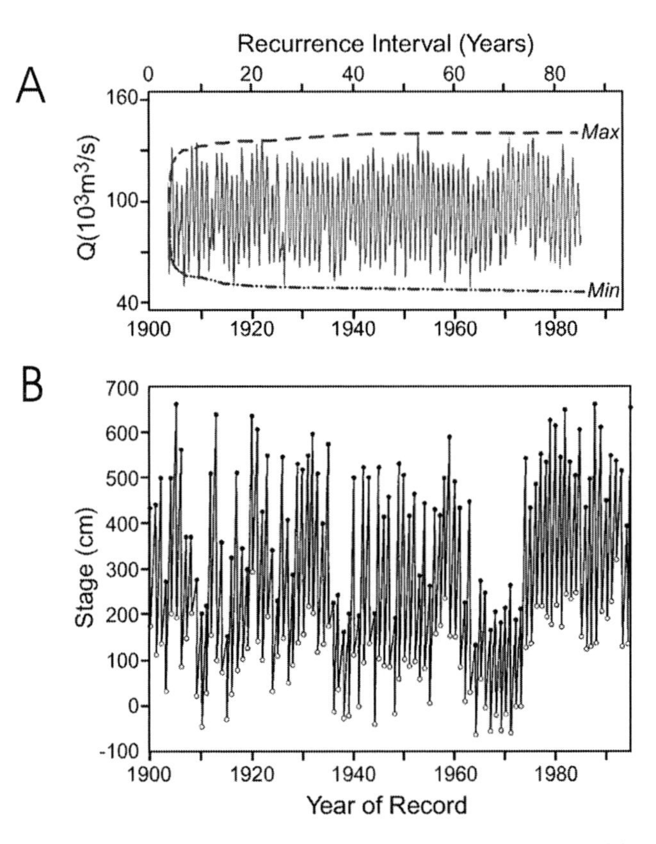

Figure 5.4 Long-term hydrographs (early 1900 to 1990s) for (A) the Amazon at Manacapurú (Richey et al., 1989, fig. 3), and (B) the Paraguay at Corumbá, Bolivia (Ponce, 1995, fig. 9), showing inter-annual variability and persistence. Data for the Amazon River are discharge and probability of recurrence for maximum and minimum annual flows. Recurrence represents the average interval in years between events equaling or exceeding a given magnitude. The stage data for the Paraguay River have not been transformed to discharge.

Table 5.1 Flow characteristics of selected South American rivers

Scale	Basin area [km²]	Country	River	Mean annual flow [m³ s⁻¹]	Annual runoff [mm]	Maximum monthly discharge [m³ s⁻¹]	Minimum monthly discharge [m³ s⁻¹]	Source
Large								
	4,620,000	Brazil	Amazonas	173,167	1182	229,000	103,000	5
	3,100,000	Argentina	Rio de la Plata	18,200	185			4
	2,300,000	Argentina	Paraná	20,663	283	23,700	17,350	4
	1,306,000	Brazil	Madeira	29,300	707	54,000	5,400	5
	1,100,000	Venezuela	Orinoco	38,000	1300	81,100	1,330	1
	1,095,000	Argentina	Paraguay	4,550	130			4
	727,900	Brazil	Tocantins	8,318	360	16,657	2,053	4
	510,800	Brazil	São Francisco	2,162	133	3,799	1,163	4
	468,000	Brazil	Negro	29,560	1990	58,000	13,200	5
	320,290	Brazil	Araguaia	6,100	601	23,000	800	4
	266,600	Colombia	Magdalena	7,018	830	11,374	1,800	2
	199,200	Colombia	Caquetá	13,180	2090			2
	166,200	Colombia	Guaviare	8,200	1560			2
Medium								
	95,000	Argentina	Negro	581	193	934	373	4
	69,000	Bolivia	Beni	2,274	1039	5,388	728	4
	63,000	Uruguay	Negro	1,139	570	3,749	285	4
	53,160	Colombia	Putumayo	6,250	3707			2
	35,700	Colombia	Atrato	4,155	3670			2
	35,000	Brazil	Taquari	732	659	4,470	92	3
	29,000	Argentina	Salado	12	13	40.5	0.5	4
	24,300	Chile	Bio Bio	1,014	1300	2,000	230	4
	22,300	Argentina	Colorado	85	120	288	31	4
Small								
	13,100	Colombia	Patia	205	491	310	105	4
	2,300	Ecuador	Zapotal	297	4072	627	64	4

Data sources: 1. Warne et al. (2002); 2. Garcia Lozano and Dister (1990); 3. Souza et al. (2002); 4. Vörösmarty et al. (1996, 1998); and 5. Richey et al. (1989).

tions of up to 1.8 standard deviations in rainy season totals in the northwest Amazon basin (the wettest region) and 0–1.2 standard deviations in the central and eastern portions of the basin. By contrast, the southwest Amazon basin, at 10°S latitude and closer to the perturbed offshore area of the east Pacific, shows an increase of up to 1.2 standard deviations in rainy season rainfall. In agreement with this pattern, figure 5.4A illustrates the trend in Amazon discharge at Manaus for much of the twentieth century, calibrated by Richey using modern discharge measurements. A more persistent, but still unexplained pattern of low- and high-flow years is exhibited by the record from the upper Paraguay River near Corumbá, Bolivia (fig. 5.4B).

5.3.2 *Seasonality*

Runoff responds to the seasonal pattern of rainfall (see chapter 3) but lags behind the rainfall record as a result of storage as groundwater and along the floodplains of large alluvial rivers. Figure 5.5 shows the mean monthly hydrographs for several of the rivers listed in table 5.1. The hydrographs are arranged to correspond generally to large (A), medium (B), and small (C) rivers. The largest rivers have discharge maxima ranging from 10,000 to 240,000 m³ s⁻¹, with marked seasonality that varies based on their locations relative to the Equator. With decreasing size, the yearly rise and fall observed for the largest rivers becomes more irregular, with the Negro in Argentina showing the greatest number of peaks.

The seasonality of precipitation over South America is controlled mainly by the annual north-south interhemispheric march of the insolation maximum and consequent land-surface heating (see chapter 3, especially fig. 3.2). This is the signal, lagged by one to three months as a result of groundwater and floodplain storage, that is reflected in the hydrographs of the largest rivers (fig. 5.5). The Tocantins River, a major Southern Hemisphere tributary of the Amazon, has a peak flow in March–April, soon after the rainfall

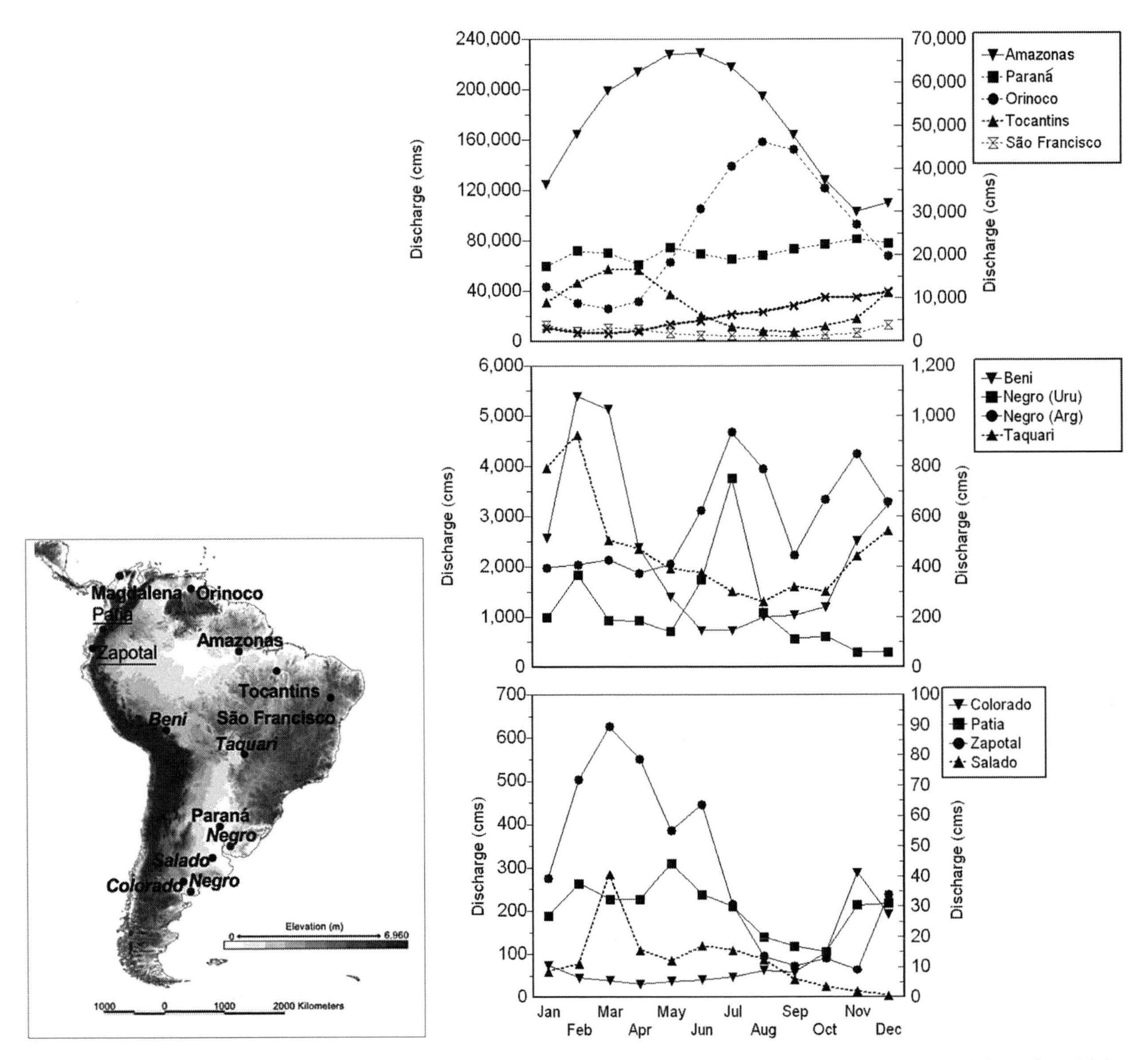

Figure 5.5 Seasonal hydrographs of mean monthly discharge for representative large, medium, and small rivers for which discharge data are available (see table 5.1). Solid lines relate to the left y-axis for the Amazonas, Beni, Negro (Uruguay), Colorado, Patia, and Zapotal rivers. Dotted lines relate to the right y-axis for the Paraná, Orinoco, Tocantins, São Francisco, Negro (Argentina), Taquari, and Salado rivers. The location map (based on USGS shaded relief map GTOPO30) shows gauging stations for which data are plotted (small rivers underlined, medium rivers in italics, and large rivers in normal font). Discharge data for the Taquari River are based on Souza et al. (2002) and for all other rivers on Vörösmarty et al. (1996, 1998).

peak, whereas the Orinoco River in the Northern Hemisphere peaks in August. The main-stem Amazon at Óbidos peaks in May–June because Southern Hemisphere tributaries contribute most of its drainage area and runoff. The regime of the Paraná is radically altered by storage in large reservoirs, which can accommodate up to 60% of the river's average annual flow.

The most influential alteration of a river regime through wetland storage occurs in the Pantanal of Mato Grosso, which delays the flow of the upper Paraguay River between Cáceres, Brazil, and Concepción, Paraguay, by about six months. From March to August, the river floods into the Pantanal, and drains out through a system of distributary channels, including the Taquari River (Souza et al., 2002), between September and February. In the central Pantanal, the river usually peaks in June, but in exceptionally high-flow years, the maximum flow is advanced to May or April, roughly in proportion to the magnitude of the flood stage; the flood also occurs earlier in exceptionally dry years when the water remains within the river banks. The smaller rivers, with less wetland storage to modify their regimes, respond more directly to the seasonality of rainfall, with for example the Salado River of southern Argentina exhibiting a March runoff peak of 25 $m^3 s^{-1}$, and the Zapotal River of Ecuador also showing a March peak of 650 $m^3 s^{-1}$.

5.3.3 Inundation

The generally low gradients (1–10 cm km^{-1}) of the continent's large rivers, the breadth of their structurally controlled alluvial valleys with extensive floodplains, inland deltas, and alluvial fans, and the limited extent of channel engineering or impoundment all favor massive overbank flows and seasonal storage of water in wetlands (Junk and Furch, 1993; Richey et al., 1989). Inundation thus occurs from the lower Paraná delta plain in Argentina to the *várzeas* of the central Amazon basin (Mertes et al., 1996), the *ciénagas* and lakes of the Magdalena valley (Garcia Lozano and Dister, 1990), the *llanos* of the western Amazon in Bolivia and Peru (Goulding et al., 2003), and the fringing floodplains, inland delta, and coastal delta of the Orinoco basin (Hamilton and Lewis, 1990). Hamilton et al. (2002) used satellite technology to monitor seasonal changes in the extent of inundation in six large wetlands in South America (fig. 5.6). Large seasonal changes in inundated area correlated well with water levels in adjacent rivers, and averaged a continent-wide total of 187,000 km^2, with a nonsynchronous maximum exceeding 500,000 km^2, although the passage of the annual flood wave exaggerates this flooded area above that which is flooded simultaneously (463,000 km^2). Maximum depths of inundation range up to 15 m. Inundation is also caused at the mouths of the many tributaries entering the major rivers through the backwater effects of large seasonal changes in water levels (e.g., Meade et al., 1991).

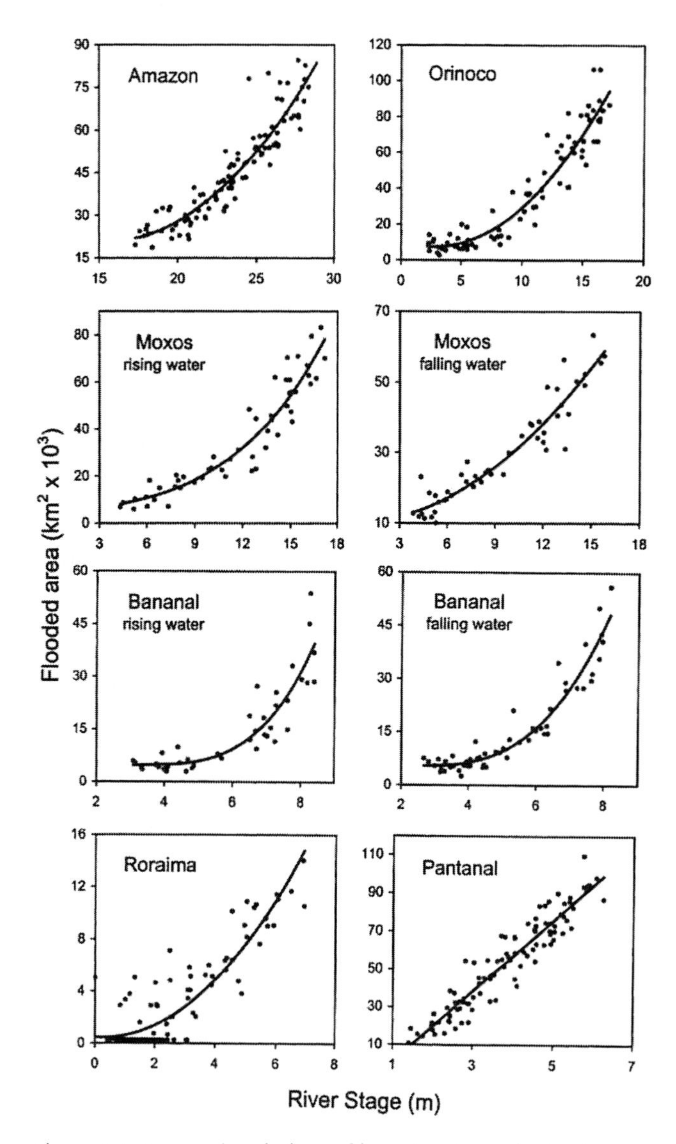

Figure 5.6 Seasonal variations of inundated area in South American wetlands (Hamilton et al., 2002, fig. 7)

5.4 Sediment Supplies and Transport

Sufficient data are available to report on sediment transport for very few South American rivers, but a useful sample of annual sediment yields per unit area of drainage basin is presented in table 5.2. As with large rivers everywhere, the quantity of sediment supplied to the rivers reflects mainly the lithology and topography of their drainage basins. Thus most sediment enters the large rivers of the continent from the mechanically weak rocks and steep topography of the Andes. The most extensive dataset presently available on sediment fluxes in South America was compiled by Bolivian and French hydrologists, and summarized by Guyot and his colleagues (Guyot, 1993; Guyot et al., 1988, 1990). Aalto et al. (2006) made a statistical summary of these data, correlating long-term annual sediment yields in the east-flowing Andean tributaries of the Madeira River in Bolivia with indices of basin-averaged hillslope steepness and of lithologic resistance to weathering and erosion. Yields range from 30 to 8000 t $km^{-2} yr^{-1}$ on igneous rocks, from 100 to 10,000 t $km^{-2} yr^{-1}$ on metasedimentary rocks, and from 200 to 18,000 t $km^{-2} yr^{-1}$ on weakly consolidated sediments. More than 70% of the annual load carried by these tributaries leaves the Andes in a three-month period from January to March.

Deposition of much of this sediment in structural basins adjacent to the Andes and in the long floodplain reaches of the major rivers alters the sediment transport patterns and limits the amount of sediment reaching the ocean. Guyot (1993) estimated that, of the annual sediment load of 550 Mt (million tonnes) leaving the Eastern Cordillera of Bolivia for the Madeira River, about 60% is trapped in the floodplains of the foreland basin, with

Table 5.2 Sediment yields of selected South American rivers

River	Station	Country	Basin area [km²]	Sediment yield [t km⁻² yr⁻¹]	Source
Magdalena	mouth	Colombia	266,600	716	Garcia Lozano and Dister (1990)
Beni	Rurrenabaque	Bolivia	68,000	3220	Guyot (1993)
Beni	Cachuela Esperanza	Bolivia	282,500	570	Guyot (1993)
Madeira	Guayaramerin	Bolivia	903,500	250	Guyot (1993)
Negro	Manaus	Brazil	691,000	10	Dunne et al. (1998)
Orinoco	Ciudad Bolivar,	Venezuela	1,100,000	140	Warne et al. 2002)
Pilcomayo	Villamontes,	Bolivia	81,000	1000	Guyot et al. (1990)
Paraguay	Paraná confluence	Paraguay	1,095,000	100	Anderson et al. (1993)
Upper Paraná	Porto São Jose	Brazil	670,000	30*	Orfeo & Stevaux (2002)
Middle Paraná	Corrientes	Argentina	1,950,000	80	Orfeo & Stevaux (2002)

*after impoundments

the amount retained ranging from 80% in the Mamore basin to 43% in the steeper Beni River with its narrower floodplain.

The main-stem Amazon carries a load of more than 1000 Mt yr^{-1}, with an irregular downstream reduction due to variations in channel gradient, valley width, and tributary input. Dunne et al. (1998) calculated sediment budgets for the various processes (bar deposition, bank erosion, and diffuse and channelized overbank deposition) exchanging sediment between the channel and the floodplain along a 2,010-km reach of the Amazon. These lateral exchanges exceed the total annual sediment load moving past Óbidos at tidewater (~1200 Mt yr^{-1}), and a further 300–400 Mt yr^{-1} of deposition was estimated to occur in a delta plain at the head of the estuary.

In the lowland Amazon basin, erosion rates on the forested Brazilian and Guiana shields are very slow. Guyot (1993) estimated a sediment yield of 16 t km^{-2} yr^{-1} for the Río Guaporé in southeast Bolivia, and Dunne et al (1998) reported a yield of 10 t km^{-2} yr^{-1} for the Río Negro in Brazil. Latrubesse and Stevaux (2002) reported a yield of 49 t km^{-2} yr^{-1} from the 364,000 km^2 Araguaia River basin, which drains the ecotone between savanna and seasonally dry forest into the Tocantins River on the eastern Brazilian Shield. Thus the Andes remain the dominant sediment source of the main-stem Amazon all the way to the Atlantic Ocean.

The interacting effects of runoff, lithology, gradient change, and to some extent impoundment are reflected in the contrasting sediment transport regimes feeding into the Río de la Plata system, between those tributaries emerging from the Andes and those draining from the more resistant rocks of the upper Paraná Highlands and the Brazilian Shield (Orfeo and Stevaux, 2002). Downstream of large dams, the upper Parana River has suspended sediment concentrations of only 6–30 mg L^{-1} (Orfeo and Stevaux, 2002), and 90–200 mg L^{-1} at the confluence with the Paraguay River (Drago and Amsler, 1988, 1998). The river's large annual discharge transports about 15 Mt yr^{-1} of measured fine washload, and a calculated sandy bed-material load of about 5 Mt yr^{-1}, the coarser component of which travels as large sand waves on the wide channel bed. The smaller Paraguay River, at its confluence with the Paraná, has an average flow of about 4,500–5,000 m^3 s^{-1} and a sediment yield of 1,188 Mt yr^{-1} of finer sediment and a sandy bed–material load of about 40 Mt yr^{-1}. Most of this load derives from the Andean tributaries: the Río Pilcomayo (81,300 km^2) with a sediment load averaging 80 Mt yr^{-1} (Guyot et al., 1990), and the Río Bermejo (16,000 km^2), where concentrations average about 6,000 mg L^{-1} (Drago and Amsler, 1988). The wave of suspended sediment enters the middle and lower Paraná weeks after the main flood wave from the upper Paraná.

5.5 Fluvial Geomorphology

In order to illustrate the variety of riverscapes of the Amazon basin, figure 5.7 presents schematically a generalized cross-section showing the distinctive fluvial geomorphology of various sectors from the headwaters to the downstream delta plain near the Xingu confluence.

On the forested Brazilian and Guiana shields (exemplified by the northeast corner of figure 5.7D), most rivers receive small sediment loads, mostly silt and clay, and little bed material, so that they flow in bedrock channels that exhibit strong structural control along fractures or faults, irregular alignments, and rapids with intervening sand bars and little if any floodplain. However, some structural basins within the shields have extensive alluvial plains with actively meandering rivers and floodplains that contain paleo-channels and various other off-channel water bodies and marshes, some of them supplied only by local runoff. Latrubesse and Stevaux (2002) described one example in the alluvial plain of the middle Araguaia, central Brazil, which drains about 200,000 km^2 of savanna that is being aggressively deforested for agriculture. An extreme example of a sediment-starved river with a bedrock-dominated

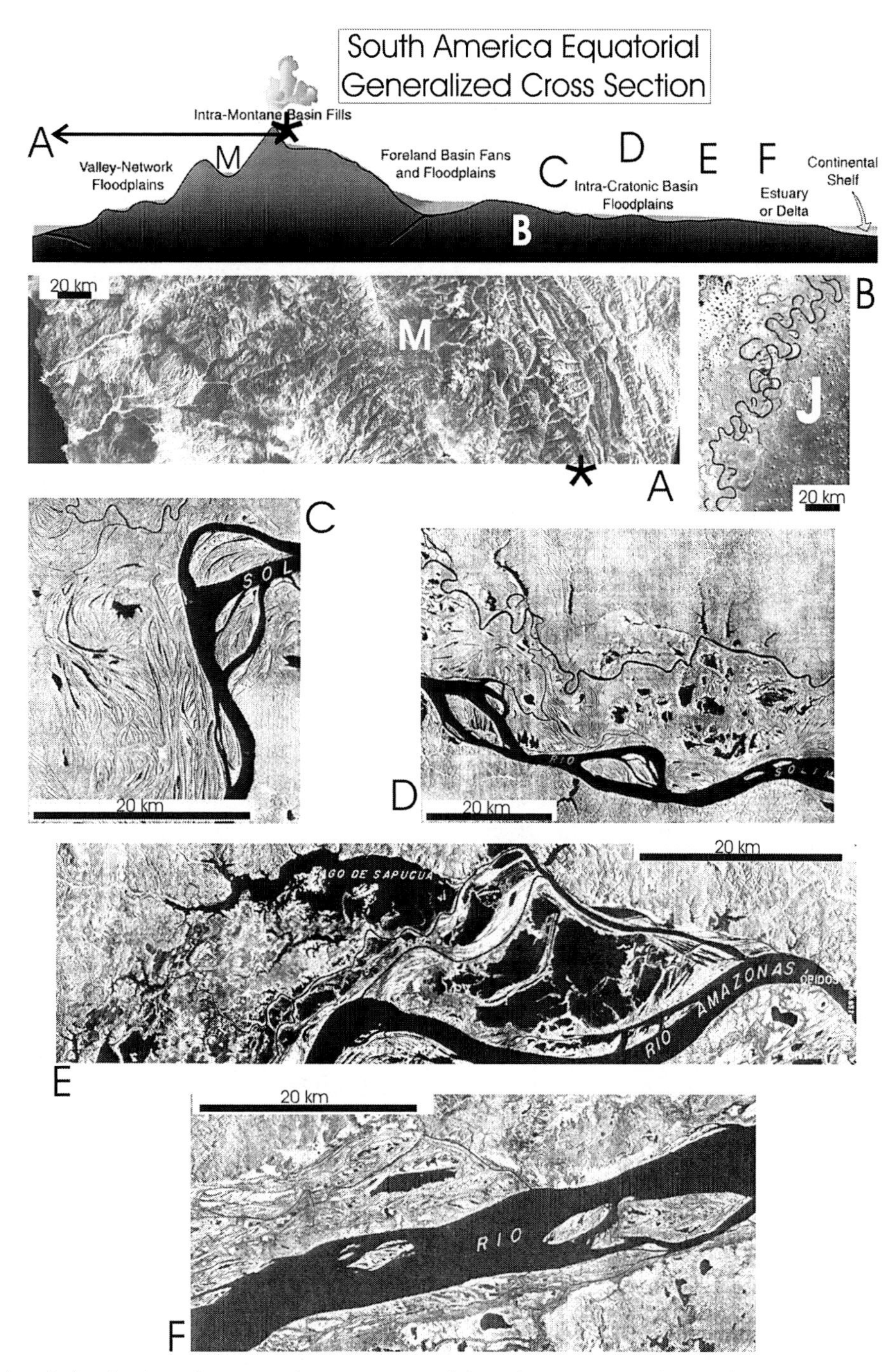

Figure 5.7 Regional variation in river character along an equatorial section across South America. A: Andean headwaters of Marañon (M) and Huallaga rivers (*), Perú (Landsat Thematic Mapper image, Band 4, 30 December 1989, centered ~9°S 78°W). B: Complex meanders of the Juruá River (J), Brazil, in an intracratonic setting just south of its confluence with the Amazon (Landsat Enhanced Thematic Mapper image, Band 4, 31 July 2001). C: Scroll-bar floodplain along upstream reaches of Amazon River near Xibeco. D: Transitional floodplain in middle reaches of Amazon River near Itapeua. E: Downstream floodplain of Amazon River near Óbidos. F: "Filled" delta-plain floodplain of Amazon River between Tapajós and Xingu confluences (figs. 5.7C–5.7F from Mertes et al., 1996).

channel is the lower Negro River in Amazonas state (Franzinelli and Igreja, 2002). The river, which exceeds 20 km in width in some reaches, flows in a half-graben between cliffs of friable sandstone and around long linear islands that are believed to be founded on up-faulted bedrock ridges. Blocks of bedrock and unconsolidated sediment also surrounded the wide bedrock channel in various attitudes and elevations. The river transports only enough sediment to maintain narrow beaches capping and lining the islands and in pockets where the valley-side alignment changes sharply.

Steep Andean rivers (fig. 5.7A) flow in narrow valleys, over bedrock and sparse gravel bars, occasionally interrupted by wider sand and gravel floodplains in intramontane basins that are also lined with gravel terraces reflecting the tectonic and climatic history of the range. When these rivers enter the Subandean basins, below the 500–m contour, they rapidly deposit sandy point and midchannel bars and crevasse-splay deposits, and overlay them with silt and clay levees and flood basin deposits. The rapid channel shifting and oxbow formations along the Beni and Mamore Rivers in the Beni Basin, for instance, create extensive and diverse habitats for fish and river dolphins (Lauzanne et al., 1990).

Larger rivers formed entirely within the Amazon lowland carry small sediment loads and have tortuous irregular planforms with sinuosities often exceeding 2.4, as exemplified by the Juruá River in western Amazonas state (fig. 5.7B; Latrubesse and Kalicki, 2002; Stölum, 1998). The rate of meander migration is generally unknown, but the Holocene floodplains are broad and contain many oxbow lakes that may be thousands of years old, since the rate of overbank sedimentation is probably limited by the small loads of these rivers.

Along the main-stem Amazon (figs. 5.7C–5.7F), deposition and channel migration have constructed an intricate valley-floor topography consisting of scroll-bar complexes, levees, and flood basins, many of them still occupied by perennial lakes. The main channel shifts at rates of <10 to >100 m yr^{-1}, depending on the degree to which the floodplain is confined between cohesive terraces. Mertes et al. (1996) described a structurally controlled alternating pattern of confined and freely meandering reaches with varying representation of alluvial features such as lakes and scroll bars (fig. 5.7C) that are scaled by three types of channel: the main Amazon channel, its occasional anabranches, and smaller floodplain channels that originate either in the main river (and therefore are sediment-rich) or in local floodplain or terrace drainage (and are sediment-starved). Latrubesse and Franzinelli (2002) emphasized the segmented alignment of the middle Amazon, which they attribute to neotectonic activity of tilted fault blocks. They also added detail and historical perspective concerning the development of the middle Amazon floodplain, differentiating between an older (late Holocene), smoother floodplain created dominantly by vertical accretion, and a later regime of lateral accretion of bars, levee complexes, channels, and lake deltas fringing the modern channel within the past millennium (fig. 5.7D).

Much of the Amazon valley downstream of Manaus was excavated of sediment as sea level fell to 120 m below the present during the Last Glacial Maximum, about 20 ka (Mertes and Dunne, in press). With the subsequent postglacial rise of sea level, sedimentation began refilling this valley, which may also still be subsiding due to continued graben dynamics (fig. 5.7E). The same lowering of base level scoured sediment from the lower reaches of each tributary entering the central and eastern Amazon, but the low sediment supply from these tributaries has not been able to keep pace with the rise of sea level and of the Amazon floodplain, leaving long bays at the mouths of all tributaries except for the sediment-rich Madeira (figs. 5.7E, 5.7F; Sioli, 1968; Tricart, 1975, 1977). At the mouth of the Amazon, a 25,000 km^2 delta plain and a subaqueous delta represent the continuing response of sedimentation to postglacial sea level rise (Nittrouer et al., 1986a, 1995).

The Orinoco River is confined to the southern edge of its valley by southeastward tilting associated with the uplift of the Venezuelan Andes and by the clastic wedge spreading from the mountains (fig. 5.8). This confinement forces the channel to cross eight bedrock outcrops that form rapids and reduce channel migration (Warne et al., 2002). The valley alluvium is thus limited to a narrow fringing floodplain with more extensive inland deltas at the mouths of left bank tributaries, such as the Apure and Arauca rivers, where deposition is induced by backwaters from the Orinoco as their flow is restricted by bedrock outcrops. Almost the entire sediment supply to the Orinoco is provided by left-bank tributaries draining from the Andes, but only after depositing much of their load on the gently sloping sedimentary wedge south and east of the mountains. The right-bank tributaries from the Guiana Shield have very low sediment loads. The Orinoco has built the continent's only large ocean-front delta, with a 22,000 km^2 plain consisting of sandy and silty fluvial deposits near its apex and southern margin, and along tributaries, grading into muddy and peaty inter-distributary plains with increasing tidal influence to the north and east (Warne et al., 2002).

In the upper Paraguay basin, between the Andes and the Brazilian Shield, the Pantanal of Mato Grosso do Sul, Brazil, comprises a series of low-gradient coalescing alluvial fans and inland deltas that cover nearly 200,000 km^2 and merge with the floodplain of the Paraguay River. Souza et al. (2002) described the downstream variation of channel form, inundation, and sedimentation along the largest of the fans, on the Taquari River. Near the fan apex, the channels are entrenched, single threaded, and sinuous, grading downstream to a multi-threaded channel more or less at the fan surface, which then merges with the Paraguay floodplain via fans, crevasses splays, and lakes. The

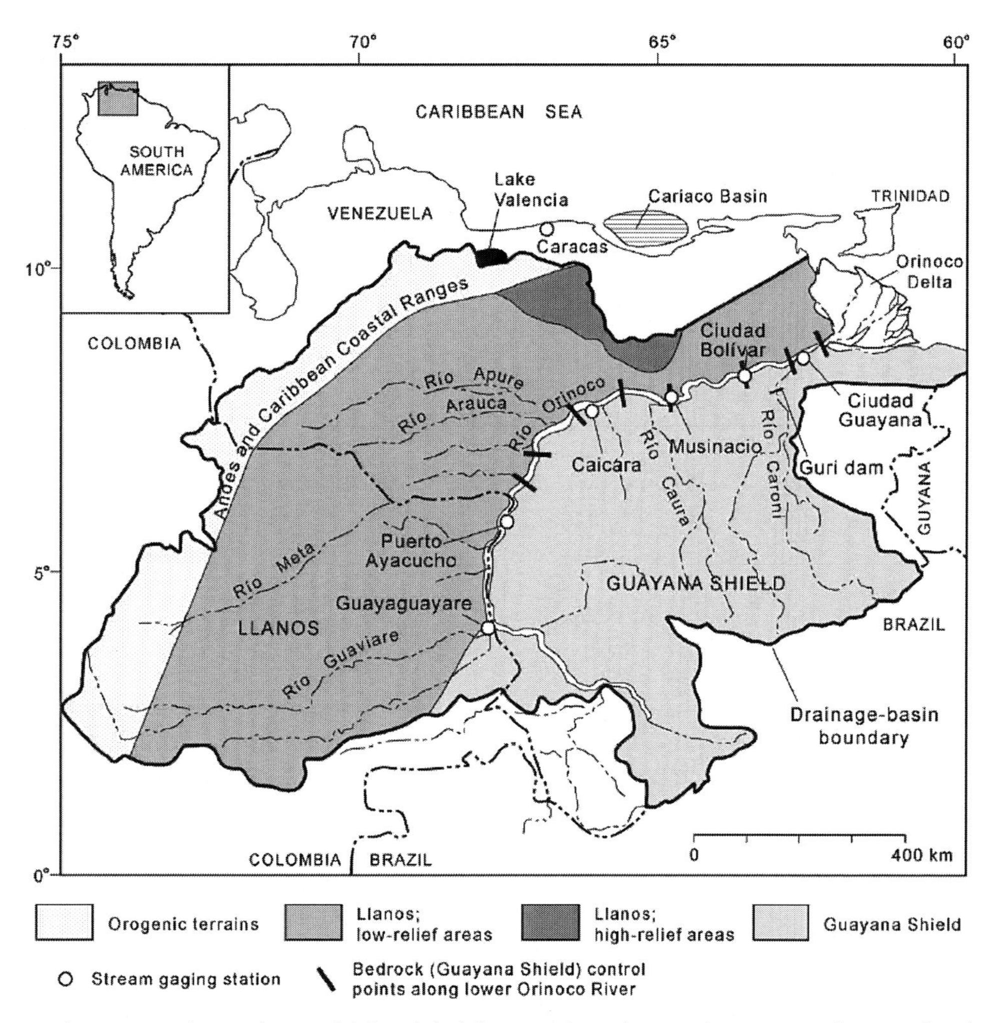

Figure 5.8 Orinoco River and delta plain (after Meade et al., 1983, in Warne et al., 2002, fig. 3)

lower Paraguay floodplain contains extensive sand-bar deposits, although significant meander activity also produces scroll bars and oxbow lakes. Older meander belts, now distant from the channel, receive only local sediment from smaller streams that are gradually resurfacing the plain (Drago, 1990).

The major channels of the upper and middle Paraná basin are exceptionally wide and shallow (width-depth ratio >100), with significant multi-thread reaches and frequent, complex sand bars, some of them eroded from older floodplain sediment and others deposited directly from the flow (Orfeo and Stevaux, 2002). Channel shifting by bend deformation occurs at rates of tens of meters per year, but chute cutoffs can displace the channel suddenly by several kilometers (Ramonell, 2002). The channel thalweg of the middle Paraná also shifts rapidly and erratically within its wide banks. As the bed narrows and widens, and scours and fills by up to 15 m in a flood season, the channel vacillates between single thread and multi-thread behavior, except where it impinges on resistant terrace materials of Neogene age (Drago, 1990). The modern floodplain consists of a complex of scroll bars, channels, lakes, lacustrine deltas, marshes, and islands, grading into a delta plain (Drago, 1990).

5.6 The Future of South American Rivers

The rivers of South America have as yet suffered relatively little disturbance from human activities, but this situation is likely to change in coming decades. The most pervasive changes are likely to arise through large impoundments, channel engineering works to improve navigation, forest and scrub clearance for agriculture, and urban and industrial pollution. Large dams have been spreading throughout the continent since the 1970s, mainly for hydroelectric power generation, and the potential for further development is large because of the generally high rates of runoff, the availability of favorable sites for large dams and reservoirs, and improvements in the efficiency of power transmission (Petts, 1990). The Paraná River is the basin most intensively developed for hydropower, with at least 20 dams on the main stem and tributaries of the upper Paraná already storing 60% of the mean annual flow. The seasonal flow pattern of the river has already been significantly altered, and together with toxic effluent from agriculture and industry, the regulation of the river-floodplain flow system appears to be reducing fish catches (Quiros, 1990). There are also major water impoundments in the São Francisco and Tocantins rivers in Brazil, and in the Andes where countries from Venezuela to Chile have constructed reservoirs in the face of high rates of sedimentation (see chapter 18) and where Bolivia, Perú, and Ecuador contain large canyons with attractive dam sites. It is likely that the alteration of flows and sediment dynamics downstream of large dams throughout the continent will change channel geometry and channel-floodplain connections.

The long rivers of the continent also provide navigation, but the frequent rapids, tortuous mobile bends, and sand bars limit the access of large modern barges. Several major dredging projects aimed at improving navigation have been initiated or proposed. The general nature of their impact is likely to be to deepen and simplify channels, thereby reducing connections with their floodplains and draining the wetlands of the major alluvial valleys. These events will likely have a major impact of aquatic and riparian ecosystems.

Deforestation for pasture and cultivation has spread widely through the continent and is likely to continue expanding through the Amazon basin for decades. Acceleration of sediment supply from deforested areas has appeared in savanna regions of the Brazilian Shield from widespread cultivation on the drier margins of Amazonia (Latrubesse and Stevaux, 2002), but sedimentation problems have not yet been reported in the wetter parts of the Amazon. However, pollution from urban and industrial development has already been recognized there (Biggs et al., 2004). Severe pollution problems also have been associated with urbanization and industrialization in more intensely developed drainage basins of the continent, notably along rivers flowing through the developed coastal regions of southeast Brazil, the Río del la Plata estuary of Argentina and Uruguay, central Chile, and the Caribbean coast (see chapter 20, Bartone, 1990; Filoso et al., 2003).

References

Aalto, R.E., T. Dunne, and J-L. Guyot, 2006. Geomorphic controls on Andean denudation rates: *Journal of Geology*, 114, 85–100.

Anderson, R.J., Jr., N. da Franca Ribeiro dos Santos, and H.F. Diaz, 1993. An analysis of flooding in the Paraná/Paraguay River basin. LATEN Dissemination Note 5, Latin American Technical Department, The World Bank.

Augustinus, P.G.E.F., 1982. Coastal changes in Suriname since 1948. In: *Future of roads and rivers in Suriname and neighboring region*. Paramaribo, Suriname, 329–338.

Bartone, C.R., 1990. Water quality and urbanization in Latin America. *Water International*, 15, 3–14.

Biggs, T.W., T. Dunne, and L.A. Martinelli, 2004. Natural controls and human impacts on stream nutrient concentrations in a deforested region of the Brazilian Amazon basin. *Biogeochemistry*, 68, 227–257.

Birkett, C., L.A.K. Mertes, T. Dunne, M.H. Costa, and M.J. Jasinski, 2002. Surface water dynamics in the Amazon Basin: Application of satellite radar altimetry. *Journal of Geophysical Research—Atmospheres*, 107, D20.

Caputo, M.V., 1984. *Stratigraphy, tectonics, paleoclimatology and paleogeography of northern basins of Brazil*. Ph.D. dissertation, University of California, Santa Barbara.

Caputo, M.V., 1991. Solimões megashear: Intraplate tectonics in northwestern Brazil. *Geology*, 19, 246–249.

Center for International Earth Science Information Network (CIESIN), 2000. *Gridded Population of the World (GPW), Version 2*. CIESIN, Palisades, NY, Columbia University; International Food Policy Research Institute (IFPRI); and World Resources Institute (WRI).

Clapperton, C., 1993. *Quaternary geology and geomorphology of South America*. Elsevier, New York.

Cox, K.G., 1989, The role of mantle plumes in the development of continental drainage patterns. *Nature*, 342, 873–877.

Damuth, J.E., and N. Kumar, 1975. Amazon Cone: Morphology, sediments, age, and growth pattern. *Geologic Society of America Bulletin*, 86, 863–878.

Drago, E.C., 1990. Geomorphology of large alluvial rivers: Lower Paraguay and middle Paraná. *Interciencia*, 15, 378–387.

Drago, E.C., and M.L. Amsler, 1988. Suspended sediment at a cross section of the Middle Paraná River: Concentration, granulometry and influence of the main tributaries. *Sediment Budgets, International Association of Hydrological Sciences*, Publication 174, 381–396.

Drago, E.C., and M.L. Amsler, 1998. Bed sediment characteristics in the Paraná and Paraguay rivers. *Water International*, 23, 174–183.

Dunne, T., L.A.K. Mertes, R.H. Meade, J.E. Richey, and B.R. Forsberg, 1998. Exchanges of sediment between the floodplain and channel of the Amazon River in Brazil. *Geological Society of America Bulletin*, 110, 450–467.

Filoso, S., L.A. Martinelli, and M.R. Williams, 2003. Land use and nitrogen export in the Piracicaba River basin, southeast Brazil. *Biogeochemistry*, 65, 275–294.

Franzinelli, E., and H. Igreja, 2002. Modem sedimentation in the lower Negro River, Amazonas State, Brazil. *Geomorphology*, 44, 259–271.

Garcia Lozano, L.C., and E. Dister, 1990. La planicie de inundacion del medio-bajo Magdalena: Restauracion y conservation de habitats. *Interciencia*, 15, 396–410.

Goulding, M., C. Cafias, B. Barthem, B.R., Forsberg, and H. Ortega, 2003. *Amazon headwaters: Rivers, wildlife, and conservation in Southeastern Peru*. Amazon Conservation Association, Lima, Peru.

Guyot, J.L., 1993. *Hydrogéochimie des Fleuves de*

l'Amazone Bolivienne. Collection Etudes & Thèses, Editions de l'ORSTOM.

Guyot, J.L., J. Bourges, R. Hoorelbecke, and M.A. Roche, 1988. Exportation des matières en suspension des Andes vers l'Amazonie par le Rio Beni, Bolivie. *Sediment Budgets, International Association of Hydrological Sciences*, Publication 174, 443–451.

Guyot, J. L., Calle, H., Cortes, J., and Pereira, M., 1990. Transport de matières dissoutes et particulaires des Andes vers le Río de La Plata par les tributaires boliviens (ríos Pilcomayo et Bermejo) du Río Paraguay. *Hydrological Sciences Journal*, 35, 653–665.

Hamilton, S.K., and W.M. Lewis, 1990. Basin morphology in relation to chemical and ecological characteristics of lakes on the Orinoco River floodplain, Venezuela. *Archiv für Hydrobiologie*, 119, 393–425.

Hamilton, S.K., S.J. Sippel, and J.M. Melack, 2002. Comparison of inundation patterns among major South American floodplains. *Journal of Geophysical Research*, 107D, 5.1–5.14.

Junk, W.J., and K. Furch, 1993. A general review of tropical South American floodplains. *Wetlands Ecology and Management*, 2, 231–238.

Kuehl, S.A., D.J. DeMaster, and C.A. Nittrouer, 1986. Nature of sediment accumulation on the Amazon continental shelf. *Continental Shelf Research*, 6, 209–225.

Kuehl, S.A., C.A. Nittrouer, and D.J. DeMaster, 1982. Modern sediment accumulation and strata formation on the Amazon continental shelf. *Marine Geology*, 49, 279–300.

Latrubesse, E.M., and E. Franzinelli, 2002. The Holocene alluvial plain of the middle Amazon River, Brazil. *Geomorphology*, 44, 241–257.

Latrubesse, E.M., and T. Kalicki, 2002. Late Quaternary palaeohydrological changes in the upper Purús Basin, southwestern Amazonia, Brazil. *Zeitschrift für Geomorphologie Supplementband*, 129, 41–59.

Latrubesse, E.M., and J.C. Stevaux, 2002. Geomorphology and environmental aspects of the Araguaia fluvial basin, Brazil. *Zeitschrift für Geomorphologie Supplementband*, 129, 109–127.

Lauzanne, L., G. Loubens, B.L. Guennec, 1990. Pesca y biologia pesquera en el Mamore medio (Region de Trinidad, Bolivia): *Interciencia*, 15, 452–459.

Lewis, W.M., Jr., S.K. Hamilton, and J.F. Saunders, III, 1989. Rivers of northern South America. In: C. Cushing, G.W. Minshall, and K. Cummins (Editors), *Ecosystems of the world, rivers*. Elsevier Science, New York, 219–256.

Manley, P.L., and R.D. Flood, 1988. Cyclic sediment deposition within the Amazon deep-sea fan. *American Association of Petroleum Geologists Bulletin*, 72, 912–925.

Meade, R.H., J.M. Rayol, S.C. Conceicão, and J.R.G. Natividade, 1991. Backwater effects in the Amazon River basin of Brazil. *Environmental Geology and Water Science*, 18, 105–114.

Meade, R.H., C.F. Nordin, Jr., D. Pérez-Hernández, A. Mejía-B, and J.M. Pérez-Godoy, 1983. Sediment and water discharge in Río Orinoco, Venezuela and Colombia. In: *2nd International Symposium on River Sedimentation*, Nanjing, China, 1134–1144.

Mertes, L.A.K., and T. Dunne, in press. Effects of tectonism, climate change, and sea-level change on the form and behavior of the modern Amazon River and its floodplain. In: A.B. Gupta (Editor), *Large Rivers*. John Wiley, Chichester.

Mertes, L.A.K., T. Dunne, and L.A. Martinelli, 1996. Channel-floodplain geomorphology along the Solimões-Amazon River, Brazil. *Geological Society of America Bulletin*, 108, 1089–1107.

Milliman, J.D., C.P. Summerhayes, H.T. Barretto, 1975. Quaternary sedimentation on the Amazon continental margin: A model. *Geological Society of America Bulletin*, 86, 610–614.

Molion, L.C.B., 1990. Climatic variability and its effects on Amazonian hydrology. Interciência, 15, 367–372.

Nittrouer, C.A., T.B. Curtin, and D.J. DeMaster, 1986a. Concentration and flux of suspended sediment on the Amazon continental shelf. *Continental Shelf Research*, 6, 151–174.

Nittrouer, C.A., S.A. Kuehl, D.J. DeMaster, and R.O. Kowsman, 1986b. The deltaic nature of Amazon shelf sedimentation. *Geological Society of America Bulletin*, 97, 444–458.

Nittrouer, C.A., S.A. Kuehl, R.W. Sternberg, A.G. Figueiredo, Jr., and L.E.C. Faria, 1995. An introduction to the geological significance of sediment transport and accumulation on the Amazon continental shelf: *Marine Geology*, 125, 177–192.

Orfeo, O., and J.C. Stevaux, 2002. Hydraulic and morphological characteristics of middle and upper reaches of the Paraná River (Argentina and Brazil). *Geomorphology*, 44, 309–322.

Petts, G.E., 1990. Regulation of large rivers: Problems and possibilities for environmentally sound river development in South America. *Interciencia*, 15, 388–395.

Ponce, V.M., 1995. *Hydrologic and environmental impact of the Paraná-Paraguay waterway on the Pantanal of Mato Grosso, Brazil*. San Diego State University, San Diego, California.

Potter, P.E., 1978. Significance and origin of big rivers: *Journal of Geology*, 86, 13–33.

Pujos, M., J. Monente, C. Latouche, and N. Maillet, 1997. Origin of sediment accumulation in the Orinoco delta and the Gulf of Paria: Amazon River sediment input. *Oceanologica Acta*, 20, 799–809.

Quiros, R., 1990. The Paraná River basin development and the changes in the lower basin fisheries. *Interciencia*, 15, 442–451.

Ramonell, C.G., M.L. Amsler, and H. Toniolo, 2002. Shifting modes of the Paraná River thalweg in its lower/middle reach. *Zeitschrift für Geomorphologie Supplementband*, 129, 129–142.

Richey, J.E., L.A.K. Mertes, T. Dunne, R. Victoria, B.R. Forsberg, A. Tancredi, and E. Oliveira, 1989. Sources and routing of the Amazon River flood wave. *Global Biogeochemical Cycles*, 3, 191–204.

Sioli, H., 1968. Hydrochemistry and geology in the Brazilian Amazon region. *Amazoniana*, 1, 267–277.

Souza, O.C. da, M.R. Araujo, L.A.K. Mertes, and J.M. Melack, 2002. Form and process along the Taquari River alluvial fan, Pantanal, Brazil. *Zeitschrift für Geomophologie Supplementband* 129, 73–107.

Sternberg, H.O'R., and R.D. Russell, 1952. Fracture patterns in the Amazon and Mississippi valleys. In: *Proceedings, 17th International Congress, Washington, D.C.*, 380–385.

Stölum, H-H., 1998. Planform geometry and dynamics of meandering rivers. *Geological Society of America Bulletin*, 110, 1485–1498.

Tricart, J., 1975. Influence des oscillations climatiques

récentes sur le modelé en Amazonie Orientale (Region de Santarém) d'aprés les images radar latéral. *Zeitschrift für Geomorphologie*, 19, 140–163.

Tricart, J., 1977. Types de lits de fluviaux en Amazonie brésilienne. *Annales de Géographie*, 473, 1–54.

Ussami, N., S. Shiraiwa, S., and J.M.L. Dominguez, 1999. Basement reactivation in a sub-Andean foreland flexural bulge: The Pantanal wetland, SW Brazil. *Tectonics*, 18, 25–39.

Vörösmarty, C.J., B. Fekete, and B.A. Tucker, 1996. *River Discharge Database, Version 1.0 (RivDIS 1.0), 0–6.* A contribution to IHP-V Theme 1, Technical Documents in Hydrology Series. UNESCO, Paris.

Vörösmarty, C.J., B. Fekete, and B.A. Tucker, 1998. *River Discharge Database, Version 1.1 (RivDIS 1.0 supplement).* Institute for the Study of Earth, Oceans, and Space, University of New Hampshire, Durham.

Warne, A.G., R.H. Meade, W.A. White, E.H. Guevera, J. Gibeaut, R.C. Smyth, A. Aslan, and T. Tremblay, 2002. Regional controls on geomorphology, hydrology, and ecosystem integrity in the Orinoco Delta, Venezuela. *Geomorphology*, 44, 273–307.

6

Flora and Vegetation

Kenneth R. Young
Paul E. Berry
Thomas T. Veblen

South America's shape, size, and geographic position, now and in the past, have acted to influence the development of diverse coverings of land surfaces with plants of different sizes, adaptations, and origins. Underlying geologic structures have been exposed to weathering regimes, thereby resulting in a multiplicity of landforms, soil types, and ecological zones (see chapters 2 and 7). The most notable large-scale features are the Andes, which curl along the western margin of the continent, and the broad swath of the Amazon lowlands in the equatorial zone. However, there are also extensive, more ancient mountain systems in the Brazilian Shield of east-central Brazil and the Guiana Shield in northern South America. The interplay of environmental factors has given rise to a panoply of vegetation types, from coastal mangroves to interior swamplands, savannas, and other grasslands, deserts, shrublands, and a wide array of dry to moist and lowland to highland forest types (fig. 6.1). The narrower southern half of South America is also complex vegetationally because of the compression of more vegetation types into a smaller area and the diverse climatic regimes associated with subtropical and temperate middle latitudes.

Alexander von Humboldt began to outline the major features of the physical geography of South America in his extensive writings that followed his travels in the early nineteenth century (von Humboldt, 1815–1832). For example, he first documented the profound influences of contemporary and historical geologic processes such as earthquakes and volcanoes, how vegetation in mountainous areas changes as elevation influences the distributions of plant species, and the effect of sea surface temperatures on atmospheric circulation and uplift and their impacts on precipitation and air temperatures (Botting, 1973; Faak and Biermann, 1986). His initial insights, in combination with modern observations (Hueck and Seibert, 1972; Cabrera and Willink, 1973; Davis et al., 1997; Lentz, 2000), still serve to frame our synthesis of the major vegetation formations of South America.

In this chapter, we relate vegetation formations to spatial gradients of soil moisture and elevation in the context of broad climatic and topographic patterns. This task is framed in the context of the three latitudinal subdivisions of South America that are associated, respectively, with tropical latitudes and the trade winds, subtropical latitudes and seasonally shifting air circulation patterns, and middle latitudes dominated by the westerlies (see chapter 3).

Charles Darwin was also a prominent nineteenth-century traveler to South America. He journeyed through these three

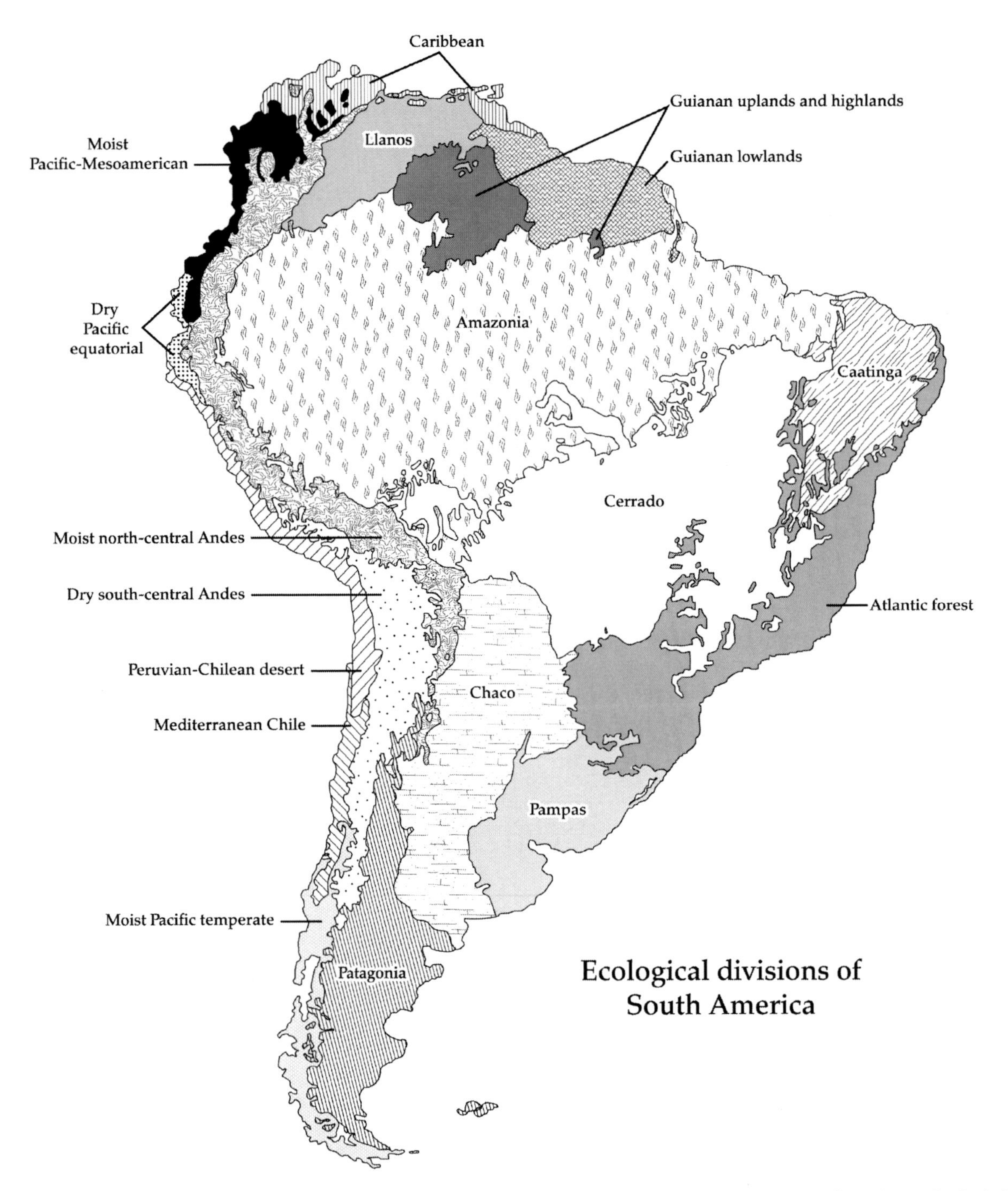

Figure 6.1 Primary ecological divisions of South America, modified from Josse et al., 2003. These areas reflect the main floristic regions and vegetation types of the continent.

latitudinal belts in the 1830s on his voyage in the *Beagle* and during excursions on land, and he used those experiences to develop his theory of the evolution of living organisms (Darwin, 1859). A variety of evidence collected on his travels served to show that plants and animals change through time or sometimes go extinct as they are exposed to natural selection and natural catastrophes. Thus, it is also important to consider historical and evolutionary changes in the biota, which can leave an imprint on what species are found in particular places. The flora of a particular place consists of a listing of all the plant species to be found there. The number of species, their respective distributions, and their adaptations to environmental conditions can be described, mapped, and quantified in a variety of ways. Explanations of those patterns, however, will often need to consider processes associated with evolution, in addition to ecological and physiological processes.

The early perspectives of Darwin and von Humboldt presaged modern attempts to understand the nature and distribution of the flora and vegetation of South America. Today we relate the modern flora to historical changes in environments and land connections that were so important to the migrations and extinctions of organisms. We also stress the role of contemporary climates and other aspects of modern environments that determine the continued success of different plant species and their shaping into communities of different structures. Thus, this chapter first considers the broad vegetation patterns of the continent, and then reviews the floristic implications of some of the major paleo-environmental legacies for subdivisions of the South American flora and vegetation. It concludes with brief comments about the rapidly changing land cover and plant and animal distributions of the continent, produced by the now dominant role of humans on large sections of the landscape. This approach provides a preface to the regional environments discussed in more detail later in this book.

There are a variety of ways in which the vegetation types of South America have been mapped and classified (Hueck and Seibert, 1972; Cabrera and Willink, 1973; Dinerstein et al., 1995; Josse et al., 2003; Eva et al., 2004). Most of these systems combine both the structural or physiognomic aspects of the vegetation itself (e.g., grassland, forest, swamp, desert), as well as the floristic or biogeographical aspects of the plants that determine the actual species composition of a particular region. To represent best the broad scale of plant formations in South America, we chose the approach developed by NatureServe (Josse et al., 2003) and depicted in their map, "Ecological Divisions of South America" (fig. 6.1). Their system further divides these 18 broad-scale units into nearly 700 *terrestrial ecological systems*, which are defined as "groups of plant community types that tend to co-occur within landscapes with similar ecological processes, substrates, and/or environmental gradients." The intent is to facilitate the recognition of these units through remote imagery, so that they may be mapped; in addition, they should be readily identifiable in the field. Although the detailed map is beyond the scope of this chapter and is still being refined by NatureServe (www.natureserve.org), it should provide an excellent tool for conservation monitoring and management in the future. Vegetation surveys in the field can measure the abundance of plants in terms of their sizes, layering, and composition along transects, typically linear sampling units that are arranged along or across important environmental gradients. In this chapter, we describe what might be found using imaginary transects placed across South America.

6.1 Tropical South America

We can begin our vegetation survey of South America with hypothetical transects radiating out from the city of Manaus, located in the heart of the Amazon basin in Brazil. Manaus was once an economic powerhouse with close links to North America and Europe during the rubber boom of the late 1880s. It now owes its development largely to factories that produce computers and other electronic consumer items. Secondarily, it is the center of an active ecotourism business that features nearby examples of the high biological diversity of the Amazon region. Surrounding the city are tropical rain forests, some growing on nutrient-poor white sands, tropical swamps dominated by palms, and Amazon waterways with blackish clear waters, muddy sediment-laden waters, and oxbow lakes (see chapter 5; Goulding et al., 1996).

6.1.1 Transect 1: From Manaus to the Southeast Coast of Brazil

A 3000-km transect beginning in Manaus and heading southeast would cross a strong physical environmental gradient, especially in terms of decreasing soil moisture (see chapter 9; Prance, 1989; Daly and Mitchell, 2000). The tall, multistory rain forest of the Amazon lowlands gradually becomes more seasonal, with an increasing proportion of deciduous trees, until a mixture of grass and low scrub known as the Cerrado occurs in places where the yearly dry season can be five or more months long (fig. 6.1). The seasonally dry southeastern portion of the Amazon basin is where the major arc of today's tropical deforestation is located (Achard et al., 2002), because the marked dry season allows people to set fires that convert forests to grasslands. A large percentage of the Cerrado biome recently has been converted to industrial soybean production (Oliveira and Marquis, 2002).

Toward the Brazilian coast near its prominent eastern bulge (fig. 6.1) are drylands known as Caatinga that were originally covered by small-leaved and often thorny shrubs several meters tall. This scrub vegetation type has been used extensively as ranchlands and, when irrigation is possible, is suitable for agriculture such as arabica coffee production. Moving southward along the coast, the coastal uplands are exposed to humid air masses from the Atlantic Ocean. These moister environments support the Atlantic rain forests, originally extending over 3,000 km from 6° to 32°S latitude, and isolated from the Amazon rain forests to the west and north. These 20–30 m tall forests are much threatened by deforestation and the pernicious influences of habitat fragmentation (Ranta et al., 1998; Galindo-Leal and de Gusmão Camara, 2003). They are in fact the epitome of the concept of threatened biological hotspots, as defined by Myers et al. (2000), in that over 75% of the habitats are strongly degraded, yet the area still harbors exceptionally high levels of endemism due to numerous species only found locally.

Close inspection of the Amazonian forests through large inventory plots that identify thousands of individual trees reveals that the dominant tree species shift from place to place (Pitman et al., 2001), undoubtedly with parallels in the Atlantic rain forests. This aspect in turn suggests that variations in modern and historical environments as well as disturbance-related phenomena form important, sometimes subtle, spatial gradients that organize the flora into different forest types or phases. Tuomisto and Ruokolainen (1994) and Duivenvoorden (1995) documented the role of relatively minor changes in topography, soil type, and soil moisture in determining different species' compositions in both the canopy and the understory layers. Junk (1989) had already pointed out that differences in flooding tolerance sort out tree species along environmental gradients going from upland to floodplain to swamp forests. Fine et al. (2004) suggested that these Amazonian environmental gradients also have consequences over evolutionary time. Not only do closely related tree species grow on different soil types in the same area in the western Amazon, but the species adapted to the more impoverished white sand soil type are apparently more protected by their leaf chemistry from herbivory by insects. Plant-animal interactions can be antagonistic, as in the case of herbivory, but also include hundreds of cases of mutualisms, for example when flowers are adapted to provide nectar to visiting pollinators, which then carry pollen to other flowers, thus allowing cross-pollination to take place.

Important site-to-site differences in species composition also characterize the Neotropical savannas, again due to relatively subtle environmental changes in conjunction with disturbances and interactions with animals. Furley (1999) described how the mosaics formed by open woodlands, arboreal grasslands, and open grasslands in the Cerrado in turn include fine-grained differences that are associated with minor shifts in edaphic conditions or are the product of disturbances such as fires. The tropical dry forests are ecotonal to these savannas and as a result can also be affected by fires, and indeed may have a grass understory. Other dry forests are only partly deciduous and may rival their rain forest neighbors in their stature and diversity. Pennington et al. (2000) documented their marginal but wide geographic distribution and used that information to suggest that these forests, in addition to the drier forms found in the Caatinga, Cerrado, Chaco, intermontane valleys of the Andes, and the dry phases of Caribbean forests, were remnants of an even wider distribution under more seasonally dry conditions in the past. The distribution of many plant groups across this entire subhumid and semi-arid belt suggests that under Pleistocene climatic episodes of major glacial expansion in the tropical Andes and at higher latitudes, such forests would have persisted over much larger areas of the tropical lowlands.

6.1.2 Transect 2: From Manaus Northward to the Atlantic and Caribbean Coasts

A 1,000–1,500-km transect extending north from Manaus would illustrate a different set of changes in the present physical environment. This hypothetical transect would cross from the Amazon rain forests into the fringes of the largely sandstone mountains of Guiana Shield and then into the Llanos of Venezuela and Colombia (fig.6.1). The Llanos are renowned for their tropical savannas, with a progression from extensive palm swamps to seasonally inundated grasslands, and then mixed savanna-shrublands similar to some of the Cerrado vegetation farther south in Brazil. Plant adaptations include mycorrhizal mutualisms between plant roots and fungi that permit survival on nutrient-poor soils. Plants found in wetlands have specialized tissues that keep their root systems aerated. Those species subjected to fire often are able to resprout, quickly replacing damaged leaves and branches and so recovering their ability to photosynthesize.

North of the Llanos in Colombia, Venezuela, and Guyana, the flora and vegetation along the coast and somewhat inland show strong affinities with the Caribbean region farther north (fig. 6.1). Much of this vegetation is deciduous forest or scrub, but there are many variations present, from deserts along the Colombian-Venezuelan border to cloud forests on the coastal mountain chains and in offshore islands like the Venezuelan Isla Margarita or nearby Trinidad.

Particularly in southern Venezuela, the sandstone-topped tepuis are a prominent feature of the landscape; these are parts of the Guiana Shield that have proven especially resistant to weathering (fig. 6.1). They serve as dramatic outposts for plants that can thrive at elevations 1,000–2,000 m above adjacent lowlands, often growing in bogs or on bare rock. Because of their isolation from other tropical uplands, the species here are unique, and the region is recognized as one of the foremost centers of endemism in the New World (Berry et al., 1995; Davis et al., 1997), especially for plant families such as Bromeliaceae, Ochnaceae, Orchidaceae, Rapateaceae, Velloziaceae, and Xyridaceae.

Givnish et al. (2000) used both DNA and morphological data to reconstruct the likely evolutionary lineages of the Rapateaceae in the tepuis and nearby lowlands of the Guiana Shield. They hypothesized a family origin 65 million years ago, subsequent colonization and evolution on individual tepuis uplands, and occasional long-distance dispersal among the uplands, including at least one event that crossed the Atlantic and resulted in the colonization of a similar environment in west Africa. Floristic uniqueness is thus due to evolution and isolation, but also can show evidence of long distance dispersal at times in the past to help explain the plants found in these unique places, called the "Lost World" by novelist Arthur Conan Doyle. Such phylogenetic reconstructions are beginning to elucidate the confounding and at times complementary roles of vicariance by plate tectonics and by dispersal (Young, 2003; Givnish and Renner, 2004; Berry and Riina, 2005).

To the east of the Guianan highlands and north of the Amazonian region lies another extensive area of lowland humid forests (fig. 6.1). These are designated the Guianan lowlands because they differ somewhat in floristic composition from central Amazonian forests, reflecting the influence of poor soils derived from the erosion of the Guiana Shield. They also occupy drainage basins that empty directly to the Atlantic Ocean rather than to the Amazon system.

6.1.3 Transect 3: From Manaus to the Andes

A third hypothetical transect from Manaus, this time extending 1,500–2,000 km to the northwest or west or southwest, would traverse some of the wettest and most diverse areas of the Amazon basin (ter Steege et al., 2003), and then extend into the Andes of Venezuela, Colombia, Ecuador, Perú, or northern Bolivia. Such a transect would cross more marked environmental gradients, with an initially gradual but then abrupt transition from the Amazon lowlands to the slopes of the Andes (see chapter 12; Luteyn and Churchill, 2000). The forest trees grade from tall rainforest specimens, with the canopy often at 25 to 35 m and emergents to 50 m, to the Andean foothills where Amazonian trees of those heights intermingle with the lower limit of Andean trees and shrubs. The tropical premontane forests of the foothills, and the belts of montane forests found sequentially up to timberline at around 3,500 m, become progressively shorter in stature with rising elevation. However, they remain dense, with understory shrubs or bamboos, and are often immersed in fogs resulting from the uplift and cooling of humid air. Trees become festooned by epiphytes, from mosses and filmy ferns to some of the several thousand species of orchids for which the region is well known.

Both von Humboldt and Darwin visited the Andes in different locations on their travels, observing how the vegetation changes along steep elevational gradients. Above the upper limits of tree growth, even up to snowline, plants are still present, where they form grasslands or open meadows of low-growing herbs or miniature shrubs. These plants are often exposed to freezing temperatures during the night, then warm temperatures during the day. Since they occur in the tropics, there is no lengthy cold season, and plants appressed to the ground are able to photosynthesize as long as solar radiation reaches them. Elsewhere in the Andes, moisture is the main limitation to plant growth, particularly in places not reached by the direct influences of the trade winds or where rain-shadow effects dry out air parcels. In these localities, the dominant vegetation is usually some kind of shrubland, dominated by heavily branched, woody plants just a few meters high. Drier places have thorny woodlands, cacti, or possibly tropical dry forests, depending on the intensity of past deforestation and the persistence of goats and sheep.

The Andes are also home to millions of farmers, working their lands for the production of field crops such as maize and potatoes, and for the grazing of livestock (Denevan, 2001). The mosaic of shrublands, seasonal grasslands, and forest patches found in these humanized landscapes still contain much diversity, in particular the high-beta diversity that signifies much species turnover from place to place. The native flora contains many endemic species, the result of internal barriers due to topography and to limited extensions of particular ecological zones, both of which add to speciation processes that result in the presence of numerous species with small ranges (Young et al., 2002). As a generalization, tropical highlands have higher levels of endemism than do tropical lowlands, even though alpha diversity, the number of species found in one place, of the highlands may never rival that found in most tropical rain forests.

6.1.4 Transects 4 and 5: From the Andes to the Pacific

The western side of the tropical Andes collectively includes the widest gamut of physical environments on the continent, where a relatively short 200-km-long transect from over 5,500 m on the highest peaks in Ecuador or Colombia would end in some of the wettest lowland tropical rain forests in the world, namely the Moist Pacific-Mesoamerican region (fig. 6.1), known more commonly as the Chocó forests. This area is renowned both for rainfall totals nearing 10 m yr^{-1} and for its high plant diversity (Gentry, 1982a, 1986). While many plants are endemic or show close evolutionary relationships to relatives on the eastern side of the Andes (Gentry, 1982b), others come from Central American groups that have dispersed along the Pacific coast from the tropical forests of lowland Panama (Jackson et al., 1996). The extensive mangrove forests in Pacific intertidal waters are considerably more diverse in tree species than those along the Caribbean and open Atlantic coasts of South America (Tomlinson, 1986; Seeliger, 1992).

A similar hypothetical elevational transect farther to the south would begin with the highest peaks in Perú or Bolivia, traverse the bare high-elevation steppes or deserts and salt pans of the dry central Andes, and then descend southwest to the Peruvian Desert or west to the Atacama Desert (fig. 6.1). These coastal deserts are among the driest places in the world, with much of the region receiving, on average, only 2–20 mm of precipitation a year. Most precipitation arrives as tiny droplets of moisture derived from the persistent fogs and low stratus clouds that move onshore over cold Pacific waters (see chapter 10). Plants here either depend on slight atmospheric moisture, or else go dormant as seeds to await infrequent rains, notably those associated with major El Niño events that occur once or twice a decade (see chapter 19).

6.2 Subtropical South America

The subtropics include mixtures of floras, vegetation types, and climatic influences from both the tropical and higher to middle latitudes. In the Southern Hemisphere's late spring and summer months (December to March), the tropical influences predominate as the Intertropical Convergence Zone shifts southward and warm moist air masses from the central Atlantic Ocean and the Amazon basin are carried far to the south (see chapters 2 and 3). The lowlands include trees from the southern Amazon basin that reach southward into northern Argentina. However, the species must also resist the cold air masses that press their way to the north in the winter months (June to September), originating with cold fronts pushed along by the westerlies and then directed to the north by the Andes (see chapter 3). This seasonal shift in air circulation patterns has been termed the South American monsoonal system (Zhou and Lau, 1998).

Similarly, the vegetation types of the subtropical Andes come from a mixture of plant species, some having their principal distributions to the north in the tropics, with neighbors that have mainly temperate distributions in mountains farther south (Cleef, 1979a, 1979b). This effect creates a regional diversity pattern of relatively high numbers of plant species in the lower latitudes of the subtropics (Zuloaga et al., 1999).

Much of Paraguay and northern Argentina is covered by the dense shrublands and woodlands of the Chaco (fig. 6.1). Plant growth is seasonal; the shrubs leaf out and maximize photosynthesis when seasonal rains begin and danger of frost has passed. Farther to the north, in eastern Bolivia and adjacent parts of Brazil, lie the wetlands of the Pantanal. This region, which receives water and groundwater recharge from seasonal rains and runoff, is covered by marshes, palm swamps, and flood-tolerant tree species. The Pantanal lies between the Chaco and Cerrado regions, and grades northward into tropical savanna and lowland rain forests (fig. 6.1).

A 1,500 km-long hypothetical transect from the Pantanal to the east would cross the scrub and forested mountains of southeast Brazil, eventually ending in the subtropical portion of the Atlantic rain forest, where numerous plant groups are found that have affinities with plants occurring either in the Andes or to north in the Amazon basin.

6.3 Temperate South America

Along the southwest coast of South America, a north-south transect of approximately 3,000 km would extend from the southern limit of the hyper-arid Atacama Desert near 30°S, cross the Mediterranean-type shrubland and woodland ecosystems of central Chile, then transition through the mixed evergreen-deciduous forests south of 36°S into the Moist Pacific temperate ecological region, and terminate in the bleak landscapes of Tierra del Fuego (fig. 6.1). West of the Andes, between 30 and 36°S (38°S in the Central Depression), are sclerophyllous shrublands, consisting of plant species with small, rigid, xeromorphic leaves adapted to hot, dry summers and wet, cool winters. Moist, moderate springs and autumns provide two seasons in which plant growth is feasible. The mostly evergreen vegetation is an adaptation to this climatic seasonality shared with other Mediterranean-type climate regions around the world, such as California, the southwestern tip of Africa, southwest Australia, and the Mediterranean basin itself. The plants of the Chilean *matorral* (shrubland) are among the most unique in South America, with almost 60% endemic and some with their closest relatives in North America, in the shrublands or deserts of California (see chapter 11). Although there is clear evidence that long-distance dispersal has operated repeatedly between the Mediterranean regions of Chile and western North America (Berry, 1995; Simpson et al., 2005), the physiognomic similarity of the vegetation of central Chile with that of central and southern California appears to be primarily due to convergent evolution under similar climates.

A hypothetical transect from the lowlands of the Mediterranean zone of Chile into the Andes would reflect the high degree of vegetation heterogeneity associated with topography and precipitation gradients. In the lowlands, which have been heavily transformed by human activities, remnant vegetation includes sclerophyllous shrublands and dry thorn woodlands. At higher elevations in the Coastal Cordillera and the Andes foothills, sites are more mesic and are dominated by evergreen sclerophyllous trees and tall shrubs. The highest elevation forests consist of winter-deciduous *Nothofagus* trees, a strictly Southern Hemisphere genus that continues to dominate forests to the southernmost limit of forest on Tierra del Fuego.

South of the Mediterranean-type ecosystems is a relatively narrow zone of deciduous *Nothofagus* forests that grades into mixed deciduous-evergreen *Nothofagus* forests at about 36°S (fig. 6.1). The former is interspersed with sclerophyllous forests and shrublands, and has been highly altered by human activities. The evergreen *Nothofagus* species are typical of areas with higher precipitation and less extreme temperatures, whereas deciduous species occur in lowland sites with hot dry summers, in subalpine forests ecotonal to Andean tundra, and on sites transitional to the dry Patagonian steppe at high latitudes (see chapter 13).

The highest precipitation in middle and high latitudes (37 to 56°S) is found in coastal and foothill locations where persistent westerly winds bring in moisture-laden air masses from the Pacific Ocean. These air masses also bring high precipitation, as rain or snow, to elevations above 800 m at these latitudes. Temperate rain forests occur over most of the area along the western side of the continent south of approximately 37°S (fig. 6.1). In southern Chile, maximum forest biomass and woody species diversity occur in the north (around 40°S) and gradually decline southward along the environmental gradient toward cooler temperatures and shorter growing seasons (see chapter 13). The floristic composition of these rain forests, specifically the abundance of endemic species and genera, reflects a long period of isolation from other major floras (Arroyo et al., 1996). It is in this region that the Neotropical elements mix with Gondwana elements shared with other Southern Hemisphere land masses. The plant species here have evolved in isolation from other temperate forests over at least the past 50 million years.

The temperate rain forests of South America, like their counterparts in the Pacific Northwest of North America, are characterized mainly by evergreen trees of great age and size, and high total biomass. In the South America, however, the evergreen trees are primarily angiosperms, whereas in North America the conifer (gymnosperm) element is overwhelmingly dominant. Temperate rain forest zones in both hemispheres are characterized by cool summer temperatures, relatively mild winters, and abundant moisture throughout the year, but owing to the smaller size of the land mass at higher latitudes in South America, temperature seasonality there is relatively muted (Veblen and Alaback, 1996).

A hypothetical west-east transect across the Andes at about 40°S would reveal shifts in land cover from tall rain forests in Chile to the low vegetation of the high Andes, pass through mesic (but less species-rich) forests on the Argentinean side of the Andes, and then transition to xeric woodlands in the Andean foothills and Patagonian plains. Farther east, the aridity of the Andean rain shadow combines with the interior location to provide for vast areas of steppe with low shrubs and bunchgrasses (see chapter 14). To the northeast, these steppes grade into the grasslands of the Pampas of east-central Argentina, renowned for their good soils and grasses, and for high plant and crop productivity. Originally, the Pampas extended to the Atlantic coast but they have been strongly transformed by humans, for both cattle and sheep ranches as well as for crop agriculture. Here, and in the steppe and temperate forest regions of both Chile and Argentina, industrial forestry has planted thousands of square kilometers of tree plantations with species from North America and Australia.

South America ends southward in the wind-swept steppes of low shrubs and grasses of Patagonia (see chapter 14), and in the bogs and forests of Tierra de Fuego (fig. 6.1). Some 7,000 km of latitudinal change separate these natural environments from the Caribbean habitats of forests and scrub along the north coast of Venezuela. Plant adaptations to mutualisms with animals and fungi, to differing degrees of stress due to moisture or nutrient limitations, and to the magnitude and predictability of natural disturbances can be identified across the continent. It was in the southern cone of South America where Darwin found fossils of extinct ground sloths and other large mammals in 1832 and began thinking about possible explanations of Earth's biotic history. Further inspired by the remarkable biota of the Galapagos Islands, where every small island had unique species of ground finches and forms of the tortoise (Moorehead, 1969), Darwin eventually developed a theory to explain changes in evolutionary lineages that result in dramatic place-to-place changes in the composition of floras and faunas (Browne, 1995, 2002).

6.4 Historical Legacies

Detailed observations on the Galapagos finches by Grant and Grant (1995) expand upon Darwin's original hypotheses and show the harsh ways in which natural selection removes individuals from populations, eliminating over time their particular gene combinations from the breeding pool. Average beak size and shape in the finches are quite responsive to the stresses in a particular year, a population-level result of these sources of differential mortality. Over much longer time periods, new species and novel adaptations appear.

The legacies of past evolutionary regimes, of selective extinctions of some but not other lineages of plants and animals, and the direct or indirect connections to other continents have left their marks on the flora of South America. Given the multiple influences and the length of time involved, deciphering these factors is a difficult and incomplete task. Nonetheless, it is possible to detect the influence of the ancient breakup of the supercontinent Pangea into its Southern Hemisphere land masses belonging to Gondwana, and then the further separation of Gondwana into the continents of today (see chapters 1, 2, and 8; Parrish, 1993). *Podocarpus* trees, for example, are an ancient lineage of conifers, with leaves similar to those contained in fossils from 100 million years ago. They are found today in the lowland and upland forests of the Southern Hemisphere continents that were once united in Gondwana: Africa, South America, Australia, New Zealand, New Guinea and others. Thus, podocarps are found throughout South America, but typically are not the dominants, rather sharing ecological space with more recently evolved tree groups, including some that originated in tropical areas to the north of the equator.

Over geologic time, the tectonic movements that raise hills and mountains may be counteracted by weathering and erosion that wear down those land surfaces into areas of relatively low relief. Thus, the rocks in the Guiana and Brazilian Shields are amalgams of ancient lands that have existed in one form or another for over a billion years (see chapter 1). These and other such places may have served in the past as sources of plant lineages that currently occupy other areas, for example those formed as the Andes began to rise during later Cenozoic time. The products of weathering and erosion can be found deposited in thick sediments at the base of the Andes, in the Amazon and Orinoco basins, and beneath the plains supporting the Pantanal wetlands and Chaco shrublands. For example, some of the modern ecological diversity in the Amazon basin is due to the presence of white sands and other unique soil types embedded into a general matrix of Oxisols and Ultisols; these sands originate from the weathering, erosion, transportation, and deposition of materials from sandstones originally found in distant uplands (see chapter 7).

The Andean orogenesis that has dominated the western margin of South America during later Cenozoic time has been the key to the development of aridity in a large sector of the southern and western parts of South America and to the evolution of the flora adapted to it. The development of the Andes has involved a complex sequence of events extending back over more than 500 million years, but the most recent phase of uplift, over the past 25–30 million years, has shut off the westward flow of moist tropical air masses from the Atlantic Ocean from the Pacific coast (see chapters 1, 2, and 3). Similarly, prior to this blockage, the eastward flow of moist Pacific air masses would have maintained a wetter climate, at least seasonally, over a much larger area of subtropical and midlatitudes in the southern part of the continent. The legacies of this Andean uplift are thus reflected in peculiar biotic distributions across the continent.

Pleistocene glaciations have also left their legacy, both on the physical landscape (see chapter 4) and on the organisms inhabiting areas under their influence. The repeated glacier advances of the Pleistocene, followed by the retreat of ice and upslope shifts of many plant and animals species, have been a particular feature of the last two or three million years. Direct impacts can be seen in the topography and soils of the Andes, certainly in the Patagonian Andes but also in the tropical and subtropical Andes above 3,000 m, with the indirect legacies of related climate changes experienced over much of the rest of the continent. However, the persistent effect of these changes on the composition and distribution of lowland plant species is controversial. Some scholars detect current distributions that have resulted from past climate change acting over large areas of the Amazon lowlands (Prance, 1982), while others question their data, methods, and conclusions (Colinvaux, 1996, 1997). Given the complexity and size of the areas involved, multiple explanations are likely to be needed (Bush, 1994; Bates et al., 1998; Ron, 2000).

6.5 Recent Changes

Flora and vegetation are affected by current ecological processes acting upon the legacies of past Earth history, including speciation, adaptations, and extinctions. Over short time periods, rapid changes in land cover and plant usage may be outside the range of those kinds and magnitudes of change experienced by plant species in their evolutionary past. If so, increased extinction rates are possible and indeed likely. This concern is one of the factors driving research interests in mapping the locations of deforestation and other land cover conversions (Eva et al., 2004), and predicting the consequences for biological diversity. In some places, the conversion might predate European colonization, given the antiquity of dense human settlement in some parts of the tropical Andes and western coastal South America (Lynch, 1980; Sandweiss et al., 1998; Shady et al., 2001), and perhaps elsewhere. Elsewhere, the growth of cities and globalization is currently producing novel land covers and usages (see chapter 20; Gwynne and Kay, 2004).

The regional differences in biota described in this chapter and exemplified elsewhere in this book speak to the importance of understanding the underlying biogeographical patterns and processes of South America's biological diversity. Knowledge of specific places is crucial, as is scientific understanding of the reasons for spatial changes, including those affecting plant cover. Land-use systems imposed by people upon those living systems create both opportunities and responsibilities (e.g., Kellman and Tackaberry, 1997). The resiliency of vegetation in relation to natural disturbances, and the environmental gradients that define the distributions and range limits of species, can be used to help predict future changes in flora and vegetation cover, given trends of change in current land use and population density. It is likely that von Humboldt and Darwin would be pleased by the further insights into physical geography that have accumulated since their respective expeditions to South America two centuries ago. They would probably be distraught, however, by the profound shifts that have occurred in the environment since their visits, shifts imposed by human activities, both deliberately and inadvertently.

References

Achard, F., H.D. Eva, H.-J. Stibig, P. Mayaux, J. Gallego, T. Richards, and J.-P. Malingreau, 2002. Determination of deforestation rates of the world's humid tropical forests. *Science*, 297, 999–1002.

Arroyo, M.T.K., L. Cavieres, A. Peñalosa, M. Riveros, and A.M. Faggi, 1996. Relaciones fitogeográficas y patrones regionales de riqueza de especies en la flora del bosque lluvioso templado de Sudamérica. In: J.J. Armesto, C. Villagran, and M.T.K. Arroyo (Editors), *Ecología de los Bosques Nativos de Chile*. Editorial Universitaria, Santiago, 71–99.

Bates, J.M., S.J. Hackett, and J. Cracraft, 1998. Area relationships in the Neotropical lowlands: An hypothesis based on raw distributions of Passerine birds. *Journal of Biogeography*, 25, 783–793.

Berry, P.E., 1995. Diversification of Onagraceae in montane areas of South America. In: S.P. Churchill, H. Balslev, E. Forero, and J. Luteyn (Editors), *Biodiversity and Conservation of Neotropical Montane Forests*. New York Botanical Garden, New York, 415–420.

Berry, P.E., O. Huber, and B.K. Holst, 1995. Floristic analysis and phytogeography. In: P.E. Berry, B.K. Holst, and K. Yatskievych (Editors), *Flora of the Venezuelan Guyana, I: Introduction*. Timber Press, Portland, Oregon, 161–192.

Berry, P.E., and R. Riina, 2005. Insights into the diversity of the Pantepui flora and the biogeographic complexity of the Guayana Shield. Proceedings of the Symposium "Plant Diversity and Complexity Patterns—Local, Regional, and Global Dimensions." *Biologiske Skrifter* 55, 145–167.

Botting, D., 1973. *Humboldt and the Cosmos*. Harper and Row, New York.

Browne, J., 1995. *Charles Darwin: Voyaging*. Knopf, New York.

Browne, J., 2002. *Charles Darwin: The Power of Place*. Princeton University Press, Princeton.

Bush, M.B., 1994. Amazonian speciation: A necessarily complex model. *Journal of Biogeography*, 21, 5–17.

Cabrera, A.L., and A. Willink, 1973. *Biogeografía de America Latina*. Organization of American States, Washington, D.C.

Cleef, A.M., 1979a. Characteristics of Neotropical paramo vegetation and its subantarctic relations. In: C. Troll and W. Lauer (Editors), Geoecological Relations Between the Southern Temperate Zone and the Tropical Mountains. *Erdwissenschaftliche Forschung*, XI. Steiner, Wiesbaden, Germany, 365–390.

Cleef, A.M., 1979b. The phytogeographical position of the neotropical vascular páramo flora with special reference to the Colombian Cordillera Oriental. In: K. Larsen and L.B. Holm-Nielsen (Editors), *Tropical Botany*. Academic Press, London, 175–184.

Colinvaux, P.A., 1996. Quaternary environmental history and forest diversity in the Neotropics. In: J.B.C. Jackson, A.F. Budd, and A.G. Coates (Editors), *Evolution and Environment in Tropical America*. University of Chicago Press, Chicago, 359–405.

Colinvaux, P.A., 1997. An arid Amazon? *Trends in Ecology and Evolution*, 12, 318–319.

Daly, D.C., and J.D. Mitchell, 2000. Lowland vegetation of tropical South America: An overview. In: D.L. Lentz (Editor), *Imperfect Balance: Landscape Transformations in the Precolumbian Americas*. Columbia University Press, New York, 391–453.

Darwin, C., 1859. *On the Origin of the Species by Means of Natural Selection, or, the Preservation of Favoured Races in the Struggle for Life*. J. Murray, London.

Davis, S.D., V.H. Heywood, O. Herrera MacBryde, and A.C. Hamilton (Editors), 1997. *Centres of Plant Diversity: A Guide and Strategy for their Conservation. Volume 3: The Americas*. WWF and IUCN, London.

Denevan, W.M., 2001. *Cultivated Landscapes of Native Amazonia and the Andes*. Oxford University Press, Oxford.

Dinerstein, E., D.M. Olson, et al., 1995. *A Conservation Assessment of the Terrestrial Ecoregions of Latin America and the Caribbean*. The World Bank in association with WWF, Washington, D.C.

Duivenvoorden, J.F., 1995. Tree species composition and rain forest-environment relationships in the middle Caquetá, Colombia, NW Amazonia. *Vegetatio*, 120, 9–113.

Eva, H.D., A. S. Belward, E.E. De Miranda, C.M. Di Bella, V. Gond, O. Huber, S. Jones, M. Sgrenzaroli, and S. Fritz, 2004. A land cover map of South America. *Global Change Biology*, 10, 731–744.

Faak, M., and K.R. Biermann, 1986. *Alexander von Humboldt: Reise auf dem Rio Magdalena, durch die Anden und Mexico*. Akademie-Verlag, Berlin.

Fine, P.V.A., I. Mesones, and P.D. Coley, 2004. Herbivores promote habitat specialization by trees in Amazonian forests. *Science*, 305, 663–665.

Furley, P.A. 1999. The nature and diversity of Neotropical savanna vegetation with particular reference to the Brazilian cerrados. *Global Ecology and Biogeography* 8, 223–241.

Galindo-Leal, C., and I. de Gusmão Câmara (Editors), 2003. *The Atlantic Forest of South America: Biodiversity Status, Threats, and Outlook*. Island Press, Washington, D.C.

Gentry, A.H., 1982a. Patterns of Neotropical plant diversity. *Evolutionary Biology*, 15, 1–84.

Gentry, A.H., 1982b. Neotropical floristic diversity: Phytogeographical connections between Central and South America, Pleistocene climatic fluctuations, or an accident of the Andean orogeny? *Annals of the Missouri Botanical Garden*, 69, 557–593.

Gentry, A.H., 1986. Species richness and floristic composition of Chocó region plant communities. *Caldasia*, 15(71–75), 1–91.

Givnish, T.J., T.M. Evans, M.L. Zjhra, T.B. Patterson, P.E. Berry, and K.J. Sytsma, 2000. Molecular evolution, adaptive radiation, and geographic diversification in the amphiatlantic family Rapateaceae: Evidence from *ndh*F sequences and morphology. *Evolution*, 54, 1915–1937.

Givnish, T.J., and S.S. Renner, 2004. Tropical intercontinental disjunctions: Gondwana breakup, immigration from the Boreotropics, and transoceanic dispersal. *International Journal of Plant Science*, 165 (Supplement 4), S1–S6.

Goulding, M., N.J. Smith, and D.J. Mahar, 1996. *Floods of Fortune: Ecology and Economy along the Amazon*. Columbia University Press, New York.

Grant, P.R., and B.R. Grant, 1995. Predicting microevolutionary responses to directional selection on heritable variation. *Evolution*, 49, 241–251.

Gwynne, R.N., and C. Kay (Editors), 2004. *Latin America Transformed: Globalization and Modernity*. 2nd edition. Arnold, London.

Hueck, K., and P. Seibert, 1972. *Vegetationskarte von Südamerika*. G. Fischer-Verlag, Stuttgart.

Jackson, J.B.C., A.F. Budd, and A.G. Coates (Editors), 1996. *Evolution and Environment in Tropical America*. University of Chicago Press, Chicago.

Josse, C., G. Navarro, P. Comer, R. Evans, D. Faber-Langendoen, M. Fellows, G. Kittel, S. Menard, M. Pyne, M. Reid, K. Schulz, K. Snow, and J. Teague, 2003. *Ecological Systems of Latin America and the Caribbean: A Working Classification of Terrestrial Systems.* NatureServe, Arlington, Virginia.

Junk, W.J., 1989. Flood tolerance and tree distribution in central Amazonian floodplains. In: L.B. Holm-Nielsen, I.C. Nielsen, and H. Balslev (Editors), *Tropical Forests, Botanical Dynamics, Speciation, and Diversity.* Academic Press, San Diego, 47–64.

Kellman, M., and R. Tackaberry, 1997. *Tropical Environments: The Functioning and Management of Tropical Ecosystems.* Routledge, London.

Lentz, D.L. (Editor), 2000. *Imperfect Balance: Landscape Transformations in the Precolumbian Americas.* Columbia University Press, New York.

Luteyn, J.L., and S.P. Churchill, 2000. Vegetation of the tropical Andes. In: D.L Lentz (Editor), *Imperfect Balance: Landscape Transformations in the Precolumbian Americas.* Columbia University Press, New York, 281–310.

Lynch, T. 1980. *Guitarrero Cave: Early Man in the Andes.* Academic Press, New York.

Moorehead, A., 1969. *Darwin and the Beagle.* Penguin Books, London.

Myers, N., R.A. Mittermeier, C.G. Mittermeier, G.A. B. da Fonseca, and J. Kent, 2000. Biodiversity hotspots for conservation priorities. *Nature* 403, 853–858.

Oliveira, P.S., and R.J. Marquis (Editors), 2002. *The Cerrados of Brazil: Ecology and Natural History of a Neotropical Savanna.* Columbia University Press, New York.

Parrish, J.T., 1993. The palaeogeography of the opening South Atlantic. In: W. George and R. Lavocat (Editors), *The African-South American Connection.* Oxford Science Publications, Clarendon Press, Oxford, 8–27.

Pennington, R.T., D.E. Prado, and C.A. Pendry, 2000. Neotropical seasonally dry forests and Quaternary vegetation changes. *Journal of Biogeography* 27, 261–273.

Pitman, N.C.A., J.W. Terborgh, M.R. Silman, P. Núñez V., D.A. Neill, C.E. Cerón, W.A. Palacios, and M. Aulestia, 2001. Dominance and distribution of tree species in upper Amazonian terra firme forests. *Ecology,* 82, 2101–2117.

Prance, G.T. (Editor), 1982. *Biological Diversity in the Tropics.* Columbia University Press, New York.

Prance, G.T., 1989. American tropical forests. In: H. Lieth and J.J.A. Werger (Editors), *Tropical Rain Forest Ecosystems, Biogeographical and Ecological Studies.* Elsevier, Amsterdam, 99–132.

Ranta, P., T. Blom, J. Niemala, E. Joensu, and M. Siitonen, 1998. The fragmented Atlantic rain forest of Brazil: Size, shape and distribution of forest fragments. *Biodiversity and Conservation,* 7, 385–403.

Ron, S.R., 2000. Biogeographic area relationships of lowland Neotropical rainforest based on raw distributions of vertebrate groups. *Biological Journal of the Linnaean Society,* 71, 379–402.

Sandweiss, D.H., H. McInnis, R.L. Burger, A. Cano, B. Ojeda, R. Paredes, M. del Carmen Sandweiss, and M.D. Glascock. 1998. Quebrada Jaguay: Early South American maritime adaptations. *Science* 281, 1830–1832.

Seeliger, U. (Editor). 1992. *Coastal Plant Communities of Latin America.* Academic Press, San Diego.

Shady, R., J. Haas, and W. Creamer, 2001. Dating Caral, a preceramic site in the Supe Valley on the central coast of Peru. *Science,* 292, 723–726.

Simpson, B.B., J.A. Tate, and A. Weeks, 2005. The biogeography of *Hoffmannseggia* (Leguminosae, Caesalpinioideae, Caesalpinieae): A tale of many travels. *Journal of Biogeography,* 32, 15–27.

Ter Steege H., N. Pitman , D. Sabatier , H. Castellanos, P. Van der Hout, D.C. Daly, M. Silveira, O. Phillips, R. Vasquez, T. Van Andel, J. Duivenvoorden, A.A. de Oliveira, R. Ek, R. Lilwah, R. Thomas, J. Van Essen, C. Baider, P. Maas, S. Mori, J. Terborgh, P.N. Vargas, H. Mogollon, and W. Morawetz, 2003. A spatial model of tree alpha-diversity and tree density for the Amazon. *Biodiversity and Conservation,* 12, 2255–2277.

Tomlinson, P.B., 1986. *The Botany of Mangroves.* Cambridge University Press, New York.

Tuomisto, H., and K. Ruokolainen, 1994. Distribution of Pteridophyta and Melastomataceae across an edaphic gradient in an Amazonian rain forest. *Journal of Vegetation Science,* 5, 25–34.

Veblen, T.T., and P.B. Alaback, 1996. A comparative review of forest dynamics and disturbance in the temperate rain forests of North and South America. In: R.G. Lawford, P.B. Alaback, and E. Fuentes (Editors), *High-Latitude Rainforests and Associated Ecosystems of the West Coasts of the Americas: Climate, Hydrology, Ecology, and Conservation.* Springer, New York, 173–213.

von Humboldt, A., 1815–1832. *Reise in die aequinoctial-gegenden des neuen continents in den jahren 1799, 1800, 1801, 1803 und 1804.* Verfasst von Alexander von Humboldt und A. Bonpland. 6 volumes, J.G. Cotta, Tubingen, Stuttgart.

Young, K.R., 2003. Genes and biogeographers: Incorporating a genetic perspective into biogeographical research. *Physical Geography,* 24, 447–466.

Young, K.R., C. Ulloa Ulloa, J. L. Luteyn, and S. Knapp, 2002. Plant evolution and endemism in Andean South America: An introduction. *Botanical Review,* 68, 4–21.

Zhou, J. and K.-M. Lau, 1998. Does a monsoon climate exist over South America? *Journal of Climate,* 11, 1020–1040.

Zuloaga, F.O., O. Morrone, and D. Rodríguez. 1999. Análisis de la biodiversidad en plantas vasculares de la Argentina. *Kurtziana,* 27, 17–167.

7

Soils

Stanley W. Buol

Soil is a physical, chemical, and biological medium at the upper surface of Earth's land areas capable of accepting plant roots and thereby enabling plants to extend their photosynthetic tissues upward and intercept radiant energy from the sun. Each day, chemical and biological activities in soil change in response to temperature and moisture dynamics. Each soil has a range of physical, chemical, and biological properties determined by inherited mineral composition and biogeochemical processes existing in a quasi-steady state of flux.

Most primary minerals in soil formed in geologic environments of high temperature and pressure. When exposed to lower temperatures and pressures, meteoric water, and organic compounds near Earth's surface, primary minerals slowly decompose in response to weathering processes. As primary minerals weather, some elements necessary for plant growth are released as inorganic ions, some reassemble to form secondary silicate and oxide clay minerals, and some elements are lost via dissolution and leaching. After prolonged or intense weathering, few minerals containing elements necessary for plant growth remain.

Weathering most often occurs in or slightly below the soil but may not be entirely related to the present soil. Material from which a present soil is formed may have been weathered in a soil environment, and eroded and deposited many times before coming to rest in its present location. Such materials are often almost devoid of nutrient-bearing minerals, and the soils formed provide scant amounts of the elements essential for plant growth. In contrast, minerals exposed to a soil environment for the first time on rapidly eroding slopes, fluvial deposits, or as volcanic ejecta succumb more rapidly to weathering and release essential elements in forms needed by plants.

Plants ingest inorganic ions and water from the soil through their roots and combine them with carbon secured as carbon dioxide from the air, and with hydrogen and oxygen from water to form organic tissues. Organic residues are added to the soil as various plant parts, insects, and animals die. Microorganisms in and above the soil then decompose the organic residues, carbon returns to the air as carbon dioxide, and essential elements contained in the organic compounds are released as inorganic ions within the soil. With the exception of carbon, nitrogen, hydrogen, oxygen, and some sulfur, all of the elements necessary for plant growth are derived from decomposition of minerals in the soil.

Some accumulation of organic compounds is a universal process in the uppermost layer of all soils. Organic carbon content of each soil exists at a level determined by the rate of accumulation via photosynthetic production of plant material and rate of organic decomposition by microbes in the soil. Organic compounds in plant and animal residues contain all the elements necessary for plant growth, but as organic compounds these elements are unavailable to plants. Also, the content of essential elements in organic residue is highly variable. Plants growing on soils formed in mineral material deficient in specific essential elements return organic residues deficient in those elements. The rate at which organic compounds decompose and release the essential elements as inorganic ions in the soil is critical to the growth of new plants. In frequently saturated soils, organic residue decomposition is slowed by lack of oxygen needed by microbes, and these soils have higher organic carbon contents than unsaturated soils. Colder soils have higher organic carbon contents than warmer soils because microbial decomposition is slowed at lower temperatures. Soils maintain higher organic carbon contents under forest shade than when cultivated and exposed to direct radiation and higher maximum diurnal temperatures for much of the year. Clearing, artificially draining, and burning are all techniques used by humans to accelerate organic decomposition and temporally enhance the availability of essential nutrients in the soil.

Processes of soil formation are controlled by the materials available and the energy available to alter those materials (Buol et al., 1997). Many properties of the mineral material from which present soils are formed were determined by ancient geologic events. Within each type of mineral material temperature and moisture conditions unique to location on the landscape and ambient vegetation influence soil properties and processes. However, the chemical fertility of soil is ultimately limited by the composition of the mineral material in which it forms.

Human societies uniquely alter natural soil-biogeochemical processes by transporting essential elements, in the form of food and fiber, from the site where food plants are grown to distant sites for consumption and waste disposal. This depletes essential nutrients that originate from soil minerals at the site where human food is grown. Although the function of soil is similar in both natural ecosystems and human food production, distinct quantitative differences are present. Most human food crops grow and mature more rapidly than native vegetation. Therefore they require rates of nutrient uptake from the soil that are several times greater than natural vegetation (Buol, 1995). Human societies are capable of enhancing rates of nutrient uptake via the application of externally produced inorganic fertilizer, thereby enabling food crop production on soils long considered too infertile for crop production. Therefore, in this chapter the role of soil is related to both undisturbed ecosystems and systems of human food production.

Many classification systems have been devised to aid communication of the evolving human understanding of soils (Buol and Sanchez, 1986). Criteria of soil texture, color, and to a lesser degree chemistry and mineralogy are used in most soil classifications. Most soil publications in South America reflect soil-classification nomenclature in the United States at the time of publication or that of the World Soil Map (FAO-UNESC, 1971), and do not identify soil temperature and moisture regimes. Soil temperature and soil-moisture regimes are closely related to climate and although they profoundly dictate many soil uses were not directly considered criteria for soil classification until the advent of Soil Taxonomy in 1960 (Soil Survey Staff, 1999). Nomenclature from both Soil Taxonomy and the World Soil Map is utilized in this chapter.

7.1 Soil-Temperature Regimes

Mean annual soil temperatures are from 2 to 4°C warmer than mean annual air temperatures. Soil-temperature regimes (STRs) are defined by two criteria, mean annual soil temperature and seasonal temperature difference determined as the mean soil temperature of June, July, and August compared to the mean soil temperature of December, January, and February (Soil Survey Staff, 1999). If the mean soil temperature differs seasonally by less than 6°C the prefix "iso" is placed before the identification of the mean annual soil temperature. In South America, almost all soils north of the Tropic of Capricorn, and along the entire extent of the Pacific coast where the ocean mollifies seasonal temperature differences, have "iso" STRs. Seasonal soil temperatures seldom have to be considered when planting food crops in soils with "iso" STRs. Cold soil temperatures seasonally limit crop growing only in the "non-iso" STRs of southern temperate latitudes (Van Wambeke, 1991).

Figure 7.1 outlines the location of STRs in South America. Complexes of isomesic, isothermic, and cryic regimes are delineated at the scale of the figure, indicating warmer temperatures at lower elevations and colder temperatures at higher elevations in the Andes.

Soils with mean annual soil temperatures of 22°C or higher are classified as hyperthermic and isohyperthermic STRs. Freezing conditions that may kill citrus trees are unlikely in these STRs. Thermic and isothermic STRs have mean annual soil temperatures of 15°C or higher but less than 22°C. Freezing conditions are seldom a problem in the isothermic STR. In the thermic STR, summers are warm enough for cotton and most grain crops but freezing conditions often occur during the winter months. Mesic and isomesic STRs have mean annual soil temperatures of 8°C or higher but less than 15°C. Freezing conditions occur seasonally in the mesic STR but summers are frost-free and warm enough for most grain crops. In the isomesic STR, soil temperatures are cool throughout the year and crop growth is slow. At high elevations, nighttime freezing is

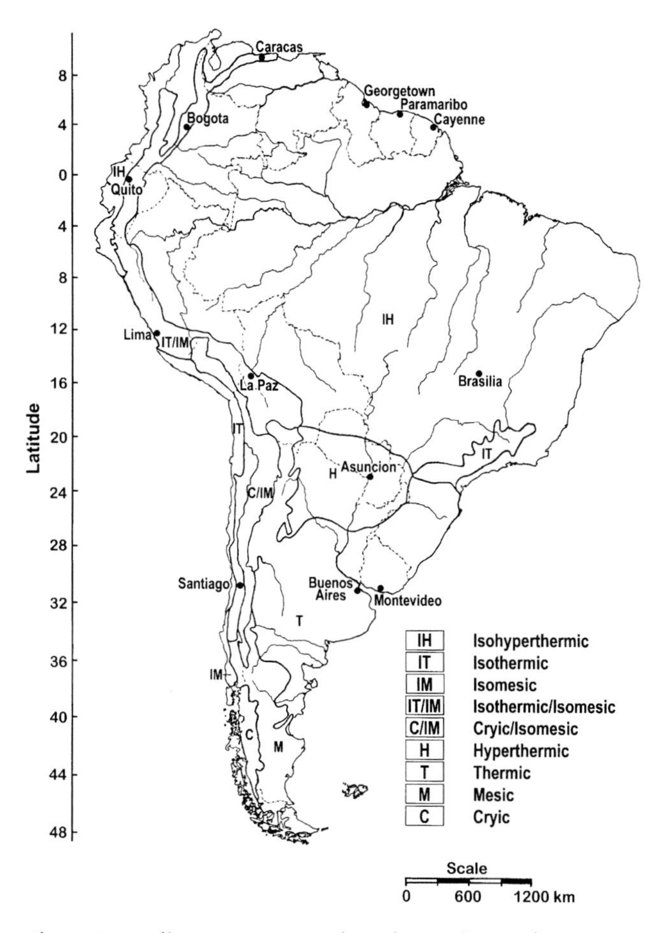

Figure 7.1 Soil-temperature regimes in South America (generalized from Van Wambeke, 1981)

common and even cold-tolerant crops like potatoes seldom survive in that part of the isomesic STR with a mean annual soil temperature below 10°C. The cryic STR has mean annual soil temperatures less than 8°C, and only limited forage is available for grazing.

7.2 Soil-Moisture Regimes

Soil-moisture regimes (SMRs) are defined in Soil Taxonomy to classify a soil's ability to supply water to plants without irrigation (Soil Survey Staff, 1999). In soils where the ground-water table is not reached by the roots of most crop plants, the SMR is determined by the average reliability and seasonal distribution of rainfall in "normal" years. "Normal" years are plus or minus one standard deviation of long-term averages. Most food crops require a reliable supply of water for at least 90 consecutive days. To calculate SMRs, the average monthly precipitation is compared to the calculated potential evapotranspiration for that month. A soil-water balance is then constructed for an average rainfall amount in the area, assuming the available water-holding capacity of the rooting zone in the soil to be five centimeters of water. The SMR is determined by the duration of time during which water is reliably available, either from average rainfall or as stored available water in the soil during a period when soil temperatures are warm enough to grow the crop.

Figure 7.2 outlines the upland soils in each of the SMRs in South America. Soils with aquic soil moisture conditions are related to local landscape relief and cannot be delineated at the scale of the figure. Each of the SMRs can be briefly summarized as follows.

The aridic SMR has less than 90 consecutive days in which plant-available water is present in the rooting zone in normal years. It is not possible to grow most food crops without irrigation. In normal years, the ustic SMR has at least 90 consecutive days when moisture is available in the root zone but more than 90 cumulative days when water is not available in the rooting zone. On soils with an ustic SMR, at least one crop can be reliably grown each year, and on some soils it is possible to grow two crops per year, but there is a seasonal dry period of 90 days or more when crop production is not possible without irrigation. The reliable dry season of the ustic SMR is a distinct advantage for weed and disease control and grain-crop harvest in isohyperthermic

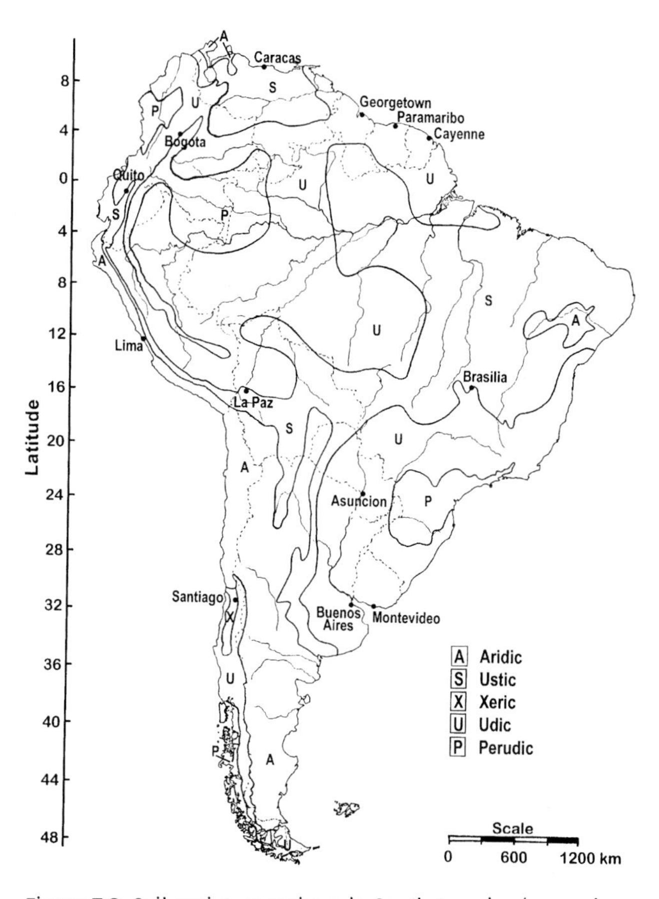

Figure 7.2 Soil-moisture regimes in South America (generalized from Van Wambeke, 1981)

and isothermic STRs (Buol and Sanchez, 1988). The udic SMR has less than 90 cumulative days when water is not available in the rooting zone in normal years. It is possible to grow food crops any time of the year without irrigation when the temperature is warm enough for that crop. In most udic SMR areas, available water is less reliable during some part of the year, and farmers often select more drought-tolerant crops or may choose not to plant during that period, but perennial plants are adequately supplied with water throughout most years. The perudic SMR has precipitation that exceeds potential evapotranspiration during every month of normal years. Although this may seem desirable, these areas present weed, insect, and disease problems, and the constantly humid conditions make it difficult to harvest mature grain crops. The xeric SMR—often referred to as a Mediterranean-type climate—has hot dry summers with cool moist winters. A small area of the xeric SMR utilized for grape and olive growing is present on the western flanks of the Andes in central Chile (see chapter 11).

The above SMRs are used to classify only the well-drained upland soils. Within each SMR area there are soils where the ground water often saturates the rooting zone. These soils are identified as having an aquic soil-moisture condition and are commonly referred to as poorly drained soils. Soils with such conditions most often occur in areas adjacent to rivers and lakes, or in broad level areas within perudic, udic, and ustic SMRs. Soils with aquic soil-moisture conditions are normally saturated with water during only a part of the year. As such, they are often seasonally utilized for subsistence food crops, but most require engineered drainage systems before commercial agricultural production is possible.

7.3 Chemical and Mineralogical Properties of Soils

Most chemical and mineralogical properties of soils are derived from the geologic material within which the soil is formed. Although some chemical and mineralogical alterations take place within soil, the initial composition of the mineral material often determines a soil's chemical composition. Among the chemical elements needed for plant and animal physiology, only carbon, oxygen, hydrogen, nitrogen, and to some extent sulfur are derived from air and water. The other life-essential elements are obtained from the minerals in the soil. Although an inadequate supply of any life-essential element limits plant growth, the most frequent limitations result from insufficient plant-available nitrogen, phosphorus, potassium, calcium or magnesium.

Practically no nitrogen is present in soil minerals. Almost all of the nitrogen in the soil is in the organic remains of plants, animals, and microbes. Small amounts of nitrogen enter the soil as ammonia and nitrate dissolved in rainwater, but most enters via fixation from the air by nitrogen-fixing microbes. Some nitrogen-fixing microbes in the soil are symbiotic and the nitrogen they extract from the air is incorporated into their legume-plant host. Other nitrogen-fixing microbes are not symbiotic and the nitrogen they extract from the air is incorporated into their cells. Nitrogen is concentrated in organic residues in the surface layers of soil. As organic residues decompose, inorganic forms of nitrogen are released into the soil solution and become available to growing plants, leach into the groundwater during periods of excessive rainfall, or return to the air as nitrogen gas during periods when the soil is saturated with water.

Phosphorus is present in only a few minerals. The iron and aluminum phosphates are extremely insoluble and do not release phosphorus rapidly enough for rapid plant growth. The release rate is so slow that soils with high iron and aluminum contents tend to absorb phosphate applied as fertilizer and decrease its availability to plants. Apatite, a rather soluble calcium phosphate mineral capable of supplying plant-available phosphorus, is a common source of phosphorus and often present in limestone.

Potassium is most often present in mica and feldspar minerals. These minerals are rather easily weathered and consequently are seldom present in materials that have been repeatedly transported and deposited on the land surface.

Calcium and magnesium are most abundant in carbonate minerals associated with limestone and some sandstone. Carbonate minerals are also relatively unstable when subjected to weathering and therefore are present only in recent sediment, limestone, and some sandstone.

Soil pH is a measure of acidity or alkalinity of water in the soil and has a direct effect on how rapidly many of the essential elements are available to growing plants. In the absence of carbonate minerals, soils are often acid in reaction, and only limited quantities of essential elements present in the soil are available for plant growth. Acid soils with pH values less than about 5.2 also have a concentration of aluminum ions that is toxic to some, but not all crop plants. To reduce or eliminate aluminum toxicity and increase the availability of the essential elements to plants growing in acid soils, additions of lime (finely ground calcium and calcium:magnesium carbonates) are desirable and often necessary for most crop plants (Sanchez, 1976).

The rate at which essential elements in the soil become available to plants is critical to understanding soil fertility. Plants extract the elements they need from the soil as inorganic ions in the soil solution. Of the total amount of each essential element in the soil, less than approximately 1% is present in an available form. The available amount of each essential element changes rapidly as the moisture content of the soil changes and also depends on the rate at which organic compounds are decomposed to release organically bound elements as available inorganic ions. Plant species differ greatly in the rate at which they need to acquire essential elements for adequate growth. The rate at which nutrients become available influences natural plant communities, as can be observed in the succession of fast-growing species immediately after a fire oxidizes a large quantity of organic material and releases the essential elements as inorganic ions available for plant uptake. As quantities of available essential elements in the soil are transferred into the plant biomass, slower-growing species invade the area, obtaining the nutrients needed from the slow decomposition of organic residue deposited by the death of the fast-growing species that first invaded the site of the fire.

Human activities in food production are directly related to the rate at which nutrients are available in the soil. Most human food crops require from 90 to 120 days to mature. During their growing period, food crops must have a rate of nutrient availability in the soil that is many times faster than required by native ecosystems. For example, a high-yielding grain crop of rice, wheat, or corn must acquire approximately as much phosphorus in 90 days as trees acquire from the same area of land in more than 20 years. Also, tree roots usually penetrate more deeply and exploit a larger volume of soil than do the roots of food crops. Therefore the concentration of available nutrient elements near the soil surface must be considerably greater to supply adequately the needs of a food crop than to support tree growth.

Humans harvest their food crops, transport them to a domicile that is some distance from the site where the crop was grown, and return only a portion of the total elements in the crop plant as residue or manure. Often the seed portion of the plant is consumed and the less nutrient-rich stems and leaves are returned to the site of growth. Often only the root system of the crop plant remains as organic residue.

Humans have developed many strategies to obtain the high rate of nutrient availability required for the production of food crops. Historically people have populated areas of soil with high levels of mineral fertility. These are commonly igneous or volcanic materials of basic mineral composition, sedimentary rocks such as limestone rich in calcium, magnesium, and phosphorus, and recent flood plains frequently renewed with deposits of material derived from fertile geologic materials and eroded surface soil. Where domestic animals are available and allowed to forage large areas of native vegetation, essential elements from that vegetation are concentrated in their excrement and collected near human settlements, enabling food crop growth on those sites.

Where the mineral composition of the soils contains only small amounts of essential elements and a large amount of slow-growing natural biomass is present, a system of food production known as "slash-and-burn" is practiced. Although some amounts of essential elements are volatilized and lost, fire is the primary method of rapidly decomposing organic material and creating a short period when a high rate of nutrient availability is present in the soil. As the biomass is burned the essential elements contained are released as inorganic ions. If there is enough biomass at least one 90-day food crop can be successfully grown from the essential elements released from organic compounds by the fire. If correctly done, burning also assures that the surface temperature of the soil becomes high enough to reduce weed competition by killing most weed seeds near the soil surface. A second and third crop is often possible before the available supply of essential elements is exported from the field as human food and the rate of nutrient availability is reduced to a point that crop yields are so low and/or weeds become such a major problem that the farmer plants no more crops. After the farmer abandons the land, it is invaded by a succession of native communities that are able to grow with lower rates of nutrient flow from the soil. After some years the vegetation can again be cut, dried, and burned to obtain a site for another brief sequence of crop plants. This method of nutrient availability management has numerous variations among different indigenous cultures (see chapter 16). Because of the long periods of time, usually between 10 and 30 years, that must be allowed for natural vegetation to accumulate sufficient quantities of essential elements needed to fertilize a food crop after burning, only low human population densities are sustained by slash-and-burn agriculture. In areas where infrastructure enables crops to be exported and essential nutrients imported as concentrated fertilizer, continuous food-crop production is practiced on even the most chemically infertile soils. Many combinations and variations of these strategies presently are in use in parts of South America.

The mineral composition of soils results not only from the initial underlying rocks but also from geologic materials that have been altered by exposure to weathering during many cycles of surface erosion and sedimentation prior to deposition in the location of the present soil. A large portion of South America encompasses land surfaces exposed to weathering, soil formation, and erosion since Precambrian time (see chapter 2). In these areas, the geologic materials from which the present soils have formed are almost devoid of minerals that contain essential elements, and soil acidity renders the low quantities present only slightly available to plants. Soils formed in such materials are not only deficient in available forms of the major plant-essential elements but also often deficient in plant-available forms of copper, zinc, boron, and sulfur. The limited amount of nutrient-poor natural biomass growing on such areas has an inadequate content of essential nutrients for slash-and-burn agriculture. Without importing essential nutrients, human habitation is often nomadic and limited to small isolated sites of somewhat more fertile soil.

Where tectonic uplift has exposed areas rich in carbonate and other nutrient-bearing rocks, the soils are relatively rich in essential elements. Other areas with relatively nutrient-rich material are those derived from volcanic activity. Sediments eroded from limestone and volcanic uplands, and rather recently deposited in lowlands, often have high to intermediate mineral fertility. If the mineral composition of the soil is relatively rich in essential elements, people are able to grow and harvest food with little or no attention to the chemical fertilization of the soil. Crop yields are usually low but reliable.

Regardless of soil's natural fertility, importing essential elements to compensate for those harvested in human food can increase crop yields. Applying animal and human wastes, wherein the essential elements have been gathered from sites distant from the food-growing area, is a partial solution in many cultures (see chapter 16). However, the concentration of essential elements in organic residue and waste materials is low, transportation is difficult and expensive, and the rate at which the residues decompose to release inorganic ions for crop uptake is slow. Where commercial infrastructure is available, concentrated inorganic fertilizers have proven to be the most economical method for supplying nutrients harvested in human food. In areas where all the soils are acid and extremely deficient in essential elements, importation of lime and fertilizer is the only method of food production possible.

The broad soil regions of South America are outlined in figure 7.3. Within each region, significant soil differences related to topographic position, geologic material, and moisture conditions are present. Such localized conditions may depart significantly from the general character of each region and form the basis for further distinctions in plant cover and land use.

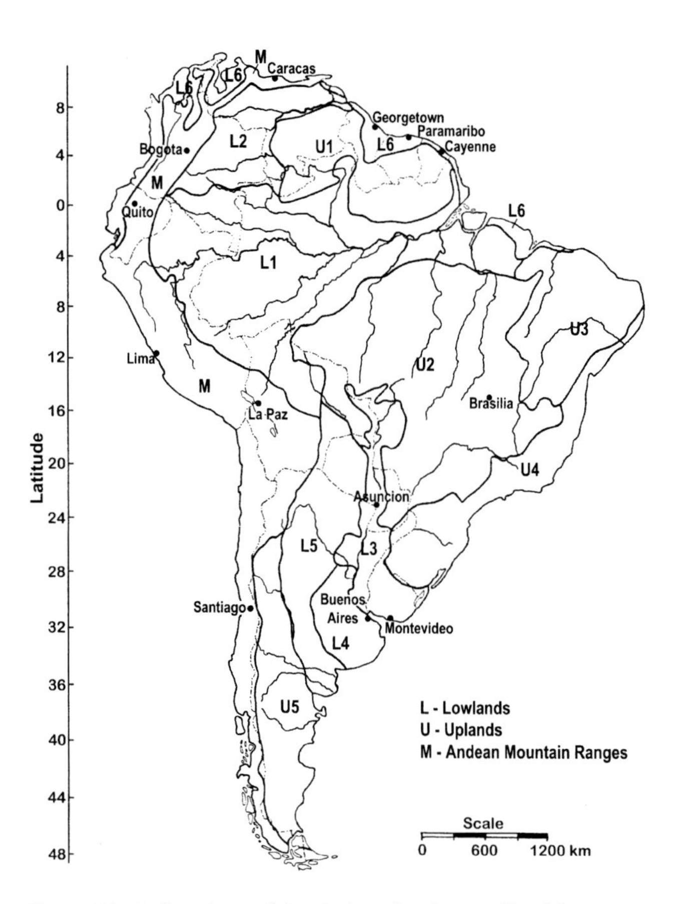

Figure 7.3 Soil regions of South America (generalized from FAO-UNESCO, 1971)

7.4 Soils of Mountainous Regions (M)

The major mountainous region in South America is the Andean chain along the west coast of the continent. Formed by repeated tectonic uplift and associated volcanic activity, the mineral composition of the soils is diverse. Some of these materials are limestones rich in carbonate, while other parts of the Andes expose granites and schists relatively poor in essential elements. Volcanic ash and basalts of various ages, including recent volcanic activity, form Andisols (Andosols).[1] Tectonic instability, steep slopes, and glaciers at the highest elevations have resulted in rapid erosion of surface materials in the Andes. Inceptisols (Cambisols) are prevalent on mountain slopes not covered by volcanic ash. When formed in basic rock, the soils are fertile Eutrudepts and Haplustepts (Eutric Cambisols), but on acid igneous rock the soils are acid, infertile Dystrudepts and Dystrustepts (Dystric Cambisols). Orographic control of moisture-laden air masses results in locally contrasting SMRs. Elevation is inversely related to STRs.

The broad intermountain valleys of the Andean Altiplano have deep soils with mineral composition that reflects the composition of adjacent mountains but is often mixed with contrasting materials and admixtures of volcanic materials. Many of these valleys have nutrient-rich Alfisols (Luvisols) and Mollisols (Phaeozems) but, where material is derived from acid igneous rocks such as granite, some contain nutrient-poor Ultisols (Nitosols). In the broad central portions of the larger valleys, the water table is near the surface and aquic soil-moisture conditions (Gleysols) are present. Organic soils, Histosols, are present in some areas near rivers and lakes. Rainfall decreases southward in the Altiplano. A udic SMR is present in the northern valleys, and a ustic SMR occurs toward the south. Elevations in most of the Altiplano are above 3,500 meters, and

1. Nomenclature of Soil Taxonomy (Soil Survey Staff, 1999) is used to identify soils and, where applicable, the corresponding units of the Soil Map of the World (FAO-UNESCO, 1971) follow in parentheses.

the STR is isomesic with extreme diurnal temperature flux. The year-round cool temperature slows the rate of crop growth, but the relatively fertile soil mineralogy of recent sediments and volcanic ash support limited agriculture and grazing.

The western slopes, valleys, and coastal plains between the equator and 32°S latitude have an aridic SMR and are dominated by Aridisols (Yermosols) and associated Torriorthents (Regosols). Over many centuries, humans have constructed irrigation systems that collect runoff water from higher elevations for use in food-crop production on portions of the coastal plain and in narrow piedmont valleys (see chapter 19). Most of the soils selected for irrigation are formed in nutrient-rich sediments and have an isohyperthermic or isothermic STR and abundant solar radiation. Conditions are more humid north of the equator where the coastal plain is broader and outlined as lowlands (L6) in figure 7.3.

South of 32°S latitude, most of the western slopes are thickly blanketed by rather acidic Andisols (Andosols) and have xeric, ustic, udic, and perudic SMRs. The STR near the coast is isomesic due to year-long temperature modification by the ocean, but inland the STRs range from mesic to cryic at higher elevations. The rugged topography is primarily used for forestry and grazing, with mixed farming and dairy operations in the warmer valleys.

The eastern slopes of the Andes north of about 25°S latitude have deep and usually infertile Ultisols (Nitosols) and Dystrudepts (Dystric Cambisols), with udic and perudic SMRs. Most areas are sparsely populated with subsistence slash-and-burn agriculture. To the south, the lower elevations of the eastern slopes have Aridisols (Yermosols) and associated soils with an aridic SMR. Low-intensity grazing is the primary land use.

7.5 Upland Soils (U)

The upland areas portrayed in figure 7.3 include the ancient Guyana, Brazilian, and Patagonian shields. The Guyana shield uplands of eastern Venezuela, the interiors of Guyana, Surinam, French Guiana, and the northern extremity of Brazil are a complex of hilly to mountainous uplands of igneous and sandstone rock surrounded by more level plateaus of soil material that has been repeatedly weathered, transported, and redeposited over geologic time (U1). Except for rare intrusions of basic rock where Alfisols (Luvisols) are present, all of the soils have low contents of essential elements, and most are Ultisols (Acrisols). Many of the sandy soils in the plateaus are Psamments (Arenosols) and the finer textured soils are Oxisols (Ferralsols). Some of the soils contain appreciable amounts of ironstone gravel and iron-enriched plinthite horizons. On steep slopes, soils tend to be shallow, namely Lithic subgroups of Inceptisols (Lithosols). Most of the area is covered with tropical forests. Limited populations practice shifting slash-and-burn cultivation for subsistence.

The Brazilian shield, centered on the watershed uplands between the Amazon and Paraná rivers, dominates central Brazil (U2). It is a geologically old land where the soils form in thick deposits of material that has been weathered through several cycles of erosion and deposition (Lepsch et al., 1977). Quartz, kaolinite, gibbsite, and iron oxides are the predominant minerals in this material. Content of essential elements is very low and the majority of the soils are Oxisols (Ferralsols).[2] The most infertile of these soils, Acrustox (Acric Ferrosols) support a sparse native vegetation of grass and low shrubs called *cerrado* (Lopes and Cox, 1977a, 1977b). Each year, wildfires seasonally burn extensive areas during the dry season of the ustic SMR. The sparse and nutrient poor natural vegetation on these soils defies even slash-and-burn attempts at utilization for human food crops. Attempts to graze the native vegetation often result in nutrient deficiencies in the cattle. With reliable infrastructure for marketing and importation of lime and fertilizer, however, large-scale modern agriculture and production of beef on fertilized pastures now flourish in parts of the area (Buol and Eswaran, 2000). On inclusions of limestone, basalt, and sediment from these more fertile materials, Eutrustox (Rhodic Ferralsols) are present.[3] These more fertile soils, often present in broad bands where ancient basalt flows followed valleys in the landscape, were naturally forested with seasonal deciduous forest. Most of the deciduous forests were removed as early European settlers sought the more fertile soils for food-crop production. Ultisols (Nitosols) predominate on less infertile sediment in the southern part of the region in Brazil and eastern Paraguay. Extensive areas of these soils, augmented by lime and fertilizer from external sources, are now being utilized for mechanized agriculture. In addition to wheat, soybeans, and other food crops, sugar cane for ethanol fuel production is grown. With fertilization, much coffee production has moved into the southern margins of the region. Large areas of Quartzipsamments (Ferralic Arenosols) formed in very sandy sediments are near the northern boundary of the shield as it merges with the Amazon lowlands.

The uplands of northeast Brazil (U3) merge with the central Brazilian uplands. Infertile Oxisols (Ferralsols) formed in sediments from the Brazilian shield are present on ancient plateaus. Psamments (Arenosols) are present on sandy deposits. In the northeast of Brazil, south of the Amazon basin, the uplands are nearly level, with some dissection. The soils formed from sedimentary rocks and associated local sediment in the valleys have a moderate

2. Oxisols (Ferrosols) are Latossolos in the Brazilian Classification system (EMBRAPA, 1981; 1999). "Eutroficos" and "Dystroficos" phases identify high base saturation and low base saturation, respectively, in the Brazilian system.

3. See note 2.

supply of essential nutrients, but moisture is a limiting factor. Alfisols (Luvisols) are common in the ustic SMR and Aridisols (Yermosols) in the aridic SMR. Vertisols (Vertisols) are present on some of the more clayey sediments derived from limestone. The soils in the river valleys are fertile Fluvents (Fluvisols), many with aquic soil conditions (Gleysols). The chemical fertility of the soils attracted European settlers to the area but, as population density increased, attempts to farm the less fertile soils resulted in low yields, further aggravated by lack of water in the ustic SMR. Lack of leaching water and improper irrigation techniques have created salinity problems in some irrigated croplands.

The udic SMR upland along the southeast coast of Brazil (U4) from Rio de Janeiro northward is hilly to mountainous. Some higher areas are blanketed by nutrient-poor Oxisols (Ferralsols). In the valleys, the soils are a mixture of Ultisols (Acrisols), formed on nutrient-poor materials, and Alfisols (Luvisols), formed on nutrient-rich material eroded from base-rich rock. Most of the native vegetation has been altered by agriculture and grazing. Some soils rich in calcium and phosphate are renowned for cocoa production. On level areas of the coastal plain, under aquic soil-moisture conditions (Gleysols), nutrient properties vary with the underlying sediment but, with proper drainage and fertilization, soils support modern agricultural production.

South of São Paulo, to the border with Uruguay, erosion has carved a landscape from basalt whose high calcium and magnesium content provides quite fertile soils. These nutrient-rich soils with udic to perudic SMRs once supported vigorous vegetative growth that contributed abundant organic matter to the soil, as shown by dark-colored surface layers. Alfisols (Luvisols) and nutrient-rich Eutrudox (Humic Ferralsols) are common. On the more clayey sediments Vertisols (Vertisols) are present, while Eutrudepts (Humic Cambisols) occupy less stable hillsides. Pasture and tree plantations have replaced coffee growing in many of the isothermic STR hills subject to infrequent frosts. Sugar cane for ethanol fuel production has become an important crop in much of the area. Argiaquolls (Mollic Gleysols) predominate in the broad valleys near the border with Uruguay, where most soils have aquic soil moisture conditions (Gleysols). The upland soils in this region have a thermic STR and udic to perudic SMRs that, coupled with a nutrient-rich mineralogy, have supported sustained agricultural production since European settlement.

East of the Andes in Argentina is a broad upland plain crossed by wide valleys (U5). The soils are formed in mineral-rich sediments and basalt, with admixtures of volcanic ash. Most materials are nutrient-rich and contain carbonates. The soils have an aridic SMR. Most soils are Aridisols (Yermosols) with inclusions of more sandy Psamments (Arenosols) and some clayey Vertisols (Vertisols). Dunes occur in some of the sandier areas, desert pavement gravels are locally widespread, and sparse grazing is the primary land use.

7.6 Lowland Soils (L)

The lowlands areas include the Amazon, Orinoco, and Paraguay River basins and broad low-lying coastal plains (figure 7.3). The mineral composition of the sediments in which the soils form reflects the composition of the uplands and mountains from which they have been derived.

The western portion of the Amazon basin (L1) has udic and perudic SMRs, while a ustic SMR with a pronounced dry season is present east of the confluence with the Rio Negro and Amazon rivers near the city of Manaus, Brazil. Sediments in the central part of the basin west of Manaus are derived from the Andes and are slightly more nutrient-rich than those to the east that are derived from the Guyana and Brazilian shields. Ultisols (Nitosols) predominate in the upper basin (Osher and Buol, 1998; Tyler et al., 1978), while Oxisols (Ferralsols) predominate in the lower portion of the basin. Although both Ultisols and Oxisols are nutrient poor, significant inclusions of more nutrient-rich material are present within the basin. Although long undetected in the poorly explored and sparsely inhabited basin, large areas of Alfisols (Luvisols and Eutric Nitosols) are formed in base-rich sediments in western Brazil primarily in the states of Rondonia and Acre (EMBRAPA, 1981). Many areas of these more chemically fertile soils have recently been cleared of forest vegetation for crop production and cattle grazing.

Some areas in these basins are sandy sediments, forming very infertile Quartzipsamments (Arenosols) where the water table is deep, and Spodosols (Podzols) where the water table is near the soil surface. It has been estimated that perhaps 10% of the Amazon basin has these kinds of soils (Cochrane et al., 1985). One major area of sandy soils is in the headwaters of the Rio Negro where it merges with the Orinoco watershed; smaller sandy areas are scattered throughout the basin, often detectable by the coffee-colored waters of the rivers that drain these watersheds.

The flood plains along the various rivers in the Amazon basin are composed of recent sediments, many quite fertile and high in montmorillonite clay, especially where derived from base-rich materials. Where the sediments are derived from siliceous materials, soils are often infertile. Most of these soils have aquic soil-moisture conditions (Gleysols). These flood plains are often quite broad and subject to flooding and stream-bank erosion from the seasonal rise and fall of river levels. With road maintenance difficult, the rivers provide routes for travel, and most human settlements are located near the rivers (see chapter 16).

In the lower Amazon basin most soils are pale yellow Oxisols (Ferralsols) and infertile. Exceptions occur where basic rock protrudes above the basin sediments. Some of these areas are known and identified, but much of the region is poorly explored due to sparse population and lack of infrastructure. Shifting slash-and-burn cultivation is the primary form of agriculture.

The Orinoco basin (L2) imperceptivity merges with the Amazon basin in an area of deep sands. The western watersheds in the Orinoco basin, south of the Meta River in Colombia, have deep infertile Oxisols (Ferralsols) and Ultisols (Nitosols). North of the Meta River and west of the Orinoco River the soils are more fertile Alfisols (Luvisols) with some Mollisols (Phaeozems) and Vertisols (Vertisols) on the more clayey sediments. Many of these areas have aquic soil conditions (Gleysols). This broad level area, known as the *llanos* of Venezuela and extending south and west into Colombia, is subject to seasonal flooding from waters discharged from the Andes. Except for slightly higher areas near the base of the Andes that are less subject to flooding, there is little agronomic use, other than grazing, because of severe seasonal flooding. The eastern portion of the Orinoco basin rises abruptly onto the Guyana uplands. Most soils in the eastern portion of the basin are infertile Ultisols (Nitosols).

The Paraná basin (L3) extends from the central Brazilian uplands south to the Rio de la Plata. The northern part of the basin is filled with infertile quartz sand and kaolinite clay sediment from the Brazilian shield. Infertile Ultisols (Acrisols) and Oxisols (Ferralsols) predominate. Many of the soils have aquic soil-moisture conditions (Gleysols). In the southern portion of the basin, admixtures of mineral-rich sediments from the Andes enter the basin and montmorillonite clay is more abundant (Troeh, 1969). Clayey Mollisols (Phaeozems), often with aquic soil conditions (Gleysols), predominate. Vertisols are present in some clayey sediments where montmorillonite is the predominate clay mineral.

The Argentine pampas (L4) is the largest area of chemically fertile soil in South America. Mollisols (Phaeozems) are present in wind-deposited loess derived from calcareous sediments rich in plagioclase, pyroxenes, hornblende, and volcanic glass. The loess material becomes finer from west to east, and Vertisols (Vertisols) predominate in the finer textured materials of old lakebeds. The STR is thermic and the SMR udic merging to ustic in the southern part of the region. Sodium-rich Natrudolls (Solonetz) are present in some areas. Aquic soil-moisture conditions (Gleysols) are common on the broad flat landscape. Famed for agricultural production and lush pastures, much of the pampas requires drainage, and the more clayey soils are often physically difficult to cultivate.

To the west of the Argentine pampas and the Paraná basin is a region dominated by sediment derived from the Andes (L5). In most of area the SMR is aridic. Saline and alkali-affected Aridisols (Kastanozems, Solonetz, and salic phases of Xerosols) are common. The clayey texture of the subsoil and the level topography make the area susceptible to flooding during infrequent rain events. Extensive grazing is the primary land use.

Coastal lowlands in Ecuador, Colombia, and the Atlantic coast of Venezuela, Guyana, Surinam, French Guiana, and Brazil have a similar yet contrasting array of soil conditions (L6). Most soils have aquic soil-moisture conditions (Gleysols) and are chemically fertile. In Colombia and Ecuador, some sediment contains appreciable quantities of volcanic ash. The Maracaibo basin has a few salt-affected Aridisols (salic phases of Xerosols) in the driest areas near the lake. As precipitation increases inland, infertile Ultisols (Acrisols), many with aquic soil-moisture conditions (Gleysols), are formed on the more stable terraces (Paredes and Buol, 1981). Small fruit plantations are irrigated in the aridic SMR. Beef and dairy pastures are located in the ustic and udic SMRs. The coastal lowlands north and south of the mouth of the Amazon River have clayey soils with aquic soil-moisture conditions (Gleysols). In some of these areas, canals have been dug to facilitate drainage and serve as waterways for transport of sugar cane and other crops. In some areas, Sulfaquents (Thionic Fluvisols) containing appreciable amounts of iron sulfide are present. These soils, commonly known as "cat-clays," become extremely acid when drained. Farther inland, areas of deep sandy Quartzipsamments (Arenosols) border the uplands. Some of these deep sandy soils are Spodosols (Podzols), where organic-rich spodic horizons are present in the subsoil. Infertility and poor water-holding capacity of the sands have precluded agricultural use.

7.7 Implications for Agriculture

The predominant soil limitation for agriculture in South America is low chemical fertility. Major portions of the soils on the continent are formed in materials that have been repeatedly weathered, transported, and redeposited several times over geologic time. Minerals containing life-essential phosphorus, potassium, calcium, and magnesium were depleted from these materials, leaving a mixture of quartz sand, kaolinite and gibbsite clays, and often rather high concentrations of iron oxide to serve as parent material for soil formation.

The most chemically infertile soils are present as uplands in the interior of the continent on and around the ancient Guyana and Brazilian shields. The Brazilian shield upland has level to rolling topography. Its soils are physically deep and have reliable moisture for one or two food crop seasons each year, and the small amount of biomass in the natural *cerrado* vegetation is frequently burned in wildfires during the dry season. Nevertheless, except for small areas where more fertile soils have formed from basalt or limestone, crop production is impossible without importation of fertilizer. Since about 1985, large-scale grain-crop farming has expanded into this area, which previously was almost devoid of inhabitants. Before crops can be grown, quantities of phosphorus must be applied and mixed into the soil to saturate the iron and aluminum oxide surfaces to the extent that sufficient phosphorus becomes available to crop plants. Carbonate, in the form of crushed limestone, must be applied to raise the pH of the soil and inactivate the exchangeable aluminum. Nitrogen fertilizer is needed for nonlegume crops. Potassium is required for high yields, and small amounts of copper, zinc, boron, and molybdenum are needed in many areas. After an initial investment is made for altering the chemical conditions, fertilizer requirements are annually no greater than in other grain-growing soils around the world. Modern soil-testing technology is utilized to determine annual fertilizer formulations and rates. Economic analyses indicate that grain production in this region competes favorably in world grain markets. Approximately 23 percent of the *cerrado* area is now developed for improved pasture and grain production and produces one-third of Brazil's total food production (Lopes, 1996). Rapid and reliable overland transportation is essential to develop and sustain commercial agriculture that relies on export of grain and import of fertilizer and fuel. In most parts of the region economical road construction is favored by physical stability of inert soil minerals, paucity of river systems that must be bridged, and gentle topography. This is not the case in most of the more rugged Guyana shield region with similar soils.

Infertile sediments from the uplands fill the lower Amazon basin and major portions of the Orinoco and upper Paraná basins. Most soils formed in these materials are Oxisols (Ferralsols). Although these soils have the potential for food-crop production if chemically improved, the many wet and unstable river floodplains make it difficult to provide the necessary transportation infrastructure. In the upper Amazon basin, Ultisols (Nitosols) are formed in slightly more fertile sediments derived from the Andes. Most food production is limited to slash-and-burn subsistence farming. Reliable overland transportation is difficult to construct westward over the Andean mountains in Perú, Ecuador, and Colombia. Significant crop and cattle production is developing in areas of western Brazil where roads from developing areas in the central Brazilian shield provide access to areas of more fertile Alfisols (Luvisols).

Along the continent's western perimeter, volcanic ash, basalt, and limestone within, or derived from, the Andes have injected more fertile mineralogy to the soils of this mountainous region. Within the Andean mountains and intermountain valleys, fertile Andisols (Andosols) and Alfisols (Luvisols) have historically supported human settlement. Most of the intermountain valleys have a cool isomesic STR that limits food crop yields. Development and maintenance of roads to population centers is expensive due to the rugged relief, earthquakes, and volcanic activity. Transportation infrastructure is seldom reliable beyond local markets.

Nutrient-rich sediments eroded from the eastern slopes of the southern Andes lie as Aridisols (Kastanozems) and other potentially fertile soils in western Argentina, but the aridic SMR creates conditions too dry for agricultural production without irrigation. In contrast, loess from similar sources in semiarid central Argentina has been deposited across the pampas around Buenos Aires, where very fertile Mollisols (Phaeozems) and Vertisols (Vertisols) are formed. Much of this traditionally productive area has aquic soil-moisture conditions that require artificial drainage. Increased yields and the continued export of nutrients in the products marketed have increased the need for fertilizer. Fertile sediments from the eastern flanks of the Andes are also spread across the nearly level western Orinoco basin in Venezuela and north of the Meta River in Colombia, but an aquic soil-moisture regime and seasonal flooding restrict land use to grazing, except at higher elevations near the western edge of basin.

The narrow coastal plains of Chile and Perú have an aridic SMR, and irrigation is required to support agricultural production. In Ecuador and Colombia, where the Pacific coastal lowlands have udic and perudic SMRs, aquic soil-moisture conditions and seasonal floods limit infrastructure maintenance, food-crop production, and grazing.

The eastern margin of the continent comprises uplands of igneous and basalt rock, intermingled with plateaus underlain by ancient sediment, and with narrow coastal valleys. In areas where they are formed in basalt, or sediment derived from basalt, the soils tend to be rather fertile Alfisols (Luvisols) and their associates. Where soils are formed from acid igneous rock or less fertile sediment, Ultisols (Acrisols) are present. The more fertile soils have long been sites of agricultural production. South of São Paulo much of the area has a thermic STR, but seasonal winter temperatures present a frost hazard to coffee and citrus production. Aridic and severe ustic SMRs limit production in northeast Brazil and southern Argentina, but perudic and udic SMRs with thermic to hyperthermic STRs are nearly ideal for rapid tree growth on the more fertile soils. Where the relief is too rugged for mechanized agriculture, commercial tree plantations of exotic trees are now replacing the natural forests once cleared for agricultural crops.

Narrow low-lying coastal plains on the north coast of the continent have relatively fertile soils. Because many of these soils have aquic soil moisture conditions (Gleysols), artificial drainage has been developed to sustain a wide range of subsistence and commercial agricultural activities.

7.8 Conclusion

Human habitation of South America is concentrated in areas of naturally fertile soils on the coastal perimeters and mountainous areas of the continent. The chemically infertile soils of the interior are not capable of sustaining food production unless essential elements are imported as fertilizer. The impressive amount of vegetative biomass growing in the interior jungles depends upon complete biocycling of essential elements and is possible only when there is little or no exportation of plant material. The silicate and oxide minerals in the soil contain and release such small amounts of essential elements that the production of fast-growing food crops is not possible, except for short periods of time after jungle biomass is burned and the essential elements it contains are released into the soil solution as inorganic ions. Modern agricultural methods are economically overcoming the fertility constraints in areas where it has been possible to establish reliable roads and make world markets accessible to commercial farmers. Areas of infertile but well-watered soils not served by reliable transportation are limited to traditional subsistence slash-and burn-agriculture. In the central Atlantic uplands, greater food production is now possible through nutrient replenishment, and cultivation is concentrating on the more level areas, with plantation forestry replacing cropland on hills once cleared of native forests. Food production in arid areas is determined by availability of irrigation water. The chemical fertility of cool to cold mountain valleys in the Andes continues to nourish indigenous peoples via traditional agriculture. Technology can do little to warm the soils and increase production in these mountains. The Andes also physically hinder infrastructure development necessary for increased food production in the warmer interior areas of those nations along the Pacific coast.

References

Buol, S.W., 1995. Sustainability of soil use. *Annual Review of Ecology and Systematics,* 26, 25–44.

Buol, S.W., and H. Eswaran, 2000. Oxisols. *Advances in Agronomy,* 68, 151–195.

Buol, S. W., F.D. Hole, R.J. McCracken, and R.J. Southard, 1997. *Soil Genesis and Classification.* 4th edition. Iowa State University Press, Ames, Iowa.

Buol, S.W., and P.A. Sanchez, 1986. Red soils in the Americas: morphology, classification and management. In: Academica Sinica (Editors), *Proceedings of the International Symposium on Red Soils.* Science Press, Beijing, and Elsevier, Amsterdam, 14–43.

Buol, S.W., and P.A. Sanchez, 1988. Soil characteristics and agronomic practices for sustainable dryland farming. In: P.W. Unger, W.R. Jordan, T.V. Snead, and R.W. Jensen (Editors), *Challenges in Dryland Agriculture: A Global Perspective.* Proceedings of the International Conference on Dryland Farming, 1988, USDA-ARS, Amarillo/Bushland, Texas, 367–370.

Cochrane, T.T., L.G. Sanchez, L.G. de Azevedo, J.A. Porras, and C.L. Garver, 1985. *Land in Tropical America* (3 volumes). Centro Internacional de Agricultura Tropical (CIAT), Apartado 6713, Cali, Colombia (in English, Spanish, and Portuguese).

EMBRAPA, 1981. *Mapa de Solos do Brasil* (Escala 1:5,000,000). Serviço Nacional de Levantamento e Conservação de Solos, Geocarta.

EMBRAPA, 1999. Sistema Brasileiro de Classificação de Solos. Serviço de Produção de Informação-SPI, Brasilia, DF.

FAO-UNESCO, 1971. *Soil Map of the World: Volume IV South America.* United Nations Educational, Scientific and Cultural Organization, Place de Fontenoy, 75 Paris-7e.

Lepsch, I.R., S.W. Buol, and R.B. Daniels, 1977. Soil-landscape relationships in the Occidental Plateau of São Paulo, Brazil: II. Soil morphology, genesis and classification. *Soil Science Society of America Journal,* 41, 109–115.

Lopes, A.S., and F.R. Cox, 1977a. A survey of the fertility status of surface soils under "Cerrado" vegetation in Brazil. *Soil Science Society of America Journal,* 41, 742–747.

Lopes, A.S., and F.R. Cox, 1977b. Cerrado vegetation in Brazil: An edaphic gradient. *Agronomy Journal,* 69, 828–831.

Lopes, A.S., 1996. Soils under cerrado: A success story in soil management. *Better Crops International,* 10 (2), 9–15.

Osher, L.J., and S.W. Buol, 1998. Relationship of soil properties to parent material and landscape position in eastern Madre de Dios, Peru. *Geoderma,* 83, 143–166.

Paredes, J.R., and S.W. Buol, 1981. Soils in an aridic, ustic, udic climosequence in the Maracaibo Lake Basin, Venezuela. *Soil Science Society of America Journal,* 45, 385–391.

Sanchez, P.A., 1976. *Properties and Management of Soils in the Tropics.* John Wiley, New York.

Soil Survey Staff, 1999. *Soil Taxonomy: A Basic System of Soil Classification for Making and Interpreting Soil Surveys.* 2nd edition. Agriculture Handbook 436, U.S. Department of Agriculture, Natural Resources Conservation Service, Washington, DC.

Troeh, F.R., 1969. Noteworthy features of Uruguayan soils. *Soil Science Society of America Proceedings,* 33, 125–128.

Tyler, E.J., S.W. Buol, and P.A. Sanchez, 1978. Genetic association of soils in the Upper Amazon Basin of Peru. *Soil Science Society of America Journal,* 42, 771–776.

Van Wambeke, A., 1981. *Calculated soil moisture and temperature regimes of South America.* SMSS Technical Monograph 2. Cornell University, Ithaca, New York, and Soil Management Support Services, USDA-SCS, Washington, DC.

Van Wambeke, A., 1991. *Soils of the Tropics: Properties and Appraisal.* McGraw-Hill, New York.

8

Zoogeography

Peter L. Meserve

South America forms the greater part of the Neotropical faunal realm, which extends northward through Central America to tropical southern Mexico. Although making up only 12% of the world's land area, South America is the richest continent for virtually all organismal groups, including vertebrates. For example, of the known 23,250 species of fish (Eschmeyer, 1998), 41% or 9,530 species are freshwater, and of these, more than 2,800 species (29%) are in South America (Moyle and Cech, 2000). A comparable level of diversity exists for amphibians and birds. Of Earth's 5,900 species of amphibians, at least 1,749 or 30% occur in South America (Duellman, 1999a, 1999b; Köhler et al., 2005; www.amphibiaweb.org). More than 3,200 (or nearly 32%) of Earth's 9,900 species of birds occur in South America (Sibley and Monroe, 1990). For reptiles and mammals, diversity is only slightly lower; at least 1,560 (19%) of 8,240 reptile species (Uetz and Etzold, 1996; www.reptile-database.org), and 1,037 (19%) of 5,416 mammal species (Nowak, 1999; Wilson and Reeder, 2005) are found in South America.

Four major geological events or features are important to understanding South America's contemporary zoogeography. The first was the breakup of Pangea, and then of Gondwana (see chapter 1). South America and Africa remained close for an extended period of the Mesozoic, and thus share important similarities in their faunas, including groups not fully evolved at the time of separation. South America also maintained connections to other Gondwanan continents, directly with Antarctica, indirectly with Australia, until the early Cenozoic. The second major feature was South America's long period of isolation in the Cenozoic, particularly from North America pending establishment of the late Pliocene land bridge after 3 Ma (million years before present). The latter resulted in "The Great American Interchange" (Webb, 1976; Marshall et al., 1982), which had profound consequences for the fauna. The third major feature of South America has been the Andes, which, in addition to modifying climate, have been a center of speciation, a dispersal route, and a barrier. The cordillera has had an overriding effect on distributions and histories of both past and current biotas on the continent. Finally, the Amazon basin continues to challenge scholars and conservationists alike as they attempt to catalog its extraordinary diversity in the face of an accelerating pace of tropical rainforest destruction. Yet underlying explanations for its diversity remain controversial.

This chapter concentrates on patterns of South American vertebrate distributions and diversity. Although comprising a small proportion of the total biota (currently 2.7% of all species; Jeffries, 1997), our current knowledge of vertebrates is significantly greater than, for example, the more than one million insects (Stork, 1999). Valuable references for nonvertebrates include Fittkau et al. (1969a, 1969b) and Bănărescu (1990, 1995) on aquatic invertebrates and selected terrestrial arthropods, Briggs (1995) on the paleogeography of some invertebrates, and Goldblatt (1993), and George and Lavocat (1993) for some insect groups and their relevance for past connections between South America and Africa.

8.1 Zoogeographic Patterns

8.1.1 Fish

There is a vast range of taxonomic diversity among South American fish. Projections of the actual number of obligatory freshwater species are much higher than the presently accepted number of 2,800 species in 32 families (Moyle and Cech, 2000). Up to 8,000 species, perhaps a quarter of all fish, may be in the Neotropics (Vari and Malabarba, 1998) and 2,000 in the Amazon basin alone (Lundberg and Chernoff, 1992). Over 400 new species are being described per year, many in South America. Ostariophysan fish make up about 28% of the world species but about 72% of the freshwater species (Briggs, 1995). Cyprinids (minnows and carp; Cyprinidae: Cypriniformes), the dominant group of ostariophysan fish elsewhere (except Australia), are entirely absent as native species in South America. In their place, the dominant families are other ostariophysan fish, including characins (order Characiformes, 14/18 families, of which 7 are endemic, comprising about 1,200 species; Buckup, 1998; Moyle and Cech, 2000), and catfish (order Siluriformes, 11/30+ families, 8 endemic, around 1,300 species; de Pinna, 1998). A third order, the electric knifefishes (Gymnotiformes, 5 families, 100+ species; Albert and Campoz-da-Paz, 1998), shows strong morphological convergence with the African knifefishes (Notopteridae) and unrelated mormyrids (Helfman et al., 1997; Moyle and Cech, 2000). Nonotophysan elements making up >10% of the ichthyofauna include the cichlids (Cichlidae: Perciformes, around 50 genera and 450 species; Kullander, 1998), pupfishes (Cyprinodontidae: Cyprinodontiformes, 80+ species) and poeciliids (Poeciliidae: Atheriniformes, 20+ species). Finally, two ancient freshwater families, the bonytongues (Osteoglossidae: Osteoglossiformes, 2 species) and South American lungfish (Lepidosirenidae: Lepidosireniformes, 1 species) are here; osteoglossids and other lungfish families are also found in Africa and Australia (Moyle and Cech, 2000).

In terms of geographic distribution, Géry (1969) identified eight faunistic regions for South American freshwater fish, namely the (1) Orinoco-Venezuelan, (2) Magdalenean, (3) Trans-Andean, (4) Andean, (5) Paranean, (6) Patagonian, (7) Guianean-Amazonian, and (8) East Brazilian (figure 8.1). Three regions (1, 2, and 7) possess <70% of the species but differ greatly in richness of the ichthyofauna. The Trans-Andean, Andean, and East Brazilian regions contain high endemicity, ranging from 26% to over 50%. In the Andes, the trend is for impoverishment of the ichthyofauna as elevation rises from 1,000 to 4,500 m, and toward the southern tip.

These patterns to some extent reflect the geological history of South American drainage patterns, involving not only uplift of the Andes but also flexuring of cratonic shields and structural arches, and changes in sea level and climate. Most diversification of the modern freshwater ichthyofauna took place over a 70-million-year period between the late Cretaceous and late Miocene. Little or none of the present diversification at the family and genus group level has occurred since then (Vari and Weitzman, 1990;

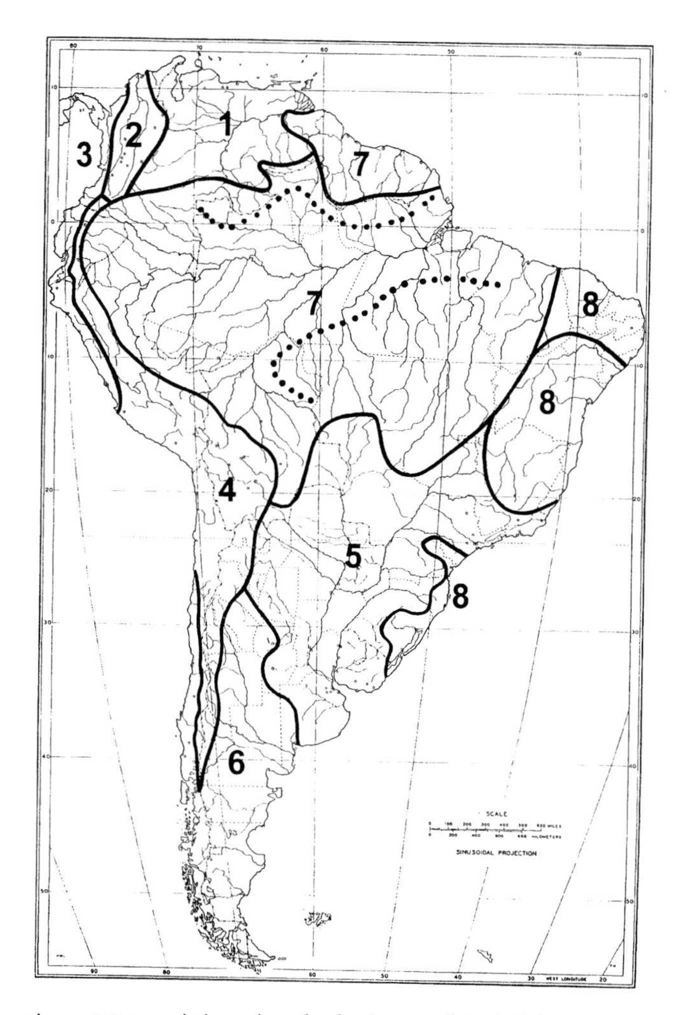

Figure 8.1 Faunistic regions for freshwater fish: 1. Orinoco-Venezuelan, 2. Magdalenean, 3. Trans-Andean, 4. Andean, 5. Paranean, 6. Patagonian, 7. Guianean-Amazonian, and 8. East Brazilian. Dotted lines delimit ancient shields. Modified from Helfman et al. (1997).

Lundberg, 1998). Thus, the ichthyofaunas of some regions now isolated by the Andes had common drainages in the more distant past and shared important elements, such as the Magdalena and Guianean-Amazonian ichthyofaunas on the Atlantic slope (Lundberg and Chernoff, 1992; Lundberg et al., 1998).

8.1.2 *Amphibians*

Earth's living amphibians include 5,873 species in three clades: caecilians (order Gymnophiona), salamanders (order Caudata), and frogs and toads (order Anura; Duellman, 1999a, 1999b; Köhler et al., 2005; www.amphibiaweb.org). Of 170 known species of caecilians, which occur in the tropics worldwide except for Madagascar and Papua-Australia, almost half are in South America. Two families (Caeciliidae and Rhinatrematidae) are represented; perhaps 95% and one family are endemic. Of the 549 known species of salamanders, only one family (Plethodontidae) is represented in South America; at least 20/24 species (83.3%) are endemic. In contrast, of the 5,154 currently recognized species of anurans, at least 1,651 species (32%) in 12 families are found in South America and of these, 1,580 species (95.7%) and four families are endemic. Five families (Leptodactylidae, Hylidae, Dendrobatidae, Bufonidae, and Centrolenidae) account for 1,341 species or 81.2% of the South American anurans. At least 96% of a minimum 1,749 amphibian species in South America are endemic, the highest such proportion for any vertebrate class or continent except for mammals in Australia. South America also has the highest density of amphibian species per unit area, 97.9 per 10^6 km^2, or twice as high as the next highest continent, North America (44.4 per 10^6 km^2); 94.4% of this density is due to anurans. Around 100 new species are being identified per year, and the documented diversity of South American amphibians increased 59% from 1979 to 1999 (Duellman, 1979a, 1999a).

With respect to the herpetofauna, Duellman (1979a) recognized 18 eco-physiographical regions in South America. This was simplified to 12 morphoclimatic domains by Ab'Saber (1977) and termed "natural biogeographic regions" by Duellman (1999b; figure 8.2). The resemblance of this latter map to the climatic maps of Hueck and Seibert (1972) is close. The most diverse regions for amphibians are the Andes (753 species, 95.0% endemic), the Amazonia-Guianan domain (335 species, 82.6% endemic), and the Atlantic Forest domain (340 species, 92.9% endemic). At the lower end of the diversity range, the Atacama, Patagonia, and Austral Temperate Forest domains have 5, 10, and 32 species, and high levels of endemism (100, 100, and 80%), respectively (Duellman, 1999b). One of the most unusual families in the latter region is the Rhinodermatidae or mouth-brooding frogs represented by only two species, which are wholly endemic to the southern temperate rainforests of Chile and Argentina (figure 8.3).

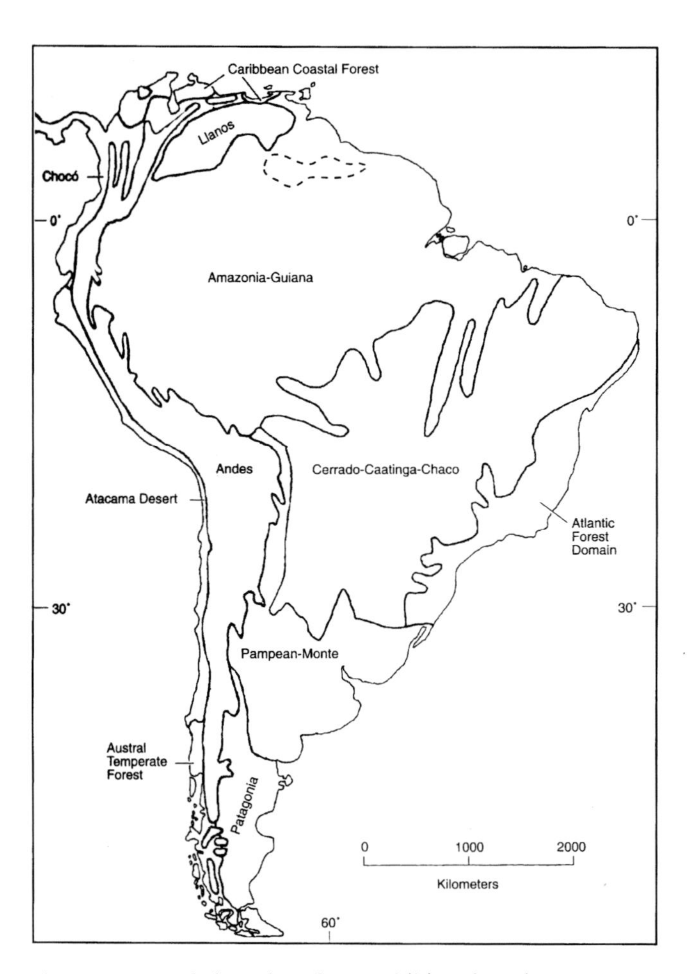

Figure 8.2 Faunistic regions for amphibians based on morphoclimatic domains recognized by Ab'Saber (1977). From Duellman (1999b).

8.1.3 *Reptiles*

Of the 8240 currently recognized species of reptiles on earth, 4,930 or 59.8% are saurians [lizards and amphibaenians (burrowing forms), 2,978 or 36.1% are serpents (snakes, including 60 marine species], and 332 are other groups [testudines (turtles), crocodilians, and tuataras (Rhynchocephalia); Uetz and Etzold, 1996; www.reptile-database.org. Of the minimum 1,560 reptile species in South America, about 53% are lizards and 43% are snakes (Bauer, 1993). Most (89%) are endemic (Dixon, 1979; Duellman, 1979a). The lower proportion of lizards relative to snakes may be due to the high proportion of tropical rain forest and other mesic habitats in South America (Duellman, 1979a; Bauer, 1993), since 48.3% and 51.4% of the reptile genera and species, respectively, are snakes

Figure 8.3 *Rhinoderma darwinii* (Darwin's frog), an endemic mouth-brooding frog, is found only in the southern temperate rainforests of Chile and Argentina. Males shelter developing embryos in vocal sacs, and tadpoles metamorphose into froglets, which emerge through the mouth, giving the appearance of parturition there (photo: P.L. Meserve).

in those habitats (vs. 42.3% and 43.0% for lizards; Dixon, 1979). In the rain forest, colubrid snake assemblages are predominantly generalized terrestrial and/or arboreal predators feeding on frogs or lizards (Cadle and Greene, 1993). Proportions of snakes reverse toward the Andes and the southern triangle (Cei, 1979; Formas, 1979). Only five snakes are found in Chile (vs. 79 species of lizards; Jaksic, 1998), the only South American country besides Uruguay lacking tropical forests. The number of described reptiles in South America increased from 1,115 to 1,560 (40%) between 1979 and 2001. Although the known overall reptile diversity of South America remains lower than that of Asia (2,080 species) and Africa (1,680 species; Uetz and Etzold, 1996; www.reptile-database.org), clearly this situation could change in the future.

Tropical rain forest regions, comprising the Chocó, Amazonian, and Atlantic forests, have about 73.4% of the genera and 49.5% of the reptile species in South America (figure 8.2). In addition to the proportions of snakes, lizards, and amphisbaenids noted above, 4.4% are turtles and 1.1% are crocodilians. Of the South American genera, 66.3% and 76.6% of lizards and snakes, respectively, occur in the tropical rain forests (Dixon, 1979; Bauer, 1993). These areas also share the highest proportion of species; as in the rest of South America, endemism is high. In contrast, the Atacama, Patagonia, and Austral Temperate Forest Domains have very low reptile diversity; the latter two regions have 33 and 6 species of lizards and amphisbaenids, 11 and 2 species of snakes, and 1 and 0 species of turtles, respectively (Cei, 1979; Formas, 1979; Meserve and Jaksic, 1991). Virtually 100% are endemic or shared only with the Andean region (Duellman, 1979b). An example of an unusual endemic reptile in central and northern Chile is the dwarf tegu (*Callopistes palluma/maculatus*), a large predaceous lizard in the widespread Western Hemispheric family Teiidae (figure 8.4).

8.1.4 Birds

With nearly one third of the world's 9,881 birds, regional species numbers in South America are astounding (Sibley and Monroe, 1990; Myers et al., 2000). Colombia alone has 1,758 species, making up 19% of the world's avifauna (Stattersfield et al., 1998), and over 300 species can be found in a 3 km^2 area of Amazonian rain forest (Terborgh et al., 1984), nearly half the number for the United States. Unlike all other continents, the South American avifauna is dominated by nonpasserines and suboscine passerines; the latter (including the tyrannine flycatchers) make up >1,000 species, more than half the passerines there (Ridgely and Tudor, 1989; Ricklefs, 2002). Of some 95 avian families present, nearly one third (31) are endemic to the region. Particularly diverse and characteristic in South America are the tinamous (Tinamidae, 43 species present/46 total species), hummingbirds (Trochilidae; 241/322 species), woodcreepers and ovenbirds (Furnariidae; 256/280 species), tapaculos (Rhinocryptidae; 28/29 species; figure 8.5), antbirds, antthrushes, antshrikes, and antwrens (Thamnophilidae and Formicariidae; 234/250 species), and cotingas and manakins [Cotingidae and Pipridae; alternatively Tyrannidae (Sibley and Ahlquist, 1990); 108/118 species] (Ridgely and Tudor, 1994; Feduccia, 1999; Rahbek and Graves 2000).

Stotz et al. (1996) recognized 15 zoogeographic regions, 34 zoogeographic subregions, and 41 principal habitats for birds in South America (figure 8.6). Cracraft (1985) distinguished 33 areas of avian endemism, which were re-

Figure 8.4 The dwarf tegu (*Callopistes palluma/maculatus*), an endemic teiid lizard, is found in arid-semiarid to mediterranean scrub habitats from northern to south-central Chile (photo: B.K. Lang).

Figure 8.5 The chucao tapaculo (*Scelorchilus rubecula*) is a common forest bird in undergrowth and edges of southern temperate rainforests of Chile and Argentina. The family Rhinocryptidae to which it belongs is almost wholly endemic to South America (photo: S.R. Morello).

duced to 20 areas by Ridgely and Tudor (1989) and amplified to 56 areas by Bibby et al. (1992). The highest species richness is in the Northern and Central Andean regions, as well as the Northern and Southern Amazonia regions (all with ≥772 species). The highest avian diversity in equatorial South America is in humid montane regions (Rahbek and Graves, 2001); 901 species (9% of the world's avifauna) occur just in the Manu Biosphere Reserve in southeastern Perú (Patterson et al., 1998). The Central South America and Northern South America regions have lower numbers of species (from 652 to 645). Regions with particularly low numbers of avian species (<250) include the Southern Andes, Patagonia, and Pampas; endemism is higher in the former two areas than the latter (i.e., 22–23 species vs. 8). The numbers of locally endemic species are highest (≥199 species) in the Northern and Central Andes plus Atlantic forests. Overall endemism (>100 species) is greatest for these areas plus Central South America. The percentage of avian species at risk is greatest (>25%) in Northern and Central Andes, Northern Amazonia, and Atlantic Forests, whereas the percentage of endemic species at risk is greatest (>60%) in the Northern Andes, Central South America, and Atlantic Forests (Stotz et al., 1996).

The effect of elevation on avian species numbers varies in different parts of the Andes. Avian species numbers peak on the eastern slope of the tropical Andes at around 1,000–1,600 m and subsequently decline above 3,600–5,000 m (Terborgh, 1977; Rahbek, 1997); in southern Perú, numbers decline up to about 2,000 m and reach an asymptote above here (Patterson et al., 1998). In interior Bolivian valleys between 1,600 and 3,800 m, species numbers peak at around 2,500 m (Kessler et al., 2001). On the western, more xeric side, they increase moderately up to 5,000 m, and are strongly affected by vegetation complexity and plant species diversity (Pearson and Ralph, 1978). Avian species numbers and endemicity are generally low in the high Andean Puna and Páramo (Vuilleumier, 1986, 1997).

From 361 to 422 nonpelagic avian species are northern migrants in the Neotropics; most migrate to Central America and the West Indies islands (DeGraaf and Rappole, 1995; Stotz et al., 1996). However, only Northern South America has more than 99 species of migrants; the Andean and Patagonian regions have less than 50 species. Whereas passerine migrants (principally warblers, flycatchers, and vireos) are concentrated in Northern and Central Amazonia, Northern South America, and Northern Andes, nonpasserines (e.g., hawks, ducks, shorebirds) are more evenly distributed throughout most non-montane regions. The principal habitats used by Neotropical migrants are scrub, montane forests, and wetlands, rather than lowland rain forests. About 250 nonpelagic species migrate northward from southern South America during the southern winter, and they occupy forest, aquatic, and open habitats in roughly equal proportions (Stotz et al., 1996).

8.1.5 Mammals

Of a minimum 1,037 terrestrial mammal species in South America, 71 are marsupials (in three orders, two endemic), comprising 25% of the world's total. Twenty-nine of the world's 30 xenarthrans (order Xenarthra; sloths, anteaters, and armadillos) occur in South America. Particularly speciose groups in South America include primates [Primates; 104 species (37% of Earth's total)], bats [Chiroptera; 219 species (22%)], and rodents [Rodentia; 522 species (25%)]. Particularly notable within the last group are the high numbers of sigmodontine (ca. 286 species; 67%) and hystricognath rodents [14 families; 180 species (72%); Nowak, 1999]. The Neotropical Region (including Central

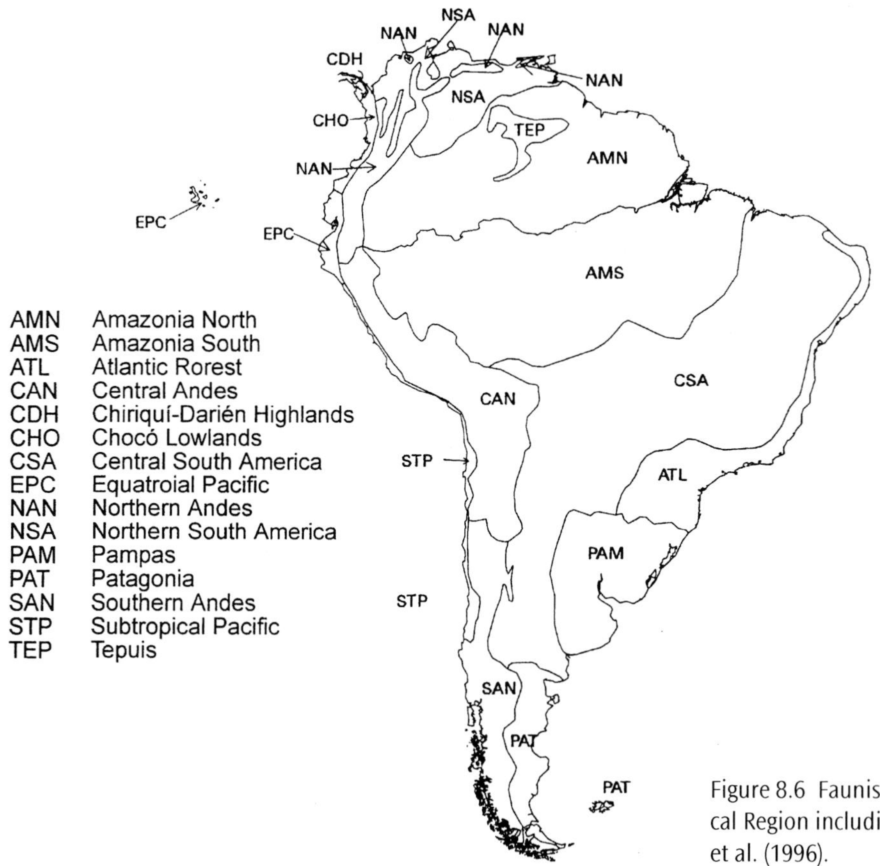

Figure 8.6 Faunistic regions for birds for the Neotropical Region including South America. Modified from Stotz et al. (1996).

America and tropical Mexico) has 1,273 species (Patterson, 1994), thus exceeding that for the next richest area, the Ethiopian Region (1,045 species). Compared to the latter, less speciose South American groups include insectivores, primates, and ungulates (Eisenberg, 1999a). Species endemism is generally high (82%) for Neotropical mammals, exceeded only by Africa (Ethiopian Region; 93%) and Australia (89%; Cole et al., 1994). At the family level, the Neotropical Region has the second-highest mammal endemicity (47%), exceeded only by the Australian Region (97%). New species are being described at a rate of about 40–45 per year, and they are predominantly smaller forms under 60–70 g (Patterson, 1994, 2000, 2001; Wilson and Reeder, 2005).

Only two zoogeographic regions have been distinguished for mammals in South America: the Brazilian extending northward into Central America, and the Patagonian (Hershkovitz, 1958, 1972; Simpson, 1969). However, faunal analyses suggest some further distinctions within these regions. Mammal-species richness is high in Amazonia; 350 species are known here, of which 205 (58.6%) are endemic. Brazil, which accounts for 48% of the area of South America and holds 62% of Amazonia, has 457 species of mammals, just under half the total known for the continent (Fonseca et al., 1999). It is the leading country in the Western Hemisphere for mammal-species richness (closely followed by Mexico, however, with 449; Ceballos and Navarro, 1991). In addition to Amazonia, five other speciose biomes of those recognized by Hueck (1972) include the Cerrado, Atlantic Forest, Gran Chaco, Caatinga, and Pantanal (in order of decreasing area) with 159, 229, 102, 124, and 113 mammal species, respectively. Endemicity ranges from 58.6% and 31.9% in Amazonia and the Atlantic Forest, respectively, to 2.4 and 0% in the Pantanal and Chaco, respectively; Amazonia and the Atlantic Forest account for 31% of all endemic South American mammals. Amazonia has the least mammalian affinity with neighboring biomes; the greatest affinity (and lower endemicity) occurs between the adjacent Pantanal and Cerrado biomes, and among the biomes separating Amazonia from the Atlantic Forest (Mares, 1992; Fonseca et al., 1999). For some taxonomic groups, species richness and levels of endemicity are particularly high. The highest numbers of species of primates (65) and a high level of endemicity (74%) are found in Amazonia; the Atlantic Forest follows with 18 species and slightly greater endemicity (78%; Rylands et al., 1997). Rodents make up about 43% of the world's mammal species, but over 50% of the mammal species in South America. Although bats make up only 21% of the South American mammalian fauna (close to the worldwide figure of 20%; Nowak, 1999), they outnumber rodents in most major tropical biomes except the Chaco (Fonseca et al. 1999, Patterson et al. 1998). Much of the increase in bat diversity appears to be due to latitude-specific rather than biome-specific effects, but there are also strong changes at the familial-level between New World temperate and tropical bat faunas (Willig and Selcer, 1989; Stevens, 2004). Nonvolant groups, including marsupials, primates, and rodents, account for most of the diversity differences among Neotropical rain forest mammal assemblages (Voss and Emmons, 1996); bats and primates are distinctly less diverse (or absent) in temperate South American assemblages. There is a higher proportion of intermediate-sized mammals (i.e., ca. 70–1000 g) in tropical forests than in South America generally, or North America, probably due to greater habitat complexity (Fonseca et al., 1999, Bakker and Kelt, 2000).

Mares (1992) reported that South American drylands have 509 mammal species and 41.5% endemicity, greater than that of Amazonia (434 species, 31.8%). However, the difference may be due largely to physiognomic classifications (Chesser and Hackett, 1992; Fonseca et al., 1999); for example, Mares grouped all nonforest habitats into "drylands," including the Llanos, Cerrado and Caatinga, Pantanal, Chaco, Páramo and Puna, and Patagonia. Inventories at four Neotropical "dryland" sites in the Llanos, Caatinga, and Pantanal indicate total mammal species richness to be 55–71 species vs. 200+ for western Amazonian sites (Voss and Emmons, 1996), and 219 species occur along an elevational forest gradient in the southeastern Andes (Patterson et al., 1998). Geographically, non-volant mammal species richness is highest in the eastern versant of the central Andes followed by the Amazon Basin and Atlantic rain forests of Brazil (Tognelli and Kelt, 2004).

Further south, mammal-species numbers decline toward Patagonia. Fewer than 37 species occur over much of southern South America and the Pacific Coastal Desert of Chile and Perú (Marquet, 1994; Eisenberg, 1999b). Beyond Amazonia, areas of higher species richness are found in montane regions and northwestern Argentina, although some rodent and marsupial taxa are largely replaced at higher elevations (Pearson and Ralph, 1978; Mares et al., 1997; Eisenberg, 1999b). Small mammal-species richness increases with elevation on the western side of the Andes next to the hyperarid Pacific Coastal/Atacama Desert. However, it declines at the highest elevations—the Altiplano and Puna plus Páramo—although endemicity is high (Chesser and Hackett, 1992). Total numbers of mammal species decline with elevation at tropical latitudes on the eastern side of the Andes, mostly due to reduced numbers of bats with elevation as numbers of nonvolant small mammal species increase (Patterson et al., 1998; Eisenberg, 1999b; Tognelli and Kelt, 2004). In the southern Valdivian Andes, small mammal-species numbers decline with elevation near the evergreen-deciduous forest-scrub transition zone (Pearson and Pearson, 1982; Patterson et al., 1989). One of most unusual mammals here, the Chilean shrew opossum (*Rhyncholestes raphanurus*, order Paucituberculata; figure 8.7) is found only in the southern temperate rain forests of Chile and Argentina. A member of the family Caenolestidae with only five species, they occur into the Andes and represent an ancient marsupial lineage that were the most abundant marsupials in the South American Miocene (Carroll, 1988).

8.2 The Role of Geological History

8.2.1 Plate Tectonics and Vertebrate Zoogeography

Throughout Paleozoic time, South America formed part of the larger landmass of Gondwana, which toward the close of the era fused with Laurussia to form the supercontinent of Pangea (see chapter 1). When Pangea broke apart during Mesozoic time, Gondwana's component parts—South America, Africa, Antarctica, Australia, and India—began to separate. These tectonic events were to have major significance for the development of South America's fauna.

While Pangea existed, a more modern vertebrate fauna began to emerge, gradually replacing the Paleozoic fauna despite the massive end-Permian extinction event (245 Ma), which eliminated up to 96% of all marine species (Raup, 1979). By the early Triassic, this vertebrate group included osteichythyan and chondrichthyian fishes, labyrinthodont amphibians and therapsid synapsids (the latter gave rise to mammals), and archosaurs (from which birds eventually arose; Sepkoski, 1990). The vertebrate

Figure 8.7 *Rhyncholestes raphanurus*, the Chilean shrew opossum (family Caenolestidae), has a restricted distribution in the southern temperate rain forests of Chile and neighboring Argentina (photo: P.L. Meserve).

tetrapod fauna was mostly cosmopolitan by mid-Triassic (240–230 Ma), and included theriodonts, therocephalians, cynodonts, dicynodonts, cotylosaurs, archosaurs, and labyrinthodonts (Cracraft, 1974; Shubin and Sues, 1991). The dinosaur lineages (theropods, sauropomorphs, ornithischians) were evident by the mid-late Triassic, although they comprised a small proportion of the tetrapod fauna. By the late Triassic, they became more dominant and widespread (Weishampel, 1990; Benton, 1993). New lineages in the late Triassic included crocodylomorphs, pterosaurs, lepidosaurs, two orders of dinosaurs, several groups of synapsids, and lissamphibians, marking the true modern vertebrate fauna (Padian, 1986). Through much of the Jurassic, the dinosaurs radiated and became widespread throughout Pangea, where, despite the beginning of the impending breakup, dinosaur faunas remained quite similar throughout Laurussia and Gondwana (Cox, 1974; Rage, 1988). Six to nine orders of mammals also appeared in the late Triassic, although these were generally small in size (Crompton and Jenkins, 1979).

Beginning in the late Jurassic (~160–145 Ma) and extending into the early Cretaceous (~145–100 Ma), the separation between Gondwana and Laurussia became complete. Although Cracraft (1974) emphasized the early distinctiveness of the Gondwanan vs. Laurussian ichthyofaunas, the ostariophysan fish appear to have evolved in the late Jurassic prior to the breakup of Pangea. Alternative theories for their monophyletic origin in South America or East Gondwana require terrestrial continental connections for the current distribution of many obligatory freshwater groups (Briggs, 1995). The most prominent nonotophysan element, the cichlids, appears to be derived from basal groups found in India and Madagascar prior to the breakup of Gondwana. Cichlids have subsequently undergone accelerated rates of molecular evolution in South America (Farias et al., 1999). Despite numerous vicariant events subsequent to the breakup of Gondwana, many similarities between South America and Africa in other nonotophysan groups, such as poeciliids, osteoglossids, and lungfish, remain (Lundberg, 1993, 1998).

Caecilians (order Gymnophiona) appear to be primarily of Gondwanan origin, whereas salamanders (Caudata) are primarily Laurussian (Cracraft, 1974; Briggs, 1995). There is dispute about the importance of the breakup of the Laurussian continents in the early-middle Jurassic to the subordinal differentiation of salamanders. Anurans, on the other hand, extend back to the Triassic and include elements found in both Gondwana and Laurussia. However, the greatest radiation was in Gondwana, with Pipidae, Leptodactylidae, Bufonidae, and Microhylidae shared in common between South American and Africa. In Africa, a number of unique elements evolved, including ranids, hyperoliids, and rhacophorids.

Patterns of reptile distribution relative to Gondwana and Laurussia are more complicated due to the absence of unique groups in either area. True lizards or lacertilians became dichotomous in the Jurassic when the two landmasses began to separate (Estes, 1983). The most primitive group, the iguanids, initially occupied Africa and South America and then dispersed to North American in the late Cretaceous. Iguanids apparently were isolated in Africa by epicontinental seas and gave rise to chamaeleonids in the west and agamids in the east; the scincomorphs radiated in Laurussia, giving rise to the teiids in North America and lacertids in Europe. Snakes probably arose from an ancestral stock within lizard groups during the early Cretaceous of the Southern Hemisphere. Gondwana appears to be the site where the more primitive boids and aniliids evolved; more progressive groups appeared later in North America (Rage, 1987).

The poor fossil record in South America provides only a fragmentary appraisal of the implications for birds of the Gondwanan-Laurussian split. Initially, their origin from theropod ancestors in the Jurassic was based on fossil material in Laurussian continents (Feduccia, 1999). However, recent discoveries of Cretaceous fossils in Madagascar demonstrate that early avian evolution was probably cosmopolitan (Forster et al., 1998). Cracraft (1974) emphasized the Southern Hemispheric distribution of ratites (ostriches, rheas, emus), and their derivation from a single common ancestor in Gondwana during the Cretaceous with subsequent vicariance. Recently, a South America origin with subsequent dispersal and vicariance involving Laurussia, Africa, and/or New Zealand was proposed (van Tuinen et al., 1998; van Tuinen and Hedges, 2001). Other avian groups that may have been influenced by the separation of Gondwana and Laurussia include Galliformes (fowl-like birds including megapodes), and Sphensiformes (penguins; Cracraft, 1974). Recent analyses support the diversification of basal lineages of neornithines (including ratites, galliforms, anseriforms, and more advanced birds such as gruiforms and passeriforms) 90 Ma or later, but prior to the Cretaceous-Tertiary boundary (van Tuinen and Hedges, 2001); a Gondwanan origin for passerines has been proposed (Cracraft, 2001; Ericson et al., 2003).

Among mammals, symmetrodonts and a distinct group of multituberculates were present in South America, thus suggesting the importance of connections with North America (Webb, 1999). The multituberculates were the most successful non-therian group, persisting for over 100 million years from the late Jurassic to the Oligocene. Two primary Cretaceous groups were the Ptilodontoidea in North America and the Taeniolabidoidea in Asia (Carroll, 1988), but multituberculates had a more cosmopolitan distribution, occurring also in the Cretaceous of Argentina and Madagascar (Krause et al., 1992, 1997). Among therian mammals, marsupials and placentals appear to have diverged from a common ancestor in the early or mid-Cretaceous (Lillegraven, 1974; Briggs, 1995; Luo et al., 2001). Earlier views suggested that this occurred in

Laurussian continents (Cracraft, 1974; Lillegraven, 1974), but ancestral therian mammals with tribosphenic molars were also present on Gondwanan continents back to the middle Jurassic and early Cretaceous (Rich et al., 1997; Flynn et al., 1999), suggesting a Gondwanan origin. Molecular evidence suggests a fundamental division for eutherian mammals between African groups and other placentals in the Cretaceous (Eizirik et al., 2001). The recent discovery of a late Cretaceous marsupial in Madagascar emphasizes their broad distribution in Gondwana (Krause, 2001). Finally, a diphyletic origin of mammals with early vicariant evolution recently was proposed; monotremes may represent the remnants of a southern Gondwanan mammal radiation during the Jurassic or early Cretaceous (Luo et al., 2001); however, extinct monotremes have only been verified in South America from the Paleocene (Pascual et al., 1992). Besides the marsupials, other groups that now appear to have been founders of the age of mammals in South America include the xenarthrans, insectivores, pantodonts, and various ungulates, many of which were shared by both North and South America in the early Tertiary (Webb, 1999).

The final separation of Africa and South America in the mid-Cretaceous was an important event. Initially, narrow seas were present; even by the Paleocene, the water barrier was perhaps no more than 600 km wide (Briggs, 1995), and volcanic islands from the Mid-Atlantic ridge probably enabled island-hopping. Recent volumes have focused on relationships between important elements of the faunas of the two continents, including freshwater fish, pipid frogs, reptiles, advanced primates, and hystricognath rodents (see George and Lavocat, 1993; Goldblatt, 1993, and references therein). Following separation, isolation between the two former Gondwanan continents led to sharp faunal divergence. On the other hand, dispersal between South America and Australia is supported by Eocene marsupial fossils in Antarctica (e.g., Woodburne and Case, 1996; Goin et al., 1999). The precursors of the Australian marsupial fauna may have come from South America prior to the separation of Australia and Antarctica around 64 Ma (Goin et al., 1999). South America continued to maintain at least an archipelagic connection to Antarctica until the late Eocene (~35 Ma; Pascual et al., 1992; Pascual, 1996).

In contrast, South America and North America may have been sufficiently close in the late Cretaceous for faunal exchanges to occur. Elements exchanged between the continents then probably included hadrosaurid dinosaurs, condylarth and marsupial mammals from the north, and boid and aniliid snakes, iguanid lizards, titanosaurid dinosaurs, and leptodactylid, bufonid, hylid, and microhylid frogs from the south (Estes, 1983; Rage, 1987, 1988). By late Cretaceous time, South America was certainly isolated from North America, but later, during late Miocene and early Pliocene times, island archipelagos permitted waif dispersal across what is now Central America (Hershkovitz, 1962; Reig, 1984; Briggs, 1995; see chapters 1 and 2).

8.2.2 The Isolation of South America and the Great American Interchange

During the rise of mammals in the Cenozoic, South America was characterized by about 60 million years of "splendid isolation" (Simpson, 1980); a similar situation may apply to suboscine bird radiations there (Ricklefs, 2002). For mammals, this isolation ended around 3 Ma with "The Great American Interchange," although some prior interchange of mammals and other vertebrates had occurred. The first stage of this interchange was in the late Mesozoic to early Cenozoic (Simpson, 1980) and included cosmopolitan mammals such as multituberculates and symmetrodonts (probable immigrants from North America), monotremes (shared with Australia), and xenarthrans, various ungulates, marsupials, insectivores, and pantodonts in South America; the latter two groups were shared between North and South America, and marsupials were cosmopolitan or immigrants from North America (Marshall, 1985; Webb, 1999). The presence of several orders of now-extinct ungulates in South America soon after this exchange in the early Cenozoic suggests a much more extensive faunal affiliation with North America than previously postulated by Simpson (1980). These included condylarths, litopterns, notoungulates, pyrotheres, astrapotheres, and xenungulates; around 17% of the mammalian genera were of North American origin (Webb, 1999). On the other hand, suboscine passerines appear to have evolved independently in South America (and Africa) since the early Cenozoic (Ricklefs, 2002).

The second stage of this interchange may have occurred in the Oligocene around 30 Ma and included dispersal of hystricognath rodents and platyrrhine primates from the north (Wood, 1985; Marshall and Sempere, 1993). Alternatively, either both groups may have originated in southern Gondwanan continents and subsequently dispersed from Africa to South America (Lavocat, 1980, 1993; Aiello, 1993; George, 1993), or vicariance may have occurred in groups present in both South America and Africa (Hershkovitz, 1972). Other groups that may have entered South America from North America then were tortoises and phyllostomid bats (Webb, 1978, 1999); somewhat later (20–25 Ma), the earliest oscine passerines (Ricklefs, 2002; Ericson et al., 2003). Unlike other stages of the interchange, this one was not reciprocated by South American mammals; approximately 29% of the mammalian genera from North America were contributed then (Webb and Marshall, 1982; Marshall and Cifelli, 1990; Webb, 1999). The complex "conveyor belt" movement of the Caribbean plate eastward permitted these "old island hoppers" to disperse into South America (Webb, 1985; Marshall and Sempere, 1993).

Although not recognized by Simpson (1980), a third stage for mammals was heralded by the arrival of "new island hoppers" via waif dispersal in the late Miocene. These included the arrival of procyonids (raccoons) in South America and ground sloths in North America. Adept swimmers, these groups probably crossed between islands characterized by savanna or lowland rain forest; around 3% of the South American mammal genera were received then (Webb and Marshall, 1982; Webb, 1985; Marshall and Cifelli, 1990; Marshall and Sempere, 1993). One particularly successful group that probably arrived in the late Miocene was sigmodontine murid rodents. Although Hershkovitz (1972) and Reig (1984, 1987) argued for a late Oligocene-early Miocene arrival, fossil records fail to show their presence in South America before the Pliocene (Patterson, 1999; Webb, 1999). On the other hand, the degree of diversification and the dominance of more primitive forms in South America do not support a Pliocene arrival (Patterson and Pascual, 1972), and it seem more likely that they arrived by island hopping in the late Miocene (Hershkovitz, 1962; Marshall, 1979; Patterson, 1999).

By far the most significant event in South American mammal history occurred around 3 Ma with the establishment of a permanent land bridge between North and South America. This was facilitated by tectonic closure of marine straits across the Central American Isthmus, and by the presence of savanna and open woodland corridors from Florida to Argentina (Webb, 1978; Webb and Marshall, 1982; Marshall and Sempere, 1993). The corridors were opened and closed several times during the late Pliocene, thus leading to a series of successional, reciprocal immigration pulses (Webb, 1976; Hoffstetter, 1986; Marshall and Sempere, 1993). Major groups from North America involved 17 families of land mammals, including camelids (camels, llamas), cervids (deer), equids (horses), Tapiridae (tapirs), tayassuids (peccaries), canids (dogs, wolves, and foxes), felids (cats), ursids (bears), mustelids (skunks and allies), and gomphotherids (now-extinct mastodonts) (Marshall et al., 1982; Webb, 1999). Fully 51% of the North American contribution to South American mammal genera was made then. An almost equal number of land mammal families extended their ranges into North America from South America including dasypodids (armadillos), Glyptodontidae (glyptodonts), Hydrochoeridae (capybaras), Erethizontidae (porcupines), and later, Toxodontidae (toxodonts), various primates, other hystricognath rodents, Choleopodidae and Bradypodidae (tree sloths), and Myrmecophagidae (anteaters). Most of these families are poorly represented or absent in North America today. The relative magnitude of the interchange is close to that expected from the relative land areas of the two continents and from equilibrium theory (Marshall et al., 1982), but explanations for the greater success of North American mammals in South America (except by groups lacking competitors there) remain controversial (Marshall et al., 1982; Webb, 1985, 1999).

Fully 75% of the South American tropical mammal fauna became extinct at the end of the Pleistocene, including many larger herbivorous and carnivorous mammals (Webb and Rancy, 1996; Martin and Steadman, 1999; see table 8.1). Xeric-adapted species suffered even higher extinction rates; disruption and elimination of arid habitats may have prevented the persistence of refugia in contrast to arid and semiarid zones of North America (Mares, 1985). In addition to climatic change (Kurtén and Anderson, 1980; Graham, 1992; Webb and Barnosky, 1989), other explanations for the reduction in South American mammals include hunting ("Pleistocene overkill"; Martin, 1984; Martin and Steadman, 1999), or other anthropogenic-related causes such as disease, pests, or parasites (MacPhee and Flemming, 1999). Although other groups such as birds (Grayson, 1977) and small mammals were affected (Kurtén and Anderson, 1980), the majority of victims were larger mammals.

The implications of the Great American Interchange for other vertebrate groups were less dramatic. For example, fish dispersal across the Central American Isthmus was exceedingly slow, especially by obligatory freshwater species (Bussing, 1985). Earlier dispersal prior to the separation of North and South America may account for some of the similarities between Central and South American ichthyofaunas. Later dispersal during the Pliocene has been predominantly from south to north; fully 95% of the Central American ichthyofauna is now of South American origin (Bussing, 1985). Only single species of two South American families have reached North America, whereas none from North America have reached South America (Moyle and Chech, 2000). Although the northern Andes in Colombia are currently a formidable barrier to dispersal by obligatory freshwater fish, fossils in the Magdalena region related to those in Guiana-Amazonia indicate that the ichthyofaunas of the two areas had pre-Miocene connections and that they originated from elements prior to the final separation of South America from Africa in the Cretaceous (Lundberg and Chernoff, 1992). Most diversification of the South American ichthyofauna occurred prior to the late Miocene, and Pliocene/Pleistocene events have played virtually no role in the present ichthyofauna at the family and genus group level (Lundberg, 1998). Similarly, most interchange by reptiles and amphibians pre-dated the Pliocene land connection. The herpetofauna of North America has been much more strongly impacted by contributions from South America than the converse. Although Central America has significant contributions of both reptiles and amphibians from North and South America, the major part of its herpetofauna is originally derived from South America (Vanzolini and Heyer, 1985; Cadle and Greene, 1993). Central America has a very high level of endemicity, similar to that for South America (Savage,

Table 8.1 Genera of large (>44 kg) extinct mammals of the late Pleistocene from South America

Order	Family	Genus
Order Xenarthra		
	Family Dasypodidae (armadillos)	
		Eutatus
		Holmesina
		Pampatherium
		Propropus
	Family Glyptodontidae (glyptodonts)	
		Chlamydotherium
		Doedicurus
		Glyptodon
		Hoplophorus
		Lomaphorus
		Neosclerocalyptus
		Neothoracophorus
		Panochthus
		Parapanochthus
		Plaxhaplous
		Sclerocalyptus
	Family Megalonychidae (ground sloths)	
		Nothropus
		Nothrotherius
		Ocnopus
		Valgipes
	Family Megatheriidae (giant ground sloths)	
		Eremotherium
		Megatherium
		Paramegatherium
	Family Mylondontidae (giant ground sloths)	
		Glossotherium
		Lestodon
		Mylodon
		Scelidodon
		Scelidotherium
Order Litopterna (litopterns)		
	Family Macraucheniidae	
		Macrauchenia
		Windhausenia
Order Notoungulata (notoungulates)		
	Family Toxodontidae	
		Mixotoxodon
		Toxodon
Order Rodentia		
	Family Hydrochoeridae (capybaras)	
		Neochoerus
	Family Octodontidae (octodontid rodents)	
		Dicolpomys
Order Carnivora		
	Family Canidae (large canids)	
		Theriodictis
	Family Felidae (saber-tooth cats)	
		Smilodon
	Family Ursidae (giant bear)	
		Arctodus
Order Proboscidea		
	Family Gomphotheriidae (mastodons)	
		Cuvieronius
		Haplomastodon
		Notiomastodon
		Stegomastodon
Order Perissodactyla		
	Family Equiidae (horses)	
		Equus
		Hippidion
		Onohippidium
Order Artiodactyla		
	Family Camelidae (camels and llamas)	
		Eulamaops
		Hemiauchenia
		Palaeolama
	Family Cervidae (deer)	
		Agalmaceros
		Charitoceros
		Morenelaphus
		Paraceros
	Family Tayassuidae (flat-head peccaries)	
		Platygonus

From Martin and Steadman (1999), and Marshall and Cifelli (1990).

1982), and the principal interchange between North and South America has been via Central American radiations of various herpetofaunal elements. Within the colubrid snakes, the dispersal of xenodontines occurred in early to mid-Cenozoic, whereas that of colubrines was more recent but pre-dated the Pliocene (Cadle and Greene, 1993).

For nonpasserine birds, 99% of the genera were restricted to North America and 97% to South America prior to the Pliocene land connection (Vuilleumier, 1985a). Only one genus *(Presbyornis)* seems to have been shared between the two continents. Following establishment of the land bridge, the proportion of genera restricted to North or South America fell to 57%. Inclusion of recent South American genera has reduced the proportion of North American endemics to 37% and that of South America to 47%. Therefore, after the initial oscine passerine invasion in the mid-Miocene, avian faunas appear to have mixed extensively through the late Pliocene and Pleistocene up to the present, unlike mammals, whose interchange was largely completed by 1 Ma (Webb and Marshall, 1982; Vuilleumier, 1985a). Non-tyrannine suboscines continue to dominate forest interiors in the Amazon, whereas oscines and tyrannine suboscines (the only successful suboscine lineage in North America) dominate forest canopies and more open habitats there (Ricklefs, 2002).

8.2.3 *The Role of the Andes in Vertebrate Distributions*

The Andes have been a dominant feature of the South American landscape since the beginning of the Cenozoic (see chapter 1). Although a low elevation proto-Cordillera existed in the late Jurassic and early Cretaceous, the greatest uplift has occurred over the past 30 Ma, particularly in the central Andes (Marshall and Sempere, 1993). In addition to altering river drainage patterns, direct climatic consequences of Andean uplift included, on the one hand, prominent rain shadows to the west between 5° and 35°S, and to the east between 28° and 38°S, and, on the other hand, recurrent corridors of vegetation on the Andean flanks, particularly of savanna biomes (Duellman, 1979b; Marshall and Sempere, 1993). Since 2.5 Ma, glacial stages have alternated with interglacial stages, strongly affecting vegetation. During glacial stages, conditions were cooler and drier on the east side of the Andes, resulting in expansion of the savannas at the expense of lowland rain forests; a savanna corridor permitted north-south dispersal of various faunal elements (Webb, 1978; Marshall and Sempere, 1993). During interglacials, conditions were warm and humid and rain forests expanded, thus isolating savanna areas and eliminating the corridor. In addition to corridors, the Andes have been a significant barrier and isolating mechanism for vertebrates. Glaciers, lakes and bogs, and zones of continuous aridity across the Cordillera have operated at various times, especially over the past 2.5 Ma, to prevent east-west or north-south dispersal (Simpson, 1979; Marshall and Sempere, 1993). Finally, the Andes have been an important factor in speciation within South America. The presence of a "stress environment" as a result of the rain shadow on the eastern flank of the southern Andes (south of 30°S), beginning around 15 Ma, appears to have strongly favored species origination by creating opportunities for local differentiation and isolation. Similarly, elevational zonation of habitats on both sides of the Andes has provided opportunities for local differentiation and specialization in response to increased ecological diversity (Marshall and Sempere, 1993).

Various groups demonstrate the importance of the Andes as a center for speciation. Although the Andean ichthyofauna is generally depauperate, even at equatorial latitudes, and dominated by torrent fishes between 1,000 and >4,000 m (Géry, 1969), the spectacular radiation of 23 species of pupfishes (*Orestias*) in Lake Titicaca at 4,750 m in the central Altiplano of Perú and Bolivia is remarkable (Moyle and Cech, 2000). *Orestias* appear to have undergone parasympatric speciation through ecological and perhaps microgeographic isolation (Villwock, 1986). Several other Andean genera are endemic but tend to be small, monotypic groups with uncertain taxonomic affinities (Bănărescu, 1995). All 753 species of amphibians found in the 27 regions of the Andes distinguished by Duellman (1999b) are anurans. More than half of South American genera (47/85) are represented, and 95.0% of the species are endemic. Only 20% of the Andean species have their major distributions in adjacent areas (Duellman, 1979b). One group of Andean frogs, *Telmatobius*, constitute about 30 species occupying lakes and streams in the highlands above 1,300 m between the equator and 29–33°S, and may be derived from southern ancestors in the recently uplifted Andes (Cei, 1986). Another anuran genus, *Eleutherodactylus* (>500 species), diversified extensively in response to repeated periods of climatic compression on the western side of the Andes in Ecuador (Lynch and Duellman, 1997). A similar process was suggested to explain areas of high biotic diversity in the Amazon Basin (Colinvaux, 1993, 1996).

Reig (1984, 1986, 1987) and Pearson (1982) supported the Andes as a major site of sigmodontine rodent speciation. More primitive sigmodontines may have become established in the northern Andes of Colombia as early as the late Oligocene or early Miocene with a second diversification center in the central Andes where present-day Perú, Chile, and Bolivia meet. Descendants of these groups then invaded the eastern lowlands; 71.6% of the sigmodontines were distributed in Andean localities (Reig, 1987). Pearson (1982) noted a high diversity of sigmodontines in the central Altiplano, and distinguished the radiation and subsequent dispersal of autochothonous forms from invasions of northern immigrants via the Central American land bridge. The oldest sigmodontines are specialized pastoral forms; more primitive tribes predominate in forested lowlands, and they may have dispersed to higher latitudes and completed their transition to pastoral life in the arid regions of the south, where they continue to make up the majority of sigmodontines (Hershkovitz, 1962; Patterson, 1999). Finally, the Andes provided a dispersal route for reinvasion of the north, and to the west of most lowland relatives. The dominance of derived, austral relatives in higher-elevational small mammal assemblages with more primitive forms in lowland rain forests continues today (Patterson, 1999).

The role of the Andes as a dispersal route is shown by savanna corridors following the establishment of the Pliocene land connection to North America; temperate-adapted mammals such as camelids and sloths probably used the Andes as a route for traversing the humid tropics (Webb and Rancy, 1996). During intervening warmer interglacial stages when the savanna corridors were replaced by rain forest, the Andes may have acted as a barrier to temperate savanna forms and hence promoted provincialism in Cis- and Trans-Andean faunas among various groups of sloths, ungulates, and glyptodonts (Webb and Rancy, 1996). Although recent evidence suggests that a savanna corridor may not have existed in the central Amazon basin (that is, tropical dry forests replaced rain forest during Quaternary climatic cycles), significant savanna corridors did exist along the Andes and on the Atlantic coast (da Silva and Bates, 2002). The Andes have also had strong effects on mammalian species ranges; the most restricted geographic ranges are in the equatorial and central portions of South America rather than in the south, where elevations are lower (Ruggiero et al., 1998). The Andes have been important in the location of major dispersal centers for terrestrial vertebrates in South America (Müller, 1973).

For birds, the effects of isolation have been analyzed using island biogeographic theory and historical evidence for the Páramo and Puna regions (e.g., Vuilleumier, 1969, 1970, 1986; Vuilleumier and Simberloff, 1980). The species diversity of birds in Páramo habitat islands of the northern Andes shows a strong positive relationship with area, although there is no relationship between endemism and area. For Páramo and Puna regions as a whole, only 166/319 (52%) species may be considered "true" Páramo/Puna birds. Of these, 248 (78%) species breed in the region whereas 71 (22%) breed elsewhere, or are accidentals/migratory from North America or southern South America. Only 48 species (19–29%) are endemic as compared with higher levels in the Amazon; affinities of the Páramo and Puna avifauna are greatest with the Patagonian region. The origins of this avifauna appear to be in the late Pliocene and Pleistocene when alternating interglacial and glacial stages were conducive to geographic isolation and speciation, respectively, in the Páramo and Puna (Vuilleumier, 1969, 1970, 1986; Vuilleumier and Simberloff, 1980).

Nowhere are the effects of isolation and geography more evident in South America than in the southern triangle (or "cone"), including central to southern Chile and Argentina. This is reflected by faunal comparisons at several spatial scales. For example, Chile stretches 4,300 km from 18° to 56°S and contains all major Neotropical biomes, except tropical forests, in an area of 483,000 km^2. Isolated to the east by the Andean crest reaching >6,000 m, and to the north by the hyperarid Pacific Coastal Desert, continental Chile has only 26 species of primary freshwater fish (Géry, 1969), 42 species of amphibians (all anurans), and 84 species of reptiles (Jaksic, 1998). There are around 380 bird species (excluding accidentals), and 101 species of nonmarine mammals (Jaksic, 1998). Caviedes and Iriarte (1989) argued that the low diversity of small mammals in central and northern Chile is due largely to its isolation from the rest of South America and a historical trend toward increasing aridity (but see Marquet, 1989; Meserve and Kelt, 1990). However, small-mammal as well as avian diversity increases at intermediate elevations in the adjacent Andean regions of northern Chile and southern Perú (Pearson and Ralph, 1978; Marquet, 1994).

Southern Patagonia is depauperate for virtually all vertebrate groups due to increasing aridity created by the Andean rainshadow, and species-area plus peninsular effects as South America narrows to the south (Eisenberg, 1999b). Specific examples of this depauperization include the southern temperate rain forests of Argentina and Chile, now isolated to the north by 1,100–1,400 km of nonforest vegetation. Contrasts with herpetofaunas in the northern coniferous and deciduous forests in the Pacific Northwest of the United States show that diversity is significantly lower in southern temperate rain forests due largely to the absence of salamanders and snakes; these make up the majority of amphibians and reptiles, respectively, in North American forests (Meserve and Jaksic, 1991; Meserve, 1996); endemicity is also high (Formas, 1979).

Southern temperate rain forests have a somewhat impoverished avifauna with a high level of species endemicity (41%; Vuilleumier, 1985b); migrants constitute a smaller proportion of the avifauna although site diversity and average numbers of individuals/census point are similar to or higher than those in Alaskan coastal forests (Jaksic and Feinsinger, 1991; Willson et al., 1996). Small mammal species numbers are significantly lower in southern temperate rain forests, mostly because of the absence of insectivores and sciurids; however, endemicity is not especially high among mammals (Mares, 1992; Meserve and Glanz, 1978, Meserve and Jaksic, 1991; Meserve, 1996).

8.2.4 Diversity of the Amazon Basin

As noted earlier, the highest diversity for some vertebrate groups may not lie in the lowland rain forests of the Amazon basin (e.g., birds: Rahbek and Graves, 2001; mammals: Mares, 1992). Yet this area continues to be the focus of much attention regarding putative explanations for organismal diversity in general. Perhaps the most prominent early hypothesis involves putative Pleistocene refugia that may have harbored high numbers of lowland tropical species during glacial stages when conditions were likely to have been more arid (Haffer, 1969, 1985, 2001; but see Colinvaux, 1993, 1996). Initial support came from some vertebrate distributions (e.g., birds: Haffer, 1969; amphibians: Lynch, 1979; Duellman, 1982; reptiles: Dixon, 1979), as well as other groups (e.g., plants, Simpson and Haffer, 1978). Although a more recent analysis suggests strong congruence in area cladograms for various vertebrate groups in the lowland tropical forests, such cladograms are dichotomous and therefore inconsistent with simultaneous isolations of biota in putative Pleistocene refugia. Using the centers of endemism proposed by Haffer (1985), anurans, lizards, passerines, and primates show several general patterns: (1) a basal separation between faunas of Cis- and Trans-Andean regions; (2) a distinct Central American clade; (3) a sister relationship between the Chocó Region and the Central American clade; (4) an Upper Amazon clade (Napo + Inambari); (5) a Guianan clade; (6) an Amazon basin clade; and (7) a sister relationship between the Atlantic Forest and the Amazon basin clade (Bates et al., 1998; Ron, 2000; Bates, 2001). A further problem with the refugia hypothesis is that, based on the age of endemic vertebrates and rates of molecular evolution, speciation occurred over a much longer time than just the past 2 to 3 million years (Cracraft and Prum, 1988; Lynch, 1988; Bush, 1994; Patton and da Silva, 2001). Putative Pleistocene refugia cannot explain current biogeographic patterns of fish diversity, since virtually all the diversification of the South American ichthyofauna occurred before the Pliocene (Lundberg, 1998). The uplift of the Andes has been an important vicariant event for speciation in various characiform fish families (Vari and Weitzman, 1990). Some terrestrial taxa in putative refugia are predominantly savanna forms (i.e., mammals; Webb and Rancy, 1996), or have adaptations for nonforest environments (i.e., amphibians; Duellman, 1982; Heyer and Maxson, 1982). Alternative explanations for Amazonian diversity include paleogeographic hypotheses with marine incursions or embayments leading to isolation of parts of Amazonia and subsequent vicariance (Bates et al., 1998; Bates, 2001; Patton and da Silva, 2001; Nores, 2004), riverine or river-refuge hypotheses with rivers acting as barriers to dispersal and gene flow (Capparella, 1991; Ayres and Clutton-Brock, 1992; Hayes and Sewlal, 2004; but see Patton et al., 1994; Gascon et al., 2000; Patton and da Silva, 2001), gradient hypotheses with parapatric speciation across steep environmental gradients (Endler, 1982; but see Patton et al., 1990; Patton and Smith, 1992), and disturbance-vicariance hypotheses with climatically unstable, peripheral regions of the Amazon sites of more allopatric speciation and greater endemicity (Bush, 1994). Within the latter explanation, Colinvaux (1993, 1996) proposed a scenario of cooler but not necessarily more arid climatic conditions during glacial stages, resulting in treeline lowering and restructuring of lowland forests by invasions of Andean taxa and their mixing with more thermophylic rain forest species. Putative "refugia" were areas of intense species interactions resulting in more speciation and endemism, thus explaining high diversity in the Amazon basin. Others have supported the role of ecological factors while cautioning against uncritical acceptance of historical explanations (e.g., Tuomisto and Ruokolainen, 1997). Hence, there is a plethora of putative explanations for the extraordinary diversity in the Amazon Basin, but no clear consensus.

8.3 Conclusions and Prognosis

Despite the Pliocene and Pleistocene extinctions, South America remains an extraordinarily diverse continent zoologically, especially for vertebrates. However, with current trends, this condition is unlikely to continue. Tropical forests cover only 6 to 7% of Earth's land surface, yet contain at least 50% of the world's species (Myers, 1988). South America holds the largest remaining blocks of intact tropical forests in the world, particularly in Amazonia; although only an estimated 2 to 12% of Amazonia had been deforested by 1992, the estimated rate of clearing was about 0.4 to 2.2% a year in the 1980s(Prance, 1993; Jeffries, 1997). The amount of tropical moist forest in some category of protected status ranges from less than 1% in the largely forested Guyanas to 15% in Venezuela (van Schaik et al., 1997).Yet in reality, most such areas are poorly protected, and virtually all are threatened by multiple human impacts including hunting, logging, grazing, mining, and fire. Further, fragmentation of the rain forest has dramatic effects on all elements of the biota by disrupting food webs, altering spatial utilization patterns, and increasing edge effects on patches of isolated forest (Lovejoy et al., 1986; Saunders et al., 1991; Laurance and Bierregaard, 1997). By 2010, it is estimated that the only large blocks of undisturbed tropical forest in South America will be in the western and northern Brazilian Amazon and the interior of the Guyanas (Raven, 1988). Amazonia is not homogeneous, but rather highly heterogeneous due to variations in drainage, soil type, and history (Foster, 1980); in addition, there are geographic regions of higher species diversity within it. Thus, the task of preserving its biodiversity will be even more difficult than simply setting aside large areas.

Recently, Myers et al. (2000) identified four biodiversity "hotspots" in South America, all outside Amazonia. Three are at tropical latitudes, namely Chocó-Darién–western Ecuador, Atlantic Coastal Forests, and Tropical Andes. The Chocó-Darién tropical forests on the west side of the Andes have only 24.2% of their original area remaining, and the isolated Brazilian Atlantic coastal forest has only 6.5 to 7.5% undisturbed (Jeffries, 1997; Myers et al., 2000). For both these forests, only 26.1 to 35.9% of the remaining area is under protected status. An estimated 3.6% of the world's endemic vertebrates are found in just these two types of forests. For Atlantic Coastal Forests, 80% of the primates are endemic to this region, and 14 of 21 species and subspecies are endangered or on the verge of extinction (Mittermeier, 1988). Another major biodiversity "hotspot" is the Tropical Andes, including submontane, montane, and cloud forests, páramo, and puna, from the central Altiplano of Perú and Bolivia to the northern Andes of Colombia and Venezuela. This area has the highest level of plant and vertebrate diversity, and endemicity, of all 25 areas compared worldwide; an estimated 45,000 species of plants and 3,389 species of vertebrates (excluding fish) are present here, and levels of endemicity are 6.7% and 5.7%, respectively. Only 25.3% of its remaining area is under some degree of protection. Five South American countries are among the top ten in the world with the highest density of endemic bird areas (Perú, Brazil, Colombia, Ecuador, and Argentina; Bibby et al., 1992). Another "hotspot" in the top 25 is the Cerrado, which is the second largest South American biome and contains the largest tropical savanna in the world. Only 20% of its area remains undisturbed and only 1.2% is in protected areas (da Silva and Bates, 2002). Although the one temperate zone "hotspot" in the top four, the central Mediterranean-type region of Chile, has lower floral and vertebrate diversity and endemism, only 10.2% of its area is under some degree of protection.

Projections of the potential loss in species are problematic. Although historical extinctions of birds and mammals in the United States have been mainly due to hunting/exploitation or species introduction, current causes of endangerment are predominantly due to habitat destruction (Flather et al., 1994). Estimates of losses of species in South America range from 2 to 25% in the next 25 years (Jeffries, 1997; Sodhi et al., 2004). Exacerbating this situation is that most South American countries continue to experience high population growth rates, increasing demand for natural resources, and strong pressure to accelerate industrial development. This in turn limits their ability to allocate significant efforts and funds to the preservation and protection of their biodiversity. Although many South American countries appear underpopulated, they often contain large areas of land unsuitable for dense populations such as in the Atacama and Monte Deserts, Patagonia, the Paraguayan Chaco, and the Altiplano, and thus human populations are clustered around urbanized areas. Finally, areas of particularly high biodiversity frequently span national boundaries, thus complicating protection efforts. Two biodiversity "hotspots" in South America, the Chocó-Darién–western Ecuadorian tropical forests and the Tropical Andes, are in areas where the annual population growth rates are 2.8 to 3.2% a year, well over the world's average growth rate of around 1.3% a year (Cincotta et al., 2000). These "hotspots" cross the territorial boundaries of 4 and 7 countries, respectively (Myers et al., 2000).

Given this situation, the prognosis for South America's biota is grim. Without quick action, we stand to lose a significant part of the world's biodiversity within our lifetime, and much of it still inadequately described. Unfortunately, there are tremendous obstacles to ensuring the preservation of South America's fauna and flora for future generations.

References

Ab'Saber, A.N., 1977. Os domínios morfoclimáticos na América do sul. Primeira Aproximacao. *Geomorfologia*, 53, 1–23.

Aiello, L.C., 1993. The origin of New World monkeys. In: W. George and R. Lavocat (Editors), *The Africa-South America Connection*. Clarendon Press, Oxford, 100–118.

Albert, J.S., and R. Campos-da-Paz, 1998. Phylogenetic systematics of Gymnotiformes with diagnoses of 58 clades: A review of the available data. In: L.R. Malabarba, R.E. Reis, R.P. Vari, Z.M.S. Lucena, and C.A.S. Lucena (Editors.), *Phylogeny and Classification of Neotropical Fishes*. EDIPUCRS, Porto Alegre, 419–446.

Ayres, J.M.C., and T.H. Clutton-Brock, 1992. River boundaries and species range size in Amazonian primates. *American Naturalist*, 140, 531–537.

Bakker, V.J., and D.A. Kelt, 2000. Scale-dependent patterns in body size distributions of Neotropical mammals. *Ecology*, 81, 3530–3547.

Bănărescu, P. 1990. *Zoogeography of Fresh Waters. 1: General Distribution and Dispersal of Freshwater Animals*. AULA-Verlag, Wiesbaden, Germany.

Bănărescu, P. 1995. *Zoogeography of Fresh Waters*. 3: Distribution and Dispersal of Freshwater Animals in Africa, Pacific Areas and South America. AULA-Verlag, Wiesbaden, Germany.

Bates, J.M., 2001. Avian diversification in Amazonia: Evidence for historical complexity and a vicariance model for a basic diversification pattern. In: I.C.G. Vieira, J.M.C. da Silva, D.C. Oren, and M.A. D'Incao (Editors), *Biological and Cultural Diversity of Amazonia*. Museu Paraense Emílio Goeldi, Belém, 119–137.

Bates, J.M., S.J. Hackett, and J. Cracraft, 1998. Area-relationships in the Neotropical lowlands: An hypothesis based on raw distributions of Passerine birds. *Journal of Biogeography*, 25, 783–793.

Bauer, A.M., 1993. African-South American relationships: a perspective from the Reptilia. In: P. Goldblatt (Editor), *Biological Relationships between Africa and South America*. Yale University Press, New Haven, 244–288.

Benton, M.J., 1993. Late Triassic extinctions and the origin of the dinosaurs. *Science*, 260, 769–770.

Bibby, C.J., N.J. Collar, M.J. Crosby, M.F. Heath, C. Imboden, T.H. Johnson, A.J. Long, A.J. Stattersfield, and S.J. Thirgood. 1992. *Putting Biodiversity on the Map: Priority Areas for Global Conservation*. International Council for Bird Preservation, Cambridge, U.K.

Briggs, J.C., 1995. *Global Biogeography*. Elsevier, Amsterdam.

Buckup, P.A., 1998. Relationships of the Characidiinae and phylogeny of characiform fishes (Teleostei: Ostariophysi). In: L.R. Malabarba, R.E. Reis, and R.P. Vari, Z.M.S. Lucena, and C.A.S. Lucena (Editors), *Phylogeny and Classification of Neotropical Fishes*. EDIPUCRS, Porto Alegre, 123–144.

Bush, M.B., 1994. Amazonian speciation: A necessarily complex model. *Journal of Biogeography*, 21, 5–17.

Bussing, W.A., 1985. Patterns of distribution of the Central American ichthyofauna. In: F.G. Stehli and S.D. Webb (Editors), *The Great American Biotic Interchange*. Plenum Press, New York, 453–473.

Cadle, J.E., and H.W. Greene, 1993. Phylogenetic patterns, biogeography, and the ecological structure of Neotropical snake assemblages. In: R.E. Ricklefs and D. Schluter (Editors), *Species Diversity in Ecological Communities. Historical and Geographic Perspectives*. University of Chicago Press, Chicago, 281–293.

Capparella, A., 1991. Neotropical avian diversity and riverine barriers. Congress of International Ornithology, 10. *Acta*, 1, 307–316.

Carroll, R.L., 1988. *Vertebrate Paleontology and Evolution*. W.H. Freeman, New York.

Caviedes, C.N., and A.W. Iriarte, 1989. Migration and distribution of rodents in central Chile since the Pleistocene: The paleogeographic evidence. *Journal of Biogeography*, 16, 181–187.

Ceballos, G., and D. Navarro L. 1991. Diversity and conservation of Mexican mammals. In: M.A. Mares and D.J. Schmidly (Editors), *Latin American Mammalogy: History, Biodiversity, and Conservation.* University of Oklahoma Press, Norman, 167–198.

Cei, J.M., 1979. The Patagonian herpetofauna. In: W.E. Duellman (Editor), *The South American Herpetofauna: Its Origin, Evolution, and Dispersal.* Monograph 7, Museum of Natural History, University of Kansas, Lawrence, 309–339.

Cei, J.M., 1986. Speciation and adaptive radiation in Andean *Telmatobius* frogs. In: F. Vuilleumier and M. Monasterio (Editors), *High Altitude Tropical Biogeography.* Oxford University Press, New York, 374–386.

Chesser, R.T., and S.J. Hackett, 1992. Mammalian diversity in South America. *Science,* 256, 1502–1504.

Cincotta, R.P., J. Wisnewski, and R. Engelman, 2000. Human population in the biodiversity hotspots. *Nature,* 404, 990–992.

Cole, F.R., D.M. Reeder, and D.E. Wilson, 1994. A synopsis of distributional patterns and the conservation of mammal species. *Journal of Mammalogy,* 75, 266–276.

Colinvaux, P. 1993. Pleistocene biogeography and diversity in tropical forests in South America. In: P. Goldblatt (Editor), *Biological Relationships between Africa and South America.* Yale University Press, New Haven, 473–499.

Colinvaux, P. 1996. Quaternary environmental history and forest diversity in the Neotropics. In: J.B.C. Jackson, A.F. Budd, and A.G. Coates (Editors), *Evolution and Environment in Tropical America.* University of Chicago Press, Chicago, 359–405.

Cox, C.B., 1974. Vertebrate paleodistribution patterns and continental drift. *Journal of Biogeography,* 1, 75–94.

Cracraft, J., 1974. Continental drift and vertebrate distribution. *Annual Review of Ecology and Systematics,* 5, 215–261.

Cracraft, J., 1985. Historical biogeography and patterns of differentiation within the South American avifauna: Areas of endemism. In: P.A. Buckley, M.S. Foster, E.S. Morton, R.S. Ridgely, and F.G. Buckley, editors. *Neotropical Ornithology.* Ornithological Monographs, 36. American Ornithological Union, Washington, D.C., 49–84.

Cracraft, J., 2001. Avian evolution, Gondwana biogeography and the Cretaceous-Tertiary mass extinction event. *Proceedings of the Royal Society of London,* B 268, 459–469.

Cracraft, J., and R.O. Prum, 1988. Patterns and processes of diversification: Speciation and historical congruence in some Neotropical birds. *Evolution,* 42, 603–620.

Crompton, A.W., and F.A. Jenkins, Jr., 1979. Origin of the mammals. In: J.A. Lillegraven, Z. Kielan-Jaworowska, and W.A. Clemens (Editors), *Mesozoic Mammals: The First Two-Thirds of Mammalian History.* University of California Press, Berkeley, 59–73.

DeGraaf, R.M., and J.H. Rappole, 1995. Neotropical Migratory Birds: Natural History, Distribution, and Population Change. Comstock Publishing/Cornell University Press, Ithaca.

Dixon, J.R., 1979. Origin and distribution of reptiles in lowland tropical rainforests of South America. In: W.E. Duellman (Editor), *The South American Herpetofauna: Its Origin, Evolution, and Dispersal.* Monograph 7. Museum of Natural History, University of Kansas, Lawrence, 217–240.

Duellman, W.E., 1979a. The South American herpetofauna: A panoramic view. In: W.E. Duellman (Editor), *The South American Herpetofauna: Its Origin, Evolution, and Dispersal.* Monograph 7. Museum of Natural History, University of Kansas, Lawrence, 1–28.

Duellman, W.E., 1979b. The herpetofauna of the Andes: Patterns of distribution, origin, differentiation, and present communities. In: W.E. Duellman (Editor), *The South American Herpetofauna: Its origin, Evolution, and Dispersal.* Monograph 7. Museum of Natural History, University of Kansas, Lawrence, 371–464.

Duellman, W.E., 1982. Quaternary climatic ecological fluctuations in the lowland tropics: frogs and forests. In: G.T. Prance (Editor), *Biological Differentiation in the Tropics.* Columbia University Press, New York, 389–402.

Duellman, W.E., (Editor), 1999a. *Patterns of Distribution of Amphibians.* Johns Hopkins University Press, Baltimore.

Duellman, W.E., 1999b. Distribution patterns of amphibians in South America. In: W.E. Duellman (Editor), *Patterns of Distribution of Amphibians.* Johns Hopkins University Press, Baltimore, 255–328.

Eisenberg, J.F., 1999a. The contemporary mammalian fauna of South America. In: J.F. Eisenberg and K.H. Redford (Editors), *Mammals of the Neotropics. 3: The Central Neotropics: Ecuador, Peru, Bolivia, Brazil.* University of Chicago Press, Chicago, 582–591.

Eisenberg, J.F., 1999b. Biodiversity reconsidered. In: J.F. Eisenberg and K.H. Redford (Editors), *Mammals of the Neotropics. 3: The Central Neotropics: Ecuador, Peru, Bolivia, Brazil.* University of Chicago Press, Chicago, 527–548.

Eizirik, E., W.J. Murphy, and S.J. O'Brien, 2001. Molecular dating and biogeography of the early placental mammal radiation. *Journal of Heredity,* 92, 212–219.

Endler, J.A., 1982. Pleistocene forest refuges: Fact or fancy? In: G.T. Prance (Editor), *Biological Differentiation in the Tropics.* Columbia University Press, New York, 641–657.

Ericson, P.G.P., M. Irestedt, and U.S. Johansson. 2003. Evolution, biogeography, and patterns of diversification in passerine birds. *Journal of Avian Biology,* 34:3–15.

Eschmeyer, W.N., (Editor), 1998. *Catalogue of Fishes.* 3 volumes. California Academy of Sciences, San Francisco.

Estes, R., 1983. The fossil record and early distribution of lizards. In: A.G.J. Rodin and K. Miyata (Editors), *Advances in Herpetology and Evolutionary Biology.* Museum of Comparative Zoology, Harvard University, Cambridge, 365–398.

Farias, I.P., G. Orti, I. Sampaio, H. Schneider, and A. Meyer, 1999. Mitochondrial DNA phylogeny of the family Cichlidae: Monophyly and fast molecular evolution of the Neotropical assemblage. *Journal of Molecular Evolution,* 48, 703–711.

Feduccia, A., 1999. *The Origin and Evolution of Birds.* 2nd Edition. Yale University Press, New Haven.

Fittkau, E.J., J. Illies, H. Klinge, G.H. Schwabe, and H. Sioli (Editors), 1969a. *Biogeography and Ecology in South America, 1.* Monographie Biologicae, 18, Dr. W. Junk, The Hague.

Fittkau, E.J., J. Illies, H. Klinge, G.H. Schwabe, and H. Sioli (Editors), 1969b. *Biogeography and Ecology in South*

America, 2. Monographie Biologicae, 19. Dr. W. Junk, The Hague.

Flather, C.H., L.A. Joyce, and C.A. Bloomgarden, 1994. *Species Endangerment Patterns in the United States.* U.S.D.A. Forest Service, General Technical Report, RM-241.

Flynn, J.J., J.M. Parrish, B. Rakotosaminanana, W.F. Simpson, and A.R. Wyss, 1999. A Middle Jurassic mammal from Madagascar. *Nature,* 401, 57–60.

Fonseca, G.A.B. da, G. Herrmann, and Y.L.R. Leite, 1999. Macrogeography of Brazilian mammals. In: J.F. Eisenberg and K.H. Redford (Editors), *Mammals of the Neotropics. 3: The Central Neotropics: Ecuador, Peru, Bolivia, Brazil.* University of Chicago Press, Chicago, 549–563.

Formas, J.R., 1979. La herpetofauna de los bosques templados de Sudamérica. In: W.E. Duellman (Editor), *The South American Herpetofauna: Its Origin, Evolution, and Dispersal.* Monograph 7. Museum of Natural History, University of Kansas, Lawrence, 341–369.

Forster, C.A., S.D. Sampson, L.M. Chiappe, and D.W. Krause, 1998. The theropod ancestry of birds: New evidence from the late Cretaceous of Madagascar. *Science,* 279, 1915–1919.

Foster, R.B., 1980. Heterogeneity and disturbance in tropical vegetation. In: M.E. Soulé and B.A. Wilcox (Editors), *Conservation Biology: An Evolutionary-Ecological Perspective.* Sinauer Associates, Sunderland, Massachusetts, 75–92.

Gascon, C., J.R. Malcolm, J.L. Patton, M.N.F. da Silva, J.P. Bogart, S.C. Loughheed, C.A. Peres, S. Nickels, and P.T. Boag, 2000. Riverine barriers and the geographic distribution of Amazonian species. *Proceedings of the National Academy of Sciences,* 97, 13672–13677.

George, W., 1993. The strange rodents of Africa and South America. In: W. George and R. Lavocat (Editors), *The Africa-South America Connection.* Clarendon Press, Oxford, 119–141.

George, W., and R. Lavocat (Editors), 1993. *The African-South America Connection.* Oxford Monographs on Biogeography, 7. Clarendon Press, Oxford,

Géry, J. 1969. The fresh-water fishes of South America. In: E.J. Fittkau, J. Illies, H. Klinge, G.H. Schwabe, and H. Sioli (Editors), *Biogeography and Ecology in South America, 2.* Monographie Biologicae, 19. Dr. W. Junk, The Hague, 824–848.

Goin, F.J., J.A. Case, M.O. Woodburne, S.F. Vizcaíno, and M.A. Reguero, 1999. New discoveries of 'oppossum-like' marsupials from Antarctica (Seymour Island, Medial Eocene). *Journal of Mammalian Evolution,* 6, 335–365.

Goldblatt, P. (Editor), 1993. *Biological Relationships between Africa and South America.* Yale University Press, New Haven.

Graham, R.W. 1992. Late Pleistocene faunal changes as a guide to understanding effects of greenhouse warming on the mammalian fauna of North America. In: R.L. Peters and T.E. Lovejoy (Editors), *Global Warming and Biological Diversity.* Yale University Press, New Haven, 76–87.

Grayson, D.K., 1977. Pleistocene avifaunas and the overkill hypothesis. *Science,* 195, 691–692.

Haffer, J., 1969. Speciation in Amazonian forest birds. *Science,* 165, 131–137.

Haffer, J. 1985. Avian zoogeography of the Neotropical lowlands. In: P.A. Buckley, M.S. Foster, E.S. Morton, R.S. Ridgely, and F.G. Buckley (Editors), *Neotropical Ornithology.* Ornithological Monograph 36. American Ornithological Union, Washington, D.C., 113–146.

Haffer, J., 2001. Hypotheses to explain the origin of species in Amazonia. In: I.C.G. Vieira, J.M.C. da Silva, D.C. Oren, and M.A. D'Incao (Editors). *Biological and Cultural Diversity of Amazonia.* Museu Paraense Emílio Goeldi, Belém, 45–118.

Hayes, F.E., and J-A.N. Sewlal, 2004. The Amazon River as a dispersal barrier to passerine birds: Effects of river width, habitat and taxonomy. *Journal of Biogeography,* 31, 1809–1818.

Helfman, G.S., B.B. Collette, and D.E. Facey, 1997. *The Diversity of Fishes.* Blackwell Science, Malden, Massachusetts.

Hershkovitz, P., 1958. A geographic classification of Neotropical mammals. *Fieldiana: Zoology,* 36, 581–620.

Hershkovitz, P., 1962. Evolution of Neotropical cricetine rodents (Muridae), with special reference to the phyllotine group. *Fieldiana: Zoology,* 46, 1–524.

Hershkovitz, P., 1972. The recent mammals of the Neotropical Region: Azoogeographic and ecological review. In: A. Keast, F.C. Erk, and B. Glass (Editors), *Evolution, Mammals, and Southern Continents.* State University of New York Press, Albany, 311–341.

Heyer, W.R., and L.R. Maxson. 1982. Distributions, relationships, and zoogeography of lowland frogs/the *Leptodactylus* complex in South America, with special reference to South America. In: G.T. Prance (Editor), *Biological Differentiation in the Tropics.* Columbia University Press, New York, 375–388.

Hoffstetter, R., 1986. High Andean mammalian faunas during the Plio-Pleistocene. In: F. Vuilleumier and M. Monasterio (Editors), *High Altitude Tropical Biogeography.* Oxford University Press, New York, 218–245.

Hueck, K. 1972. *Las Florestas da América do Sul.* Editora Polígono, São Paulo.

Hueck, K., and P. Seibert, 1972. *Vegetationskarte von Süd Amerika.* Gustav Fischer Verlag, Stuttgart, Germany.

Jaksic, F.M. ,1998. *Ecología de los Vertebrados de Chile.* Second edition. Ediciones Universidad Católica de Chile, Santiago.

Jaksic, F.M., and P. Feinsinger, 1991. Bird assemblages in temperate forests of North and South America: A comparison of diversity, dynamics, guild structure, and resource use. *Revista Chilena de Historia Natural,* 64, 491–10.

Jeffries, M.J., 1997. *Biodiversity and Conservation.* Routledge, London.

Kessler, M., S.K. Herzog, J. Fjeldså, and K. Bach, 2001. Species richness and endemism of plant and bird communities along two gradients of elevation, humidity and land use in the Bolivian Andes. *Diversity and Distribution,* 7, 61–77.

Köhler, J., D.R. Vieites, R.M. Bonett, F.H. García, F. Glaw, D. Steinke, and M. Vences, 2005. New amphibians and global conservation: A boost in species discoveries in a highly endangered vertebrate group. *BioScience,* 55, 693–696.

Krause, D.W., 2001. Fossil molar from a Madagascan marsupial. *Nature,* 412, 497–498.

Krause, D.W., Z. Kielan-Jaworoswska, and J.F. Bonaparte, 1992. *Ferugliotherium* Bonaparte, the first known multituberculate from South America. *Journal of Vertebrate Paleontology,* 12, 351–376.

Krause, D.W., G.V.R. Prasad, W. von Koenigswald, A.

Sahni, and F.E. Grine, 1997. Cosmopolitanism among Gondwanan Late Cretaceous mammals. *Nature*, 390, 504–507.

Kullander, S.O., 1998. A phylogeny and classification of the South American Cichlidae (Teleostei: Perciformes). In: L.R. Malabarba, R.E. Reis, R.P. Vari, Z.M.S. Lucena, and C.A.S. Lucena (Editors), *Phylogeny and Classification of Neotropical Fishes.* EDIPUCRS, Porto Alegre, 461–498.

Kurtén, B., and E. Anderson, 1980. *Pleistocene Mammals of the North America.* Columbia University Press, New York.

Laurance, W.F., and Bierregaard, R.O., Jr. (Editors), 1997. *Tropical Forest Remnants: Ecology, Management, and Conservation of Fragmented Communities.* University of Chicago Press, Chicago.

Lavocat, R., 1980. The implications of rodent paleontology and biogeography to the geographical sources and origins of the platyrrhine primates. In: R.L. Ciochon and A.B. Chiarelli (Editors), *Evolutionary Biology of the New World Monkeys and Continental Drift.* Plenum Press, New York, 93–102.

Lavocat, R., 1993. Conclusions. In: W. George and R. Lavocat (Editors), *The Africa-South America Connection.* Clarendon Press, Oxford, 142–150.

Lillegraven, J.A., 1974. Biogeographical considerations of the marsupial-placental dichotomy. *Annual Review of Ecology and Systematics,* 5, 263–283.

Lovejoy, T.E., R.O. Bierregaard, Jr., A.B. Rylands, J.R. Malcolm, C.E. Quintiles, L.H. Harper, K.S. Brown, Jr., A.H. Powell, G.V.N. Powell, H.O.R. Schubart, and M.B. Hays, 1986. Edge and other effects of isolation on Amazon forest fragments. In: M.A. Soulé (Editor), *Conservation Biology: The Science of Scarcity and Diversity.* Sinauer Associates, Sunderland, Massachusetts, 257–285.

Lundberg, J.G., 1993. African-South American freshwater clades and continental drift: Problems with a paradigm. In: P. Goldblatt (Editor), *Biological Relationships between Africa and South America.* Yale University Press, New Haven, 156–198.

Lundberg, J.G., 1998. The temporal context for the diversification of Neotropical fishes. In: L.R. Malabarba, R.E. Reis, and R.P. Vari, Z.M.S. Lucena, and C.A.S. Lucena (Editors). *Phylogeny and Classification of Neotropical Fishes.* EDIPUCRS, Porto Alegre, 49–68.

Lundberg, J.G., and B. Chernoff, 1992. A Miocene fossil of the Amazonian fish *Arapaima* (Teleostei, Arapaimidae) from the Magdalena River Region of Colombia—biogeographic and evolutionary implications. *Biotropica*, 24, 2–14.

Lundberg, J.G., L.G. Marshall, J. Guerrero, B. Horton, M.C.S.L. Malabarba, and F. Wesselingh. 1998. The stage for Neotropical fish diversification: A history of tropical South American rivers. In: L.R. Malabarba, R.E. Reis, R.P. Vari, Z.M.S. Lucena, and C.A.S. Lucena (Editors), *Phylogeny and Classification of Neotropical fishes.* EDIPUCRS, Porto Alegre, 13–48.

Luo, Z-X., R.L. Cifelli, and Z. Kielan-Jaworowska, 2001. Dual origin of tribosphenic mammals. *Nature*, 409, 53–57.

Lynch, J.D., 1979. The amphibians of the lowland tropical forests. In: W.E. Duellman (Editor), *The South American Herpetofauna: Its Origin, Evolution, and Dispersal.* Monograph 7. Museum of Natural History, University of Kansas, Lawrence, 189–215.

Lynch, J.D., 1988. Refugia. In: A.A. Myers and P.S. Giller (Editors), *Analytical Biogeography.* Chapman and Hall, London, 311–342.

Lynch, J.D., and W.E. Duellman, 1997. Frogs of the genus *Eleutherodactylus* (Leptodactylidae) in western Ecuador. *Systematics, Ecology, and Biogeography.* Special Publication 23. Museum of Natural History, University of Kansas, Lawrence.

MacPhee, R.D.E., and C. Flemming, 1999. Requiem æternam: The last five hundred years of mammalian species extinctions. In: R.D.E. MacPhee (Editor), *Extinctions in Near Time: Causes, Contexts, and Consequences.* Kluwer Academic, New York, 333–371.

Mares, M.A., 1985. Mammal faunas of xeric habitats and the Great American Interchange. In: F.G. Stehli and S.D. Webb (Editors), *The Great American Biotic Interchange.* Plenum Press, New York, 489–520.

Mares, M.A., 1992. Neotropical mammals and the myth of Amazonian biodiversity. *Science*, 255, 976–979.

Mares, M.A., R.A. Ojeda, J.K. Braun, and R.M. Barquez. 1997. Systematics, distribution, and ecology of the mammals of Catamarca Province, Argentina. In: T.L. Yates, W.L. Gannon and D.E. Wilson (Editors), *Life among the Muses: Papers in Honor of James S. Findley.* Museum of Southwest Biology, University of New Mexico Press, Albuquerque, 89–141.

Marquet, P.A., 1989. Paleobiogeography of South American cricetid rodents: A critique to Caviedes & Iriarte. *Revista Chilena de Historia Natural*, 62, 193–197.

Marquet, P.A., 1994. Diversity of small mammals in the Pacific Coastal Desert of Peru, Chile and in the adjacent Andean area: Biogeography and community structure. *Australian Journal of Zoology*, 42,527–542.

Marshall, L.G., 1979. A model for paleobiogeography of South American cricetine rodents. *Paleobiology*, 5, 126–132.

Marshall, L.G., 1985. Geochronology and land-mammal biochronology of the transamerican faunal interchange. In: F.G. Stehli and S.D. Webb (Editors), *The Great American Biotic Interchange.* Plenum Press, New York, 49–85.

Marshall, L.G., and R.L. Cifelli, 1990. Analysis of changing diversity patterns in Cenozoic land mammal age faunas, South America. *Palaeovertebrata*, 19, 169–210.

Marshall, L.G., and T. Sempere, 1993. Evolution of the Neotropical Cenozoic land mammal fauna in its geochronologic, stratigraphic, and tectonic context. In: P. Goldblatt (Editor), *Biological Relationships between Africa and South America.* Yale University Press, New Haven, 329–392.

Marshall, L.G., S.D. Webb, J.J. Sepkoski, Jr., and D.M. Raup, 1982. Mammalian evolution and the great American interchange. *Science*, 215, 1351–1357.

Martin, P.S., 1984. *Quaternary Extinctions: A Prehistoric Revolution.* University of Arizona Press, Tucson.

Martin, P.S., and D.W. Steadman, 1999. Prehistoric extinctions on islands and continents. In: R.D.E. MacPhee (Editor), *Extinctions in Near Time: Causes, Contexts, and Consequences.* Kluwer Academic, New York, 17–55.

Meserve, P.L., 1996. Patterns of terrestrial vertebrate diversity in New World temperate rainforests. In: R.G. Lawford, P.B. Alaback, and E. Fuentes (Editors), *High-latitude Rainforests and Associated Ecosystems of the West Coast of the Americas.* Springer, New York, 214–227.

Meserve, P.L., and W.E. Glanz, 1978. Geographical ecology of small mammals in the northern Chilean arid zone. *Journal of Biogeography*, 5, 135–148.
Meserve, P.L., and F.M. Jaksic, 1991. Terrestrial vertebrates of temperate rainforests of North and South America. *Revista Chilena de History Natural*, 64, 511–535.
Meserve, P.L., and D.A. Kelt, 1990. The role of aridity and isolation on central Chilean small mammals: A reply to Caviedes & Iriarte (1989). *Journal of Biogeography*, 17, 681–684.
Mittermeier, R.A., 1988. Primate diversity and the tropical forest: case studies from Brazil and Madagascar and the importance of the megadiversity countries. In: E.O. Wilson (editor), *Biodiversity*. National Academy Press, Washington, D.C., 145–154.
Moyle, P.B., and Cech, J.J., Jr., 2000. *Fishes: An Introduction to Ichthyology*. 4th edition. Prentice Hall, Upper Saddle River, New Jersey.
Müller, P., 1973. The dispersal centres of terrestrial vertebrates in the Neotropical realm. *Biogographica* 2. Dr. W. Junk, The Hague.
Myers, N., 1988. Tropical forests and their species: going, going . . . ? In: E.O. Wilson (Editor), *Biodiversity*. National Academy Press, Washington D.C., 28–35.
Myers, N., R.A Mittermeier, C.G. Mittermeier, G.A.B. da Fonseca, and J. Kent, 2000. Biodiversity hotspots for conservation priorities. *Nature*, 403, 853–858.
Nores, M., 2004. The implications of Tertiary and Quaternary sea level rise events for avian distribution patterns in the lowlands of northern South America. *Journal of Biogeography*, 13, 149–161.
Nowak, R.M., 1999. *Walker's Mammals of the World*. 6th edition. Johns Hopkins University Press, Baltimore.
Padian, K., 1986. Introduction. In: K. Padian (Editor), *The Beginning of the Age of Dinosaurs*. Cambridge University Press, Cambridge, 1–7.
Pascual, R., 1996. Late Cretaceous-Recent land mammals: An approach to South American geobiotic evolution. *Mastozoología Neotropical*, 3, 133–152.
Pascual, R., M. Archer, E.O. Jaureguizar, J.L. Prado, H. Godthelp, and S.J. Hand, 1992. First discovery of monotremes in South America. *Nature*, 356, 704–705.
Patterson, B., and R. Pascual, 1972. The fossil mammal fauna of South America. In: A. Keast, F.C. Erk, and B. Glass (Editors), *Evolution, Mammals, and Southern Continents*. State University of New York Press, Albany, 247–310.
Patterson, B.D., 1994. Accumulating knowledge on the dimensions of biodiversity: Systematic perspectives on Neotropical mammals. *Biodiversity Letters*, 2, 79–86.
Patterson, B.D., 1999. Contingency and determinism in mammalian biogeography: The role of history. *Journal of Mammalogy*, 80, 345–360.
Patterson, B.D., 2000. Patterns and trends in the discovery of new Neotropical mammals. *Diversity and Distribution*, 6, 145–151.
Patterson, B.D., 2001. Fathoming tropical biodiversity: The continuing discovery of Neotropical mammals. *Diversity and Distribution*, 7, 191–196.
Patterson, B.D., P.L. Meserve, and B.K. Lang, 1989. Distribution and abundance of small mammals along an elevational transect in temperate rainforests of Chile. *Journal of Mammalogy*, 70, 67–78.
Patterson, B.D., D.F. Stotz, S. Solari, J.W. Fitzpatrick, and V. Pacheco, 1998. Contrasting patterns of elevational zonation for birds and mammals in the Andes of southeastern Peru. *Journal of Biogeography*, 25, 593–607.
Patton, J.L., and M.N.F. da Silva, 2001. Molecular phylogenetics and the diversification of Amazonian mammals. In: I.C.G. Vieira, J.M.C. da Silva, D.C. Oren, and M.A. D'Incao (Editors), *Biological and Cultural Diversity of Amazonia*. Museu Paraense Emílio Goeldi, Belém, 139–164.
Patton, J.L., and M.F. Smith, 1992. MtDNA phylogeny of Andean mice: A test of diversification across ecological gradients. *Evolution*, 46, 174–183.
Patton, J.L., M.N.F. da Silva, and J.R. Malcolm, 1994. Gene genealogy and differentiation among arboreal spiny rats (Rodentia: Echimyidae) of the Amazon Basin: A test of the riverine barrier hypothesis. *Evolution*, 48, 1314–1323.
Patton, J.L., P. Myers, and M.F. Smith, 1990. Vicariant versus gradient models of diversification: The small mammal fauna of eastern Andean slopes of Peru. In: G. Peters and R. Hutterer (Editors), *Vertebrates in the Tropics*. Alexander Koenig Zoological Research Institute and Zoological Museum, Bonn, 355–371.
Pearson, O.P., 1982. Distribución de pequeños mamíferos en el Altiplano y los desiertos de Peru. In: P.J. Salinas (Editor), *Zoología Neotropical*, Merída, Venezuela, 263–284.
Pearson, O.P., and A.K. Pearson, 1982. Ecology and biogeography of the southern rainforests of Argentina. In: M.A. Mares (Editor), *Mammalian Biology in South America*. Pymatuning Laboratory of Ecology, University of Pittsburgh, Linesville, 129–142.
Pearson, O.P., and C.P. Ralph, 1978. The diversity and abundance of vertebrates along an altitudinal gradient in Peru. *Memorias del Museo de Historia Natural Javier Prado*, 18, 1–97.
Pinna, M.C.C. de, 1998. Phylogenetic relationships of Neotropical Siluriformes (Teleostei: Ostariophysi): Historical overview and synthesis of hypotheses. In: L.R. Malabarba, R.E. Reis, R.P. Vari, Z.M.S. Lucena, and C.A.S. Lucena (Editors), *Phylogeny and Classification of Neotropical Fishes*. EDIPUCRS, Porto Alegre, 279–330.
Prance, G.T., 1993. The Amazon: paradise lost? In: L. Kaufman and K. Mallory (Editors), *The Last Extinction*. 2nd edition. MIT Press, Cambridge, 69–114.
Rahbek, C., 1997. The relationship among area, elevation, and regional species richness in Neotropical birds. *American Naturalist*, 149, 875–902.
Rahbek, C., and G.R. Graves, 2000. Detection of macro-ecological patterns in South American hummingbirds is affected by spatial scale. *Proceedings of the Royal Society of London*, B, 267, 2259–2265.
Rahbek, C., and G.R. Graves, 2001. Multiscale assessment of patterns of avian species richness. *Proceedings of the National Academy of Sciences*, 98, 4534–4539.
Rage, J.-C., 1987. Fossil history. In: R.A. Seigel, J.T. Collins, and S.S. Novak (Editors). *Snakes: Ecology and Evolutionary Biology*. MacMillan, New York, 51–76.
Rage, J.-C., 1988. Gondwana, Tethys and terrestrial vertebrates during the Mesozoic and Cenozoic. In: M.G. Audley-Charles and A. Hallam (Editors), *Gondwana and Tethys*. Geological Society Special Paper 37, 255–272.
Raup, D.M., 1979. Size of the Permo-Triassic bottleneck and its evolutionary implications. *Science*, 206, 217–218.

Raven, P.H., 1988. Our diminishing tropical forests. In: E.O. Wilson (Editor), *Biodiversity.* National Academy Press, Washington, D.C., 119–122.

Reig, O.A., 1984. Geographic distribution and evolutionary history of South American muroids, Cricetidae: Sigmodontinae). *Revista Brasilera de Genética*, 7, 333–365.

Reig, O.A., 1986. Diversity patterns and differentiation of high Andean rodents. In: F. Vuilleumier and M. Monasterio (Editors), *High-altitude Tropical Biogeography.* Oxford University Press, New York, 404–439.

Reig, O.A., 1987. An assessment of the systematics and evolution of the Akodontini, with the description of new fossil species of *Akodon* (Cricetidae: Sigmodontinae). In: B. D. Patterson and R. M. Timm (Editors), Studies in Neotropical Mammalogy: Essays in Honor of Philip Hershkovitz. *Fieldiana: Zoology*, 39, Chicago, 347–399.

Rich, T.H., P. Vickers-Rich, A. Constantine, T.F. Flannery, L. Kool, and N. van Klaveren, 1997. A tribosphenic mammal from the Mesozoic of Australia. *Science,* 278, 1438–1442.

Ricklefs, R.E., 2002. Splendid isolation: Historical ecology of South American passerine fauna. *Journal of Avian Biology*, 33, 207–211.

Ridgely, R.S., and G. Tudor, 1989. *The Birds of South America. 1: The Oscine Passerines.* University of Texas Press, Austin.

Ridgely, R.S., and G. Tudor, 1994. *The Birds of South America. II: The Suboscine Passerines.* University of Texas Press, Austin.

Ron, S.R., 2000. Biogeographic area relationships of lowland Neotropical rainforest based on raw distributions of vertebrate groups. *Biological Journal of the Linnaean Society*, 71, 379–402.

Ruggiero, A., J.H. Lawton, and T.M. Blackburn, 1998. The geographic ranges of mammalian species in South America: Spatial patterns in environmental resistance and anisotropy. *Journal of Biogeography*, 25, 1093–1103.

Rylands, A.B., R.A. Mittermeier, and E. Rodríquez-Luna, 1997. Conservation of Neotropical primates: Threatened species and an analysis of primate diversity by country and region. *Folia Primatologica*, 68, 134–160.

Saunders, D.A., R.J. Hobbs, and C.R. Margules, 1991. Biological consequences of ecosystem fragmentation: A review. *Conservation Biology*, 5, 18–32.

Savage, J.M., 1982. The enigma of the Central American herpetofauna: Dispersals or vicariance? *Annals of the Missouri Botanical Garden*, 69, 464–547.

Sepkoski, J.J., Jr., 1990. Evolutionary faunas. In: D.E.G. Briggs and P.R. Crowther (Editors), *Palaeobiology: A Synthesis.* Blackwell Scientific, Oxford, 37–41.

Shubin, N.H., and H.-D. Sues, 1991. Biogeography of early Mesozoic continental tetrapods: Patterns and implications. *Paleobiology*, 17, 214–230.

Sibley, C.G., and J.E. Ahlquist, 1990. *Phylogeny and Classification of Birds: A Study of Molecular Evolution.* Yale University Press, New Haven.

Sibley, C.G., and B.L. Monroe, Jr., 1990. *Distribution and Taxonomy of Birds of the World.* Yale University Press, New Haven.

Silva, J.M.C. da, and J.M. Bates, 2002. Biogeographic patterns and conservation in the South American Cerrado: A tropical savanna hotspot. *BioScience*, 52, 225–233.

Simpson, B.B., 1979. Quaternary biogeography of the high montane regions of South America. In: W.E. Duellman (Editor), *The South American Herpetofauna: Its Origin, Evolution, and Dispersal.* Monograph 7. Museum of Natural History, University of Kansas, Lawrence, 157–188.

Simpson, B.B., and J. Haffer, 1978. Speciation patterns in the Amazonian forest biota. *Annual Review of Ecology and Systematics*, 9, 497–518.

Simpson, G.G., 1969. South American mammals. In: Fittkau, E.J., J. Illies, H. Klinge, G.H. Schwabe, and H. Sioli (Editors), *Biogeography and Ecology in South America, 2.* Monographie Biologicae, 19. Dr. W. Junk, The Hague, 879–909.

Simpson, G.G., 1980. *Splendid Isolation: The Curious History of South American Mammals.* Yale University Press, New Haven.

Sodhi, N.S., L.H. Liow, and F.A. Bazzaz, 2004. Avian extinctions from tropical and subtropical forests. *Annual Review of Ecology and Systematics*, 35, 323–345.

Stattersfield, A.J., M.J. Crosby, A.J. Long, and D.C. Wege, 1998. *Endemic Bird Areas of the World: Priorities for Biodiversity Conservation.* BirdLife International, Cambridge.

Stevens, R.D., 2004. Untangling latitudinal richness gradients at higher taxonomic levels: Familial perspectives on the diversity of New World bat communities. *Journal of Biogeography*, 31, 665–674.

Stork, N.E. 1999. The magnitude of global biodiversity and its decline. In: J. Cracraft and F.T. Grifo (Editors), *The Living Planet in Crisis: Biodiversity Science and Policy.* Columbia University Press, New York, 3–32.

Stotz, D.F., J.F. Fitzpatrick, T.A. Parker, III, and D.K. Moskovits, 1996. *Neotropical Birds: Ecology and Conservation.* University of Chicago Press, Chicago.

Terborgh, J.W., 1977. Bird species diversity on an Andean elevational gradient. *Ecology*, 58, 1007–1019.

Terborgh, J.W., J.W. Fitzpatrick, and L.H. Emmons, 1984. Annotated checklist of bird and mammal species of Cocha Cashu Biological Station, Manu National Park, Peru. *Fieldiana: Zoology*, 2, 1–29.

Tognelli, M.F., and D.A. Kelt. 2004. Analysis of determinants of mammalian species richness in South America using spatial autoregressive models. *Ecography*, 27, 427–436.

Tuomisto, H., and K. Ruokolainen, 1997. The role of ecological knowledge in explaining biogeography and biodiversity in Amazonia. *Biodiversity and Conservation*, 6, 347–357.

Uetz, P., and T. Etzold, 1996. The EMBL/EBI reptile database. *Herpetological Review*, 27, 174–175.

van Tuinen, M., C.G. Sibley, and S.B. Hedges, 1998. Phylogeny and biogeography of ratite birds inferred from DNA sequences of the mitochondrial ribosomal genes. *Molecular Biology and Evolution*, 15, 370–376.

van Tuinen, M., and S.B. Hedges, 2001. Calibration of avian molecular clocks. *Molecular Biology and Evolution*, 18, 206–213.

Van Schaik, C.P., J. Terborgh, and B. Dugelby, 1997. The silent crisis: The state of rain forest nature reserves. In: R. Kramer, C. van Schaik, and J. Johnson (Editors), *Last Stand: Protected Areas and the Defense of Tropical Biodiversity.* Oxford University Press, New York, 64–89.

Vanzolini, P.E., and W.R. Heyer, 1985. The American herpetofauna and the interchange. In: F.G. Stehli and S.D. Webb (Editors), *The Great American Biotic Interchange.* Plenum Press, New York, 475–487.

Vari, R.P., and L.R. Malabarba, 1998. Neotropical ichthyology: An overview. In: L.R. Malabarba, R.E. Reis, R.P. Vari, Z.M.S. Lucena, and C.A.S. Lucena (Editors), *Phylogeny and Classification of Neotropical Fishes.* EDIPUCRS, Porto Alegre, 1–11.

Vari, R.P., and S.H. Weitzman, 1990. A review of the phylogenetic biogeography of the freshwater fishes of South America. In: G. Peters and R. Hutterer (Editors), *Vertebrates in the Tropics.* Alexander Koenig Zoological Research Institute and Zoological Museum, Bonn, 381–393.

Villwock, W., 1986. Speciation and adaptive radiation in Andean *Orestias* fishes. In: F. Vuilleumier and M. Monasterio (Editors), *High-altitude Tropical Biogeography.* Oxford University Press, New York, 387–403.

Voss, R.S., and L.H. Emmons, 1996. Mammalian diversity in Neotropical lowland rainforests: A preliminary assessment. *Bulletin of the American Museum of Natural History,* 230, 1–115.

Vuilleumier, F. 1969. Pleistocene speciation in birds living in the high tropical Andes. *Nature,* 223, 1179–1180.

Vuilleumier, F., 1970. Insular biogeography in continental regions. I: The northern Andes of South America. *American Naturalist,* 104, 373–388.

Vuilleumier, F., 1985a. Fossil and recent avifaunas and the interamerican interchange. In: F.G. Stehli and S.D. Webb (Editors), *The Great American Biotic Interchange.* Plenum Press, New York, 387–424.

Vuilleumier, F. 1985b. Forest birds of Patagonia: Ecological geography, speciation, endemism, and faunal history. In: P.A. Buckley, M.S. Foster, E.S. Morton, R.S. Ridgely, and F.G. Buckley (Editors), *Neotropical Ornithology.* Ornithological Monographs, 36. American Ornithological Union, Washington, D.C., 255–304.

Vuilleumier, F., 1986. Origins of the tropical avifaunas of the high Andes. In: F. Vuilleumier and M. Monasterio (Editors), *High-altitude Tropical Biogeography.* Oxford University Press, New York, 586–622.

Vuilleumier, F., 1997. How many bird species inhabit the puna desert of the high Andes of South America? *Global Ecology and Biogeography Letters,* 6, 149–153.

Vuilleumier, F., and D. Simberloff, 1980. Ecology versus history as determinants of patchy and insular distributions in high Andean birds. *Evolutionary Biology,* 12, 235–379.

Webb, S.D., 1976. Mammalian faunal dynamics of the Great American Interchange. *Paleobiology,* 2, 220–234.

Webb, S.D., 1978. A history of savanna vertebrates in the New World. II: South America and the Great American Interchange. *Annual Review of Ecology and Systematics,* 9, 393–426.

Webb, S.D. 1985. Late Cenozoic dispersals between the Americas. In: F.G. Stehli and S.D. Webb (Editors), *The Great American Biotic Interchange.* Plenum Press, New York, 357–386.

Webb, S.D., 1999. Isolation and interchange. A deep history of South American mammals. In: J.F. Eisenberg and K.H. Redford (Editors), *Mammals of the Neotropics. 3: The Central Neotropics: Ecuador, Peru, Bolivia, Brazil.* University of Chicago Press, Chicago, 13–19.

Webb, S.D., and A.D. Barnosky, 1989. Faunal dynamics of Pleistocene mammals. *Annual Review of Earth and Planetary Sciences,* 17, 413–438.

Webb, S.D., and L.G. Marshall, 1982. Historical biogeography of Recent South American land mammals. In: M.A. Mares and H.H. Genoways (Editors), *Mammalian Biology in South America.* Pymatuning Laboratory of Ecology, University of Pittsburgh, Linesville, 39–52.

Webb, S. D., and A. Rancy, 1996. Late Cenozoic evolution of the Neotropical mammal fauna. In: J.B.C. Jackson, A.F. Budd, and A.G. Coates (Editors), *Evolution and Environment in Tropical America.* University of Chicago Press, Chicago, 335–358.

Weishampel, D.B. 1990. Dinosaurian distribution. In: D.B. Weishampel, P. Dodson, and H. Osmólska (Editors), *The Dinosauria.* University of California Press, Berkeley, 63–139.

Willig, M.R., and K.W. Selcer, 1989. Bat species density gradients in the New World: A statistical assessment. *Journal of Biogeography,* 16, 189–195.

Willson, M.F., T.L. De Santo, C. Sabag, and J.J. Armesto, 1996. Avian communities in temperate rainforests of North and South America. In: R.G. Lawford, P.B. Alaback, and E. Fuentes (Editors), *High-latitude Rainforests and Associated Ecosystems of the West Coast of the Americas.* Springer, New York, 228–247.

Wilson, D.E., and D. M. Reeder, 2005. *Mammal Species of the World: a Taxonomic and Geographic Reference.* 3rd edition. Johns Hopkins University Press, Baltimore.

Wood, A.E. 1985. Northern waif primates and rodents. In: F.G. Stehli and S.D. Webb (Editors), *The Great American Biotic Interchange.* Plenum Press, New York, 267–282.

Woodburne, M.O., and J.A. Case, 1996. Dispersal, vicariance, and the late Cretaceous to early Tertiary land mammal biogeography from South America to Australia. *Journal of Mammalian Evolution,* 3, 121–161.

II

REGIONAL ENVIRONMENTS

9

Tropical Forests of the Lowlands

Peter A. Furley

Most of South America lies within the tropics, and lowland tropical ecosystems make up the majority of its landscapes. Although there is great concern for the Amazon ecosystem, the largest of the world's tropical forests, there are many other fascinating and in some cases more endangered types of lowland forest. Such forests may be defined as lying below 1,000 m above sea level, although it is difficult to set arbitrary limits (Hartshorn, 2001). The two main lowland moist evergreen forests are the *Hylea* (a term coined by Alexander von Humboldt to denote rain forests of the Amazon Basin) and the much smaller *Chocó* forest on the Pacific coast between Panama and Ecuador. Two related yet distinctive types of forest are the *Mata Atlântica* or Atlantic moist evergreen forest and the *Mata Decidua* or dry deciduous forest, including the *caatinga* woodland, which is both deciduous and xerophytic (Rizzini et al., 1988). The latter two formations are among the most threatened of all South American forests.

Lowland forests vary from dense and multilayered to open and single-layered, from evergreen to deciduous, and from flooded or semi-aquatic to near-arid. Tree heights range from 30 to 40 m with emergent trees reaching over 50 m, to forests where the tallest trees barely attain 20 m (Harcourt and Sayer, 1996; Solorzano, 2001). However, because of its extent and importance, Amazonia will form the principal focus of this chapter.

9.1 The Amazon Lowland Forests

Amazonia covers a vast area (>6×10^6 km^2) and contains some 60% of the world's remaining tropical forest. The Amazon and Orinoco basins influence not only regional climates and air masses, but also atmospheric circulation patterns both north and south of the Equator. The sheer size and diversity of Amazonia exhausts a normal repertoire of grandiose adjectives (fig. 9.1). The Amazon may or may not be the longest river in the world but it is by far the greatest in terms of discharge, sending around one fifth of the world's fresh water carried by rivers to the oceans (see chapter 5; Eden, 1990; Sioli, 1984). The drainage basin is twice as large as any other of the world's catchments. The second and third rivers by volume are tributaries of the Amazon (the rivers Negro and Madeira), forming a vast water network, which can expand in the wet season by nearly 80,000 km^2. This annual flux of water sets up a complex set of plant-animal-water interrelationships unrivalled anywhere in the world.

Nine countries share Amazonia, with the largest proportion (around two thirds) occupied by Brazil, while the next largest, Perú, covers 8.6%. Lowland forest is mostly found below 1,000 m in altitude. However, the forest has been defined in several ways and may include areas outside the Amazon drainage (for instance *Amazônia Legal,* the legal

Figure 9.1 Comparative size: Western Europe overlain across Amazonia (modified from Ratter 1990)

or political unit of Amazonia as it is known in Brazil). The figures for forested and nonforested areas are only estimates, based mostly on satellite imagery, and require constant updating as more accurate sensors provide new data and deforestation gathers pace (table 9.1).

Because of its size and variety, Amazonia contains a vast number of plant, animal, and microbial species. It is not, however, uniformly carpeted by tall, moist evergreen forest (*rain forest*), as often imagined. For example, about 20% of *Amazônia Legal* in Brazil is not strictly forest (~700,000 km^2 out of 3.5 × 10^6 km^2), comprising instead numerous patches of more open woodland, wet grassland, and woody savanna (fig. 9.2). Local variations are usually determined by the incidence of flooding, topography, and drainage and changes in the depth and nature of soils, although the dominant factor is climate. Environmental and ecological changes over the Quaternary have also greatly influenced the nature and distribution of present-day vegetation.

9.1.1 Amazonian Environments

Climate and Hydrology Most of Amazonia lies between 5°N and 10°S of the Equator. In its central river section, the duration of daylight remains nearly constant throughout the year, although cloud cover limits insolation. The greater part of Amazonia falls within the humid tropical zone with relatively constant temperatures (mean annual temperatures of 23–27°C) and a diurnal range of ~8–10°C (Molion, 1987), although there are episodic extremes (Nieuwolt, 1977; Marengo and Nobre, 2001). The only exception to temperature uniformity is the occurrence of cold spells, with air masses from southern polar regions occasionally penetrating Amazonia (see chapter 3). Climate is principally affected by seasonal changes in the Intertropical Convergence Zone (ITCZ). From April to July the ITCZ moves northward across Amazonia and the surface trade winds transfer moisture-laden Atlantic air westward across the region (see figs. 2.2, 3.2). From August to March, the ITCZ loops south into western Brazil and Bolivia and draws the trade winds southward, often bringing heavy rain to the central Amazon Basin. Despite the relative constancy of the easterly trade winds, the lowland subregions experience distinctly different local climates, influenced by factors such as topography, distance from the sea, and the local water balance (fig. 9.3a). Furthermore, as air masses move toward the west they undergo a series of cyclical convectional movements that affect the water balance (fig. 9.3b; Salati 1985, 1987). When these air masses finally reach the Andean foothills, they are forced to rise and in doing so cause heavy (orographic) precipitation. Rainfall

Table 9.1 Estimated areas (million ha) of the main vegetation types in Amazonia

	Amazonia	Total forest area	*Terra firma* forest	Wetland forest	Total non-forest area	Dry savanna	Herbaceous wetland	Cleared/ cultivated land (end-1970s)
Bolivia	52.7	36.6	30.5	6.1	16.1	0.6	14.5	1.0
Brazil	497.8	388.2	348.3	39.9	109.6	80.1	12.1	17.4
Colombia	45.7	42.0	38.1	3.9	3.7	1.3	—	2.4
Ecuador	9.3	8.4	7.6	0.8	0.9	—	—	0.9
Fr. Guiana	9.7	9.6	9.4	0.2	0.1	trace	0. 1	trace
Guyana	21.4	18.6	18.1	0.5	2.8	2.1	—	0.7
Perú	65.1	60.3	51.1	9.2	4.8	—	0.1	4.7
Surinam	16.3	15.0	13.5	1.5	1.3	0.1	0.5	0.7
Venezuela	38.6	34.1	30.7	3.4	4.5	3.8	0.4	0.3
Total	756.6	612.8	547.3	65.5	143.8	88.0	27.7	28.1

Source: Based on Eden (1990).

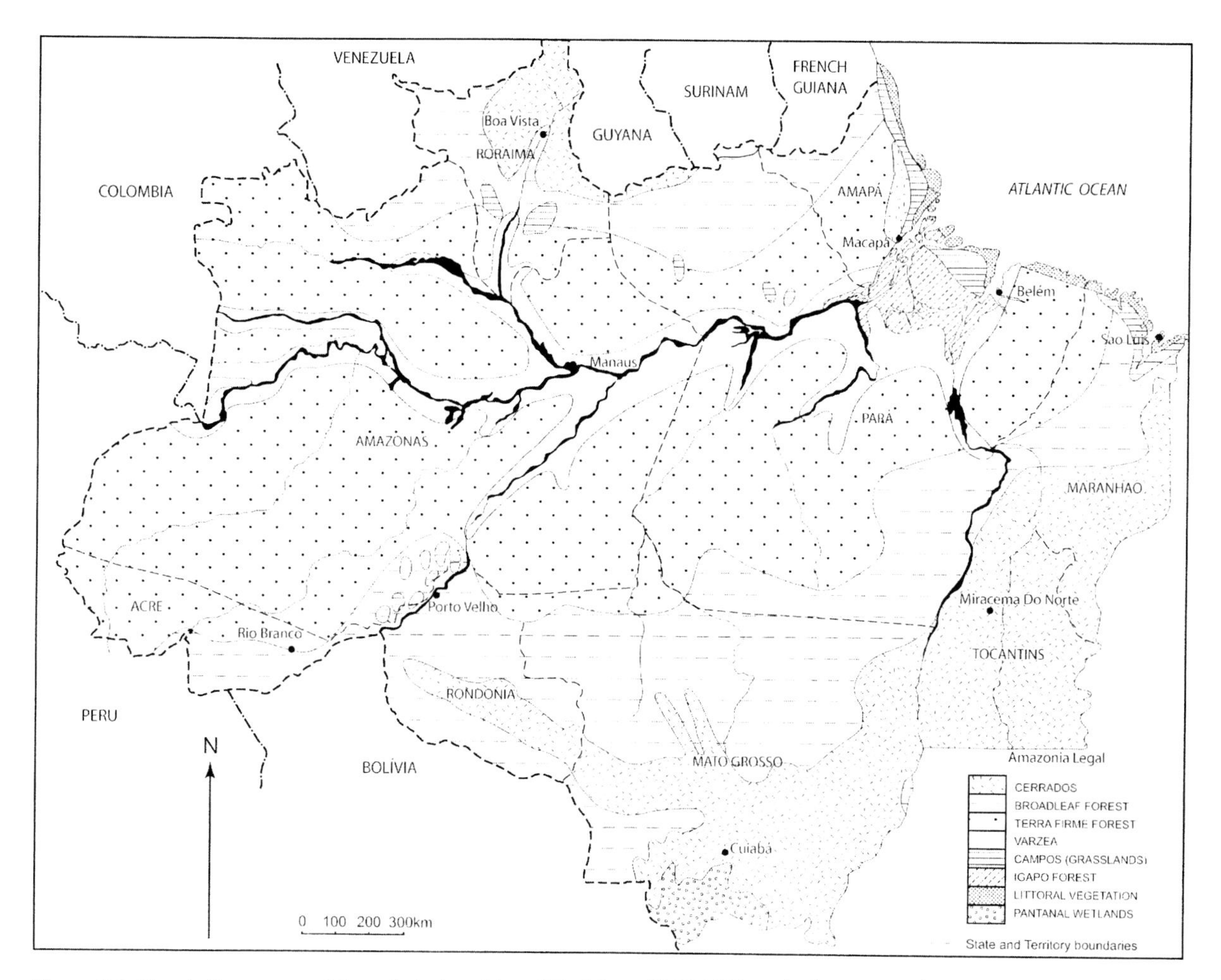

Figure 9.2 Vegetation zones of Amazônia Legal (modified from IBGE 1993 and other sources)

varies considerably over the basin (fig. 9.3c), and variability in any one place is significant from year to year (15–20% of the mean). This can result in moisture deficiencies in areas vulnerable to seasonal swings and high evapotranspiration. Even a short dry spell of 10 to 15 days can cause water deficits (average losses of around 4 mm d^{-1}), and trees near Manaus have been seen to lose some of their leaves in this short time. Rainfall of <100 mm in a month can lead to water deficit. The volume of canopy leaves in a single mature forest tree is immense, and it is easy to imagine the huge transpiration and evaporation losses that occur under clear skies. Rainfall intensity can also cause problems of compaction and erosion, especially in cleared areas where the soils are exposed. Furthermore, at any specific locality, such as a natural gap in the forest, microclimatic differences assume greater significance (fig. 9.3d).

The Amazon River system undergoes marked seasonal changes, which strongly affect the biology and human activities. There is a distinct wet season, whose timing varies throughout the basin, when flooding and powerful flows dominate the river. During floods, boats pass more easily along tributary waterways, but these in turn become obstructed during drier periods, when the river level may drop significantly, for example by 6–7 m at Iquitos and by 8–12 m or more at Manaus. The timing of peak floods varies according to tributary location, and flood discharge and variability now appear to be increasing, largely caused by deforestation in the catchments (Smith, 1981, 1999; Goulding et al., 1996). The main Amazon River, or Solimões as it is known upstream from Manaus, and western tributaries such as the Madeira and Purus, are referred to as *white water* rivers, reflecting their distinctive load of suspended and dissolved sediment (neutral to alkaline pH values of 6.2–7.2) derived from the Andes. In contrast, other rivers such as the Río Negro are *black*, because they drain old crystalline rocks with low relief and less erosion, and consequently carry little solid mineral waste but a great deal of dissolved or colloidal organic matter and tannins (pH 3.8–4.9). Intermediate *clear water* rivers have also been identified, with pH values 4.5–7.8 (Junk and Furch, 1985; Sioli, 1984). Satellite-based microwave data reveal the area and periodicity of flooding (e.g., EOS, 2001). The maximum area of inundation has been calculated at around 77,000 km^2 along the Amazon mainstream, representing about 30% of the central

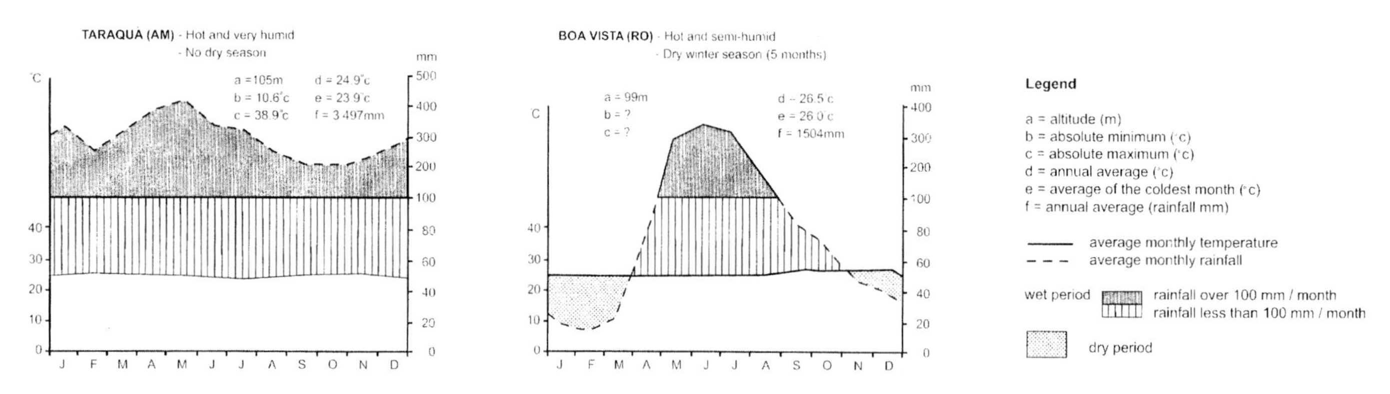

Figure 9.3a Climograms for contrasting sites in Amazonia (one seasonal and one with marked seasonality (compiled from Salati, 1985; Salati, 1987;IBGE, 1977)

part of the basin, and as much as 87,000 km² in wet savannas in Bolivia, 58,000 km² in the Ilha de Bananal area (Mato Grosso-Goias), and 16,000 km² in Roraima.

Geology and Geomorphology The east-flowing mainstream of the Amazon River divides Amazonia into broad regions. A 3,000-km long alluvial plain, mostly below 300 m, separates the Guiana Shield to the north from the Brazilian Shield to the south, with the Andes rising in the far west (fig. 9.4). The alluvial plain is 600–800 km wide in the west but narrows to less than 300 km toward the Atlantic Ocean. The river's discharge forces sediment-laden fresh water well out into the ocean, forming a sediment plume readily observed on satellite images and contributing to littoral budgets and the deepwater Amazon cone (see chapters 1, 5, and 15). The Amazon's many distributaries leave an island, Marajó, the size of Switzerland in its estuary.

The Guiana Shield and the Guaporé portion of the Brazilian Shield are surface expressions of Precambrian crystalline rocks forming the large Amazonian craton, connected at depth beneath the deep alluvial fill of the Amazon Basin (see chapter 1, fig. 1.2). The Brazilian Shield also encompasses the São Francisco and São Luis cratons, Neoproterozoic orogenic belts, and later sedimentary cover rocks. This is an oldland comprising deeply weathered rocks and ancient planation surfaces, most clearly seen in eastern Brazil (Bigarella and Ferreira, 1985). Overlying the Guiana Shield are the remains of thick Proterozoic sandstones that form spectacular *tepuis* (table mountains) and inselbergs, rising to 2,810 m on Mount Roraima at the border between Venezuela, Brazil, and Guyana. Later Mesozoic sandstones overlie both shield areas and produce distinctive hilly features, such as the Serra do Roncador in Mato Grosso.

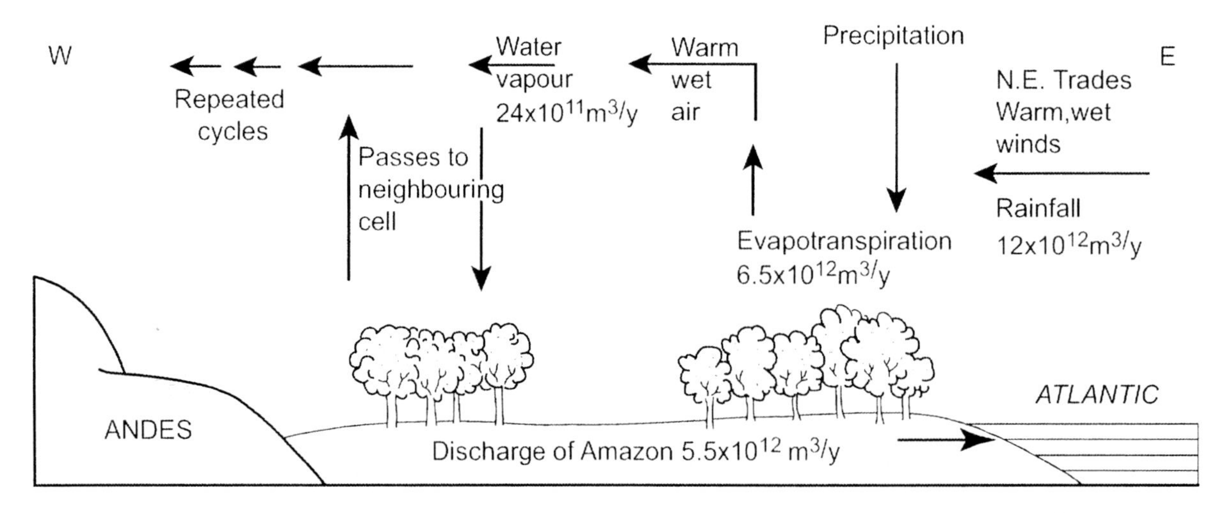

Figure 9.3b Water balance and convectional cycles across Amazonia, showing the influence of evapotranspiration on water availability (modified from Salati, 1985; Furley, 1986). Salati (1985) distinguished between primary water vapor (carried by winds from the ocean) and secondary water vapor (derived from transpiration and evaporation). In most parts of the world the latter is unimportant relative to the former, but in tropical moist forests such as those of Amazonia they are of similar magnitude.

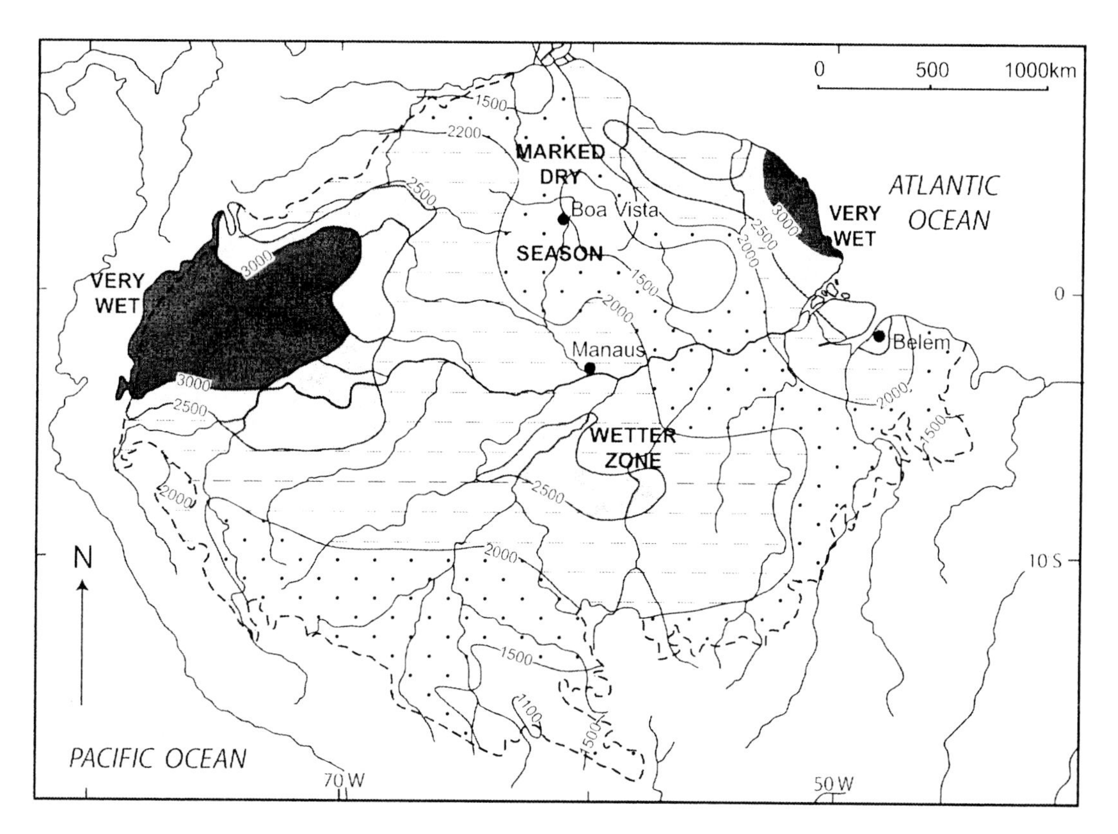

Figure 9.3c Rainfall of Amazonia, showing the influence of dry periods even at the equator. Sections of the moist forest are much more vulnerable to water deficits than might be thought, particularly in the sensitive lower-rainfall areas of the central Amazon in the face of deforestation (modified from Salati, 1985; IBGE, 1977).

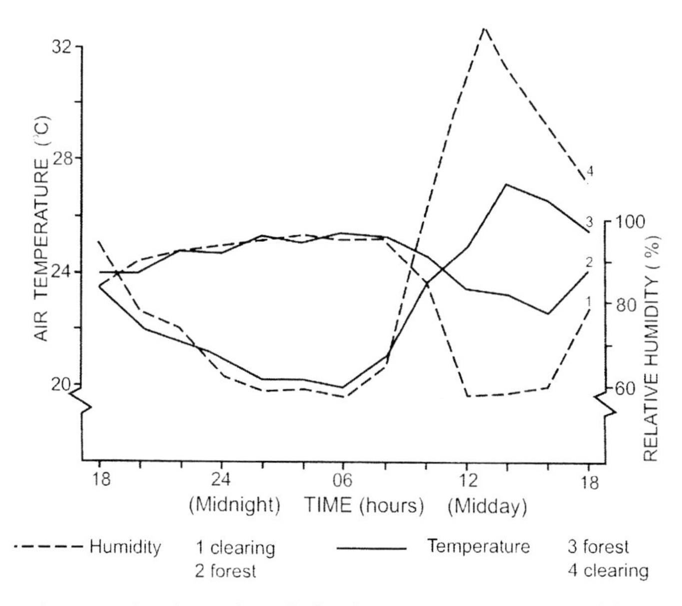

Figure 9.3d Schematic variation in temperature and humidity between undisturbed forest and a clearing. Cleared areas and natural gaps in the forest are characterized by a rapid rise of temperature and a marked drop in relative humidity, and thus greater desiccation, increased organic decomposition, and more susceptibility to erosion and soil degradation.

The Amazonian lowlands are made up of sediments of varying age. Until sometime in the Neogene, these were deposited from westward-flowing rivers, but the rise of the Andes obstructed former outlets and the drainage system reversed (Eden 1990; Bigarella and Ferreira, 1985). Most of the basin, the central part of which may have remained a fresh water sea into the Pleistocene, is filled with late Cenozoic deposits, although there are outcrops of older sediments around the periphery of the shields. The valley deposits consist mostly of Plio-Pleistocene clay (Belterra Clay) forming a 10–20-m thick terrace-like surface, together with later Pleistocene terrace deposits and Holocene flood plain alluvium. White sands occur in localized patches throughout the basin, supporting a distinctive type of vegetation (*caatinga*) over very poor soils. The Brazilian alluvial areas, which form regularly or periodically flooded wetlands, are known widely as *várzeas* (Junk and Furch, 1985) and are distinguished from the dry land, known as *terra firme* on older, well-drained sediments and residual deposits.

Soils In those parts of the world where high temperatures and heavy rainfall promote rapid chemical weathering and leaching, plants need to devise mechanisms to combat nutrient losses from the soil. It is a curious paradox that the most luxuriant forests on Earth grow over some of the poor-

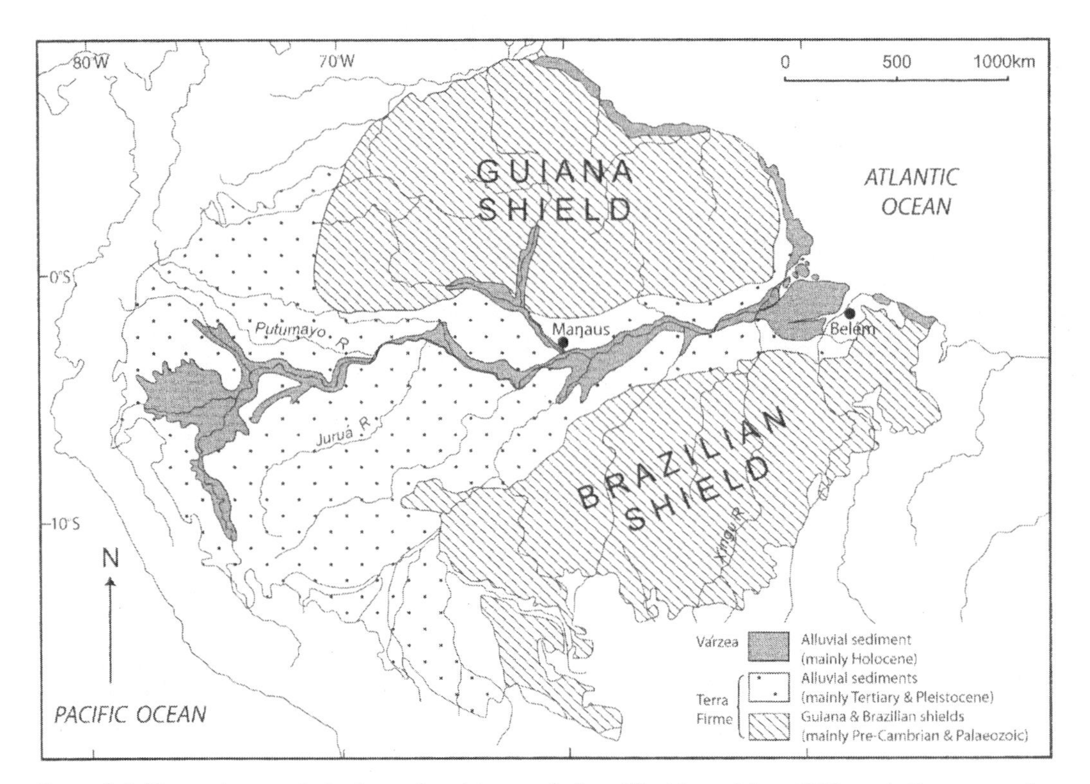

Figure 9.4 The geology and physiography of Amazonia (modified from Eden, 1990; and other sources)

est soils, and this is a reflection of the efficiency of the nutrient cycling processes. Vital plant nutrients would be lost from the system without buffering mechanisms. Nutrients released by leaf-litter decomposition or brought in by rainfall are trapped by the dense, fine network of roots and fungal hyphae (mycorrhizae) with their huge absorbent surface area (Salati et al., 1986; Jordan, 1982, 1985a, 1985b; Herrera, 1985). Mechanisms for nutrient conservation are therefore crucial for plants and may be classified into four groups (fig. 9.5). So effective are these processes that waters draining into black-water streams from oligotrophic or nutrient-poor forests may contain practically no exchangeable anions (such as forms of phosphorus or nitrogen) or exchangeable cations (such as calcium, magnesium, and potassium).

A major advance in understanding the nature and distribution of soils in the Brazilian Amazon was the completion of Projeto Radambrasil (Furley, 1986), a synthesis of which was also presented by CIAT (Cochrane et al., 1985). Around 75% of the Amazon basin is covered by deep acidic, nutrient-poor soils, specifically Oxisols and Ultisols (fig. 9.6; Soil Survey Staff, 1999). Water-affected soils, mostly hydromorphic gleys or poorly drained Entisols and Inceptisols, cover about 14% of the basin and are typical of the alluvial *várzeas* and swampy ground affected by high ground-water levels (fig. 9.6). There are patches of extremely infertile, usually sandy, acidic soil (Psamments), but also scattered patches of moderately fertile, well-drained soils (especially sub-order Udalfs, Fluvents and Tropepts, Soil Survey Staff 1999), which collectively total ~6% of the basin (Furley, 1990; Jordan, 1985b; and Sombroek, 1979).

Tracts of more fertile soil are limited in extent but are very important in influencing plant composition as well as possibilities for agricultural development. The alluvial *várzea* soils from white water streams provide the most extensive better soils, though constrained by annual floods. Limited tracts of more fertile soil are found throughout the Amazon, usually associated with outcrops of more favorable parent materials. Projeto Radambrasil identified two areas in the Brazilian Amazon where igneous intrusions furnish base-rich parent materials at the surface (in southwest Amazonia in Acre-Rondônia, and south of the lower Amazon in parts of Pará). More limited patches of black or dark-colored soils (*terra preta*) occur in association with earlier indigenous settlements usually located on bluffs above flood plains (see chapter 16; McCann et al., 2001). They develop from repeated organic additions and burning, which produces "black carbon" (Glasser et al. 2001). This resists decomposition, attracts cations to exchange surfaces, and allows high fertility to persist over long periods. Intensive raised- and drained-field wetlands similar to those known in Central America (*chinampas*; Gliessman, 1990) also occur in Amazon wetlands, generating quite different types of soil fertility (see chapter 16; Furley, 2005)

Environmental and Ecological Change Over Time It is still a popular misconception that tropical forests are ancient vegetation formations whose biological complexity reflects a long period of evolution. Although some species diversity may indeed result from niche specialization and require a long time span to evolve, it is now known that most of the tropical forests of the world have been strongly affected by climate change, notably changes in temperature (if only a few degrees) and moisture availability during the climatic oscillations of the Quaternary (Whitmore and Prance, 1992). Parts of the tropical forest may be ancient but have been altered in size and shape through time as a result of these changes, particularly at the forest margins.

The history of Amazonian flora is far from well understood and large-scale extrapolations have to be made from a limited number of detailed studies. Most of the evidence comes from palynological analyses of wetland sites (where pollen is better preserved), but there is additional evidence from other proxy measurements such as old arid landscape deposits, diatom indicators of water chemistry, and preserved fragments of vegetation, especially wood remains, all of which need to be matched against known dates, usually from ^{14}C measurements. In addition, present distributions of diverse organisms, from plants to lizards and butterflies, provide valuable clues.

During pre-glacial times there were complex interactions between forest and savanna, and the boundary between these and other formations has changed considerably over time (Behling and Hooghiemstra, 2001; Behling, 1998; Furley et al., 1992). Some records indicate that there was grassland where today there is forest. Others show repeated alternations between forest and non-arboreal vegetation or more open woody conditions. During successive glacial maxima of the Pleistocene, the savannas, broadly defined, must have expanded under markedly drier conditions with a consequent reduction in the forest cover (Behling and Hooghiemstra, 2001; Behling, 2002). The idea that the forest area declined and retreated to a number of refuges (fig. 9.7) is generally accepted, although their location, details of their composition, and explanations of their origin are hotly argued (Colinvaux, 1997, 1996; Bush 1994). Over the past 18,000 years, since the Last Glacial Maximum, there have been changes in temperature and rainfall associated with latitudinal migration of the ITCZ, and changes in pressure systems over the south Atlantic and Antarctic, which have combined to affect the forest cover (Marchant et al., 2000). Today, at a number of Neotropical sites, the forest is advancing into the savanna as a result of slow climatic change and despite widespread clearances and destruction (Ratter 1992).

9.1.2 Amazonian Vegetation

The Amazon-Orinoco lowlands contain a complex mosaic of ecosystems, from tall moist evergreen forest to open grassy savanna *(cerrado/llanos)* and scrub *(campina)*. The plant associations can be examined from several points of

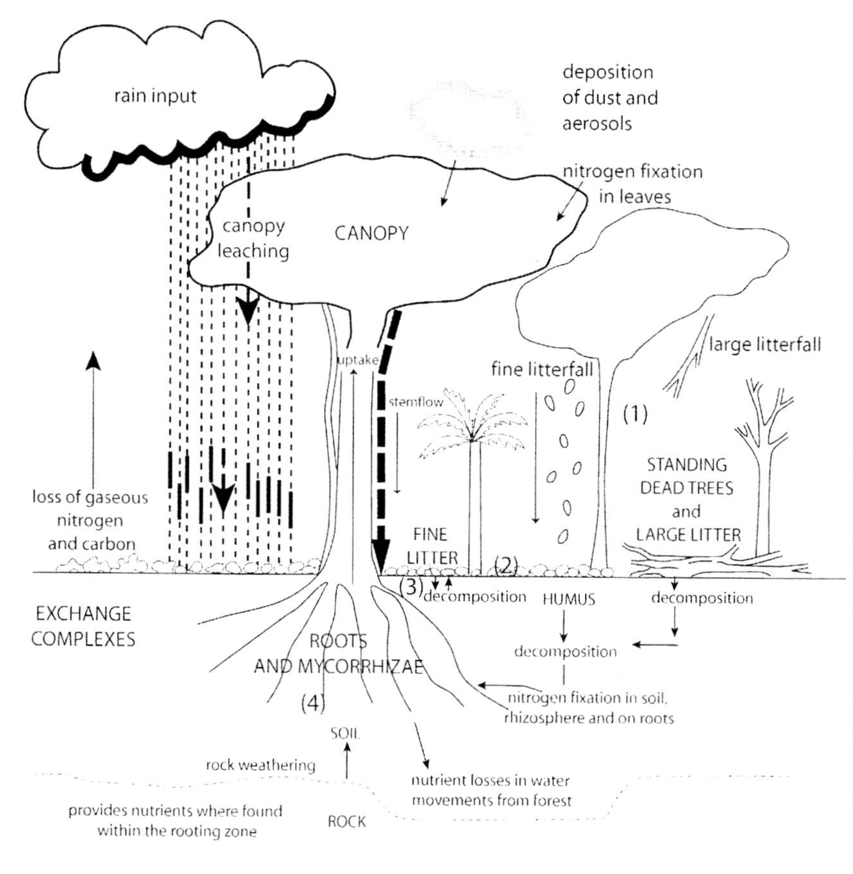

Figure 9.5 Nutrient cycling and nutrient conservation in Amazonian forests: (1) The protective and absorptive role of the above-ground biomass (which reduces rainfall impact and erosion, and retains moisture and nutrients; (2) the protective and absorptive nature of the litter layer (which acts as a protective mulch, also retaining moisture and nutrients); (3) the absorptive nature of the root mat (where fine root extension has been shown to be a rapid and effective scavenger of available nutrients); (4) nutrient conservation by mycorrhizae, which cover a very large volume of the surface soil horizons and trap nutrients.

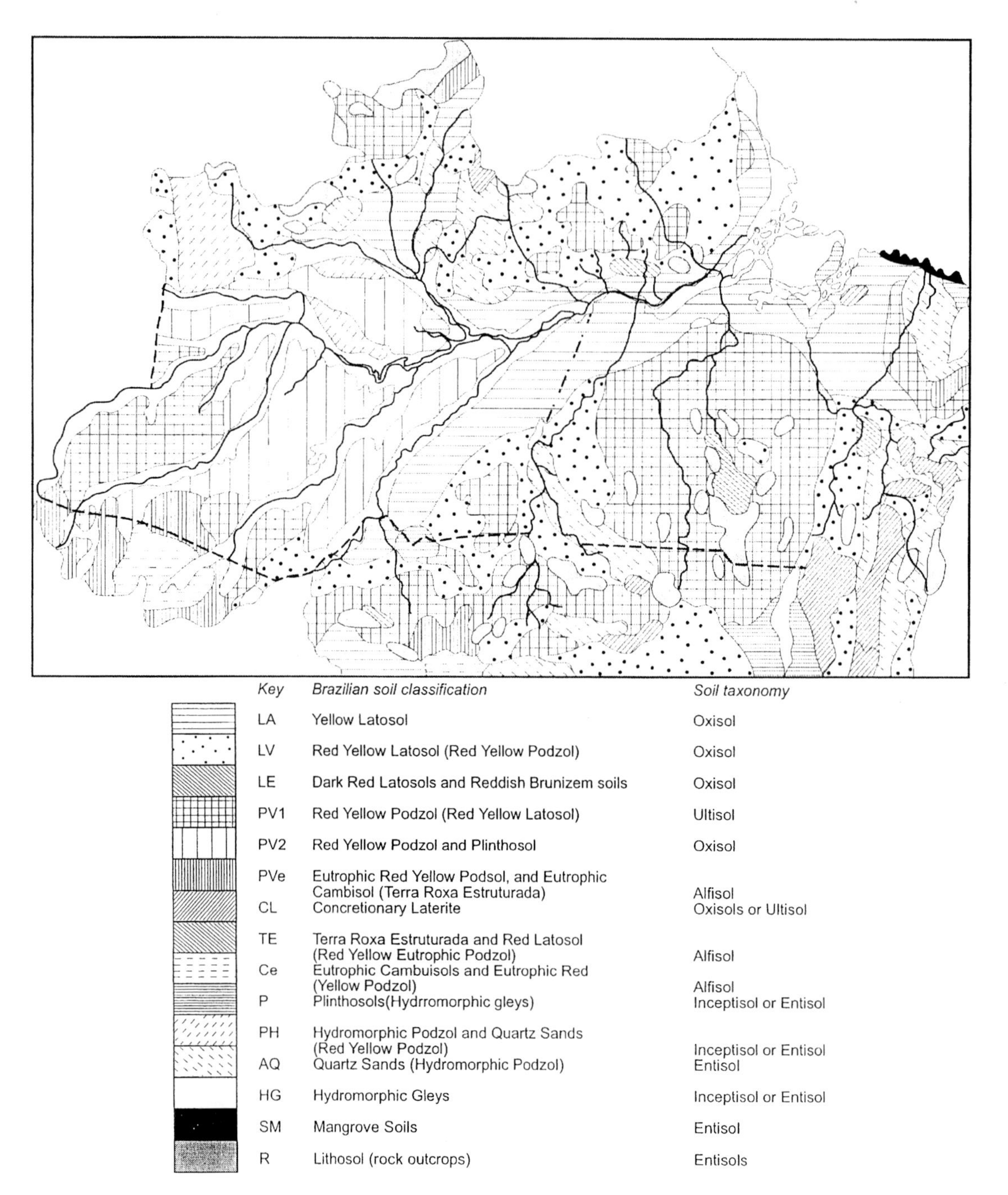

Figure 9.6 Soils of Amazonia (based on Furley, 1990, from several sources)

view, such as their life zones that reflect broad environmental controls (Holdridge, 1972), or their structure and physiognomy, or species composition. Local variations are determined by factors such as the incidence of flooding, topography, and drainage, or soil depth and character. This account follows Prance (2001) in using species composition together with structure to explain the constituent ecosystems.

The first effective overview of the whole of Amazonia came from Side-Looking Airborne Radar surveys (SLAR) of Projeto Radambrasil in the 1970s and 1980s. Despite its reconnaissance nature (published at a scale of 1:250,000), the project was a monumental achievement that gave a quantitative measure of the vegetation as well as geology, geomorphology, soils, climate, and land-use potential of the region.

Forests of Nonflooded Land (Terra firme) *Moist evergreen or rain forest* is the most extensive formation, covering about half of the Amazon region, and typically found on *terra firme*, the firm or solid ground above high

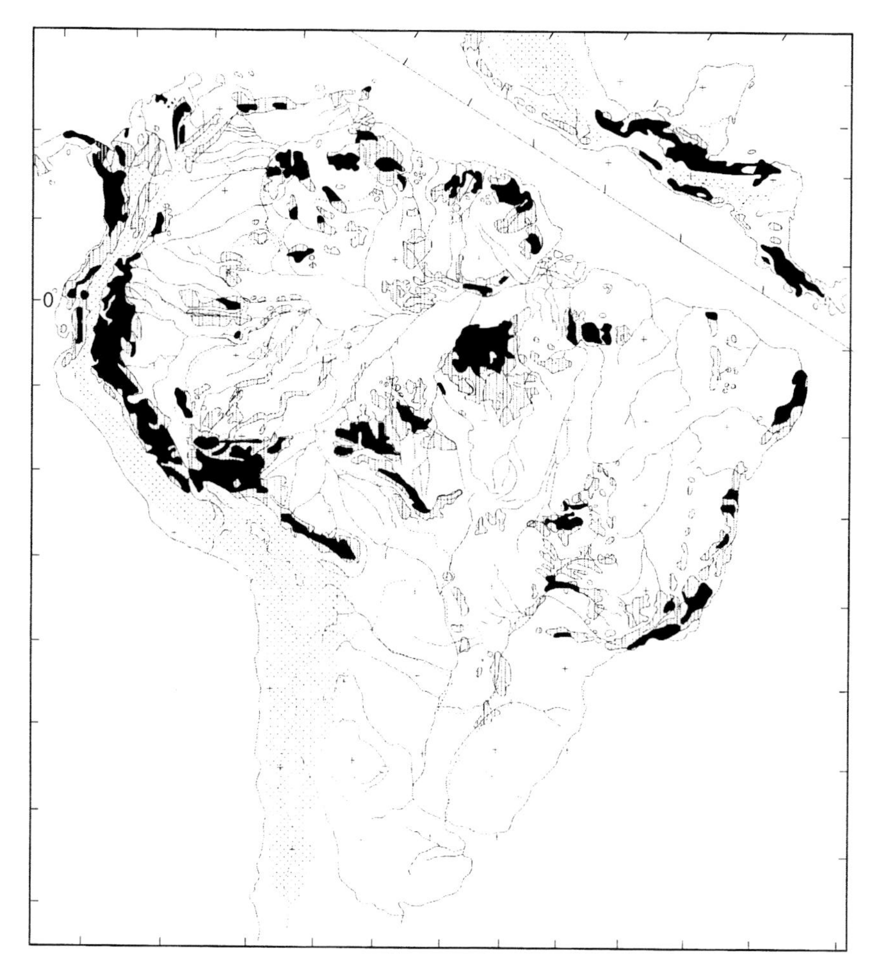

Figure 9.7 Suggested forest refuges in the Amazon (modified from Whitmore and Prance, 1992)

water mark. The forest is frequently stratified, although such layers are often difficult to visualise from the ground (Whitmore, 1998; Longman and Jenik, 1987; Richards, 1996). Single trees or emergents protrude above the main canopy. The main canopy is often so dense that a view from one of the few towers, placed in the forest to observe vertical changes in the vegetation and its microclimate, suggests that one could walk out across the apparent pavement of branches and leaves. Below this layer are subcanopy strata consisting of trees and woody plants such as palms, cacao, and wild banana, and thin spindly saplings of canopy trees waiting for a gap to appear before growing up into the forest. There is a sparse shrub and ground layer except in canopy gaps, which may occupy as much as 20% of the forest area and where exuberant low vegetation occurs. In dense forest it is typically dark and relatively easy to traverse on foot, since there is little ground vegetation. Lianas and epiphytes are common in the higher canopy of the forest. Lianas can be dominant and form dense thickets. Strangler plants, principally figs (*Ficus* spp.), are widespread. They start life as epiphytes and send down aerial roots along the surface of the host trunk. These eventually reach the soil and develop into a stranglehold, killing the host tree. Thus the host is supplanted by the strangler, which forms a tree in its own right without having to start as a seedling on the dark forest floor. The trunks and bark of forest trees are usually thin, since there is no need to protect against cold or fire. Several species exhibit cauliflory—flowering and fruiting from trunks or major branches rather than from slender shoots. Leaves at the top of the canopy, exposed to strong sunlight, heavy rain, and wind, are often hard and small (microphyllous), and angled to limit light incidence. The leaves of midcanopy trees are more commonly of medium size in more shaded conditions. Near the ground, where light is restricted, energy is captured in soft large leaves (macrophyllous) angled to catch the maximum light. Many leaves have mechanisms for shedding excess water from their surfaces (e.g., drip tips), which may help prevent the buildup of leaf epiphytes (which could block light and inhibit photosynthesis).

Some emergent trees may exceed 50 m in height, such as the Brazil nut (*Bertholletia excelsa*), but the closed canopy is mostly 25 to 35 m high. The dense cover inhibits light penetration to the floor, where plants are adapted to low light levels—usually 1–3% of the light incident at the top of the forest. Much of the light at ground level is in the form of sun flecks (Mulkey et al., 1996). Where light increases, for example through tree fall, some pioneer shrubs and trees grow remarkably quickly (e.g., *Cecropia* spp.). Despite their height and bulk, often accentuated by buttressing, trees generally have shallow rooting systems, although in some circumstances (such as areas vulnerable to drought), deep roots occur (Nepstad et al., 1994). They develop a thick dense root mat near the surface rather than taproots or axial roots because virtually all nutrients are close to the surface and they do not normally need moisture-seeking deeper roots. Where there is shallow rooting, the trees can be easily toppled in storms (or by bulldozer) and the buttresses and stilt roots may reflect an attempt to provide greater stability.

The diversity of plants in Amazonia varies greatly. Occasionally plants may form nearly pure stands of one species, such as the monospecific Peltogyne forest occurring in parts of northeast Amazonia (Milliken and Ratter, 1998), and some palm communities, but mostly species are well dispersed in an apparently random pattern. The numbers of species are amazing (Gentry, 1990). In the western Amazon, Valencia et al. (1994) found 307 tree species of 10 cm trunk diameter at breast height and about 500 woody species in a single hectare. However, over a large part of central Amazonia, there are typically far fewer tree species per hectare (60–70), but since different species occur in successive plots the total aggregate of vascular plants has been estimated at 30,000 species or more (Prance, 2001).

The vegetation of Amazonia is closely intertwined with a wealth of vertebrate and invertebrate life. Typical of the myriads of organisms making up the food web of the forest are the ants. Many plants are myrmecophiles or ant-adapted, such as the pioneer *Cecropia* trees. Some plants provide shelter and exude nutritious food (nectar) in return for an ant guard against herbivorous grazers or competing plants. Pollination and dispersal mechanisms, like so many processes in the life of the forest, are achieved through complex links with animals and insects or other organisms.

Amazonian campina (also known as *catinga* or *campinarana*) is a type of vegetation typical of infertile sandy soils derived from the erosion of sandstones surrounding the basin. The soils are frequently white, reflecting extreme weathering and leaching in a porous matrix. Some river terraces have reworked coarse-textured alluvium derived from these sandstones and have similar characteristics, as do the *restingas* of coastal sand dunes. The plants have remarkable adaptations to overcome the severe nutrient deficiencies with extremely tight nutrient cycle mechanisms.

The resulting plant formations vary from woody savanna-like formations to low forest up to around 18 m high. The trees have smaller diameters than those of moist evergreen forest, with more tortuous branching, and may not form a continuous canopy. The increased light reaching the ground surface favors a richer ground flora with numerous endemics, mostly members of the families Bromeliaceae, Marantaceae, and Rutaceae. There are also numerous epiphytic plants, including many ferns (pteridophytes) on the branches. Patches of *campina* vegetation are scattered throughout Amazonia and colonisation of these often distant, sandy sites relies on long-range dispersal by wind, birds, and bats. This contrasts with the more typical moist evergreen forest, where trees tend to produce heavy fruits and seeds and may have different dispersal strategies such as gravity fall and movement along the ground by rodents and other animals.

Dry semi-deciduous and deciduous forests typically extend outward from the evergreen and semi-evergreen formations of central Amazonia into more seasonal climates and/or over more fertile soils (see later). Although they are transitional to *cerrados* in many respects, smaller patches may be found throughout the Amazon, usually identified by indicator trees of more mesotrophic (i.e., more fertile) conditions (Milliken and Ratter, 1998; Ratter et al., 1978). Many species are common to Amazonian *terra firme* forests (e.g., species of the genera *Enterolobium*, *Parkia*, *Schizolobium*, and *Simaoruba*), but there are other species that are absent from these areas, such as the mahogany (*Swietenia macrophylla*).

Transitional Forests At the forest-savanna boundaries, the strongly seasonal climate affects the composition and structure of the forest (Furley et al., 1992). Although the boundaries may be sharp, particularly where they are affected by local factors such as slope, drainage, soils, or fire, many are diffuse and the vegetation formations grade into one another over a hundred kilometers or more (Ratter, 1992). With favorable moisture conditions, moist evergreen forest extends into areas typically covered by savanna. The reverse process is found where variants of savanna or *campina* extend well into the forest on dry, freely drained, nutrient-poor soils. Prance (2001) recognizes three principal types of more specialized transitional forest.

Palm forests are typically composed of babassu (*Attalea speciosa*), and they dominate much of eastern Amazonia. Trees are commonly 10–25 m high, are widely spaced, and often produce a mono-specific stand. There are few endemics. The high fire-resistance of babassu is believed to account for its abundance, and this is probably accentuated by local people, who leave the palms while clearing forest (partly for shade, shelter, nuts, and other uses).

Liana forest is open with scattered trees, but trees can be densely covered in lianas (forming thickets known as Mata de Cipó in Brazil). The lianas belong mainly to the families Bignoniaceae, Dilleniaceae, Leguminosae, and Malpighiaceae. The trees are similar to those found in moist forest but with less diversity and smaller structure, leaves, and biomass. The origin of this forest is often unclear but may also be related to past human disturbance.

Bamboo forests are more open and found frequently in southwest Amazonia. The bamboos are locally abundant in the upper canopy and lower strata, and can reach as much as 30 m in height. They tend to be composed of species of the same genera found widely throughout the Amazonian forests (e.g., *Merostachys*).

Floodplain and Wetland Forests Since around 10% of Amazonia is affected by periodic flooding, the floodplain forests are both extensive and vital to the life of the region. The forests may be flooded up to 10–15 m deep for 5–7 months each year in land close to the rivers. Plants need to be adapted to survive, and the highly unusual character of the forest is illustrated by fish and dolphins swimming among the tree trunks in the wet season. Much of the animal and micro-organic life migrates annually up to the higher parts of the canopy above high water or away from flooded areas onto higher ground. Farther away from the rivers, trees have less time under inundation, and the forest character approaches that of well-drained *terra firme* (Goulding, 1980; Goulding et al., 1996).

Várzea forests are temporarily flooded by mainly white water rivers and develop over more fertile floodplain alluvium than is found in the typical *terra firme* forest. Tidal *várzeas* extend far upriver and contain similar plant associations. Generally, *várzea* forests are more diverse in the upper than in the lower Amazon, and they have a physiognomy similar to that of moist evergreen forest but with fewer tree species. The trees are commonly buttressed or have stilt-roots typical of shallow-rooted woody plants. There are abundant lianas and a rich understory with well-known herbaceous families such as Marantaceae and Heliconiaceae, which feature in many garden centers beyond the tropics. Typical trees include the widespread cotton tree (*Ceiba pentandra*), rubber (*Hevea brasiliensis*), and some virolas (e.g., *Virola surinamensis*). Where extensive tracts of *várzea* forest occur, wet grasslands or meadows (*canarana*) are often found.

Igapó forests are permanently flooded forests along the courses of black-water rivers, such as the Río Negro, and clear-water rivers such as the Xingu, Tocantins, and Tapajos. In the wet season only the tree canopies may be seen above water, and the plants and wildlife are highly adapted to pronounced shifts in water level. Fish often disperse seeds and fruits. In general, trees are lower in height and more scattered than in *várzeas* or *terra firme* forests. Many trees are confined to *igapó* forests, such as palms of the genus *Leopoldinia*. In some cases where clear and black waters intermingle, the influence of both types of forest can be found.

Aquatic and semi-aquatic vegetation occurs in the many scattered lakes, such as oxbows and abandoned river courses, within the wetland forest. Some large lakes exist, such as Lago Manacupuru in the Brazilian state of Amazonas. These shallow water bodies vary in size during a year but typically contain aquatic macrophytes, such as the giant lily (*Victoria amazonica*) and water hyacinth. There are numerous free-floating species, and the force of water often carries mats of grassy vegetation downstream. There are also small patches of permanent swamp with very low tree diversity over humic gley soils. A few stands of palm occur (typically *Mauritia flexuosa* and *Euterpe*). These species indicate the presence of subsurface water throughout Amazonia and central Brazil, and are also found in gallery forests and seepages beyond the forest. Sedges and floating aquatic plants are common wherever shallow pools and semistagnant waters occur.

Mangrove forests occur along much of the tropical open coast, and extend upstream as riverside or deltaic formations in brackish water (see chapter 15). Typical plants show distinctive adaptations to their saline and waterlogged environment. Some trees have adventitious or random roots, which descend from the trunk and anchor themselves in the muddy substrate of shallow waters. Many mangrove species have stilt roots branching from the trunk that provide support in seasonally high-energy conditions. This often gives a tangled appearance to the vegetation. Unlike those of southeast Asia, South American mangroves have few tree species. Most characteristic are red mangrove (*Rhizophora mangle*), which favors the saltiest and most exposed seaward sites, black mangrove (*Avicennia* spp.), and white mangrove (*Laguncularia racemosa*). Other typical plants include *Hibiscus*, *Acrostichum* (mangrove fern), algae, and grasses. Many more species occur at the inner fringes of the forest where soils become leached of their salt content. With time, mangroves can form substantial forests inland and where they become isolated from the sea. Despite being species poor in terms of plants, these mangroves have a high productivity and biomass. Furthermore, their wildlife is often very rich with numerous birds and invertebrates (such as crabs), though they are rarely restricted to mangrove. Raccoons are frequent in Brazil but larger land animals are usually not seen. Crustaceans are very common and caiman (crocodilians related to alligators), manatees, and other larger animals frequent the rich warm waterways. Mangroves are also very important in various stages of the life cycle of marine animals, notably as a breeding ground and nursery for many species of fish.

Savannas Tropical grasslands and woody savannas are widespread within the Amazon and Orinoco basins. They occur in patches within the tropical rain forest of the Amazon lowlands and more extensively farther south, in the *cerrados* that stretch across the Brazilian hills and plateaus between major forested valleys, in the Guaporé basin of eastern Bolivia, and in the Chaco farther south (Oliveira and Marquis, 2002; Mistry, 2000; Furley, 1999). Grasslands also characterize the extensive *llanos* between the Orinoco lowland rain forest and the Venezuelan and Colombian

Andes, as well as the Rupununi high country of the Guiana Shield. They are broadly divided into dry savannas and wet (hydrologic) savannas, and also include gallery woodlands along water courses (fig. 6.1, 9.2; Sarmiento 1992, 1984).

Montane Fringing Ecosystems Many outcrops of crystalline rock around the Amazon basin provide hilly land that supports transitional ecosystems. The low *brejo* hills of northern Brazil, for example, attract and retain cloud, and the moisture released allows the development of tall forest on hilltops. Relatively small changes in slope and drainage can result in very different zones of vegetation. The most dramatic landscapes are those of the sandstone *tepuis* of the Roraima region straddling the Brazil-Venezuela border, where slopes are often clothed in forest but the summits support totally different vegetation formations, although some like the Cerro Guiaquinima in Venezuela have a forest cover. Summit floras are often highly endemic, with each summit, such as Mt. Neblina (3,045 m), having a unique group of species but closely related to those of nearby *tepuis*. Rocky ground with thin soils frequently has an open type of vegetation (*campo rupestre* in Brazil) and looks superficially like savanna but differs floristically and physiognomically. Typical cerrados species are not found but there are representatives of the genera *Byrsonima*, *Clusia*, *Vellozia*, and a number of bromeliads, eriocaulads, and lichens.

Secondary Forests As a result of human disturbance, much of the Amazon is now covered by secondary forest. Around 15% of the primeval forest has been cleared and a possible further 15% disturbed. Human activity has been significant, since the earliest occupation of the region and indeed the composition of the eastern Amazonian forests may be the result of indigenous activity over thousands of years (see chapter 16; Posey et al., 2005). The clearances of the past few decades have altered great swathes of forest, particularly in the southern, more accessible parts of Amazonia. Unlike the indigenous populations who utilized all phases of forest succession, modern day clearance has tended to clear-cut large tracts, forming *capoeiras* (areas of secondary vegetation often in the early weedy stages of plant succession), although there is still extensive selected logging deep within the forest. Secondary growth is also typical of natural gaps in the forest with characteristic pioneer plants such as *Cecropia*, *Byrsonima* or *Vismia* spp.

9.1.3 Amazonian Wildlife

There are several groups of vertebrate and invertebrate animals that are characteristic of Amazonia (see chapter 8). In addition there are hosts of microorganisms whose diversity is only just being uncovered. The silky anteater and two-toed sloth are typical endemics of Amazonia, which also contains 14 of the 16 Neotropical primates (the largest number in the world), marsupials such as opossums, and predator cats such as ocelots and jaguars. There are numerous other animals such as wild dogs, deer, two species of peccary, coatis, kinkajous, and smaller mammals, rodents, and reptiles. Birdlife is particularly rich, colourful, and noisy (such as toucans, parrots, and macaws), and ranges in size from minute hummingbirds to large harpy eagles and giant jaboro storks. There are also interesting groups such as the hoatzins, with claws on the edges of their wings.

The aquatic fauna deserve special mention because of the unique nature of the Amazon River system with the most diverse freshwater fish fauna in the world (Goulding et al. 1996). Estimates of fish species range from 2,500 to 3,000, although only around 1,700 species have been fully described. Understanding fish ecology provides a useful signpost for the survival of other aquatic animals, as well as having direct economic importance. An extraordinary aspect of aquatic life is the way in which fishes (such as the meter-long tambaqui) and their predators (such as the bôto dolphins) extend their range into dense forest during the flood season. There are numerous examples of links between plants (furnishing seeds, fruits, and organic debris) and the life cycle of fishes. The Amazon estuary is particularly productive because of accumulated organic matter, primary production by phytoplankton, and the extensive tidal forests. Ecosystem disturbances have an impact on seasonal shifts and long-distance migrations up and down the river systems. Some catfish, for instance, migrate 2,000 to 3,000 km from the estuary, where they breed, to the headwaters. Little information is available on the geographic range and limits of commercially valuable species, although spawning seems to be confined to sediment-rich waters. Floating meadows and open-water plankton (algae) are also vital for young fishes. Since the main food sources for Amazonian fish are in the floodplain forests, and in the aquatic herbaceous and phytoplankton communities, any disturbance to these systems may have long-term and widespread impacts and disrupt the complex food web (Goulding 1980; Goulding et al., 1996).

9.1.4 Biogeography and Development

The environmental and ecological legacy of Amazonia is threatened by deforestation and inappropriate development. Although large areas of forest in the Amazon and Orinoco basins have changed little over the past century, the southern and eastern margins of Amazonia, and the Pacific coasts of Ecuador and Colombia, have suffered devastating deforestation in recent decades, while the Atlantic Forest was heavily depleted before 1900 (Harcourt and Sayer, 1996). Rapid development has led to the destruction of over half a million square kilometers of forest in Brazil alone over the past 25 years, leading to an increase in the atmospheric carbon flux (Houghton et al., 2000). Nevertheless, satellite imagery indicates that less than 20% of the forest has been cleared at present, although the rate of destruction continues unabated.

Forest clearance has many indirect repercussions in addition to the direct loss of trees and the communities of plants and animals dependent upon them. Occurrences of clear-cutting and burning can be recognized by remote sensing from satellites, but selective logging can take place beneath the upper canopy and is less easy to detect and quantify. The latter may affect an area equal to that of felling (Nepstad et al., 1999). Fire has become more widespread as a result of increasing spells of dry years (ascribed to El Niño effects, see chapter 19), and may penetrate forest areas from adjacent, more flammable savannas and cleared land. Fire and accelerated organic decomposition increase carbon loss to the atmosphere, as well as producing carbon monoxide, nitrogen oxides, ozone, and aerosol particles. On the other hand, intact forest appears overall to be a carbon sink (Houghton et al., 2000; Grace et al., 1995). Repeated fires following logging can cause long-term impoverishment of soils and inhibit regeneration, leading to further carbon loss and biogeochemical impoverishment. Furthermore, watercourses become contaminated with dissolved elements and suspended sediments eroded from the newly exposed catchments and consequently affect the extraordinarily rich aquatic life.

There are many reasons for forest clearance, and it is sometimes difficult to trace the root causes as opposed to the more obvious groups of people or mechanisms responsible for clearing. The most obvious cause is the expansion of agriculture and ranching, either for extensive commercial farming (such as cattle or soy) or for small-scale, mostly subsistence, farming. The impact in southern Amazonia is particularly dramatic as shown by satellite imagery, which allows land-cover changes to be merged with agricultural census data (EOS, 2001; fig. 9.8).

It is unreasonable to expect Amazonia to escape development completely, even given its immense biological and environmental value (Clüsener-Godt and Sachs, 1995). However, the history of Amazonia has been littered with economic spurts (such as the rubber boom) and stagnation, with relatively little evidence of sustainable development however that is defined.

9.2 The Mata Atlântica (Atlantic Forest)

The *Mata Atlântica* forms one of the world's major biological hotspots, with a remarkably high proportion of endemic organisms (Olson and Dinerstein, 1998; Biotropica, 2000). It may contain as much as 7% of the world's plant and animal species (Quammen, 1996). Plant diversity is often equal to or sometimes greater than many Amazonian forests. There are an estimated 2,000 or more species of trees and shrubs, and there is increasing concern over the fate of the rich and varied wildlife. Over 100 plant species have endangered status (IUCN, 1986), including at least one important hardwood, *Caesalpinia echinata* (Mori, 1989). Many of the animals that live in these habitats are simi-

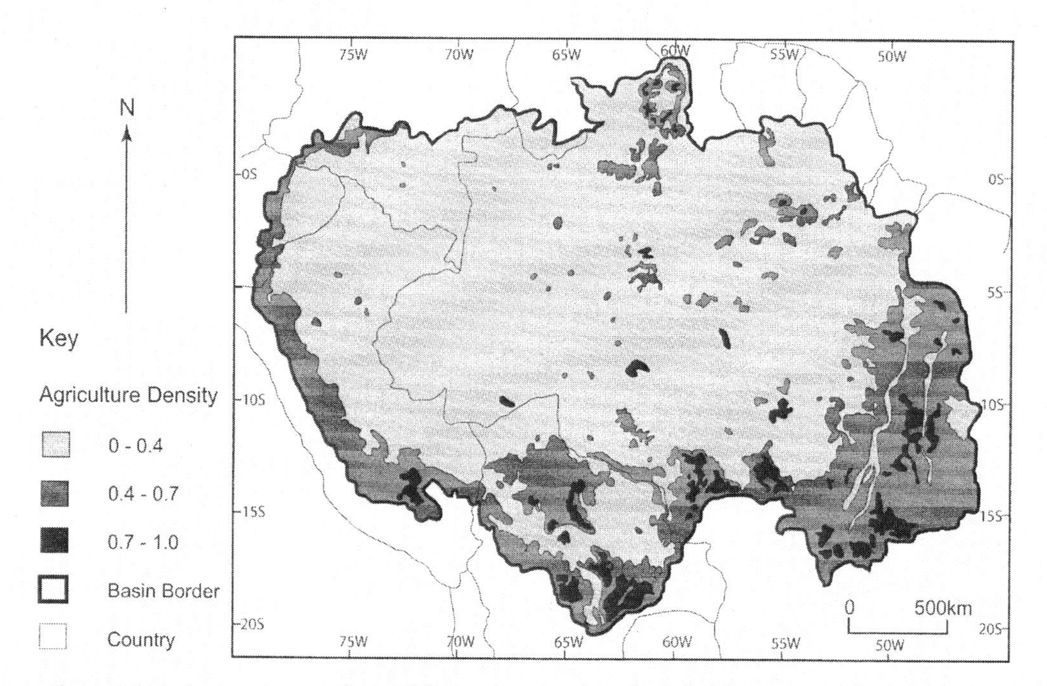

Figure 9.8 Agricultural expansion and forest clearance in Amazonia (after EOS, 2001)

larly threatened. From an original area of perhaps 1.3×10^6 km^2, progressive clearances have fragmented the forest to around 98,000 km^2 (Morellato and Haddad, 2000), 7.5% of its former extent, and sheltering 171 of Brazil's 202 officially endangered species (Brown and Brown, 1992). The remaining forest is frequently located in inaccessible and disconnected pockets, and much of the area still forested is secondary. It is therefore one of the most threatened of all tropical forests, and many initiatives have been set in place to retain what is left and restore connections between forest patches.

The Atlantic Forest forms a very distinctive landscape associated with the coastal mountains and escarpments running southward for 4,000 km along the eastern margin of Brazil, from 5°S at Cape São Roque to 30°S at the Rio Taquari (IBGE, 1993). Although it extends marginally into Uruguay, Argentina, and Paraguay to the south, it is essentially a Brazilian formation. The main core of the forest lies in the Serra do Mar and Serra de Mantiqueira in the states of São Paulo, Minas Gerais, Rio de Janeiro, and Espirito Santo (fig. 9.9). The rugged mountains, steep slopes, and deeply incised valleys, together with the incidence of heavy rains and high humidity, have left an acidic, leached, and stony soil cover, where rooting depth depends largely on gradient and position on the slope. Many soils are inherently unstable and mass movement is commonplace, leaving a mosaic of gaps and successional phases in the forest.

At its greatest former extent, the forest covered much of the narrow coastal plain and ranged inland to merge with the *caatinga* thorn forest and steppe savanna to the north, the *cerrado* and dry semi-deciduous woodland to the west, and the deciduous and mixed pine forest to the south. Today it is confined mostly to montane and submontane environments rising to around 2,700 m but, because of its paleoenvironmental range and lowland connections, can justifiably be included in this account of tropical lowland forests.

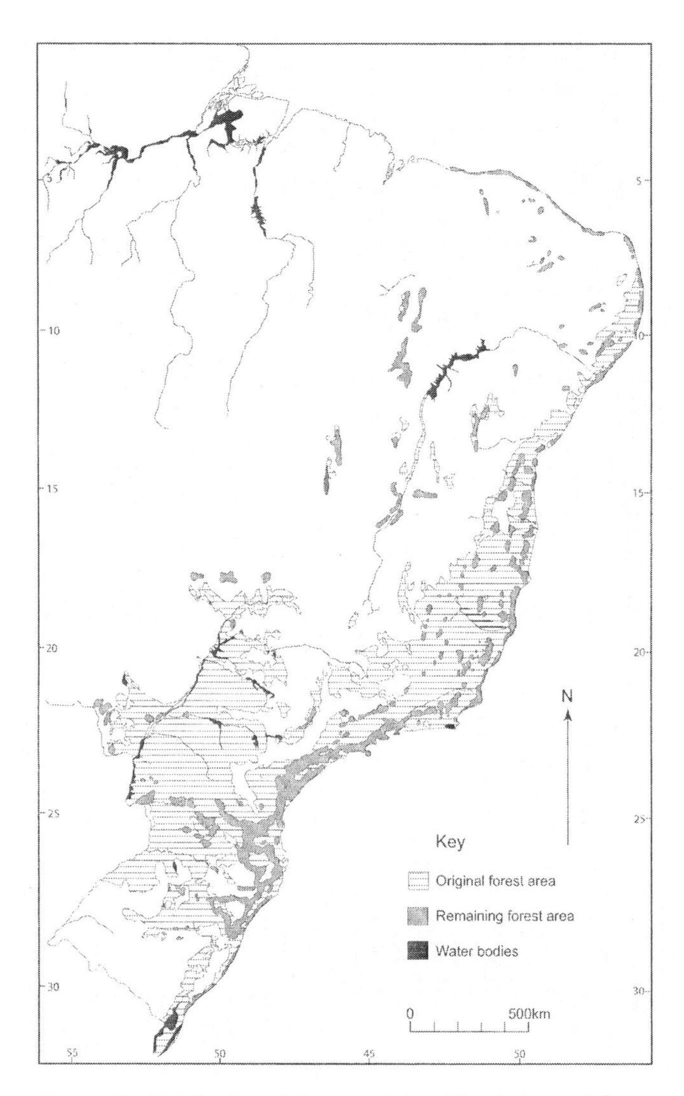

Figure 9.9 Distribution of the remaining Atlantic Forest (after Bohrer, 1998)

9.2.1 Biogeographical Links

The main differences between the Atlantic and lowland Amazon forests relate to altitude and average temperature. Annual temperatures in the Atlantic Forest average between 14 and 21°C, reaching a maximum of ~35°C and a minimum mostly >1°C, though occasionally lower farther south. Not surprising, the more mountainous areas are cooler and occasionally cold. Rainfall regimes are similar between parts of the Atlantic Forest and the Amazon. For example, the central Amazon, between Manaus and Santarém experiences an annual rainfall of ~2,000 mm, and approaches 3,000 mm in the western Amazon lowlands. In the Atlantic Forest, the Teresópolis-Pétrópolis area of Rio Janeiro state has 2,000 to 2,200 mm annually, and parts of the Serra dos Orgãos receive over 3,000 mm. Clearly the topography of the Atlantic Forest region plays a major role with, for example, an altitudinal effect on temperature, a wind-channeling influence, and a differentiating role in rainfall and humidity forming a barrier between east (wetter) and west (drier). The geology and soils are also different. The Amazon forest is supported by soils derived mostly from Cenozoic sediments, whereas the Atlantic forest has mostly evolved over ancient crystalline granite and gneiss (Rizzini et al., 1988).

The floristic composition of the Atlantic Forest bears many points of resemblance with that of the Amazon and this has led to speculation on possible paleoenvironmental links between them. The similarities are greatest at a family level, although there are differences in terms of dominance (Mori et al., 1983). Although richer in epiphytes and tree ferns, the Atlantic Forest has fewer palms and lianas. At a species level, differences between the two formations are more pronounced. At present there are marked climatic and topographic barriers but there are, arguably, older links through the network of gallery forests across the *cerrado* and through the patchwork of semideciduous or deciduous forests reaching through interior Brazil to the dry northeast (Oliveira Filho and Ratter, 1995). These formations have been postulated as linking systems during the retreat of evergreen forest associated with glacial stages of the Pleistocene (Pennington et al., 2000). However, despite a recent surge in research, many gaps in the data remain, thus making it difficult to be sure about past distributions (Behling and Hooghiemstra, 2001).

9.2.2 The Nature of the Forest

The Atlantic Forest is spectacular, though best seen from a distance, since the rugged terrain and extreme moisture make observation difficult at the forest-floor level. At different times of the year, synchronous flowering of species provides vivid splashes of color such as the violet Tibouchina or yellow flowers of *Cassia* species. Although forest structure and composition are known only in detail from a few sites (e.g., Por, 1992), they are extremely rich. The vegetation may be described as montane (>1500 m), lower montane, and submontane moist evergreen forest. However, transitional formations with semideciduous, deciduous, *Araucaria* pine, and other ecosystems are sometimes included in the morphoclimatic Atlantic-Paranaensis biogeographical province (Cabrera and Willink, 1973). The western boundary is more difficult to define because of wide transitional zones, fragmentation, and deforestation.

The vegetation of upper montane formations varies from east to west and with increasing altitude. Above 1,500 m on eastern slopes, there is a less dense form of forest. This reaches 15 m in height and contains numerous shrubs with drier, twisted tree trunks and irregular branching. Lianas and epiphytes become less frequent whereas mosses and lichens increase in proportion and drape branches in foliose, filamentous, and pendulous forms, sometimes up to a meter or more in length (e.g., "Spanish Moss," a lichen, *Usnea*). Above 1,800 m, the forest becomes high montane and loses its evergreen arboreal character to shrubby and grassy forms of vegetation. Lower montane formations occur typically between 800 and 1,700 m above sea level. Trees may attain heights of 20 to 30 m, with some emergents reaching to over 40 m. The largest trees are often *Carinia estrellensis* (Cecythidaceae) with numerous palms (e.g., *Euterpe edulis*, the source of the edible palm heart), conifers (e.g., *Podocarpus* spp.), and many ornamental species. Bamboos and tree ferns (Cyatheaceae) are common, the latter often reaching 5 m and characteristic of the mountain environment (Gentry, 1982). Favored by the high humidity, the forest is typified by an abundance of epiphytes, of which over 1,000 species may exist in the Serra dos Orgãos alone. Myrtaceae are often the most important tree family, with Lauraceae becoming frequent to the south. There are numerous endemic plant species, and some have a very restricted geographic range. It is estimated that 53% of tree species of the Atlantic Forest and up to 70% of its total plant species may be classified as endemic (Mori et al., 1981; Gentry, 1992). The remarkable species richness may be explained, in part, by the changing character of the plant cover over time, with long periods of isolation. There have also been numerous natural disturbances, such as fires and storms, which have generated greater diversity (intermediate disturbance theory). Fragmentation has also led to local species equilibrium (island biogeography theory, Laurance et al., 1998). In a sense there is perpetual disequilibrium, reflecting the sensitive environment and extreme events.

The distinctive submontane forest covers slopes from 300 to 800 m, reflecting the deeper colluvial soils and more marked dry season. Greater ease of access has meant that much of the remaining forest is secondary and shrubby in character. Proximity to or remoteness from the sea affects the plant cover. The understory plants of both montane and lower montane areas contain giant herbs and reflect variations in local environments. This emphasizes the importance of microscale and the value of topographic drainage to an understanding of the relationships among physiography, dynamic processes, nutrient fluxes, and related phenological patterns. Trees in the lower montane forest tend to be shorter (12–25 m), with greater spacing between individuals than in montane locations. Palms are less common and the lower strata are drier, with typical species such as *Piptadenia macrocarpa*, *Persea cordata*, *Ocotea rigida*, and species of *Cedrella* (cedar). Although the forest has been largely cleared from the coastal plains and lower foothills, patches still exist (Rizzini et al., 1988).

Possibly the most publicized of the wildlife are the mammals, among which the tamarins figure prominently. These small endearing primates have high public appeal and have been used as a symbol to draw attention to the plight of their habitats in general (Bierregaard et al., 1997). The highly isolated endemic populations make them very vulnerable to extinction (Rylands, et al., 1997). The lion tamarins (genus *Leontopithecus*) are found today in four disjunct areas, each with a different species having a slightly different set of resource requirements.

9.2.3 Threats to the Mata Atlântica and Conservation

Despite deforestation and settlement from the beginning of the sixteenth century, there has been surprisingly little information on the status of the Atlantic Forest until recently (Dean, 1995; Por, 1992). The forest may have covered as much as 12% of Brazil a century or so ago (Viana and Tabanez, 1996). More recently, the forest has become particularly vulnerable to human impacts because it is situated so close to some of the largest and fastest growing population centers of South America. The progressive destruction of the environment has had a marked effect on the coastal plain and now threatens the montane areas inland, although there are great differences in clearance rates across the region (Viana et al., 1997).

The forests have been exploited for selective and clear logging, cutting for fuelwood and pulp, clearance for agriculture, livestock farming, settlements, hunting of wildlife (for food and for the pet trade), and for the cultivation of fast-growing tree species such as eucalyptus and pine. Heywood and Stuart (1992) have defined the pressures of fragmentation as demographic, genetic, ecological, and accidental. Demographic pressure isolates populations of plants and animals, and species such as top predators no longer have sufficient resources to sustain their needs (Pinto and Rylands, 1997; Laurance et al., 1998). Isolation also results in genetic changes, although it has been argued that such pressures are part of the natural process of speciation and that many animals are in any case resilient. Ecological impacts relate mainly to edge effects and interactions with surrounding habitats, since the size and shape of fragments are influential in accelerating or diminishing these impacts and the problem is compounded by the penetration of tracks and roads. Finally, fragmentation results in an increase in accidental damage as a result of a greater incidence of fires, greater hunting pressures, and increased pollution.

Several attempts have been made to protect the remaining Atlantic Forest. Both the Brazilian government and active NGOs (such as SOS Atlantic Forest, Ranta et al., 1998) have raised awareness of the problems and contributed to a better understanding of the resource. Brazil's oldest national park was established in the Atlantic Forest in 1939, and there are also two World Heritage Sites and an International Biosphere Reserve declared by the United Nations, as well as other conservation units at state and municipal levels. Nevertheless, the usual problems of financial support, lack of information, and lack of qualified personnel mean that only a tiny fraction of the forest is secure (~2% according to Por, 1992). The latest thinking points to a concept of conserving whole landscapes that are usually made up of a mosaic including primary and secondary forest and a mix of agroforestry, agriculture, ranching, and settlement. There clearly needs to be a long-term benefit that is obvious to local people to offset the short-term benefits of clearance. Publicizing flagship species, such as the Golden Lion Tamarin, highlights the issues and provides a model for other conservation efforts in fragmented landscapes.

9.3 The Seasonally Dry Tropical Forests

Dry forests in tropical areas tend to be overshadowed by the publicity given to moist forests (Bellefontaine et al., 2000), yet they are highly distinctive, greatly threatened by development, and have very little protection or conservation. Most tropical ecosystems are to some extent seasonally stressed by drought, and this has been recognized from some of the earliest work in plant geography. Forests experiencing longer spells of dry conditions cover many parts of South America, including northeast Brazil (the drier *cerrados* and *caatinga* formations), the Chaco region of Paraguay, Bolivia, and Argentina (Prado and Gibbs, 1993), tracts along the Caribbean coasts of Colombia and Venezuela, the dry valleys of the Andes (Pennington et al., 2000), and scattered patches, often associated with calcareous outcrops from northeast Brazil to the southwest Chaco (fig. 9.10; Ratter et al., 1978, Prado, 2000).

9.3.1 Dry Tropical Forest Terminology

Dry forest is a confusing term and could refer to both climatic and vegetational concepts. FAO terminology groups forests into closed forest, open woodland, and savanna (>10% tree cover). Broadly therefore, there are a variety of dry forest formations that have been given a confusing number of names. This chapter recognizes dry deciduous forest, open woodland, and woody savannas, and steppe-like formations of the *chaco* and *caatinga*. There are close biological links between most of these groups within the present-day distributions, although they are often widely separated. The current spatial patterns represent simply a moment in time in the constant dynamic flux of vegetation formations (Furley et al., 1992).

Dry deciduous forest is relatively rare today. It is mostly closed (i.e., no substantial openings in the canopy cover), some 15–20 m in height, with a tree component that loses most of its leaves in the dry season. The understory is sparse with scattered evergreen and deciduous shrubs and a patchy herbaceous cover in the gaps. Mesotrophic dry forest is a subdivision of the dry deciduous forest and tends to persist, as the name suggests, in areas of more fertile soil. In these patches, where trees remain today mainly because of inaccessibility and sharp limestone relief, there are moderate pH values (often alkaline) and medium to high nutrient levels. A number of trees are characteristic of these more favourable conditions, such as *Anadenanthera colubrina*, *Aspidosperma subincanum*, *Cedrela fissilis*, *Storculia stricta*, and *Tabebuia impetiginosa*.

Open woodland generally contains smaller trees. Although the canopy cover may be fairly continuous (~40–60% according to Bellefontaine et al., 2000), light is able to penetrate to the ground, resulting in an irregular grassy ground cover. The ratio of woody to herbaceous biomass has sometimes been used to subdivide these woodlands.

Savanna woodland (*cerradão*) in the savannas has many similarities to the dry forests. A broad distinction can be made, however, between the xeromorphic, fire-tolerant vegetation typical of *cerrados* with dystrophic soils (acidic

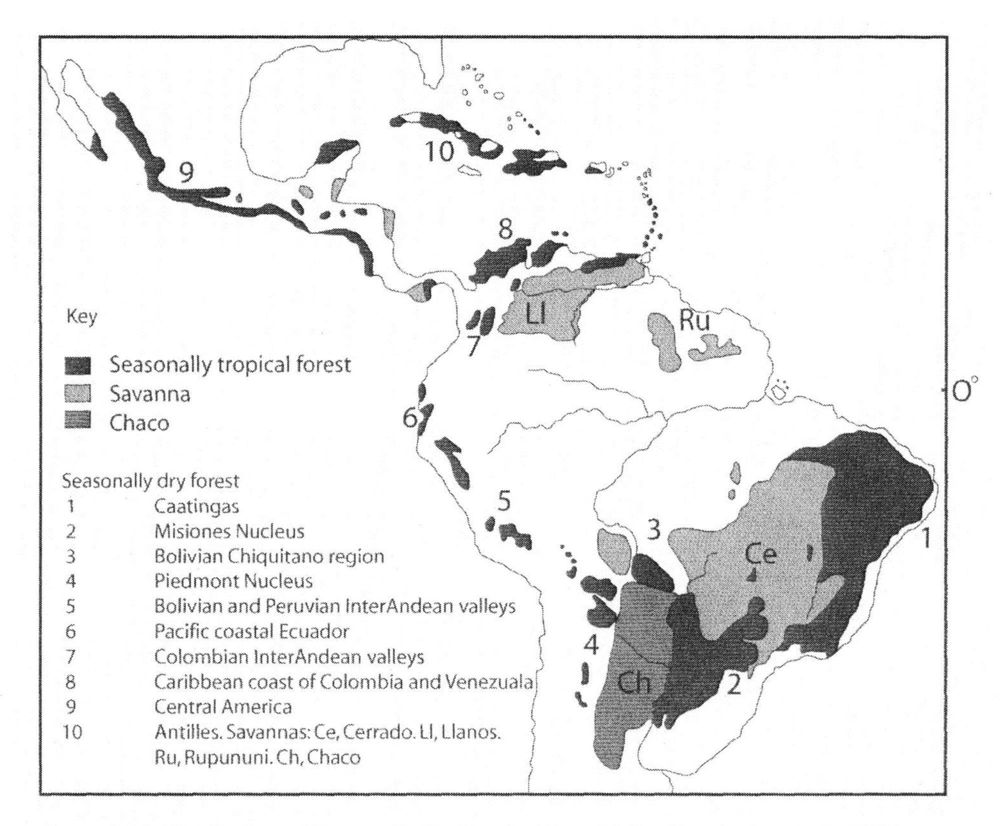

Figure 9.10 Distribution of Seasonally Dry Tropical Forest (after Pennington et al., 2000)

and nutrient poor), and the tree-dominated, mesotrophic ecosystems of the tropical dry forests that suffer far less frequent burning.

Finally, the *chaco* and *caatinga* formations represent the dry extremes of forest or woodland in South America. They have a very individual character of low, dense thickets with numerous thorny species. The woodland often appears lifeless for much of the year, although the drier forests can show an increase in evergreen and succulent species. Flowering and fruiting are strongly seasonal. For this study, the *caatinga* formation is discussed separately in section 9.4.

9.3.2 Dry Forest Environments

The dry forests are primarily a reflection of climate, though they are strongly influenced by the occurrence of more fertile soils. They are found mostly in regions where annual rainfall ranges from around 300 to 1,200 mm, typically with 5 to 10 dry months. Gentry (1995) considered seasonally dry tropical forest to occur where annual rainfall is less than 1,600 mm, with at least 5 or 6 months receiving less than 100 mm (table 9.2). These figures should be taken as guidelines, since there are exceptions and much depends on the precise definition of the vegetation and on local environmental differences. In particular the water storage capacity of soils and soil-moisture availability are critical factors that affect plant-community variation but for which there is usually little information (Sarmiento, 1992, 1984). The balance is usually regulated by human intervention with or without fire disturbance.

9.3.3 Biogeographical Links

There is substantial evidence of historic and prehistoric connections between the present day disjunct formations. Dry forest elements in the vegetation have been identified from the middle to late Eocene (Graham and Dilcher, 1995), and more widespread connections may have existed in the past between currently separate tracts of forest. For example, Prado and Gibbs (1993) found some 40 unrelated species in 10 disjunct remnants. These species do not occur in the moist forest or savanna vegetation that now surrounds them. Pennington et al. (2000) suggest that there may be over 60 further species with similar characteristics. Although many of the families and genera of Neotropical dry forests are shared with lowland moist forest, this relationship applies less at species level. It might be concluded therefore that species have evolved later in response to fragmentation and isolation. Prado and Gibbs (1993) suggest that the present-day distributions represent the remains of a once extensive and largely

Table 9.2 Climate and vegetation characteristics of dry and wet forests

	Wet forests	Dry forests
Annual rainfall (mm)	>2000	500–2000
Seasonality	slight	strong
Number of tree species (ha^{-1})	50–200	35–90
Canopy height (m)	20–84	10–40
Leaf area index (m^2/m^2)	5–8	3–7
Ground vegetation cover	low	low–high
Basal area of trees (m^2 ha^{-1})	20–75	17–40
Plant biomass (t ha^{-1})	270–1200	80–320
Root biomass as percentage of total (%)	<5–33	8–50
Net primary productivity (t ha^{-1} yr^{-1})	13–28	8–21

Adapted from Murphy and Lugo (1986).

continuous formation, which may have reached its maximum extent between 18,000 and 12,000 years ago, and was coincident with the contraction of moist forest. The floristic composition of *chaco* is however rather different, despite a similar appearance and is believed by Pennington et al. (2000) to be a subtropical extension of an essentially temperate type of vegetation with links to Andean formations.

9.3.4 Nature of the Vegetation

Dry tropical forest is defined here in a general sense and ranges from tall forest in moister sites to cactus scrub in some of the driest (Murphy and Lugo, 1995). Leguminosae and Bignoniaceae families dominate the woody floras, with strong representation from other families such as Anacardiaceae, Capparidaceae, Euphorbiaceae, Flacourtiaceae, Myrtaceae, Rubiaceae, and Sapindaceae. The most common tree genus is *Tabebuia,* and other prominent genera include *Anadenanthera, Casearia, Trichilia, Erythroxylum, Capparis, Bursera,* and *Acacia.* At the dry end of the spectrum, cacti are especially dominant in the understory, and sometimes in the tree strata, and are highly visible. The sparse ground cover usually comprises a few grasses and members of the families Bromeliaceae and Compositae, along with Malvaceae and Marantaceae.

Generally, dry forests have fewer plant species and lower biomass than moister forests, but the most diverse dry forests are not necessarily the wettest. Their centers of endemism are closer in latitude to the tropics than the Equator. There is considerable diversity of plant life forms, notably in terms of structure and physiology. Conspicuous flowers and wind-dispersed seeds are more frequent than in more moist forest. Shade-loving plants and organisms become stressed in more open conditions and with greater penetration of light, heat, and wind. On the whole there are fewer epiphytes and more vines (though fewer lianas) than in moist forests. For several features, such as the number of trees or basal area, dry forests differ little from moist forests (Gentry, 1995), but taking a general perspective across the whole of South America and Mesoamerica, there are major differences between the different continental tracts of dry forest.

Since deciduous forests often occur on more fertile soils, they have attracted agricultural development from the earliest days of colonization, leading to very extensive clearance and disturbance. For example, Janzen (1988) estimates that less than 2% of such forests are still intact along the Pacific coast of Central America, and agricultural colonists in Brazil have long targeted them. Forest exploitation has taken several forms in addition to clearance for agriculture and ranching. Many plant and animal products have been traditionally extracted, though mostly for local or regional use. In particular, dry forest timbers are hard and durable and have traditionally supplied high-quality wood. Local people have also used a variety of species for food, fuel, or medicine. As a result of this history of spontaneous exploitation and destruction, little remains of the original forests and few of the remnants have any form of protection.

9.4 The Caatinga Woodlands

In South American lowland forests, the extremely dry end of the vegetation spectrum is exemplified by *caatinga*, an indigenous term for "white forest," referring to the dusty bleached appearance of the vegetation during the dry season. The *caatinga* is a seasonally dry forest, as discussed in the previous section, but is given special emphasis here because of its unusual properties. The *caatinga* is a type of dry forest characteristic of northeast Brazil, but it reaches south to northwest Minas Gerais and northern Espirito Santo, and covers nearly one million km^2, or around 11% of Brazil (Rizzini et al., 1988; Sampaio, 1995). The forest is generally a low, dense thorn-thicket dominated by deciduous and xerophytic trees. Although predominantly woody, there are numerous prickly and succulent cacti, many bromeliads, and a patchy covering of annual herbs and grasses on the forest floor that flourish in the short wet season.

9.4.1 The Caatinga Environment

The *caatinga* stretches over low plains and plateaus lying mostly 300 to 600 m above sea level. The soils have developed over a range of crystalline and sedimentary rocks, sometimes sandy, sometimes clay-rich, but usually shallow and frequently interspersed with rocky outcrops. The sandy but fertile substrates tend to favor *caatinga* species (Andrade Lima, 1981). There may be nine months or more of rainless days and a few months of torrential but irregular rain, which can cause severe erosion. Several years may pass with virtually no rainfall, followed by intense storms that can cause a great deal of damage to crops and plants. The long, dry periods and high rates of evapotranspiration result in salts being brought to the surface by capillary action, giving saline soils that again inhibit agriculture and plant growth. The water table is usually at considerable depth, encouraging plants with long tap roots or those that can withstand water deficits (e.g., cacti) or can complete their life cycle during brief wet spells (e.g. ephemeral forbs). The onset of rain triggers a dramatic greening of the landscape.

9.4.2 Biogeographical links

Geological and biological evidence suggests that, during past climatic fluctuations, the *caatinga* formation advanced on and retreated from surrounding moister vegetation formations (Andrade Lima, 1981). The remnant hilltop humid forest (*brejo*), surrounded by *caatinga*, is testimony to these changes. Some species suggest older connections, such as *Anadenanthera colubrina* (found also in the *chaco*), or *Cavanillesia arborea* (also widespread in Acre in the southwest Brazilian Amazon). At a generic level the links are more frequent. To the west there is a transition into the *cerrado*, with the ecotonal vegetation termed "deciduous cerrado" (Andrade-Lima, 1981).

9.4.3 Structure and Composition of the Vegetation

Plant life has adapted to these extreme conditions and closely reflects the availability of water. The *caatinga* can be subdivided phytogeographically into an *agreste* type and a *sertão* type. The *agreste* type lies closer to the sea and therefore in a more humid environment. Here the vegetation is taller and denser, often on deeper soils (e.g. the Apodi plateau in Ceara). The trees may be 10 m or more in height with thick, upright trunks often dominated by a few species such as *Erythrina velutina*.

The second type is the hot dry *sertão*, a term that is used both for the dry vegetation type and as a general name for the isolated arid landscapes of the interior of northeast Brazil. Here the vegetation may be made up of stunted shrubs (cacti and bromeliads), only 2–3 m high, giving a sparse covering to rocky areas and shallow soils, or of small trees up to 5–6 m high, with a dense shrubby ground cover. The cacti are often tree-like in stature and reach 4–6 m in height. In broad terms, the plants may be viewed as permanent or periodic. The former are often long-lived and most shed their leaves in the dry season. The latter include grasses and herbs that die down during the dry season but recover during the rainy season or wet spells. Water storage adaptations are found in many *caatinga* species. Some have swollen trunks (such as the Bombaceous trees *Chorisia glaziovii* or *Cavanillesia arborea*), and some have underground storage tubers (e.g., *Spondias tuberosa*). A few species are capable of trapping water from humid air, notably the Bromeliaceaous *Tillandsia* species, which hang in festoons from the branches of trees and absorb moisture through their scales.

The arboreal vegetation displays a number of typical features. Frequently there is profuse branching and woodiness generating hard spiny branches. These are extremely sharp and justify the thick leather clothing worn by *vaqueiros* (cowboys), even covering their horses. The trees often bear medium or large sized leaves (such as *Auxemma oncocalyx,* the "pau-branco"; or *Cavenillenia arborea*, the "barriguda"), rather than the small hard leathery leaves so often typical of xeric vegetation. In the wet season such trees resemble the Atlantic Forest rather than xerophytic forest, but they are less strongly deciduous. Most of the time the *caatinga* looks like a semi-arid dry forest and only a relatively few true xerophytes retain their hard sclerophyllous, thorny, and leathery leaves (e.g., *Zizyphus*, *Capparis,* and *Maytenus* genera).

Although the dominants in *caatinga* are trees, the shrub layer is frequently the most eye-catching. Succulents such as Euphorbiaceae are very common but thick spiny cacti are perhaps the most spectacular plants, with exotic shapes and colors. Typical of this group is *Pilosocereus gounellei*, resembling candelabra with its hard erect branches. Some, such as the tall branched mandacaru (*Cereus jamacaru*), produce berries that afford a dry season source of water for grazing animals. Cacti are sometimes planted to give a water source in extreme conditions and some grazing for cattle. There are also stinging shrubs and small trees that contain a caustic liquid to dissuade grazers. Nettle-like plants belonging to the families Euphorbiaceae and Loasaceae are common in the ground layer. Overall the most common families in the *caatinga* are Leguminosae, Malvaceae, Verbenaceae, and Tiliaceae, with bromeliads being particularly common in the ground layer. There are few plant genera that are endemic (such as *Neoglaziovia*) and many of the plants also occur in other ecosystems surrounding the *caatinga*. However, the xerophytes make an exception, and over 60% of them may be exclusive to the region.

9.4.5 Wildlife

Little is known of the *caatinga* fauna, which is believed to have lost many species as a result of fires and grazing pressures. Such disturbances aggravate an environment already sensitive to periodic droughts and in many ways marginal to plant growth and animal life. Many animals are nocturnal and burrow, and most also occur in adjacent ecosystems such as the *cerrado*.

9.4.6 Development

Much of the *caatinga* has been heavily disturbed. Despite its inhospitable climate, frequent droughts, and lack of water, the *sertão* has been traditionally used for extensive ranching. Farms tend to be large and the carrying capacity for cattle may be as low as 1 head per 10 ha. There are also a number of useful native fruits such as the mandacaru (*Cereus jamacaru*), ico (*Capparis*), and umbu (*Spondias tuberosa*), although the most prized are not native to the region. A number of valuable woods come from the caatinga, some very hard such as barauna (*Schinopsis brasiliensis*), but because of the low height of the trees, they have not in recent times been utilized for timber. Many other plants have traditional uses, for instance *Manihot glaziovii* for latex. Excessive destruction of the ecosystem has enhanced the natural dryness of the environment and led to initiatives to prevent erosion and desertification, mainly through reducing the numbers of cattle and by stabilizing the vegetation. In sum, it is a semi-arid landscape suffering from episodic droughts and a general shortage of water, and the evocative terms *sertão* and *sertanistas* (those who live there) epitomize the hard lifestyle of the settlers, similar in many ways to the concept of Outback in Australia.

9.5 Conclusion

The Neotropical forest ecosystems are of immense importance from the perspective of their diverse biology but at the same time are perceived quite differently in terms of their resource potential for development. They are the home for a large proportion of the world's plant, animal, and microbial species, with exceedingly complex interactions between organisms and the environment. Their history has been one of frequently intense and constantly dynamic change. In addition to vital functions in trapping solar energy and their exuberant biological productivity (acting as a crucial carbon dioxide sink), they also moderate the flow of water across the land, limiting soil erosion and protecting watersheds. Thoughtless deforestation accompanied by short-term land management destroys these environmental services and puts at risk the unique lifeforms and resources that currently exist. If sustainable development of tropical forest really does mean addressing present justifiable demands "without compromising the ability of future generations to meet their own needs" (World Commission, 1987), then a worldwide effort is required to understand the systems better and to manage them more effectively.

References

Andrade-Lima, D. de, 1981. The caatingas dominium. *Revista Brasileira Botânica*, 4, 149–163.

Behling, H., 1998. Late Quaternary vegetation and climatic changes in Brazil. *Review of Palaeobotany and Palynology*, 99, 143–156.

Behling, H, 2002. South and south-east Brazilian grasslands during Late Quaternary times: A synthesis. *Paleogeography Palaeoclimatology Palaeoecology*, 177, 19–27.

Behling, H., and H. Hooghiemstra, 2001. Neotropical savanna environments in space and time: Late Quaternary interhemispheric comparisons. In: V. Markgraf (Editor), *Interhemispheric Climate Linkages*, Academic Press, London, 307–323.

Bellefontaine, R., et al., 2000. *Management of Natural Forest in Dry Tropical Zones*. FAO, Rome.

Bierregaard, R.O., et al., 1997. Key priorities for the study of fragmented tropical ecosystems. In: W.F. Laurance and R.O. Bierregaard (Editors), *Tropical Forest Remnants*, University of Chicago Press, Chicago.

Bigarella, J.J, and A.M.M. Ferreira, 1985. Amazonian geology and the Pleistocene and Cenozoic environments and palaeoclimates. In: G.T. Prance and T.E. Lovejoy (Editors), *Key Environments: Amazonia*, Oxford University Press, Oxford, 49–71.

Biotropica, 2000. *The Brazilian Atlantic Forest*. Special Issue 32.

Bohrer, C. 1998. Ecology and biogeography of an Atlantic montane forest in south-eastern Brazil. Ph.D. dissertation. Unversity of Edinburgh, Edinburgh.

Brown, K.S., and G.G. Brown, 1992. Habitat alteration and species loss in Brazilian forests. In: T.C. Whitmore and J.A. Sayer (Editors), *Tropical Deforestation and Species Extinction*. Chapman & Hall, London.

Bush, M. B. 1994. Amazonian speciation: A necessarily complex model. *Journal of Biogeography* 21, 5–17.

Cabrera, A.L., and A. Willink, 1973. *Biogeografia de America Latina*. OEA, Washington, D.C.

Cochrane, T., P.A. Sanchez, L.C. Azevedo, and L.C. Garve, 1985. *Land in Tropical America, 1–3*. CIAT (Centro de Agricultura Tropical), Cali, Colombia.

Colinvaux, P. A., 1996. Quaternary environmental history and forest diversity in the Neotropics. In: J.B.C. Jackson, A.F. Budd, and A.G. Coates (Editors), *Evolution and environment in Tropical America*. University of Chicago Press, Chicago, 359–405.

Colinvaux, P.A., 1997. An arid Amazon? *Trends in Ecology and Evolution*, 12, 318–319.

Clüsener-Godt, M., and I. Sachs (Editor), 1995. *Brazilian Perspectives on Sustainable Development of the Amazon Region*. UNESCO/Parthenon, Carnforth, UK.

Dean, W., 1995. *With Broadax and Firebrand: The Destruction of the Brazilian Atlantic Forest*. University of California Press, Berkeley.

Eden, M.J., 1990. *Ecology and Land Management in Amazonia*. Belhaven Press, London.

EOS, 2001. Answers sought to big questions about the Amazon region. *Transactions, American Geophysical Union*, 82, 405–406.

Furley, P.A., 1986. Radar surveys for resource evaluation in Brazil: An illustration from Rondonia. In: M.T. Eden and J.T. Parry (Editors), *Remote Sensing and Tropical Land Management*, Wiley, Chichester.

Furley, P.A., 1990. The nature and sustainability of Brazilian Amazon soils. In: D. Goodman and A. Hall (Editors), *The Future of Amazonia*, Macmillan, London, 309–359.
Furley, P.A., 1999. The nature and diversity of Neotropical savanna vegetation with particular reference to the Brazilian cerrados. *Global Ecology and Biogeography*, 8, 223–241.
Furley, P.A., 2005. Fragility and resilience in Amazonian soils. In: D.A. Posey, M. Ballick, and W. Capraro (Editors), *Human Impacts on the Amazon: The Role of Traditional Ecological Knowledge in Conservation and Development*, Colombia University Press, New York.
Furley, P.A., J. Proctor, and J.A. Ratter (Editors), 1992. *Nature and Dynamics of Forest-Savanna Boundaries*. Chapman & Hall, London.
Gentry, A.H., 1982. Patterns of Neotropical plant species diversity. *Evolutionary Biology,* 5, 1–84.
Gentry, A.H., 1990. *Four Neotropical Forests*. Yale University Press, New Haven.
Gentry, A.H., 1992. Tropical forest diversity: Distribution patterns and their conservational significance. *Oikos*, 63, 19–28.
Gentry, A.H., 1995. Diversity and floristic composition of Neotropical dry forests. In: S.H. Bullock, H.A. Mooney, and E. Medina (Editors), *Seasonally Dry Tropical Forests*, Cambridge University Press, Cambridge, 146–194.
Glasser, B., G. Guggenberger, L. Haumaier, W. Zech, 2001. Persistence of soil organic matter in archaeological soils (*terra preta*) of the Brazilian Amazon region. In: R.M. Rees, B.C. Ball, C.D. Campbell, and C.A. Watson (Editors), *Sustainable Management of Soil Organic Matter*. CAB International, Wallingford, UK, 190–194.
Gliessman, S.R., 1990. *Agroecology: Researching the Basis for Sustainable Development*. Springer, New York.
Goulding, M., 1980. *The Fishes and the Forest: Explorations in Amazonian Natural History*. University of California Press, Los Angeles.
Goulding, M., N.J.H. Smith, and D.J. Mahar, 1996. *Floods of Fortune: Ecology and Economy Along the Amazon*. Columbia University Press, New York.
Grace, J., J. Lloyd, J. McIntyre, A. C. Miranda, P. Meir, H. S. Miranda, C. Nobre, J. Moncrieff, J. Massheder, Y. Malhi, I. Wright, J. Gash 1995. Carbon dioxide uptake by an undisturbed tropical forest in southwest Amazonia, 1992–1993. *Science,* 270, 778–780.
Graham, A., and D. Dilcher, 1995. The Cenozoic record of tropical dry forest in northern Central America and the southern United States. In: S.H. Bullock, H.A. Mooney, and E. Medina (Editors), *Seasonally Dry Tropical Forests*, Cambridge University Press, Cambridge, 124–145.
Harcourt, C.S., and J.A. Sayer (Editors), 1996. *The Conservation Atlas of Tropical Forests: the Americas*. Simon and Schuster, New York.
Hartshorn, G., 2001. Tropical forest ecosystems. In: S.A. Levin (Editor), *Encyclopedia of Biodiversity*, 5, Academic Press, New York, 701–710.
Herrera, R., 1985. Nutrient cycling in Amazonian forests. In: G.T. Prance and T.E. Lovejoy (Editors), *Key Environments: Amazonia*, Pergamon Press, Oxford, 95–108.
Heywood, V.H., and S.N. Stuart, 1992. Species extinctions in tropical forests. In: T.C. Whitmore and J.A. Sayer (Editors), *Tropical Deforestation and Species Extinction*, Chapman & Hall, London, 90–117.
Holdridge, G.R., 1972. *Forest Environments in Tropical Life zones: A Pilot Study*. Pergamon Press, Oxford.
Houghton, R.A., D.L. Skole, C.A. Nobre, J.L. Hackler, K.T. Lawrence, and W.H. Chomentowski, 2000. Annual fluxes of carbon from deforestation and regrowth in the Brazilian Amazon. *Nature*, 403, 301–304.
IBGE, 1977. *Geografia do Brasil. 1: Região Norte*. Brazilian Institute of Geography and Statistics, Rio de Janeiro.
IBGE, 1993. *Map of the Vegetation of Brazil* (1:5,000,000 scale). Brazilian Institute of Geography and Statistics, Rio de Janeiro.
IUCN, 1986. *The Plant CITES Red Data Book*. International Union for the Conservation of Nature, Gland, Switzerland.
Janzen, D., 1988. Tropical dry forests: The most endangered major tropical ecosystem. In: E.O. Wilson (Editor), *Biodiversity*. National Academy Press, Washington, D.C., 130–137.
Jordan, C.F., 1982. The nutrient balance of an Amazonian rainforest. *Ecology,* 63, 647–54.
Jordan, C.F., 1985a. *Nutrient Cycling in Tropical Forest Ecosystems*. Wiley, Chichester.
Jordan, C.F., 1985b. Soils of the Amazon forest. In: G.T. Prance and T.E. Lovejoy (Editors), *Key Environments: Amazonia*. Pergamon Press, Oxford, 83–94.
Junk, W. J., and K. Furch, 1985. The physical and chemical properties of Amazonian waters and their relationships with biota. In: G.T. Prance and T.E. Lovejoy (Editors), *Key Environments: Amazonia*. Oxford: Pergamon Press, Oxford, 3–17.
Laurance, W.F., L.V. Ferreira, de M. Rankin, and S.G. Laurance, 1998. Rain forest fragmentation and the dynamics of Amazonian tree communities. *Ecology*, 79, 2032–2040.
Longman, K.A., and J. Jenik, 1987. *Tropical Forest and Its Environment*. Longman, London.
Marchant, R., et al., 2000. Pollen-based biome reconstructions for Latin America at 0, 6000 and 18000 radiocarbon years. *Journal of Biogeography*, BIOME 6000 Special Issue.
Marengo, J.A., and C.A. Nobre, 2001. General characteristics and variability of climate in the Amazon Basin and its links to the global climate system. In: M.E. McClain, R. Victoria, and J.E. Richey (Editors), *The Biogeochemistry of the Amazon Basin*, Oxford University Press, New York.
McCann, J.M., W.I. Woods, and D.W. Meyer, 2001. Organic matter and anthrosols in Amazonia: Interpreting the Amerindian legacy. In: R.M. Rees, B.C. Ball, C.D. Campbell, and C.A. Watson (Editors), *Sustainable Management of Soil Organic Matter*. CAB International, Wallingford, UK, 190–194.
Milliken, W., and J.A. Ratter (Editors), 1998. *The Biodiversity and Environment of an Amazonian Rainforest*. Wiley, Chichester.
Mistry, J., 2000. *World Savannas*. Prentice Hall, Harlow, UK.
Molion, L.C.B., 1987. Micrometeorology of an Amazonian rain forest. In: G.E. Dickinson (Editor), *The Geophysiology of Amazonia*. Wiley, New York, 255–272.
Morellato, L.P.C., and C.F.B. Haddad, 2000. Introduction. Special Issue: The Brazilian Atlantic Forest. *Biotropica*, 32, 786–792.

Mori, S.A., 1989. Eastern extra-Amazonian Brazil. In: D.C. Campbell and H.D. Hammond (Editors), *Floristic Inventory of Tropical Countries,* New Botanical Gardens, New York, 428–448.
Mori, S.A., B.M. Boom, G.T. Prance, 1981. Distribution patterns and conservation of eastern Brazilian coastal forest species. *Brittonia,* 33, 233–245.
Mori, S.A., B.M. Boom, A.M. de Carvalho, and D.S. dos Santos, 1983. Southern Bahía moist forest. *Botanical Review,* 49, 158–232.
Mulkey, S.S., R.L. Chazdon, and A.P. Smith (Editors), 1996. *Tropical Forest Plant Ecophysiology.* Chapman & Hall, New York.
Murphy, P., and A.E. Lugo, 1986. Ecology of tropical dry forest. *Annual Review of Ecology and Systematics,* 17, 67–88.
Murphy, P., and A.E. Lugo, 1995. Dry forests of Central America and the Caribbean. In: S.H. Bullock, H.A. Mooney, and E. Medina (Editors), *Seasonally Dry Tropical Forests.* Cambridge University, Press, Cambridge, 146–194.
Nepstad, D. C., et al., 1999. Large scale impoverishment of Amazonian forests by logging and fire. *Nature,* 398, 505–508.
Nepstad, D.C., de C.R. Carvalho, and S. Vieira, 1994. The role of deep roots in the hydrological and carbon cycles of Amazonian forest pastures. *Nature,* 372(6507), 666.
Nieuwolt, S., 1977. *Tropical Climatology.* Wiley, London.
Oliveira-Filho, A.T., and J.A. Ratter, 1995. A study of the origin of central Brazilian forests by the analysis of plant species distribution patterns. *Edinburgh Journal of Botany,* 52, 141–194.
Oliveira, P.S., and R.J. Marquis, 2002. *The Cerrados of Brazil.* Columbia University Press, New York.
Olson, D. M., and E. Dinerstein, 1998. The global 200: A representation approach to conserving the Earth's most valuable ecoregions. *Conservation Biology,* 12, 503–515.
Pennington, T.R., D.E. Prado, and C.A. Pendry, 2000. Neotropical seasonally dry forests and Quaternary vegetation changes. *Journal of Biogeography,* 27, 261–273.
Pinto, L.P.D S., and A.B. Rylands, 1997. Geographic distribution of the Golden Headed Lion Tamarin, *Leonopithecus chrysomelas*: Implications for its management and conservation. *Folia Primatologia,* 68, 161–180.
Por, F. D., 1992. *Sooretama: The Atlantic Rain Forest of Brazil.* SPB Academic Publishers.
Posey, D.A., M. Balick, and W. Capraro (Editors), 2005. *Human Impacts on Amazonia: The Role of Traditional Knowledge in Conservation and Development.* Columbia University Press, New York.
Prado, D.E., 2000. Seasonally dry forests of tropical South America: From forgotten ecosystems to a new phytogeographic unit. *Edinburgh Journal of Botany,* 57:437–461.
Prado, D.E., and P.E. Gibbs, 1993. Patterns of species distributions in the dry seasonal forests of South America. *Annals of the Missouri Botanical Garden,* 80, 902–907.
Prance, G. T., 2001. Amazon ecosystems. In: S.A. Levin (Editor), *Encyclopedia of Biodiversity,* 1. Academic Press, New York, 145–157.
Prance, G.T., and T.E. Lovejoy, 1985. *Key Environments: Amazonia.* Pergamon Press, Oxford.
Quammen, D., 1996. *The Song of the Dodo—Island Biogeography in an Age of Extinction.* Pimlico, London.
Ranta, P., T. Blom, J. Nameable, E. Joensu, and M. Siitonen, 1998. The fragmented Atlantic rain forest of Brazil: Size, shape and distribution of forest fragments. *Biodiversity and Conservation,* 7, 385–403.
Ratter, J.A., 1990. House of Commons Environment Committee Report. H.M.S.O., London.
Ratter, J.A., 1992. Transitions between cerrado and forest vegetation in Brazil. In: P.A. Furley, J. Proctor, and J.A. Ratter (Editors), *Nature and Dynamics of Forest-Savanna Boundaries.* Chapman & Hall, London, 417–430.
Ratter, J. A., G.P. Askew, R.F. Montgomery, and D.R. Gifford, 1978. Observations on the vegetation of northeastern Mato Grosso II. Forest and soils of the Rio Suiá-Missu area. *Proceedings of the Royal Society of London. Series B, Biological Sciences,* 203, 191–208
Richards, P. W., 1996. *The Tropical Rain Forest.* 2nd edition. Cambridge University Press, Cambridge.
Rizzini, C.T., A.F. Coimbra-Filho, and A. Houaiss (Editors), 1988. *Ecosistemas Brasileiros—Brazilian Ecosystems.* Editora Index, Rio de Janeiro.
Rylands, A.B., R.A. Mittermeier, and E. Rodriguez-Luna, 1997. Conservation of Neotropical primates: Threatened species and an analysis of primate diversity by country and region. *Folia Primatologia,* 68, 134–160.
Salati, E. 1985. The climatology and hydrology of Amazonia. In: G.T. Prance and T.E. Lovejoy (Editors), *Key Environments: Amazonia,* Pergamon Press, Oxford, 18–48.
Salati, E., 1987. The forest and the hydrological cycle. In: G.E. Dickinson (Editor), *The Geophysiology of Amazonia,* Wiley, New York, 273–296.
Salati, E., P.B. Vose, and T.E. Lovejoy, 1986. Amazon rainfall, potential effects of deforestation and plans for future research. In: G.T. Prance (Editor), *Tropical Rain Forests and the World Atmosphere.* Westview Press, Boulder, 61–74.
Sampaio, E.V.S.B., 1995. Overview of the Brazilian caatinga. In: S.H. Bullock, H.A. Mooney, and E. Medina (Editors), *Seasonally Dry Tropical Forests.* Cambridge University Press, Cambridge, 35–63.
Sarmiento, G., 1984. *The Ecology of Neotropical Savannas.* Harvard University Press, Cambridge, Massachusetts.
Sarmiento, G., 1992. A conceptual model relating environmental factors and vegetation formations in the lowlands of tropical South America. In: P.A. Furley, J. Proctor, and J.A. Ratter (Editors), *Nature and Dynamics of Forest-Savanna Boundaries,* London: Chapman & Hall, London, 583–602.
Sioli, H. (Editor), 1984. *The Amazon: Limnology, and Landscape Ecology of a Mighty Tropical River and Its Basin.* Junk, Dordrecht.
Smith, N., 1981. *Men, Fishes and the Amazon.* New York: Academic Press, New York.
Smith, N.J.H., 1999. *The Amazon River Forest: A Natural History of Plants, Animals and People.* Oxford University Press, New York.
Soil Survey Staff, 1999. *Soil Taxonomy: A Basic System of Soil Classification for Making and Interpreting Soil Surveys.* 2nd edition. Agriculture Handbook 436, U.S. Department of Agriculture, Natural Resources Conservation Service, Washington, D.C.
Solorzano, C.L.A., 2001. Ecosystems of South America. In: S. A. Levin (Editor), *Encyclopedia of Biodiversity,* 5, 327–359. Academic Press, New York.
Sombroek, W.G., 1979. *Soils of the Amazon Region.* International Soils Museum, Wageningen.
Valencia, R., H. Balslev, and G. Paz y Mino, 1994. High tree

alpha diversity in Amazonian Ecuador. *Biodiversity and Conservation*, 2, 21–28.
Viana, V. M., and A.J.J. Tabanez, 1996. Biology and conservation of forest fragments in the Brazilian Atlantic moist forest. In: J. Schelhas and R. Greenberg (Editors), *Forest Patches in Tropical Landscapes.* Island Press, Washington, D.C., 151–161.
Viana, V.M., A.J.J. Tabanez, and J.L.F. Batista, 1997. Dynamics and restoration of forest fragments in the Brazilian Atlantic moist forest. In W.F. Laurance and R.O Bierregard (Editors), *Tropical Forest Remnants*, University of Chicago Press, Chicago, 351–365.
Whitmore, T.C., 1998. *An Introduction to Tropical Rain Forests.* Oxford University Press, Oxford.
Whitmore, T.C., and G.T. Prance (Editors), 1992. *Biogeography and Quaternary History in Tropical America.* Clarendon Press, Oxford.
World Commission on Environment and Development (Brundtland Report), 1987. *Our Common Future.* Oxford University Press, Oxford.

10

Arid and Semi-Arid Ecosystems

P.W. Rundel
P.E. Villagra
M.O. Dillon
S. Roig-Juñent
G. Debandi

Arid and semi-arid ecosystems in South America are best illustrated by two desert regions, the Peruvian and Atacama Deserts of the Pacific coast and the Monte Desert of central Argentina. The *caatinga* of northeast Brazil is often described as semi-arid, but mostly receives 500–750 mm of annual rainfall and is better regarded as dry savanna. Small areas of Venezuela and Colombia near the Caribbean coast, and nearby offshore islands, support desert-like vegetation with arborescent cacti, *Prosopis*, and *Capparis*, but generally receive up to 500 mm annual rainfall. Substrate conditions, as much or more than climate, determine the desert-like structure and composition of these communities, and thus they are not discussed further here. Extensive areas of Patagonian steppe also have semi-arid conditions, as discussed in chapter 14.

10.1 The Peruvian and Atacama Deserts

10.1.1 Regional Delineation, Boundaries, and Transitions

The Peruvian and Atacama Deserts form a continuous belt along the west coast of South America, extending 3,500 km from near the northern border of Perú (5°S) to north-central Chile near La Serena (29°55'S), where the Mediterranean-type climate regime becomes dominant (fig. 10.1). The eastward extent of the Peruvian and Atacama Deserts is strongly truncated where either the coastal ranges or Andean Cordillera rise steeply from the Pacific coast and, as a biogeographic unit, the desert zone may extend from 20 to 100 km or more inland. A calculation of the area covered by these deserts depends in part on how this eastern margin is defined. Thus the Peruvian Desert covers between 80,000 and 144,000 km^2, while the Atacama Desert of Chile extends over about 128,000 km^2 if the barren lower slopes of the Andes are included. Actual vegetated landscapes are far smaller and for the *lomas* of Perú change dramatically between years depending on rainfall. Only about 12,000 km^2 of the Atacama contain perennial plant communities, largely in the southern portion known as the *Norte Chico* but also including a narrow coastal belt of *lomas* extending northward almost to Antofagasta and the *Prosopis* woodlands of the Pampa del Tamarugal. The vegetated areas of the coastal *lomas* of Perú and Chile together probably do not exceed 4,000 km^2 as a maximum following El Niño rains.

The continuous nature of this desert has led many to apply the name Atacama Desert to the entire coastal region.

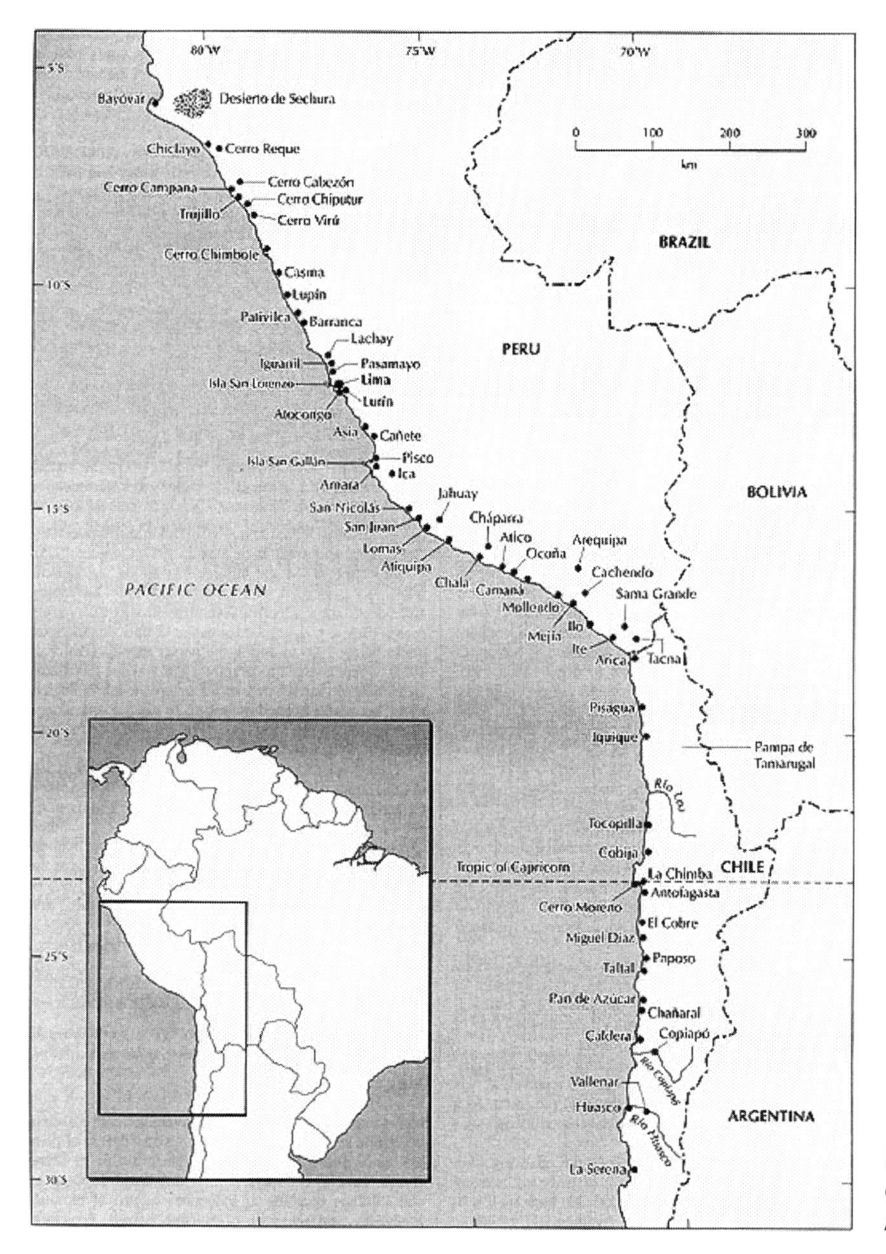

Figure 10.1 Location of loma sites and other features in the Peruvian and Atacama Deserts.

This terminology, however, has political implications and thus it is better to distinguish between the Peruvian Desert to the north and the Atacama Desert to the south. This distinction is not inappropriate since the present political boundary between Perú and Chile also forms a boundary between northern subtropical biota and southern Mediterranean-type biota.

Rainfall is very limited along this coastal area and much of the distribution of vegetation is correlated with moisture inputs from fog arising over the cold Humboldt Current. Strong thermal inversions, however, restrict the upper limit of fog to around 1,000 m and, not surprising, the eastward extent of faunal and floral elements of the Peruvian and Atacama Deserts is largely limited to areas below this elevation. Scattered vegetation at higher elevations is mostly linked to available groundwater, and major portions of these desert regions entirely lack vegetation cover.

Various concepts have been used to define the geographic limits of the Peruvian and Atacama Deserts. Biogeographers have largely focused on the coastal portions of the desert region where plant cover is present, ignoring higher, barren areas on the slopes of the Andes and coastal ranges. Climatically speaking, however, the desert zone extends well up the slopes of the Andes. Arid cactus shrublands at elevations of 2,200–4,000 m on the western slopes of the Andes in southern Perú and northern Chile have sometimes been referred to as Atacama communities (e.g. Betancourt et al., 2000). These communities are, however, more closely related to the *puna* ecosystems of the Altiplano (see chapter 12) and are not elements of the true Atacama.

The eastern limits of the arid zone as a climatic unit extend inland up to elevations of ~1,000 m in Perú, where the *lomas* vegetation is replaced by a zone termed the *serranía esteparia* that has biogeographic affinities with the higher *puna* (Brack and Mendiola, 2000). Coastal desert vegetation in Chile is likewise restricted to below ~1000 m, the upper limit of coastal fogs. The Atacama is virtually devoid of plant life from this limit up to 2,000–2,200 m, where rainfall becomes sufficient for *puna* elements to reach their lower elevational limits. South of Antofagasta (~24°S) this transition alters. Here the summer rainfall regime of the high elevation *puna* changes southward to an arid winter rainfall regime, in the northern part of which the barren Atacama Desert climatic region can extend up to 3,500 m or more.

The northern limit of the Peruvian Desert occurs at about 4°S, near the border with Ecuador. Here higher and more predictable rainfall produces a change to a subtropical scrub ecosystem. The southern limit of the Atacama Desert is commonly defined just north of La Serena (~30°S). The level of rainfall here corresponds well with desert-to-Mediterranean-type transitions in both Baja California and the Namaqualand region of South Africa. The floristic transition around La Serena is marked by the southern limit of many desert species and a change to matorral floristic elements (see chapter 11).

10.1.2 Climate

The hyper-aridity of the Peruvian and Atacama Deserts results from the combined presence of the Humboldt Current, bringing cold Antarctic waters northward along the west coast, and the high Andean Cordillera to the east (Houston and Hartley, 2003). The cold surface water temperatures pull moisture out of air masses before they reach land, while the high mountains effectively separate the coastal areas from any influence of subtropical convective events that bring summer rains to the Amazon Basin and eastern slopes of the Andes. Moreover, onshore winds from the Humboldt Current meet a strong zone of atmospheric subsidence formed by the stable South Pacific anticyclone, producing a strong temperature inversion along the coast (see chapters 2 and 3). The result is a mild and uniform climatic zone along the coast with the formation of thick stratus cloud, especially in winter, extending up to about 1,000 m in elevation (Rundel et al., 1991).

The origin of hyper-arid conditions in these coastal deserts is thought to date back 13–15 Ma (million years before present) to the middle Miocene, when the ancestral Humboldt Current was present and the Andes had reached no more than half their present height (Alpers and Brimhall, 1988). Other geologic evidence suggests intermittently arid climates extending back to as far as 25 Ma (Mortimer, 1973; Mortimer and Saric, 1975; Dunai et al., 2005) and broadly arid conditions for much longer (Hartley et al., 2005). A contrary view suggests an age for hyper-aridity of no more than 3 Ma, when fluvial deposits in the central valley of northern Chile became curtailed (Hartley and Chong, 2002). As the Andes have increased in height during the Cenozoic, the barrier against humid air masses from the east has become more effective and the role of the South Pacific anticyclone stronger (see chapter 2). At the same time, more constant southwesterly winds along the north Chilean coast would have increased marine upwelling and reinforced hyper-aridity.

Where coastal terrain is relatively flat, the fog zone diffuses well inland. However, where mountains rise abruptly from the coast, this fog zone concentrates into a relatively narrow zone against the slopes and produces a strong biological effect from fog drip (Rundel et al., 1991). These coastal fogs are referred to as *garua* in Perú and *camanchaca* in Chile. Their moisture allows for the formation of distinctive biological communities termed *lomas*.

There are three notable aspects of the climate regime of the Peruvian and Atacama Deserts, namely extreme hyper-aridity, the impact of high rainfall associated with strong El Niño events, and the remarkably constant temperature regime along the coast (Miller, 1976; Johnson, 1976; Rundel et al., 1991). The hyper-aridity of these deserts exceeds that of any other desert region in the world. Rainfall ranges from low to functionally nonexistent over extensive areas. Along the coast of Perú, mean annual rainfall commonly ranges from less than 2 mm to about 15 mm. Arica (18°28'S) in northern Chile averages less than 1 mm and may experience 10 years or more without measurable rain. Mean annual rainfall then increases slowly southward until it exceeds 100 mm at the southern margin of the Atacama. Mean values have little meaning, however, because of the high variability of rainfall between years. This variability, seen in higher coefficients of variation in interannual rainfall than in desert regions elsewhere, manifests itself in both short-term variations in rainfall and multi-decadal patterns of extreme drought. Antofagasta averaged about 10 mm annually in the first half of the twentieth century, but this dropped to only about 1 mm from 1945 to 1990. Mean annual rainfall begins to increase around Taltal (25°25'S) and Chañaral (26°20'S) to about 10–25 mm (depending on the period of record), and rises to 130 mm at the southern margin of the desert region near La Serena.

El Niño-Southern Oscillation (ENSO) conditions may have a strong impact on rainfall patterns for the Peruvian and Atacama Deserts, manifested in differing degrees and forms from north to south (see chapter 19; Caviedes, 1975). The coast of northern Perú may experience extremely intense rainfall during strong El Niño events. Chicama

(7°51'S), with a mean annual rainfall of 4 mm, was deluged by 394 mm in the strong event of 1925–1926, while Lima (12°S), with a mean of 46 mm, received 1,524 mm (Goudie and Wilkinson, 1977). Increased rainfall moved into northern Chile as far south as Antofagasta. Normally, however, strong ENSO impacts do not directly extend beyond northern Perú. The unusually strong El Niño event of 1982–1983 produced similar torrential rains in northern Perú (Goldberg and Tisnado, 1987), but these did not extend south into central Perú. The coast of northern Chile is also impacted by irregular ENSO conditions, but in contrast to Perú, these are generally linked to teleconnections with storms moving northward from central Chile. The moderate El Niño events of 1987 and 1991, for example, brought some of the heaviest rains recorded to coastal areas south of Antofagasta, but no rain farther north. The 1997–1998 event brought increased rainfall as far north as Iquique (Muñoz-Schick et al., 2001).

Qualitative historical records document the frequencies and relative intensities of four centuries of ENSO events (Quinn et al., 1987). The Holocene history of ENSO events, however, continues to be a subject of some debate, particularly for the period from 8 ka to 5 ka. It has been suggested that ENSO conditions were infrequent during this period and that a more savanna-like climate existed (Sandweiss et al., 1996), or at least fogs were more intense along the coast (Fontugne et al., 1999). This is the period when human settlements grew rapidly in size and complexity along the coast of Perú (Rollins et al., 1986). Other soil and sediment records suggest continued aridity throughout the Pleistocene and early and mid-Holocene (Ortlieb et al., 1996b; Keefer et al., 1998; Wells and Noller, 1998). Studies of flood deposits in northern Perú provide a 3,500–year chronology of extreme El Niño-induced flood events at intervals of about 1,000 years (Wells, 1990).

Combined effects of the cold Humboldt Current and coastal temperature inversion produce a unique pattern of temperature stability along the entire 3,500 km of coastal desert. Nowhere else in tropical latitudes of the world are temperatures so unchanging across a broad latitudinal gradient. The mean January maximum temperature along the coast is about 28°C at the northern limit of the Peruvian Desert and drops gradually to about 24°C at the southern limit of the Atacama Desert (Rundel et al., 1991). Mean July minimum temperatures gradually decline from about 14 to 9°C over the same latitudinal gradient. Relative humidity is high along this entire coast, particularly in the fog zone around 300–800 m in elevation. Seasonal and diurnal changes in temperature are also small. Lima, for example, has monthly mean maximum temperatures that vary only from 26°C in January to 19°C in July and August. Monthly mean minimum temperatures range from 20°C in February and March to 15°C from July through September. Below this fog zone, however, dew point temperatures are virtually never reached and fog condensation does not occur (Mooney et al., 1980; Rundel et al., 1991).

10.1.3 Geomorphology and Soils

The geomorphology of the Peruvian and Atacama deserts is strongly influenced by the ongoing subduction of the Nazca plate below the South American plate (see chapter 1). Numerous studies have documented vertical deformation and varying uplift rates along this coast for later Cenozoic time. The north-central coast of Perú from 6 to 14°S, which appears to have been stable or even subsiding since Pliocene time, is the only exception to this pattern. Elsewhere, Quaternary rates of uplift range from about 100–240 mm 10^{-3} years in northern Chile to as high as 740 mm 10^{-3} years in southern Perú (Devries, 1988; Macharé and Ortlieb, 1992; Ortlieb et al., 1996a, 1996b, 1996c).

Peruvian Desert The coast of northwest Perú is backed by a broad plain extending from near Casma (9°28'S) and broadening to the border with Ecuador (3°20'S). Small sections have been uplifted but only a few areas rise above 1,000 m (Robinson, 1964). The Sechura Desert, which extends inland for over 100 km to the foothills of the Andes in the northern portion of this region (4°20'–6°50'S), is marked by arid sandy plains and active dunes (fig. 10.2; Barclay, 1917; Silbitol and Ericksen, 1953; Kinzl, 1958; Broggi, 1961). Although the name Sechura Desert has sometimes been applied to the entire Peruvian Desert, this use is inaccurate.

Southward, from Chiclayo (6°50'S) to Trujillo (8°05'S) the coastal plain narrows to 20–50 km and is dominated by large alluvial fans formed by several dry river valleys (e.g., Reque, Saña, Jequetepeque, Chacama, and Moche valleys). The region has deep soils developed in fluvial deposits of Andean origin that, with irrigation, form some of the richest agricultural lands in Perú. The topography is flat with a few isolated mountains of late Cretaceous and early Cenozoic rocks near the coast (e.g., Cerro Reque, Cerro Cabezón, Cerro Campana [fig. 10.3], Cerro Cabras, and Cerro Negro) (Carranza-R., 1996). These mountains seldom reach 600 m but seasonal vegetation does develop on west-facing slopes. Between Trujillo and Lima, the flat coastal plain continues to narrow, with a few mountains such as Cerro Mongon (9°36'S).

Between Lima (12°S) and Pisco (13°50'S), the coastal plain disappears and the landscape is dominated by the foothills of the Western Cordillera (Cordillera Occidental) of the Andes and the many river valleys draining these mountains. The city of Lima lies at the mouth of the valley where the Rímac and Chillón rivers merge. These rivers fall steeply from 4,800 m to sea level over a distance of only 120 km, and during late Miocene time cut deep valleys into the coastal plain that was then uplifting rapidly (le Roux et al., 2000). Coastal cliffs near Lima rise to over 80 m (fig. 10.4). Steep coastal ridges with flights of marine

Figure 10.2 Sechura desert in northern Perú (photo: M.O. Dillon)

terraces continue south of the Paracas Peninsula near Pisco to Ica. The coastal plain farther inland lies largely below 300 m, but some terraces reach 700 m (Robinson, 1964). Scattered barchan fields also occur (Hastenrath, 1987; Gay, 1999). The coastal zone of southern Perú, from Ocoña (16°30'S) to the Chilean border (18°24'S), is characterized by a 20- to 30-km-wide plain broken by steep ridges extending out from the Andes. The low topography here allows coastal fogs to penetrate farther inland, thereby diffusing much of the effect of their moisture.

Atacama Desert The sharp change in direction of the desert coastline at the Arica bend reflects changing late Cenozoic relations between the converging South American and Nazca plates (see chapter 1). One expression of this has been the uplift of the Coastal Cordillera (Cordillera de la Costa) in Chile. This range extends along the coast for more than 800 km, from Arica (18°25'S) to south of Taltal (25°25'S), often running close to the sea with narrow emergent wave-cut terraces. The cordillera averages about 700 m in elevation but rises above 2,000 m near Antofagasta. The most recent phase of uplift, which began in the Miocene, created a barrier to westward drainage from the rising Andes to the east. Erosional debris from the Andes was thus trapped in the Central Valley, where it accumulated to depths of over 1,000 m (Ericksen, 1981; Mortimer, 1972, 1980; Mortimer and Saric, 1975). This debris then overtopped the Coastal Cordillera locally, allowing rivers to cut deep gorges across the range to the ocean. Some of these gorges are now more than 800 m deep near the coast but relatively shallow in their paths across the Central Valley. The Coastal Cordillera plunges very steeply to the coast from elevations of 1,000 m or more between Arica and Iquique such that virtually no marine terraces are preserved

Figure 10.3 Cerro Campana, 20 km north of Trujillo, northern Perú (photo: M.O. Dillon)

Figure 10.4 Absolute desert and along the coast near Lima, Perú (photo: M.O. Dillon)

(Ortlieb et al., 1996c). In many places south of Iquique there is a narrow coastal plain, 1–3 km wide, that extends south beyond Chañaral. This plain broadens to the north of Antofagasta, at the structural block of the Mejillones Peninsula where Cerro Moreno rises to 1,150 m.

The continental shelf is less than 10 km wide off northern Chile. Alluvium carried to the coast by rare floods is either swept into the deep Perú-Chile Trench or transported northward by strong littoral currents. There have been many interpretations of the origin of this narrow shelf. For example, the present coastal cliffs may represent a modified fault scarp that has retreated inland since Miocene time through wave benching influenced by Pleistocene sea-level changes (Paskoff, 1980). Others contend that tectonic movement has simply raised a steep coastal slope formed by long-term marine erosion along a subsiding coastline (Mortimer and Saric, 1972).

The Coastal Cordillera in northern Chile slopes gently eastward to the Central Valley whose elevation around 1,000 m reflects Neogene sedimentation. The hyper-aridity of this valley since Miocene time has prevented river channels from cutting through the Coastal Cordillera near Antofagasta. Indeed, surface rivers are entirely absent in the Central Valley between Iquique and Antofagasta, with the exception of the Río Loa, which collects water in a 100-km northward reach along the valley before cutting seaward through the coast range north of Tocopilla (Abele, 1988). South of Antofagasta, from north of Paposo to Chañaral, deep gorges cut through the Coastal Cordillera but do not link with permanent river channels across the Central Valley. South of Chañaral, major seasonal river channels include the Río Copiapó, Río Huasco, and Río Elqui.

In varying forms, the Central Valley extends more than 1,000 km from south of Arica to Chañaral and is characterized by undulating topography and small precordilleras separating internal drainage basins, many of which contain dry lakes or *salares* (Eriksen, 1981). The broad Pampa de Tamarugal inland from Iquique is the largest of the basins. Fluvial deposits here show that running water had a significant geomorphic role during the Pleistocene (Börgel, 1973). High surface runoff into the Pampa from the Andes still occurs occasionally today, forming shallow lakes that may last for months. Records from the nineteenth and twentieth centuries indicate that such floods occur at intervals of 15–20 years (Billinghurst, 1886; Bowman, 1924; Brüggen, 1936). Other data suggest high rainfall events in the Río Loa drainage at frequencies of about 70 years (CEPAL, 1960). This region is known for its abundance of saline minerals and, beginning in the mid-nineteenth century, sodium nitrate was extensively mined here for use as fertilizer and in the manufacture of explosives. Although some mining continues today, the major economic impact of this mineral declined rapidly following the development of synthetic nitrates early in the twentieth century.

10.1.4 Vegetation

Broad descriptions of the vegetation patterns and floristics of the *lomas* formations of the Peruvian and Atacama Deserts have been published by Rauh (1985), Rundel et al. (1991), and Dillon et al. (2003). Broad vegetation/floristic regions of northern Chile have been mapped by Quintanilla (1988) and Gajardo (1994).

Plant life is largely found in two types of communities: riparian and *lomas*. Riparian vegetation is restricted to river valleys draining from the Andes where flows are sufficient to reach the coast. Valleys fed by such water supported some of the earliest population centers along the Peruvian coast. These riverine habitats, present along all of the major rivers reaching the coast of Perú, are now much disturbed and generally under intense agricultural use. Nonagricultural areas are invariably high in introduced alien species that have naturalized. In northern Chile, only the Río Loa maintains a flow to the coast. Problems of salinity, trace metal pollutants, and human impacts have largely eliminated natural plant communities from this riverine habitat. The dry canyons of the Camarones, San Jose, and Lluta rivers in the barren far north of Chile do contain enough subsurface flow for the survival of local riparian communities.

More interesting are the *lomas* communities. These are largely restricted to small fog oases from 200 to 1,500 m in elevation where the combined effects of onshore moisture-laden fogs and steep topography produce sufficient moisture condensation to allow for plant growth. The *lomas* collectively have a relatively large flora of over 1,300 species, many endemic and restricted to these fog zones.

Peruvian Desert Large expanses of the Sechura Desert support little or no vegetation. Low hummocky inland dunes are generally stabilized by woody species, while coastal dunes are stabilized by both salt grass and woody species (Weberbauer, 1911; Ferreyra, 1979). Extensive mangrove formations are also present along the coast (Clüsener and Breckle, 1987). River valleys flowing through the Sechura Desert once supported lush riparian tree thickets, but these have now been largely cleared for agriculture (Ferreyra, 1983). Farther inland, semi-arid communities of dwarf trees, columnar cacti, and terrestrial bromeliads are well developed on the slopes of the Andes (Weberbauer, 1936; Ferreyra, 1960; Dillon et al., 1995).

The northernmost limit of floristic elements of the *lomas* is found at Cerro Reque (6°52'S) near Chiclayo, but this area does not receive sufficient moisture to develop *lomas* communities every year (Rundel et al., 1991). The northern extent of well-developed *lomas* vegetation is better seen northwest of Trujillo at Cerro Campana (7°58'S) and Cerro Cabezón (7°54'S). Here there is a diverse community of shrubs and herbaceous perennials above 400 m in the zone of maximum fog input, and a lower belt from 200 to 400 m dominated by a *Tillandsia* zone with a mixture of cacti and prostrate shrubs (Rundel et al., 1991).

The coastal hills between Chiclayo and Lachay near Lima have a relatively poor development of *lomas* vegetation. Discrete mountains on the coastal plain with slopes exposed to fog generally develop sparse *lomas* communities of shrubs and herbaceous species at elevations of 200–600 m, but species richness is low (Sagástegui et al., 1988). Examples of such communities occur at Cerro Prieto (7°59'S), Cerro Cabras (8°03'S), Cerro Chiputur (8°10'S), Cerro Negro (8°18'S), and Lomas de Virú (8°19'S). Closer to Lima the luxuriance of *lomas* increases, for example at Lomas de Lupin (10°33'S), Chimbote (9°04'S), and Casma (9°28'S).

The most famous of the Peruvian *lomas* are those at Lachay, 60 km north of Lima, where two distinctive plant zones have been described (fig. 10.5; Ferreyra, 1953; Rundel et al., 1991). The lower zone, between 100 and 300 m elevation, is dominated by cyanobacteria, especially *Nostoc*, and a diverse array of foliose and fruticose lichens. Vascular plant diversity is low here, but a belt with rich vascular plant communities begins at about 300 m elevation. Dry rocky slopes are dominated by terrestrial bromeliads. Moist hillsides are covered by a species-rich assemblage of low shrubs and herbaceous perennials. Canyons and valleys in this *lomas* zone support relatively dense stands of small trees. The significance of fog condensation on the water relations of woody vegetation at Lachay has been studied in some detail (Ellenberg, 1959).

Lomas formations that once occurred on the coastal slopes around Lima have been heavily impacted by human activities over more than four centuries and present poor communities today. Examples of these are the *lomas* of Cajamarquilla, Cerro Agustino, and Manzano. The three guano islands of San Lorenzo, San Gallan, and Veijas are sufficiently high to support sparse *lomas* communities (Rundel et al., 1991).

Figure 10.5 Lomas de Lachay, Perú, 1984 (photo: M.O. Dillon)

South of Lima, the frequency and intensity of coastal fogs decreases such that the species richness of the *lomas* decreases significantly. The Lomas de Lurín (12°17'S), Chorrillos (12°10'S), Atocongo (12°08'S), and Amancaes (12°01'S) support moderate diversities of 30–80 species each, with several typically Andean species also present (Rundel et al., 1991; Cuya and Sánchez, 1991). The Lomas de Amana (14°4'S) and other coastal plant communities around Ica and Pisco have been described generally (Craig and Psuty, 1968; ONERN, 1971a, 1971b, 1971c). Extensive stands of mesquite, acacia, and other phreatophytes occur along the channels of the Pisco and Ica rivers. Here, the most developed *lomas* communities here have scattered shrubs in a matrix of terrestrial *Tillandsia,* forming terraced stripes across the landscape (fig. 10.6A). A biotic origin for these stripes has been suggested (Broggi, 1957), but others have attributed them to soil creep and slumping (Lustig, 1968). A belt of columnar cactus occurs above the *lomas* communities.

The Lomas de Atiquipa (15°48'S) and Chala (15°53'S) form a continuous zone of fog vegetation broken only by a broad river valley. These received early study by Peruvian botanists because of their rich species diversity, with 120 species known from the former (Rundel et al., 1991). The Lomas de Ocoña (16°30's) north of Camaná and other *lomas* south of the city (16°35'S), each with about 80 species, and the Lomas de Mollendo (16°55'S) 20 km north of Mejia (Péfaur, 1982), with nearly 100 species, also support rich floras (Rundel et al., 1991).

South of Mejia (17°07'S) in southern Perú, *lomas* formations are less well developed (Ferreyra, 1961; Rundel et al., 1991). The plant cover of *lomas* northeast of Tacna is largely dominated by herbaceous species, sufficient to allow grazing by sheep and goats in favorable years. The strong El Niño event of 1982–1983 brought enough moisture to this region for limited agriculture to be possible. Large colonies of *Tillandsia* cover sand dunes in fog areas too dry for rooted vascular plants. The southernmost populations of terrestrial *Tillandsia* of the *lomas* occur northwest of Tacna in the Pampa de La Yarada, where there are large stands of these terrestrial epiphytes with scattered cactus mounds. Beyond the *lomas* communities, vegetation is largely restricted to scattered stands of trees and shrubs, and halophytic herbs that grow along dry streambeds fed by occasional runoff from the Andes.

Atacama Desert From Arica to south of Antofagasta, vegetation is extremely sparse along the coast and species richness is low. Near Iquique, plant cover is virtually absent as the weak fog zone limits plant growth to scattered occurrences in steep canyons near the coast. A narrow band of terrestrial Tillandsia occurs around 1,000 m at the crest of coastal ranges east of the city (Rundel et al., 1998). The total flora of the Iquique area amounts to only 69 highly localized native species (Muñoz-Schick et al., 2001), and many of the herbaceous species among these may be absent for decades until light rains occur. Slightly larger flo-

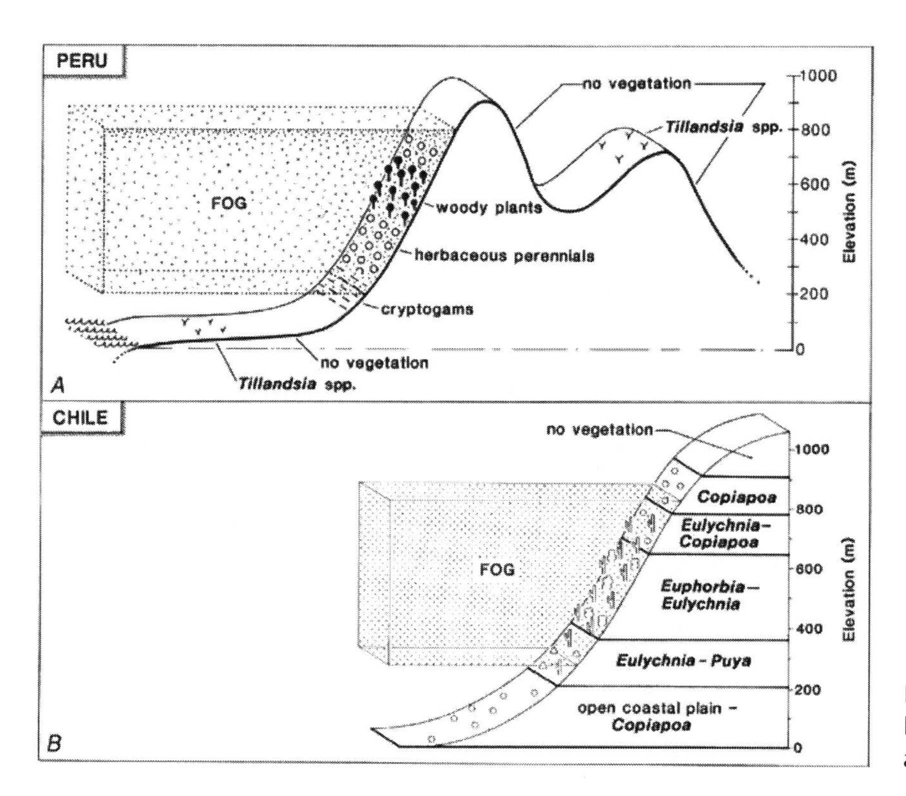

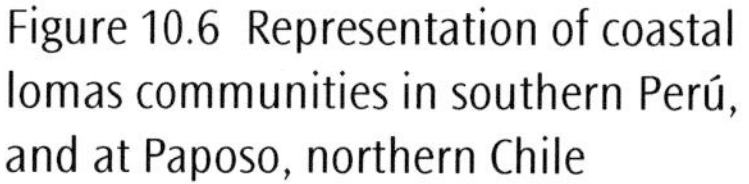
Figure 10.6 Representation of coastal lomas communities in southern Perú, and at Paposo, northern Chile

ras, again largely restricted to a few favorable canyon sites along the coast, are present at Tocopilla and Antofagasta.

Scattered vegetation is present along the Río Loa as it reaches the coast between Iquique and Antofagasta, an area impacted by human activities for thousands of years. *Pluchea absinthioides* and *Distichlis spicata* form characteristic communities where water reaches the surface (Gutiérrez et al., 1999). High levels of minerals such as lithium and boron are present in the Río Loa and in plant tissues along its course.

From south of Antofagasta to Chañaral, vegetation is restricted to the fog zone along the western slopes of the Coastal Cordillera and the narrow coastal plain. This zone exhibits the highest plant endemism of the Atacama Desert. The greatest species diversity and most luxuriant vegetation occur between about 300 and 800 m, where the escarpment is relatively steep and fog condensation is maximized (fig. 10.6B; Rundel and Mahu, 1998). The dominant plants are species of *Eulychnia*, an arborescent cactus, and the shrubby *Euphorbia lactiflua*. Vegetation above and below this elevation is much more open and lower in diversity, and succulents such as the cactus *Copiapoa* are common. The upper limit of plant occurrence occurs around 1,000 m, where fog is no longer present. The classic site for fertile islands of fog-zone vegetation in northern Chile occurs at Paposo, where more than 250 species of vascular plants occur (fig. 10.7).

A rich flora is also present south of Paposo in the Quebrada de Taltal and the Parque Nacional Pan de Azúcar (Rundel et al., 1996). Mesoclimatic patterns control vegetation zonation within the latter and, although the fog zone in this park has sufficient growth and productivity to support remnant populations of guanaco, the arid lower areas below the fog zone contain more total numbers of plant species despite low coverage (Rundel et al., 1996). This park is notable for a high diversity and dominance of species of the cactus genera *Copiapoa* and *Neoporteria* and for unusual modes of drought adaptation (Ehleringer et al., 1980, 1981; Rundel et al., 1980).

Rainfall is low and fog relatively rare in the *Norte Chico* south of Chañaral but sufficiently predictable to favor open communities of scattered shrubs, not just along the coast but also in the interior (Muñoz-Schick, 1985). Years with good rainfall, often associated with El Niño events, produce a remarkable flowering of annuals, geophytes, and herbaceous perennials known locally as the *desierto florido*. The flora and vegetation structure of elements of the *Norte Chico*, and the sharp interannual variations in herbaceous plant cover, have been detailed (Armesto et al., 1993; Armesto and Vidiella, 1993; Gutiérrez et al., 1993; Vidiella et al., 1999).

10.1.5 Biodiversity and Biogeography

Floristics The composition and community patterns of distribution of the Peruvian and Atacama deserts demonstrate a long history of extreme aridity, supporting the geologic evidence for a Miocene origin for the desert climate. Within the *lomas*, the important families are the same that predominate in many warm desert regions—Aizoaceae, Apiaceae, Asteraceae, Boraginaceae, Brassicaceae, Cactaceae, Fabaceae, Malvaceae, Portulaceae, and Solanaceae. The family Malesherbiaceae is largely restricted to this desert belt and adjacent arid slopes of the Andes.

Endemism is significant at the generic level with two groups of endemic genera along the coast. One group, centered at 15–18°S in Perú, is home to *Islaya* (Cactaceae), *Weberbauriella* (Fabaceae), and *Mathewsia* and *Dictyophragmus* (Brassicaceae). The Chilean group, centered from 24°14'–26°21'S, has *Domekoa* and *Gymnophyton* (Apiaceae), *Gypothamnium* and *Oxyphyllum* (Asteraceae), *Dinemandra* (Malpigheaceae), and *Copiapoa* and *Eulychnia* (Cactaceae). The level of endemism at the species level for the desert region is about 45%, but higher in some families. The Solanaceae, for example, are represented by 130 spe-

Figure 10.7 Lomas north of Paposo, northern Chile (photo: M.O. Dillon)

cies in the *lomas* formations, of which 105 are endemic (Dillon, 2005). Within the genus *Nolana*, nearly all of the 86 species are considered endemic to either the desert coast or adjacent Andean habitats (Tago-Nakawaza and Dillon, 1999). More locally, species endemism has been estimated at 42% for the Peruvian *lomas*, with the highest levels occurring in central Perú, where 62% of *lomas* species are endemic (Müller, 1985).

Much as with the floras of island archipelagos, the *lomas* can be considered as islands of favorable habitat in a sea of extreme desert. While the extent and location of the *lomas* have clearly altered during the Quaternary, in response to changes in sea level and the related fog zone, the influence of habitat isolation is significant in the development of individual *lomas* floras. The floras of individual *lomas* sites reflect a combination of dispersal history from neighboring and inland communities, as well as disjunctions from North American deserts.

Relatively complete checklists of the floras of the Peruvian and Atacama deserts are now available (Dillon et al., 2003). These lists reinforce the relatively strong biogeographic boundary between the two continuous coastal desert regions. Current estimates of the flora of the Peruvian *lomas* include 800 species, distributed in 357 genera and 85 families. The Atacama Desert has a flora of nearly 650 species, distributed in 248 genera and 85 families. Only about 115 (12%) of the roughly 1,350 vascular plant species known from the combined desert region occur on both sides of the Peruvian-Chilean border. When widespread weeds are removed from the analysis, less than 6% of the native species are shared by the two regions. The floristic separation across the border is strongly seen in the Cactaceae, where the three most important genera of coastal desert cacti in Chile (*Copiapoa, Eulychnia*, and *Neoporteria*) all fail to cross into Perú, while the four common genera of coastal Peruvian cacti (*Neoraimondia, Loxanthocereus, Haageocereus* and *Islaya*) do not occur along the coast south of Arica (Rauh, 1985). Similarly, only four of the 70 species of *Nolana* in the coastal desert occur in both Perú and Chile (Dillon 2005; Tago-Nakazawa and Dillon, 1999).

In general terms, the origin of the *lomas* flora in Perú shows strong subtropical connections, while the flora of the Chilean *lomas* show more affinities to the floras of central Chile and the adjacent Andes. The *lomas* flora of northern Perú, for example, demonstrates a particularly strong influence of less xeric species with Andean origins. It is only here that such families as the Pteridaceae, Orchidaceae, Passifloraceae, Piperaceae, and Begoniaceae enter the coastal flora. Terrestrial species of *Tillandsia*, a dominant aspect of the coastal vegetation of southern Perú, have very limited distribution and diversity in northern Chile (Rundel and Dillon, 1998).

Vertebrates Certainly the extreme aridity of the Peruvian and Atacama Deserts limits the diversity and abundance of vertebrate species. Among mammals, rodents form the most diverse group, as is the case in other South American drylands (Jaksic et al., 1999; Ojeda et al., 2000). Guanacos (*Lama guanacoe*) have also been historically present in many *lomas* from central Perú southward, including Cerro Moreno near Antofagasta. The only other native ruminant is the white tailed deer (*Odocoileus virginianus*), which rarely enters Peruvian *lomas*. Other large mammals along the coastal desert are fox species, including the widespread *Pseudolapex culpaeus* and *P. griseus*, and the Peruvian endemic *P. sechurae*, while pumas (*Felis concolor*) occasionally venture into the coastal deserts to prey or scavenge on marine mammals.

Ecological studies of mammals in coastal Perú are limited. Only 24 mammal species are native to the Peruvian *lomas*, including two marsupials, six rodents, and five carnivores (Brack and Mendiola, 2000). In the Chilean fog desert of Pan de Azúcar, the small mammal fauna consists of only three rodents and one marsupial (Rundel et al., 1996). Notably present in this park, however, is a small relict population of guanacos that survive on plants in the fog zone in the coastal *lomas*. Small mammal populations in the semi-arid *Norte Chico* of Chile are more diverse and respond sharply in population numbers to the effects of cyclical El Niño climatic conditions (Meserve et al., 1996, 1999, 2001; Jaksic et al., 1997). Complex interactions between small mammals, their food plants, and predators exist in this area (Contreras et al., 1993; Meserve et al., 1993, 1996; Gutiérrez et al., 1997).

Bird diversity and population size is surprisingly large along much of the arid Pacific coast. Notable among these are diverse assemblages of marine and pelagic birds that nest on protected offshore islands. Many islands off central Perú and northern Chile were the sites of hugely valuable guano mines in the nineteenth century. Raptors are diverse in terrestrial habitats, including Andean condors, which feed on the carrion of marine mammals. The bird fauna of terrestrial *lomas* include 71 species (Brack and Mendiola, 2000).

The diversity of reptiles and amphibians in the coastal deserts is relatively low. As with other biotic groups, there are strong biogeographic differences between the Peruvian and Atacama desert regions. Desert lizards in the widespread genus *Liolaemus* possess special adaptations in behavior and physiological thermoregulation to survive in the Atacama (Labra et al., 2001). Unusual among lizards is the iguanid *Tropidurus atacamensis,* which consumes marine algae along the littoral zone and excretes excess salt through special nasal glands.

Arthropods There have been few studies of invertebrate biology in the Peruvian and Atacama Deserts. The most detailed data come from research on Coleoptera (Jerez, 2000) and Lepidoptera (Peña and Ugarte, 1996). Ant diversity is low (Heatwole, 1996).

Biogeography The coastal deserts of Perú and Chile have been classified as the *Provincia del Desierto* within the *Dominio Andino-Patagónico* (Cabrera and Willink, 1980). This biogeographic classification roughly parallels that of Takhtajan et al. (1986), who defined the coastal deserts north of 25°S as the Central Andean Province of the Andean phytogeographic region. Although as many as six phytogeographic provinces have been suggested for the Peruvian Desert alone (Galan de Mera et al., 1997), our studies have suggested a simpler classification into four primary subregions: the savanna-like Sechura Desert north of 7°S, the northern Peruvian subregion from 7 to 13°S, the southern Peruvian subregion from 13 to 18°S, and the Chilean subregion from 20 to 28°S (Rundel et al., 1991). These subregions exhibit high distributional fidelity with minimal overlap in their floristic composition, suggesting independent evolutionary histories. Plant communities from northern Perú share only about 100 plant species with southern Perú, and only 15 species with northern Chile. The coastal desert at 18–20°S lacks the topography to allow for *lomas* vegetation and thus has limited genetic exchange across this boundary.

Recent phylogenetic reconstructions of several *lomas* lineages have allowed a better understanding of the evolutionary history of these biogeographic regions (Morrone and Crisci, 1995). In three cases—*Malesherbia*/Malesherbiaceae (Gengler-Nowak et al., 2002), *Polyachyrus*/Asteraceae (Katinas and Crisci, 2000), and *Nolana*/Solananceae (Tago-Nakazawa and Dillon, 1999; Dillon 2005)—the southern Peruvian flora appears to have radiated from original introductions from Chile, perhaps as early as the Pliocene. *Nolana*, a genus with species distributed throughout the Peruvian and Atacama deserts and in virtually every *lomas* site, has a unique assemblage of endemics within each subregion with virtually no overlap. Distinct and deeply rooted Chilean and Peruvian clades suggest independent radiations (Riddle, 1996). Northern Perú has two endemics, southern Perú has 31 endemics, and northern Chile has 33 endemics. The only four species spanning southern Perú and northern Chile occur in the 17–20°S latitude range near their boundaries. There is evidence for ancient *Nolana* introduction to Perú from Chile that produced a major evolutionary radiation of closely related Peruvian endemics. Similar biogeographic patterns have been reported in animal groups (Morronne, 1994; Kelt et al., 2000; Ojeda et al., 2000).

10.1.6 Impacts, Threats, and Conservation

Human Impacts Threats to the biodiversity of coastal desert habitats of Perú and Chile come from both direct and indirect human influences. Large urban centers, from Lima to Arica, Iquique, and Antofagasta, are all rapidly expanding over coastal terraces and impacting plant communities. All population centers must look for groundwater resources in local watersheds as well as distant water basins in the Central Valley of Chile or in the Andes. Groundwater resources of the Pampa de Tamarugal are heavily tapped to bring water to Iquique, and a lowering of the water table has killed stands of *Prosopis tamarugo* in scattered areas (Mooney, 1980a, 1980b).

Historical records indicate the valleys of both the Río Copiapó and Río Huasco once supported dense stands of phreatophytic trees. Indeed, the city of Copiapó was once known as San Francisco de la Selva because of the extensive stands of trees in its river valley (Klohn, 1972). These trees are now gone and agricultural activities cover much of the valley. Urbanization and agricultural activities have drawn down the water table of the Río Copiapó and Río Huasco, eliminating surface flow and promoting problems of downstream salinization.

Mining activities and associated air pollution from copper smelters have also had a significant impact on scattered areas of the coastal desert. These impacts have not been studied in detail but may explain some of the declining vegetation vigor at Paposo in recent decades. Epiphytic lichen communities once abundant in the *lomas* of northern Chile have also undergone dramatic recent declines in vigor and diversity. (Follmann, 1995), but it is unclear if this change has resulted from human impacts on air quality or from expanded drought.

Protected Areas Despite the rare habitats and significant levels of endemism, relatively few areas of high biodiversity along the coastal fog desert of the Peruvian and Atacama Deserts enjoy protected conservation status (D'Achille, 1996; CONAF, 2001). The small Reserva Nacional de Lachay near Lima preserves a diverse but degraded *lomas* community. Much larger is the Reserva Nacional de Paracas near Pisco, which conserves about 335,000 ha of coastal habitat as well as a large marine area. It was designed largely to protect wetlands and marine habitats important to migratory birds. Four sites along the Peruvian coast have received designation as international RAMSAR sites because of their significance for wetland bird populations. These are the mangrove lagoons of the Santuario Nacional Los Manglares de Tumbes near the border with Ecuador (2,972 ha), the Zona Reservada Los Pantanos de Villanear Lima (396 ha), the Reserva Nacional de Paracas, and the Santuario Nacional Lagunas de Mejía (691 ha).

Protected areas of ecological significance in northern Chile include the large Parque Nacional Pan de Azucar along the coast between Taltal and Chañaral (Rundel et al., 1996). This park protects 43,679 ha of significant habitat including breeding habitat for marine mammals and sea birds, as well as intact ecosystems of coastal fog-zone communities containing guanacos and other wildlife. The area of Paposo to the north, which harbors the richest diversity of fog-zone plant species along the northern coast of Chile, is under partial conservation management as the Proyecto Reserva Nacional Paposo, but mining activities and associated pollution have limited the effectiveness of this process. The Parque Nacional Llanos de Challe, 17 km north of Vallenar near the southern margin of the Atacama Desert, protects 45,708 ha of diverse desert shrubland and cactus habitat, as well as coastal penguin rookeries.

10.2 The Monte

10.2.1 *Regional Delineation, Boundaries, and Transitions*

The Monte Biogeographical Province extends over 460,000 km^2 in Argentina, from 24°35'S in Quebrada del Toro (Salta Province) to 44°20'S in Chubut; and from 69°50'W at the foot of the Andes to 62°54'W on the Atlantic coast (fig. 10.8). It consists of plains, *bolsones* (mountain-girt basins), foothills, alluvial fans, and plateaus. Climatically, it is an arid and semi-arid region, with mean annual rainfall ranging from 30 to 350 mm, and mean temperature from 13° to 15.5°C (Morello, 1958). A shrub steppe dominated by species of Zygophyllaceae is the typical plant formation, while open woodlands of several species of *Prosopis* occur where groundwater is accessible (Morello, 1958; Cabrera, 1976).

The boundaries of the Monte Desert were delineated by Morello (1958) through chorological and ecological criteria. Although these limits have not been redefined at a macro-regional scale, local studies have rectified them in some areas (Cabrera, 1976; F. Roig, 1998). The Monte Desert maintains broad ecotones with the Chaco and Espinal biomes to the east, with the Patagonian desert to the south, and with the Pre-puna and Puna biomes to the west.

The transition between the Monte and the Chaco and Espinal regions is governed by a rainfall gradient. Monte elements are gradually replaced by Chaco elements, accompanied by an increase in structural complexity, species richness, and total vegetation cover (Morello, 1958; Cabido et al., 1993). The transition between the Monte and Patagonia is gradual, especially in areas with no geomorphologic discontinuities, and corresponds to the boundary between the southernmost influence of Atlantic air masses and the northernmost influence of Pacific air masses. This produces a gradual change in both thermal and hydrological regimes as temperature decreases and winter rainfall increases southward. Year-to-year fluctuations generate ecotonal areas where Monte and Patagonian elements coexist (F. Roig, 1998). These ecotonal areas begin at an elevation of 200 m in the south and rise northward to 1450 m (F. Roig, 1998). The boundary between the Monte and Pre-puna occurs between elevations of 2,000 and 3,400 m, depending on latitude (Cabrera, 1976). However, taking floristic affinities into account, some authors consider the Pre-puna a Monte district, or an ecotone between the Monte and Puna (Morello, 1958; Mares et al., 1985).

10.2.2 *Climate*

The latitudinal extension of the Monte determines a climatic gradient involving temperature, rainfall seasonality, and predominant winds. According to the United Nations Environmental Programme (1992), the Monte is defined bioclimatically as semiarid to arid. Mean annual rainfall varies between 30 and 350 mm, and the ratio between precipitation and potential evapotranspiration ranges between 0.05 and 0.5, indicating a water deficit for the whole area. Consequently, water shortage is the main limiting factor in the Monte. The largest drought nucleus is located in the central Monte, between southern La Rioja and northern Mendoza, and in the Calingasta-Iglesia Valley, with annual rainfall less than 80 mm. Most precipitation occurs in localized heavy downpours. A major characteristic is high variability between years (coefficient of variation of mean rainfall ranging between 40% and 70%) (Le Houérou, 1999).

The Monte shows a gradient in rainfall seasonality, with marked summer rainfall in the north (70% of the total), whereas precipitation is more evenly distributed throughout the year in the south (only 19% in summer). Long dry periods may extend over seven months in the northern Monte but rarely occur in the south. Figure 10.9 shows mean annual and monthly precipitation for different localities of the Monte, following the latitudinal gradient. Differences between summer and winter temperatures, as well as diurnal ranges, emphasize the region's continentality. The thermal gradient is about 6°C in mean temperature between the northern and southern areas. Annual absolute maximum temperature is 48°C, and the absolute minimum is –17°C. Elevation strongly influences temperature in areas at similar latitudes (Le Houérou, 1999).

Latitudinal variations also occur in the predominant wind direction. Westerly winds predominate year round in the southern Monte, while in the northern Monte winds from the south and east prevail. Warm dry föhn winds from the west, locally called *zonda*, occur across the Monte as air masses that have cooled adiabatically and lost moisture on rising against the west-facing slopes of the Andes, gain heat rapidly on descending the eastern slopes.

10.2.3 *Geomorphology and Soils*

Morello's (1958) review of the geomorphology of the Monte has since been updated with local studies (Abraham and Rodríguez Martinez, 2000). From 24°35'S to 27°S, the Monte occupies longitudinal valleys and nearby mountain slopes. South of 27°S, it comprises mainly *bolsones* and intermontane valleys of the Precordillera and Sierras Pampeanas. Beyond 32°S, it occupies the western sector

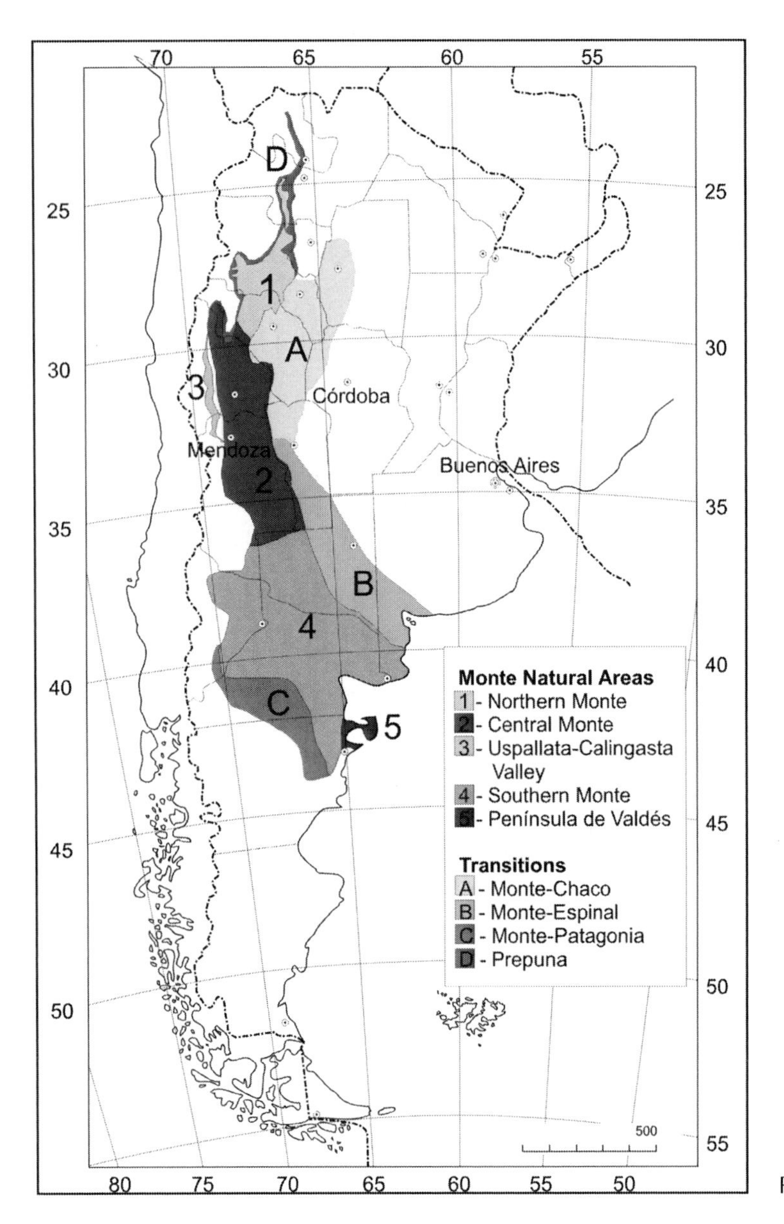

Figure 10.8 Monte regions and transitions, Argentina

of the plain between the Precordillera and Sierras Pampeanas. South of the Río Colorado (38°S), plateaus form the characteristic landscape.

The bolsones and intermontane valleys occupy a long narrow zone in which individual bolsones are isolated by intervening mountain chains. Although variable in shape and size, the general arrangement of the bolsones is consistent: a flat central area flanked by gently sloping alluvial fans and eroded mountain slopes. Farther south, the sedimentary plains slope smoothly eastward from the Andes and the Precordillera to the Sierras Pampeanas. These plains are underlain by clastic and carbonate materials of continental origin. The undulating plateaus of the south are covered with gravels, locally eroded, and central circular or elliptical depressions supporting marshes and salares. Throughout the Monte, piedmonts form transitional belts 10–50 km wide between the mountain ranges and basin floors (Garleff, 1987).

Minor landforms occur throughout the Monte. *Bardas* are steep rugged plateau edges typical of Neuquén, Río Negro, and Chubut provinces. *Huayquerías* are badlands formed in Cenozoic sediments whose lithosols are poor in organic matter but rich in clays that on weathering form impermeable muds during rain events. Large areas of W–E or NW–SE sand ridges, produced by aeolian reworking of Holocene sediments, are found near Cafayate (Salta), in Campo del Arenal and Belén (Catamarca), near Vinchina and Guandacol (La Rioja), and in the Médanos Grandes desert (San Juan) (fig. 10.10). *Barriales*, formed from fine material deposited on basin floors, are hardpans with polygonal

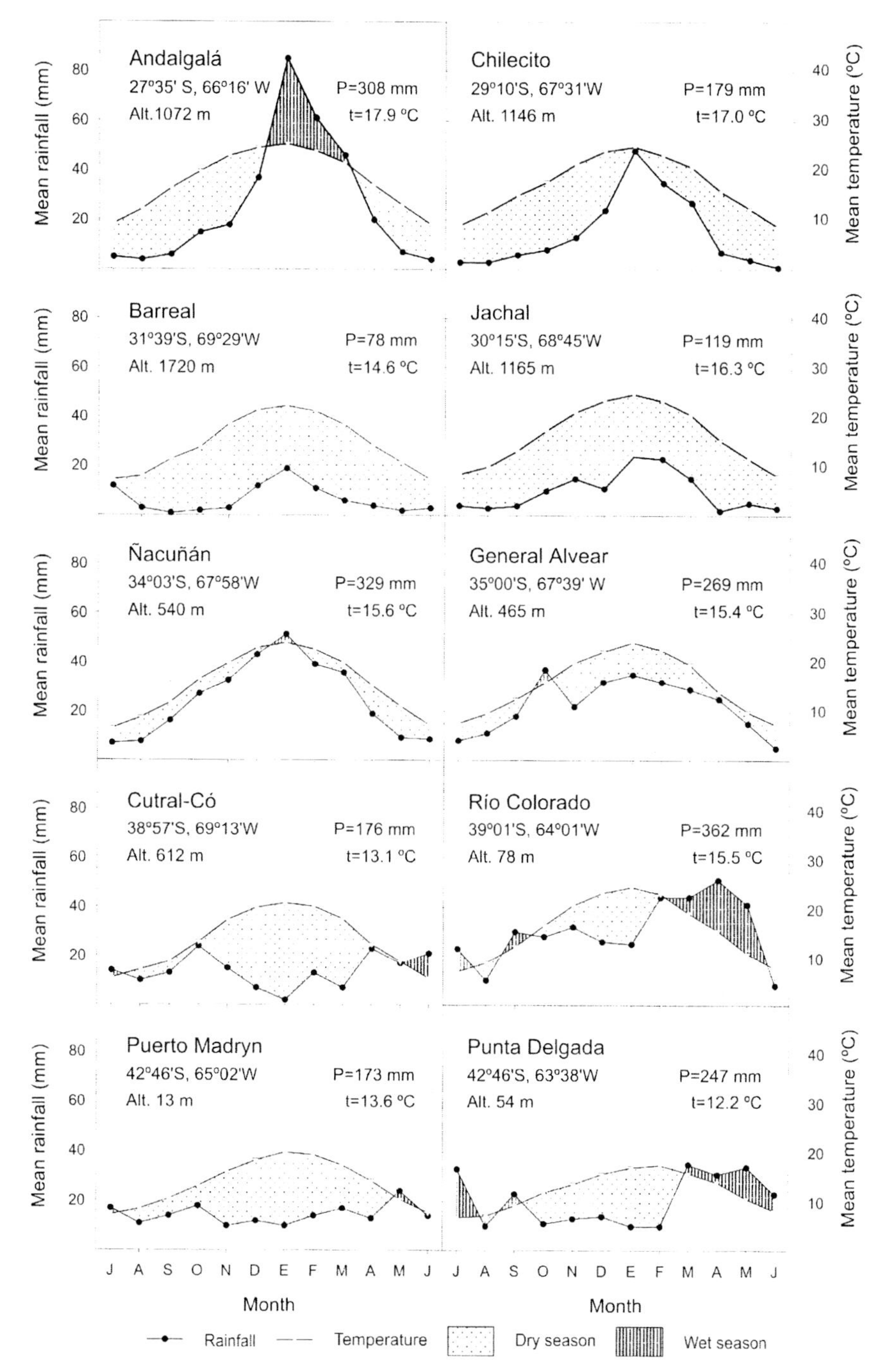

Figure 10.9 Annual and monthly temperature and rainfall at ten climate stations in the Monte

cracks at the end of the dry season, but are transformed into muddy swamps during the rainy season. They are widely distributed throughout the Monte, but most important in Neuquén, Río Negro, and the Calingasta-Rodeo valley. *Salares* form where salts precipitate in internal drainage basins, as at Salar de Pipanaco in Catamarca Province.

The soils of the Monte are mostly Entisols, which predominate from Valles Calchaquíes to southern La Pampa province, and Aridisols, which dominate from southern Mendoza and La Pampa to the transition with Patagonia (INTA/CIRN, 1990). Entisols include sandy Torripsamments and Torrifluvents, derived from aeolian and fluvial deposits. Aridisols include widespread carbonate-rich Calcids, notably in southern Mendoza, and clay-rich Argids farther south (del Valle, 1998; Abraham and Rodríguez Martínez, 2000).

The few permanent rivers that flow across the Monte are allochthonous, since they are fed by rainfall and snowmelt in Andean headwaters. Variations in discharge can be great. For example, the Río Durazno, with headwaters in the

Figure 10.10 Sand-dune chains in northeastern Mendoza Province, central Monte (photo: P.E. Villagra)

Sierra de Famatina, has a summer flow of 400 to 600 l s^{-1} but during the rainy season flows can reach 6000 l s^{-1}. Whereas mountain rivers rarely have well-developed floodplains, when these rivers reach the plains they can meander extensively, resulting in widespread fluvial sedimentation. An example is the Río Mendoza 5-km-wide meander belt in the Travesía de Guanacache (Abraham and Rodríguez Martínez, 2000). These mountain rivers are the main source for groundwater recharge in the Monte. Water tables are highest close to rivers but deepen or disappear with increasing distance from the riverbanks.

10.2.4 Biogeography

Biogeographic History The Monte is a transitional region, comprising a vast area of contact between tropical and cold temperate biotas. Historical biogeography has provided some explanations about the origin of its flora and fauna. South American deserts including the Monte are very old, as indicated by paleoendemic genera well adapted to desert conditions (S. Roig-Juñent and Flores, 2001; S. Roig-Juñent et al., 2001). These paleoendemics are mixed with other neoendemic taxa that have speciated more recently. Thus, the Monte biota has multiple origins, in both time and space. Most genera are of Neotropical origin, followed by groups with Patagonian affinities (Ringuelet, 1961; Müller, 1973).

Although the early Cenozoic was a humid interval in South America, paleogeographic data suggest that a restricted semiarid area existed in west-central Argentina during that time (Volkheimer, 1971; Axelrod et al., 1991). The subsequent elevation of the Andes led to the expansion of arid and semi-arid conditions in the Monte to the east, which favored the formation and expansion of xeric shrubby steppes (Axelrod et al., 1991). Desertification was further enhanced by the continued uplift of the Sierras Pampeanas along the western border of the Chaco plains, creating a barrier against winds coming from the northeast (Pascual and Ortiz Jaureguizar, 1990).

There are superficial biogeographic relations today between the Monte and North American deserts, based principally on the distribution of important woody plant genera such as *Prosopis*, *Larrea*, and *Cercidium*. Their presence in North America represents a long-distance dispersal from an origin in South America rather than a parallel evolution of groups from a common source. The distribution of vertebrate animal species between the Monte and Sonora is not vicariant, and the shared genera have species distributed all over South and Central America (Willink, 1991).

Biogeographic Regions Endemic assemblages of reptilian, anuran, and insect species allow the identification of five natural areas within the Monte Desert (fig. 10.8, S. Roig-Juñent et al., 2001).

The northern Monte is an elongated area extending from Salta to northern La Rioja province. It is formed by three longitudinal valleys at the eastern boundary of the Puna (Calchaquí, Quilmes, and Santa María). This area also includes Bolsón de Pipanaco, Tinogasta in Catamarca, and Famatina in La Rioja. The northern area is nearly completely enclosed by mountain chains to 28°40'S, and partially separated from the central part of the Monte by the Sierra de Velazco and Sierra de Famatina in La Rioja where the Monte area is very narrow. This area, which includes the northern and Catamarqueño-Riojano districts defined by Morello (1955), shares several insect taxa and floristic elements with the Chaco to the east. Several endemic genera occur only in this area, mainly in the valleys of Catamarca, Tucumán, and Salta.

The central Monte extends from La Rioja province to southern Mendoza. It is a wide strip of land (almost 200 km from east to west), with isolated mountain chains such as the Precordillera and Pie de Palo. It is in broad contact with the Chaco to the northeast and the Espinal to the east. This area has several elements that belong to both the Neotropical and Patagonian biotas. The southern part is connected with the southern Monte by a 300 km long belt, without a clear biogeographical barrier, and with elements from both the southern and central Monte. This area roughly coincides with the central district defined by Morello (1955).

The Uspallata-Calingasta Valley is a small area within the Monte that stretches from northern San Juan province to northern Mendoza. Formed by two longitudinal valleys that run 300 km from north to south between the Andes and the Precordillera, it is the most arid region of the Monte with sparse vegetation.

The southern Monte, stretching from southern Mendoza to east-central Chubut, constitutes the largest natural area within the Monte, extending from the foothills of the Andes to the Atlantic coast. This area, the southern district of Morales (1955), is dominated by a low steppe of woody plants and is rich in endemic species of several Patagonian genera.

The Valdés Peninsula is the smallest natural area of the Monte, confined to the peninsula and the nearby Chubut littoral. It is equivalent to the Atlantic Shrub district of the Monte (F. Roig, 1998). The western limit is determined by precipitation (160–250 mm). It is occupied by steppes of small-leaved woody plants, strongly influenced by its maritime climate. Others may view this area as an ecotone between the Monte and Patagonia, or as a part of the latter (Soriano, 1956; 1985), but the majority of its floristic elements belong to the Monte (F. Roig, 1998).

10.2.5 Vegetation

Physiognomically, the Monte is a mosaic of two types of vegetation: shrubby steppes, which are the dominant type, and open woodlands of *Prosopis,* which occur exclusively where ground water can be tapped. Floristically, the Monte is characterized by the dominance of species of the Zygophyllaceae, with *Larrea, Bulnesia,* and *Plectocarpha* forming significant components of the shrub steppes. There are other shrub steppes whose presence is determined by edaphic factors, constituting associations of species adapted to azonal conditions with saline, sandy, or waterlogged soils.

Morello (1958) described the most important plant associations in the Monte and analyzed them ecologically, while Cabrera (1976) summarized the main associations. Detailed studies of plant communities are available in several works that describe the vegetation of different zones of the Monte at the local scale. (e.g. Hunziker, 1952; Vervoorst, 1954; F. Roig, 1972, 1976, 1985, 1989, 1998; Solbrig et al., 1977; González Loyarte et al., 1990; Cabido et al., 1993; León et al., 1998; Villagra and Roig 1999; Martínez Carretero 2000).

Larrea Shrublands *Larrea* shrublands, or *jarillales*, are the dominant communities in the Monte and give the area a physiognomic unity. They extend across the Monte from sea level to 3,000 m, and occupy low plains, plateaus, alluvial fans, and *bardas*, as well as the plains of Patagonia. They prefer sandy or loamy soils, and do not tolerate high salinity. The shrub layer, strongly dominated by *Larrea* species, ranges between 1.5 and 2 m in height, with varying cover determined by climatic conditions. *Larrea* density varies from 30 plants ha^{-1} in the driest zones to over 500 plants ha^{-1} in the wettest.

Larrea cuneifolia and *L. divaricata* share dominance in most parts of the Monte, accompanied by other shrubs such as *Montea aphylla, Bougainvillea spinosa, Zuccagnia punctata, Prosopidastrum globosum, Bulnesia retama,* and *Cercidium praecox*, and by an herbaceous layer dominated by *Pappophorum caespitosum, Trichloris crinita,* and *Setaria mendocina* (fig. 10.11).

Larrea species behave differently depending on the area. In the northern Monte, *L. cuneifolia* colonizes the drier bajadas, while *L. divaricata* behaves as a riparian species along washes (Ezcurra et al., 1991). In the central Monte, *L. divaricata* dominates in sandy or loamy soils and reaches higher altitudes, while *L. cuneifolia* dominates soils with higher clay content. Here also *L. nitida* can be found forming a third altitudinal range where colder and moister conditions prevail, or in depressions at lower elevations where both cold air and water tend to accumulate (Morello, 1956; F. Roig, 1972; Ezcurra et al., 1991). *Larrea nitida* is abundant in northern Patagonia and the southern Monte, but

Figure 10.11 *Larrea cuneifolia* shrubland (*jarillal*) in the Bolsón de Pipanaco, northern Monte. The Sierra de Ambato rises in the distance (photo: P.E. Villagra).

also expands northward and rises in altitude along the west side of the Monte to around 28°S. In this species, latitude is inversely related to altitude. *Larrea ameghinoi* occupies the southern Monte and reaches to 37°S in Neuquén. It is usually found on wind-swept sites such as beaches and sandy ridges (Ezcurra et al., 1991). Finally, *L. nitida* is limited to an associate role in the *Chuquiraga avellanedae* steppe in the Península de Valdés. In the *jarillales*, some accompanying species are common throughout the Monte, but there is overlap with neighboring areas. For example, the southern Monte includes species of Patagonian origin not found in other areas.

Prosopis Woodlands Open *Prosopis* woodlands, or *algarrobales*, occupy basin floors and lower alluvial fans (Morello, 1958). The climatic boundary for this woodland between 29° and 33°S is at a limit near 350 mm of annual rainfall. With less than 350 mm, *algarrobales* can occur only where supplementary groundwater exists.

Dominant *Prosopis* trees can reach 15 m in height in the northern Monte and 7 m in the central Monte. Crowns are not completely closed, and the understory is typically formed by one or two shrub strata and an herbaceous layer (fig. 10.12). The tree layer can be monospecific or composed of several species, *P. flexuosa* and *P. chilensis* being the more common. These species are sympatric in the northern and north-central Monte, but at a local scale they occupy different habitats. *Prosopis chilensis* is found along washes while *P. flexuosa* occurs in more upland areas. The density of *Prosopis* woodlands varies from as low as 20 trees ha^{-1} in some places of the central Monte to 300 trees ha^{-1} in Valles Calchaquíes in the northern Monte.

In the northern Monte, woodlands reach their greatest height and density. Examples are the woodlands of *P. flexuosa*, 6–7 m in height, between the Sierras de Jagüé and Umango, and the Sierra de Famatina. In the Valles Calchaquíes there is a mixed woodland of *P. flexuosa* and *P. alba*. Vervoorst (1954) described a woodland of *P. flexuosa, P. alba, Bulnesia retama, and Cercidium praecox* at Tilciao (Catamarca). In the Bolsón de Pipanaco, *P. flexuosa* forms dense rings around the salt flat with a thick understory dominated by *Suaeda divaricata*.

In the central Monte, *P. flexuosa* is the dominant species, and from 33°S southward it is the only tree species of this genus. *Prosopis flexuosa* occupies the lowlands of San Juan and Mendoza up to an altitude of 1,200 m, in sites with groundwater available throughout the year. This species can tolerate high salinity of 10,000 to 20,000 μmhos cm^{-1} (F. Roig, 1985). In Bolsón de Mogna (San Juan), there are woodlands of *P. flexuosa and P. chilensis*. In Mendoza, the tree layer is composed of *P. flexuosa*, the shrub layer includes *Larrea divaricata, Capparis atamisquea, Condalia microphylla, Lycium tenuispinosum, Mimosa ephedroides*, and *Suaeda divaricata*, and the herbaceous layer is dominated by *Trichloris crinita* and *Pappophorum caespitosum* (Villagra et al., 2000).

Figure 10.12 *Prosopis flexuosa* woodland in the Bolsón de Pipanaco, northern Monte showing individuals 15 m high and coverage over 50% (photo: P.E. Villagra).

In the southern Monte, *P. flexuosa* takes on a shrub-like morphology and forms mixed shrublands with *L. divaricata*. In the Península de Valdés and Uspallata-Calingasta areas, no woodlands have *Prosopis* species.

Halophytic Communities Three communities of saline environments, with composition controlled by salt concentration and water availability, can be distinguished (Morello, 1958). These are the vegetation of salt flats (*salinas*), saltpeter beds (*salitrales*), and muddy areas (*barriales*). Salt flats occupy the interior of basins throughout the Monte area, where plants form vegetation belts or rings. The inner ring is an open shrubland of *Heterostachys ritteriana* and *Plectocarpha tetracantha*, which may be accompanied by the salt grass *Distichlis scoparia*. The second ring is typically dominated by *Allenrolfea vaginata*, associated with *Cyclolepis genistoides, Atriplex vulgatissima, Prosopis alpataco*, and *P. strombulifera*. The third and outer ring is a dense shrubland of *Suaeda divaricata*, associated mainly with species of saltbush, *Atriplex*. Beyond the outer ring, there may be woodlands of *P. flexuosa* (salt tolerant) mixed with halophytes, as in the Bolsón de Pipanaco.

In alkaline saltpeter beds occurring on stream banks, there are grasslands of *Distichlis spicata* and *Scirpus olneyi* accompanied by *Nitrophila australis* and *Tessaria absinthioides*. Saline saltpeter beds support *Sporobolus pyramidatus* and *S. phleoides* associated with *Scirpus asper* and *Juncus acutus*.

In the southern Monte, muddy areas are covered by concentric vegetation rings of *Atriplex roseum, Bassia hyssopholia, Halophytium ameghinoi*, and *Suaeda divaricata. Atriplex lampa* and *Suaeda divaricata* constitute the outermost and widest of these belts, growing as patches and covering large areas in Mendoza, Neuquén, Río Negro, and Chubut.

Sand-dune Communities Sand dunes occupy discontinuous but sometimes extensive areas of the Monte, as at Campo del Arenal and Belén (Catamarca). Some are actively expanding, as in Cafayate (Salta), while others form discontinuous strips along rivers. There vegetation involves a contrast between low and taller woody plants, and a dense carpet of annuals and geophytes appears before the first summer rains. The shrub stratum is dominated by *Trichomaria usillo, Mimosa ephedroides, Prosopis argentina, Larrea divaricata, Senna aphylla, Ephedra boelkei* (endemic to the central Monte), *Prosopidastrum globosum*, and *Bulnesia retama*. The most important grasses are *Sporobolus rigens, Panicum urvilleanum, Chondrosum barbatum, Aristida mendocina*, and *A. adsencionis*. Annuals are a diverse group whose abundance varies greatly in space and time. Species of *Gomphrena, Verbesina, Portulaca, Heliotropium, Sclerophyllax, Ibicella*, and *Glandularia* are the most common plants in this group.

Floodplains For Monte floodplains, the frequency and magnitude of inundation determine the floristic composition, structure, and function of plant communities and regulate their dynamics. For example, in the Río Mendoza, floodplain communities are replaced as distance from the watercourse increases. Thus, an initial community of pioneer species is replaced by a shrubland of *Baccharis salicifolia*, associated with *Tessaria absinthioides, Lycium tenuispinosum*, and *Tamarix gallica*, followed by a shrubland of *Prosopis alpataco* and *Capparis atamisquea* accompanied by *Lycium tenuispinosum*, and finally, by the regional woodlands of *Prosopis flexuosa, Larrea divaricata*, and *Bulnesia retama*. The riverbanks support a grassland of Cortaderia rudiuscula, accompanied by *Baccharis salicifolia* and *Tessaria* absinthioides, and a creeping layer of Cynodon dactylon and Equisetum bogotense (Villagra and Roig, 1999).

Similar associations occur in other areas of the Monte. These include shrublands of *Larrea nitida* and woodlands of *Discaria trinervis* in the Uspallata Valley (Martínez Carretero, 2000), and gallery woodlands of *Salix humboldtiana* in the perennial rivers of the central and southern Monte. Hydrophytes such as *Juncus balticus, Acaena magellanicus, Trifolium repens, Scirpus, Carex gayana, Eleocharis*, and *Hydrocotile* are often found in riverbeds.

Rocky Slope Vegetation Communities dominated by Cactaceae and Bromeliaceae flourish on steep rocky hillsides, constituting an area with mixed elements of Monte and Puna, and have often been included within the Monte (Morello, 1958; Mares et al., 1985). However, Cabrera (1976), Cabrera and Willink (1980), and F. Roig (1989) viewed them as part of the Prepuna (or cardonal).

Key Ecosystem Processes Water availability is the major factor determining the productivity of the Monte desert, triggering the new flush of leaves in most perennial woody plants and the germination of annual herbs and grasses. Other factors, such as soil texture and temperature, generally act by regulating water availability. Nutrient availability seems to be limiting only in years where rainfall exceeds 400 mm/year during the growing season (Guevara et al., 2000).

The growth of grasses is opportunistic and can be triggered by adequate soil moisture after periods of inter- and intra-seasonal dormancy, and whenever temperature conditions are suitable (Seligman et al., 1992). Useful rainfall events—those able to initiate some growth—were determined as those higher than 0.1 PET (potential evapotranspiration) on a weekly basis. In the central Monte, this represents rainfall events of about 5 mm (Guevara et al., 1997). In contrast, *P. flexuosa*, as a phreatophytic species, has continuously available water and is therefore somewhat independent of rainfall (Cavagnaro and Passera, 1993).

In arid lands, patchy vegetation patterns appear to determine ecosystem functioning (Aguiar and Sala, 1999). Dominant woody plants cause changes in microclimate and soil properties by attenuating harsh temperatures and solar radiation, and by modifying substrate characteristics and the availability of water and nutrients (Gutiérrez et al., 1993). Through these environmental changes, shrubs and trees can affect species composition, phenology, productivity, and biomass allocation of its understory, as well as the establishment and spatial distribution of species.

The presence of *Prosopis* modifies the spatial pattern of plant species in the shrub and herbaceous layers, as well as soil chemistry, generating spatial heterogeneity at different scales. Differences in floristic composition and soil conditions are observed between exposed and under-canopy areas, and also between different microhabitats beneath *Prosopis* canopies. These differences are the result of species replacement rather than species enrichment and contribute to ecosystem diversity (Mares et al., 1977; Rossi and Villagra, 2003).

10.2.6 Biodiversity

The Monte has been traditionally viewed as an impoverished Chaco. Many genera present in the Monte also occur in the Chaco but with lower species richness (Stange et al., 1976). Distribution patterns of 1,500 Chaco and Monte insect species confirm that the latter has lower species richness than the former (Roig-Juñent et al., 2001), showing a trend of decreasing biodiversity with aridity. However, the reverse appears to be true when comparing endemic taxa. Overall endemism in the Monte is 35% for insect species and 11% for insect genera, whereas in the Chaco it is 29% and 5%, respectively. This shows a trend of increasing levels of endemism with greater aridity. Among vertebrates, reptiles show a similar level of endemism,

~30%, while mammals and birds show 21% and 12%, respectively. The presence of numerous species and genera exclusive to the Monte justifies regarding this area as an independent evolutionary center, wherein one third of the biodiversity of insects and reptiles originated.

Vertebrates The vertebrate fauna of the Monte has been related to elements of the Pampas and Chaco, although some reptiles show Patagonian relationships (Dabbene, 1910; Cabrera and Yepes, 1940; Ringuelet, 1961; Fittkau, 1969; V. Roig, 1972; Müller, 1973).

There are few mammal species in the Monte (60). Rodents show the highest diversity with 30 species, nine of which and two genera (*Tympanoctomys* and *Octomys*) are endemic to the region (Redford and Eisenberg, 1992). Cuises like *Microcavia australis* and *Galea musteloides* (Caviidae) are very common, while the Patagonian cavy (*Dolichotis patagonum*, Caviidae), the vizcacha (*Lagostomus maximus*, Chinchillidae), and tuco-tucos (*Ctenomys* species, Ctenomyidae) are also typical. Carnivores follow in importance with 13 species. These include several species of the genus *Felis*, three fox species (*Pseudolapex*), and mustelids of the genera *Conepatus*, *Galictis*, and *Lyncodon*. Edentates like hairy armadillo (*Chaetopharactus villosus* and *C. vellerosus*), the pichi (*Zaedyus pichiy*), and the pichiciego (*Chlamydophorus truncatus*) constitute another typical group, the last being endemic to the Monte.

Birds are the most diverse group of vertebrates in the Monte (Roig and Contreras, 1975), with a total 235 species (Olrog, 1963), four of them and 25 subspecies being endemic. The ñandú (*Rhea americana*) and several tinamous of the genera *Nothura* and *Eudromia* are among the biggest birds. Passerines are the most diverse group.

Of a total of 63 reptilian species, 19 are endemic to the Monte (Cei, 1986). *Liolaemus* is the most characteristic genus, with 19 species present, with 10 of these endemic. Other typical lizards are matuastos (*Leiosaurus* species), *Mabuya dorsivittata*, and the iguana colorada (*Tupinambis rufescens*) (V. Roig, 1972; Roig and Contreras, 1975). Many snakes such as *Leptotyphlops borrichianus*, *Philodrias burmeisteri*, and species of the genera *Tomodon*, *Chlorosoma*, *Elapomorphus*, and *Pseudoboa* are widely distributed in the Monte. *Pseudotomodon* is an endemic genus. Two turtle species are also found, *Chelonoidis chilensis*, and the endemic *C. donosobarrosoi*, which is restricted to the southern Monte.

Amphibians are very scarce in the Monte and there are no endemic species. This group includes the common toad *Bufo arenarum mendocinum*, ranging up to 1,500 m, the escuerzos (*Odontophrynus occidentalis*), and *Ceratophrys cranwellii*, the frogs *Pleurodema nebulosa*, *P. bufonina*, and *Leptodactylus* species, and several species of *Telmatobius* (Cei and Roig, 1974; Cei, 1980).

Arthropods Arthropods constitute the most diverse group of animals in the Monte. The region contains at least 49 endemic genera of insects (Roig-Juñent et al., 2001). There are also a high number of endemic solfugids (windscorpions) such as the genera *Procleobis*, *Uspallata*, *Mortola*, *Patagonolpuga*, *Nothopuga*, *Syndaesia*, and *Valdesia*. Studies of Monte arthropods and some taxonomic revisions have greatly enhanced knowledge of this fauna (Porter, 1975; Terán, 1973; Stange et al., 1976; Willink, 1991; Roig-Juñent et al., 2001). The most diverse insect orders in the Monte, as in other deserts, are Coleoptera and Hymenoptera. Within the Coleoptera the richest families are Tenebrionidae and Scarabeidae, with a high proportion of endemic genera and species. Within Hymenoptera, the richest superfamilies are Apoidea and Vespoidea.

10.2.7 Biological Processes

Foliage production is essential for the growth and reproduction of highly diverse groups of phytophagous insects and herbivorous mammals. Mammalian herbivores can be very conspicuous, like the guanaco (*Lama guanicoe*), which feed mainly on grasses and forbs (Puig et al., 1996; 1997). Medium-sized herbivorous mammals (4–6 kg) include the mara and the vizcacha that prefer open habitats and can serve as indicators of degradation processes. Tucotucos are important herbivores that live underground and seem to be generalists in their diet preferences (Puig et al., 1999). Cavies and tuco-tucos have been reported as important causes of mortality of shrubs such as *Larrea cuneifolia*, *Lycium* species, *Atriplex lampa*, and *Geoffroea decorticans*, due to their browsing habits in the central Monte (Borruel et al., 1998; Tognelli et al., 1999). Small mammals (less than 100 g) mainly have omnivorous habits, although the rodent *Graomys griseoflavus* shows a more folivorous diet (Campos and Ojeda, 1997).

Herbivorous insects are very diverse in the Monte. *Schistocerca americana* (Orthoptera) breeds were formerly a major plague that devastated both natural and cultivated plants (Claver and Kufner, 1989). *Acromyrmex lobicornis* is the most common species of leaf-cutting ants; these ants are able to consume as much fresh biomass as grazing cattle in the Monte, although they have little specific diet overlap with cattle (Claver, 2000). Perennial vegetation in the Monte suffers from a high incidence of monophagous insects, and as result the phytophagous insect communities differ greatly between woody perennials in the central Monte (Debandi, 1999).

Granivory may play a significant role in ecosystem function, being the primary cause for the loss of grass (Marone et al., 2000b) and tree seeds (Villagra et al., 2002). Small mammals, ants, and birds are the most important granivores in desert environments. It was once thought that granivory in South American deserts was depressed, mainly owing to the lack of specialized granivorous rodents (Mares and Rosenzweig, 1978; Morton, 1985), but recent studies have found that seed removal rates in the Monte are comparable to those in other deserts, except for those in North America, where seed removal is exceptionally high (Lopez de Casenave et al., 1998; Marone et al., 2000a).

Another important biological process is pollination. As in other desert environments, the main pollen vectors in the Monte are solitary bees. Flower visitors, like vertebrates and social bees, are expected to be poor in deserts due to low water availability and the short period of blooming. The pollination environment is also highly variable in space and time. In years when rains are more or less evenly distributed in the spring, there is a good flower supply for the diverse fauna of pollinators. In contrast, years with scattered spring rains may decouple blooming and pollinator emergence, an asynchrony that seems frequent in the Monte and may be a cause for generalization in the pollination system of many woody plants (Simpson et al., 1977; Debandi et al., 2002). *Larrea* shrubs provide an excellent example of this, since they have a high diversity of bee visitors in both North and South American deserts, and the dominance in taxonomic composition varies from site to site within each desert, as well as within microhabitats, within and between seasons, and during different parts of the day (Hurd and Linsley, 1975, Simpson et al., 1977, Minckley et al., 1999; 2000, Rossi et al., 1999).

10.2.8 Human Impacts, Threats, and Conservation

Human Impacts Two forms of socioeconomic organization coexist in the Monte, one with a market economy based on irrigated oases and the other with a subsistence economy based in the arid areas (Ministerio de Ambientes y Obras Públicas, 1997). Some 90% of the population lives in irrigated oases located in valleys crossed by perennial rivers, but these oases occupy only 3% of the Monte. The inhabitants are mostly engaged in intensive agriculture, especially viticulture, and the growing of fruit and olive trees, and related local industries.

In the arid lands near aquifers there are small settlements, *puestos*, controlled by families with ancient Hispanic-indigenous roots (called *puesteros*). Their principal activities are goat and cattle breeding, as well as wood extraction and the making of handicrafts. Land ownership is usually unclear, and the puesteros are legally regarded as mere occupants, although historically they are the true landowners.

Thus, in a highly fragile environment, competition for water is one of the most serious environmental issues concerning oasis-desert relationships, since deserts do not receive surface water because it is mostly destined for irrigation and human consumption.

Desertification is another important issue in the Monte, as vast fluvial-aeolian plains are subjected to overgrazing and deforestation, and steep slopes undergo erosion. These systems have slow capacity for natural recovery due to both the intrinsically harsh environmental conditions and currently severe degradation.

From an economic viewpoint, the *algarrobal* has been the most important plant community, the source of subsistence for human communities over many centuries and still heavily used by local people (Abraham and Prieto, 1999). Woodlands have provided people and domestic livestock with shade, firewood, vineyard posts, timber, and food. Economically, the algarrobal has been exploited nonsustainably without adjusting the extraction rate to the recovery rate of the natural resources. This has led to severe desertification and consequently to the impoverishment of the local people. Other plant associations have also been exploited and degraded. In San Juan province, for example, the geographic range of *Bulnesia retama* has been reduced from 4 million to 400,000 hectares as a consequence of harvesting.

Protected Areas Protected areas of the Monte have been created with the primary purposes of protecting landscapes (Valle de La Luna and Talampaya), tree populations (Nacuñán and Telteca), or some particular species or habitat (Lihue Calel). All 16 reserves within the Monte, representing only 1.52 % of the total area, are located in the central and southern Monte (Roig-Juñent and Claver, 1999). These protected areas are not enough to ensure biodiversity conservation, and since they are far from one another they function as islands within an increasingly degraded landscape. Only three reserves (Valle Fértil, Ischigualasto, and Talampaya) are near enough to each other to allow biological interchange.

As conservation of all areas and all species is not feasible, efficient methodological tools should be used for making decisions on priorities so as to ensure as much conservation of biodiversity as possible and to identify new conservation areas in the Monte. Applying traditional priorization criteria, such as species richness or endemism, is difficult because knowledge of species richness and distribution in the Monte is still very limited, even for the currently protected areas. Several methods proposed recently incorporate phylogenetic information to set conservation priorities (Vane-Wright et al., 1991; Humphries et al., 1991). Using this approach, Roig-Juñent et al. (2001) have proposed conservation priorities for natural areas in the Monte and Chaco. In order of importance, areas were ranked as follows: (1) northern Monte; (2) Chaco; (3) central Monte; (4) southern Monte; (5) Valdés Peninsula; and (6) Uspallata-Calingasta Valley. Although these results show that the northern Monte is the most important area for conservation, there are paradoxically no reserves in this area.

10.3 Conclusions

Although not as well studied as their North American counterparts, the coastal Peruvian and Atacama deserts and interior Monte Desert provide important geographic models for studies of arid-land climatology, geomorphology, hydrology, and biogeography. There are many parallels between ecosystem processes and biodiversity in the Sonoran Desert of the southwest United States and Mexico and those of the Monte that deserve new study. Moreover, the hyper-aridity of the Peruvian and Atacama deserts makes these regions important in understanding how organisms can adapt to such extreme conditions. With increasing international focus on global change and on processes of human impact that increase desertification, there will undoubtedly be expanded interest in these arid regions.

References

Abele, G., 1988. Geomorphological west-east section through the north Chilean Andes near Antofagasta. In: C. Bahlburg, C. Breitkreuz, and P. Giese (Editors), *The Southern Central Andes.* Springer-Verlag, Berlin, 153–168.

Abraham, E. M., and F. Rodríguez Martínez (Editors), 2000. *Argentina. Recursos y problemas ambientales de la zona árida.* Provincias de Mendoza, San Juan y La Rioja. Junta de Gobierno de Andalucía–Universidades y Centros de Investigación de la Región Andina Argentina, Mendoza.

Abraham, E.M., and M.R. Prieto, 1999. Vitivinicultura y desertificación en Mendoza. In: B. García Martínez (Editor), *Estudios de historia y ambiente en América: Argentina, Bolivia, México, Paraguay.* IPGH–Colegio de México. México, 109–135.

Aguiar, M.R., and O.E. Sala, 1999. Patch structure, dynamics and implications for the functioning of arid ecosystems. *Trends in Ecology and Evolution,* 14, 273–277.

Alpers, C.N., and G.H. Brimhall, 1988. Middle Miocene climatic change in the Atacama Desert, northern Chile: Evidence from supergene mineralization at La Escondida. *Geological Society of America Bulletin,* 100, 1640–1656.

Armesto, J.J., and P.E. Vidiella, 1993. Plant life-forms and biogeographic relations of the flora of Lagunillas (30°S) in the fog-free Pacific coastal desert. *Annals of the Missouri Botanical Garden,* 80, 499–511.

Armesto, J.J., P.E. Vidiella, and J.R. Gutierrez, 1993. Plant communities of the fog-free coastal desert of Chile: Plant strategies in a fluctuating environment. *Revista Chilena de Historia Natural,* 66, 271–282.

Axelrod, D.I., M.T. Kalin Arroyo, and P.H. Raven, 1991. Historical development of temperate vegetation in the Americas. *Revista Chilena de Historia Natural,* 64, 413–446.

Barclay, W.S., 1917. Sand dunes in the Peruvian desert. *Geographical Journal,* 49, 53–56.

Betancourt, J.L., C. Latorre, J.A. Rech, J. Quade, and K.A. Rylander, 2000. A 22,000–year record of monsoonal precipitation from northern Chile's Atacama Desert. *Science,* 289, 1542–1546.

Billinghurst, G.E., 1886. *Estudio sobre la geografía de Tarapacá.* "El Progresso," Santiago.

Börgel, R., 1973. The coastal desert of Chile. In: D.H.K. Amiran and A.W. Wilson (Editors), *Coastal Deserts: Their Natural and Human Environments.* University of Arizona Press, Tucson, 111–114.

Borruel, N., C.M. Campos, S.M. Giannoni, S.M., and C.E. Borghi, 1998. Effect of herbivorous rodents (cavies and tuco-tucos) on a shrub community in the Monte Desert, Argentina. *Journal of Arid Environments,* 39, 33–37.

Bowman, I., 1924. *Desert Trails of Atacama.* American Geographical Society, New York.

Brack, A., and C. Mendiola, 2000. *Ecología del Perú.* Editorial Bruño, Lima.

Broggi, J.A., 1957. Las terracitas fitogénicas de las lomas de la costa del Perú. *Boletin de la Sociedad Geológica del Perú,* 32, 51–64.

Broggi, J.A., 1961. Las ciclópeas dunas compuestas de la costa peruana, su origen y significación climática. *Boletin de la Sociedad Geológica del Perú,* 36, 61–66.

Brüggen, J., 1936. El agua subterránean en la Pampa de Tamarugal y morfología general de Tarapacá. *Revista Chilena de Historia Natural,* 80, 111–166.

Cabido, M., C. González, and S. Díaz, 1993. Vegetation changes along a precipitation gradient in central Argentina. *Vegetatio,* 109, 5–14.

Cabrera, A.L., 1976. Regiones Fitogeográficas Argentinas. In: W.F. Kugler (Editor.) *Enciclopedia Argentina de Agricultura y Jardinería.* Buenos Aires. Vol. 2, Fascículo 1, 85.

Cabrera, A.L., and A. Willink, 1980. *Biogeografía de America Latina.* Organización de los Estados Americanos (Washington, D.C.), Serie Biología, Monografía 13, 1–122.

Cabrera, A., and J. Yepes, 1940. *Mamíferos Sudamericanos.* Compañia Argentina de Editores, Buenos Aires.

Campos, C.M., and R.A. Ojeda, 1997. Dispersal and germination of *Prosopis flexuosa* (Fabaceae) seeds by desert mammals in Argentina. *Journal of Arid Environments,* 35, 707–714.

Carranza-R., L., 1996. Lomas del Cerro Campana: Estudio geológico y geomorfológico. *Arnaldoa,* 4, 95–101.

Cavagnaro, J.B., and C.B. Passera, 1993. Relaciones hídricas de *Prosopis flexuosa* (algarrobo dulce) en el Monte, Argentina. pp. 73–78, In: IADIZA, *Contribuciones Mendocinas a la Quinta Reunión de Regional para América Latina y el Caribe de la Red de Forestación del CIID.* Conservación y Mejoramiento de Especies del Género Prosopis. Mendoza, Argentina.

Caviedes, C.L., 1975. Secas and El Niño: Two simultaneous climatic hazards in South America. *Proceedings of the Association of American Geographers,* 7, 44–49.

Cei, J.M., and V.G. Roig, 1974. Fauna y ecosistemas del oeste arido argentino. II Anfibios de la provincia de Mendoza. *Deserta,* 4, 141–146.

Cei, J.M., 1980. Amphibians of Argentina. *Italian Journal of Zoology,* 2, 1–609.

Cei, J. M., 1986. Reptiles del centro, centro-oeste y sur de la Argentina. Herpetofauna de las zonas áridas y semiáridas. *Monografie del Museo Regionale di Science Naturali di Torino,* 4, 1–527.

CEPAL, 1960. *Recursos hidráulicos de America latina.* United Nations, Mexico City.

Claver, S., 2000. *Ecología de* Acromyrmex lobicornis *(E.) (Hymenoptera: Formicidae) en la Reserva de Biosfera*

de Ñacuñán, Provincia Biogeográfica del Monte. Preferencia de hábitat, abundancia de colonias y patrones de actividad. Tesis Doctoral. Universidad Nacional de La Plata.La Plata.
Claver, S., and M.B. Kufner, 1989. La fauna de invertebrados de zonas áridas y sus relaciones con el sobrepastoreo y la desertificación. In: IADIZA, *Detección y control de la desertificación.* UNEP, Mendoza, 287–293.
Clüsener, M., and S.W. Breckle, 1987. Reasons for the limitation of mangrove along the west coast of northern Peru. *Vegetatio,* 68, 173–177.
CONAF, 2001. *Guia de Parques Nacionales y Areas Silvestres Potegídas de Chile.* Corporación Nacional Forestal, Santiago.
Contreras, L.C., J.R. Gutiérrez, V. Valverde, and G.W. Cox, 1993. Ecological relevance of subterranean herbivorous rodents in semiarid coastal Chile. *Revista Chilena de História Natural,* 66, 357–368.
Craig, A.K., and N.P. Psuty, 1968. The Paracas Papers, 1(2): Reconnaissance report. *Florida Atlantic University, Department of Geography Occasional Publication* 1, Boca Raton.
Cuya, O., and S. Sánchez, 1991. Flor de Amancaes: lomas que deben conservarse. *Boletin, Lima,* 76, 59–64.
Dabbene, R., 1910. Ornitología Argentina. Catálogo sistemático y descriptivo de las aves de la República Argentina. Parte II. Distribución geográfica de las Aves en el territorio argentino. *Anales del Museo Nacional de Historia Natural, Buenos Aires,* 18, 169–513.
D'Achille, B., 1996. *Kuntursuyo: El Territorio del Cóndor* (2nd ed.) Ediciones Peisa, Lima.
Debandi, G., 1999. *Dinámica de la comunidad de artrópodos asociados a* Larrea *(Zygophyllaceae).* Tesis Doctoral. Universidad Nacional de La Plata.
Debandi, G., B.E. Rossi, J. Araníbar, J.A. Ambrosetti, and I. Peralta, 2002. Breeding system of *Bulnesia retama* (Gill. ex Hook.) Gris. (Zygophyllaceae) in the central Monte Desert (Mendoza, Argentina). *Journal of Arid Environments,* 51, 141–152.
del Valle, H.F., 1998. Patagonian soils: A regional synthesis. *Ecología Austral,* 8, 103–123.
Devries, T.J., 1988. The geology of late Cenozoic marine terraces (*tablazos*) in northwestern Peru. *Journal of South American Earth Sciences,* 1, 121–136.
Dillon, M.O., 2006. The Solanaceae of the *lomas* formations of coastal Peru and Chile. *Annals of the Missouri Botanical Garden* (in press).
Dillon, M.O., 2005. Andean Botanical Information System (6). URL: <http://www.sacha.org>.
Dillon, M.O., M. Nakawaka, and S. Leiva, 2003. The *lomas* formations of coastal Peru: Composition and biogeographic history. In: J. Haas and M.O. Dillon (Editors), El Niño in Perú: Biology and Culture over 10,000 years. Fieldiana: Botany, 1524, 1–9.
Dillon, M.O., A. Sagástegui, I. Sánchez V., S. Llastas, and N.C. Hensold, 1995. Floristic inventory and biogeographic analysis of montane forests in northwestern Peru, pp. 251–270. In: S.P. Churchill, H. Balslev, E. Forero, and J.L. Luteyn (Editors), *Biodiversity and Conservation of Neotropical Montane Forests.* The New York Botanical Garden, New York.
Dunai, T.J., G.A. González Lopez, and J. Juez-Larré, 2005. Oligocene-Miocene age of aridity in the Atacama Desert revealed by exposure dating of erosion-sensitive landforms. *Geology,* 33, 321–324.
Ehleringer, J.R., H.A. Mooney, S.L. Gulmon, and P.W. Rundel, 1980. Orientation and its consequences for *Copiapoa* (Cactaceae) in the Atacama Desert. *Oecologia,* 45, 63–67.
Ehleringer, J., H.A. Mooney, S.L. Gulmon, and P.W. Rundel, 1981. Parallel evolution of leaf pubescence in *Encelia* in arid lands of North and South America. *Oecologia,* 48, 38–41.
Ellenberg, H., 1959. Über den Wasserhaushalt tropischer Nebeloasen in der Küstenwüste Perus. *Bericht uber das Geobotanische Forschungsinstitut Rübel,* 1958, 47–74.
Ericksen, G.E., 1981. Geology and origin of the Chilean nitrate deposits. *U.S. Geological Survey Professional Paper,* 1188, 1–37.
Ezcurra, E., C. Montana, and S. Arizaga, 1991. Architecture, light interception, and distribution of *Larrea* species in the Monte Desert, Argentina. *Ecology,* 72, 23–34.
Ferreyra, R., 1953. Comunidades de vegetales de algunas lomas costaneras del Perú. *Estación Experimental Agrícola "La Molina," Boletin,* 53, 1–88.
Ferreyra, R., 1960. Algunos aspectos fitogeográficos del Perú. *Revue del Institut Geográfica de la Universidad Nacional Mayor San Marcos (Lima),* 6, 41–88.
Ferreyra, R., 1979. El algarrobal y manglar de la costa norte del Perú. *Boletin, Lima,* 1, 12–18.
Ferreyra, R., 1983. Los tipos de vegetación de la costa peruana. *Anales del Jardín Botanico de Madrid,* 40, 241–256.
Fittkau, E.J. 1969. The fauna of South America. In: E.J. Fittkau, J. Illies, H. Klinge, G. H. Schwabe, and H. Sioli (Editors), *Biogeography and Ecology in South America, 2.* W. Junk, The Hague, 624–650.
Follmann, G., 1995. On the impoverishment of the lichen flora and the retrogression of the lichen vegetation in coastal central and northern Chile during the last decades. *Cryptogamic Botany,* 5, 224–231.
Fontugne, M., P. Usselmann, D. Lavallée, M. Julien, and C. Hatté, 1999. El Niño variability in the coastal desert of southern Peru during the mid-Holocene. *Quaternary Research,* 52, 171–179.
Gajardo, R., 1994. *La Vegetación Natural de Chile: Clasificación y Distribución Geográphica.* Editorial Universitaria, Santiago.
Galán de Mera, A., J.A. Vicente, J.A. Lucas, and A. Probanza, 1997. Phytogeographical sectoring of the Peruvian coast. *Global Ecology and Biogeography Letters,* 6, 349–367.
Garleff, K., 1987. El piedemonte andino de la Argentina. In: IADIZA, *Detección y control de la desertificación.* UNEP, Mendoza, 133–138.
Gay, S.P., 1999. Observations regarding the movement of barchan sand dunes in the Nazca to Tacna area of southern Peru. *Geomorphology,* 27, 279–293.
Gengler-Nowak, K., 2002. Reconstruction of the biogeographic history of Malesherbiaceae. *Botanical Review,* 68, 171–188.
Goldberg, R.A., and G. Tisnado, 1987. Characteristics of extreme rainfall events in northwestern Peru during the 1982–1983 El Niño period. *Journal of Geophysical Research,* 92, 14225–14241.
González Loyarte, M., E. Martínez Carretero, and F.A. Roig, 1990. Forest of *Prosopis flexuosa* var. *flexuosa* (Leguminosae) in the NE of Mendoza, I: Structure and Dynamism in the Area of the "Telteca Natural Reserve." *Documents Phytosociologiques, n.s.,* 12, 285–289.

Goudie, A., and J. Wilkinson, 1977. *The Warm Desert Environment.* Cambridge University Press, Cambridge.

Guevara, J.C., J.B. Cavagnaro, O.R, Estevez, H.N. Le Houérou, C.R. Stasi, 1997. Productivity, management and development problems in the arid rangelands of the central Mendoza plains (Argentina). *Journal of Arid Environments,* 35, 575–600.

Guevara, J.C., C.R. Stasi, O.R. Estévez, O.R., and H.N. Le Houérou, 2000. N and P fertilization on rangeland production in Midwest Argentina. *Journal of Range Management,* 53, 410–414.

Gutiérrez, J.R., F. Lopez-Cortes, and P.A. Marquet, 1999. Vegetation in an altitudinal gradient along the Río Loa in the Atacama Desert of northern Chile. *Journal of Arid Environments,* 40, 383–399.

Gutiérrez, J.R., P.L. Meserve, L.C. Contreras, H. Vásquez, and F.M. Jaksic, 1993. Spatial distribution of soil nutrients and ephemeral plants beneath and outside the canopy of *Porlieria chilensis* shrubs (Zygophyllaceae) in arid coastal Chile. *Oecologia,* 95, 347–352.

Gutiérrez, J.R., P.L. Meserve, F.M. Jaksic, L.C. Contreras, S. Herrera, and H. Vasquez, 1993. Structure and dynamics of vegetation in a Chilean arid thornscrub community. *Acta Oecologica,* 14, 271–285.

Gutiérrez, J.R., P.L. Meserve, S. Herrera, L.C. Contreras, and F.M. Jaksic, 1997. Effects of small mammals and vertebrate predators on vegetation in the Chilean semiarid zone. *Oecologia,* 109, 398–406.

Hartley, A.J., and G. Chong, 2002. Late Pliocene age for the Atacama Desert: Implications of the desertification of western South America. *Geology,* 30, 43–46.

Hartley, A.J., G. Chong, J. Houston, and A.E. Mather, 2005. 150 million years of climatic stability: Evidence from the Atacama Desert, northern Chile. *Journal of the Geological Society,* 162, 421–424.

Hastenrath, S.L., 1987. The barchan dunes of southern Peru revisited. *Zeitschrift fur Geomorphologie,* 31, 300–311.

Heatwole, H., 1996. Ant assemblages at their dry limits: The northern Atacama Desert, Peru, and the Chott El Djerid, Tunisia. *Journal of Arid Environments,* 33, 449–456.

Houston, J., and Hartley, A.J. 2003. The central Andean west-slope rainshadow and its potential contribution to the origin of hyper-aridity in the Atacama Desert. *International Journal of Climatology,* 23, 1453–1464.

Humphries, C.J., R.I. Vane-Wright, and P.H. Williams, 1991. Biodiversity reserves: Setting new priorities for the conservation of wildlife. *Parks,* 2, 34–38.

Hunziker, J.H., 1952. Las comunidades vegetales de la cordillera de La Rioja. *Revista de Investigaciones Agrícolas,* 6, 167–196.

Hurd, J.P.D., and E. G. Linsley, 1975. The principal *Larrea* bees of the southwestern United States (Hymenoptera: Apoidea). *Smithsonian Contribution to Zoology,* 193, 1–74.

INTA/CIRN, 1990. *Atlas de suelos de la República Argentina.* Escala 1:500.000 y 1:1.000.000. Secretaría de Agricultura Ganadería y Pesca. Proyecto PNUD ARG. 85/019. Buenos Aires.

Jaksic, F.M., S.I. Silva, P.L. Meserve, and J.R. Gutiérrez, 1997. A long-term study of vertebrate predator responses to an El Niño (ENSO) disturbance in western South America. *Oikos,* 78, 341–354.

Jaksic, F.M., J.C. Torres-Mura, C. Cornelius, and P.A. Marquet, 1999. Small mammals of the Atacama Desert (Chile). *Journal of Arid Environments,* 42, 129–135.

Jerez, V., 2000. Diversity and geographic distribution patterns of coleopteran insects in desert ecosystems of the Antofagasta region, Chile. *Revista Chilena de Historia Natural,* 73, 79–92.

Johnson, A.M., 1976. The climate of Peru, Bolivia and Ecuador. In: W. Schwerdtfeger (Editor), *Climates of Central and South America.* World Survey of Climatology, 12. Elsevier, Amsterdam, 147–218.

Johnston, I.M., 1929. Papers on the flora of northern Chile. *Contributions of the Gray Herbarium,* 4, 1–172.

Katinas, L., and J.V. Crisci. 2000. Cladistic and biogeographic analysis of the genera *Moscharia* and *Polyachyrus* (Asteraceae, Mutisieae). *Systematic Botany,* 25, 33–46.

Keefer, D.K., S.D. de France, M.E. Moseley, J.B. Richardson, D.R. Satterlee, and A. Day-Lewis, 1998. Early maritime economy and El Niño events at Quebrada Tacahuay, Peru. *Science,* 281, 1833–1835.

Kelt, D.A., P.A. Marquet, and J.H. Brown, 2000. Geographical ecology of South American desert small mammals: Consequences of observations at local and regional scales. *Global Ecology and Biogeography Letters,* 9, 219–223.

Kinzl, H., 1958. Die Dünen in der Küstenlandschaft von Peru. *Mitteilungen der Geologischen Gesellschaft in Wien,* 100, 5–17.

Klohn, W., 1972. *Hidrogafía de las Zonas Deserticas de Chile.* United Nations Development Program, Santiago.

Labra, A., M. Soto-Gamboa, and F. Bozinovic, 2001. Behavioral and physiological thermoregulation of Atacama desert-dwelling *Liolaemus* lizards. *Ecoscience,* 8, 413–420.

Le Houérou, H.N., 1999. *Estudios e Investigaciones ecológicas de las zonas áridas y semiáridas de Argentina.* Informe Interno. IADIZA, Mendoza.

León, R.J.C., D. Bran, M.B. Collantes, J.M. Paruelo, and A. Soriano, 1998. Grandes unidades de vegetación de la Patagonia extra-andina. *Ecología Austral,* 8, 125–144.

le Roux, J.P., C. Tavares, and F. Alayza, 2000. Sedimentology of the Rimac-Chillon alluvial fan at Lima, Peru, as related to Plio-Pleistocene sea-level changes, glacial cycles and tectonics. *Journal of South American Earth Sciences,* 13, 499–510.

Lopez de Casenave, J., V.R. Cueto, and L. Marone, 1998. Granivory in the Monte desert, Argentina: Is it less intense than in other arid zones of the world. *Global Ecology and Biogeography,* 7, 197–204.

Lustig, L.K. 1968. Geomorphology and surface hydrology of desert environments. In: W.G. McGinnies, B.L. Goldman, and P. Paylor (Editors), *Deserts of the World: An Appraisal of Research into their Physical and Biological Environments.* University of Arizona Press, Tucson, 93–283.

Macharé, J., and L. Ortlieb, 1992. Plio-Quaternary vertical motions and the subduction of the Nazca Ridge, central coast of Peru. *Tectonophysics,* 205, 97–108.

Mares, M.A., F.A. Enders, J.M. Kingsolver, J.L. Neff, and B.B. Simpson, 1977. *Prosopis* as a niche component. In: B.B. Simpson (Editor), *Mesquite: Its Biology in Two Desert Scrub Ecosystems.* Dowden, Hutchinson and Ross, Stroudsburg, 123–149.

Mares, M.A., J. Morello, and G. Goldstein, 1985. The Monte Desert and other subtropical semi-arid biomes of Argentina, with comments on their relation to North American arid areas. In: M. Evenari, I. Noy-Meir, and

D.W. Goodall, (Editors), *Hot Deserts and Arid Shrublands, 12A*. Elsevier. Amsterdam, 203–236.

Mares, M.A., and M.L. Rosenzweig, 1978. Granivory in North and South American deserts: rodents, birds and ants. *Ecology,* 59, 235–241.

Marone, L., J. Casenave, and V. Cueto, 2000a. Granivory in southern South American Deserts: Conceptual issues and current evidence. *BioScience,* 50, 123–132.

Marone, L., M. Horno, and R. Gonzalez del Solar, 2000b. Post-dispersal fate of seeds in the Monte desert of Argentina: Patterns of germination in successive wet and dry years. *Journal of Ecology,* 88, 940–949.

Martínez Carretero, E. 2000. Vegetación de los Andes centrales de la Argentina. El Valle de Uspallata, Mendoza. *Boletín de la Sociedad Argentina de Botánica,* 34, 127–148.

Meserve, P.L., J.R. Gutiérrez, L.C. Contreras, and F.M. Jaksic, 1993. Role of biotic interactions in a semiarid scrub community in north-central Chile: A long-term ecological experiment. *Revista Chilena de Historia Natural,* 66, 225–241.

Meserve, P.L., J.R. Gutiérrez, J.A. Yunger, L.C. Contreras, and F.M. Jaksic, 1996. Role of biotic interactions in a small mammal assemblage in semiarid Chile. *Ecology,* 77, 133–148.

Meserve, P.L., W.B. Milstead, J.R. Gutiérrez, and F.M. Jaksic, 1999. The interplay of biotic and abiotic factors in a semiarid Chilean mammal assemblage: Results of a long-term experiment. *Oikos,* 85, 364–372.

Meserve, P.L., W.B. Milstead, and J.R. Gutiérrez, J.R. 2001. Results of a food addition experiment in a north-central Chile small mammal assemblage: evidence for the role of "bottom-up" factors. *Oikos,* 94, 548–556.

Miller, A., 1976. The climate of Chile. In: W. Schwerdtfeger (Editor), *Climates of Central and South America.* World Survey of Climatology, 12. Elsevier, Amsterdam, 113–145.

Minckley, R.L., J.H. Cane, and L. Kervin, 2000. Origins and ecological consequences of pollen specialization among desert bees. *Proceedings of the Royal Society of London B,* 267, 265–271.

Minckley, R.L., J.H. Cane, L. Kervin, and T.H. Roulston, 1999. Spatial predictability and resource specialization of bees (Hymenoptera: Apoidea) at a superabundant, widespread resource. *Biological Journal of the Linnaean Society,* 67, 119–147.

Ministerio de Ambientes y Obras Públicas, 1997. *Informe Ambiental.* Gobierno de Mendoza, Mendoza.

Mooney, H.A., S.L. Gulmon, J.R. Ehleringer, and P.W. Rundel, 1980a. Atmospheric water uptake by an Atacama Desert shrub. *Science,* 209, 693–694.

Mooney, H.A., S.L. Gulmon, P.W. Rundel, and J.R. Ehleringer, 1980b. Further observations on the water relations of *Prosopis tamarugo* of the northern Atacama Desert. *Oecologia,* 44, 177–180.

Morello, J., 1955. Estudios Botánicos en las Regiones Aridas de la Argentina. 2. Transpiración de los arbustos resinoso de follaje permanente en el Monte. *Revue Agronómica del Noroeste Argentina,* 1, 385–524.

Morello, J., 1956. Estudios Botánicos en las Regiones Aridas de la Argentina. 3. Reacciones de las plantas a los movimientos de suelos en Neuquén extra-andino. *Revue Agronómica del Noroeste Argentina,* 2, 79–152.

Morello, J., 1958. La Provincia Fitogeográfica del Monte. *Opera Lilloana,* 2, 5–115.

Morello, J., 1984. *Perfil Ecológico de Sudamérica.* Instituto de Cooperación Iberoamericana.

Morronne, J.J., 1994. Distributional patterns of species of Rhytirrhinini (Coleoptera: Curculionidae) and the historical relationships of the Andean provinces. *Global Ecology and Biogeography Letters,* 4, 188–194.

Morronne, J.J., and J.V. Crisci, 1995. Historical biogeography: Introduction to methods. *Annual Review of Ecology and Systematics,* 26, 373–401.

Mortimer, C., 1973. The Cenozoic history of the southern Atacama Desert. *Journal of the Geological Society,* 129, 505–526.

Mortimer, C., 1980. Drainage evolution of the Atacama Desert of northernmost Chile. *Revista Geológica de Chile,* 11, 3–28.

Mortimer, C., and N. Saric, 1972. Landform evolution in the coastal region of Tarapacá Province, Chile. *Revue de Geomorphologie Dynamique,* 4, 162–170.

Mortimer, C., and N. Saric, 1975. Cenozoic studies in northernmost Chile. *Geologische Rundschau,* 64, 395–420.

Morton, S. R., 1985. Granivory in arid regions: Comparison of Australia with North and South America. *Ecology,* 66, 1859–1866.

Müller, G.K., 1985. Zur floristischen Analyse der peruanische, Loma-Vegetation. *Flora,* 176, 153–165.

Müller, P., 1973. *The Dispersal Centres of Terrestrial Vertebrates in the Neotropical Realm.* W. Junk, The Hague.

Muñoz-Schick, M., 1985. *Flores del Norte Chico.* Direción de Bibliotecas, Archivos y Museos, La Serena, Chile.

Muñoz-Schick, M., R. Pinto, A. Mesa, and A. Moreira-Muñoz, 2001. Fog oases during the El Nino-Southern Oscillation 1997–1998, in the coastal hills south of Iquique, Tarapaca region, Chile. *Revista Chilena de Historia Natural,* 74, 389–405.

Ojeda, R.A., P.G. Bledinger, and R. Brandl, 2000. Mammals in South American drylands: Faunal similarity and trophic structure. *Global Ecology and Biogeography Letters,* 9, 115–123.

Olrog, C.C., 1963. Lista y distribución de las aves argentinas. *Opera Lilloana,* 9, 1–377.

ONERN, 1971a. *Inventario, evaluación y uso racional de los recursos naturales de la costa. Cuenca del río Grande (Nazca).* Oficina Nacional de Evaluación de Recursos Naturales, Lima.

ONERN, 1971b. *Inventario, evaluación y uso racional de los recursos naturales de la costa. Cuenca del río Ica.* Oficina Nacional de Evaluación de Recursos, Lima.

ONERN, 1971c. *Inventario, evaluación y uso racional de los recursos naturales de la costa. Cuenca del río Pisco.* Oficina Nacional de Evaluación de Recursos, Lima.

Ortlieb, L., A. Diaz, and N. Guzman, 1996a. A warm interglacial episode during the oxygen isotope stage 11 in northern Chile. *Quaternary Science Reviews,* 15, 857–871.

Ortlieb, L., C. Zazo, J.L. Goy, C. Dabrio, and J. Macharé, 1996b. Pampa del palo: An anomalous composite marine terrace on the uprising coast of southern Peru. *Journal of South American Earth Sciences,* 9, 367–379.

Ortlieb, L., C. Zazo, J.L. Goy, C. Hillaire-Marcel, B. Ghaleb, and L. Cournoyer, 1996c. Coastal deformation and sea-level changes in the northern Chile subduction area (23°S) during the last 330 ky. *Quaternary Science Reviews,* 15, 819–831.

Pascual, R., and E. Ortiz Jaureguizar, 1990. Evolving climates and mammal faunas in Cenozoic South America. *Journal of Human Evolution,* 19, 23–60.

Paskoff, R., 1980. Late Cenozoic crustal movements and sea-level variations in coastal areas of northern Chile. In: N.A. Mürner (Editor), *Earth Rheology, Isostasy and Eustasy*. Wiley, New York, 487–495.

Péfaur, J.E., 1982. Dynamics of plant communities in the lomas of southern Peru. *Vegetatio,* 49, 163–171.

Peña, L.E., and A.J. Ugarte, 1996. *Las Mariposas de Chile.* Editorial Universitaria, Santiago.

Porter, C.C., 1975. Relaciones zoogeográficas y origen de la fauna de Ichneumonidae (Hymenoptera) en la provincia biogeográfica del Monte del noroeste argentino. *Acta Zoológica Lilloana,* 31(15), 175–252.

Puig, S., M. Rosi, M.I. Cona, V.G. Roig, and S. Monge, 1999. Diet of a piedmont population of *Ctenomys mendocinus* (Rodentia, Ctenomyidae): Seasonal patterns and variations according sex and relative age. *Acta Theriologica,* 44, 15–27.

Puig, S., F. Videla, and M.I. Cona, 1997. Diet and abundance of the guanaco (*Lama guanicoe* Müller 1776) in four habitats of northern Patagonia, Argentina. *Journal of Arid Environments,* 36, 343–357.

Puig, S., F. Videla, S. Monge, and V.G. Roig, 1996. Seasonal variations in guanaco diet (*Lama guanicoe* Müller 1776) and food availability in northern Patagonia, Argentina. *Journal of Arid Environments,* 34, 215–224.

Quinn, W.H., V.T. Neal, and S.E. Antúnez de Mayolo, 1987. El Niño occurrences over the past four and a half centuries. *Journal of Geophysical Research—Oceans,* 92,14,449–14,461.

Quintanilla, V.G., 1988. Fitogeografia y cartografía de la vegetación de Chile arido. *Contibuciones Científicas y Tecnológicas, Arena Geociencias VI,* 82, 1–27.

Rauh, W., 1985. The Peruvian-Chilean deserts. In: M. Evenari, I. Noy-Meir, and D.W. Goodall (Editors), *Hot Deserts and Arid Shrublands (12A).* Elsevier, Amsterdam, 239–267.

Redford, K.H., and J.E. Eisenberg, 1992. *Mammals of the Neotropics: The Southern Cone (2).* University of Chicago Press, Chicago.

Riddle, B.R., 1996. The molecular phylogenetic bridge between deep and shallow history in continental biota. *Trends in Ecology and Evolution,* 11, 207–211.

Ringuelet, R., 1961. Rasgos fundamentales de la zoogeografía de La Argentina. *Physis,* 22, 151–188.

Robinson, D.A., 1964. *Peru in Four Dimensions.* American Studies Press, Lima.

Roig, F.A., 1972. Bosquejo fisonómico de la vegetación de la provincia de Mendoza. *Boletín de la Sociedad Argentina de Botánica,* XIII(Suplemento), 49–80.

Roig, F.A., 1976. Las comunidades vegetales del piedemonte de la precordillera de Mendoza. *Ecosur,* 3, 1–45.

Roig, F.A., 1985. Arboles y bosques de la región árida centro oeste de la Argentina (Provincias de Mendoza y San Juan) y sus posibilidades silvícolas. In: *Forestación en Zonas Aridas y Semiáridas. Segundo Encuentro Regional CIID. América Latina y el Caribe.* Santiago, 145–188.

Roig, F.A., 1989. Ensayo de detección y control de la desertificación en el W de la Ciudad de Mendoza, desde el punto de vista de la vegetación. In: IADIZA, *Detección y control de la desertificación.* IADIZA-UNEP. Mendoza, 196–232.

Roig, F.A., 1998. La vegetación de la Patagonia. *Flora Patagónica,* INTA Colección Científica Tomo VIII (I).

Roig, V.G., 1972. Esbozo general del poblamiento animal en la Provincia de Mendoza. *Boletín de la Sociedad Argentina de Botánica,* 13(Suplemento), 81–88.

Roig, V.G., and J.R. Contreras, 1975. Aportes ecologicos para la biogeografia de la provincia de Mendoza. *Ecosur,* 2, 185–217.

Roig-Juñent, S.A. and S. Claver, 1999. La entomofauna del monte y su conservación en las áreas naturales protegidas. *Revista de la Sociedad Entomológica Argentina,* 58, 117–127.

Roig-Juñent, S., and G. Flores, 2001. Historia biogeográfica de las áreas áridas de América del Sur Austral. In: J. Llorente Bousquets and J.J. Morrone (Editors), *Introducción a la Biogeografía en Latinoamérica: Teorías, conceptos, métodos y aplicaciones.* Las Prensas de Ciencias, Facultad de Ciencias, UNAM. México D.F., 257–266.

Roig-Juñent, S., G. Flores, S. Claver, G. Debandi, and A. Marvaldi, 2001. Monte Desert (Argentina): Insect biodiversity and natural areas. *Journal of Arid Environments,* 47, 77–94.

Rollins, H.B., J.B. Richardson, and D.H. Sandweiss, 1986. The birth of El Niño: Geoarchaeological evidence and implications. *Geoarchaeology,* 1, 3–15.

Rossi, B.E., G.O. Debandi, I.E. Peralta, and E. Martinez. 1999. Comparative phenology and floral patterns in *Larrea* species (Zygophyllaceae) in the Monte desert (Mendoza, Argentina). *Journal of Arid Environments,* 43, 213–226.

Rossi, B.E., and P.E. Villagra, 2003. Effects of *Prosopis flexuosa* on soil properties and the spatial pattern of understory species in arid Argentina. *Journal of Vegetation Science,* 14, 543–550.

Rundel, P.W., and M.O. Dillon. 1998. Evolutionary patterns, habitat selection, and speciation of Bromeliaceae in the loma formations of the Peruvian and Atacama Desert regions. *Plant Systematics and Evolution,* 212, 261–278.

Rundel, P.W., M. O. Dillon, and B. Palma, 1996. The vegetation and flora of Pan de Azúcar National Park in the Atacama Desert of northern Chile. *Gayana, Botanica,* 53, 295–315.

Rundel, P.W., M.O. Dillon, B. Palma, H.A. Mooney, S.L. Gulmon, and J.R. Ehleringer, 1991. The phytogeography and ecology of the coastal Atacama and Peruvian deserts. *Aliso,* 13, 1–50.

Rundel, P.W., J.R. Ehleringer, S.L. Gulmon, and H.A. Mooney, 1980. Patterns of drought response in leaf succulent shrubs of the coastal Atacama Desert in northern Chile. *Oecologia,* 46, 196–200.

Rundel, P.W., and M. Mahu, 1976. Community structure and diversity in a coastal fog desert in northern Chile. *Flora,* 165, 493–505.

Rundel, P.W., B. Palma, M.O. Dillon, M.R. Sharifi, E.T. Nilsen, and K. Boonpragob, 1997. *Tillandsia landbeckii* in the coastal Atacama Desert of northern Chile. *Revista Chilena de Historia Natural,* 70, 341–348.

Sagástegui, A., J. Mostacero, and S. López, 1988. Fitoecología del Cerro Campana. *Boletin de la So.iead Botanica, La Libertad,* 14, 1–47.

Sandweiss, D.H., J.B. Richardson, E.J. Reitz, H.B. Rollins, and K.A. Maasch, 1996. Geoarcheological evidence from Peru for a 5,000 year B.P. onset on El Niño. *Science,* 273, 1531–1533.

Seligman, N.G., J.B. Cavagnaro, and M.E. Horno, 1992. Simulation of defoliation effects on primary produc-

tion of a warm-season, semiarid perennial-species grassland. *Ecological Modelling*, 60, 45–61.
Shmida, A., 1985. Biogeography of the desert flora. In: M. Evenari, I. Noy-Meir, and D.W. Goodall (Editors), *Hot Deserts and Arid Shrublands, 12A.* Elsevier, Amsterdam, 23–77.
Silbitol, R.H., and Ericksen, G.E., 1953. Some desert features of northwest central Peru. *Boltin de la Sociedad Geologica, Perú*, 26, 225–246.
Simpson, B.B., J.L. Neff, and A.R. Moldings, 1977. Reproductive Systems of *Larrea*. In: T.J. Mabry, J. Hunziker, and D.R. DiFeo Jr. (Editors), *Creosote Bush: Biology and Chemistry of Larrea in New World Deserts.* Dowden, Hutchinson and Ross, Stroudsburg.
Solbrig, O., M.A. Barbour, J. Cross, G. Goldstein, C.H. Lowe, J. Morello, and T.W. Yang, 1977. The strategies and community patterns of desert plants. In: G.H. Orians and O.T. Solbrig, (Editors), *Convergent Evolution in Warm Deserts.* Dowden, Hutchinson and Ross, Stroudsburg, 67–106.
Soriano, A. 1956. Los distritos florísticos de la Provincia Patagónica. *Revista de Investigaciones Agrícolas*, 10, 323–335.
Soriano, A., 1985. Deserts and semi-deserts of Patagonia. In: M. Evenary, I. Noy-Meir, and D.W. Goodall (Editors), *Hot Deserts and Arid Shrublands (A).* Elsevier, Amsterdam, 423–460.
Stange, L. A., A. L. Teran, and A. Willink, 1976. Entomofauna de la provincia biogeográfica del Monte. *Acta Zoológica Lilloana*, 32, 73–120.
Tago-Nakazawa, M., and M.O. Dillon. 1999. Biogeografía y evolución en el clado *Nolana* (Solaneae-Solanaceae). *Arnaldoa*, 6, 81–116.
Takhtajan, A., T.J. Crovello, and A. Cronquist, 1986. *Floristic Regions of the World.* University of California Press, Berkeley.
Terán, A., 1973. Entomofauna del Dominio Subandino. I. Las cochinillas (Hom. Coccoidea) de *Larrea divaricata* y *L. cuneifolia* (Zygophyllaceae). *Acta Zoológica Lilloana*, 30, 190–206.
Tognelli, M., C. Borghi, and C. Campos, 1999. Effect of gnawing by *Microcavia australis* (Rodentia, Cavidae) on *Geoffroea decorticans* (Leguminosae) plants. *Journal of Arid Environments*, 41, 79–85.
United Nations Environment Programme, 1992. *World Atlas of Desertification.* Edward Arnold. London.
Vane-Wright, R.I., C.J. Humphries, and P.H. Williams, 1991. What to protect? Systematics and the agony of choice. *Biological Conservation*, 55, 235–254.
Vervoorst, F. 1954. *Observaciones ecológicas y fitosociológicas en el bosque de algarrobo del Pilciao, Catamarca.* Doctoral Thesis. Universidad de Buenos Aires.
Vidiella, P.E., J.J. Armesto, and J.R. Gutierrez, 1999. Vegetation changes and sequential flowering after rain in the southern Atacama Desert. *Journal of Arid Environments*, 43, 449–458.
Villagra, P.E., M.A. Cony, N.G. Mantován, B.E. Rossi, M.M. González Loyarte, R. Villalba, and L Marone, 2004. Ecología y manejo de los algarrobles de la Provincia Fitogeográfica del Monte. In: M.F. Arturi, J.L. Frangi, and J.F. Goya (Editors), *Ecología y Manejo de Bosques Nativos de Argentina.* Editorial Universidad Nacional de la Plata.
Villagra, P.E., L. Marone, and M.A. Cony, 2002. Mechanism affecting the fate of *Prosopis flexuosa* seeds during secondary dispersal en the Monte desert. *Austral Ecology*, 27, 416–421.
Villagra, P.E., and F.A. Roig, 1999. Vegetación de las márgenes de inundación del Río Mendoza en su zona de divagación (Mendoza, Argentina). *Kurtziana*, 27, 309–317.
Volkheimer, W., 1971. Aspectos paleoclimatológicos del Terciario argentino. *Revista del Mueso Argentino de Ciencias Naturales (Paleontología)*, 18, 172–190.
Weberbauer, A., 1911. *Die Pflanzenwelt der peruanischen Anden.* Vegetation der Erde 12. Englemann, Leipzig.
Weberbauer, A., 1936. Phytogeography of the Peruvian Andes. *Field Museum of Natural History, Botanical Series*, 13, 13–81.
Weberbauer, A., 1945. *El mundo vegetal de los Andes peruanos.* Ministerio de Agricultura, Dirección de Agricultura, Estación Experimental Agrícola de La Molina, Lima.
Wells, L.E., 1990. Holocene history of the El Niño phenomenon as recorded in flood sediments of northern coastal Peru. *Geology*, 18, 1134–1137.
Wells, L.E., and J.S. Noller, 1997. Determining the early history of El Niño. *Science*, 276, 966.
Willink, A., 1991. Contribucion a la Zoogeografia de insectos argentinos. *Boletin Academico Nacional de Ciencias, Cordoba*, 59, 125–147.

11

The Mediterranean Environment of Central Chile

Juan J. Armesto
Mary T. K. Arroyo
Luis F. Hinojosa

The Mediterranean-type environment of South America, broadly defined as the continental area characterized by winter rainfall and summer drought, is confined to a narrow band about 1,000 km long on the western side of the Andes in north-central Chile (Arroyo et al., 1995, 1999). Although much has been written about the climate, vegetation, and landscapes of this part of Chile, and comparisons have been drawn with California and other Mediterranean-type regions of the world (Parsons, 1976; Mooney, 1977; Rundel, 1981; Arroyo et al., 1995), a modern synthesis of information on the physical setting, regional biota, and historical development of ecosystems in central Chile has not been attempted. This chapter is intended to provide such an integrated picture, emphasizing those aspects most peculiar to the region.

Since the earlier floristic work on the Chilean *matorral* (e.g. Mooney, 1977), the name given to the vegetation of central Chile, there is now a much greater appreciation of the geographic isolation and high levels of biological diversity and endemism in this region of South America (Arroyo and Cavieres, 1997; Villagrán, 1995; Arroyo et al., 1995, 1999). Because of the great richness and singularity of its terrestrial flora, this area of the continent is considered to be one of the world's 25 hotspots in which to conserve global biodiversity (Arroyo et al., 1999; Myers et al., 2000). An analysis of the main features of the Mediterranean environment in South America should therefore address the causes of such high floristic richness, the nature of current threats to biodiversity, and the prospects for its conservation in the long-term. A discussion of conservation concerns closes the present chapter (but see also: Arroyo and Cavieres, 1997; and Arroyo et al., 1999). In view of the vast literature on the biota and physical setting of central Chile, this chapter adopts a selective approach, from a biogeographic perspective, of what we consider to be the most remarkable historical, physical, and ecological features of this environment, which in turn may explain its extraordinary richness in plants and animals.

11.1 The Physical Character of the Chilean Mediterranean-Type Ecosystems

Mediterranean-type ecosystems occupy a narrow band along the western margin of South America, from 30 to 36°S in central Chile (fig.11.1). These ecosystems represent the transition between one of the driest deserts in the world, the southern Atacama Desert, north of 28°S, and the mixed deciduous-evergreen temperate forests, which occur south of

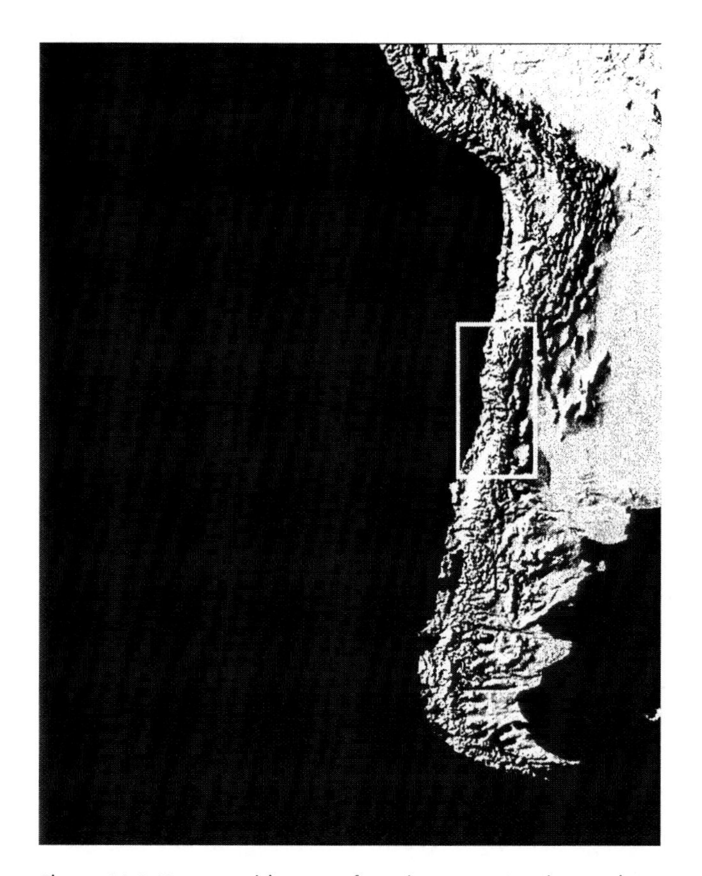

Figure 11.1 Topographic map of southwestern South America showing the approximate extent of the Mediterranean-climate region (box), 30–36°S, and the position of the massive Andean Cordillera, which intercepts winds flowing from both east and west, and locks the position of the anticyclone over the Pacific, thus configuring the "arid diagonal"

36°S (fig. 11.2). Mediterranean ecosystems in central Chile are therefore a highly heterogeneous vegetation mosaic. The major vegetation types are dry xerophytic thorn scrub dominated by deciduous shrubs and succulents; mesic communities dominated by evergreen sclerophyllous trees in the coastal and Andean foothills; and forests dominated by winter-deciduous trees in the south of the Mediterranean area. Within the latitudinal range of the Mediterranean-climate ecosystems, annual precipitation varies from less than 200 to 700 mm (Rundel, 1981).

Topographically, this ecologically complex region is characterized by the presence of two parallel mountain ranges oriented from north to south, the Coastal Cordillera and the Andes (fig. 11.1). These ranges are separated by a narrow tectonic basin, the 80–100-km wide Central Depression, which is filled with a large volume of rock waste derived mainly from the weathering and erosion of the Andes. Extensive accumulations of ignimbrites also occur, derived from the activity of Andean volcanoes during Pleistocene time. Soils in the Central Depression are derived primarily from these sedimentary and volcanic materials. By contrast, soils in the Coastal Cordillera are developed on highly weathered metamorphic parent materials of Jurassic age. These soils have a high organic matter and clay content that makes them particularly susceptible to runoff erosion after the loss of plant cover.

Mountain chains with an east-west orientation *(angosturas)* interrupt the Central Depression both to the north and south of the Mediterranean-climate area, connecting the Andes with the Coastal Cordillera. The latter range reaches a maximum elevation of 2,200 m above sea level, thus allowing for the development of an alpine vegetation belt, above the sclerophyllous vegetation zone, in a way similar to the Andes. Several important river systems originating in the Andes cross the Central Depression in spacious and fertile valleys. From north to south these include the Limari, Aconcagua, Maipo, and Maule valleys. These river systems have been completely transformed by a long history of agricultural activity and presently sustain a major portion of the human population of Chile.

Climatic characterizations of the Chilean Mediterranean region have been provided by Miller et al. (1977), di Castri and Hajek (1975), and Arroyo et al. (1995), among others. A summary of salient features of this climatic regime is presented here, emphasizing those aspects that have received less attention in previous work. At the latitude of Santiago (33°S), rainstorms occur sporadically during the cooler months of the year (April to September), but drought may extend for as long as six months during the spring and summer (October to March). The single cool rainy season is a distinctive feature of north-central Chile, as it is of Mediterranean-type environments elsewhere. The monthly distribution and total amount of precipitation in a year, two of the most important variables affecting the local vegetation, vary greatly between years, although the former has received less emphasis.

This variable climatic regime can be explained by seasonal changes in the strength and latitudinal position of the South Pacific anticyclone, a high-pressure center located around 30°S off the Pacific coast of southern South America (see chapter 3; also Aceituno, 1988). During the austral summer, this high-pressure center occupies a broad latitudinal range off South America and blocks the westerly flow of humid air masses moving across the Pacific Ocean, thereby leaving central Chile completely dry. During the austral winter, however, the anticyclone weakens and may often be displaced equatorward, thus allowing for cyclonic storms and related frontal systems to progress toward central Chile. Because of the dominant role played by the South Pacific anticyclone in maintaining the seasonal rainfall pattern over central Chile, the atmospheric and oceanic conditions that affect the strength and latitudinal position of this high pressure system can greatly alter the amount and distribution of precipitation in the Chilean Mediterranean region. Large inter-annual variabil-

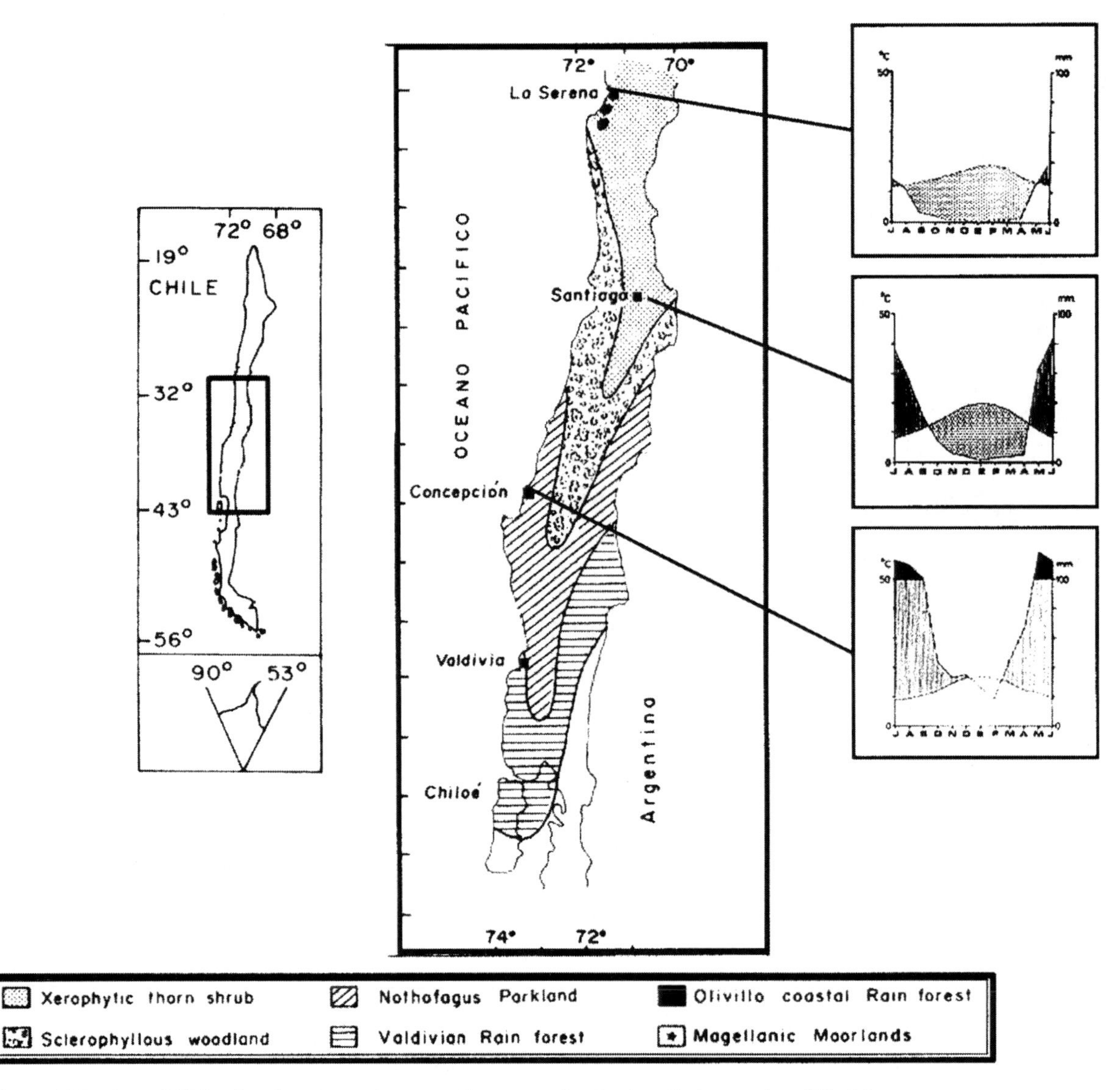

Figure 11.2 Central Chile showing the main vegetation zones for the late Holocene (modified from Villagrán, 1995). The Mediterranean-climate region includes xerophytic thorn scrub, Olivillo coastal rain forest, sclerophyllous woodland, and the northern zone of the *Nothofagus* parkland (Maulino forest). Climatic diagrams represent the latitudinal variation within the range of the Mediterranean environment.

ity in precipitation, associated with changes in anticyclone strength and position, is an intrinsic feature of the region.

In recent decades, a major component of the interannual variability in precipitation has been linked to El Niño–Southern Oscillation (ENSO) (Aceituno, 1988; Holmgren et al., 2001). During El Niño years, weakening of the anticyclone and decreased upwelling of ocean waters along the west coast of South America may cause large increases in total rainfall across the entire Mediterranean-climate region of Chile. In contrast, during longer La Niña episodes, due to the strengthening of the anticyclone, rainfall decreases well below the average, producing extremely dry conditions, particularly in the northern part of the Mediterranean region. These effects extend also to the wetter southern limit of the region, causing long droughts with major ecological consequences (see chapter 19; also Holmgren et al., 2001). In recent years, the often dramatic environmental impacts of the long droughts associated with La Niña have been understood more thoroughly by the public because of marked reductions in irrigation water and potable water supplies, and the loss of generating capacity at hydroelectric power plants that depend on artificial dams placed across the main river systems of central Chile.

The climatic regime of central Chile differs in some important aspects from other Mediterranean ecosystems of the world because of the complete absence of summer rainfall and associated thunderstorm activity. This is a feature of great significance for understanding the disturbance regime that has historically affected Chilean Mediterranean ecosystems. It accounts for the negligible importance of natural fire as a shaper of vegetation structure and dynamics in central Chile. Summer rains, derived from an easterly source, are prevented from reaching Mediterranean Chile by the massive physical barrier of the Andes, where many peaks rise above 5,000 m, including the highest, Aconcagua (6,962 m; fig. 11.1). The rain-shadow effect of the Andes is responsible for maintaining the hyperarid conditions of the Atacama Desert in northern Chile (Arroyo et al., 1988), but it also isolates the Mediterranean region from the easterly flow of warm moist air that brings summer rainfall to the subtropical eastern side of the cordillera. Lightning storms that could start wildfires are rare in the Chilean Mediterranean region and when they do occur, occasionally in summer, they are generally confined to elevations above 3,000 m in the Andes (e.g., Arroyo et al., 1981), above the altitudinal limit of sclerophyllous vegetation.

Despite its narrowness, the Mediterranean region of Chile is characterized by pronounced environmental gradients from east to west. Particularly important, although often ignored, is the strong rain-shadow effect of the Coastal Cordillera. This range intercepts westerly wind flows of moist oceanic air, causing a substantial drop in precipitation in the Central Depression, an effect that has not been properly quantified but is evident from the comparison of rainfall records at climate stations both east and west of the range (e.g., Rundel, 1981). Consequently, habitats in the Central Depression are drier than the slopes of the Coastal Cordillera and the Andes. The drier vegetation zones in the north of the Mediterranean region are thus projected southward to higher latitudes along the Central Depression, giving each zone a V-shape on figure 11.2.

An additional moisture source for vegetation in the Coastal Cordillera derives from cloud-water condensation at altitudes above 600 m. A permanent fog-zone, triggered by a stable temperature inversion above the cold Humboldt Current (Rundel et al., 1991), forms at this altitude, supplementing rainfall, especially in the drier part of the Mediterranean-climate region where normal rainfall is insufficient to maintain vegetation (Kummerow, 1966; Rundel and Mahu, 1976). From 33 to 30°S, as annual rainfall declines to less than 200 mm, isolated patches of rainforest are maintained by fogs in the windward slopes of the coastal mountains (Rundel, 1981). Fog condensation in tree canopies may play a significant, but understudied, role in enhancing moisture and nutrient availability for sclerophyllous vegetation, especially through the long dry summers.

11.2 History of Semiarid Environments in Western South America

The history of the Mediterranean-type climate and vegetation of central Chile is connected with the development of the "arid diagonal" of South America (see chapter 2; also Hinojosa and Villagrán, 1997). The progress of aridity along the western margin of South America is coupled with the rise of the Andes, which interrupted the flow of moist air to southern Perú and northern Chile from tropical sources farther east, and with the beginning of the equatorward flow of cold Antarctic waters in the South Pacific. In this section, we summarize the current knowledge of the events that led to the onset of the Mediterranean-type climate in central Chile.

The Andean orogenesis has dominated the western margin of South America during the latter part of the Cenozoic Era. This massive geophysical transformation progressed through a series of tectonic uplifts that became more intense during the Neogene (Reynolds et al., 1990). Geologic dating of magmatism and exposed rocks across the Argentina-Chile border assigns the folding and shortening of the continental crust in central Chile to the last 15 Ma (million years before present), during which the Andes rose to elevations ranging from 3,000 m to more than 6,000 m (Reynolds et al., 1990). The final uplift of the Frontal Cordillera and Precordillera, dated to the Miocene between 15 and 8 Ma, progressively shut off the easterly flow of air masses originating in the tropics and subtropics farther east, leading to the establishment of the present Mediterranean climate in central Chile with its single rainy season. Prior to these events, tropical air masses flowed freely from east to west across South America, maintaining a wetter subtropical climate in central Chile, possibly with two rainy seasons (Hinojosa and Villagrán, 2005). The existence of a subtropical fossil flora of Miocene age, containing the ancestors of certain floristic components of the Chilean sclerophyllous vegetation (e.g., Lauraceae), has been documented in the Andean foothills near Santiago (Hinojosa and Villagrán, 1997). The character of this fossil assemblage supports the existence of a different climatic regime, with lower mountain ranges, in the middle Miocene, and suggests that the Mediterranean-type climate of central Chile did not become established until the late Neogene.

Another factor contributing to the desiccation of the climate along the western margin of South America was the establishment of the cold Humboldt Current, a major feature of the present oceanic circulation. The equatorward flow of cold polar waters began in earnest after the breakup of Gondwana. The separation of Antarctica, first from Africa, then from Australia, and later from southern South America, generated a massive westerly (zonal) circulation of water around Antarctica that reduced the poleward (meridional) flow of warm tropical waters (see chapter 2). The Antarctic Circumpolar Current did not fully form until the final opening of the Drake Passage after 35 Ma, but once it became established, Antarctic ice sheets expanded and the cooling of the proto-Humboldt Current was enhanced (Zachos et al., 2001). These events led to the development of a steeper temperature gradient from tropical to high latitudes and probably to the present positioning and strength of the South Pacific anticyclone that determines the summer drought in central Chile.

The evolution of the morphological and physiological mechanisms that allowed the local flora, of tropical ancestry, to endure the long summer droughts characteristic of the Mediterranean-type climate must, therefore, have occurred during the Neogene. The ability to tolerate seasonal desiccation became the primary determinant of species survival during the Quaternary Period, as seasonal droughts in central Chile became more frequent and intense, especially during interglacial stages (Villagrán, 1995).

11.3 Mediterranean Vegetation Types

The mountainous topography of central Chile, with up to 6,000 m of relief, contrasting radiation and moisture regimes, strong rain-shadow effects, and a mosaic of soil types and nutrient supplies, generates pronounced environmental gradients that have stimulated the evolutionary differentiation of the biota (Armesto and Martínez, 1978; Rundel, 1981; Rozzi et al., 1989). High floristic richness and a diversity of plant communities are a consequence of this environmental heterogeneity (Arroyo et al., 1993, 1995). In addition, strong climatic variability resulting from varying frequencies of ENSO events during the Quaternary (Villagrán, 1995; Holmgren et al., 2001) must have promoted genetic variability among local populations. The Mediterranean region is also a depository of ancient tropical lineages that found refuge in coastal valleys from the drying trend initiated by Neogene uplift of the Andes (Troncoso et al., 1980; Villagrán and Armesto, 1980; Arroyo et al., 1995; Hinojosa et al., 2006). Also, as a consequence of the repeated cooling and warming cycles of the Quaternary, the regional vegetation has been exposed to wetter or drier periods relative to present conditions (Villagrán, 1995). In response to these cycles, xerophytic and cool-temperate forest taxa have successively expanded and contracted their ranges in central Chile at different times in the past, as documented from pollen studies (Heusser, 1983; Villagrán, 1995). These floral migrations must have contributed greatly to increased species richness and the heterogeneity of the vegetation mosaic within the Mediterranean zone (Villagrán, 1995, Arroyo et al., 1995). Andean uplift also provided opportunities for colonization and differentiation of local alpine floras, adding further to the floristic richness and diversity of vegetation types in the region.

In addition, the strong influence of insect and bird pollinators in the evolution of plant-reproductive strategies in Mediterranean-type ecosystems (Arroyo and Uslar, 1993; Arroyo et al., 1993, 1995) and the role of animals in the dispersal of seeds of sclerophyllous tree species (Hoffmann and Armesto 1995) are important conditions favoring genetic variability and speciation in biotic communities within the region. High levels of self-incompatibility have been documented for the woody species in montane sclerophyllous vegetation (Arroyo and Uslar, 1993).

Arroyo et al. (1995) recognized a greater diversity of vegetation types in the Mediterranean-climate region of central Chile than previously described in the context of comparative studies of analogous ecosystems in Chile and California (e.g., Parsons, 1976; Mooney, 1977; Rundel, 1981). In the following account of the vegetation of central Chile, the general floristic scheme proposed by Arroyo et al. (1995) will be followed for describing the main vegetation types, although names given here to some vegetation units may differ. A few vegetation types not listed by Arroyo et al. are also included (table 11.1). We discuss each vegetation type in terms of its relative importance regarding floristic richness, endemism, distribution, and natural history. Gaps in knowledge are indicated. The conservation status of Mediterranean ecosystems in Chile will be considered later, although a brief mention of the history of these plant formations is made here. Finally we discuss succinctly the main floristic affinities of the vegetation of central Chile with other dry plant formations found in South America.

11.3.1 Relict Coastal Forests

Olivillo Forests A widely distributed vegetation type in central Chile is the coastal forest dominated by olivillo, *Aextoxicon punctatum*, a tree from the endemic monotypic family Aextoxicaceae (figs. 11.2, 11.3). These forests, described in detail by Muñoz and Pisano (1947), Villagrán and Armesto (1980), and Pérez and Villagrán (1985), have an important number of species of climbers and epiphytes, including narrow endemics such as *Peperomia coquimbensis* (Piperaceae). Olivillo forests are found as far north as the latitude of La Serena (30°S) in the semiarid region of Chile, immersed in a matrix of xerophytic vegetation. The northernmost extensions of Olivillo forests are known as the relict forests of Fray Jorge and Talinay (Troncoso et al., 1980), and occur as isolated forest fragments on the upper slopes of the coastal mountains, under the direct influence of maritime fogs. Eco-physiological studies have shown that fog condensation is the main source of water and nutrients for these isolated relict forests (Kummerow, 1966; P.E. Vidiella and T.E. Dawson, unpublished). Other isolated patches of Olivillo forest are scattered across coastal mountaintops between Fray Jorge and the coast, from where these forests extend more or less continuously southward at lower elevations, especially on slopes and valleys facing the ocean (Villagrán and Armesto; 1980, Pérez and Villagrán, 1985).

The fragmented distribution of the flora of Olivillo forests in Chile has intrigued biogeographers for many years (e.g., Skottsberg, 1948; Schmithüsen, 1956; Muñoz and

Table 11.1 Major plant formations in central Chile, modified from Arroyo et al. (1995) and Rundel (1981)

Plant formation	Approximate latitudinal range (°S) in central Chile	Commonly associated woody taxa
Olivillo forests	Fragmented 30–33 Continuous 33–36	*Aextoxicon punctatum, Myrceugenia correifolia, Raphithamnus spinosus, Drimys winteri, Sarmienta repens, Mitraria coccinea*
Nothofagus montane forest	Fragmented 33–36	*N. obliqua, Ribes punctatum, Lomatia hirsuta*
Maulino forest	35–36	*N. alessandrii, N. glauca, N. dombeyi, Gomortega keule, Pitavia punctata, Cryptocarya alba*
Sclerophyllous matorral	32–36	*Cryptocarya alba, Quillaja saponaria, Lithrea caustica, Peumus boldus, Proustia pyrifolia, Kageneckia oblonga*
Summer-deciduous matorral with succulents	30–34	*Retanilla trinervia, Flourensia thurifera, Trichocereus, Eulychnia, Puya* spp.
Acacia caven savanna	30–36	*Prosopis chilensis, Cestrum parqui, Trevoa trinervis*
Swamp forests	Scattered patches	*Myrceugenia exsucca, R. spinosus, D. winteri, Boquila trifoliolata*
Palm forests	33–34	*Jubaea chilensis, Cryptocarya alba, Drimys winteri, Schinus latifolius*
Coastal matorral	30–36	*Lithrea caustica, Bahia ambrosioides, Schinus latifolius, Escallonia pulverulenta, Senna* spp.
Andean montane woodland	Discontinuous 32–36	*Kageneckia angustifolia, Schinus montanus, Austrocedrus chilensis*

Figure 11.3 Olivillo forests in Fray Jorge National Park (30°S) (photo: J.J. Armesto)

Pisano, 1947; Villagrán and Armesto, 1980; Troncoso et al., 1980). Many plant species found in the northernmost relicts (e.g., *Sarmienta repens* and *Mitraria coccinea*, Gesneracieae, *Azara microphylla*, Celastraceae) also occur as components of the Valdivian evergreen rainforests, with a disjunction of nearly 900 km (Villagrán and Armesto, 1980). Current interpretations of the phylogenetic relations, South American distribution, and fossil record of the main relict taxa suggest that these floristic assemblages may be descendants of subtropical rainforests that occupied central Chile in the mid-Cenozoic, before the onset of events that shaped the Mediterranean-type climate (Troncoso et al., 1980; Villagrán, 1990; Hinojosa and Villagrán, 1997). They are consequently unrelated to equatorward advances of the southern temperate forests during colder and wetter intervals of the Pleistocene, as has sometimes been argued (e.g., Solbrig et al., 1977; Rundel, 1981). However, southern Chilean temperate forests are ancestrally related to these relict communities, thus explaining the considerable number of shared taxa. The future of these northern forest fragments, now designated as protected areas, remains uncertain in the face of past human disturbance by logging and fire, and present climatic desiccation. Understanding the ability of trees to capture moisture from cloud water, especially from the perspective of water supply to growing seedlings in the understory, and the climatic factors affecting the frequency of coastal fogs, are critical issues for the survival of these relicts.

Nothofagus Forests *Nothofagus* species (Nothofagaceae) occur in the Mediterranean region of central Chile at its southern boundary with the temperate rainforest region, and also as a chain of small fragmented populations above 1,000 m on the highest peaks of the Coastal Cordillera—El Roble, La Campana, and Altos de Cantillana (fig. 11.2). These forest fragments represent the northernmost populations of *Nothofagus* in South America (Casassa, 1985) and are isolated from the continuous distribution of *Nothofagus obliqua*, a deciduous tree broadly distributed across the Central Depression as well as on coastal and Andean foothills of south-central Chile . At the southern margin of the Mediterranean region, several narrowly distributed species of *Nothofagus* occur in mixed stands with both sclerophyllous species from central Chile and evergreen broad-leaved trees from the Valdivian temperate rainforest, a vegetation known locally as the *Maulino* forest (San Martin and Donoso, 1996). Two important narrow endemics of the *Maulino* forest are *Nothofagus glauca* and *N. alessandrii*, both deciduous. Sheltered ravines are dominated by the evergreen *N. dombeyi*, which is a characteristic tree species of the Valdivian rainforests farther south. Other narrow endemic tree species found in the coastal ranges near the southern limit of the Mediterranean region are *Gomortega keule* (Gomortegaceae) and *Pitavia punctata* (Rutaceae). The high local endemism probably reflects the conservative character of coastal range habitats, compared to those of the Central Depression and Andean slopes (Hinojosa and Villagrán, 1997; Villagrán et al., 1998).

The disjunct northern populations of *Nothofagus obliqua* have been interpreted as remnants from a more continuous distribution that existed at lower elevations during colder stages of the Pleistocene (Villagrán, 1990). The cooler and wetter climate of the last cold stage would have facilitated the equatorward expansion of the *Nothofagus* woodlands in central Chile, an idea supported by fossil pollen assemblages from Laguna Tagua-Tagua, 100 km south of Santiago (Heusser, 1983).

11.3.2 Sclerophyllous Shrublands (Matorral) and Thorn Scrub

Evergreen sclerophyllous vegetation extends more or less continuously across both slopes of the Coastal Cordillera and into the foothills of the Andes, but it is more scattered in river valleys crossing the Central Depression (figs. 11.2, 11.4). It is the most common plant formation in central Chile, dominated by evergreen trees and shrubs, such as *Cryptocarya alba* (Lauraceae), *Lithrea caustica* (Anacardiaceae), *Quillaja saponaria* (Rosaceae), *Maytenus boaria* (Celastraceae), *Kageneckia oblonga* (Rosaceae), and *Peumus boldus* (Monimiaceae). The matorral commonly exhibits a patchy spatial structure with open spaces between shrub clumps, especially on the Andean foothills (Fuentes et al., 1986). However, in deep creeks and along permanent water courses, as well as on steep south-facing slopes in the coastal range, it may develop a continuous canopy, 8–12 m tall. Some tree species, such as *Beilschmiedia miersii* (Monimiaceae), *Drimys winteri* (Winteraceae), *Luma chequen* (Myrtaceae), *Citronella mucronata*, and *Persea lingue* (Lauraceae), may be restricted to such habitats (Armesto and Martínez, 1978; Rundel, 1981; Arroyo et al., 1995). On ocean-facing slopes of the coastal range, influenced by moisture-laden marine air, sclerophyllous trees such as *Cryptocarya alba* grow together with *Aextoxicon punctatum* (see above) and may be covered by a profuse growth of climbers, such as *Proustia pyrifolia* (Asteraceae), *Bomarea salsilla* (Amaryllidaceae), and *Lardizabala biternata*, (Lardizabalaceae) (fig. 11.5).

On dry north-facing slopes, or on frequently disturbed sites (Armesto and Martínez, 1978; Armesto and Gutiérrez, 1978), the sclerophyllous matorral is frequently replaced by a xerophytic thorn scrub, with a combination of deciduous shrubs, such as *Trevoa trinervis* (Rhamnaceae), *Flouresia thurifera* (Asteraceae), and *Colliguaja odorifera* (Euphorbiaceae), and often includes succulent species such as *Puya* spp. (Bromeliaceae) and columnar cacti, *Echinopsis* and *Eulychnia* spp., especially on steep rocky slopes (fig. 11.6).

There seems to be a fragile balance in the regional plant cover of central Chile between dry xerophytic vegetation

Figure 11.4 Coastal sclerophyllous shrubland (photo: J.J. Armesto)

and mesic sclerophyllous tree communities. This balance has historically depended on climatic cycles between conditions that are cool and wet, and those that are hot and dry, which in turn may be linked to varying frequencies of ENSO events in the South Pacific Ocean during the Holocene (Villagrán, 1995). Large-scale regeneration of sclerophyllous trees is normally limited by drought stress and the lack of seed banks (Fuentes et al., 1986; Jiménez and Armesto, 1992), except in unusually wet periods or along water courses. It is likely that the wet phases of ENSO in central Chile are linked to significant regeneration of sclerophyllous tree species, especially in the drier habitats. Owing to recurrent anthropogenic disturbance through fire and permanent grazing pressure, however, vegetation cover may be presently shifting regionally toward dominance by dry xerophytic species (Armesto and Gutiérrez, 1978), despite a regional climatic trend towards increasing rainfall in the late Holocene (see below). Fossil pollen assemblages suggest that poleward expansions of the xerophytic thorn scrub and exclusion of sclerophyllous trees from large areas in the Central Depression occurred periodically during hot dry periods in the early Holocene (Heusser, 1983; Villa-Martínez and Villagrán, 1997; Villagrán, 1995).

Figure 11.5 *Bomarea salisilla* (Amaryllidaceae), a vine endemic to coastal sclerophyllous mattoral (photo: J.J. Armesto)

11.3.3 *Acacia caven* Savanna

Xerophytic open woodlands, with the physiognomic aspect of a savanna (Fuentes et al., 1990), are widespread along the Central Depression in the Mediterranean-climate region, extending beyond the limit with the southern temperate forest region. These woodlands are dominated by leguminous trees, mainly *Acacia caven*, occasionally forming mixed stands with *Prosopis chilensis*, and have a dense herbaceous cover composed almost entirely of introduced European annual herbs (e.g., *Erodium* spp.) and grasses, which are typically associated with grazing pastures. Scattered patches of native perennials (*Loasa* spp.) and geophytes, such as *Pasithaea coerulea*, *Alstroemeria* spp., and *Leucocoryne* spp., and grazing-resistant *Cestrum parqui* (Solanaceae), can be found within the range of this dry plant formation. The same geophytes and perennials can be found

Figure 11.6 Xerophytric thorn scrub (photo: J.J. Armesto)

in patches of sclerophyllous vegetation in coastal and Andean foothills. Some discussion has centered around the historical character of this dry formation (Arroyo et al., 1995). The high cover of introduced grasses under *Acacia* trees, and the resprouting ability of dominant shrubs, suggests that this vegetation type has been shaped by human impact, especially through cattle grazing and fire. In some cases, the invasion of the *Acacia caven* savanna by sclerophyllous species following the exclusion of fire and cattle has been postulated (Fuentes et al., 1986; Armesto and Pickett, 1985). However, it is also possible that this formation dominated the Central Depression before intense human impact, as a consequence of the rain-shadow effect of the Coastal Cordillera, and that at least part of its present distribution may have been associated with dry periods during the Holocene. This hypothesis remains to be tested. Whatever the case, this dry formation is expanding due to intense woodland clearing in the Central Depression and the dispersal of *Acacia caven* seeds by introduced cattle in recent times (Armesto and Gutiérrez, 1978).

11.3.4 Palm Forests

A narrowly distributed, and frequently overlooked, component of the mosaic of vegetation types in coastal areas of central Chile is the mixed sclerophyllous forest with numerically abundant populations of the tall endemic palm, *Jubaea chilensis* (fig. 11.7). These evergreen forests occur mainly on slopes and deep canyons of the Coastal Cordillera, and very likely profit from the moisture-laden marine air and the limited temperature oscillation in these habitats (Arroyo et al., 1995). The dominance by palms is another manifestation of the persistence of woody tropical lineages in maritime locations in central Chile. The physiognomic dominance of palms (up to 20 m tall) over the evergreen sclerophyllous canopy makes these local forests especially picturesque. Their distribution is presently limited to two small protected areas in central Chile, with scattered remnant populations in several coastal outposts around Valparaíso, and as far north as Illapel, suggesting that mixed palm forests may have been more widespread in the past. Palms have been intensely exploited for the production of syrup used for food and beverages since colonial times. Unfortunately, harvesting the syrup requires killing the palm trees that must be 30–40 years old to be of use (see later discussion on conservation). In addition, human-set fire probably contributed to the

Figure 11.7 Palm forests in La Campana National Park, central Chile (photo: J.J. Armesto)

demise of the palm from many sites, as the palms lack the resprouting ability of other sclerophyllous shrubs. Adult palms often survive ground fires because of their size, but regenerating juveniles are most certainly killed by fire and are conspicuously absent from most remnant populations. Propagation in tree nurseries and fire protection programs in the wild can help to prevent further population decline.

11.3.5 Swamp Forests

Another often overlooked, but ecologically important component of the Chilean Mediterranean vegetation are the swamp forests (*bosques pantanosos*), which are scattered at various locations, generally in the narrow coastal plains between the Pacific Ocean and the Coastal Cordillera (fig. 11.8; Ramírez et al., 1996). Several woody species are characteristically or exclusively found in these swamp communities, particularly *Myrecugenia exsucca* (Myrtaceae), *Blepharocalyx cruckshanksii* (Myrtaceae), and vines such as *Boquila trifoliolata* (Lardizabalaceae). The origin of these coastal swamp forests has been studied recently by Villagrán and Varela (1990), Villa-Martínez and Villagrán (1997), and Maldonado (1999). Their results indicate that most swamp forests originated no earlier than the last 5,000 years, as a result of factors that contributed to raise the water table in small coastal basins, leading to sediment accumulation and later invasion by trees. These factors involved the blocking of local streams by littoral dune systems, a climatic trend toward wetter conditions in the late Holocene, and recent expansion of evergreen sclerophyllous species (Villa-Martínez and Villagrán, 1997). Swamps are thus living witnesses to recent climatic fluctuations in central Chile, possibly associated with changes in the frequency of ENSO events (Villagrán 1995). They are unrelated to relict Olivillo forests mentioned earlier, except that they share some species with avian-dispersed fleshy fruits (e.g., *Drimys winteri*) that may be recent colonizers of these wetlands.

11.3.6 Coastal Matorral

The coastal plain is underdeveloped in most of central Chile. However, in some areas marine terraces formed by Pleistocene sea-level changes and tectonic events, often covered by fossil dunes, have provided open spaces for the development of coastal plant formations that include sclerophyllous shrubs (e.g., *Lithrea caustica, Schinus latifolius*, both Anacardiaceae), together with typical littoral plants (e.g., *Bahia ambrosioides*, Asteraceae; *Heliotropium chilensis*, Scrophulariaceae), and including some narrow endemics (e.g., *Lucuma valparidisiaca, Lobelia salicifolia,* Lobeliaceae, *Myrcianthes coquimbensis*, Myrtaceae). A number of geophytes that flower massively in wet years enrich the coastal matorral formation, including *Alstroemeria* spp., *Leucocoryne* spp., *Hippeastrum* spp., *Pasithea coerulea*, and some orchid species.

11.3.7 Andean Montane Woodland

Finally, above 1,500 m in the Andean mountains, an upland sclerophyllous woodland is found that differs in composition from the mesic communities in the foothills of the Andes (Hoffmann and Hoffmann, 1982; Arroyo et al., 1995) and from the upland *Nothofagus* communities of the

Figure 11.8 Swamp forest (Quintero, central Chile) (photo: J.J. Armesto)

Coastal Cordillera. This vegetation is dominated by *Kageneckia angustifolia* (Rosaceae), which is restricted to this altitudinal belt and constitutes a discontinuous treeline (Arroyo and Uslar, 1993). Above treeline, at 2300 m in central Chile, the montane woodland gives way to a low subalpine shrubland (Arroyo et al., 1981). Scattered patches of the conifer *Austrocedrus chilensis* (Cupressaceae) occur near the treeline at various locations between 32 and 36°S (Rundel, 1981). The distribution of *Austrocedrus* becomes more continuous on the eastern side of the Andes south of 36°S. The presence in central Chile of these relict stands of *Austrocedrus*, which rely strongly on their sprouting ability for reproduction and currently lack seedling regeneration, is another indication that the vegetation of central Chile is very sensitive to climatic variability over large temporal scales (C. Le-Quesne and J.C. Aravena, personal communication). These long-lived populations are probably remnants of stands that became established in wetter periods of the Holocene.

11.4 Floristic Affinities of Chilean Matorral with Other Dry Plant Formations

The Andean uplift fragmented the rich subtropical floras that prospered in the north-central regions of Chile and Argentina, extending to Bolivia and southern Brazil, during mid-Cenozoic time (Hinojosa and Villagrán, 1997). This process would have been gradual, producing the early segregation of more mesic taxa, followed by the development of a still continuous xerophytic flora. The present Mediterranean flora of central Chile must have originated from this transitional dry vegetation.

Based on physiognomic and floristic attributes, Sarmiento (1975) recognized 13 arid plant formations in South America, including the semiarid formations of north-central Chile. The area of Mediterranean vegetation analyzed by Sarmiento (1975) is equivalent to the summer-deciduous matorral, with succulents or *Acacia caven* savanna, as defined above (table 11.1), both subtypes of Mediterranean vegetation. The following genera are cited by Sarmiento as belonging to the semiarid formation of north-central Chile: *Geoffroea*, *Prosopis*, *Proustia*, *Puya*, and *Trichocereus*. Considering floristic similarities at the genus level (Sorensen's similarity coefficient) among these 13 arid plant formations, as reported by Sarmiento (1975), we constructed the dendrogram shown in figure 11.9. The Mediterranean vegetation of north-central Chile is weakly related to the driest vegetation types within South America, even those associated with the Pacific coastal desert and the occidental Andes, which are geographically contiguous. In contrast, it has stronger floristic affinities (43%) with two formations found on the eastern side of the Andes (the Monte and Andean Prepuna). These relationships suggest that these dry vegetation formations were connected in past geologic times (Cabrera and Willink, 1980; Hinojosa and Villagrán, 1997). The breakup of the dry trans-andean flora is probably related to the time of maximum Andean uplift. According to Gregory-Wodzicki (2000), the central Andes reached half of their present altitude by 10.7 Ma. This elevation was probably sufficient for the establishment of semiarid environments, but the floristic connection across the Andes could have been maintained for some time. The shift to much drier conditions in northern Chile and Argentina is indicated by the replacement of C3 by C4 grasses in northwest Argentina, recorded at 7 Ma (Latorre et al., 1997). This may be the

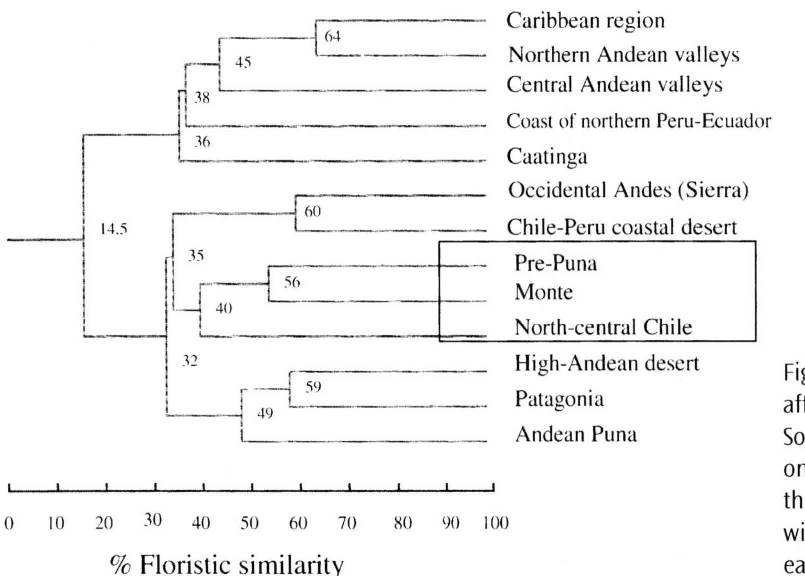

Figure 11.9 Dendrogram representing floristic affinities among dry plant formations within South America (data from Sarmiento 1975, based on woody species only). The closest affinities of the Mediterranean flora (north-central Chile) are with the Pre-Puna and Monte vegetation of the eastern side of the Andes (box).

minimum age for the initial separation of the dry plant formations on both sides of the Andes and the subsequent isolation of the Mediterranean vegetation in South America.

11.5 Human Impacts on the Chilean Mediterranean Ecosystem

Mediterranean-type ecosystems in South America, as elsewhere in the world, have long been modified by human activities. Because of the large population contained within its relatively small area, human impacts have been probably more pervasive in central Chile than in many other parts of South America (Arroyo et al., 1995). Greater human impact on the sclerophyllous vegetation of central Chile may be expected from the fact that humans frequently use uncontrolled fire to open areas for agriculture and settlement. The effects of fire on local ecosystems would have been greater in central Chile than in other Mediterranean-climate areas where wildfire has been a regular component of the disturbance regime at least through the Pleistocene. For example, the Mediterranean-type ecosystems of California exhibit a higher species richness of annuals (Arroyo et al., 1995) and larger seed banks of woody species (Jiménez and Armesto, 1992) than those of central Chile, thus providing indirect evidence that Californian ecosystems may be more resilient to fire disturbance. Presently, as discussed below, several factors associated with a long human presence influence vegetation dynamics and the distribution of plant communities in central Chile.

11.5.1 Succession and Disturbance

Successional recovery of sclerophyllous vegetation in many open areas of central Chile, including abandoned farmland, is a slow process (Fuentes et al., 1986; Armesto et al., 1995). Slow vegetation change may be a consequence of environmental conditions that are limiting for seedling recruitment. These include grazing by introduced herbivores, such as rabbits and goats, especially in open spaces between shrub patches, nd physiological stress related to the long summer droughts (Fuentes et al., 1986; Armesto et al., 1995). Although climatic trends may significantly influence the probability of successful establishment and expansion of shrublands into open, less favorable habitats (Arroyo et al., 1993), several conditions derived from human impact may prevent the regeneration of matorral shrubs and trees. The most important factor is the notable increase in the frequency and extent of human-set fire (Armesto et al., 1995), which is directly related to the growth of human settlements in central Chile. Fire-recurrence cycles of 20 years or less, as controlled by rates of fuel accumulation from resprouting species, are not uncommon in many areas of central Chile, especially in the foothills of the two mountain ranges and in the Central Depression. Most likely, fires led to negative selection of species that are unable to resprout from rootstocks and must result in widespread mortality of young plants. Succession may also be impaired by limited seed dispersal from population sources (Jiménez and Armesto, 1992). Many sclerophyllous shrub species bear fleshy fruits dispersed by animals (Hoffmann and Armesto, 1995) and require the presence of perching trees to be recruited on open areas (Armesto and Pickett, 1985). Consequently, the loss of tree cover from large areas of central Chile may have a negative feedback on immigration via seed dispersal.

11.5.2 Invasion

A second relevant impact of humans on Mediterranean-climate regions is the introduction of exotic species of plants and animals (Sala et al., 2000). The original landscapes of central Chile have been greatly transformed by the deliberate introduction of vertebrate species such as goats and rabbits (whose range expanded from southern Chile). Rabbits and goats are major predators of shrub seedlings, thereby limiting population regeneration in open areas (Fuentes et al., 1983, Fuentes and Muñoz, 1995). It is likely that herbivore pressure in the Mediterranean region is presently much greater than before Europeans arrived in Chile. Natural herbivores, such as rodents, probably had a more limited impact on vegetation due to their control by native predators, although the latter are now much reduced in number (Jaksic, 1997). Little is known about the past effects of wild populations of guanacos (*Lama guanicoe*) on the vegetation of central Chile, from where these animals are long gone.

Furthermore, the large-scale planting of *Eucalyptus* spp. and *Pinus radiata* for forestry purposes has profoundly changed the landscape of central Chile, especially along the coast. In addition, many weedy species have accidentally spread through central Chile, along roadsides, making this region the richest in the country in terms of numbers of exotic species, with more than 400 recorded taxa (Arroyo and Cavieres, 1997). Ecosystems in the foothills of the two mountain ranges and the Central Depression, subjected to intense grazing pressure by cattle, have been heavily invaded by annuals, originally rare in central Chile (Arroyo et al., 1995). Among the most common introduced annuals are *Erodium* spp., *Bromus* spp., and *Avena* spp. The ecological impact of exotic weeds on sclerophyllous vegetation has been less studied than the effects of introduced herbivorous mammals, which are well documented (Fuentes and Muñoz, 1995). In the Coastal Cordillera, extensive areas of species-rich sclerophyllous vegetation, with a native herbaceous flora, have been replaced by a closed canopy of pines and eucalyptus, leading to considerable habitat loss. One of the most dramatic cases is the near demise of the Maulino forest as the coastal mountains in the southern parts of the Mediterranean region have become almost completely covered by a blanket of forest plantations (Bustamante and Castor, 1998).

11.5.3 *Hydrologic Problems*

Two effects associated with recurrent fire and logging for firewood are significant in relation to the water supply to sclerophyllous vegetation. On the one hand, the loss of forest cover and the exposure of bare ground have increased the amount of surface runoff, leading to massive losses of organic soil and nutrients to downstream ecosystems. These losses can affect the ability of sclerophyllous vegetation to recover from disturbance, slowing down succession. Disturbance effects may be enhanced further by the extreme variability of precipitation in this climatic regime, leading to serious runoff and erosion problems when burning or logging of native vegetation, or plantations, coincide with wet years. On the other hand, frequent anthropogenic fire and logging appear to have changed the physiognomic structure of sclerophyllous vegetation everywhere, converting tall forests into low-stature shrublands, with many stems sprouting from the base of each shrub. The loss of stature of sclerophyllous vegetation may be critical, especially in upland areas of the Coastal Cordillera where interception of maritime fogs contributes importantly to the ecosystem's water supply. Consequently, a long history of chronic disturbance to coastal vegetation has disrupted the hydrologic cycle over many years, with a negative feedback on plant cover and tree regeneration, causing further desiccation and loss of forest habitat.

11.6 Prospects for Conservation and Restoration of Degraded Lands

The conservation of Mediterranean-type ecosystems in central Chile, as in other Mediterranean regions of the world, is a serious challenge because of the long history of human occupation and the present relevance of the areas for agriculture, industry, and human habitation (Armesto et al., 1998). The challenge is greater in Chile because less than 5% of the areas currently protected by the state are found within the Mediterranean-climate region, which in turn contains about 50% of the vascular plant species endemic to the Chilean flora (Arroyo et al., 1995). In addition, about two-thirds of the vegetation types in central Chile are poorly represented or absent from the national system of protected areas, which is also flawed by a serious lack of attention to the herbaceous flora (Arroyo et al., 1995). For the benefit of land managers, and as a reference for more quantitative assessments in the future, a qualitative ranking of the present conservation status of the major plant formations in central Chile is given in table 11.2. This assessment shows that most plant formations in central Chile are represented in only one or two conservation areas (national parks, or the less restrictive forest reserves), and that the majority of these originally narrowly distributed communities currently have a greatly restricted and/or fragmented range due to human exploitation or past disturbance. Consequently, we consider these ecosystems either threatened or endangered (table 11.2).

Considering the high demand on the land, there is little chance that the number of protected areas in central Chile could be greatly increased through direct government efforts. Nevertheless, public or private initiatives to increase

Table 11.2 Representation in protected areas and qualitative conservation status of major plant formations in central Chile

Plant formation	Representation in protected areas (number)*	Conservation status **
Olivillo forests	P (1) only in the north	Threatened
Nothofagus montane forest	P (1)	Endangered
Maulino forest	R (2)	Endangered
Sclerophyllous matorral	P (1), R (1)	Threatened
Summer-deciduous matorral with succulents	P (1), R (1)	No danger
Acacia caven savanna	R (1)	No danger
Swamp forests	(1) Municipal park	Endangered
Palm forests	P (1), R (1)	Threatened
Coastal matorral	P (1)	Threatened
Andean montane woodland	P (1)	Endangered

*P = national parks, R = national forest reserves, M = national monuments

**Endangered = greatly restricted in distribution due to human exploitation or disturbance, only few, small patches left. Threatened = restricted distribution due to human exploitation or disturbance, several large patches still persist. No danger = broadly distributed in central Chile.

the extent of protected land should be strongly encouraged, especially given the limited protection of critical remnant sites that help sustain the elevated species richness and endemism of this region. Even small areas, if set aside for preservation, may contain an unusually high number of species and should have a significant conservation value (Arroyo et al., 1995). A high priority in the agenda for the conservation of biodiversity in central Chile must be placed on the regulation of land use, particularly in restricting the use of fire. Restoration programs based on plantations of native shrubs and trees should be favored over the expansion of exotic tree plantations in coastal areas, especially considering the flammability of pine and eucalyptus plantations and the potential loss of some of the most valuable habitats for the native flora and fauna (Arroyo et al., 1999, Arroyo and Cavieres, 1997). Restoration efforts, although expensive, must consider the need to exclude exotic herbivores, such as cattle, goats, and rabbits, and keep in mind the large inter-annual variability in precipitation, which may limit the survival probability of young trees. Nevertheless, restoration programs should have added economic value because of their potential to reduce runoff and erosion rates, and the risk of destructive landslides that periodically affect rural and urban landscapes in central Chile, causing significant losses of human life and property.

References

Aceituno, P., 1988. On the functioning of the southern oscillation in the Southern American sector. Part I. Surface climate. *Monthly Weather Review*, 116, 505–524.

Armesto, J.J., and J. Gutiérrez, 1978. El efecto del fuego en la estructura de la vegetación de Chile central. Anales del Museo de Historia Natural de Valparaíso 11, 43–48.

Armesto, J.J., and J.A. Martínez, 1978. Relations between vegetation structure and slope aspect in the Mediterranean region of Chile. *Journal of Ecology*, 66, 881–889.

Armesto, J.J., and S.T.A. Pickett, 1985. A mechanistic approach to the study of succession in the Chilean matorral. *Revista Chilena de Historia Natural*, 58, 9–17.

Armesto, J.J., P.E. Vidiella, and H.E. Jiménez, 1995. Evaluating causes and mechanisms of succession in the Mediterranean regions of Chile and California. In: M.T.K. Arroyo, P.H. Zedler, and M.D. Fox (Editors), *Ecology and Biogeography of Mediterranean Ecosystems in Chile, California, and Australia.* Springer-Verlag, New York, 418–434.

Armesto, J.J., R. Rozzi, C. Smith-Ramirez, and M.T.K. Arroyo, 1998. Conservation targets in South American temperate forests. *Science*, 282, 1271–1272.

Arroyo, M.T K., J. J. Armesto, and C. Villagrán, 1981. Plant phenological patterns in the high Andean Cordillera of central Chile. *Journal of Ecology*, 69, 205–223.

Arroyo, M.T.K., and L. Cavieres, 1997. The Mediterranean-type climate flora of central Chile—What do we know and how can we assure its protection? *Noticiero de Biología* (Chile), 5, 48–56.

Arroyo, M.T.K., F. Squeo, J J. Armesto, and C. Villagrán, 1988. Effects of aridity on plant diversity in the northern Chilean Andes: Results of a natural experiment. *Annals of the Missouri Botanical Garden*, 75, 55–78.

Arroyo, M.T.K., J.J. Armesto, F. Squeo, and J. Gutiérrez, 1993. Global change: The flora and vegetation of Chile. In: H. Mooney, E. Fuentes, and B. Kronberg (Editors), *Earth-System Responses to Global Change: Contrasts between North and South America.* Academic Press, New York, 239–264.

Arroyo, M.T.K., L. Cavieres, C. Marticorena, and M. Muñoz-Schick, 1995. Convergence in the Mediterranean floras in central Chile and California: Insights from comparative biogeography. In: M.T.K. Arroyo, P.H. Zedler, and M.D. Fox (Editors), *Ecology and Biogeography of Mediterranean Ecosystems in Chile, California, and Australia.* Springer-Verlag, New York, 43–88.

Arroyo, M.T.K., R. Rozzi, J.A. Simonetti, and M. Salaberry, 1999. Central Chile. In: R.A. Mittermeier, N. Myers, and C.G. Mittermeier (Editors), *Hotspots: Earth's Biologically Richest and Most Endangered Terrestrial Ecoregions,* CEMEX, Mexico City, 161–171.

Arroyo, M.T.K., and P. Uslar, 1993. Breeding systems in a temperate Mediterranean-type climate montane sclerophyllous forest in central Chile. *Botanical Journal of the Linnean Society*, 111, 83–102.

Bustamante, R.O., and C. Castor, 1998. The decline of an endangered temperate ecosystem: The ruíl (*Nothofagus alessandrii*) forest in central Chile. *Biodiversity and Conservation*, 7, 1607–1626.

Cabrera, A.L., and A. Willink, 1980. *Biogeografía de América Latina*, Second edition. Secretaría General, OEA, Washington, D.C.

Casassa, I., 1985. *Estructura de edades de poblaciones de Nothofagus obliqua en un gradiente climático latitudinal de Chile.* M.Sc. Thesis, University of Chile, Santiago, Chile.

Castri, F. di, and E.R. Hajek, 1975. *Bioclimatología de Chile.* Ediciones de la Vicerectoría Académica. Universidad Católica de Chile, Santiago.

Fuentes, E. R., and M. Muñoz, 1995. The human role in changing landscapes in central Chile: Implications for intercontinental comparisons. In: M.T.K. Arroyo, P.H. Zedler, and M.D. Fox (Editors), *Ecology and Biogeography of Mediterranean Ecosystems in Chile, California, and Australia.* Springer-Verlag, New York, 401–407.

Fuentes, E.R., R. Avilés, and A. Segura, 1990. The natural vegetation of a heavily man-transformed landscape: The savanna of central Chile. *Interciencia*, 15, 293–295.

Fuentes, E.R., A.J. Hoffmann, A. Poiani, and M.C. Alliende, 1986. Vegetation change in large clearings: Patterns in the Chilean matorral. *Oecologia*, 68, 358–366.

Fuentes, E.R., F.M. Jaksic, and J. Simonetti, 1983. European rabbits vs. native rodents in central Chile: Effects on shrub seedlings. *Oecologia,* 58, 411–414.

Gregory-Wodzicki, K.M., 2000. Uplift history of the Central and Northern Andes: A review. *Geological Society of America Bulletin*, 112, 1091–1105.

Heusser, C J., 1983. Quaternary pollen record from Laguna Tagua Tagua, Chile. *Science,* 219, 1429–1432.

Hinojosa, L.F., and C. Villagrán, 1997. Historia de los bosques del sur de Sudamérica: I. Antecedentes paleobotánicos, geológicos y climáticos del Terciario del cono sur de América. *Revista Chilena de Historia Natural*, 7, 225–239.

Hinojosa, L.F., and C. Villagrán, 2005. Did South American Mixed Paleofloras evolve under thermal equability or in the absence of an effective Andean barrier during the Cenozoic? *Palaeogeography, Palaeoclimatology, Palaeoecology*, 217, 1–23.

Hinojosa, L.F., J. J. Armesto, and C. Villagrán, 2006. Are Chilean coastal forests Cenozoic relicts? Evidence from foliar physiognomy, paleoclimate and phytogeography. *Journal of Biogeography*, 33, 331–341.

Hoffmann, A.J., and A.E. Hoffmann, 1982. Altitudinal ranges of phanerophytes and chamaephytes in central Chile. *Vegetatio*, 48, 151–163.

Hoffmann, A.J., and J.J. Armesto, 1995. Convergence vs non-convergence in seed dispersal modes in the Mediterranean-climate vegetation of Chile, California, and Australia. In: M.T.K. Arroyo, P.H. Zedler, and M.D. Fox (Editors), *Ecology and Biogeography of Mediterranean Ecosystems in Chile, California, and Australia.* Springer-Verlag, Berlin, 289–310.

Holmgren, M., M.E. Ezcurra, J.R. Gutiérrez, and G.M.J. Mohre, 2001. El Niño effects on the dynamics of terrestrial ecosystems. *Trends in Ecology and Evolution*, 16, 89–84.

Jaksic, F.M., 1997. Ecología de los vertebrados de Chile. Ediciones Universidad Católica de Chile, Santiago.

Jiménez, H.E., and J.J. Armesto, 1992. Soil seed bank of disturbed sites in the Chilean matorral: Its importance in early secondary succession. *Journal of Vegetation Science*, 3, 579–586.

Kummerow J. 1966. Aporte al conocimiento de las condiciones climáticas del bosque de Fray Jorge. Boletín técnico. Santiago:Universidad de Chile, Facultad de Agronomía, 21–24.

Latorre, C., J. Quade, and W. McIntosh, 1997. The expansion of C4 grasses and global change in the late Miocene: Stable isotope evidence from the Americas. *Earth and Planetary Science Letters*, 146, 83–96.

Maldonado, A., 1999. *Historia de los bosques pantanosos de la costa de Los Vilos (IV Región, Chile) durante el Holoceno medio y tardío.* M.Sc. Thesis. University of Chile, Santiago.

Miller, P.C., D.E. Bradbury, E. Hajek, V. LaMarche, and N.J.W. Thrower, 1977. Past and present environment. In: H.A. Mooney (Editor), *Convergent Evolution in Chile and California: Mediterranean Climate Ecosystems.* Dowden, Hutchinson and Ross, Stroudsburg.

Mooney, H.A. (Editor), 1977. *Convergent evolution in Chile and California: Mediterranean Climate Ecosystems.* Dowden, Hutchinson and Ross, Stroudsburg.

Muñoz, C., and E. Pisano, 1947. Estudio de la vegetación y la flora de los Parques Nacionales de Fray Jorge y Talinay. *Agricultura Técnica* (Chile), 7, 71–190.

Myers, N., R.A. Mittermeier, G.A.B. d. Fonseca, and J. Kent, 2000. Biodiversity hotspots for conservation priorities. *Nature*, 40, 853–858.

Parsons, D.J., 1976. Vegetation structure in the Mediterranean climate scrub communities of California and Chile. *Journal of Ecology*, 64, 435–447.

Pérez, C., and C. Villagrán, 1985. Distribución de abundancias de especies de bosques relictos de la zona mediterránea de Chile. *Revista Chilena de Historia Natural*, 58, 157–170.

Ramirez, C., C. San Martin, and J. San Martin, 1996. Estructura florística de los bosques pantanosos de Chile central. In: J. Armesto, M. T. Kalin-Arroyo, C. Villagrán. Ecología del Bosque Nativo en Chile. Editorial Universitaria, Santiago, 215–234.

Reynolds, J. H., T.E. Jordan, N.M. Johnson, J.F. Damanti, and K.D. Tabutt, 1990. Neogene deformation of the flat-subduction segment of the Argentine-Chilean Andes: Magnetostratigraphic constraints from Las Juntas, La Rioja province, Argentina. *Geological Society of America Bulletin*, 102, 1607–1622

Rozzi, R., J.D. Molina, and P. Miranda, 1989. Microclima y períodos de floración en laderas de exposición ecuatorial y polar en los Andes de Chile central. *Revista Chilena de Historia Natural*, 62, 74–85.

Rundel, P.W. 1981. The matorral zone of central Chile. In: F. di Castri, D.W. Goodall, and R.L. Specht (Editors), *Mediterranean-type Shrublands.* Elsevier, Amsterdam, 175–201.

Rundel, P.W., and M. Mahu, 1976. Community structure and diversity of a coastal fog zone in northern Chile. *Flora*, 165, 493–505.

Rundel, P.W., M.O. Dillon, B. Palma, H.A. Mooney, S.L. Gulmon, and J.R. Ehrleringer, 1991. The phytogeography and ecology of the coastal Atacama and Peruvian deserts. *Aliso*, 13, 1–49.

Sala, O.E., F.S. Chapin, J.J. Armesto, E. Barlow, J. Bloomfield, R. Dirzo, E. Huber-Sanwald, L. Huenneke, R.B. Jackson, A. Kinzing, R. Leemans, D.M. Lodge, H.A. Mooney, M. Oesterheld, N.L. Poff, M.T. Sykes, B.H. Walker, M. Walker, and D.W. Hall, 2000. Global biodiversity scenarios for the year 2100. *Science*, 287, 1770–1774.

San Martín, J., and C. Donoso, 1996. Estructura florística e impacto antrópico en el bosque Maulino de Chile. In: J J. Armesto, C. Villagrán, and M.T.K. Arroyo (Editors), *Ecología de los Bosques Nativos de Chile.* Editorial Universitaria, Santiago, 153–168.

Sarmiento, G., 1975. The dry plant formations of South America and their floristic connections. *Journal of Biogeography*, 2, 233–251.

Schmithüsen, J., 1956. Die räumliche Ordnung der chilenischen Vegetation. *Bonner Geographische Abhandlungen*, 17, 1–89.

Skottsberg, C., 1948. Apuntes sobre la flora y vegetación de Fray Jorge (Coquimbo, Chile). *Acta Horti Goobers*, 18, 90–184.

Solbrig, O.T., M.L. Cody, E.R. Fuentes, W. Glanz, J.H. Hunt, and A.R. Moldenke, 1977. The origin of the biota. In: H.A. Mooney (Editor), *Convergent Evolution in Chile and California: Mediterranean Climate Ecosystems.* Dowden, Hutchinson and Ross, Stroudsburg, 13–26.

Troncoso, A., C. Villagrán, and M. Muñoz, 1980. Una nueva hipótesis acerca del origen y edad del bosque de Fray Jorge (Coquimbo, Chile). *Boletín del Museo Nacional de Historia Natural, Chile*, 37, 117–152.

Villagrán, C., 1990. Glacial climates and their effects on the history of vegetation of Chile: A synthesis based on palynological evidence from Isla de Chiloé. *Review of Paleobotany and Palynology*, 6, 17–24.

Villagrán, C., 1995. Quaternary history of the Mediterranean vegetation of Chile. In: M.T.K. Arroyo, P.H. Zedler, and M.D. Fox (Editors), *Ecology and Biogeography of Mediterranean Ecosystems in Chile, California and Australia.* Springer Verlag, Berlin, 3–20.

Villagrán, C., and J.J. Armesto, 1980. Relaciones florísticas entre las comunidades de relictuales del norte Chico y la zona central con el bosque del sur de Chile. *Boletín del Museo Nacional de Historia Natural, Chile*, 37, 87–101.
Villagrán, C., and J. Varela, 1990. Palynological evidence for increased aridity of the central Chilean coast during the Holocene. *Quaternary Research*, 34, 198–207.
Villagrán, C., C. Le-Quesne, J.C. Aravena, H. Jiménez, and F. Hinojosa, 1998. El rol de los cambios de clima del Cuaternario en la distribución actual de la vegetación de Chile central-sur. *Bamberger Geographische Schriften*, 15, 227–242.
Villa-Martínez, R., and C. Villagrán, 1997. Historia de la vegetación de los bosques pantanosos de la costa de Chile central durante el Holoceno medio y tardío. *Revista Chilena de Historia Natural*, 70, 391–401.
Zachos, J., M.Pagani, L. Sloan, E. Thomas, and K. Billups. 2001. Trends, rhythms and aberrations in global climate 65 Ma to present. *Science*, 292, 686–693.

12

Tropical and Subtropical Landscapes of the Andes

Kenneth R. Young
Blanca León
Peter M. Jørgensen
Carmen Ulloa Ulloa

The Andes represent Earth's longest mountain system and include some of the world's highest peaks. The rugged relief found above 1,000 m elevation produces strong environmental gradients tied to dramatic changes in temperature, moisture, and atmospheric pressure. These physical factors provide the background to understanding Andean landforms and land cover. In this chapter, we review these factors and patterns, and the complicating influences of geology and human land use, for the tropical and subtropical portions of the Andes, above 1,000 m and from 11°N to 24°S, in Venezuela, Colombia, Ecuador, Perú, Bolivia, and northernmost Argentina and Chile (fig. 12.1).

The tropical Andes are recognized as one of the most important regions in the world from the viewpoint of biodiversity conservation (Myers et al., 2000; Brooks et al., 2002). They are home to ancient human settlements and early civilizations (Burger, 1992; Bruhns, 1994; Dillehay, 1999), and large indigenous populations (Maybury-Lewis, 2002) living in some of the highest permanent settlements in the world. As a result, a better understanding of the physical geography of this complex region is important for sustainable development initiatives and other global environmental concerns.

Historically important overviews have been written for this region by von Humboldt (1807), Troll (1931), and Ellenberg (1958). Country-level studies include those for Venezuela (Monasterio, 1980), Colombia (Cuatrecasas, 1958; Rangel, 2000), Ecuador (Whymper, 1896; Acosta Solís, 1968; Jørgensen and León-Yánez, 1999), Perú (Weberbauer, 1945; Young and León, 2001), and Bolivia (Navarro and Maldonado, 2002). Luteyn (1999) has assembled information on the plants of the high elevations of the northern Andes, Luteyn and Churchill (2000) have examined the plant communities of the tropical Andes, and Kappelle and Brown (2001) have provided descriptive accounts of the montane forests. Inspiring chronicles can be found in Steele (1964) and Botting (1973). In this chapter, we first describe the relationships among the physical environments and natural landscapes of the tropical and subtropical Andes. We then discuss the natural vegetation types to be found, as typified by the forests, shrublands, grasslands, high Andean types, and wetlands. Finally, we summarize key aspects of the role

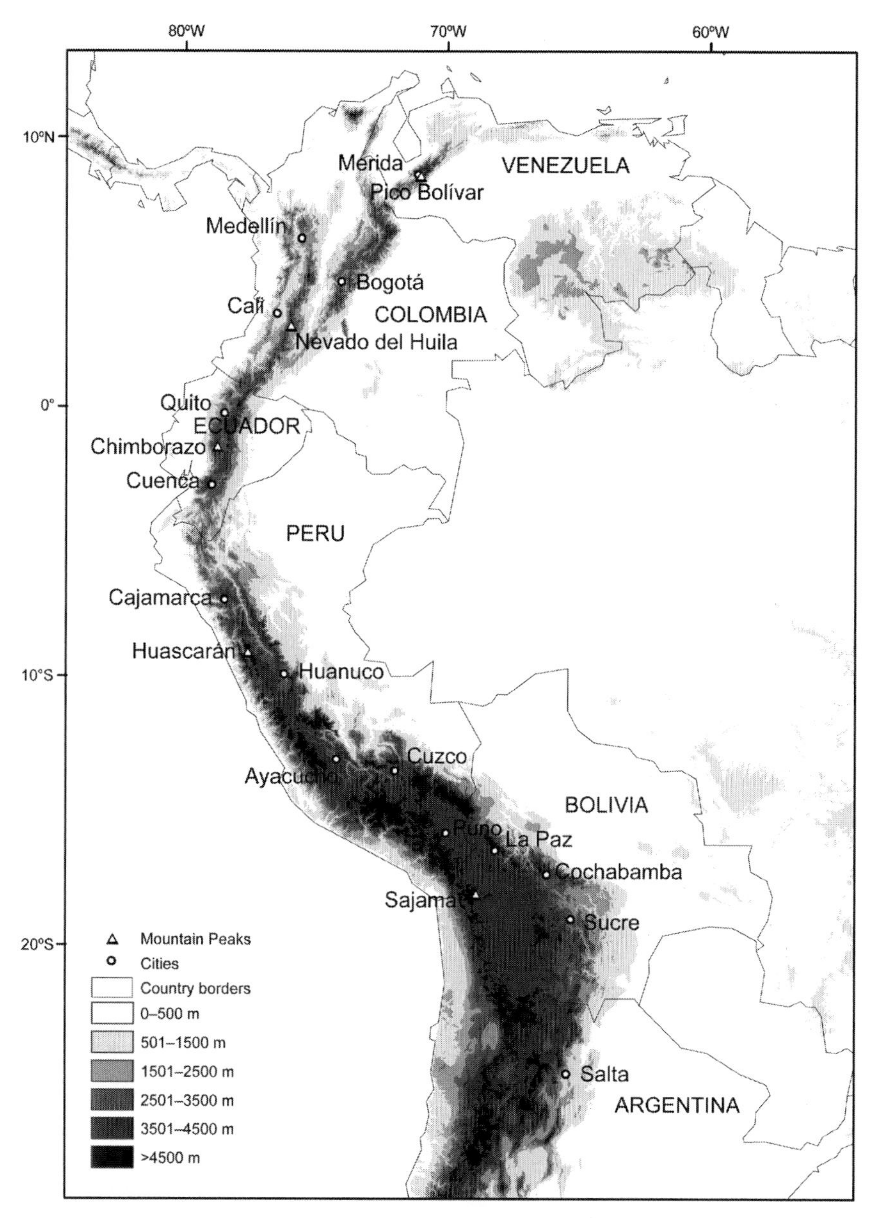

Figure 12.1 Topographic relief found among the tropical and subtropical Andean countries

of historical biogeography and human influences on and within those landscapes.

12.1 Physical Environments and Biophysical Controls on Landscapes

Landscapes are shaped by processes that originate in global, regional, and local contexts. Thus, particular types of landforms and land covers are influenced by processes acting both at different spatial scales and over time periods ranging from a few seconds, for example for a wind burst, to millions of years for some geological and evolutionary processes.

It was about 85 Ma (million years ago) that the South American tectonic plate began to interact with the Nazca plate, involving the subduction of the latter and creating the first components of the present Andes in central and southern South America (see chapter 1; Parrish, 1993; Taylor, 1995). Africa and South America had fully separated somewhat earlier, resulting in the final opening of the Atlantic Ocean and the start of almost 100 million years of partial biogeographical isolation for South America, to be broken only some 3 million years ago by land connections to North America through what is now Panama (Jackson et al., 1996). The northern Andes began to rise later than the central and southern Andes and are influenced by a complex interplay between the South American, Caribbean, Cocos, and Nazca plates. There were highlands in northern South America by about 55 million years ago, although, as elsewhere in the Andes, they include rocks of many ages (Riller et al., 2001; Pratt et al., 2005).

Geological processes of uplift, volcanism, and denudation have continued to shape the Andes over the past 100 million years. Macroscale feedbacks between climatic setting and tectonic and denudational processes have also played important roles in establishing Andean relief (Montgomery et al., 2001; Harris and Mix, 2002). However, much of the final uplift responsible for the elevational belts found today only occurred during late Pliocene and Pleistocene time. This timing means that the plants and animals found at the highest elevations occupy relatively recent ecological zones that previously did not exist. In a variety of lineages, this has resulted in a trend toward specialization for species that can tolerate the harsh climatic conditions found there. For example, the most recently evolved species of *Polylepis* (Rosaceae) are distributed at the highest elevations occupied by that genus of trees (Simpson, 1979). This also appears true for some species of *Centropogon* (Stein, 1992) and of *Fuchsia* (Berry, 1995).

The current topography (fig. 12.1) consists of one to several generally north-south trending mountain ranges broken by both parallel and east-west oriented valleys that serve as dispersal barriers to mountain organisms and as habitat for those species found in the intermontane valleys. The connectivity from one place to another in the Andes has varied over geological time and must have had profound influences on the evolution of the biota.

In northwestern South America, the Andes show a fanlike system of cordilleras. Venezuela's Andes consists of one prominent cordillera that reaches 5,007 m on Pico Bolívar and runs from northeast to southwest before connecting with the Eastern Cordillera of Colombia. There are three large cordilleras in Colombia, reaching 5,750 m at Nevado del Huila, making this the tropical country with the largest mountainous area in South America. These cordilleras approach each other southward and narrow into two ranges. The Andes in Ecuador are characterized by two large cordilleras separated by numerous interconnected intermontane valleys. The highest point is Volcano Chimborazo (6,310 m). In southern Ecuador and northern Perú, the cordilleras are less well defined because deep valleys crisscross the mountains and many ridges are as low at 2,500 to 3,000 m. The lowest point of the continental divide, around 2,000 m, is found here. In northern Perú, there are two and at times three cordilleras, and the highest peak is Huascarán (6,768 m). The mountains farther south are unusual in that they bracket a large, relatively flat plateau, the Altiplano, located at 3,800 to 4,100 m in elevation. Here, the north-central Andes embrace the widest part of this mountain system. The Altiplano continues into northwest Bolivia, where it is bounded by a large Eastern Cordillera that descends eastward into Amazonia, and a Western Cordillera whose highest peak, Nevado Sajama (6,542 m), lies close to the border with Chile. Northernmost Argentina shares the converging cordilleras and narrowing intermontane basins of the south-central Andes with northern Chile (see chapter 1).

12.1.1 Atmospheric Circulation

The climates of the tropical Andes are principally influenced by interactions between topography and movements of the Intertropical Convergence Zone (ITCZ) and the trade winds, although there have been large shifts in climatic circulation during late Cenozoic time (see chapters 2, 3; Markgraf, 2001). Today, during the high-sun season in the Northern Hemisphere, from June to August, the ITCZ is positioned over parts of Central America, the Caribbean, and northernmost South America. Trade winds move moisture-laden air from the Atlantic Ocean in a westerly direction north of the equator and in a southwesterly direction farther south. Where these winds meet steep mountains slopes, the cordilleras can be drenched by rains and immersed in clouds caused by orographic uplift. For example, the Chocó portion of western Colombia is thought to receive more than 10 m of annual rainfall. Very different are the intermontane valleys and mountain slopes facing away from the trade winds or where winds are altered in their trajectories by the mountains. In those locations, climates are likely to be seasonally or permanently dry owing to rain-shadow effects. The cold water currents off the coasts of Perú and northern Chile, aided by subsiding air and the influence of the Andes, exacerbate these patterns by producing stable air masses that create linearly extensive coastal deserts.

The subtropical portions of the Andes experience varying influences of easterly, westerly, and northerly air flows (Vera et al., 2002; Rutllant and Garreaud, 2004; Siqueira and Toledo Machado, 2004; Wang and Fu, 2004). In the austral summer (December–February), low-level winds from the Atlantic cross the Amazon basin and bring significant moisture, cloud, and rain to the Andes. In the austral winter (June–August), air flows are commonly westerly and conditions in the Andes are dry and cloud free. These circulation patterns have been described as monsoonal (Jones and Carvalho, 2002; Gan et al., 2004). In addition, during the austral winter, cold air masses may move from southern Argentina into the subtropics, at both low and high elevations. Thus, the dry season in the subtropics may also experience much cooler temperatures than the wet season. The increased seasonality in temperature is what most distinguishes the climate regimes of the tropical from the subtropical Andes.

12.1.2 Macroclimatic Patterns

Given that the Andes are high mountains, many of the physical environmental changes to be found are associated with elevation (Körner, 1999). Higher elevations have lower air pressures and cooler mean temperatures. In the tropical Andes, different elevational belts have relatively constant mean temperatures but pronounced seasonality in precipitation owing to the shifting effects of the ITCZ and moisture-bearing winds. In contrast, the subtropical Andes experience distinct seasonality in both temperature and precipitation.

Slopes that face winds from the tropical oceans or the Amazon basin are humid or perhumid (very humid). Annual rainfall ranges from 1,000 to above 3,500 mm. Typical mean temperatures at 1,000 m are ~21°C, often cooling at night due to downslope winds, and also during the day with significant cloud buildup caused by orographic and convectional uplift. Mean temperatures at 2,000 m are ~16°, and this altitude is often near the elevation where persistent cloud banks can form, subjecting plants and animals to frequent fog and high relative humidity. At 3,000 m, average temperatures are ~11° and at 4,000 m are ~5°. The mean annual zero degree isotherm occurs at about 5,200 m.

Tropical slopes sheltered from frequent moist air masses are drier, relative humidity is lower, and fog is less important, often only occurring during the rainy season. However, temperature lapse rates appear to be similar (~0.5°C/100 m), and mean annual temperatures in relation to elevation are roughly the same as on the humid and perhumid slopes. Most deep intermontane valleys have long dry seasons or at least seasonally dry environments, although they may also have small enclaves with more humid microclimates.

Mean annual temperatures are lower at similar elevations in the subtropical Andes. Not only is there more seasonality in temperature, but the landforms conspire to expose large areas to freezing temperatures and local aridity. The Altiplano of southern Perú and western Bolivia experiences subfreezing temperatures at night during the dry season, while conditions in southwest Bolivia form some of the world's largest salt flats (Mpodozis et al., 2005).

12.1.3 Diurnal Variation

While mean temperatures in the Andes may be almost constant throughout the year, intra-daily variations increase with elevation and decreasing humidity and cloud cover. On the equator, it is not uncommon to have nightly temperatures as low as 6°C at 2,800 m elevation, while the daily maximum may reach 24°. There is a loss of longwave radiation at night and in the shade during the day, and a gain of shortwave radiation in the sun during the day. This effect is more pronounced at higher elevations and often results in clear cool nights, sunny mornings with rapid warming, and cloud formation possibly leading to rain in the afternoons. The high Andean climate has been described as a tropical climate with four seasons in one day.

12.1.4 Inter-Annual Variation

A conspicuous factor in the year-to-year behavior of climate regimes in the Andes is the influence of El Niño–Southern Oscillation (ENSO) events, triggered by changes in sea temperatures in the Pacific Ocean and their influence on rainfall (see chapter 19; Caviedes, 2001; Glantz, 2001). Owing to the direct or indirect effects of El Niño, which recurs irregularly between three to eight years, some portions of the Andes receive more rainfall than normal, while other portions receive less (Garreaud and Aceituno, 2001; Vera et al., 2004).

The ENSO impact is direct in southern Ecuador and northern to central Perú. As surface waters in the eastern Pacific Ocean warm during El Niño years, warm waters from the north reach farther south than normal, and evaporation and condensation of air rising from the ocean surface lead to massive thunderstorms. Areas that are typically arid to semi-arid may receive the equivalent precipitation of several years of rainfall in only a few months. Further, because atmospheric pressure systems and circulation patterns are altered, other portions of the Andes are indirectly connected (teleconnnected) to ENSO influences. Southern Perú and the subtropical Andes are drier than normal during El Niño events. Northern Ecuador sees increased rainfall, but not nearly to the level of areas farther south and west. The effect is further moderated northward in Colombia and Venezuela, and much of the Amazon basin sees lower seasonal rainfall.

12.1.5 Landforms and Geomorphic Processes

The geomorphic processes that shape Andean landforms originate from the interaction of climatic regimes with the raw materials provided by tectonism and volcanism (see chapter 2). Above 3,500 m, most Andean landscapes were shaped directly by glacial or periglacial conditions during the Pleistocene (see chapter 4; Clapperton, 1993; Smith et al., 2005). Glacier ice sculpted bedrock, formed cirques and U-shaped valleys, and deposited eroded materials downslope as moraines, while fluvioglacial processes formed outwash plains and winds redistributed fine debris as loess. In drier areas, despite low temperatures, glaciers were smaller or nonexistent. In areas lacking glaciers, freeze-thaw processes heaved soils while solifluction moved lobes of debris downslope. During Pleistocene cold stages, areas influenced by glacial and periglacial processes would have been three or more times larger than they are today.

In the humid and perhumid Andes of today, mass wasting and stream flows are the dominant geomorphic processes (Trauth et al., 2000). Steep slopes are shaped by mass movement ranging from persistent creep to episodic rockfalls and landslides (Wilcke et al., 2003). These processes are exacerbated when mountain sides are undercut by rivers swollen by rains during the wet season, and by road building, which may generate sequential landslides (Young, 1994).

Volcanic landscapes are shaped by similar processes, but also include special features such as lava flows, deep ash deposits, crater lakes, and lahars—flows of volcanic debris often augmented by melting glacier ice. These kinds of landscapes are found in the Northern Volcanic Zone of southern Colombia and northern Ecuador, and in the Central Volcanic Zone of southern Perú and northern Chile (see chapter 2). These landscapes are also known for natural disasters triggered by volcanic eruptions and earthquakes. For example, the 1985 eruption of Nevado del Ruiz (5,321 m) in Colombia generated lahars that brought widespread death and destruction downslope, and swept away the town of Armero. Frequent eruptions of Ecuador's active volcanoes spread ash over wide areas, disrupting the lives and livelihoods of thousands of people. An earthquake in Perú in 1970 released ice, water, and debris from the west side of Huascarán, killing most inhabitants in the towns of Yungay and Ranrahirca.

12.1.6 Soils and Edaphic Processes

Whereas both geology and climate influence the chemistry and physical structure of the overlying soil, no easy generalizations can be made about Andean soils because of the wide range of rocks and climatic conditions that occur throughout the mountains. Suffice it to say that outcrops of granite, andesite, limestone, and metamorphic rock are widespread, each providing very different mineral conditions for soil development. The specific soil properties that most affect living organisms are derived from a site's topography, moisture and temperature regimes, and the influence of the biota related to those climatic regimes (see chapter 7; Birkeland, 1999).

12.2 Natural Vegetation Types

The plants found growing in a particular landscape constitute the vegetation cover of that area. The species may segregate themselves spatially, thus forming several distinct vegetation types. As a generalization, the sharper the difference in underlying abiotic environmental factors, the more likely this is the case. For example, vegetation in wetlands is quite different in terms of species composition and dominance than on sites with better-drained soils, although these may be spatially contiguous. More subtle environmental gradients are typically associated with indistinct or interdigitated ecological transition zones (ecotones). The result displayed by the vegetation would be a more gradual change in the dominant species and plant life-forms. Disturbances may interact with these edaphic patterns to create abrupt spatial changes, but the exact location of this boundary may change through time. For example, its location may be set or controlled by shifts in moisture or by occasional fires that remove or damage woody plant species, giving at least temporary assistance to the shrubs and to the graminoids that can resprout from their bases, including the grasses and sedges.

Collectively, these sets of biophysical processes create the general patterns of potential vegetation types discussed here. In addition, actual vegetation types present may differ from what would be expected for a particular climate, elevation, and soil type, if there has been a recent disturbance or other alteration, including the effects of human land-use practices.

12.2.1 Montane Forests

Forests are defined by the presence and dominance of tall woody plant species. Several kinds of montane forests occur along the tropical and subtropical Andes in response to differences in precipitation, seasonality, latitude, and altitude. The largest expanses are found in the wettest environments, particularly between 1,000 and 3,500 m in places exposed to rising humid air masses, notably along the steep eastern slopes of the Andes from Venezuela to northern Argentina, and along the western slopes of the mountains from Venezuela into Perú (fig. 12.2).

Most Andean forests found in other areas are fragmented, found especially in ravines, rocky sites, and wetter slopes of places that otherwise are covered by shrublands, grasslands, or agricultural landscapes. Some of this habitat fragmentation is inherent, caused by the lack of connectivity of the mountainous terrain (Young, 1998). In other situations, fragmentation has been caused or amplified by the conversion of forests into grazing lands and croplands. These forest remnants are often very useful to local people for extraction of forest products, in addition to serving important hydrological functions for watersheds.

Vegetation and Flora The Andean forests are among the most speciose types of vegetation to be found in the tropical Andes (Churchill et al., 1995). There are floristic differences from north to south, and from wetter to drier slopes. In the wetter areas, important families include Lauraceae, Melastomataceae, Myrtaceae, Rubiaceae, and Solanaceae. In drier locations, the shift is to the Fabaceae and Asteraceae. As shown by Gentry (1995) for woody plants and Gentry and Dodson (1987) for epiphytes, the more humid the environment the more plant species are to be found. Dominant tree genera in these forests are *Clusia*, *Miconia*, *Ocotea*, *Oreopanax*, and *Weinmannia*, and shrubs of *Palicourea*, *Psychotria*, *Saurauia*, and *Viburnum*. Factoring in elevation, the relationship would be that the higher the altitude, the lower overall plant diversity.

Figure 12.2 Cloud forests are found on mountain slopes exposed to cooling and uplifting air masses that produce fog and rain (photo: K.R. Young).

Endemism, however, is not necessarily a correlate to diversity. In particular, the drier and higher sites in the Andes, especially those found in isolated intermontane valleys, often have relatively large numbers of species with restricted ranges (Kessler, 2002a, 2002b).

The humid to perhumid forests found at 1,000 m can be as diverse and complex as forests found in humid tropical lowlands. Speciose woody families include Annonaceae, Arecaceae, Burseraceae, Euphorbiaceae, Fabaceae, Lauraceae, Melastomataceae, Meliaceae, Moraceae, and Rubiaceae. At slightly higher elevations, diversity begins to decline, perhaps due to cold-air drainage from above that prevents temperature-sensitive plant species from surviving. These foothill or premontane forests include mixtures of species found in lowlands, some species that grow all the way up into upper montane forests, and a set of endemic species that are altitudinal specialists. On gentle slopes, these forests can be 25 to 30 m tall, with emergents even taller. The steeper the slope, the smaller the stature of the forest. For example, steep ridges often have shrubs or small trees even in mountainous areas otherwise covered by dense forests. The dwarfed vegetation may be caused by shallow soils, exposure to persistent winds, or perhaps slow recovery following disturbances.

Moist lower montane forests are to be found at middle elevations, centered around 2,500 m in the Andes. These forests can also reach heights of 20 m, with emergents to 35 m or even more. Common trees are from genera such as *Brunellia*, *Ceroxylon*, *Hieronyma*, *Meliosma*, *Miconia*, *Myrcianthes*, *Saurauia*, and *Sapium*. Typically, the understories are densely filled with shrubs and herbs in the Araceae, Bromeliaceae, Ericaceae, Gesneriaceae, and Piperaceae. In more seasonal environments, with a more prolonged dry season, human imprints become much more noticeable and often these landscapes have been historically or even prehistorically deforested. Forest remnants can be found in ravines, while alder (*Alnus acuminata*) trees typically dominate along river courses.

The moist upper montane forests extend in humid sites from 2,500 m up to about 3,500 or even 3,900 m elevation. Generally these forests are 5 to 20 m tall, with shorter forests at treeline and in places where soils are shallow or disturbances have altered the vegetation in the past. Mean annual temperatures are 5 to 6°C (Körner and Paulsen, 2004). Tree genera such as *Axinaea*, *Cervantesia*, *Columellia*, *Drimys*, *Escallonia*, *Gaiadendron*, *Hedyosmum*, *Ilex*, *Miconia*, *Myrsine*, *Styrax*, *Symplocos*, and *Vallea* are characteristic. Understories are typically filled with scandent bamboos, most commonly from the genus *Chusquea*, with vines of *Clematis*, *Jungia*, *Mutisia*, and *Passiflora*, and shrubs especially of the genus *Piper*, but many of the woody genera above contain species that also grow as shrubs (Young, 1991, 1993). The highest elevation forests in the Andes are often rather open woodlands with trees of the genera *Buddleja*, *Gynoxys*, and *Polylepis*. They can be found to 4,700 m in fairly dry sites in western Bolivia and often at 3,800 to 4,200 m from southern Perú to Venezuela. Morales et al. (2004) showed that seasonal rainfall controlled tree growth at high elevations in subtropical northwestern Argentina.

Intermontane valleys cut through the Andes, often creating rain shadows and thus dry climates. Some of the largest valleys contain tropical dry forests, which can be 20 to 30 m tall, seasonally deciduous forests dominated by Anacardiaceae, Bombacaceae, Fabaceae, Meliaceae, and Sapotaceae (Bridgewater et al., 2003; Linares-Palomino, 2004). Most of the valleys, however, have only small areas of forest today, with the remainder dominated by shrubs such as *Dodonaea viscosa* and *Euphorbia laurifolia, Acacia* trees, or even cacti.

Species Adaptations Plant species of true montane cloud forests must be adapted to nearly constant high humidity and resulting poor soils, often highly acidic and lacking in nitrogen. They often have relatively thick leaves and short stature (Grubb, 1977).

All moist Andean sites are exposed to slow soil creep, in addition to slope failures in the form of debris flows, landslides, and rockfalls. Some plants are specialized in dispersing to and colonizing these sites (Stern, 1995), or in adjusting to slopes by the growth of curved stems (Young and León, 1990). Some of these same species also grow on river gravels stranded as fast currents cut new channels through their valleys. In environments where moisture is seasonally limiting, some plant species are deciduous and shut down growth during the dry season.

Most plant species have mutualisms with other species, typically fungi and animals. Many plants disperse their seeds with aid of birds, bats, and other small mammals, while pollination is effected through bees, moths and butterflies, hummingbirds, and bats (e.g., Muchhala and Jarrín-V., 2002). Pollination and seed dispersal by wind are relatively rare.

12.2.2 Shrublands

Shrublands are among the most widespread vegetation types in the tropical and subtropical Andes, although they are probably the least known in terms of generalization owing to their great spatial heterogeneity. They are dominated by woody plant species, usually no taller than 3 to 5 m in height (fig. 12.3). Shrublands are found on slopes, talus, or flat areas and they also have been described as part of the pre-puna, puna, or páramo communities (Villagrán et al., 1981; Rundel et al., 2003). They tend to be found in places where actual evapotranspiration is greater than precipitation, a climatic regime often without forests, except near rivers. However, they also tend to occupy coarse and rocky soils, while grasslands tend to be found on finer grained soils.

Shrublands can also be ecotonal in nature, with gradual shifts into either grasslands or montane forests. The largest areas of shrublands occupy the flanks of interandean valleys, and in many cases have a xeric component with the dominance of shrubby and spiny plants, succulent plants such as bromeliads, some euphorbs and Cactaceae, and an array of ephemeral herbs. It is in this type of vegetation that grasses using the C4 photosynthetic pathway can be more frequent than in any other community. There are also areas of scrub found in places with steep slopes and rock walls. These habitats share many xeric conditions due to poor soil development, high solar radiation, and often a rain-shadow effect.

Vegetation and Flora Shrublands include much plant diversity, in terms of number of species and rates of ende-

Figure 12.3 Shrublands are dominated by multiple stemmed woody plants and are frequently used by local people for the grazing of livestock and the extraction of firewood (photo: K.R. Young).

mism. Several families important at the continental scale maintain their dominance here, including Asteraceae, Campanulaceae, Ericaceae, Melastomataceae, Rosaceae, Rubiaceae, and Solanaceae (e.g., Ulloa Ulloa and Jørgensen, 1993; Jørgensen and León-Yánez, 1999). Plant composition varies with altitude, latitude, and location of the Andean flanks. In many drier areas, such as in intermontane valleys and western slopes of the central Andes, Asteraceae, Cactaceae, and Fabaceae are important components. At higher elevations, important genera of typically smaller shrubs are *Barnadesia*, *Bejaria*, *Berberis*, *Brachyotum*, *Centropogon*, *Chuquiraga*, *Escallonia*, *Fuchsia*, *Ribes*, and *Valeriana*.

Species Adaptations By definition, the dominant plant species have multiple woody stems. Many are species that are able to resprout following damage or disturbance. This resiliency is one reason that shrubs often come to dominate in areas that are subjected to fires, firewood extraction, or livestock grazing (e.g., Young and Keating, 2001). Other plants are adapted to water stresses through morphological and physiological adaptations, as shown for example with the involuted leaves of some of the *Chuquiraga* species (Ezcurra, 2002), by deciduousness, and by succulency among the Cactaceae, Euphorbiacaeae, and Piperaceae. Most plant species are pollinated by insects or hummingbirds, and have seeds dispersed by fruit-eating birds and timed to germinate with the start of the rainy season. The Asteraceae are wind dispersed.

12.2.3 Grasslands

Open spaces dominated by herbaceous plants, usually grasses or graminoids (i.e. grass-like plants such as sedges) occupy a wide variety of altitudes in the tropical Andes. The grasslands found on well-drained sites are discussed here. Grassland-dominated landscapes are usually associated with high elevations, but are also found on lower slopes, notably where precipitation is strongly seasonal or where other processes prevent encroachment by shrubs. At the highest elevations, grasslands are replaced by High Andean vegetation types. The grasslands have been called steppes by Weberbauer (1945) and Schnell (1987), but also include extensive areas in places known as páramo (Cabrera, 1980; Luteyn and Churchill, 2000). Some grasslands found in areas with annual precipitation above 1,200 mm might result from deliberate burning.

Evidence of the presence of grasslands at high altitudes unequivocally indicates that they were established by the early Pliocene, with a probable earlier occurrence at lower altitudes (Jacobs et al., 1999). Thus, past changes in tectonics and climate shaped the floristic composition of Andean grasslands (e.g., Boom et al., 2001).

Vegetation and Flora Tropical and subtropical grasslands are species-rich, nonwoody plant vegetation types in the Andes. For the northern Andes, a preliminary account, including several types of vegetation above 2,500 m, listed approximately 3,000 species (Luteyn, 1999). Collectively, there are probably 3,500 species in the Andean grasslands.

Although grasses and sedges are the most dominant families in grasslands, plant composition varies with altitude, and several other plant families form part of the biological richness of grassland ecosystems. Between 3,000 and 4,500 m, where grasslands are 20 to 150 cm tall, the most important families are Apiaceae, Asteraceae, Brassicaceae, Caryophyllaceae, Gentianaceae, Lycopodiaceae, Orobanchaceae, Poaceae, Rosaceae, and Scrophulariaceae; these families include most of the dominant or more common plant species. At lower altitudes, besides Asteraceae, Brassicaceae, Poaceae, and Scrophulariaceae, important families shift to include Fabaceae, Verbenaceae, and Solanaceae.

At the genus level, there are over 500 genera inhabiting grasslands (Sklenář et al., 2005). Attempts to understand the origin of Andean plant genera show that in general there is a tendency for the presence and dominance of tropical elements, especially from the Andes, with a clear decline of this dominance southward to more temperate elements (Gutte, 1992; Simpson and Todzia, 1990). Floristic insights into the origin and evolution of Andean floras clearly show the importance of past cold-climate events (Simpson, 1975), and the importance of colonization from neighboring lower-elevation plant communities (Young et al., 2002). Important genera, species rich and conspicuous, include *Agrostis*, *Bromus*, *Calamagrostis*, *Deyeuxia*, *Festuca*, and *Poa* in the Poaceae; *Baccharis*, *Bidens*, *Coreopsis*, *Diplostephium*, *Gamochaeta*, *Hieracium*, *Hypochaeris*, and *Senecio* in the Asteraceae; *Bartsia* in the Orobanchaceae; *Calceolaria* of the Calceolariaceae; *Cerastium*, *Drymaria*, and *Stellaria* of the Caryophyllaceae; and *Lysipomia* in the Campanulaceae.

The predominant life forms in grasslands are tussock or cespitose grass/graminoid. Overgrazing may change the aspect of these grasslands to shorter vegetation with more open barren areas (e.g. Podwojewski et al., 2002). This short vegetation is especially conspicuous in areas with well-defined rain seasonality, while in more humid conditions overgrazing is manifested by the presence of plant communities dominated by forbs of the Asteraceae, Ericaceae, Fabaceae, and Rosaceae. Heavily overgrazed and xeric areas in the central Andes can also be dominated by low cushion-forming plants of the Cactaceae, mostly *Opuntia*, and other unpalatable species (Ellenberg, 1979; Smith, 1994). Some grasslands are dominated by several other types of life forms that have been associated with high Andean plant communities, such as rosettes, chamephytes, and creeping hemicriptophytes (Cabrera, 1968). Tussock grasses behave as protective "nursing" sites for shorter plants, similar to what has been documented for temperate Andean sites (e.g. Arroyo et al., 2003).

Most grassland species are herbaceous, but semi-woody and low woody habits are also found dispersed or forming localized plant communities. These interspersed woody elements are mostly shrubs inhabiting drier localities, especially among boulders or rocky talus. Woody elements belong mostly to Asteraceae genera, such as *Diplostephium, Gynoxys,* and *Pentacalia,* but species of *Valeriana* (Valerianaceae) and *Monnina* (Polygalaceae) are not infrequent. In areas with higher humidity, Ericaceae genera are co-dominant elements (Cleef, 1980; van der Hammen and Cleef, 1986; Luteyn, 2002). Rocky areas where woody species can be found also host communities dominated by the grasses of *Cortaderia* spp. (e.g. Weberbauer, 1945).

In riverine environments of the central Andes, at middle elevations (2,000 to 3,500 m), grasslands are found in narrow bands dominated by tall conspicuous grasses reaching 3 m heights in the native *Gynerium sagittatum* and the exotic *Arundo donax.* Often these are mixed plant communities with shrubland and riverine forest elements.

In the northern Andes, and in areas with higher precipitation on the eastern side of the central Andes, thickets of bamboo species, mostly of the genus *Chusquea,* appear frequently on slopes and near forested areas (Cleef, 1980; Rangel et al., 1997). Toward the western Andes, in southern Perú and northern Chile, where grasslands exclusively occupy the arid high plateaus, tussock grasses are dominant in species-poor communities (e.g. Villagrán et al., 1981). Hemiparasitic species are also present at the base of tussocks or of scattered shrubs; these include several species of Orobanchaceae, like *Bartsia* and *Castilleja.*

Species Adaptations Under drier conditions, especially at lower elevations and higher latitudes, there is a tendency for more annual or biennial life cycles among herbaceous plants. These plant communities are usually present in the altitudinal belt most associated with agriculture. This disturbance shapes those plant communities, especially favoring invasive species (e.g. Gutierrez et al., 1998). In general, under more humid conditions with less precipitation seasonality, perennial herbaceous species dominate. In most high Andean grasslands, needle-type leaves are conspicuous, while scattered shrubs have thick, small, and usually acuminate leaves (Sarmiento, 1986; Leuschner, 2000).

An important factor shaping grassland communities is related to soil chemistry, texture, and development. Gutte (1985) described different plant communities for grasslands in central Perú in relation to depth, texture, and soil properties, especially those associated with water retention. For example, he showed *Calamagrotis vicunarum* occupying low flat terrains with soil depths usually up to 25 cm while *C. rigida* (as *C. antonianae*) was found on slopes with soil depths up to 20 cm and mixed with rocks. For the drier slopes of southern Perú and northern Chile, loamy sand soils are often found above 3,000 m (e.g. Gutierrez et al. 1998), which, in conjunction with strong rain seasonality and aridity, contribute to a more scattered and species-poor grassland. The presence of toxic minerals, like selenium, appears also to add to the dominance of some species, as in the case of *Astragalus garbancillo* in northern Perú (Becker et al., 1989). In an area with a high amount of salt, Gutte and Müller (1985) documented the presence of communities dominated by the succulent genus *Salicornia.*

Overgrazing undoubtedly plays a role, alone or in conjunction with other physical and biological factors, in shaping grassland plant composition. The dominance of low grasses, weedy species, and the increment in C4 taxa might be the result of overgrazing and other forms of human interference. For temperate Andean grasslands, however, overgrazing appears not to interfere with other biological interactions like colonization of endomycorrhiza (Lugo and González Meza, 2003). Invasive species can play an important role in some grasslands along the Andes. Although probably less than 5% of the total flora is exotic, in some places exotics can become dominants, for example, the exotic grass *Pennisetum clandestinum.* In addition, some native species become weedy in response to disturbance, especially due to overgrazing and fire. Many areas are dominated by low herbs, such as *Werneria nubigena* and *Lachemilla* spp., or woody plants, such as *Margyricarpus, Senecio,* and *Tetraglochin* species. These plants are usually not consumed by livestock and thus tend to spread easily, mostly through a combination of seed output and vegetative growth.

12.2.4 High Andes

Above 4,500 m and below the elevations of permanent snow or ice, Andean landscapes are often characterized by rocky substrates and relatively sparse vegetation types. Many of these landscapes are not contiguous, forming island-like habitats along or between mountain ranges.

Biophysical environmental conditions are harsh because of cool temperatures and strong winds. Precipitation and air humidity are relatively low. As in other Andean landscapes, the total amount of precipitation decreases from north to south, varying from 950 to 700 mm/yr for the northern to central Andes (Gutte, 1987; Monasterio, 1986), with a marked decline to 300 mm and even as low as 100 mm/yr in the western parts of the subtropical Andes. Daily extreme changes in temperature, and especially the presence of ice-needles, produce localized soil cryoturbation. The ecological importance of cryoturbation and associated morphological processes undoubtedly shapes plant composition and life forms (Monasterio, 1986; Pérez, 1987).

Vegetation and Flora Plant communities in the high Andes have been categorized as belonging to the super-paramo, desert paramo, jalca, puna, or subnival strata (Cabrera, 1980; Cleef, 1980; Monasterio, 1986; Smith, 1994). It is this environment where rosette growth forms dominate the landscape, especially giant *Espeletia*, *Lupinus* (fig. 12.4), and *Puya* species.

Low-growing, 2 to 40 cm tall, cushion-forming plants occur in both xeric and wetter areas (Smith, 1994). In noninundated and rocky areas, cushion plants form the last conspicuous perennial plants, in many cases the only ones capable of growing above the altitudinal limit of tree-forming species. Little is known about the ecology of their establishment and the biotic features that allow them to withstand the harsh conditions at high elevations. Three families include most cushion-forming plants: Apiaceae, Asteraceae, and Caryophyllaceae.

There are probably fewer than 1,000 species above 4,500 m along the tropical and subtropical Andes. Locally, high Andean vegetation is species poor, and plant communities may include only a dozen or fewer species (e.g. Gutte, 1987). Most elements of the high Andes are recent arrivals with probable evolutionary links to taxa from lower altitudes, and to lineages with a temperate origin or amphitropical distribution (e.g. Mummenhoff et al., 2001; Tate and Simpson, 2003). Important species-rich families are Apiaceae, Asteraceae, Brassicaceae, Caryophyllaceae, Gentianaceae, Malvaceae, and Poaceae (Gutte 1987; Jørgensen and León-Yánez 1999). Species-rich genera are *Azorella*, *Baccharis*, *Draba*, *Gentianella*, *Lachemilla*, *Nototriche*, *Senecio*, and *Valeriana*. *Calamagrostis* and *Festuca*, in addition to *Poa*, are important grass genera.

Figure 12.4 This endemic high Andean lupine (*Lupinus weberbaueri*) is adapted to the harsh environmental conditions found above 4,000 m elevation (photo: K.R. Young).

Species Adaptations High Andean vegetation is dominated by perennial species, and thus also includes plant communities dominated by woody species, some of which, like *Polylepis*, form woodlands (Velez et al., 1998), or others, like *Pentacalia*, that may develop into more open shrublands among rocks. These areas harbor probably more species than any other community at this elevation. Most plant species are perennials of several different life forms (Rundel et al., 1994). Besides rosette plants, other forms are cushion-like, woody chamaeophytes, and cespitose hemicryptophytes (Gutte, 1987), and a few tussock-forming grasses.

Monasterio (1986) described the microclimatic conditions associated with the tall rosette-forming *Espeletia*, favoring the establishment of other plant species, especially low hemicryptophytes. For a temperate Andean area, Arroyo et al. (2003) showed that 87% of those species studied have a potential for establishing permanent seed banks.

Plants located in rock outcrops include most of the rosettes and shrubs. Crevices facilitate the survival of many species, probably because of protection against sudden changes in temperature, wind erosion, and herbivory. At this elevation, rock outcrops occur on sites varying from 5 to 40° or more in slope. *Puya raimondii* from the central Andes of Perú and Bolivia occurs here. Crevices appear to be important sites for the survival of its seeds and seedlings (Vadillo et al., 2004). It appears that most of the colonizing species found on moraines and retreating glaciers belong to widespread families, but from genera or lineages restricted to the Andes, as in the case of Brassicaceae and *Draba*, Apiaceae and *Azorella*, and Juncaceae. Other important groups of organisms include lichens and bryophytes. Some assemblages appear on rocks accompanied by low rosette herbs, mostly of the Asteraceae (Rangel et al., 1997). Berry and Calvo (1989) showed that the high elevation *Espeletia* species were wind pollinated, while those species found only below 4,000 m were pollinated by insects. They also found that seed set in a high Andean orchid was limited by infrequent insect visits (Berry and Calvo, 1991).

12.2.5 *Wetlands*

Wetland vegetation is defined here as those plant communities or plant covers found in areas with water-saturated soils, often associated with margins of lakes and rivers.

They are azonal in the sense that they can be found at an array of elevations. Similar to other kinds of vegetation in the Andes, their location, origin, and biota respond to the diverse physical conditions (slope, precipitation, bedrock) found there, and the result is also affected by an interaction with historical and present human uses.

Differences in the number and origin of lacustrine wetlands are associated with altitude. For example, most lakes in Ecuador are found above 3,000 m, and are either volcanic or glacial in origin, mostly oligotrophic, and species poor (Terneus, 2000). At lower elevations, the number of lakes diminishes. Water-table changes also affect the presence of wetland vegetation. Irrigation, water control, and glacier retreat may impact wetlands by diversions and drainage.

Vegetation and Flora Wetland floras include species that are strict aquatics and others tolerant of saturated soils. Strictly aquatic plants probably include fewer than 500 species for the tropical and subtropical Andes. Facultative aquatics are more diverse and may exceed 1,000 species. For all wetland floras, important families are Asteraceae, Crassulaceae, Cyperaceae, Isoetaceae, Juncaceae, Poaceae, Potamogetonaceae, and Ranunculaceae (León and Young, 1996).

Margins of lakes and ponds include communities of emergent aquatics belonging mostly to Cyperaceae and Juncaceae. At high to middle elevations, *Schoenoplectus californicus*, *S. tatora*, and *Typha domingensis* are frequently found. They usually form monospecific communities (*totorales*) or are associated with small aquatic herbs such as *Callitriche* and the fern *Azolla* (Rangel et al., 1997). These habitats have a long history of use (Erickson, 2000), and human intervention might play a role in controlling their dynamics through continuous harvest (Macía and Balslev, 2000). Exotic species become important elements in wetlands dominated by facultative species. *Nasturtium aquaticum* can be found at different elevations. Other native species may behave like weedy taxa, notably *Azolla* spp., *Cardamine bonariensis*, and *Ranunculus trichophyllus*. Special communities within the wetlands realm are found on wetter Andean flanks, where *Sphagnum* moss may be a dominant. Some of these assemblages may include shrubs of Asteraceae, rosette species of the genus *Puya*, and the fern *Blechnum*, creeping herbs, climbers, and bamboos.

Species Adaptations Floating and submerged species are less diverse, and at different altitudes can be dominated by Potamogetonaceae or Ranunculaceae. Some small herbs such as *Callitriche* and *Elatine* species, and *Crassula venezuelensis*, can also be found along margins where there is sediment. Most wetland assemblages at high elevation have rosette-forming species, like species of *Chrysactinium*, *Hieracium*, *Lupinus*, and *Plantago*. Some grasses are interspersed, mostly of the genus *Calamagrostis*. The main constraint for submerged aquatic vegetation in the Andes is thought to come from carbon limitations (Keeley, 1998).

12.3 Human Influences

The Andes have been an important center of domestication. Andean landscapes support vegetation types from which native species have been selected and manipulated genetically beginning thousands of years ago, but with modern repercussions (e.g., Raimondi and Camadro, 2003; Rodríguez-Burrouzo et al., 2003). This region provided the world with the tomato (*Solanum esculentum*), peppers (*Capsicum* spp.), and the potato (*Solanum tuberosum*). When the Spaniards arrived, Andean societies cultivated a wide array of plant species, but the subsequent introduction of European agricultural methods and crops often relegated native species to local consumption. Only a few became of worldwide importance (National Research Council, 1989). Grasslands house most Andean plant species of economic value, although not necessarily exclusively linked to the evolution and speciation of these species (e.g. Emshwiller, 2002). Over 100 plant species are used as fodder for animals.

Above 3,000 m, the most important tuber is the potato, which actually consists of a complex based on four species (*Solanum goniocalyx*, *S. phureja*, *S. stenotomum*, and *S. tuberosum*). Lesser known tuber species include *Arracacia xanthorrhiza* (aracacha), *Oxalis tuberosa* (oca), *Tropaeolum tuberosum* (mashua), and *Ullucus tuberosus* (melloco or ulluco). Among the grains, *Chenopodium quinoa* (quinua) has recently arrived in international markets. Other grains consumed in the region are *Chenopodium pallidicaule* (cañihua) and *Amaranthus caudatus* (kiwicha). From the legumes, there is the domesticated lupine (*Lupinus mutabilis*). Land covers of many high Andean landscapes have been transformed in order to support fields for these crops, in addition to serving as grazing.

The montane forests and valleys were localities for Andean fruit trees such as the tree tomato or tomatillo (*Solanum betaceaum*), montane papayas (*Carica* spp.), lucuma (*Pouteria lucuma*), various species of the legume genus *Inga*, and shrubs such as naranjilla (*Solanum quitoense*), cherimoya (*Annona cherimoya*), sweet cucumber (*Solanum muricatum*), goldenberry or uvillla (*Physalis peruviana*), and no fewer than ten species of passion fruits in the vine genus *Passiflora*. Some of the plants mentioned above have made their way into international markets through cultivation in New Zealand and California.

Many plants are used in traditional medicines, among them some lycopods of the genus *Huperzia* and some Andean cheilanthoids (e.g., Thomas and Winterhalder, 1976; Murillo, 1983). The most famous medicine coming out from the Andes is quinine, used for centuries to control malaria. Originally extracted from the bark of the cloud forest tree genus *Cinchona*, the alkaloid is nowadays produced synthetically. The coca plant (*Erythroxylum coca*) grows on the warm Andean slopes and the leaves are socially chewed or drunk as tea (mate de coca), especially in Perú and Bolivia, as a stimulant and to help with the effects of high-altitude sickness.

Most forest, shrubland, grassland, and wetland vegetation types in the Andes have been modified by humans for millennia (fig. 12.5; Gade, 1999; Mayer, 2002). Frequently, this has been through burning, forest clearance, drainage, introduction of exotic species, or extraction for firewood or medicinals. For example, in the Lake Titicaca basin, humans have occupied and extensively used the lake and surrounding areas for at least 8,000 years, and some uses include the change of the surrounding land for agriculture, which affects water levels (Erickson, 2000). The origin of Andean pastoralism is closely linked to the use and modification of high Andean plant assemblages. Today, wetlands and grasslands in the high plateaus continue to be the main source for forage, especially during the dry season (Moreau and Le Toan, 2003).

12.4 Conclusions

High place-to-place variation in species composition, as shown here for plants (e.g., Svenning, 2001, Sklenár and Jørgensen, 1999), is also known for Andean invertebrates (Brehm et al., 2003; González and Watling, 2003; Kořínek and Villalobos, 2003), amphibians (Duellman, 1999), birds (Fjeldså et al., 1999; Poulsen and Krabbe, 1998; Ruggiero, 2001), and mammals (Leo, 1995; Voss, 2003). This phenomenon is related to topographic barriers, species endemism, and species that are habitat or elevational specialists (see chapter 8; Young, 1995; Young et al., 2002). As such, spatial variation in species composition should be expected to be a defining characteristic of Andean biogeography, in addition to the differences with changes in elevation, soils, and effective moisture regimes that affect the vegetation types, as exemplified by Burns and Noaki (2004) for a Neotropical genus of birds.

Owing to past connections between continents, plant and animal lineages either have arrived in the region from elsewhere (Givnish et al., 2004; Pennington and Dick, 2004; Sanmartín and Ronquist, 2004), or they have evolved *in situ* with the formation of the Andes. Relatively recent colonists, such as alder, have not speciated since their arrival from Central America. Andean intra-regional connectivity and dispersal barriers are behind many of the floristic and faunistic differences that have developed over time and from place to place. Each mountain ridge and valley constitutes a barrier for one or another species. The Huancabamba in northern Perú may be the biggest barrier (Weigend, 2002; Sancho, 2004), but other large

Figure 12.5 Inhabited Andean landscape in north-central Perú showing planted *Eucalyptus* trees and agricultural fields amid splendid topography (photo: K.R. Young)

intermontane valleys also play such a role. Climate change has also left a mark on the biota and current assemblages (Paduano et al., 2003; Hooghiemstra and Van der Hammen, 2004; Chepstow-Lusty et al., 2005).

Mountainous and high-elevation landscapes can be conceived of as a series of vertical ecological continua (Seastedt et al., 2004) that affect climate, soils, vegetation, and ecosystem processes. Climate change and tectonic shifts have altered those continua for centuries, millennia, and millions of years. The development of Andean landscapes can be explained spatially in relation to environmental gradients and over long periods with reference to their history and evolutionary processes. The surprisingly high human populations that occupy middle- to high-elevation tropical landscapes in the Andes distinguish these areas from landscapes found at similar elevation elsewhere in the world (Cohen and Small, 1998). Thus, human influences must be considered in explanations of the variations in land cover across the tropical and subtropical Andes (Gade, 1999; D'Altroy, 2000; Denevan, 2001).

Much attention now focuses on relatively recent land-cover changes due to human impacts on tropical lowlands (Lambin et al., 2003). However, the anthropogenic changes affecting tropical uplands, such as the Andes, typically predate by centuries these lowland changes (e.g., Achard et al., 2002). There are likely many lessons for the better understanding of Earth's surface to be found among these long-inhabited landscapes and their landscape-forming processes. Similarly, the biological diversity and the implications of future environmental change of tropical uplands have sometimes been overlooked (Chapin et al., 2001). These reasons emphasize the urgency for considering Andean landscapes and similar places worldwide. This task needs to be undertaken with due respect for the underlying causes and consequences of the spatial heterogeneity and dynamisms that have shaped these landscapes and their biota.

References

Achard, F., H.D. Eva, H-J. Stibig, P. Mayaux, J. Gallego, T. Richards, and J-P. Malingreau, 2002. Determination of deforestation rates of the world's humid tropical forests. *Science*, 297, 999–1002.

Acosta Solís, M., 1968. *Las Divisiones Fitogeográficas y las Formaciones Geobotánicas del Ecuador.* Casa de la Cultura Ecuatoriana, Quito.

Arroyo, M.T.K., L.A. Cavieres, A. Peñaloza, and M.A. Arroyo-Kalin, 2003. Positive associations between the cushion plant *Azorella monantha* (Apiaceae) and alpine plant species in the Chilena Patagonian Andes. *Plant Ecology*, 169, 121–129.

Becker, B., F.M., Terrones H., and M.E. Tapia, 1989. *Los Pastizales y Producción Forrajera en la Sierra de Cajamarca. Proyecto Piloto de Ecosistemas Andinos.* Cajamarca, Peru.

Berry, P.E., 1995. Diversification of Onagraceae in montane areas of South America. In: S.P. Churchill, H. Balslev, E. Forero, and J.L. Luteyn (Editors), *Biodiversity and Conservation of Neotropical Montane Forests.* New York Botanical Garden, New York, 415–420.

Berry, P.E., and R.N. Calvo, 1989. Wind pollination, self-incompatibility, and altitudinal shifts in pollination systems in the high Andean genus *Espeletia* (Asteraceae). *American Journal of Botany*, 76, 1602–1614.

Berry, P.E., and R.N. Calvo, 1991. Pollinator limitations and position dependent fruit-set in the high Andean orchid *Myrosmodes cochleare* (Orchidaceae). *Plant Systematics and Evolution*, 174, 93–101.

Birkeland, P.W., 1999. *Soils and Geomorphology.* 3rd edition. Oxford University Press, New York.

Boom, A., G. Mora, A.M. Cleef, and H. Hooghiemstra, 2001. High altitude C4 grasslands in the northern Andes: Relicts from glacial conditions? *Review of Palaeobotany and Palynology*, 115, 147–160.

Botting, D., 1973. *Humboldt and the Cosmos.* Harper and Row, New York.

Brehm, G., D. Süssenbach, and K. Fiedler, 2003. Unique elevational diversity patterns of geometrid moths in an Andean montane rainforest. *Ecography*, 26, 456–466.

Bridgewater, S., R.T. Pennington, C.A. Reynel, A. Daza, and T.D. Pennington, 2003. A preliminary floristic and phytogeographic analysis of the woody flora of seasonally dry forests in northern Peru. *Candollea*, 58, 129–148.

Brooks, T.M., R.E. Mittermeier, C.G. Mittermeier, G.A.B. da Fonseca, A.B. Rylands, W.R. Konstant, P. Flick, J. Pilgrim, S. Oldfield, G. Magin, and C. Hilton-Taylor, 2002. Habitat loss and extinction in the hotspots of biodiversity. *Conservation Biology*, 16, 909–923.

Bruhns, K. O., 1994. *Ancient South America.* Cambridge University Press, Cambridge.

Burger, R.L., 1992. *Chavin and the Origins of Andean Civilization.* Thames and Hudson, London.

Burns, K.J., and K. Naoki, 2004. Molecular phylogenetics and biogeography of Neotropical tanagers in the genus *Tangara. Molecular Phylogenetics and Evolution*, 32, 838–854.

Cabrera, A.L., 1968. Ecología Vegetal de la Puna. *Colloquium Geographicum*, 9, 91–115.

Cabrera, A.L., 1980. *Biogeografía de América Latina.* Serie de Biología. Organización de Estados Americanos. Monografía 13, 1–122.

Caviedes, C.N., 2001. El Niño in history: Storming through the ages. *University Press of Florida*, Gainesville, Florida.

Chapin III, F.S., O.E. Sala, and E. Huber-Sannwald (Editors), 2001. *Global Biodiversity in a Changing Environment. Scenarios for the 21st Century.* Springer, New York.

Chepstow-Lusty, A., M.B. Bush, M.R. Frogley, P.A. Baker, S.C. Fritz, and J. Aronson, 2005. Vegetation and climate change on the Bolivian Altiplano between 108,000 and 18,000 years ago. *Quaternary Research,* 63, 90–98.

Churchill, S.P., H. Balslev, E. Forero, and J.L. Luteyn (Editors), 1995. *Biodiversity and Conservation of Neotropical Montane Forests.* New York Botanical Garden, New York.

Clapperton, C., 1993. *Quaternary Geology and Geomorphology of South America.* Elsevier, Amsterdam.

Cleef, A.M., 1980. La vegetación del paramo neotropical y

sus lazos australo-antarticos. *Revista del Instituto Geográfico "Agustín Codazzi,"* 7, 68–86.
Cohen, J.E., and C. Small, 1998. Hypsographic demography: The distribution of human population by altitude. *Proceedings of the National Academy of Science*, 95, 14009–14014.
Cuatrecasas, J., 1958. Aspectos de la vegetación natural de Colombia. *Revista Académica Colombiana de Ciencias Exactas,* 10(40), 221–264.
D'Altroy, T.N., 2000. Andean land use at the cusp of history. In: D.L. Lentz (Editor), *Imperfect Balance: Landscape Transformations in the Precolumbian Americas.* Columbia University Press, New York, 357–390.
Denevan, W.M., 2001. *Cultivated Landscapes of Native Amazonia and the Andes.* Oxford University Press, Oxford.
Dillehay, T.D., 1999. The Late Pleistocene cultures of South America. *Evolutionary Anthropology*, 7(6), 206–216.
Duellman, W.E., 1999. Distribution patterns of amphibians in South America. In: W.E. Duellman (Editor), *Patterns of Distribution of Amphibians.* Johns Hopkins University Press, Baltimore, 255–328.
Ellenberg, H., 1958. Wald oder Steppe? Die natürliche Pflanzendecke der Anden Perus I, II. *Umschau Wissenschaftliche Technologie*, 21, 645–648; 22, 679–681. 1, 5
Ellenberg, H., 1979. Man's influence on tropical mountain ecosystems in South America. *Journal of Ecology*, 67, 401–416.
Emshwiller, E., 2002. Biogeography of the *Oxalis tuberosa* alliance. *Botanical Review*, 68, 128–152.
Erickson, C. L. 2000. The Lake Titicaca Basin: A Precolumbian built landscape. In: D.L. Lentz (Editor), *Imperfect Balance: Landscape Transformations in the Precolumbian Americas.* Columbia University Press, New York, 311–356.
Ezcurra, C., 2002. Phylogeny, morphology, and biogeography of *Chuquiraga*, an Andean-Patagonian genus of Asteraceae-Barnadesioidea. *Botanical Review*, 68, 153–170.
Fjeldså, J., E. Lambins, and B. Mertens, 1999. Correlation between endemism and local ecoclimatic stability documented by comparing Andean bird distributions and remotely sensed land surface data. *Ecography*, 22, 63–78.
Gade, D.W., 1999. *Nature and Culture in the Andes.* University of Wisconsin Press, Madison.
Gan, M.A., V.E. Kousky, and C.F. Ropelewski, 2004. The South American Monsoon Circulation and its relationship to rainfall over west-central Brazil. *Journal of Climate*, 17, 47–66.
Garreaud, R.D., and P. Aceituno, 2001. Interannual rainfall variability over the South American Altiplano. *Journal of Climate,* 14, 2779–2789.
Gentry, A.H., 1995. Patterns of diversity and floristic composition in Neotropical montane forests. In: S.P. Churchill, H. Balslev, E. Forero, and J.L. Luteyn (Editors), *Biodiversity and Conservation of Neotropical Montane Forests.* New York Botanical Garden, New York, 103–126.
Gentry, A.H., and C.H. Dodson, 1987. Diversity and biogeography of Neotropical vascular epiphytes. *Annals of the Missouri Botanical Garden,* 74, 205–233.
Givnish, T.J., K.C. Millam, T.M. Evans, J.C. Hall, J.C. Pires, P.E. Berry, and K.J. Sytsma, 2004. Ancient vicariance or recent long-distance dispersal? Inferences about phylogeny and South American-African disjunctions in Rapateaceae and Bromeliaceae based on *ndh*F sequence data. *International Journal of Plant Science*, 165 (Supplement 4), S35–S54.
Glantz, M.H., 2001. *Currents of Change: Impacts of El Niño and La Niña on Climate and Society.* Cambridge University Press, Cambridge.
González, E.R., and L. Watling, 2003. Two new species of *Hyalella* from Lake Titicaca, and redescriptions of four others in the genus (Crustacea: Amphipoda). *Hydrobiologia*, 497, 181–204.
Grubb, P.J., 1977. Control of forest growth and distribution on wet tropical mountains, with special reference to mineral nutrition. *Annual Review of Ecology and Systematics*, 8, 83–107.
Gutierrez, J.R., F. López-Cortes, and P.A. Marquet, 1998. Vegetation on an altitudinal gradient along the Río Loa in the Atacama Desert of northern Chile. *Journal of Arid Environments*, 40, 383–399.
Gutte, P., 1985. Beitrag zur Kenntnis zentralperuanischer Pflanzengesellschaften IV. Die grasreiche Vegetatio der alpinen Stufe. *Wissenschaftliche Zeitschrift der Karl-Marx-Universitat Leipzig, Mathematisch-Naturwissenschaftliche Reiche*, 34(4), 357–401.
Gutte, P., 1987. Beitrag zur Kenntnis zentralperuanischer Pflanzengesellschaften V. Die Vegetation der subnivalen Stufe. *Feddes Repertorium*, 98(7–8), 447–460.
Gutte, P., 1992. Die Herkunft hochandiner zentralperuanischer Gattungen-Versuch einer Florenanalyse. *Feddes Repertorium*, 103(3–4), 209–214.
Gutte, P., and G.K. Müller, 1985. Salzpflanzengesellschaften bei Cusco, Peru. *Wissenschaftliche Zeitschrift der Karl-Marx-Universitat Leipzig, Mathematisch-Naturwissenschaftliche Reiche*, 34(4), 402–409.
Harris, S.E., and A.C. Mix, 2002. Climate and tectonic influences on continental erosion of tropical South America, 0–13 Ma. *Geology*, 30, 447–450.
Hooghiemstra, H., and T. Van der Hammen, 2004. Quaternary Ice-Age dynamics in the Colombian Andes: Developing an understanding of our legacy. *Philosophical Transactions of the Royal Society of London* B 359, 173–181.
Humboldt, F.W.H.A. von, 1807. *Essai sur la Géographie des Plantes, Acompagné d'un Tableau Physique des Régions Equinoxiales, Fondé sur des Mesures Exécutées, depuis le Dixième Degré de Latitude Boréale jusqu' au Dixième Degré de Latitude Australe, pendant les Années 1799, 1800, 1801, 1802 et 1803.* Levrault et Schoell, Paris.
Jackson, J.B.C., A.F. Budd, and A.G. Coates (Editors), 1996. *Evolution and Environment in Tropical America.* University of Chicago Press, Chicago.
Jacobs, B.F., J.D. Kingston, and L.L. Jacobs, 1999. The origin of grass-dominated ecosystems. *Annals of the Missouri Botanical Garden*, 86, 590–643.
Jørgensen, P.M., and S. León-Yánez (Editors). 1999. Catalogue of the Vascular Plants of Ecuador. *Monographs in Systematic Botany from the Missouri Botanical Garden,* 75, 1–1181.
Jones, C., and L.M.V. Carvalho, 2002. Active and break phases in the South American monsoon system. *Journal of Climate*, 15, 905–914.
Kappelle, M., and A.D. Brown (Editors), 2001. *Bosques Nublados del Neotrópico.* INBio, Santo Domingo de Heredia, Costa Rica.

Keeley, J.E., 1998. CAM photosynthesis in submerged aquatic plants. *Botanical Review*, 64(2), 121–175.
Kessler, M., 2002a. The elevational gradient of Andean plant endemism: Varying influences of taxon-specific traits and topography at different taxonomic levels. *Journal of Biogeography*, 29, 1159–1165.
Kessler, M., 2002b. Environmental patterns and ecological correlates of range size among bromeliad communities of Andean forests in Bolivia. *Botanical Review*, 68, 100–127.
Kořínek, V., and L. Villalobos, 2003. Two South American endemic species of *Daphnia* from high Andean lakes. *Hydrobiologia*, 490, 107–123.
Körner, C., 1999. *Alpine Plant Life: Functional Plant Ecology of High Mountain Ecosystems.* Springer, Berlin.
Körner, C., and J. Paulsen, 2004. A world-wide study of high altitude treeline temperatures. *Journal of Biogeography*, 31, 713–732.
Lambin, E.F., H.J. Geist, and E. Lepers, 2003. Dynamics of land-use and land-cover change in tropical regions. *Annual Review of Environment and Resources*, 28, 205–241.
Leo, M., 1995. The importance of tropical montane cloud forest for preserving vertebrate endemism in Peru: The Rio Abiseo National Park as a case. In: L.S. Hamilton, J.O. Juvik, and F.N. Scatena (Editors), *Tropical Montane Cloud Forests: Proceedings of an International Symposium.* Springer Verlag, New York, 198–205.
León, B., and K.R. Young, 1996. Aquatic plants of Peru: Diversity, distribution, and conservation. *Biodiversity and Conservation*, 5, 1169–1190.
Leuschner, C., 2000. Are high elevations in tropical mountains arid environments for plants? *Ecology*, 81, 1425–1436.
Linares-Palomino, R., 2004. Los bosques tropicales estacionalmente secos. II: Fitogeografía y composición florística. *Arnaldoa,* 11 (1), 103–138.
Lugo, M.A., and M.E. González Maza, 2003. Arbuscular mycorrhizal fungi in a mountain grassland. II: Seasonal variation of colonization studied, along with its relation to grazing and metabolic host type. *Mycologia*, 95(3), 407–415.
Luteyn, J.L., 1999. Páramos: A Checklist of Plant Diversity, Geographical Distribution, and Botanical Literature. *Memoirs of The New York Botanical Garden* 84, 1–278.
Luteyn, J.L., 2002. Diversity, adaptation, and endemism in Neotropical Ericaceae: Biogeographical patterns in the Vaccinieae. *Botanical Review*, 68, 55–99.
Luteyn, J.L., and S.P. Churchill, 2000. Vegetation of the tropical Andes: An overview. In: D.L. Lentz (Editor), *Imperfect Balance: Landscape Transformations in the Precolumbian Americas.* Columbia University Press, New York, 281–310.
Macía, M.J., and H. Balslev. 2000. Use and management of totora (*Schoenoplectus californicus*, Cyperaceae) in Ecuador. *Economic Botany*, 54(1), 82–89.
Markgraf, V. (Editor), 2001. *Interhemispheric Climate Linkages.* Academic Press, San Diego.
Maybury-Lewis, D. (Editor), 2002. *The Politics of Ethnicity: Indigenous Peoples in Latin American States.* Harvard University Press and David Rockefeller Center for Latin American Studies, Cambridge.
Mayer, E., 2002. *The Articulated Peasant: Household Economies in the Andes.* Westview Press, Boulder, Colorado.
Monasterio, M., (Editor), 1980. *Estudios Ecológicos en los Páramos Andinos.* Ediciones de la Universidad de Los Andes, Mérida, Venezuela.
Monasterio, M., 1986. Adaptive strategies of *Espeletia* in the Andean desert paramo. In: F. Vuilleumier and M. Monasterio (Editors), *High Altitude Tropical Biogeography.* Oxford University Press, New York, 49–80.
Montgomery, D.R., G. Balco, and S.D. Willett, 2001. Climate, tectonics, and the morphology of the Andes. *Geology*, 29, 579–582.
Morales, M.S., R. Villalba, H.R. Grau, and L. Paolini. 2004. Rainfall-controlled tree growth in high-elevation subtropical treelines. *Ecology*, 85, 3080–3089.
Moreau, S., and T. Le Toan, 2003. Biomass quantification of Andean wetland forages using ERS satellite SAR data for optimizing livestock management. *Remote Sensing of Environment*, 84, 477–492.
Mpodozis, C., C. Arriagada, M. Basso, P. Roperch, P. Cobbold, and M. Reich, 2005. Late Mesozoic to Paleogene stratigraphy of the Salar de Atacama basin, Antofagasta, northern Chile: Implications for the tectonic evolution of the central Andes. *Tectonophysics*, 399, 125–154.
Muchhala, N., and P. Jarrín-V., 2002. Flower visitation by bats in cloud forests of western Ecuador. *Biotropica*, 34, 387–395.
Mummenhoff, K., H. Brüggemann, and J.L. Bowman, 2001. Chloroplast DNA phylogeny and biogeography of *Lepidium* (Brassicaceae). *American Journal of Botany*, 88, 2051–2063.
Murillo, M.T., 1983. Usos de los Helechos. *Biblioteca José Jerónimo Triana*, 5, 1–156.
Myers, N., R.A. Mittermeier, C.G. da Fonseca, A.B. Gustavo, and J. Kent, 2000. Biodiversity hotspots for conservation priorities. *Nature*, 403, 853–858.
National Research Council, 1989. *Lost Crops of the Incas.* National Academy Press, Washington, DC.
Navarro, G., and M. Maldonado, 2002. *Geografía Ecológica de Bolivia: Vegetación y Ambientes Acuáticos.* Centro de Ecología Simón I. Patiño, Cochabamba, Bolivia.
Paduano, G.M., M. Bush, B.A. Baker, S.C. Fritz, and G.O. Seltzer, 2003. A vegetation and fire history of Lake Titicaca since the Last Glacial Maximum. *Palaeogeography, Palaeoclimatology, Palaeoecology,* 194, 259–279.
Parrish, J.T., 1993. The palaeogeography of the opening South Atlantic. In: W. George and R. Lavocat (Editors), *The African-South American Connection.* Clarendon Press, Oxford, 8–27.
Pennington, R.T., and C.W. Dick, 2004. The role of immigrants in the assembly of the South American rainforest tree flora. *Philosophical Transactions of the Royal Society of London*, B 359, 1611–1622.
Pérez, F. L., 1987. Needle-ice activity and the distribution of stem-rosette species in a Venezuelan paramo. *Arctic and Alpine Research*, 19, 135–153.
Podwojewski, P., J. Poulernard, T. Zambrana, and R. Hofstede, 2002. Overgrazing effects on vegetation cover and properties of volcanic ash soil in the paramo de Llangahua and La Esperanza (Tungurahua, Ecuador). *Soil Use and Management*, 18(1), 45–55.
Poulsen, B.O., and N. Krabbe, 1998. Avifaunal diversity of five high-altitude cloud forests on the Andean western slope of Ecuador: Testing a rapid assessment method. *Journal of Biogeography*, 25, 83–93.
Pratt, W.T., P. Duque, and M. Ponce, 2005. An autochthonous geological model for the eastern Andes of Ecuador. *Tectonophysics*, 399, 251–278.

Raimondi, J.P., and E.L. Camadro 2003. Crossability relationships between the common potato, *Solanum tuberosum* spp. *tuberosum*, and its wild diploid relatives *S. kurtzianaum* and *S. ruiz-lealii*. *Genetic Resources and Crop Evolution*, 50, 307–314.

Rangel, J.O. (Editor), 2000. *Colombia Diversidad Biótica III: La Región de Vida Paramuna de Colombia.* Universidad Nacional de Colombia, Bogotá, Colombia.

Rangel, J.O., P.D. Lowy, M. Aguilar, and A. Garzón, 1997. Tipos de vegetación en Colombia. In: J.O. Rangel, P.D. Lowy, M. Aguilar (Editors), *Colombia Diversidad Biótica II: Tipos de Vegetación en Colombia.* Instituto de Ciencias Naturales, Universidad Nacional de Colombia; Instituto de Hidrología, Meteorología y Estudios Ambientales, IDEAM, Ministerio Del Medio Ambiente. Santa Fé de Bogotá, Colombia, 89–391.

Riller, U., I. Petrinovic, J. Ramelow, M. Strecker, and O. Oncken, 2001. Late Cenozoic tectonism, collapse caldera, and plateau formation in the central Andes. *Earth and Planetary Science Letters*, 188, 299–311.

Rodríguez-Burruezo, A., J. Prohens, and F.Nuez, 2003. Wild relatives can contribute to the improvement of fruit quality in pepino (*Solanum muricatum*). *Euphytica*, 129, 31–318.

Ruggiero, A., 2001. Size and shape of the geographical ranges of Andean passerine birds: Spatial patterns in environmental resistance and anisotropy. *Journal of Biogeography*, 28, 1281–1294.

Rundel, P.W., A.C. Gibson, G.S. Midgley, S.J.E. Wand, B. Palma, C. Kleier, and J. Lambrinos, 2003. Ecological and ecophysiological patterns in a pre-altiplano shrubland of the Andean Cordillera in northern Chile. *Plant Ecology*, 169, 179–193.

Rundel, P.W., F.C. Meinzer, and A.P. Smith, 1994. Tropical alpine ecology: Progress and priorities. In: P.W. Rundel, A.P. Smith, F.C. Meinzer (Editors), *Tropical Alpine Environments.* Cambridge University Press, New York, 355–363.

Rutllant, J. A., and R. D. Garreaud, 2004. Episodes of strong flow down the western slope of the subtropical Andes. *Monthly Weather Review*, 132, 611–622.

Sancho, G., 2004. Phylogenetic relationships in the genus *Onoseris* (Asteraceae, Mutisieae) inferred from morphology. *Systematic Botany*, 29, 432–447.

Sanmartín, I., and F. Ronquist, 2004. Southern hemisphere biogeography inferred by event-based models: Plant versus animal patterns. *Systematic Biology*, 53, 216–243.

Sarmiento, G., 1986. Ecological features of climate in high tropical mountains. In: F. Vuilleumier and M. Monasterio (Editors), *High Altitude Tropical Biogeography.* Oxford University Press, New York, 11–45.

Schnell, R., 1987. *La Flore et la Végétation de l'Amérique Tropicale. Tome 2: Es Formations Xériques; Le Peuplemnt des Montagnes; Le Végétation Aquatique et Littorale; Les Plantes Utiles.* Masson, Paris.

Seastedt, T.R., W.D. Bowman, T.N. Caine, D. McKnight, A. Townsend, and M.W. Williams, 2004. The landscape continuum: A model for high-elevation ecosystems. *Bioscience*, 54, 111–121.

Simpson, B., 1975. Pleistocene changes in the flora of the high tropical Andes. *Paleobiology*, 1, 273–294.

Simpson, B.B., 1979. A revision of the genus *Polylepis* (Rosaceae: Sanguisorbeae). *Smithsonian Contributions to Botany*, 43, 1–62.

Simpson, B.B., and C. Todzia, 1990. Patterns and processes in the development of the high Andean flora. *American Journal of Botany*, 77, 1419–1432.

Siqueira, J.R., and L.A. Toledo Machado, 2004. Influence of the frontal systems on the day-to-day convection variability over South America. *Journal of Climate*, 17, 1754–1766.

Sklenář, P., J.L. Luteyn, C. Ulloa Ulloa, P.M. Jørgensen, and M.O. Dillon, 2005. *Flora Genérica de los Páramos—Guía Ilustrada de las Plantas Vasculares.* The New York Botanical Garden Press, New York.

Sklenář, P., and P.M. Jørgensen, 1999. Distribution patterns of páramo plants in Ecuador. *Journal of Biogeography*, 26, 681–692.

Smith, A.P., 1994. Introduction to tropical alpine vegetation. In: P.W. Rundel, A.P. Smith, and F.C. Meinzer (Editors), *Tropical Alpine Environments.* Cambridge University Press, Cambridge, 1–19.

Smith, J.A., G.O. Seltzer, D.L. Farber, D.T. Rodbell, and R.C. Finkel, 2005. Early local Last Glacial Maximum in the tropical Andes. *Science*, 308, 678–681.

Steele, A.R., 1964. *Flowers for the King: The Expedition of Ruiz and Pavon and the Flora of Peru.* Duke University Press, Durham.

Stein, B.A., 1992. Sicklebill hummingbirds, ants, and flowers. *Bioscience*, 42, 27–33.

Stern, M.J., 1995. Vegetation recovery on earthquake-triggered landslide sites in the Ecuadorian Andes. In: S.P. Churchill, H. Balslev, E. Forero, and J.L. Luteyn (Editors), *Biodiversity and Conservation of Neotropical Montane Forests.* New York Botanical Garden, New York, 207–220.

Svenning, J.-C., 2001. Environmental heterogeneity, recruitment limitation and the mesoscale distribution of palms in a tropical montane rain forest (Maquipucuna, Ecuador). *Journal of Tropical Ecology*, 17, 97–113.

Tate, J.A., and B.B. Simpson, 2003. Paraphyly of *Tarasa* (Malvaceae) and diverse origins of the polyploidy species. *Systematic Botany*, 28, 723–737.

Taylor, D.W., 1995. Cretaceous to Tertiary geologic and angiosperm paleobiogeographic history of the Andes. In: S.P. Churchill, H. Balslev, E. Forero, and J.L. Luteyn (Editors), *Biodiversity and Conservation of Neotropical Montane Forests.* New York Botanical Garden, New York, 3–9.

Terneus, E., 2002. Aquatic plant communities of Ecuadorian páramo lakes. In: A. Freire Fierro and D. A. Neill (Editors), *Memorias del Tercer Congreso Ecuatoriano de Botánica.* Funbotánica, Quito, 88–103.

Thomas, B.R., and B.P. Winterhalder. 1976. Physical and biotic environment of southern highland Peru. In: P.T. Baker and M.A. Little (Editors), *Man in the Andes*, US/IBP Synthesis Series, Stroudsburg, 21–59.

Trauth, M.H., R.A. Alonso, K.R. Haselton, R.L. Hermanns, and M.R. Strecker, 2000. Climate change and mass movements in the NW Argentine Andes. *Earth and Planetary Science Letters*, 179, 243–256.

Troll, C., 1931 [1932]. *Die Landschaftsgürtel der Tropischen Anden.* Verh. Wiss. Abh. 24. Deutsch.

Ulloa Ulloa, C., and P. M. Jørgensen, 1993. *Arboles y Arbustos de los Andes de Ecuador.* AAU Reports, 30, 1–264.

van der Hammen, T., and A.M. Cleef, 1986. Development of the high Andean páramo flora and vegetation. In: F. Vuilleumier and M. Monasterio (Editors), *High Altitude Tropical Biogeography.* Oxford University Press, New York, 153–201.

Vadillo, G., M. Suni, and A. Cano, 2004. Viabilidad y germinación de semillas de *Puya raimondii* Harms (Bromeliaceae). *Revista Peruana de Biología*, 11, 71–78.

Velez, V., J. Cavelier, and B. Devia, 1998. Ecological traits of the tropical treeline species *Polylepis quadrijuga* (Rosaceae) in the Andes of Colombia. *Journal of Tropical Ecology*, 15, 771–787.

Vera, C.S., G. Silvestre, V. Barros, and A. Carril, 2004. Differences in El Niño response over the Southern Hemisphere. *Journal of Climate*, 17, 1741–1753.

Vera, C.S., P.K. Vigliarolo, and E.H. Berbera, 2002. Cold season synoptic-scale waves over subtropical South America. *Monthly Weather Review*, 130, 684–699.

Villagrán, C., J.J. Arrest, and M.T. Kalin Arroyo, 1981. Vegetation in a high Andean transect between Turi and Cerro León in northern Chile. *Vegetatio*, 48, 3–16.

Voss, R.S., 2003. A new species of *Thomasomys* (Rodentia: Muridae) from eastern Ecuador, with remarks on mammalian diversity and biogeography in the Cordillera Oriental. *American Museum Novitates*, 3421, 1–47.

Wang, H., and R. Fu, 2004. Influence of cross-Andes flow on the South American low-level jet. *Journal of Climate*, 17, 1247–1262.

Weberbauer, A., 1945. *El Mundo Vegetal de los Andes Peruanos*. Ministerio de Agricultura, Lima, Peru.

Weigend, M., 2002. Observations on the biogeography of the Amotape-Huancabamba zone in northern Peru. *Botanical Review*, 68, 38–54.

Whymper, E., 1896. *Travels amongst the Great Andes of the Equator*. Scribner, New York.

Wilcke, W., H. Valladarez, R. Stoyan, S. Yasin, C. Valarezo, and W. Zech, 2003. Soil properties on a chronosequence of landslides in montane rain forest, Ecuador. *Catena*, 53, 79–95.

Young, K.R., 1991. Natural history of an understory bamboo (*Chusquea* sp.) in a tropical timberline forest. *Biotopica*, 23, 542–554.

Young, K.R., 1993. Woody and scandent plants on the edges of an Andean timberline. *Bulletin of the Torrey Botanical Club*, 120, 1–18.

Young, K.R., 1994. Roads and the environmental degradation of tropical montane forests. *Conservation Biology*, 8, 972–976.

Young, K.R., 1995. Biogeographical paradigms useful for the study of tropical montane forests and their biota. In: S.P. Churchill, H. Balslev, E. Forero, and J. L. Luteyn (Editors), *Biodiversity and Conservation of Neotropical Montane Forests*. New York Botanical Garden, New York, 79–87.

Young, K.R., 1998. Deforestation in landscapes with humid forests in the central Andes: Patterns and processes. In: K. S. Zimmerer and K.R. Young (Editors), *Nature's Geography: New Lessons for Conservation in Developing Countries*. University of Wisconsin Press, Madison, 75–99.

Young, K.R., and P.L. Keating, 2001. Remnant forests of Volcán Cotacachi, northern Ecuador. *Arctic, Antarctic, and Alpine Research*, 33, 165–172.

Young, K.R., and B. León, 1990. Curvature of woody plants on the slopes of a timberline montane forest in Peru. *Physical Geography*, 11, 66–74.

Young, K.R., and B. León, 2001. Perú. In: M. Kappelle and A. D. Brown (Editors), *Bosques Nublados del Neotrópico*. INBio, Heredia, Costa Rica, 549–580.

Young, K.R., C. Ulloa Ulloa, J.L. Luteyn, and S. Knapp, 2002. Plant evolution and endemism in Andean South America: An Introduction. *Botanical Review*, 68, 4–21.

13

Temperate Forests of the Southern Andean Region

Thomas T. Veblen

Although most of the continent of South America is characterized by tropical vegetation, south of the tropic of Capricorn there is a full range of temperate-latitude vegetation types including Mediterranean-type sclerophyll shrublands, grasslands, steppe, xeric woodlands, deciduous forests, and temperate rain forests (fig. 13.1). Southward along the west coast of South America the vast Atacama desert gives way to the Mediterranean-type shrublands and woodlands of central Chile (see chapter 11), and then to increasingly wet forests all the way to Tierra del Fuego at 55°S. To the east of the Andes, these forests are bordered by the vast Patagonian steppe of bunch grasses and short shrubs (see chapter 14). The focus of this chapter is on the region of temperate forests occurring along the western side of the southernmost part of South America, south of 33°S. The forests of the southern Andean region, including the coastal mountains as well as the Andes, are presently surrounded by physiognomically and taxonomically distinct vegetation types and have long been isolated from other forest regions. Although small in comparison with the extent of temperate forests of the Northern Hemisphere, this region is one of the largest areas of temperate forest in the Southern Hemisphere and is rich in endemic species.

For readers familiar with temperate forests of the Northern Hemisphere, it is difficult to place the temperate forests of southern South America into a comparable ecological framework owing both to important differences in the histories of the biotas and to contrasts between the broad climatic patterns of the two hemispheres. There is no forest biome in the Southern Hemisphere that is comparable to the boreal forests of the high latitudes of the Northern Hemisphere. The boreal forests of the latter are dominated by evergreen conifers of needle-leaved trees, mostly in the Pinaceae family, and occur in an extremely continental climate. In contrast, at high latitudes in southern South America, forests are dominated mostly by broadleaved trees such as the southern beech genus (*Nothofagus*). Evergreen conifers with needle or scale-leaves (from families other than the Pinaceae) are a relatively minor component of these forests. Furthermore, the relatively narrow width of South America at high latitudes results in a maritime climate with mild temperature ranges compared to similar latitudes in the Northern Hemisphere. Thus, the southern temperate forests do not fit into pre-conceived notions of vegetation patterns derived from experiences in the Northern Hemisphere. The strongest parallels between the Northern and Southern Hemisphere temperate forests are for the coastal forests of the Pacific Northwest (including coastal Alaska) and the high latitudes of the Pacific coast of Chile (Veblen and Alaback, 1996). But even for these climatically similar

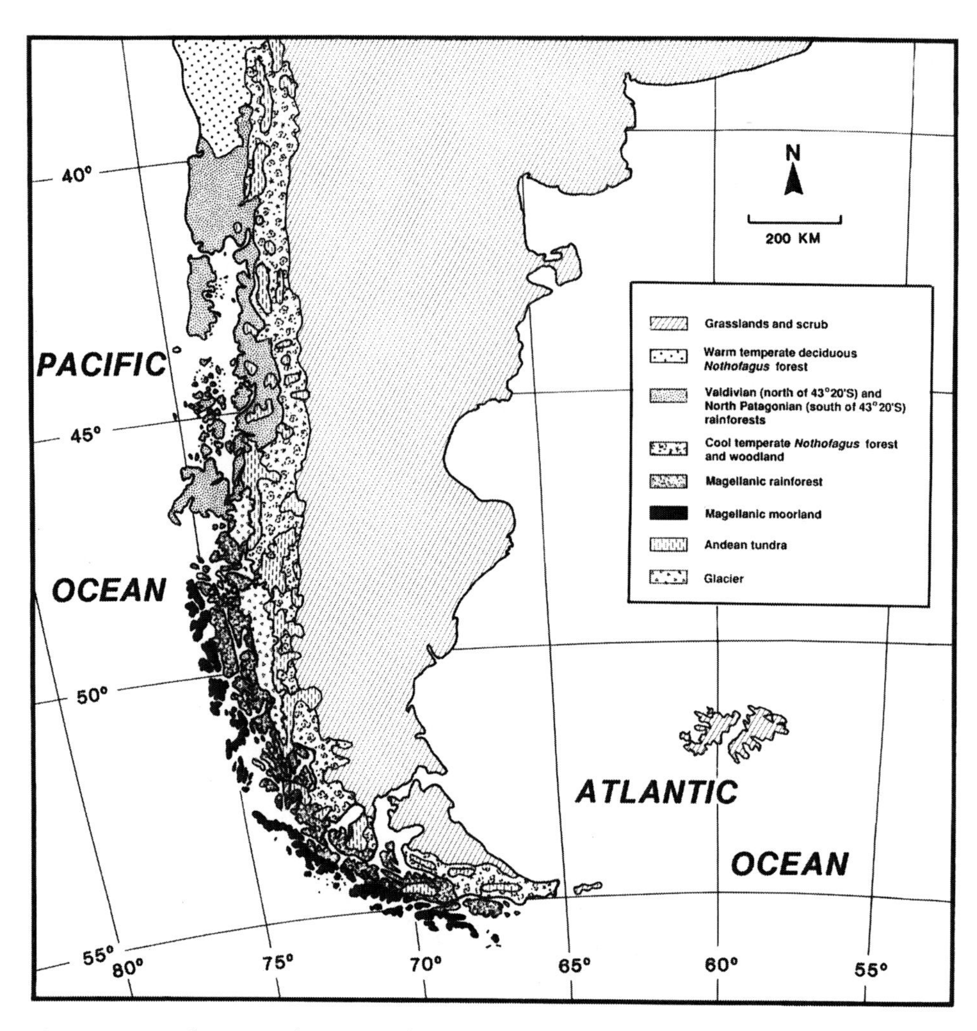

Figure 13.1. Major vegetation zones of southern South America

regions, there are major differences in forest physiognomies that reflect the different biogeographic histories of the two regions. The key biogeographic theme that helps us understand these global vegetation patterns is the interplay between history and modern environments. The migration and dispersal of plants over geological time interact with regional climates and other aspects of the modern environment to govern which plant groups survive and dominate today.

This chapter emphasizes the dynamic nature of the temperate forest ecosystems of South America, both over millennial time scales in terms of their biogeographic history and over centennial time scales in response to natural and human-caused disturbances. A recurrent theme is the constant interplay of climatic variation with natural ecological disturbances and the activities of humans that have and will continue to shape the vegetation patterns of this region. The chapter starts with three background sections, summarizing the broad-scale physical geography and vegetation patterns of the region, and then continues with a more detailed account of the patterns and processes of vegetation change.

13.1 Physical Background

13.1.1 Geology, Physiography, and Soils

The Andes of southern Chile and southwestern Argentina are a major determinant of climate and vegetation patterns in southern South America. From the Mediterranean region of central Chile southward to 42°S, a west-to-east transect across southern Chile and southwestern Argentina comprises the Coastal Cordillera, the Central Depression, the Andes, and the Patagonian plateaus and plains. South of 42°S, the Coastal Cordillera breaks into numerous islands, including large Chiloé Island and the Chonos Archipelago, the Central Depression becomes submerged, and the Andes rise directly from a maze of coastal fjords. Here, the drainage divide between Pacific and Atlantic streams lies east of the main Andes range, whereas in central Chile the two coincide. South of the Gulf of Peñas (47°S), there is no feature similar to the Central Depression and the coastal archipelagos merge progressively with the main Andes (Hervé et al., 2000).

The structural framework of this region derives from the prolonged collision of the South American continental plate with oceanic plates beneath the Pacific Ocean farther west. The subduction of the Nazca plate, particularly over the past 50 million years, has led to massive uplift of the Andes accompanied by widespread volcanic activity, processes that continue today (see chapter 1). South of the Chile Rise, which reaches the coast beneath the Taitao Peninsula and Gulf of Peñas, the Antarctic plate is now subducting beneath the continent, while in the extreme south the Fuegan Andes are influenced by the Scotia plate.

The Coastal Cordillera is underlain by some of the oldest rocks in the region, namely Precambrian and Paleozoic metamorphic rocks. Although an important influence on climate, this range rarely exceeds 1,000 m in elevation. The Central Depression is a linear structure filled with late Cenozoic glacial, fluvioglacial, aeolian, and alluvial deposits, and with volcanic ash. The Andes comprise a metamorphic basement complex, igneous and metamorphic rocks of the Patagonian batholith, and Mesozoic and Cenozoic sedimentary and volcanic rocks (Veit and Garleff, 1996; Hervé et al., 2000). The Patagonian Andes reach 4,050 m above sea level in Mt. San Valentin, inland from the Gulf of Peñas, and also contain several active volcanoes.

The mountainous relief of this region is due partly to late Cenozoic plate interactions and partly to repeated glaciation. Pleistocene ice caps formerly extended from Tierra del Fuego northward to beyond 40°S, and glacier ice covered eastern Chiloé Island and reached almost to the foothills of the Coastal Cordillera. This region thus supports a legacy of glaciated valleys, moraine-dammed lakes, fjords, bare rock, and shallow soils, and many glaciated surfaces have been covered by extensive volcanic deposits of recent origin (see chapters 2 and 3). Where glaciers were absent, periglacial conditions involving frost heaving and solifluction were widespread, not only in the Andes but above 800 m in the Coastal Cordillera. Today, though much smaller, ice fields still cover sizeable areas of the Patagonian Andes between 47 and 52°S, and reach tidewater to the south and west, and Patagonian lakes to the east.

Soil development in the southern Andes has been strongly influenced by volcanism that has resulted in volcanic ash and derived aeolian deposits often several meters thick (Veit and Garleff, 1996; see also chapter 7). The most widespread soils in the southern Andes on both the Chilean and Argentine side of the Andes are Andosols derived from recent, deep volcanic ash deposits. Locally, they are called *trumaos*, and are known for their high porosities and low bulk densities. They are typically rich in allophane, a noncrystalline mineral that contributes to their susceptibility to mass movement.

Aquepts are common in the southern Andes in areas of low relief where recent volcanic ash lies on glaciofluvial sediments (Veit and Garleff, 1996). These soils are characterized by poor drainage, high organic content, and typically have an iron-oxide-rich layer where the volcanic-aeolian deposits contact the glaciofluvial deposits. Aquepts occur extensively in the Central Depression of south-central Chile, where they are known as *ñadis*, and in southern Patagonia in areas of relatively recent glaciofluvial deposits.

Red-clay and brown-clay soils are widespread on old volcanic ashes in the southern Andes, as well as on metamorphic rocks in the coastal mountains. There are many areas where topography results in poor drainage and associated soil features, such as gley layers and accumulations of organic matter. Spodosols (podzols) are relatively rare in south-central Chile, except on old metamorphic rocks in the coastal mountains, but occur more extensively at cooler high latitudes in southern Patagonia and on Tierra del Fuego where rainfall is high.

13.1.2 Climate

The southern Andes are almost continuously bathed by the persistent mid-latitude westerlies of the Southern Hemisphere (see chapter 3). Orographic uplift by the Coastal Cordillera and the Andes results in mean annual precipitation of from 3,000 to over 5,m000 mm on the windward slopes from around 40°S to Tierra del Fuego. Most winter precipitation is in the form of snow at higher elevations, for example above 800 m at 40°S, and declines to nearly sea level at 55°S. The coastal mountains produce a minor rain-shadow effect on the Central Depression of south-central Chile at 40°S, where mean annual precipitation is typically less than 2,000 mm. East of the Andes, the rain-shadow effect becomes very pronounced: mean annual precipitation declines from around 3,000 mm to less than 800 mm over a west-to-east distance of only 50 km.

Climatic gradients are generally more gradual along the latitudinal gradient. Mean annual temperature declines from north to south, but decreasing continentality buffers higher latitudes from temperature extremes. The west coast maritime climate of South America south of 37°S is characterized by a mild temperature range and high annual precipitation (fig. 13.2). Along the west coast as far south as 42°S, Mediterranean-type precipitation seasonality is associated with the summer presence of a subtropical high pressure cell in the southeastern Pacific. Year-to-year variation in the intensity and latitude of this cell is the major determinant of annual variation in precipitation in the mid-latitudes of the southern Andes (see chapter 3). Farther south, the seasonal distribution of precipitation is relatively uniform, with stormy conditions prevailing year round. In the far south, at 52 to 55°S, the influence of the circum-Antarctic

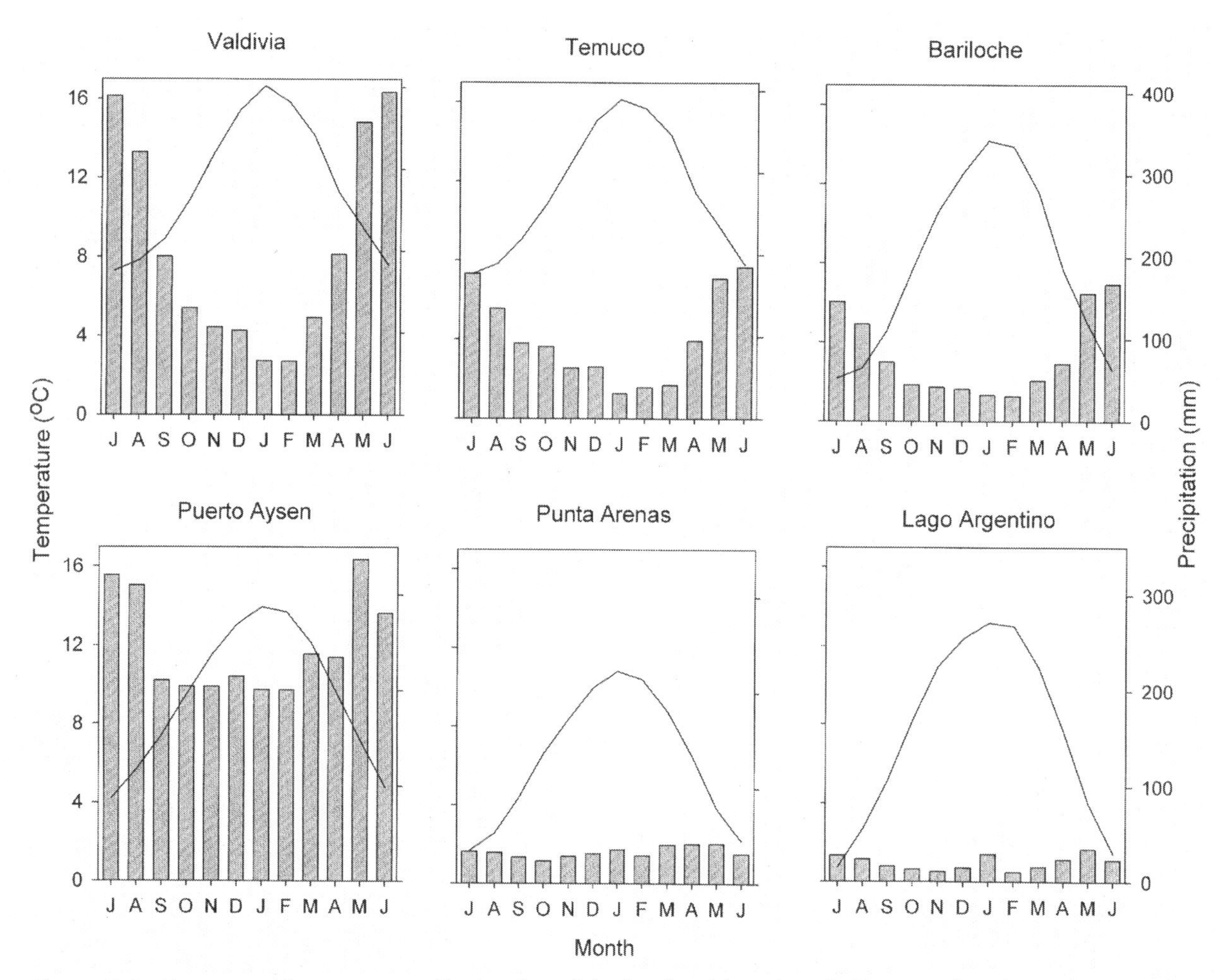

Figure 13.2. Mean monthly temperature (line) and precipitation (bars) for selected climate stations in southern South America. Details on the stations are given in table 13.1.

low pressure trough becomes more evident, and cool windy conditions prevail for most of the year.

13.2 Regional Vegetation Patterns

The broad-scale vegetation patterns of southern Chile and southwestern Argentina mirror the north-to-south and west-to-east climatic gradients. Bordering the evergreen sclerophyll shrublands and woodlands of central Chile is the district of warm temperate deciduous *Nothofagus* forest. Owing to deforestation and conversion of the land to plantations of exotic conifers, only relatively small patches of *N. glauca* and *N. obliqua* remain (see chapter 11, Donoso, 1996). Temperate rain forests prevail south of 37°45'S (Veblen et al., 1996). The Valdivian rain forest occurs from 37°45' to 43°20'S, the North Patagonian rain forest from 43°20' to 47°30'S, and the Magellanic rain forest south of 47°30'S (fig. 13.1, table 13.1). Although these rain forests occur mainly in Chile, on the windward side of the Andes and in the coastal range, small patches also occur in Argentina where precipitation is high on the eastern slope of the Andes.

As mean annual temperatures and the length of the growing season decline along this north-to-south gradient, so do species richness and total biomass in these rain forests. Thus, the Valdivian rain forest constitutes the center of maximum forest biomass and arboreal species richness (Arroyo et al., 1996). Although comprising mainly evergreen broadleaved trees and evergreen conifers, the Valdivian rain forest district also includes the deciduous *Nothofagus obliqua* and *N. alpina* (= *N. nervosa*; table 13.2). In contrast, the Patagonian and Magellanic rain forests are purely evergreen (mainly broadleaved trees with a small conifer component). Common dominants of the Patagonian rain forests are the evergreen *Nothofagus* (*N. dombeyi, N. nitida,* and/or *N. betuloides*), which typically occur in stands associated with fewer than five or six other angiosperm or conifer tree species. The most characteristic tree of the Magellanic rain forests is *Nothofagus betuloides,* which often forms monotypic stands or co-occurs with just a few other tree species (table 13.2). The deciduous *N. pumilio* and *N. antarctica* also occur within all three rain-

Table 13.1 Details of climate stations used in figure 13.2

Station	Vegetation zone	Lat. and long.	Elevation (m)	Duration of record	Mean annual temperature (°C)	Mean annual precipitation (mm)
Temuco	Valdivian rainforest	38°45'S 72°38'W	114	1951–1998	11.5	2,444
Valdivia	Valdivian rainforest	39°48'S 73°14'W	20	1941–1998	11.6	1175
Bariloche	Steppe and woodland ecotone	41°09'S 71°16'W	825	1905–1997	7.9	970
Puerto Aysen	Patagonian rainforest	45°24'S 72°42'W	20	1961–1995	5.6	2956
Lago Argentino	Steppe and woodland ecotone	50°20'S 72°18'W	223	1961–1998	7.5	224
Punta Arenas	Steppe and woodland ecotone	53°00'S 70°51'W	20	1914–1905	6.4	400

forest districts, in areas transitional to subalpine forests (*N. pumilio*), or on poorly drained sites (*N. antarctica*). Bamboos (*Chusquea* spp.) typically dominate the rain-forest understories in the north but are absent south of 48°S.

A particularly impressive component of the Valdivian and Patagonian rain forests is the evergreen conifer *Fitzroya cupressoides* (Cupressaceae), the largest and longest-lived conifer in South America (Veblen et al., 1995). *Fitzroya cupressoides* reaches up to 5 m in diameter and 50 m in height, and attains ages greater than 3,000 years. Its distribution consists of discontinuous populations from 39°50' to 42°30'S in the Coastal Cordillera and Andes of Chile, and less extensive stands in Argentina from 41 to 43°S. Generally, it occurs in areas of highest precipitation from 100 m to 1,200 m in elevation, typically on nutrient-poor soils that often are poorly drained, and where there is less competition from other tree species.

Other evergreen conifers are also important components of the rain forests. *Pilgerodendron uviferum* (Cupressaceae) occurs from 39°30'S, in the Valdivian rain forest, to 55°30'S in Magellanic rain forest and bogs on Tierra del Fuego. It occurs in poorly drained areas, often in association with *Fitzroya* in the north, and as the only tree in lowland bogs or in coastal forests of *Nothofagus betuloides* and *Drimys winteri* in the far south. The conifers *Podocarpus nubigena* and *Saxegothaea conspicua* (both in the Podocarpaceae) are widespread in the Valdivian and Patagonian rain forests and reach heights >25 m and diameters >2 m. Other conifers with more restricted distributions include the dwarf *Lepidothamnus fonkii* in bogs, and *Podocarpus saligna* typically in riparian habitats in the Valdivian rain forest.

Parallel to the strong west-to-east decline in precipitation is a gradient from temperate rain forests, through cool temperate *Nothofagus* forests and xeric woodlands, to the Patagonian steppe of bunchgrasses and shrubs (fig. 13.1). Cool temperate *Nothofagus* forests and woodlands occur from 37°30'S southward to around 55°S on Tierra del Fuego, and include subalpine Andean forests as well as drier forests eastward in the rain shadow of the Andes (Veblen et al., 1995). These forests and woodlands occur most extensively on the Argentinean side of the Andes and include stands dominated by evergreen *N. dombeyi* at mesic mid-elevation sites, and by deciduous *N. pumilio* and *N. antarctica* stands at high elevation and/or xeric sites. In the northern part of the extensive cool temperate *Nothofagus* forests, the evergreen conifers *Araucaria araucana* (north of 40°20'S) and *Austrocedrus chilensis* (north of 44°S) occur along the mesic forest to xeric woodland portion of this trans-Andean gradient. In Chile, the xeric conifer *Prumnopitys andina* forms small stands in dry habitats near the northern limit of the Valdivian rain forest. South of 47°S, this strong west-to-east precipitation gradient is also evident, but here the conifers are absent from the xeric eastern habitats. The far western district (west of the Magellanic rain forest), is known as Magellanic moorland and is characterized by scattered *N. betuloides*, bog, and ericaceous heath. This is an area of typically waterlogged soils and nutrient-poor substrates such as the diorite of the Chilean archipelago.

13.3 Historical Biogeography

13.3.1 Origins of the Modern Flora

The origin of the temperate forest flora of southern South America reflects Cretaceous and Cenozoic geologic histories that gradually isolated the region from North American (and indirectly Eurasian) and Gondwanan floras. The major geologic events that are important to understanding the contemporary flora and fauna of southern South America are described in more detail in chapters 1, 2, and 8. They include the breakup of Pangea, and then Gondwana. South America's long isolation from North America in the Cenozoic resulted in a strong differentiation between the temperate floras of North and South America. Yet the Andean Cordillera also provided an important dispersal route for some temperate taxa shared

Table 13.2 Common tree species of the major forest zones of southern Chile and southwestern Argentina

	Valdivian rain forest	Patagonian rain forest	Magellanic rain forest	Cool temperate Nothofagus forests and woodlands
*Nothofagus alpina**	x			
*Nothofagus obliqua**	x			
Laurelia sempervirens	x			
Amomyrtus meli	x			
Crinodendron hookerianum	x			
Eucryphia cordifolia	x			
Persea lingue	x			
Aextoxicon punctatum	x			
Gevuina avellana	x			
Fitzroya cupressoides#	x			
Luma apiculata	x			
Podocarpus saligna	x			
Nothofagus nitida	x	x		
Laureliopsis philippiana	x	x		
Amomyrtus luma	x	x		
Pseudopanax laetevirens	x	x		
Chusquea bamboos	x	x		
Caldcluvia paniculata	x	x		
Saxegothaea conspicua#	x	x		
Weinmannia trichosperma	x	x		
Nothofagus dombeyi	x	x		x
Podocarpus nubigena#	x	x		
Maytenus magellanica	x	x	x	
Drimys winteri	x	x	x	
Pilgerodendron uviferum#	x	x	x	
Embothrium coccineum	x	x	x	
Nothofagus betuloides	x	x	x	x
*Nothofagus antarctica**	x	x	x	x
*Nothofagus pumilio**	x	x	x	x
Araucaria araucana				x
Austrocedrus chilensis				x
Lomatia hirsuta				x

Deciduous broadleaved species are indicated by * and evergreen conifers by #. All others are evergreen broadleaved trees.

between the two continents, as did long-distance dispersal by birds.

An overriding influence on the temperate forest flora of South America has been the long-standing relationship of the continent with Antarctica and other Southern Hemisphere (Gondwanan) land masses throughout the Cretaceous and early Cenozoic (Hill and Dettmann, 1996). An earlier Cenozoic connection across Antarctica, then much warmer, facilitated the migration of elements from Australia into South America. In addition, though relatively rare, long-distance dispersal over water barriers from the southwest Pacific to South America when the continents were much closer may have contributed to the strong floristic similarities between the floras of now separated Gondwanan lands (Hill and Dettmann, 1996).

The critical period of Cretaceous and early Cenozoic dispersals among Gondwanan lands is strongly reflected in the modern flora of the temperate forests of southern South America. Eighteen genera of woody species in southern South America are shared with other Gondwana land masses such as Australia, New Zealand, New Guinea, and New Caledonia. These include many of the most dominant tree and shrub species such as *Araucaria, Aristotelia, Caldcluvia, Discaria, Eucryphia, Gevuina, Laurelia, Lomatia, Nothofagus, Podocarpus, Pseudopanax, Prumnopitys*, and *Weinmannia* (Arroyo et al., 1996). Many dominant tree species now placed in distinct genera in the southwest Pacific and in South America clearly had common ancestors and today are strikingly similar (e.g., *Laurelia* of the Valdivian rain forest and *Atherosperma* of southeast Australia; *Austrocedrus* of South America and *Libocedrus* of New Zealand). Overall, 21 woody angiosperm genera of the temperate forests of South America are considered to have a Gondwanan linkage (Arroyo et al., 1996). Furthermore, all seven genera of southern South American gymnosperms evolved in Gondwana, so that at least one third of the woody genera of the temperate forest zone of South America are of Gondwanan origin (Arroyo et al., 1996).

There is also a substantial component (19 woody genera) of Neotropical origin (including tree genera such as *Dasyphyllum*, *Drimys*, and *Crinodendron*) (Arroyo et al., 1996), but these elements are much less dominant physiognomically than the Gondwanan elements.

13.3.2 *Impacts of Quaternary Environmental Changes*

The major changes in land connections during late Mesozoic and earlier Cenozoic time are strongly reflected in the taxonomic affinities of the modern flora of the temperate latitudes of southern South America. Although land connections to the region have not varied significantly since Pliocene time, the fluctuating climates of the past two million years, the Pleistocene and Holocene epochs, have left a strong imprint on the modern pattern of vegetation. Repeated alternations between glacial and interglacial conditions affected the southern Andean region and resulted in latitudinal and elevational shifts of vegetation types (Markgraf et al., 1996). During glacial times in the area south of 42°S, ice covered most of the land from 120 m below sea level on the Chilean coast across the Andes to the foothills bordering the Patagonian steppe. The area of forest in this region was substantially reduced but probably persisted in relatively small, ice-free refuges.

In addition to the direct local impacts of glacier advance and recession, the associated climatic variations had major influences on regional vegetation pattern (Markgraf et al., 1996). For the area south of 43°S latitude, fossil pollen records from both onshore and marine sites suggest that glacial intervals were generally arid, with steppe-scrub vegetation dominating, and that interglacials were more mesic with greater dominance by forests (Markgraf et al., 1996). South of 51°S, after deglaciation, conditions were dry and steppe was widespread at 12.5 Ka (thousands of years before present). *Nothofagus* forests were present in only relatively small pockets, where apparently they had survived in glacial refugia. Sedimentary charcoal and pollen records indicate that fires were frequent during the early Holocene (11.7 to 5.5 Ka) and that vegetation was in greater flux (Huber and Markgraf, 2003). Paleo-Indian hunters potentially were an important ignition source for these fires, but clearly the warmer, drier climate favored the spread of fire. In comparison with the modern period, the vegetation consisted of more open *Nothofagus* woodlands during this time of increased fire frequency (Huber and Markgraf, 2003). By 8 ka, *Nothofagus* forests became extensive, and fire frequency declined. During the mid-Holocene (6 to 5 Ka), *Nothofagus* forests declined as more xeric shrubs advanced, indicating a period of decreased precipitation (Markgraf et al., 1996). Conditions became more mesic after 5 Ka, and by 3 Ka modern conditions of cooler winters and warmer, drier summers prevailed. Between 5.5 and 4.0 Ka, charcoal records indicate that fires were substantially less frequent and were associated with other environmental indicators, suggesting a prolonged regional increase in effective moisture (Huber and Markgraf, 2003). After 1600 A.D., fire frequency and European weed taxa increase abruptly in the far south (Huber and Markgraf, 2003). These are likely to have been the result of European settlement in the coastal areas and the escape of livestock into the interior.

At latitudes 42 to 51°S, steppe vegetation was dominant during the late-glacial interval (~14 Ka) but small amounts of *Nothofagus* pollen indicate the presence of forest in restricted glacial refugia. Even during full glacial times, small forest refugia occurred despite the coverage of most of the landscape by ice (Markgraf et al., 1996). For example, the geographical distribution of genetic variation in the conifer *Fitzroya cupressoides* indicates that its modern populations originated from multiple glacial-refugia as opposed to a post-glacial migration from a more northerly ice-free area (Premoli et al., 2000). The treeless vegetation on the Chilean side of the Andes was replaced by *Nothofagus* woodland and forest shortly after 13 Ka, or substantially earlier than forest spread in latitudes south of 51°S. In the early Holocene, sedimentary charcoal records also indicate an early and widespread importance of fire (Heusser, 1987, 1994), which may have been associated with humans. In contrast, on the Argentine side of the Andes, transition from steppe to *Nothofagus* forest did not occur until 10 Kka (i.e., the same time as at high latitude).

During the last millennium, glacier advances and retreats document regional-scale climatic variation in the southern Andes. South of 40°S, glaciers retreated during a warming from 1080 to 1250 A.D., approximately coincident with the Medieval Warm Epoch of the Northern Hemisphere. That retreat was followed by a cool, moist trend that peaked about 1340–1640, overlapping with the Little Ice Age of the Northern Hemisphere (Villalba, 1994).

Tree-ring records are relatively abundant for the southern Andean region and document important trends in temperature over the past several centuries. Tree-ring records from *Nothofagus pumilio*, which grows at high elevations, extend as far back as 1640 A.D. with adequate spatial coverage and sample sizes for reliable reconstruction of past precipitation and temperatures (Villalba et al., 1998, 2003). The mean annual temperatures reconstructed for the Andes at 37 to 55°S for the period 1900 to 1990 are 0.53 to 0.86°C above the mean for 1640–1899 A.D. The rate of temperature increase from 1850 to 1920 was the highest over the past 360 years (Villalba et al., 2003). Instrumental records indicate that there is significant variation in temperature and precipitation trends for different subregions within the southern Andean region. For example, at high latitudes along the coast of Chile there has been a marked cooling since the early 1960s, whereas at midlatitudes on the eastern side of the Andes temperatures have risen markedly since the mid-1970s (fig. 13.3; Daniels and Veblen, 2000).

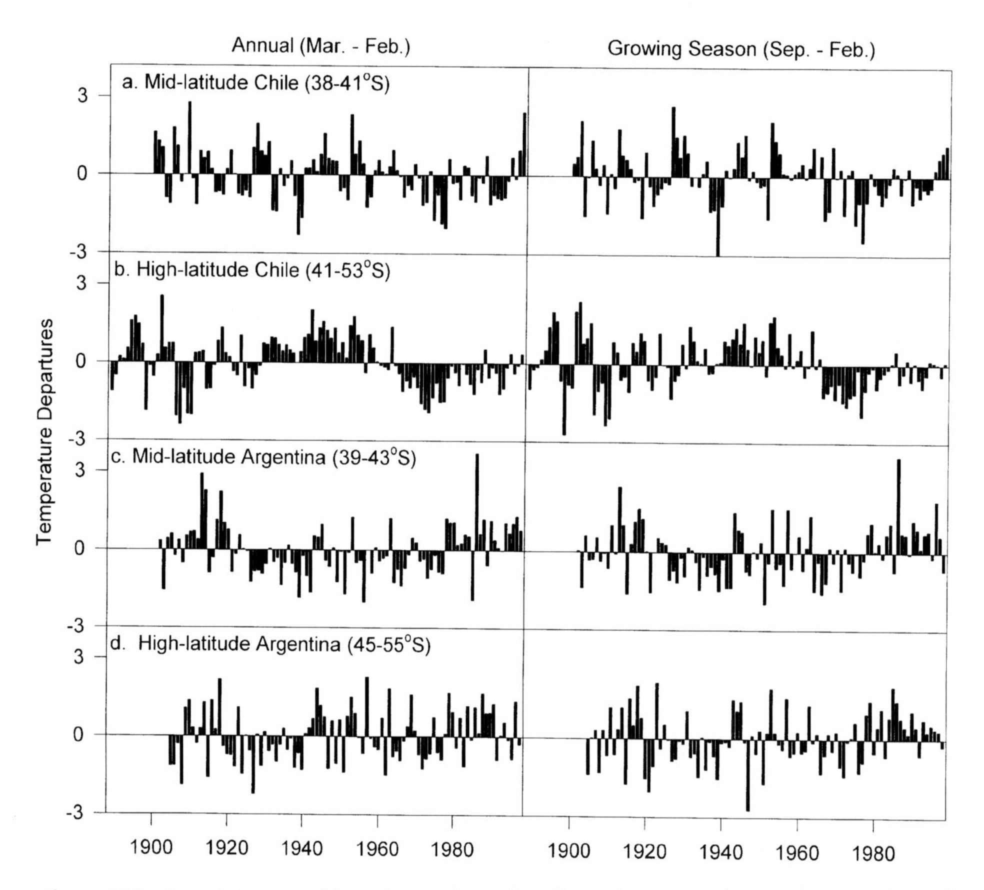

Figure 13.3. Trends in annual (March to February) and growing season (September to February) temperature departures (standard deviations) by region for southern South America. For each region, multiple climate stations were used. Note the post-1960 cooling trend for high-latitude Chile in contrast to the post-1976 warming trend for the other regions, especially for mid-latitude Argentina (northern Patagonia). From Daniels and Veblen (2000).

Thus, at a broad spatial scale and over the past several hundred years, the region as a whole exhibits a warming trend, but there have been contrasting patterns for different subregions at decadal time scales. These decadal-scale changes in climate have had major influences on the vegetation patterns of the southern Andes, as discussed below in relation to fire ecology.

13.4 Forest Dynamics

13.4.1 *Patterns and Processes of Forest Dynamics*

Patterns of stand development in South American *Nothofagus* forests result from both coarse-scale disturbances producing whole-stand replacement patterns, and small treefalls resulting in fine-scale gap dynamics. Over extensive areas in the Valdivian rain forest district (37°45' to 43°20'S), relatively old (>300 years) forests are dominated by emergent trees of shade-intolerant species such as *Nothofagus* spp., *Weinmannia trichosperma,* and *Eucryphia cordifolia* (Veblen et al., 1996). In undisturbed stands, young or small trees of these species are absent or scarce. In contrast, shade-tolerant tree species (such as *Laurelia sempervirens, L. philippiana, Persea lingue, Aextoxicon punctatum,* and numerous myrtaceous trees) are typically abundant and occur as all-aged populations. In the absence of natural disturbance, there is a gradual successional trend toward dominance by these shade-tolerant species. However, in landscapes with a high frequency of coarse-scale disturbances, this successional trend is not completed and the shade-intolerant trees (especially *Nothofagus*) remain dominant in the oldest stands (Veblen and Ashton, 1978, Veblen et al., 1996). Thus, over extensive areas the frequency of disturbances such as landslides, floods, wind storms, and fire permit the dominance of the forest by shade-intolerant species that, in the absence of coarse-scale disturbance, would be mostly replaced by shade-tolerant species.

The dependence of *Nothofagus* on coarse-scale disturbance is best documented for the Valdivian rain forest district but also has been described in the Patagonian rain forest district (43°20' to 47°30'S; Innes, 1992). In landscapes characterized by lower rates of coarse-scale disturbance, forests exhibit greater dominance by shade-tolerant trees and appear to be closer to equilibrium in species composition (Veblen et al., 1996). In the coastal mountains of the Chilean Lake District, for example, natural disturbances by landslides and volcanism are markedly less than in the Andes, and forest dominance by shade-intolerant *Nothofagus* species is less. Even in the Andes, at higher latitudes or elevations, and on drier sites with few or no shade-tolerant species, the tendency for *Nothofagus* to be replaced successionally is necessarily less or absent.

Coarse-scale disturbance plays an analogous role in permitting the shade-intolerant conifer *Fitzroya cupressoides* to regenerate at sites where otherwise more shade-tolerant tree species would successionally replace it. This long-lived conifer is adapted to regenerate on open sites disturbed by landslides, floods, or fire (Veblen et al., 1995). At sites of relatively favorable habitat where many of the shade-tolerant tree species of the Valdivian or Patagonian rain forests also grow, the occasional occurrence of these disturbances appears to be critical in creating regeneration opportunities for *Fitzroya*. Although the shade-intolerant seedlings of this species are typically rare or absent beneath a closed tree canopy, they establish in abundance on bare sites created by disturbance, especially where the climate is wet. Despite its growth in a rain-forest environment, this species is adapted to survive fire and to regenerate following fire. The very thick bark (>10 cm thick on large trees) of *Fitzroya* allows it often to survive intense crown fires that kill all the competing species of thin-barked trees. Thus, infrequent drought and severe fires that may occur at intervals of 100 or more years at rain-forest sites are important in permitting *Fitzroya* to maintain its dominance despite the competitive superiority of the associated shade-tolerant tree species. At sites where the habitat is unfavorable for most tree species, such as at water-logged sites and on nutrient poor soils, tree canopies are less dense and the regeneration of *Fitzroya* is less dependent on coarse-scale disturbances (Veblen et al., 1995). Where site quality is high, this poor competitor depends on coarse-scale disturbance for its regeneration; where site quality is lower, it can regenerate in relatively small openings created by the death of an individual canopy tree.

The recognition that in certain habitats the regeneration of shade-intolerant tree species such as *Nothofagus* species and *Fitzroya cupressoides* is dependent on infrequent coarse-scale disturbance has been important in understanding forest structures that previously were puzzling or sometimes incorrectly attributed to recent climate change (Veblen et al., 1995). However, the dependence of these shade-intolerant tree species on coarse-scale disturbance is highly variable across the landscape, and even within the same habitat the fortuitous availability of a safe site for seedling establishment may permit these shade-intolerant species to regenerate occasionally beneath relative small canopy openings.

13.4.2 *Chusquea* Bamboos as Keystone Species

Bamboos of the Neotropical genus *Chusquea* dominate the understories of most of the Valdivian and Patagonian rain forests as far south as 47°S latitude, and function as keystone species. As keystone species, they exert influences on ecosystem processes and landscape patterns that far exceed the size of these plants. There are at least five bamboo species that occur commonly in the understories of these forests that can impede tree regeneration through their competitive influences on tree seedlings, especially seedlings of shade-intolerant species such as *Nothofagus* species. The most common *Chusquea* species are *C. quila* and *C. culeou,* which occur at low (<600 m) and middle elevations (500–900 m), respectively. *Chusquea quila* is a climbing bamboo with multibranched stems allowing it to reach heights of over 20 m in the tree canopy. *Chusquea culeou* is an erect, clump-forming bamboo with unbranched culms reaching over 7 m in height. Both bamboo species can form dense thickets beneath relatively open tree canopies, and occur in even greater densities where there is no tree overstory. These species can tolerate semishade in a forest understory but are rare or absent beneath the densest tree canopies. Both of these bamboo species spread vegetatively, allowing them gradually to proliferate as disturbances such as natural treefalls or logging increase light levels in a forest stand.

The degree to which *Chusquea* species impede the regeneration of tree species is highly variable, depending on the factors that control the size and abundance of the bamboos and the relative shade-tolerance of the tree species. The highest densities of *Chusquea* bamboos occur beneath the relatively open canopies of stands dominated exclusively by shade-tolerant tree species such as *Nothofagus dombeyi* and *N. pumilio*. In such stands, relatively little tree regeneration can occur without either a large-scale disturbance such as fire, blowdown, or a snow avalanche, or the death of some or all of the bamboo clones.

Most of the *Chusquea* bamboo species follow a monocarpic reproductive pattern in which the plant lives for several or more decades, flowers, produces seed, and then dies (Veblen, 1982). Because flowering and seeding occur synchronously among large percentages (20 to >90%) of the clones over areas of 100s or 1,000s of square kilometers, the life cycle of the bamboo is a major determinant of vegetation dynamics and related ecosystem processes. For example, in 2000–2001, for the first time in 60 years, a massive flowering and withering of *Chusquea culeou* occurred in Argentina and Chile from 38 to 40°S latitude. In some forest types, the death and withering of the bamboo allows the accelerated growth of tree seedlings previously established beneath the bamboo cover. Thus, a bamboo flowering event has a long-lasting impact on forest structure. The abundance of seeds produced in a flowering event of *Chusquea* results in an explosion of the seed-eating rodent populations—including both native rodents and introduced rats. The irruption of the rodent population results in turn in a sharp population increase of predator animals such as foxes and raptors. The large rodent populations consume agricultural crops and infest human habitations so that a flowering event is viewed as a calamity from the human perspective. Furthermore, the large rodent population also carries the hanta virus, which breaks out among the human population due to increased probability of human and rodent contact. The outbreaks of hanta virus in human populations, in turn, have devastating impacts on local tourism industries when frightened tourists from the metropolitan districts of Chile and Argentina avoid the Lake Districts of both countries. Potentially, the greatest impact of a *Chusquea* flowering event on landscape structure is through the massive amount of fuels produced for subsequent fires. When flowering events are followed by severe drought, extensive fires may occur even in the wettest areas of rain forest. During the twentieth century, for example, periods of peak forest burning in mesic *Nothofagus dombeyi* forests in northern Patagonia followed massive bamboo flowering events around 1912 and in 1940–1941 (Veblen et al., 2003).

13.4.3 The Roles of Natural Disturbances in Forest Dynamics

Coarse-scale ecological disturbances are particularly important sources of the patterns of forest structure and composition in the southern Andean region. These include disturbances of natural origin, such as earthquake-triggered landslides, volcanic activity, and wind storms; disturbances of purely human origin, such as forest clearing; and disturbances that can have either a natural or human origin, such as fire.

Disturbances Related to Tectonic Activity Situated near the intersection between four major crustal plates, the southern Andean region is an area of great volcanic and tectonic activity. On the wet western side of the Andes, in the temperate rain forests of Chile, strong earthquakes such as the 1960 Valdivian earthquake may in a single event trigger thousands of slope failures (e.g., rockfalls, landslides, and debris flows) in the Andes (Veblen and Ashton, 1978). Earthquakes similar in magnitude to that of the 1960 event have affected the Chilean Lake District (40°S) seven times since 1520. In addition to mass movement triggered by earthquakes, many other slope failures are associated with prolonged and torrential rainfall in this region of glacially steepened slopes. Bare sites created by mass movement are favorable to the establishment of the shade-intolerant species (e.g. *Nothofagus* spp., *Weinmannia trichosperma, Fitzroya cupressoides*) that otherwise tend to be replaced by the numerous shade-tolerant tree species of the Andean rain forests (Veblen and Ashton, 1978, Veblen et al., 1996). Whether caused by earthquakes or precipitation events, landslides are a relatively frequent occurrence in the southern Andes, permitting shade-intolerant tree species to regenerate and dominate stands for hundreds of years after the disturbance event. On the Argentine side of the Andes, the 1960 earthquake triggered few landslides, but intense shaking on unstable debris fans caused tree mortality and affected tree growth rates (Veblen et al., 1996).

Volcanism is another regionally extensive disturbance throughout the southern Andes that creates bare surfaces suitable, in some cases, for tree establishment (Veblen et al., 1996). At a regional scale, vast areas (10^5 km^2) have been buried by volcanic ash several times during the Holocene. In the past several centuries, deposition of volcanic ash over hundreds of square kilometers has provided establishment opportunities for shade-intolerant tree species such as *Nothofagus*.

Disturbance by Wind The mid- to high-latitudes of the southern Andean region are well known for strong winds, and blowdowns are a common disturbance throughout the temperate rain-forest zone. For example, in the Valdivian and Patagonian rain forests, the fall of the 50-m-tall *Nothofagus dombeyi* often creates canopy openings greater than 1,000 m^2 (Veblen et al., 1996). Most treefalls occur during winter storms when these evergreen broadleaved trees are heavily laden with snow, which tends to be warm and wet in this maritime climate. Among forest types, the subantarctic *Nothofagus pumilio* forests are the most affected by blowdown. Although the structure of many stands reflects the importance of relatively small treefall gaps (openings of 100 to 200 m^2), the structure of this forest type is often described as a mosaic of even-aged patches, including patch sizes well over one hectare (Veblen et al., 1996; Rebertus et al., 1997). These structures arise as responses to both treefalls of small groups of old canopy trees and to coarse-scale exogenous disturbance. For example, following disturbance by fire, avalanche, mass movement, or blowdown, relatively large, even-aged patches may occur (Veblen et al., 1996). Beneath an even-aged cohort there may be little or no successful regeneration of *N. pumilio* until the stand is 200 or more years old. Since these cohorts can cover areas of several hectares, this often gives the impression of a nonregenerating forest, but where either the senescence and death of large canopy trees occurs or the stand is disturbed by exogenous disturbance (most commonly strong winds), regeneration occurs.

At high latitudes (south of 52°S), blowdowns are a particularly important disturbance in the zone of the subantarctic *Nothofagus* forests (Veblen et al., 1996). For example, in a 10–km^2 study area on Tierra del Fuego, areas exposed to strong winds have stand structures and stand ages showing a cycle in which the average time between blowdowns is only 145 years (Rebertus et al., 1997). Storm damage is greatest on upper leeward slopes and in valleys and sideslopes parallel to storm tracks.

Fire Throughout the southern Andean region, fire is the most spatially pervasive disturbance, but its frequency and ecological effects are highly variable, depending on the regional climate and local forest type. Along the steep east-to-west precipitation gradient from xeric woodlands of *Austrocedrus* east of the Andes in northern Patagonia to the temperate rain forests in Chile, the frequency of natural fires declines dramatically (Kitzberger et al., 1997; Lara et al., 2003). Fires in this region are ignited both by humans (see below) and by lightning. Although in many areas of the southern Andes, lightning storms are rare in most years, the infrequent occurrence of lightning-ignited fires are of major ecological importance in these forests. Even in the wet *Fitzroya* forests of Argentina and coastal Chile, where over thousands of square kilometers of forest lightning-ignited fires may only occur a few times in a century, the effects on forest structure last for centuries.

In northern Patagonia, convective storms, which are only frequent during years when humid subtropical air masses penetrate anomalously southward into the region, are associated with years of warm summers and of unusual lightning strikes (see chapter 3; Kitzberger and Veblen, 2003). Although lightning is less frequent on the Chilean side of the Andes at midlatitudes, rare lightning-ignited fires do occur, as they did in the *Araucaria* forest region in 2002. Many tree and shrub species at midlatitudes in the southern Andes are well adapted to resist or recover after fire, notably the thick-barked *Araucaria,* which can resist even many intense crown fires. When its canopy is killed by fire, it resprouts from buds at the base of the trunk, on lateral roots, and on the branches (Veblen et al., 1995). The percentage of woody species that resprout after fire reaches a peak at the ecotone between dry forests and the Patagonian steppe. Even within the wettest forests, the thick-barked *Fitzroya* and the resprouting behavior of *Chusquea* bamboos assure regeneration after fire.

Some of the dominant *Nothofagus* species (e.g., *N. antarctica, N. obliqua,* and *N. alpina*) resprout after being burned or severely damaged by other disturbances such as partial burial by mass movement (Veblen et al., 1996). Other *Nothofagus* species have traits that allow their populations to recover quickly after most types of disturbance that create open sites, including fire, even though such traits are unlikely to have been selected primarily by fires. For example, *Nothofagus dombeyi* is effective at dispersing its light, winged seeds over distances of at least twice the tree height so that following a fire there will be adequate seed supply at least to the edge of severe burns. Where fires do not consume all the canopy trees, leaving dispersed seed trees, the abundant seed production and ability of *N. dombeyi* to grow rapidly under open conditions results in extensive areas of even-aged post-fire stands in northern Patagonia. Under favorable site conditions and seed sources, *N. pumilio* also can form dense, even-aged, post-fire stands. However, there are many areas where the regeneration *N. pumilio* is absent or poor after fire. Potential explanations for the failure of post-fire regeneration of *N. pumilio* include soil changes associated with severe fires, exposure of sites to drying winds, herbivory by introduced animals, and unfavorable climate.

Other Natural Disturbances Large areas of the southern Andes are currently covered with glaciers or have been recently deglaciated. Thus, there are many habitats recently exposed by ice retreat or disturbed by glacial deposition and associated flooding. *Nothofagus* spp. commonly colonize the bare sites created by these disturbances (Veblen et al., 1996).

Snow avalanches are important disturbances in the forests of *Nothofagus pumilio* that occur near alpine treeline. Relatively infrequent stand-replacing avalanches typically result in the development of even-aged stands of this species. Where avalanches occur more often, they create a pattern of chronic disturbance that maintains a shrubby form of *Nothofagus* species.

13.4.4 Human Influences on Forest Dynamics

Humans have significantly changed forest patterns in the southern Andean region through logging and conversion of forest land to agricultural land use, and in more subtle ways through the effects of introduced animals and alteration of the fire regime.

Human Impacts on Fire Regimes Indigenous peoples in the southern Andean region are known to have set fires and undoubtedly increased the frequency of fire over what it would have been if the sole source of ignition was from lightning. However, fire spread is so dependent on weather that it is not certain if the increased frequency of fires due to humans would have resulted in major ecological changes in comparison with the natural fire regime (Veblen et al., 1999). For particular habitats such as the forest/steppe ecotone of northern Patagonia, where indigenous land use was concentrated, it is likely that human-set fires had a significant ecological impact. On the other hand, in rain forests in remote parts of the Andes the dependence of fire spread on rare droughts implies that human-set fires would not have had the opportunity to spread except on very rare occasions. Fires were set by indigenous peoples for a variety of purposes, including communication and hunting, and in certain habitats and under appropriate weather conditions some of these fires became large enough to have an impact on the landscape. Early explorers in northern Patagonia describe native peoples using fires to hunt rheas (the ostrich-like birds of the steppe) and guanacos (the American camelid) in the steppe adjacent to forests (Veblen et al., 1996). They also described large areas of burned rain forest in the eighteenth and nineteenth centuries, which they attributed to ignition by humans. Similarly, on the Chilean side of the Andes, nineteenth century fires were also believed to have been set by native peoples as a warfare tactic. Fires were commonly set in the nineteenth and twentieth centuries by native peoples in *Araucaria* forests in both Chile and Argentina in order to clear the undergrowth and facilitate the collection of nutritious *Araucaria* seeds (Veblen et al., 2003).

In northern Patagonia, in *Austrocedrus*-dominated woodlands, the fire-scar record shows a substantial increase in fire frequency in the mid-1800s that coincides both with decadal-scale climatic variation conducive to fire (Kitzberger et al., 1997) and with increased use of the land by native peoples (Veblen et al., 1999). The mid-1800s were a time of massive white settlement in southern Chile, which resulted in the expulsion of the native inhabitants from Chile and their migration to northern Patagonia. This increase in fire frequency occurs primarily in the xeric vegetation types, which was the habitat of the guanacos, the principal game animal. Fire frequencies again increase sharply in the 1890s in association with white settlement in northern Patagonia. In this case, however, the most dramatic increases are in the mesic *Nothofagus* forests, which the white colonists burned extensively in attempts to create cattle pasture, attempts that ultimately failed. This is reflected by the dominance today of the mesic zone by 80 to 100-year-old post-fire forests (Veblen et al., 1996).

In southern Chile, in comparison with northern Patagonia, there are few tree species that can record fire scars, and consequently less is known about the long-term history of fire. However, documentary sources record vast areas of the Valdivian rain forest being burned during the European colonization of the Central Depression in the mid-1800s (Perez, 1958). Farther south in Chile, massive forest burning primarily motivated by the desire to create cattle pasture began in the early 1900s and continued into the 1950s.

Although for the most part modern humans have increased fire occurrence, in some areas humans may have decreased fire occurrence. At the forest-steppe ecotone in the *Austrocedrus* woodlands of northern Patagonia, for example, the tree-ring record of fires shows a substantial decline in fire occurrence in the twentieth century (Kitzberger et al., 1997; Veblen et al., 1999). Some of this decline may be due to the cessation of intentional burning due to the demise of the native population or due to the reduction of grass fuels by introduced livestock. Much of the post-1940 decline is due, however, to the active suppression of fires by land managers. The decline in fire occurrence at the forest-steppe ecotone coincides with the beginning of a substantial increase in tree establishment and survival in formerly more open habitats (Veblen et al., 2003). Although other factors such as climatic variation and livestock impacts influence this dramatic tree invasion of the steppe, without fire suppression tree seedling survival would not be possible.

Introduced Animals and Plants The forest region of the southern Andes has been fundamentally altered by human introductions, both purposeful and accidental, of exotic plants and animals. Livestock arrived in southern Chile with the earliest Spaniards in the sixteenth century, and from coastal settlements gradually spread into the hinterlands (Veblen et al., 1996). Even from as far away as the Rio de la Plata estuary of eastern Argentina, livestock are believed to have spread across the interior of the continent to the Andes as early as the eighteenth century. Thus, by the late eighteenth century, horses, sheep, and cattle had been adopted by native peoples living on the eastern side of the Andes. From the late nineteenth century onward, livestock have been widespread in the forests of the southern Andean region, and only the most remote areas have been spared their effects. During the early twentieth century, exotic deer (red deer and fallow deer) were introduced to provide a game animal for hunters (Veblen et al., 1996).

Introduced livestock and cervids have greatly affected the vegetation of the southern Andes (Ramirez et al., 1981; Veblen et al., 1996). They have impeded post-fire recovery at many sites, and they may have had a significant impact on fuel quality and quantity. Although the major tree species are relatively resistant to browsing once they reach about 2 m in height, exceptionally heavy cattle pressure during early post-fire recovery can locally impede tree regeneration. Thus, forest burning followed by heavy livestock pressure sometimes results in establishment of herbaceous turfs (with abundant exotic species) or shrublands of spiny shrubs and dwarfed trees (Veblen et al., 2003). In forests with understories dominated by *Chusquea culeou*, livestock greatly reduce the size and cover of the bamboos so that fuel loads and heights are markedly less. The overall impact of livestock in northern Patagonia appears to be two-fold: a generalized decrease in fine fuel quantity in grasslands, and a shift toward greater dominance by shrubs following forest burning.

Other introduced wild animals that have important ecological impacts in the southern Andes include the European hare (*Lepus capensis*), wild boars (*Sus scrofa*), rabbits (*Oryctolagus cuniculus*), and beavers (*Castor canadensis*). European hares occur throughout the region and, where their populations are high, have damaging effects on tree-seedling survival. This is especially a problem in areas of *Nothofagus pumilio* forest. Wild boars alter grassland sites and forest floors through their rooting activities. Wild rabbits were introduced into central Chile and to Tierra del Fuego in the late nineteenth century, and from these north and south entry points are gradually spreading throughout the southern Andean region. They are particularly common in open vegetation types where their grazing has a significant impact on biomass (Jaksic and Yáñez, 1983). Beavers were introduced to Tierra del Fuego in the 1940s, where they have killed extensive areas of forests through their dam-building activities.

Introduced plant species have naturalized and in some habitats have become an important component of the vegetation. There are more than 300 exotic vascular plant species that have naturalized in northern Patagonia (Rapoport and Brión, 1991). Exotic species are particularly common in habitats severely disturbed by livestock and logging, and may have significantly altered natural fuel patterns and the capacity of the native vegetation to respond to fire (Veblen et al., 2003). The European gorse (*Ulex europaeus*) is now a major component of roadside landscapes in south-central Chile, and the European rose (*Rosa rubiginosa*) is widespread in disturbed habitats. In northern Patagonia, the European broom *(Sarothamnus scoparius)* is common along roadsides and is highly flammable. Similarly, Douglas fir *(Pseudotsuga menziesii)* has naturalized from timber and ornamental plantings and is a common invader along trails and abandoned logging roads in the mesic *Nothofagus dombeyi* forest.

13.5 Recent Changes in the Native Forest Resource

The forest cover of the southern Andean region has been greatly altered in character and extent during the past century of widespread European settlement. Climatic variation and human activities have interacted synergistically to shape the modern forested landscape of southern Chile and Argentina. In addition to the relatively obvious impacts of humans on the forested landscape through forest clearance and plantation of exotic timber species (discussed below), humans have affected the native forest landscape through altered fire regimes and the effects of introduced animals. However, these more subtle human influences are highly dependent on climatic variation. For example, in northern Patagonia, Argentina, tree-ring and instrumental climatic records show that the 1910s, 1940s, and post-1976 were warmer and drier than the long-term average climatic conditions (since about 1900 for instrumental records; since about 1600 for tree-ring records; Villalba et al., 2003). It is precisely during these warm dry periods that rapid changes in the forested landscape have resulted from widespread and severe fires, sometimes followed by herbivory from livestock (Veblen et al., 2003). During these periods of rapid landscape change, burned and grazed forests have been replaced by shrublands, which only very slowly succeed back toward a forest cover. Once a site is converted from tall forest to shrubland in northern Patagonia, the structure of the fuels makes it much more susceptible to fire. Thus, there is a positive feedback that results in increased probability of subsequent fires once a site is burned initially. Under a future, probably warmer, climate (consistent with the warming trend that has affected much of Patagonia since the mid-1970s) and of increased human-set fires, it is likely that the rate of fire-induced conversion from forest to tall shrubland will increase.

Less subtle forest conversion and degradation, due to logging and replacement of the native forest by plantations of exotic trees, are a great threat to the native forests of the southern Andes. Destruction of native forests in the southern Andean region is occurring at an alarming rate and poses a major challenge to resource managers and conservationists. In southern Chile, the destruction of the native forest began on a small scale with Spanish colonization in the sixteenth century and forest clearance around the coastal outposts (Lara et al., 1996). As early as 1599, the exploitation of *Fitzroya cupressoides* achieved great commercial importance in southern Chile because of the use of this decay-resistant wood in naval construction (Veblen et al., 1995). In the mid-nineteenth century, the lowlands of the Chilean Lake District were settled by European colonists who converted vast areas of native forest to agriculture. Other than the post-1960 period of rapid conversion of Neotropical forests, this late nineteenth-century deforestation of the lowlands of south-central Chile stands out as one of the most rapid and widespread deforestation events in the history of Latin America. From the late nineteenth century through to the 1950s, large areas of native forest, especially *Nothofagus* forests, were burned and cleared farther south in Chile and in Argentina in attempts to create livestock pasture. Thus, a large percentage of the native forests of the southern Andean region has been converted to crop or relatively unproductive livestock pasture without any significant use of the timber resource.

Timber utilization in southern Chile, although initiated on a small scale early in the Spanish colonial period, was not a major land use until after about 1912, when railroads linked the Chilean Lake District to markets in central Chile. Completion of trunk railroad lines in the 1930s accelerated timber utilization, which continues to result in native forest degradation or elimination over large areas of southern Chile. Forest utilization during this period consisted largely of repeated cycles of selective logging of the best-formed individuals of the most valuable tree species. Because most of the preferred timber species, such as *Nothofagus* species, are shade-intolerant and require large canopy openings for successful regeneration, selective logging has generally not resulted in adequate regeneration of the desirable timber species. On the other hand, stands of fast-growing *Nothofagus* have developed where forests were burned or cut in larger patches, and these stands can be managed effectively on a sustained-yield basis. However, most exploitation of the native forest of southern Chile continues to be unsustainable, and in many cases consists of complete replacement of the native forest by plantations of exotic conifers (Lara et al., 1996).

In Chile, beginning in 1974, national forest policy provided economic incentives for establishing plantations of exotic tree species, such that the area of plantations grew from 290,000 ha in 1974 to 1.45 million ha by 1990, by which time Chile could boast the largest area planted to Monterey pine (*Pinus radiata*) of any country in the world (Lara and Veblen, 1992). Although this rapid growth of plantation forestry has often been regarded positively from a limited economic perspective, there are many social and environmental consequences that are negative (Lara and Veblen, 1992). The plantations of exotic trees expanded mostly at the expense of the native forest in Chile, including many young stands of considerable future timber value. For example, between 1978 and 1987, nearly one third of the native forest area of the coastal range from roughly 36 to 38°S was converted to plantation forest. Detrimental effects of this massive conversion from diverse native forests dominated mainly by broadleaved trees (angiosperms) to monocultures of exotic conifers include (1) loss of plant species diversity; (2) habitat loss and decreased diversity and population sizes for native wildlife; (3) loss of valuable native timber sources; (4) soil degradation due to the acidifying effects of the conifers; (5) loss of many non-timber forest products (edible fruits, forage, and shelter for livestock, medicinal plants, fuelwood) that local human populations previously utilized; (6) increased fire hazard due to the continuous extent of highly flammable conifer stands; and (7) decreased water yield and quality for rural and urban populations. In addition, these industrial forests have primarily benefited large companies and land owners, whereas small property owners were often coerced to sell their lands and migrate to urban areas in search of employment. Lack of capital, credit, and technical assistance has largely excluded small landowners from industrial forestry, while the highly mechanized industrial forestry sector has not provided adequate new employment opportunities (Lara and Veblen, 1992).

In northern Patagonia, Argentina, the plantation of fast-growing conifers from North America, including ponderosa pine (*Pinus ponderosa*), lodgepole pine (*Pinus contorta*), and Douglas-fir (*Pseudotsuga menziesii*), constitutes a rapidly growing land use. The rate of plantation is estimated at 10,000 ha of new plantations per year, and it has been suggested that there are 700,000 to 2,000,000 ha in Patagonia that could be converted to plantations. Plantations are established in the steppe, as well as in different types of native Andean forest. The typical habitat in which exotic conifers are being planted in Patagonia is at the ecotone between the Patagonian steppe and relatively xeric woodlands of native trees such as the conifer *Austrocedrus chilensis*. Extensive areas of native vegetation have been replaced by plantations of exotic trees resulting in a decline in suitable habitat for much of the native wildlife. Such plantations tend to be more favorable for the establishment of weedy exotic plants than for native plants, further reducing the native biodiversity of the region.

Although the initial motive for planting exotic conifers in the ecotone of the Patagonian steppe and Andean forests was timber production, international and national policies to mitigate carbon dioxide releases into the atmosphere are increasingly important sources of motivation. Widespread tree planting today is partially fueled by the belief that it constitutes long-term carbon sequestration from the atmosphere. However, in the case of northern Patagonia, the plantations are located in environments that, due to the regional climate, are highly prone to wildfire, which will return the carbon to the atmosphere. Furthermore, little is known about how much carbon is stored below ground by the native vegetation in comparison to carbon storage by planted conifers.

The landscape of temperate forests of the southern Andean regions has been shaped by a long history of natural and anthropogenic influences. In recent decades, anthropogenic factors such as timber exploitation and conversion to plantations of introduced trees have been the main causes of the loss of the native forests. The pressure on native forests from these factors is unlikely to abate in the foreseeable future. Although the native forests of the southern Andes obviously contain far fewer species than Neotropical forests farther north, they are of exceptional biodiversity value because of the large number of endemic genera and of families represented by a single genus or a few genera. Likewise, from a multiple-use resource perspective, the native forests of the southern Andes provide a wealth of ecosystem services and products from wildlife habitat and watershed protection to production of high-quality wood. Thus, there are strong justifications for developing strategies of managing these forests for sustained yields and to avoid their conversion to monocultures of exotic species.

References

Arroyo, M.T.K., M. Riveros, A. Peñaloza, L. Cavieres, and A.M. Faggi, 1996. Phytogeographic relationships and regional richness patterns of the cool temperate rainforest flora of southern South America. In: R.G. Lawford, P. Alaback, and E.R. Fuentes (Editors), *High Latitude Rain Forests and Associated Ecosystems of the Western Coast of the Americas: Climate, Hydrology, Ecology and Conservation.* Springer-Verlag, New York, 134–172.

Daniels, L.D., and T.T. Veblen, 2000. ENSO effects on temperature and precipitation of the Patagonian-Andean region: Implications for biogeography. *Physical Geography,* 21, 223–243.

Donoso, C., 1996. Ecology of *Nothofagus* forests in central Chile. In: T.T. Veblen, R.S. Hill, and J. Read (Editors),

Ecology and Biogeography of Nothofagus Forests. Yale University Press, New Haven, 271–292.
Hervé, F., A. Demant, V.A. Ramos, R.J. Pankhurst, and M. Suárez, 2000. The Southern Andes. In: U.G. Cordani, E.J. Milani, A.T. Filho, and D.A. Campos (Editors), *Tectonic Evolution of South America.* 31st International Geological Congress, Rio de Janeiro, 605–634.
Heusser, C. J., 1987. Fire history of Fuego-Patagonia. *Quaternary of South America and Antarctic Peninsula* 5,93–109.
Heusser, C.J., 1994. Paleoindians and fire during the late Quaternary in southern South America. *Revista Chilena de Historia Natural* 67,435–442.
Hill, R.S., and M.E. Dettmann, 1996. Origin and diversification of the genus *Nothofagus.* In: T.T. Veblen, R.S. Hill, and J. Read (Editors), *Ecology and Biogeography of Nothofagus Forests.* Yale University Press, New Haven, 11–24.
Huber, U. and V. Markgraf, 2003. Holocene fire frequency and climate change at Rio Rubens bog, southern Patagonia. In: T.T. Veblen, W.L. Baker, G. Montenegro and T.W. Swetnam, (Editors), *Fire Regimes and Climatic Change in Temperate Ecosystems of the Western Americas.* Springer-Verlag, New York, 357–380.
Innes, J.L., 1992. Structure of evergreen temperate rain forest on the Taitao Peninsula, southern Chile. *Journal of Biogeography,* 19, 555–562.
Jaksic, F.M., and J.L. Yáñez, 1983. Rabbit and fox introductions in Tierra del Fuego: History and assessment of the attempts at biological control of the rabbit infestation. *Biological Conservation,* 26, 145–156.
Kitzberger, T., and T.T. Veblen, 2003. Influences of climate on fire in northern Patagonia, Argentina. In: T.T. Veblen, W.L. Baker, G. Montenegro, and T.W. Swetnam (Editors). *Fire Regimes and Climatic Change in Temperate Ecosystems of the Western Americas.* Springer-Verlag, New York, 296–321.
Kitzberger, T., T.T. Veblen, and R. Villalba, 1997. Climatic influences on fire regimes along a rainforest-to-xeric woodland gradient in northern Patagonia, Argentina. *Journal of Biogeography,* 24, 35–47.
Lara, A., C. Donoso, and J.C. Aravena, 1996. La conservación del bosque nativo de Chile: Problemas y desafios. In: J.J. Armesto, C. Villagran, and M.K. Arroyo (Editors), *Ecología de los Bosques Nativos de Chile.* Editorial Universitaria, Santiago, 335–362.
Lara, A., A. Wolodarsky-Franke, J.C. Aravena, M. Cortés, S. Fraver, and F. Silla, 2003. Fire regimes and forest dynamics in the lake region of south-central Chile. In: T.T. Veblen, W.L. Baker, G. Montenegro, and T.W. Swetnam (Editors), *Fire Regimes and Climatic Change in Temperate Ecosystems of the Western Americas.* Springer-Verlag, New York, 322–342.
Lara, A., and T.T. Veblen, 1992. Forest plantations in Chile: A successful model? In: A. Mather (Editor), *Afforestation: Policies, Planning and Progress.* Belhaven Press, London, 118–139.
Markgraf, V., E. Romero, and C. Villagrán, 1996. History and paleoecology of South American *Nothofagus* forests. In: T.T. Veblen, R.S. Hill, and J. Read (Editors), *Ecology and Biogeography of Nothofagus Forests.* Yale University Press, New Haven, 354–386.
Premoli, A.C., T. Kitzberger, and T. Veblen, 2000. Isozyme variation and recent biogeographical history of the long-lived conifer *Fitzroya cupressoides. Journal of Biogeography,* 27, 251–260.
Ramírez, C., R. Godoy, W. Eldridge, and N. Pacheco, 1981. Impacto ecológico del ciervo rojo sobre el bosque de Olivillo en Osorno, Chile. *Anales del Museo de Historia Natural,* 14, 197–212.
Rapoport, E.H., and C. Brión, 1991. Malezas exóticas y plantas escapadas de cultivo en el noroeste patagónico: Segunda aproximación. *Cuadernos de Alternatura* (Bariloche), 1, 1–19.
Rebertus, A.J., T. Kitzberger, T.T. Veblen, and L.M. Roovers, 1997. Blowdown history and landscape patterns in the Andes of Tierra del Fuego, Argentina. *Ecology,* 78, 678–692.
Veblen, T.T., 1982. Growth patterns of *Chusquea* bamboos in the understory of Chilean *Nothofagus* forests and their influences in forest dynamics. *Bulletin of the Torrey Botanical Club,* 109, 474–487.
Veblen, T.T., and P.B. Alaback, 1996. A comparative review of forest dynamics and disturbance in the temperate rain forests of North and South America. In: R.G. Lawford, P. Alaback, and E.R. Fuentes (Editors), *High Latitude Rain Forests and Associated Ecosystems of the West Coast of the Americas: Climate, Hydrology, Ecology and Conservation.* Springer-Verlag, New York, 173–213.
Veblen, T.T., and D.H. Ashton, 1978. Catastrophic influences on the vegetation of the Valdivian Andes. *Vegetatio,* 36, 149–167.
Veblen, T.T., B.R. Burns, T. Kitzberger, A. Lara, and R. Villalba, 1995. The ecology of the conifers of southern South America. In: N. Enright and R. Hill (Editors), *Ecology of the Southern Conifers.* Melbourne University Press, Melbourne, Australia, 120–155.
Veblen, T.T., C. Donoso, T. Kitzberger, and A.J. Rebertus, 1996. Ecology of southern Chilean and Argentinean *Nothofagus* forests. In: T.T. Veblen, R.S. Hill, and J. Read (Editors), *Ecology and Biogeography of Nothofagus Forests.* Yale University Press, New Haven, 293–353.
Veblen, T.T., T. Kitzberger, R. Villalba, and J. Denevan, 1999. Fire history in northern Patagonia: The roles of humans and climatic variation. *Ecological Monographs,* 69, 47–67.
Veblen, T.T., T. Kitzberger, E. Raffaele, and D.C. Lorenz, 2003. Fire history and vegetation change in northern Patagonia, Argentina. In: T.T. Veblen, W.L. Baker, G. Montenegro, and T.W. Swetnam (Editors), *Fire Regimes and Climatic Change in Temperate Ecosystems of the Western Americas.* Springer-Verlag, New York, 265–295.
Veit, H., and K. Garleff, 1996. Evolución del paisaje cuaternario y los suelos en Chile Central-Sur. In: J.J. Armesto, C. Villagrán, M.K. Arroyo (Editors), *Ecología de los Bosques Nativos de Chile.* Editorial Universitaria, Santiago, Chile, 29–49.
Villalba, R., 1994. Tree-ring and glacial evidence for the Medieval Warm Epoch and the Little Ice Age in southern South America. *Climate Change,* 26, 183–197.
Villalba, R., E.R. Cook, G.C. Jacoby, R.D. D'Arrigo, T.T. Veblen, and P.D. Jones, 1998. Tree-ring based reconstructions of precipitation in Patagonia since A.D. 1600. *Holocene,* 8, 659–674.
Villalba, R., A. Lara, J.A. Boninsegna, M. Masiokas, S. Delgado, J.C. Aravena, F.A. Roig, A. Schmelter, A. Wolodarsky, and A. Ripalta, 2003. Large-scale temperature changes across the Southern Andes: 20th-century variations in the context of the past 400 years. *Climatic Change,* 59, 177–232.

14

The Grasslands and Steppes of Patagonia and the Río de la Plata Plains

José María Paruelo
Estebán G. Jobbágy
Martín Oesterheld
Rodolfo A. Golluscio
Martín R. Aguiar

The Patagonian steppes and the Río de la Plata grasslands occupy a vast proportion of the plains, plateaus, and hills of southern South America, and are characterized by the almost absolute absence of trees. Prairies and steppes (grass and low shrubs) are the dominant physiognomic types, and forests are restricted to some riparian corridors. Savannas become important only in the ecotones of these regions, whereas meadows may be locally important under particular topographic or edaphic conditions. The Río de la Plata grasslands (RPG), one of the most important grassland regions in the world, extend between 28°S and 38°S latitude, covering about 700,000 km^2 of eastern Argentina, Uruguay, and southern Brazil (fig. 14.1). The boundaries of these grasslands include the Atlantic coastline to the east, dry temperate forests to the south and west, and subtropical humid forests to the north. Woody vegetation within the region is restricted to small areas near water bodies, such as the gallery forests along the large Paraná and Uruguay rivers and their tributary streams. The Patagonian steppes occupy the southern tip of the continent from approximately 40°S, and are framed by the Andes to the west and the Atlantic coast to the east and south and cover more than 800,000 km^2 of Chile and Argentina. Toward the west, the region displays a sharp ecotone with the subantarctic forests, whereas to the north it grades into a broad zone of Monte scrublands in central Argentina. The RPG and the Patagonian steppes are separated by a wide strip of woody vegetation, the Monte and Espinal phytogeographic units (see chapter 10; Cabrera and Willkins, 1973).

In this chapter, we describe the heterogeneity and main characteristics of the dominant ecosystems of the Patagonian steppes and the RPG, focusing on environmental controls and human-induced changes. Although numerous criteria have been applied to describe the internal heterogeneity of both regions, we emphasize here the structural and functional attributes of vegetation as integrators of climate, physiography, and land use.

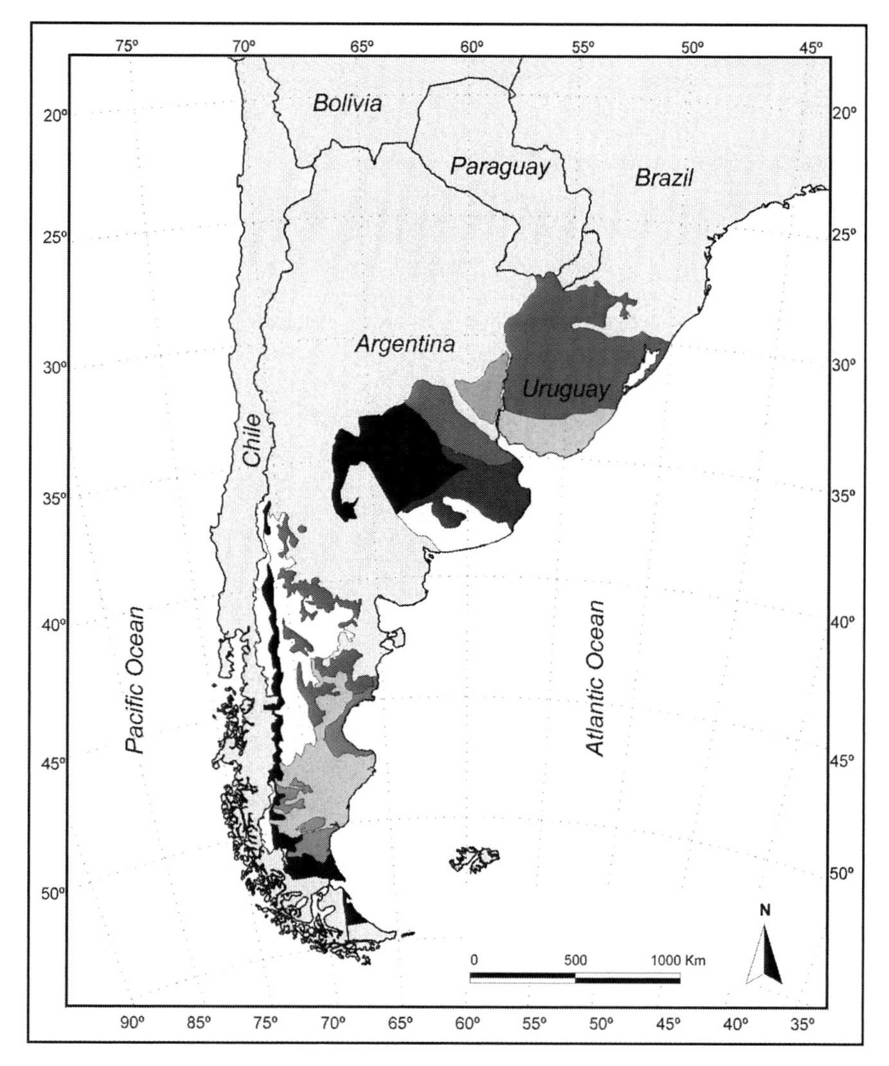

Figure 14.1 Río de la Plata grasslands and Patagonian steppes, based on the boundaries defined by Soriano (1991) and León et al. (1998)

14.1 Climate, Physiography, and Soils

14.1.1 General Climate

A key climatic feature of southern South America and other temperate regions of the Southern Hemisphere that distinguishes them from their Northern Hemisphere counterparts is the relatively low thermal amplitude caused by the low ratio of land mass to ocean in this portion of the world. This contrast may control the differences between the Southern and Northern Hemispheres in the structure and functioning of their respective prairies and steppes (Paruelo et al., 1995, 1998a). Both Patagonia and the RPG lie between the semi-permanent anticyclones of the South Pacific and South Atlantic Oceans (centered around 30°S) and the subpolar low-pressure belts around 60°S (see chapter 3; Prohaska, 1976, Paruelo et al., 1998b). During the southern summer, the RPG are also influenced by a low-pressure center developed over northern Argentina (25°S), probably associated with the Intertropical Convergence Zone (Frere et al., 1978). Although both areas can be classified as temperate, the RPG experience subtropical influences toward their northern boundary.

The western sector of the South Atlantic anticyclone conveys most of the humidity received by the RPG. Warm humid air masses generated within this anticyclone interact on their way toward the region with dry polar air masses coming from the southwest that travel over the dry areas of Patagonia. Such interaction generates most of the frontal precipitation over the area. The lack of major orographic features that could alter the course of these air masses results in a gentle NE-SW precipitation gradient that ranges from 1,500 mm/yr in southern Brazil to 600 mm/yr in central Argentina (fig. 14.2). Conditions change from humid to semiarid across this gradient, leading to important changes in vegetation and land uses (Soriano 1991). While the more continental areas toward the north and west of the RPG have rainfall maxima in spring and fall and a minimum in winter (fig. 14.2), heavy winter rainstorms associated with cyclogenesis over the central part of the area and strong southeasterly winds (*sudestada*) balance rainfall seasonality toward the south and east of the region (Prohaska, 1976). In the northeastern portion of the region (southern Brazil and eastern Uruguay), the Atlantic anticyclone determines an increase in winter precipitation. Mean annual temperatures increase northward, from 14ºC to 19ºC. Although winters are relatively mild and snowfall is infrequent or nonexistent, frosts are common and may extend well into spring and fall in the southern part of this region.

Pacific air masses have an overriding influence on the Patagonian climate. Strong, constant westerly winds dominate the region. The seasonal displacement of the low- and high-pressure systems and the equatorward ocean current determine the seasonal pattern of precipitation in Patagonia. During winter, the higher intensity of the subpolar low, the northward displacement of the Pacific high, and higher ocean temperatures relative to the continent determine an increase in precipitation over the region. The result is a clear winter-distribution pattern of precipitation over most of the area (fig. 14.2). In Patagonia, 46% of precipitation falls in winter (Jobbágy et al., 1995). Where the influence of Atlantic air masses has some importance (toward Patagonia's northern and southern extremes), precipitation is more evenly distributed over the year. The north-south orientation of the Andes on the western border of Patagonia plays a crucial role in determining the latter's climate. The Patagonian steppes are located in the rain shadow of mountains that impose an important barrier to humid air masses coming from the Pacific Ocean. Humidity is released on the western slopes of the Andes and air masses entering the Patagonian steppe become warmer and drier through adiabatic processes as they descend the eastern slopes.

The characteristics outlined above result in a strong west-east gradient of precipitation across the region (Barros and Mattio, 1979), with total annual precipitation decreasing exponentially east of the Andes. For the areas that are not directly influenced by the Atlantic, more than 90% of the mean annual precipitation variation over the region is accounted for by the distance from the Andes (Jobbágy et al., 1995). Within a precipitation gradient that may reach more than 4,000 mm/yr in the western subantarctic forests, the Patagonian steppes occur where precipitation levels are less than 600 mm/yr in the north and less than 350 mm in the south, with most of the intervening region receiving less than 200 mm (Paruelo et al., 1998b). Winter distribution of precipitation results in a strong water deficit in summer (Paruelo and Sala, 1995; Paruelo et al., 2000). According to the bioclimatic classification of Le Houreou (1996) (based on the ratio of potential evapotranspiration to mean annual precipitation), more than 55% of Patagonia is arid or hyperarid and only 9% subhumid (Paruelo et al., 1998b). As precipitation increases westward, the summer deficit decreases and has a later onset in the season, an aspect that is mirrored by vegetation phenology (Jobbágy et al., 2002).

Thermal belts in Patagonia succeed one another from northeast to southwest, following the effects of increasing latitude and altitude. Mean annual temperature ranges from 12°C along the northern margins of the region to 3°C in Tierra del Fuego. The strong westerly winds that blow over Patagonia reduce the perceived temperature (wind chill), on average 4.2°C. The wind-chill effect is more pronounced in summer, generating the cool (or even cold) summers that characterize the Patagonian climate (Coronato, 1993).

The Patagonia and RPG regions have their closest climatic analogues in North America in the intermountain zone in the western United States and in the humid portion of the tallgrass priairie, respectively (Paruelo et al., 1995). Patagonia and the Intermountain West both experience relatively low mean annual precipitation (150–500 mm MAP) and temperature (0 to 12°C MAT) (Adler et al., 2006). Most of the Río de la Plata grasslands occur in areas characterized by much higher precipitation (>1,000 mm MAP) and temperatures (15–20°C MAT), and a lack of clear seasonality in precipitation. Areas displaying such climates in North America correspond to the transition between grasslands and forests. The lack of such a transition in South America, and the presence of grasslands under climatic conditions that would appear to support forest vegetation based on relationships observed in the Northern Hemisphere, have puzzled plant geographers and ecologists for at least a century (Parodi, 1942; Ellenberg, 1962; Walter, 1968).

14.1.2 Landforms, Geologic History, and Soil Gradients

Perhaps the most significant event shaping the landscapes that host the Rio de la Plata grasslands and the Patagonian steppes has been the uplift of the Andes, the most recent phase of which began in middle Miocene time, around 14 million years ago (see chapter 1). In a region of strong westerly winds, this massive barrier has modified the climate leeward of the mountains, creating conditions favorable to desiccating winds and desertification. Meanwhile, rivers and winds have transported large quantities of sediment from the eroding mountains toward the plains and plateaus farther east. There, aided by local alluvial activity and soil-forming processes, these sediments have come to shape the distinctive landscapes of the Patagonian steppes, the Pampas, and parts of the Campos regions in the Río de la Plata plains (Teruggi, 1957; Zarate, 2003).

Most of the Río de la Plata grasslands occur over a vast plain, the Pampas, formed by thick Quaternary loess deposits that have experienced varying degrees of local reworking. Exceptions to this general pattern are a few isolated uplands in Argentina, where Precambrian-Paleozoic crystalline rocks have not been fully buried by sediment, and

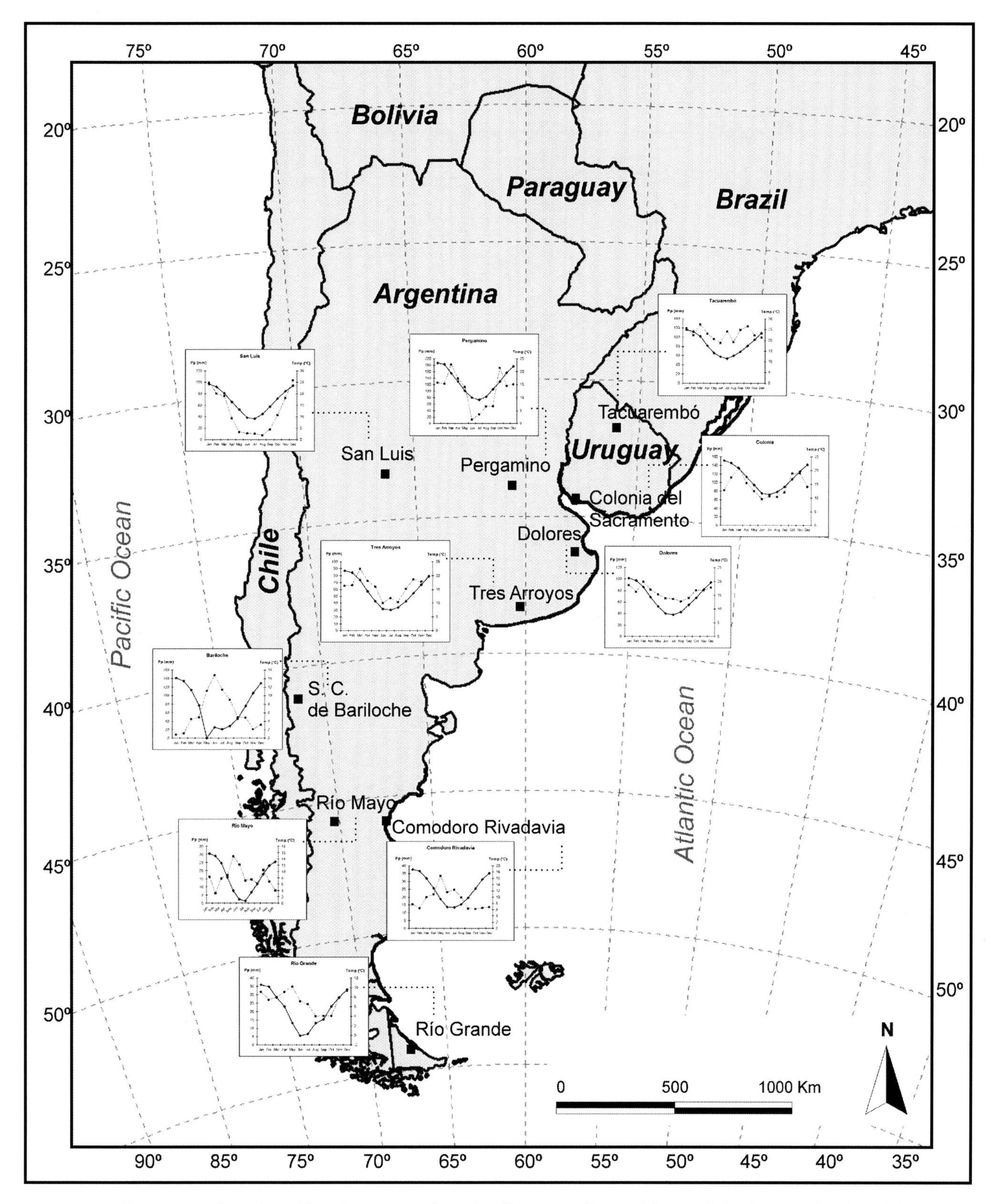

Figure 14.2 Climatograms for selected locations across the main climatic gradients of the Río de la Plata grasslands and the Patagonian steppes

most of the Uruguayan and Brazilian portions of the region, where a diverse array of rocks such as Precambrian granite, Carboniferous sandstone, and Jurassic basalt is exposed to surface and soil-forming processes.

The dominant loessic sediments of the Pampas are characterized by their unconsolidated nature, silt to fine-sand texture typical of wind-transported material, and large contributions of volcanic material. Most primary loess has been subsequently reworked by winds and streams, and by pedogenesis (Zarate, 2003). In general, these sediments are younger and coarser towards the west, where past and present climates have been more xeric and dust sources less remote. Towards the east, these loess mantles are thinner and older, becoming restricted to hill tops and fluvial terraces to the east of the Uruguay River (Panario and Gutierrez, 1999). The relatively young and flat aeolian landscape of the Pampas constrains surface-water transport, generating shallow groundwater tables under present humid conditions. This translates into a dense network of shallow and stagnant water bodies that expand over a significant proportion of the territory during wet years (Tricart, 1973). This hydrological context has made flooding and salt accumulation key driving forces for soil and ecosystem development.

Along a west-east gradient across the Pampas, soils developed on Quaternary sediment shift from mollic Entisols with incipient horizon differentiation to well-developed Mollisols with high clay accumulation in subsurface horizons (INTA, 1989). In general, these soils have high organic matter content and base saturation, posing few or no constraints for agriculture (see chapter 7). Low topographic sites host sodic and/or hydromorphic soils (natric and aquic Mollisols to Alfisols), which dominate the landscape matrix in the flattest portions of the region. Similar soil-landscape situations have been described for the plains in Hungary (Toth and Jozefaciuk, 2002) and eastern Siberia (Bazilievich, 1970). In the Uruguayan and Brazilian Campos, a more complex array of parent materials and topographic forms has resulted in higher soil heterogeneity. While well developed Mollisols similar to those of the eastern Pampas are frequent here, widespread shallow soils over rock outcrops (Entisols) and profiles with poor base-saturation are a distinctive feature (Alfisols, Inceptisols, and Oxisols in the northern edge of the region) (Duran, 1991).

Plateaus or *mesetas* are among the most characteristic features of the Patagonian steppe, decreasing in elevation from the eastern foothills of the Andes towards the Atlantic coast. Volcanic forms and sierras occur within these plateaus in northern Patagonia and central Patagonia, respectively. An earlier Cenozoic warm moist period in Patagonia, suggested by fossil evidence and large coal deposits, ended when uplift of the Andes in the Miocene imposed a barrier to the transport of moisture from the Pacific Ocean by the prevailing westerly winds (Volkheimer, 1983). Major glaciations of the Andes probably began in the Pliocene and were broadly contemporary with volcanic activity, resulting in interbedded tills and basalts (Mercer, 1976; Strelin et al., 1999). A complex interaction among Andean glaciation, volcanism, crustal adjustments, and downstream glaciofluvial and fluvial activity has produced Patagonia's present landscape of plateaus and terraces, one of whose most striking features, as noted by early naturalists like Darwin, is the widespread occurrence of pebbles or *rodados patagónicos* (Darwin, 1842; Strelin et al., 1999). Glacier expansion during Pleistocene cold stages reached its maximum around 170,000 years ago but was confined to the Andes and their foothills (Clapperton, 1994). However, its indirect influence farther east is seen in the legacy of glaciofluvial processes on the Patagonian steppe and of loess on the Pampas.

With few exceptions, the drainage network of Patagonia consists of several west-east running rivers that drain the humid slopes of the Andes and cross the steppes on their way to the Atlantic, maintaining riparian systems across the whole region. In addition to riparian environments, the Patagonian landscape hosts a profuse network of flood meadows or *mallines* associated with groundwater discharge and low-order water courses (Iriondo et al., 1974). In the western steppes, these units can occupy as much as 5% of the landscape, providing a crucial resource to wildlife and domestic herbivores (Paruelo et al., 2004). *Mallines* are supported on water drained from the steppes, resulting mainly from winter precipitation (Paruelo et al., 2000).

Glacial detritus and volcanic materials represent the major soil parent materials of the Patagonian steppe. Mollic soils are common in the more humid and cold plateaus of the western steppes, dominated by grasses in their pristine condition. Toward the east, under drier and warmer climates, Aridisols and Entisols dominate the landscape and thick cemented calcic horizons are widespread (del Valle, 1998). Rounded pebbles and gravels associated with glaciofluvial processes are characteristic of the Patagonian steppe soils and responsible for the formation of extensive desert pavements where wind erosion has been able to remove finer materials. Buried soil horizons are common throughout the area, reflecting the influence of past climatic conditions on pedogenesis. In flooded meadows, soils tend to be rich in organic matter, fitting the definition of Mollisols with hydromorphic and/or sodic characteristics, depending on their hydrological balance.

14.2 Vegetation

14.2.1 Main Vegetation Units of the Río de la Plata Grasslands

The major internal subdivisions of the RPG were defined and described by Rolando León (in Soriano 1991), based on geomorphic, hydrologic, and edaphic features and their linkages with natural vegetation and land use (fig. 14.3).

Such characterization of the heterogeneity of the area summarizes and integrates information of the flora at both regional (Vervoorst, 1967; Cabrera and Willkins, 1973) and local scales (León et al., 1979; Collantes et al., 1981; Lewis et al., 1985; Faggi, 1986; Burkart et al., 1998, 1990; Batista et al., 1988; Cantero and León, 1999; Perelman et al., 2001). We describe here the natural vegetation of these regions based mainly on that seminal description.

The four northern subdivisions share several features in common, particularly hilly relief and the existence of a well-defined drainage system (fig. 14.3). The Rolling Pampa is now covered largely by annual crops, with a predominance of soybean. Although its original vegetation can only be inferred from a few isolated remnants of unploughed vegetation and from early, but postcolonial, botanical descriptions, it is clear that these prairies were formerly dominated by tussock grasses that covered most of the ground. Dominants comprise several warm-season (C_4) and cool-season (C_3) grasses in approximately similar proportion. The most common genera among the grasses are *Stipa*, *Piptochaetium*, *Paspalum*, and *Bothriochloa*. Shrubs are little represented, but in some places, probably as a result of disturbance, one of several species of Compositae *(Baccharis and Eupatorium)* may become locally dominant. Small dicots grow in between grasses and are favored by grazing. The Southern Campos are floristically very similar to the Rolling Pampa, but the influence of agriculture has been lower, probably due to relatively shallow soils (fig. 14.4). The Mesopotamic Pampa, located between the Paraná and Uruguay Rivers, is similar in relief to the two previous units, but its soils have a finer texture. Vegetation is also similar, but is enriched in subtropical elements (warm-season grasses such as *Axonopus* and *Schizachyrium*). Cool-season grasses are still present, but rarely become dominant. Finally, the Northern Campos are structurally similar to the other northern subregions, but here species composition becomes even more enriched in subtropical species (*Andropogon*). Soriano et al. (1991) distinguished several features within this subregion, such as the strong impact of grazing on species composition, the response of vegetation heterogeneity to topography, and the flat areas or *malezal* subject to extended flooding west of the Uruguay River. Phytogeographical analysis of that portion of the Northern Campos associated with basaltic substrate has revealed a mosaic of three communities (hydrophytic and mesophytic prairies, and grass steppes), the spatial distribution of which is mainly related to soil depth (Lezama et al., 2004).

The three southern subdivisions of the RPG, all in Argentina, comprise the Flooding Pampa, the Inner Pampa,

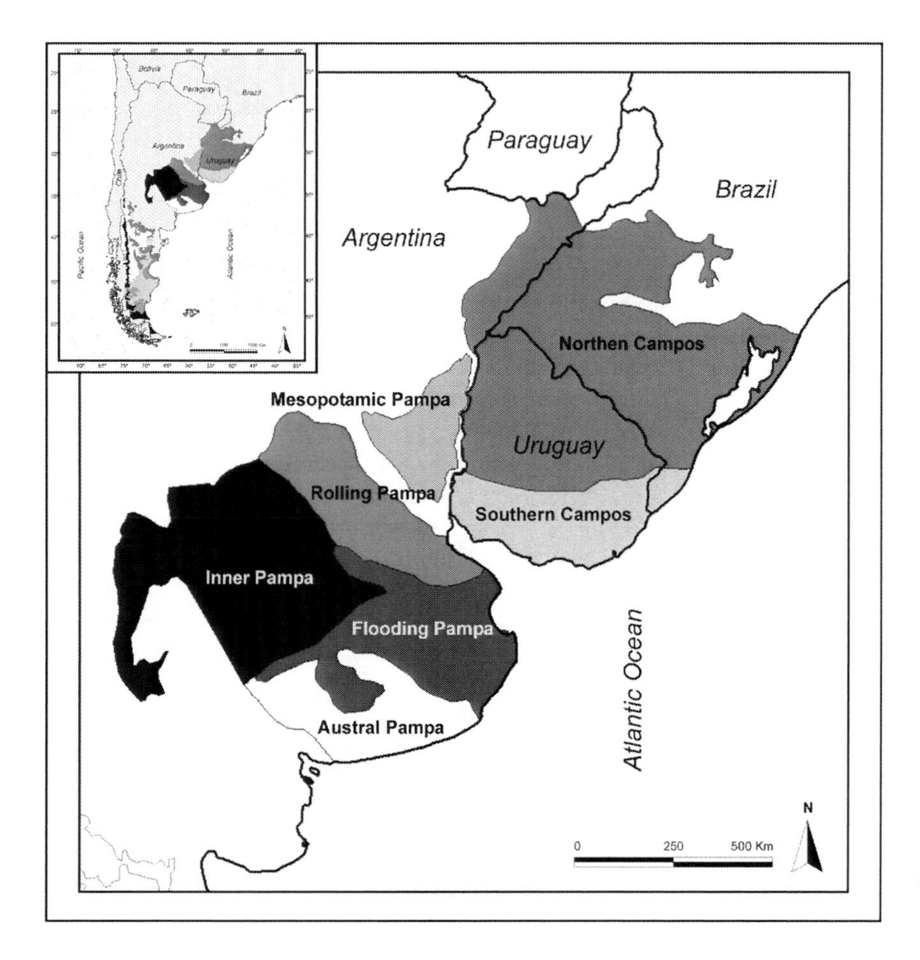

Figure 14.3 Main subdivisions of the Rio de la Plata grasslands, as defined by R.J.C. León in Soriano (1991)

Figure 14.4 Grazed prairie of the Southern Campos of the Río de la Plata grasslands (photo: J.M. Paruelo)

and the Austral Pampa (fig. 14.3). The Flooding Pampa is composed of depressions associated with the Salado and Laprida river basins in Buenos Aires Province. Extensive areas with virtually no slope give the impression of homogeneity, but a closer look reveals an intrinsic mosaic of plant communities. A large proportion of the variation of plant species composition is included in small 0.1–10 km^2 areas. Minimal topographic variations and patches with high soil salinity are responsible for this pattern. Five vegetation units are associated with these environmental variations (Perelman et al., 2001). Three of them correspond to a topographic gradient, from small, slightly elevated hills that are never flooded (mesophytic meadows, with *Stipa trichotoma, S. charruana, Diodia dasycephala, Sida rhombifolia*), to flat areas occasionally flooded (humid mesophytic meadows, with *Mentha pulegium, Leontodon taraxacoides*), to extended lowlands regularly flooded for several months (humid prairies, with *Leersia hexandra, Paspalidium paludivagum*). The other two vegetation units correspond to halophytic steppes and humid halophytic steppes (with *Distichlis spicata, D. scoparia, Sporobolus pyramidatus, Sida leprosa, Salicornia*), which occupy soils with sodic salinity frequently influenced by flooding. As a result of the strong constraints generated by the flooding regime and the high salinity of some soils, agriculture is restricted to areas where mesophytic meadows should be found.

The Inner Pampa to the west also lacks an obvious drainage system, but soils have a coarser texture and, as a result, are not extensively flooded. They have been widely converted to agriculture. With less precipitation than the other two subregions, vegetation is less dense and hardly reaches a 100% cover. According to the history of grazing by livestock, these grasslands may be dominated by different native grass (*Sorghastrum pellitum, Elionurus muticus*) and perennial dicot species, or even by exotic weeds (Soriano, 1991).

The Austral Pampa in the south is associated with the ancient rocks and later sedimentary covers of the Sierra Tandilia and Sierra de la Ventana, and their slopes towards the Atlantic Ocean (Soriano, 1991). These hills rarely exceed 1,000 m and, except for rocky outcrops, the high fertility of their soils has facilitated the conversion of most of this subregion to extensive agriculture, with a predominance of wheat. The vegetation replaced by these crops was dominated by several tussock, cool-season grasses (*Stipa, Piptochaetium*), some common to other subregions. In contrast, the rocky areas have a distinct composition of grasses and dense areas dominated by shrubs.

14.2.2 Main Vegetation Units of Patagonia

The grasslands and steppes of Patagonia are very heterogeneous, both physiognomically and floristically. This high heterogeneity contradicts the common perception of Patagonia as a vast desert at the southern end of the world. Vegetation types range from semi-deserts to humid prairies with a large variety of shrub and grass steppes in between. This vegetation heterogeneity reflects the constraints imposed by the climatic, topographic, and edaphic features of the region. In the following paragraphs we describe the internal heterogeneity of the typical vegetation units (i.e., phytogeographical districts) of the Patagonian Phytogeographical Province (Soriano, 1956, see details in León et al., 1998) (fig. 14.5).

Grass steppes characterize the most humid districts of the region, the Subandean and Magellanic Districts, which are dominated by grasses of the genus *Festuca* (*F. pallescens* and *F. gracillima*, respectively), accompanied by several other grasses, highly preferred by native and exotic herbivores, and sometimes by shrubs (fig. 14.6). In the Subandean District (fig. 14.4), shrubs *(Mulinum spinosum, Senecio filaginoides, and Acaena splendens)* seem to be indicative of degradation by grazing (León and Aguiar, 1985; Bertiller et al., 1995). In the Magellanic District, shrubs are common constituents of the grass steppe, differing between the most xeric soils *(Nardophyllum bryoides,)* and the most humid and acidic soils *(Chilliotrichum diffusum and Empetrum rubrum)* (Collantes et al., 1999).

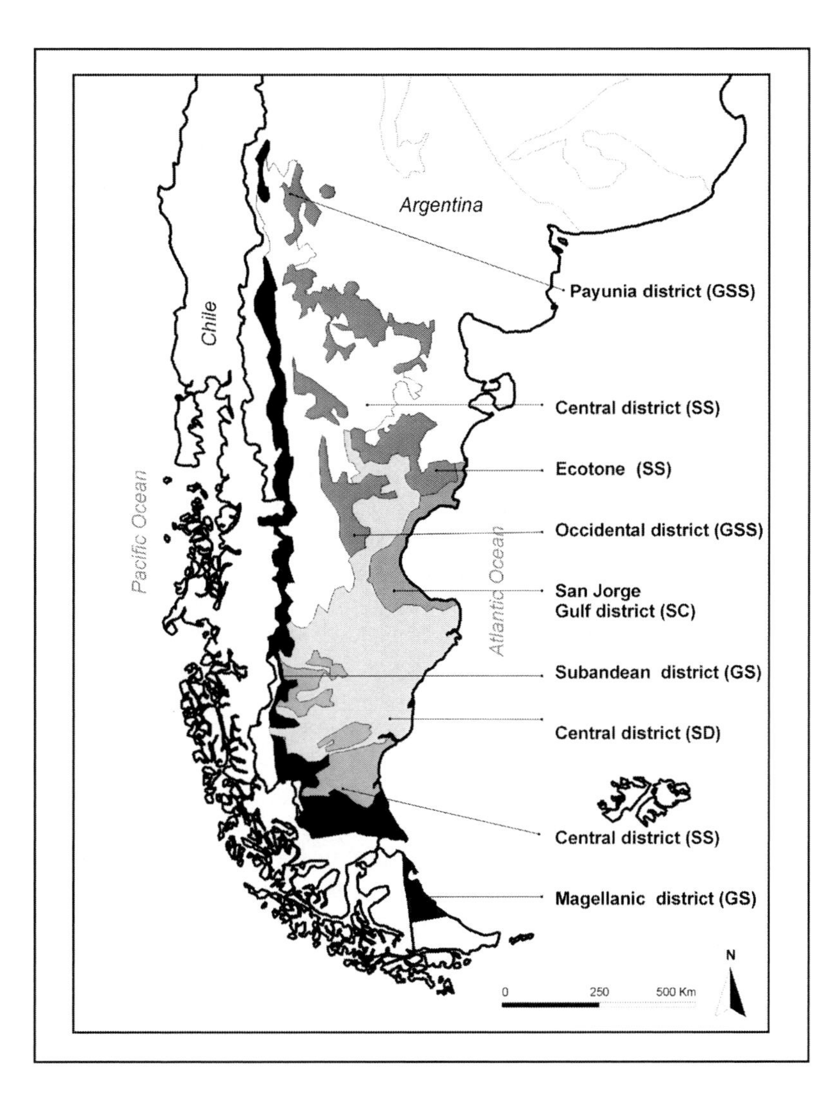

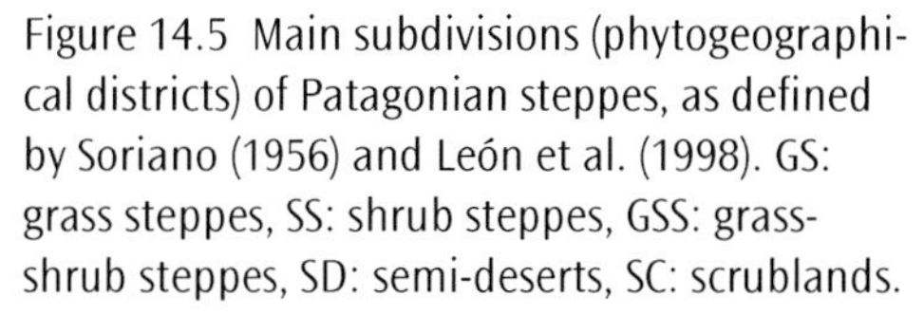

Figure 14.5 Main subdivisions (phytogeographical districts) of Patagonian steppes, as defined by Soriano (1956) and León et al. (1998). GS: grass steppes, SS: shrub steppes, GSS: grass-shrub steppes, SD: semi-deserts, SC: scrublands.

Figure 14.6 Grass steppe of Festuca pallescens in western Patagonia (Subandean District) (photo: J.M. Paruelo)

East of the Subandean District, shrub-grass steppes occupy the semiarid environments of the Occidental District and Payunia District. The vegetation of the Occidental District (fig.14.5) is a shrub-grass steppe dominated by the shrubs *Mulinum spinosum*, *Senecio filaginoides*, and *Adesmia volkmanni*, and the grasses *Stipa speciosa, S. humilis*, and *Poa ligularis* (fig. 14.7; Golluscio et al., 1982). As in the Subandean District, the north-south extent of the Occidental District only modifies the accompanying species. In contrast, the Payunia District is an intricate mosaic of shrub steppes whose dominant species vary according to edaphic substrate, elevation, and topography. Patagonian shrub species, as commonly found in the Occidental District, alternate with shrub species typical of the neighboring Monte Phytogeographical Province (see chapter 10).

Shrub-grass steppes also dominate most of the San Jorge Gulf District, where the local climate is strongly influenced by sea breezes coming off the Atlantic Ocean and by the humidity delivered by westerly winds passing over the large Musters and Colhue Huapi lakes (Coronato, 1996). This district includes two different vegetation units that are not distinguishable at the spatial resolution of figure 14.5. The slopes of the Montemayor, Pampa del Castillo, and Pampa de Salamanca plateaus are occupied by scrublands dominated by *Colliguaya integerrima* and *Trevoa patagonica,* two phanerophytes up to 2 m tall. In contrast, the plateau surfaces are occupied by shrub-grass steppes physiognomically similar to those of the Occidental District (Soriano, 1956; Bertiller et al., 1981).

The most arid environments of the Patagonian Province are included within the Central District (fig. 14.5). As precipitation is uniformly scarce, vegetation varies according to topography, soils, and temperature. The most extensive vegetation unit is a semi-desert dominated by the dwarf shrubs *Nassauvia glomerulosa, N. ulicina,* and *Chuquiraga aurea.* Plant height (<30 cm) and cover (<30%) are very low in these communities, which are often located on soils with heavy clay pans close to the surface and a highly unfavorable water balance. The rest of the district is occupied by shrub steppes, all of them growing on deeper and/or coarser soils, but each dominated by a different suite of shrub species. The most important of these shrub steppes are dominated by *Chuquiraga avellanedae* (<50 cm height; <50% cover), on the plateaus located at the northeast end of the district; by *Colliguaya integerrima* (150 cm height, 65% cover) on the basaltic hills located at the northwest end; by *Nardophyllum obtusifolium* (60 cm height; 50% cover) on the hills located in the west-center; and by *Junellia tridens* (70 cm height, 60% cover) in the plateaus located at the south end of the district (León et al., 1998).

The vegetation of the Patagonian districts displays heterogeneity of a finer grain in association with altitude, slope, and exposure (Jobbágy et al., 1996, Paruelo et al., 2004). Prairies and meadows (*mallínes*) are distributed throughout the region and are generally associated with rivers, creeks, valley bottoms, or local springs, where the high water availability generates a completely different physiognomic type (fig. 14.8). The plant cover in these systems is often 100%, and mesophytic grasses (*Poa*

Figure 14.7 Shrub-grass steppe of the Occidental District of Patagonia (photo: N. Fernández)

Figure 14.8 Meadow (mallín) surrounded by shrub steppes in Patagonia (photo: N. Fernández)

pratensis, *Deschampsia flexuosa*, etc.), rushes (*Juncus balticus*), and sedges (*Carex* spp.) dominate. Most of the species currently present in the *mallines* are cosmopolitan or exotic, mainly from Europe.

14.2.3 Ecosystems Functioning Gradients

The environmental heterogeneity and the gradients in vegetation structure described for both the RPG and Patagonia lead to profound changes in ecosystems functioning, which is the exchange of matter and energy between biota and the abiotic environment. Net primary production (NPP) and its seasonal dynamics have been identified as a key ecosystem attribute able to integrate different aspects of ecosystem functioning. Satellite data provide a simple and powerful alternative for describing regional patterns of NPP from the normalized difference vegetation index (NDVI), a spectral index derived from reflectance in the red and infrared bands of the electromagnetic spectrum (Running et al., 2000). Paruelo and Lauenroth (1995, 1998) showed that three attributes of the annual curve of the NDVI capture the main characteristics of the seasonal dynamics of carbon gains: the NDVI integral (NDVI-I), the relative range of NDVI (RREL), and the date of maximum NDVI (DMAX). Moreover, these three attributes of the NDVI curve have been used to define ecosystem functional types (EFT) for the temperate areas of South America (Paruelo et al., 2001). EFT are areas presenting a similar functioning as defined by the NDVI-I, the RREL, and the DMAX.

In the RPG, net primary production levels, as suggested by remote sensing proxies, decrease from NE to SW following the dominant precipitation gradient (fig. 14.9, top left). Such changes in NDVI-I translate into twofold differences in above-ground net primary production (ANPP) between the extremes (Oesterheld et al., 1998). In Patagonia, net primary production patterns deduced from NDVI-I follow a W-E gradient axis that is also associated with the dominant precipitation gradient. Both areas clearly reflect the strong control exerted by precipitation on the carbon uptake of grassland and shrubland worldwide (Lauenroth, 1979; Sala et al., 1988). In Patagonia, the climate gradient not only affects total carbon uptake but also its interannual variability, which increases with aridity (Jobbágy et al., 2002). At a finer spatial scale, topography, landscape configuration, and land use also play an important role in controlling net primary production patterns. In Patagonia, Paruelo et al. (2004) found that the diversity of vegetation units explained a significant portion of the spatial variability of ANPP not accounted for by climate: the more diverse a landscape, the higher the ANPP.

For the RPG, 80% of the variance of the NDVI-I of uncropped areas is explained by rainfall and 9% by potential evapotranspiration. Land-use/land-cover changes (mainly agriculture expansion) have a relatively small impact on NDVI-I (less than 15%) (Guerschman et al., 2003).

In Patagonia, the seasonality of carbon uptake, as described by the NDVI relative range (RREL), shows a similar spatial pattern as NDVI-I (fig. 14.9, top right): the relative difference between maximum and minimum production through the year increases as MAP increases. Such behavior results mainly from changes in maximum productivity, which are associated with precipitation. NPP minima are

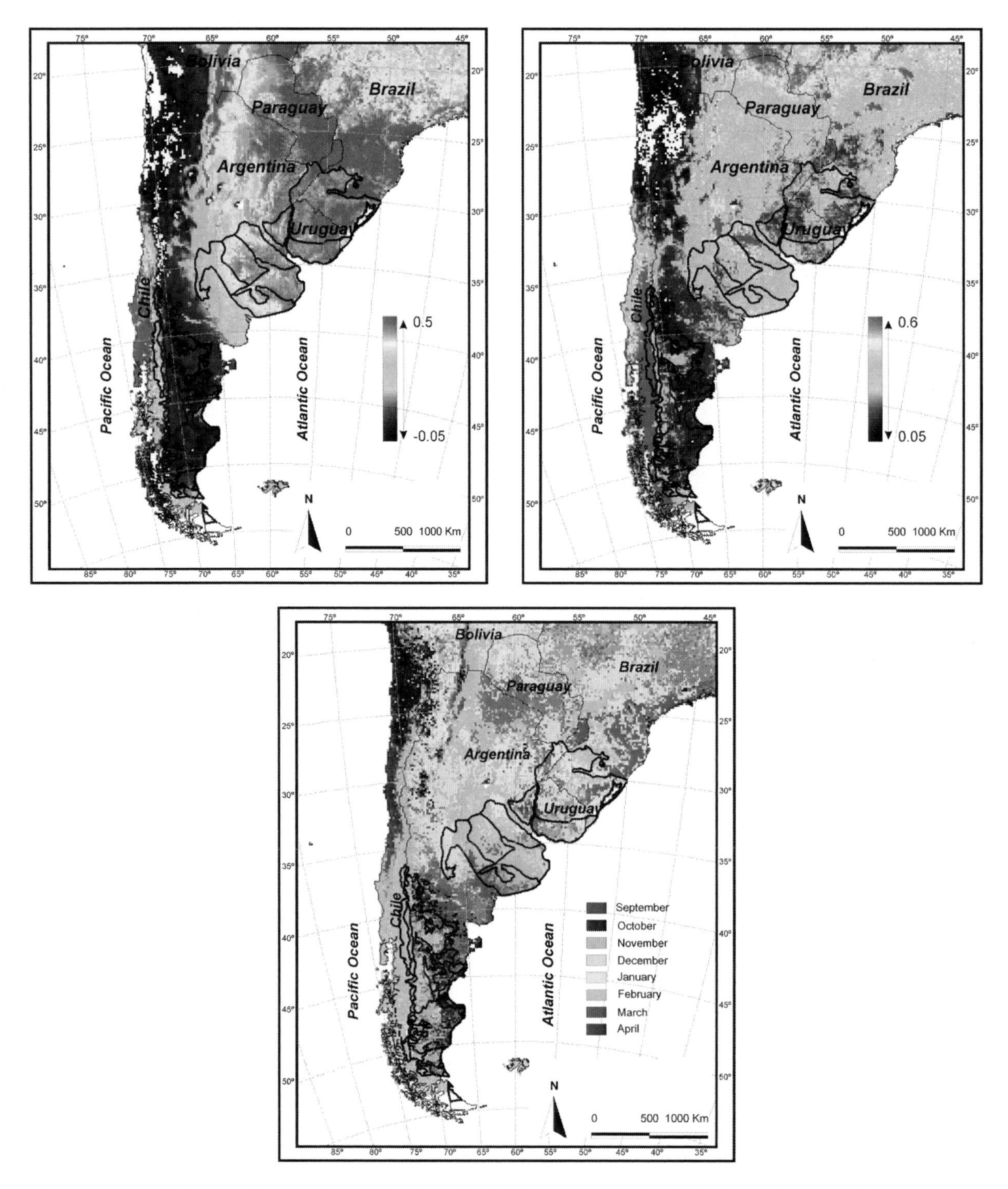

Figure 14.9 Top left: Annual integration of the normalized difference vegetation index (NDVI-I), a spectral index used as a surrogate of net primary production (NPP). Top right: Relative annual range of NDVI (maximum minus minimum NDVI relative to NDVI-I), a descriptor of NPP seasonality. Bottom: Date of maximum NDVI. Data were derived from the Pathfinder AVHRR Land database (NASA) and are averages over the 1981–2000 period. Figures based on Paruelo et al. (2001). See color insert.

independent of precipitation and are associated with thermal restrictions on plant growth during winter. In the RPG, the patterns of NDVI-I and RREL differ. Two factors contribute to such differences. First, the environmental controls of spatial variation in uncropped areas differ between attributes. Instead of MAP, as in the case of the NDVI-I, RREL is associated mainly with changes in potential evapotranspiration and in rainfall seasonality. Second, land use has a strong impact on seasonality: annual crops increase RREL up to 80% (Guerschman et al., 2003).

An additional description of the seasonality of C uptake is provided by the date of maximum NDVI (DMAX) (fig. 14.9, bottom). In Patagonia, the production peak (as described by the NDVI) is concentrated in the period November–January, with a few exceptions along the coast. In the western portion of Patagonia, the date of maximum NDVI is delayed as MAP increases (Jobbágy et al., 2002). In the RPG, the pattern is more complicated and heterogeneity is quite high: peak NDVI may occur between October and April. Though a substantial portion of the heterogeneity (73%) of the date of maximum NDVI in areas with low human impact on vegetation is explained by environmental variables, land-use/land-cover change has an overriding effect. Cropping modifies the date of peak NPP by up to 150 days (Guerschman et al., 2003).

These regional patterns of primary productivity are matched by strong patterns of livestock biomass and, as a result, the proportion of ANPP consumed by herbivores. Livestock biomass across rangelands of Argentina and Uruguay (largely from Patagonia and the RPG) increases more than proportionally as ANPP increases (Oesterheld et al., 1992). This livestock pressure is 10 times greater than that observed in natural ecosystems with similar primary productivity (McNaughton et al., 1989). The logarithmic increase of livestock biomass with ANPP or its surrogate, NDVI-I, may result from a higher forage quality in areas with high ANPP (Oesterheld et al., 1998). In addition, it may be influenced by the negative correlation between NDVI-I and both its seasonal and its interannual variability: more productive areas are also those with more reliable levels of ANPP, which allows ranchers to set livestock densities closer to the average carrying capacity (Oesterheld et al., 1998).

In a study based on correlative models that describe the climatic controls of grassland structure and functioning at regional scales, Paruelo et al. (1998b) analyzed the similarity of structural and functional characteristics of temperate grassland and shrubland ecosystems of North and South America. The evaluation included models that describe the regional distribution of plant functional types (C_3 and C_4 grasses and shrubs), soil organic carbon, and the annual amount and seasonality of above-ground net primary production. North and South America offer a unique opportunity to test the generality of regional models. As outlined above, temperate zones of North and South America show important climatic similarities (Paruelo et al.,1995). However, the flora and fauna of both continents are partially unrelated from an evolutionary viewpoint; North America belongs to the Holarctic realm and South America to the Neotropical realm (Udvardy, 1975). The estimates derived from the models were compared against observed data collected in South America, independent from those used to generate the models in North America. The results support the notion that, in climatically similar regions, structural and functional attributes, such as plant functional-type composition, soil organic carbon, and above-ground net primary productivity, have similar environmental controls, independent of the evolutionary history of the regions. Adler et al. (2005) went a step further by performing comparative experiments on the response of plant traits to grazing across analogous climatic gradients in sagebrush steppe of the United States and in the Patagonian steppe of Argentina. They found that poor-quality graminoids make the Patagonian steppe particularly resistant to overgrazing. Dominance by such species would be a consequence of the relatively long evolutionary history of grazing in Patagonia, where generalist herbivores would have exerted stronger selective pressures (Lauenroth, 1998), a hypothesis contrary to previous assumptions (Milchunas et al., 1988).

14.3 Changes in Land Use and Land Cover: Agriculture, Afforestation, and Grazing

Human transformation of this portion of South America started recently compared to temperate areas of the Old World or even of North America. The replacement of natural vegetation by crops in the RPG had its onset at the beginning of the twentieth century, with increasing European immigration. Irrigation and fertilization did not expand as widely as in the Great Plains, with the former being very rare and the latter only becoming more important over the past two decades (Hall et al., 1991).

Land-cover changes provide one of the clearest examples of the contrasts between the developed north and the developing south. Whereas the northern temperate regions are seeing a spontaneous expansion of forests over land that had been devoted to crop production (Dong et al., 2003), in the southern regions annual crops are taking over native ecosystems and new land-cover shifts are emerging—such as grassland afforestation. Superimposed on these land-cover changes are differing patterns of nitrogen deposition, acid rain, and other biogeochemical alterations that widen the divergence between southern and northern ecosystems.

Three main land-use changes can be identified in the RPG: replacement of grasslands by annual and forage crops, replacement of grasslands by tree-plantations, and intensification of grazing in native prairies. Areas receiving less than 500 mm/yr of precipitation had almost no agriculture. In areas with higher than 500 mm/yr, the proportion of the area under irrigation is lower than 0.045 and represents less than 7% of the total cropped area (Guerschman et al., 2003). The spatial variability in cropped area is associated with soil restrictions. For example, salinity and soil drainage constrain agriculture in the Flooding Pampa, whereas soil depth becomes limiting in parts of the Uruguayan Campos. Agriculture has strongly impacted the Rolling, Inner, and Austral Pampas in such a way that the natural vegetation is unknown and probably poorly represented in what might be considered relict areas. In meadows and the humid western grass steppes of Patagonia, sporadic agriculture may also play an important role as a disturbance agent.

As shown above, land-use change (particularly crop expansion) has had a significant effect on the seasonality of carbon gains and almost no effect on the total amount of carbon fixed (Guerschman et al., 2003). Agriculture basically modifies the shape of the seasonal curve of primary production but not the area beneath the curve. An increase in croplands also reduces the interannual variability of the seasonal patterns of carbon gains (Guerschman et al., 2003).

Shifts in vegetation cover can have strong and often devastating effects on ecosystems and societies, as illustrated by cases across the globe of massive replacement of native forests with pastures. Southern South America is experiencing an opposite—and less well understood—type of land-use shift: that of the conversion of native grasslands to tree plantations. Afforestation of some of the most productive native grasslands of the continent is already rapid and will probably be reinforced by the prospective carbon sequestration market. During the 1970s and 1980s, federal laws promoting afforestation with fast-growing tree species spurred the expansion of tree plantations over vast areas of Uruguay and Argentina. In coming decades, forestry probably will be the only subsidized land use in these countries. In the early years, afforestation projects were funded by national investment and were often integrated with local industrial processing (Jobbágy et al., 2005). But in the 1990s, more and more multinational companies became the major tree-planters and channeled most of the production to overseas industries. Highly productive grasslands in which crop production was usually not feasible (rocky or sandy soils, steep slopes, etc.) supported most of these plantations, shifting vast territories and their towns from a ranching to a forestry-based economy (Jobbágy et al., 2005). Afforestation has had a large and varied ecological impact. Soils under eucalypt plantations have been acidiûed to levels similar to those associated with acid rain in industrialized areas of the Northern Hemisphere (Jobbágy and Jackson, 2003). This acidification is probably due to high rates of calcium cycling following tree establishment because trees use and cycle more calcium than grasses. In addition, the establishment of tree plantations has had striking effects on evapotranspiration, increasing it up to 80% (Nosetto et al., 2005). Depending on the hydraulic properties of the soils and aquifers, tree plantation can also affect groundwater hydrology and salt dynamics, resulting in groundwater consumption of up to 300 mm/yr and large accumulations of salts in soils and aquifers (Jobbágy and Jackson, 2004; Nosetto et al., 2005).

Grazing is the main cause of vegetation changes in Patagonia and non-agricultural areas of the Pampas. Grazing effects depend on three broad categories of factors: grazing history, physical environment, and plant functional types available in the potential biota. Many large herbivores grazed in the region until the late Pleistocene extinctions around 10,000 years ago, which coincided with a large-scale reduction in grass species (see chapter 8; Markgraf, 1985). Afterward, the guanaco (*Lama guanicoe*) and the Pampas deer (Venado de las Pampas) *(Ozotoceros bezoarticus)* remained the main large herbivores, accompanied by one of the two ostrich species (Lesser Rhea, *Pterocnemia pennata* in Patagonia, and Greater Rhea, *Rhea americana*, in the RPG) (Bucher, 1987). Domestic herbivores were introduced in the sixteenth and early twentieth centuries in the Pampas and Patagonia, respectively. A key difference between natural and domestic herbivory is selectivity. Natural herbivores are allowed to select their diet at different spatial scales, from landscapes to individual plants. In contrast, domestic grazers are confined by fences and other husbandry practices. As we move from southern Patagonia to the Pampas there is change from sheep to cattle husbandry. The relative importance of sheep and cattle determines the grazing regime (i.e., intensity and selectivity) (Oesterheld et al., 1992).

As discussed above, the region encompasses a gradient of temperature and precipitation, although precipitation seems to be the main determinant of primary production and potential plant functional types, which in turn determine the response of vegetation to grazing. How do the grasslands and steppes of the RPP and Patagonia respond to grazing? We focus here on the effects of grazing on ecosystem types that exhibit the greatest vegetation heterogeneity and for which data are available.

Grazing has been perceived to be the main agent of desertification in Patagonia (Soriano and Movia, 1986; Ares et al., 1990). However the impact of grazing varies widely among vegetation units, as illustrated from the shrub steppes of the Occidental District of Patagonia and the grass steppes of the Subandean District of Patagonia. The grass-shrub steppes of the Occidental District (fig. 14.5) (45°S, 70°W) show in general no major changes in vegetation physiognomy due to grazing. The steppes have kept a structure defined by two strata: shrubs and tussock grasses. The relative abundance of perennial and annual species does not change. The main changes induced by grazing are floristic (Perelman et al., 1997). Tussocks and shrubs are represented in the group of decreasers as well as in the increasers. In contrast, the grass steppes of Subandean District (45°S, 71°W) have experienced dramatic physiognomic changes due to grazing. As grazing impact increases, the dominant tussock grass species *(Festuca pallescens)* decreases and total plant cover decreases from 85% to less than 40%. Shrub encroachment is the final stage of grazing degradation of the grass steppes formerly dominated by *Festuca* (León and Aguiar, 1985; Bertiller et al., 1995). *Mulinum spinosum* or *Acaena splendens* become the dominant shrub species. Such changes reduce primary production (Paruelo et al., 2004) and modify water dynamics and herbivore biomass (Aguiar et al., 1996). In both vegetation units plant diversity is higher in ungrazed areas.

Many studies identify domestic herbivore grazing as a key factor in shaping the structure and functioning of the RPG (e.g., Sala et al., 1986; Rusch and Oesterheld, 1997; Lavado and Taboada, 1985; Chaneton et al., 1996). In the Uruguayan Campos, for example, grazing increases the relative abundance of summer (C_4) grasses, which in turn determine changes in the seasonal dynamics of primary production, moving the peak of production from spring to summer (Altesor et al., 2005). However, the response of vegetation to herbivory varies among subregions. Even though grazing increased species richness both in the Flooding Pampa (36°S, 58°W) and in the Uruguayan Campos (34°S, 50°W), in the first case the increase was accounted for by exotic cool season forbs, whereas in the second case the plant functional types that increased were prostrate grasses (Chaneton et al., 2002; Rush and Oesterheld, 1997; Sala et al., 1986; Altesor et al., 1998; Rodríguez et al., 2003).

Grazing may either increase or decrease above-ground net primary production (ANPP). Rusch and Oesterheld (1997) found in the Flooding Pampa that ungrazed areas showed higher ANPP than grazed plots, even if the initial biomass was the same. Altesor et al. (2005) found that grazed areas produced 51% more than the ungrazed paired situation. However, when the initial biomass was the same, ungrazed areas were 29% more productive. Such differences may be associated with the effects of the structural changes promoted by grazing (species and plant functional type composition, biomass vertical distribution) on the resource level (water, nutrients and light). By removing or avoiding the accumulation of senescent material, grazing may increase radiation interception. As far as light is the limiting factor, grazing would increase ANPP. Long-term grazing significantly reduced ANPP in the grass-shrub steppes of Patagonia (Adler et al., 2005).

Although most of the rangelands in Patagonia and the RPG include woody plants, grazing did not necessarily promote their increase. The only clear case of shrub encroachment is found in the *Festuca pallescens* steppes in western Patagonia (see above). Aguiar and Sala (1998) proposed that woody species will be promoted only in those communities lacking grasses able to avoid grazing either because they are prostrate or unpalatable.

Some other disturbances may affect the grasslands and steppes of southern South America: fire and oil industry activities. In the RPG, fire is now restricted to the Campos subdivisions and small patches in the Mesopotamic Pampa and the Flooding Pampa. Fires are usually set by ranchers to eliminate low-quality forage and favor the regrowth of a more palatable and nutritious biomass. Agricultural use has significantly reduced fire frequency in the RPG (Di Bella et al., 2006). Summer fire can be an important disturbance agent in Patagonia, where accumulation of dry dead biomass is common. However, as electric storms are rare in Patagonia, fires are usually anthropogenic, associated with railway lines or human settlements. Fire frequency increases toward the boundary with the Monte scrublands. Oil extraction activities are the most intensive disturbance in Patagonia, though restricted in extent. Grazing affects almost all the region, but nowhere has it completely eliminated plant cover. Oil exploration and extraction cause extremely severe and irreversible damage in focal areas because they remove all vegetation cover, and often entire soil layers. Agriculture and fire have intermediate effects between oil extraction and grazing, both in terms of the extent and intensity of disturbance.

14.4 Conclusion

The plains, plateaus, and hills of Patagonia and the Río de la Plata host some of the largest extensions of semi-natural rangelands on Earth. The initial appearance of human populations in these areas was relatively late compared to most other continents, and population densities remained low until the late nineteenth-century period of European settlement. Patagonia is the last region of the Americas to be reached by humans entering via the Bering Strait, and also one of the last regions to be colonized by Europeans. Domestic livestock were introduced, depending on the area, between 100 and 400 years ago. Agriculture occupied a relatively small fraction of the area, even in the most humid zones.

At the onset of the twenty-first century, however, rapid changes are taking place, both in Patagonia and the Rio de la Plata region. Natural resources in Patagonia have come under intense pressure due to extractive activities (mainly oil exploitation), and sheep grazing has generated desertification in many foci across this ecologically fragile region. Impacts of sheep on this landscape have become more extensive during the past decade due to a reduction in wool prices, the lack of productive alternative land uses, and the absence of an environmental policy from federal and state agencies and governments. Poor understanding of the heterogeneity of Patagonian ecosystems and how they respond to environmental stresses and disturbances has also contributed to the degradation processes. In the Río de la Plata grasslands several factors, including the prices of commodities (mainly soybean) and the expansion of low-tillage techniques, are promoting a fast expansion of agriculture over natural and semi-natural grasslands. While producing significant economic benefits from the exports generated, land use in Patagonia and the Rio de la Plata plains is modifying the structure and functioning of ecosystems in a way that compromises their ability to provide basic ecosystem services. The extremely small area under governmental protection aggravates the threat to these unique ecosystems.

References

Adler, P.B., M.F. Garbulsky, J.M. Paruelo, and W. K. Lauenroth, 2006. Do abiotic differences explain contrasting graminoid functional traits in sagebrush steppe, USA and Patagonian steppe, Argentina? *Journal of Arid Environments*, 65, 62-82.

Adler, P., D. Milchunas, O. Sala, I. Burke, and W. Lauenroth, 2005. Plant traits and ecosystem grazing effects: Comparison of U.S. sagebrush steppe and Patagonian steppe. *Ecological Applications*, 15, 774–792.

Aguiar, M.R., and O.E. Sala, 1998. Interaction among grasses, shrubs, and herbivores in Patagonian grass-shrub steppes. *Ecología Austral*, 8, 201–210.

Aguiar, M.R., J.M. Paruelo, O.E. Sala, and W.K. Lauenroth, 1996. Ecosystem consequences of plant functional types changes in a semiarid Patagonian steppe. *Journal of Vegetation Science*, 7, 381–390.

Altesor, A., E. Dilandro, H. May, and E. Ezcurra, 1998. Long-term species changes in a Uruguayan grassland. *Journal of Vegetation Science*, 9, 173–180.

Altesor, A., M. Oesterheld, E. Leoni, F. Lezama, and C. Rodríguez, 2005. Effect of grazing exclosure on community structure and productivity of a temperate grassland. *Plant Ecology*, 179, 83-91.

Ares, J., A. Beeskow, M. Bertiller, M. Rostagno, M. Irrisarri, J. Anchorena, G., Defosse, and C. Merino, 1990. Structural and dynamics characteristics of overgrazed lands of northern Patagonia, Argentina. In: *Managed Grasslands*, Elsevier Science Publishers, Amsterdam, 149–175.

Barros, V., and H. Mattio, 1979. Campos de precipitación de la provincial de Chubut (1931–1960). *Geoacta*, 10, 175–192.

Batista, W.B., R. León, and S. Perelman, 1988. Las comunidades vegetales de un pastizal natural de la región de Laprida, Provincia de Buenos Aires, Argentina. *Phytocoenologia*, 16, 465–480.

Bazilievich, N.I., 1970. *The Geochemistry of Soda Soils.* Israel Program for Scientific Translation, Jerusalem.

Bertiller, M.B., A. Beeskow, and P. Irisarri, 1981. *Caracteres fisonómicos y florísticos de la vegetación del Chubut. 1: Sierra de San Bernardo.* Contribución 40. Centro Nacional Patagónico, CONICET, Puerto Madryn, Argentina.

Bertiller, M.B., N. Elissalde, M. Rostagno, and G. Defosse, 1995. Environmental patterns and plant distribution along a precipitation gradient in western Patagonia. *Journal of Arid Environments*, 29, 85–97.

Bucher, E., 1987. Herbivory in arid and semi-arid regions of Argentina. *Revista Chilena de Historia Natural*, 60, 265–273.

Burkart, S.E., R. León, and C.P. Movia, 1990. Inventario Fitosociológico del Pastizal de la Depresión del Salado (Provincia de Buenos Aires.) en un área representativa de sus principales ambientes. *Darwiniana*, 30, 27–69.

Burkart, S.E., R. León, S. Perelman, and M. Agnusdei, 1998. The grasslands of the flooding pampas (Argentina): Floristic heterogeneity of natural communities of the southern Rio Salado basin. *Coenoses*, 13, 17–27.

Cabrera, A., and A. Wilkins, 1973. *Biogeografia de America Latina.* Editorial OEA, Washington DC.

Cantero, J.J., and R.J.C. León, 1999. The vegetation of salt marshes in central Argentina. *Beiträge zur Biologie der Pflanzen*, 71, 203–242.

Chaneton, E., J. Lemcoff, and R. Lavado, 1996. Nitrogen and phosphorus cycling in grazed and ungrazed plots in a temperate subhumid grassland in Argentina. *Journal of Applied Ecology*, 33, 291–302.

Chaneton, E.J., S.B. Perelman, M. Omacini, and R.J.C. León, 2002. Grazing, environmental heterogeneity, and alien plant invasions in temperate Pampa grasslands. *Biological Invasions*, 4, 7–24.

Clapperton, C.M., 1994. The Quaternary glaciation of Chile: A review. *Revista Chilena de Historia Natural, 67*, 369–383.

Collantes, M.B., M. Kade, and A. Puerto, 1981. Empleo de técnicas de análisis factorial y análisis diferencial en el estudio de un área de pastizales de la Depresión del Río Salado (Provincia de Buenos Aires). *Estudio Ecológica*, 1, 89–107.

Collantes, M.B., J. Anchorena, and A.M. Cingolani, 1999. The steppes of Tierra del Fuego: Floristic and growth form patterns controlled by soil fertility and moisture. *Plant Ecology*, 140, 61–75.

Coronato, F., 1996. Influencia climática de los lagos Musters y Colhue-Huapi. *Meteorológica*, 21, 65–72.

Darwin C., 1842. On the distribution of the erratic boulders and on the contemporaneous unstratified deposits of South America. *Transactions of the Geological Society, London*, 6, 415–431.

del Valle, H., 1998. Patagonian soils: A regional synthesis. *Ecología Austral*, 8, 103–124.

Di Bella, C.M., E.G. Jobbágy, J.M. Paruelo, and S. Pinnock, 2006. Environmental and land use controls of fire density in South America. *Global Ecology and Biogeography*,15, 192–199.

Dong, J.R., R. Jaufmann, R. Myneni, C. Tucker, P.Kauppi, J.Liski, W. Buermann, V. Alexeyev, and M. Hughes, 2003. Remote sensing estimates of boreal and temperate forest woody biomass: Carbon pools, sources, and sinks. *Remote Sensing of Environment*, 84, 393–410.

Duran, A., 1991. *Los Suelos del Uruguay.* Hemisferio Sur, Montevideo.

Ellenberg, H., 1962. Wald in der Pampa Argentiniens? *Veroff Geobotanical Institute ETH* (Switzerland), 37, 39–56.

Faggi, A., 1986. Mapa de vegetación de Alsina, Provincia de Buenos Aires. *Parodiana*, 4, 381–400.

Frére, M., J. Riks, and J. Rea, 1978. *Estudio Agroclimático de la Zona Andina.* Technical Note 161, World Meterorological Organization. Geneva.

Golluscio, R.A., R.J.C. León, and S.B. Perelman, (1982). Caracterización fitosociológica de la estepa del oeste del Chubut: Su relación con el gradiente ambiental.

Boletín de la Sociedad Argentina de Botánica, 21, 299–324.

Guerschman J.P, J.M. Paruelo, O. Sala, and I. Burke, 2003. Land use in temperate Argentina: Environmental controls and impact on ecosystem functioning. *Ecological Applications*, 13, 616–628.

Hall, A.J., C. Rebella, C. Ghersa, and J. Culot, 1991. Field crop systems of the Pampas. In: C.J. Pearson (Editor), *Field Crop Ecosystems, Ecosystems of the World 19*, Elsevier, Amsterdam, 413–450.

INTA, 1989. *Mapa de Suelos de la Provincia de Buenos Aires (Escala 1:500.000).* Instituto Nacional de Tecnología Agropecuaria, Buenos Aires.

Iriondo M., J. Orellana, and J. Neiff, 1974. Sobre el concepto de mallín cordillerano. *Revista de la Asociación de Ciencias Naturales del Litoral*, 5, 45–52.

Jobbágy, E.G., J.M. Paruelo, and R.J.C. León, 1995. Estimación de la precipitación y de su variabilidad interanual a partir de información geográfica en el NW de Patagonia, Argentina. *Ecología Austral*, 5, 47–53.

Jobbágy, E.G., J.M. Paruelo, and R.J.C. León, 1996. Vegetation heterogeneity and diversity in flat and mountain landscapes of Patagonia (Argentina). *Journal of Vegetation Science*, 7, 599–608.

Jobbágy, E.G., O. Sala, and J.M. Paruelo, 2002. Patterns and controls of primary production in the Patagonian steppe: A remote sensing approach. *Ecology*, 83, 307–319.

Jobbágy, E.G., and R. Jackson, 2003. Patterns and mechanisms of soil acidiûcation in the conversion of grasslands to forests. *Biogeochemistry*, 54, 205–229.

Jobbágy, E.G., and R. Jackson, 2004. Groundwater use and salinization with grassland afforestation. *Global Change Biology*, in press.

Jobbágy, E.G, J.M. Paruelo, G. Piñeiro, D. Piñeiro, M. Carámbula, V. Morena, V. Sarli, and A. Altesor, 2005. Climate and land-use controls on ecosystem functioning: Challenges and insights from the South. In: *Annual Report 2003-2004, InterAmerican Institute for Global Change (IAI)*, San Jose dos Campos, Brazil, 47–61.

Lauenroth, W.K., 1979. Grassland primary production: North American grasslands in perspective. In N.R. French (Editor), *Perspectives in Grassland Ecology.* Ecological Studies 32, Springer-Verlag, New York, 3–24.

Lauenroth, W., 1998. Guanacos, spiny shrubs, and evolutionary history of grazing in the Patagonian steppe *Ecología Austral*, 8, 211–216.

Lavado, R.S., and MA. Taboada, 1985. Influencia del pastoreo sobre algunas propiedades químicas de un natracuol de la pampa deprimida. *Ciencia del Suelo*, 3, 102–108.

Le Houérou, H., 1996. Climate change, drought and desertification. *Journal of Arid Environments*, 34, 133–185.

León, R.J.C., and M.R. Aguiar, 1985. El deterioro por uso pasturil en estepas herbáceas patagónicas. *Phytocoenologia*, 13, 181–196.

León, R J.C., S. Burkart, and C. Movia, 1979. *Relevamiento Fitosociológico del Pastizal del Norte de la Depresión del Salado.* Serie Fitogeográfica 17, INTA, Buenos Aires.

León, R.J.C., D. Bran, M. Collantes, J.M. Paruelo, and A. Soriano, 1998. Grandes unidades de vegetación de la Patagonia extra andina. *Ecología Austral*, 8, 125–144.

Lewis J.P., M.B. Collantes, E.F. Pire, N.J. Carnevale, S.I. Boccanelli, S.L. Stofella, and D.E. Prado, 1985. Floristic groups and plant communities of southeastern Santa Fe, Argentina. *Vegetatio*, 60, 67–90.

Lezama, F., S. Baeza, A. Altesor, J. Paruelo, G. Piñeiro, and R. León, 2004. Distribución Espacial de las Comunidades de Pastizal en la Región Basáltica (Uruguay). *XX Reunión Del Grupo Técnico Regional del Cono Sur en Mejoramiento y Utilización de los Recursos Forrajeros del Area Tropical y Subtropical—Grupo Campos*, Salto, Uruguay.

Markgraf, V., 1985. Late Pleistocene faunal extinctions in Southern Patagonia. *Science*, 228, 1110–1112.

McNaughton, S. J., M. Oesterheld, D. Frank, and K. Williams, 1989. Ecosystem-level patterns of primary productivity and herbivory in terrestrial habitats. *Nature*, 341, 142–144.

Mercer, J.H., 1976. Glacial history of southernmost South America. *Quaternary Research*, 6, 125–166.

Milchunas, D.G., O.E. Sala, and W.K. Lauenroth, 1988. A generalized model of the effects of grazing by large herbivores on grassland community structure. *American Naturalist*, 132, 87–106.

Nosetto, M.D., E. Jobbágy, and J.M. Paruelo, 2005. Land use change and water losses: The case of grassland afforestation across a soil textural gradient in central Argentina. *Global Change Biology*, 11: 1–17.

Oesterheld, M., C. Di Bella, and K. Herdiles, 1998. Relation between NOAA-AVHRR satellite data and stocking rate of rangelands. *Ecological Applications*, 8, 207–212.

Oesterheld, M., O. Sala, and S.J. McNaughton, 1992. Effect of animal husbandry on herbivore-carrying capacity at a regional scale. *Nature*, 356, 234–236.

Panario, D., and O. Gutierrez, 1999. The continental Uruguayan Cenozoic: An overview. *Quaternary International*, 62, 75–84

Parodi, L., 1942. ¿Porqué no existen bosques naturales en la llanura bonaerense si los árboles crecen en ella cuando se los cultiva? *Agronomía 30*, Buenos Aires.

Paruelo, J.M., and O. Sala, 1995. Water losses in the Patagonian steppe: A modelling approach. *Ecology*, 76, 510–520.

Paruelo J.M., and W. Lauenroth, 1995. Regional patterns of NDVI in North American shrublands and grasslands. *Ecology*, 76, 1888–1898.

Paruelo J.M., and W. Lauenroth, 1998. Interannual variability of the NDVI curves and their climatic controls in North American shrublands and grasslands. *Journal of Biogeography*, 25, 721–733.

Paruelo, J.M., O. Sala, and A. Beltrán, 2000. Long-term dynamics of water and carbon in semi-arid ecosystems: A gradient analysis in the Patagonia steppe. *Plant Ecology*, 150, 133–143.

Paruelo, J.M., E. Jobbágy, and O. Sala, 2001. Current distribution of ecosystem functional types in temperate South America. *Ecosystems*, 4, 683–698.

Paruelo, J.M., A. Beltrán, O. Sala, E. Jobbágy, and R.A. Golluscio, 1998a. The climate of Patagonia general patterns and controls on biotic processes. *Ecología Austral*, 8, 85–104.

Paruelo, J.M., E. Jobbágy, O. Sala, W. Lauenroth, and I. Burke, 1998b. Functional and structural convergence of temperate grassland and shrubland ecosystems. *Ecological Applications*, 8, 194–206.

Paruelo, J.M., W.K. Lauenroth, H. Epstein, I. Burke, M.R. Aguiar, and O.E. Sala, 1995. Regional climatic similarities in the temperate zones of North and South America. *Journal of Biogeography*, 22, 2689–2699.

Paruelo, J.M., M. Oesterheld, C. Di Bella, M. Arzadum, J. Lafontaine, M. Cahuepe, and C. Rebella, 2000. A calibration to estimate primary production of subhumid rangelands from remotely sensed data. *Applied Vegetation Science*, 3, 189–195.

Paruelo, J.M., R. Golluscio, J. Guerschman, A. Cesa, V. Jouve, and M. Garbulsky, 2004. Regional scale relationships between ecosystem structure and functioning: The case of the Patagonian steppes. *Global Ecology and Biogeography*, 13, 385–395.

Perelman, S.B., R. León, and J. Bussacca, 1997. Floristic changes related to grazing intensity in a Patagonian shrub steppe. *Ecography*, 20, 400–406.

Perelman, S., R. León, and M. Oesterheld, 2001. Cross-scale vegetation patterns of Flooding Pampa grasslands. *Journal of Ecology*, 89, 562–577.

Prohaska, F., 1976. The climate of Argentina, Paraguay and Uruguay. In: E. Schwerdtfeger (Editor), *Climate of Central and South America*. World Survey of Climatology, Elsevier, Amsterdam, 57–69.

Rodríguez, C., E. Leoni, F. Lezama, and A. Altesor, 2003. Temporal trends in species composition and plant traits in natural grasslands of Uruguay. *Journal of Vegetation Science*, 14, 433–440.

Rosengurtt, B., B. R. Arrillaga de Maffei, and P. Izaguirre de Artucio, 1970. *Gramíneas Uruguayas*. Montevideo.

Running, S.W., P.E. Thornton, R.R. Nemani, and J.M. Glassy, 2000. Global terrestrial gross and net primary productivity from the Earth Observing System. In: O. Sala, R. Jackson and H. Mooney (Editors), *Methods in Ecosystem Science*, Springer-Verlag, New York, 44–57.

Rusch, G.M., and M. Oesterheld, 1997. Relationship between productivity, and species and functional group diversity in grazed and non-grazed Pampas grasslands. *Oikos*, 78, 519–526.

Sala, O. E., 1988. The effect of herbivory on vegetation structure. In M.J.A. Werger, P.J.M. van der Aart, H.J. During, and J.T.A. Verboeven (Editors), *Plant Form and Vegetation Structure*, Academic Publishing, The Hague, 317–330.

Sala, O. E., M. Oesterheld, R.J.C. León, and A. Soriano, 1986. Grazing effects upon plant community structure in subhumid grasslands of Argentina. *Vegetatio*, 67, 27–32.

Sala, O.E., W. Parton, L. Joyce, and W. Lauenroth, 1988. Primary Production of the central grassland region of the United States. *Ecology*, 69, 40–45.

Soriano, A., 1956. Los distritos florísticos de la Provincia Patagónica. *Revista Investigaciones Agropecuarias*, 10, 323–347.

Soriano, A., 1991. Río de La Plata Grasslands. In: R.T. Coupland (Editor), *Ecosystems of the World: 8A: Natural Grasslands. Introduction and Western Hemisphere.* Elsevier, Amsterdam, 367–407.

Soriano, A., and C. Movia, 1986. Erosión y desertización en la Patagonia. *Interciencia*, 11, 77–83.

Strelin, J.A., G. Reb, R. Kellerc, and E. Malagninod, 1999. New evidence concerning the Plio-Pleistocene landscape evolution of southern Santa Cruz region. *Journal of South American Earth Sciences*, 12,333–341.

Teruggi, M.E. 1957. The nature and origin of Argentine loess Journal of Sedimentary Research 27, 322-332.

Tóth, T., and G. Jozefaciuk, 2002. Physicochemical properties of a solonetzic toposequence. *Geoderma*, 106, 137–159.

Tricart, J. L. 1973, *Geomorfología de la Pampa Deprimida: Base para los Estudios Edafológicos y Agronómicos. Colección Científico*, 12, Instituto Nacional de Tecnología Agropecuaria, Buenos Aires

Udvardy, M.D.F., 1975. *A Classification of the Biogeographical Provinces of the World.* IUCN Occasional Paper 18, Morges, Switzerland.Vervoorst, F.B., 1967. *Las Comunidades Vegetales de la Depresión del Salado (Provincia de Buenos Aires).* Serie Fitogeográfica 7, INTA. Buenos Aires.

Volkheimer, W., 1983. Geology of extra-andean Patagonia. In. N. West (Editor), *Deserts and Semideserts of Patagonia. Temperate Deserts and Semideserts.* Elsevier, Amsterdam, 425–429.

Walter, H., 1968. *Die Vegetation der Erde in Okophysiologischer Betrachtung.* Gustav Fischer Verlag, Stuttgart.

Zarate, M.A., 2003. Loess of southern South America. *Quaternary Science Reviews*, 22, 1987-2006.

15

Ocean Coasts and Continental Shelves

José Araya-Vergara

Suess (1900) provided the first scientific treatment of the South American coast from a tectonic perspective when he distinguished between the Atlantic and Pacific structural styles on opposite sides of the continent. Inman and Nordstrom (1971) later complemented this approach by relating these styles to the concepts of plate tectonics that had emerged during the 1960s. Useful keys to understanding South American coastal processes and sediment supplies were then offered by Davies (1977) and Potter (1994), respectively, while regional accounts of South American coastal landforms were made by specialists in books edited by Bird and Schwartz (1985) and Schwartz (2005). Clapperton (1993) reviewed Quaternary coastal morphogenesis. Coastal sites of scientific importance and historical coastline changes were discussed by Bird and Koike (1981) and Bird (1985).

This chapter focuses on the principal factors involved in coastal evolution and morphogenesis, describes key regional landforms, and proposes a new analytical perspective for South America's coasts by introducing a hierarchical system within coastal groups.

15.1 Coastal Origins and Evolution

The main coastline of South America is approximately 31,100 km long, of which 10,400 km face the Pacific Ocean, 16,700 the open Atlantic Ocean, and the remaining 4,000 km the more sheltered Caribbean Sea (table 15.1). Of the total length, approximately two-thirds lie within the tropics, ensuring that physical and ecological responses to ocean-atmosphere circulation systems involving the Intertropical Convergence Zone dominate these coasts. The remaining one third of the coast beyond the tropics is dominated during part or all of the year by temperate westerly conditions, which become increasingly cool and stormy toward the continent's southern tip. The origins of the present coast reflect the tectonic forces that have affected the South American plate over the past 200 million years, augmented by relative sea-level changes associated with changing global (eustatic) ocean volume and regional (isostatic) crustal adjustments.

15.1.1 *The Rupture of Gondwana and Postpartum Evolution*

The Atlantic coast of South America owes its broad outline to the separation of the continent from neighboring parts of Gondwana that began more than 200 Ma (million years ago). The Pacific and Caribbean coasts have a more complex history, related to the progressive interaction of the westward-moving South American plate with four oceanic plates with which it has come into contact (see chapter 1). Of these, the Nazca plate meets the South American plate along most of the Pacific coast, but the Caribbean and Cocos plates to the

Table 15.1 Character of South American coastline by tectonic group

Tectonic group and climate zone	Length km	Length %	Edge type and wave approach
Pacific	10,400	33.4	Collision; SW high-energy swells
Colombian-Ecuadorian (wet tropical)	3,800	12.2	Collision; oblique and parallel
Peruvian (dry tropical)	2,300	7.4	Collision; transverse
North-central Chilean (dry subtropical-temperate)	2,500	8.0	Collision; oblique
Patagonian	1,800	5.8	Collision and transcurrent; transverse and parallel
Atlantic	16,700	53.7	Trailing; SE swells and NE tradewind waves
Argentinian (dry to wet temperate)	5,700	18.3	Trailing; transverse and oblique
South Brazilian (wet tropical) Uruguay-Cabo São Roque	5,200	16.7	Trailing; transverse and oblique
North Brazilian (wet tropical) Cabo São Roque-Paria	5,800	18.7	Trailing; transverse and oblique
Caribbean (mostly dry tropical)	4,000	12.9	Transcurrent; protected sea, E-NE trade-wind waves
Total	31,100	100.0	

north, and the Antarctic and Scotia plates to the south, are regionally important (see fig. 15.1, also fig. 1.1).

The Atlantic coasts of South America and Africa have been inherited from the breakup of Gondwana that began in the far north in late Triassic time (~210 Ma) and in the far south in the early Jurassic (~180 Ma), but separation was not completed until the mid-Cretaceous (~100 Ma) (see chapter 1). Today, there is still a reasonably good fit between these continents at depths of around 1000 m on the continental slope, but the fit is not perfect because of subsequent crustal adjustments, and erosion and deposition (Bullard et al., 1965). The tensional stresses involved in the breakup of Gondwana occurred along intracontinental fracture zones, leading to the formation of rift systems and sedimentary basins that still influence the coasts and offshore bathymetry of both Brazil and adjacent parts of Africa (Urien and Martins, 1999; Orme, 2005). The end-product of these tectonic changes allowed Inman and Nordstrom (1971) to designate the Atlantic margin of South America as a trailing-edge coast in that it has trailed behind the continent's leading edge in moving away from the Mid-Atlantic Ridge. As such, this coast strongly reflects the large structures, the Precambrian cratons of the Guiana and Brazilian shields and the Paleozoic platform covers of the Patagonian plateaus, that were formerly part of the continuous Gondwana land mass.

In contrast, following Inman and Nordstrom (1971), the Pacific margin is a collision-edge coast where the South American plate meets with and overrides the Nazca plate, causing massive orogenic activity along the leading edge of the former while the latter subducts beneath the resulting Andes. This is a mountainous coast whose regional character reflects a distinction between flat-slab subduction, for example along the coast of northern Perú, and steeply angled subduction with resulting volcanic activity, for example along the coast of southern Perú and northern Chile (see chapter 2; Gutschner et al., 2000). This collision-edge coast extends southward beyond the Chile Rise but here it is the Antarctic plate that is subducting beneath the South American and Scotia plates, as reflected in the arcuate structures of the west Patagonian coast and Tierra del Fuego (Stern et al., 1986; de Witt, 1977).

The northern and southern margins of South America exhibit a third tectonic style, that of transcurrent interplate contact. In the north, triggered by motion between the Nazca and Cocos plates northeast of the Galapagos spreading center (Naumann and Geist, 2000; Marshall and Fisher, 2000), the Caribbean plate is shearing eastward along the north coasts of Colombia and Venezuela, and overriding the South American plate along its southern margin. This is the Caribbean straggler shearing edge (fig. 15.1; Taboada et al., 2000). In the south, the Scotia straggler shearing edge characterizes the interaction between the west-moving South American plate and the east-moving Scotia plate (Barker and Burrell, 1977; Parra et al., 1984).

15.1.2 Relative Sea-Level Change

As indicators of past sea levels, the present elevations of marine terraces of Pleistocene and Holocene age offer a measure of the relative importance of global eustatic forcing and regional tectonic deformation to sea-level change. Between three and seven principal terrace levels are commonly found along the South American coast between present sea level and about 300 m in elevation, sometimes higher (Clapperton, 1993).These are arranged in table 15.2 according to their marine oxygen isotope signatures (Chappell and Shackleton, 1986). Following convention, oxygen isotope stages defined by odd numbers (1, 3, 5. . . .) are related to interglacial stages

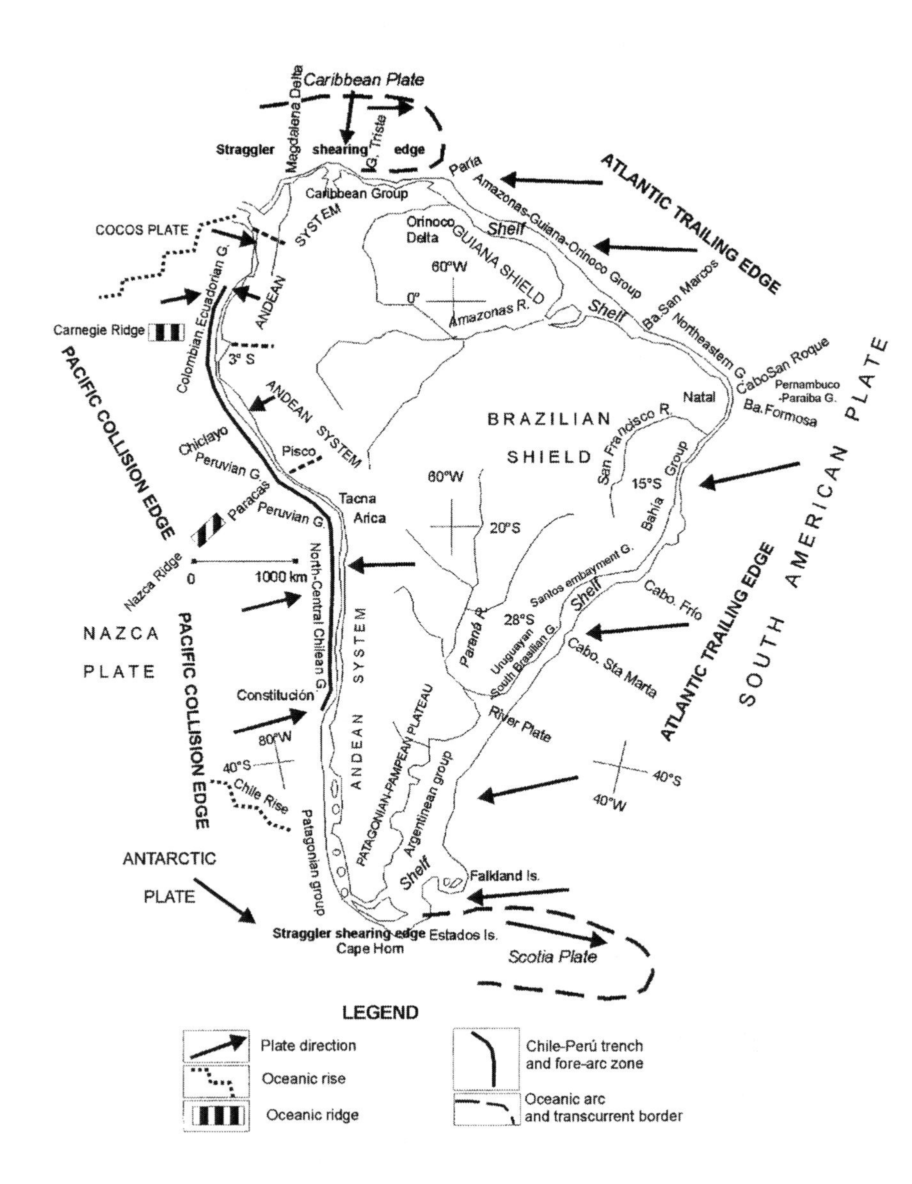

Figure 15.1 Tectonic elements and coastal groups

when eustatic sea levels are thought to have been similar to the present. The table, which excludes terraces lacking isotopic information, usually those above 200 m, thus presents a measure of the relative tectonic deformation of the coastal zone since middle Pleistocene time.

The earliest terraces, those of OIS 11 related to a middle Pleistocene marine transgression, are highest along the coast of Perú, and somewhat lower in north-central Chile. No data are available beyond the Pacific coast. Terraces of OIS 9, a later middle Pleistocene transgression, though lower, show a similar pattern of deformation along the Pacific coast, reaching 145 m in Perú and 45 m in Chile (Radke, 1987), but rising to only 15–25 m along the Atlantic coast. Although lower again, similar behavior is observed for terraces from OIS 7, a late middle Pleistocene event. Terraces of OIS 5e, the major transgression of the Last Interglacial when global warming raised the eustatic sea level to about 6 m above the present, are again higher along the coast of Perú and decline both north and south from there. Terraces of OIS 5a, a lesser oscillation during the eustatic regression associated with the onset of the Last Glaciation (*sensu lato*), are similarly deformed, while those of OIS 3, an interstadial during the Last Glaciation, have so far been found only in Perú. That terraces from OIS 5a and 3 should even appear is interesting because eustatic sea levels during these stages were below those of the present.

Terraces of OIS 1, related to the maximum level attained by Holocene seas, show little difference between the Pacific and Atlantic coasts but do rise above mean Holocene levels in central Chile and Argentina. The simple flat-earth model of Clark and Bloom (1979), which incorporates isostatic factors, predicts a modest emergence of 2–4 m over the past 5,000 years for the South American continent, a value supported by data from the Atlantic coast but less than that from Patagonia or central Chile where tectonic uplift has continued (Auer, 1970).

These data in table 15.2 reveal an elevational asymmetry between the higher Pacific and lower Atlantic marine

Table 15.2 Elevation of Pleistocene and Holocene marine terraces relative to tectonic group and oxygen-isotope stages

	Marine oxygen isotope stage						
Tectonic group	1	3	5a	5e	7	9	11
Pacific							
Colombian-Ecuadorian				10–35			
Peruvian		0–4	20–40	70	100	145	210
North-central Chilean	1–33		4–15	10–36	25–50	35–45	65–100
Patagonian	5–8						
Atlantic							
Argentinean	2–11		3–12	8–30	23–35	15–18	
South Brazilian	3–5		8–10	~15	20–25		
Santos embayment	2–5					25	
Bahía	3–5					25	
Pernambuco-Paraiba	1–4						
Northeast Brazil	4						
Guianan	0–2			2–7			
Caribbean							
Venezuelan	1–8			7–10		11–15	

Numbers (meters above mean sea level) refer to elevations of individual terraces, or range of elevations of terrace sequences, related to specific OIS stages.

Source: Elevations based on data in Clapperton (1993), with oxygen isotope stages after Chappell and Shackleton (1986).

terraces of similar age, and a latitudinal asymmetry between Pacific terraces, which are higher in the north, and Atlantic terraces, which are higher in the south, in Argentina. Brazilian terraces are little deformed by tectonic activity.

Thus the coast of South-America appears tectonically warped, as modeled in figure 15.2. The greatest distortion occurs in Perú, a possible result of the subduction of the Nazca Ridge (Le Roux et al., 2000). High levels of deformation occur throughout the Pacific collision edge, with modest distortion in the plateaus of the Argentinean trailing edge, and least distortion along the coasts fronting the Brazilian and Guiana shields. Nevertheless, the model is a simplification. Within the generalized pattern of distortion, there is tectonic segmentation, notably along the Pacific coast where terraces of the same age occur at different altitudes (table 15.2). Even so, terrace elevations from different oxygen isotope stages suggest that trends in deformation rates have remained similar throughout later Pleistocene and Holocene time.

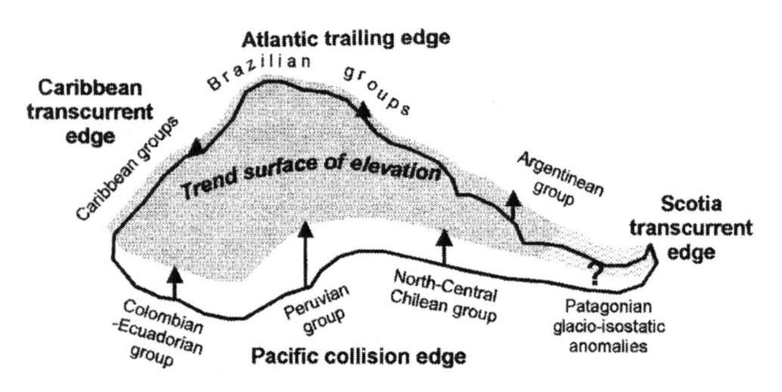

Figure 15.2 Simplified model of elevation trend surface for marine terraces (not valid for Patagonia owing to glacio-isostatic anomalies)

The morphology of the South American continental shelf provides further evidence of sea-level changes triggered by interacting eustatic and tectonic forces (fig. 15.3). Although there is as yet no comprehensive interpretation of shelf and slope bathymetry, available data indicate that the shelf edge descends to depths exceeding 100 m, or close to the accepted maximum regression of sea level to around -130 m during the Last Glaciation (OIS 2). Submarine valleys on the shelf lead to canyons that reflect extensions of terrestrial drainage basins when sea levels were lower.

The Pacific shelf is clearly much narrower than the Atlantic shelf (table 15.3). It is narrowest off southern Perú and northern Chile, where the Chile-Perú subduction trench comes nearest to shore south of the Nazca Ridge, significantly in the area of greatest marine-terrace uplift. The shelf is wider farther north, between the Nazca and Carnegie ridges, and also off southern Chile.

The Atlantic shelf reaches its maximum width off the Amazon estuary and again southward from the Río de la Plata, partly a reflection of massive fluvial sediment discharge to the ocean during late Cenozoic time, augmented in the south by fluvioglacial sediment from the Patagonian Andes (Potter, 1994). The broad shelf and Malvinas

Figure 15.3 Morphoclimatic indicators along the South American coast. Sand families modified after Potter (1994).

(Falkland) Plateau off Patagonia, however, also reflect submerged portions of the former Gondwana land mass. In contrast, the shelf is exceptionally narrow south of Cabo São Roque, between Paraíba and Pernambuco, where the coast's youngest tectonic basin reflects the late separation from Africa in mid-Cretaceous time. The Atlantic shelf is thus not easily explained by simple plate-trailing dynamics but must also accommodate rift systems and subsiding blocks of former continental crust.

15.2 Coastal Processes

15.2.1 Climatic Factors

Climatic factors influence the South American coast in two main ways: through the behavior of river regimes and the supply of sediment from morphoclimatic zones farther inland, and through their direct influence on coastal ecology.

With respect to sediment supply, Potter (1994) recognizes two great families of sands as representatives of morphoclimatic zones within the continent, namely the Brazilian and Andean families (fig. 15.3). In sands of the Brazilian family, which reach the shore between Paria and the Río de la Plata, quartz is the predominant mineral because of its provenance from the tropical weathering residues of the cratons forming the Brazilian and Guiana shields (see chapters 2 and 7). In contrast, sands of the Andean family are rich in volcaniclastic and metamorphic minerals derived from the complex orogenic systems and volcanic centers of the Andes. Andean sands prevail throughout the Pacific coast but also reach Caribbean and Argentinean shores. In addition to these sources, carbonate materials derived from coral reefs and similar sources are abundant along tropical Atlantic coasts, less so on tropical Pacific shores, and scarce farther south.

Table 15.3 Mean widths of continental shelf by regional group

Pacific	Width (km)	Atlantic and Caribbean	Width (km)
Colombian-Ecuadorian	27	Caribbean	25
North Peruvian (Chiclayo-Pisco)	66	North Brazilian (Paria-Natal) (300 km off Amazon)	100
South Peruvian (Pisco-Tacna)	13		
		Central Brazilian (Natal-Ilhéus)	25
North Chilean (Arica-Constitución)	18	South Brazilian-Argentinian (500 km at Islas Malvinas)	245
South Chilean and Patagonian	38		

Morphoclimatic factors are also expressed in the basin regimes of rivers supplying these sediments to the coast. Many rivers reaching the Atlantic coast drain exceptionally large tropical and subtropical basins, such as those of the Amazon and Paraná-Paraguay systems, but the response time between entrainment of sediment and its eventual arrival at the shore is very slow. Much sediment is stored in river floodplains for long periods and, on reaching the coast, is dominated by large quantities of clay and silt rather than sand (see chapter 5). In contrast, rivers descending the Andes to the Pacific drain relatively small basins whose climates range from humid tropical in the north, through semi-arid to arid in the center, to cool temperate and even periglacial in the south. Aided by steep gradients, these basins have faster response times and supply much sand and gravel to the shore. Despite this contrast, however, the sediment load reaching Atlantic shores is greater, and depositional landforms are thus better developed there.

The direct influence of coastal climate may be illustrated by comparing tropical and extra-tropical coasts. The tropical coasts of Pacific Colombia and Ecuador, and of the Caribbean and Brazilian sectors, have rain forests, mangrove swamps, coral reefs, and abundant sediment. Depositional processes along the Atlantic coast favor deltaic and estuarine sedimentation, and the growth of spits and barrier beaches across coastal lagoons. Such features are smaller along the more mountainous Caribbean and Pacific coasts. Although favored by abundant mud, the distribution of mangroves also reflects the influence of climatic and marine factors; their southern limit reaches almost 29°S along the Atlantic coast but, in the presence of cool ocean waters, only ~3°S along the Pacific coast (see chapters 9 and 10; Bird and Ramos, 1985).

The remaining environments of the Pacific coast, and of the Atlantic coast south of the Río de la Plata, lack the characteristic features of tropical coasts described above, with biogenic reefs and mangrove swamps notably absent. The coasts of Perú and northern Chile, though not strictly extratropical, share the desiccating effects of cool ocean waters and anticyclonic circulation systems (see chapter 3). Desert conditions prevail, streams are ephemeral or intermittent, sediment supplies are sporadic but may increase during El Niño wet events, and extensive beaches and associated dunes are scarce (see chapters 10 and 19). Desert conditions gradually disappear farther south until, in north-central Chile, winter rains and stream flows introduce more Andean sediment, and sandy coasts become more common (see chapter 11). In the wet temperate climate of southern Chile, however, forested landscapes reduce sediment yields to the shore and sandy shorelines are fewer. In the far south, in the cold wet climates of western Patagonia, forests come close to the shore in the inner fjords while coastal tundra occurs in the outer archipelagoes. Ice-scoured landscapes inherited from Pleistocene glaciations, and from some glaciers that still reach tidewater, limit the amount of sediment available for beach construction. The east coast of Patagonia is mostly rocky, augmented by coarse sands and gravels that have been transported across the semi-arid steppes from the Andes (see chapter 14). Farther north in Argentina, a wetter, more temperate climate and massive inputs of sediment from the Paraná-Paraguay drainage system, derived from both Andean and Brazilian sources, are reflected in the Río de la Plata estuary. Sandy shores are important here, and littoral drift carries sand northward to form barrier beaches along the shores of Uruguay and southernmost Brazil.

15.2.2 Oceanic Factors

Davies (1977) recognized four wave environments for the South America coast, related to the dominant wave fetch zones in the adjacent oceans (fig. 15.4). The *West Coast Swell Environment* extends along the Pacific coast from Central America southward to ~40°S. Predominant swells approach this coast from the southwest, owing to forcing effects of cyclonic activity on the surface of the Southern Ocean. These are high-energy waves because of the persistently large fetches and the vastness of the propagation zone. The *East Coast Swell Environment* extends along the Atlantic coast from the Guianas to ~40°S. From Guyana to Cabo São Roque, swells are masked by dominant NE wave systems forced by the trade winds. In contrast, from Cabo São Roque southward to ~40°S, SE swells predominate but energy levels are lower than along the Pacific coastline. The *Storm Wave Environment* dominates the Patagonian coast, a reflection of intense and persistent cyclonic activity and strong winds over the Southern Ocean and adjacent parts

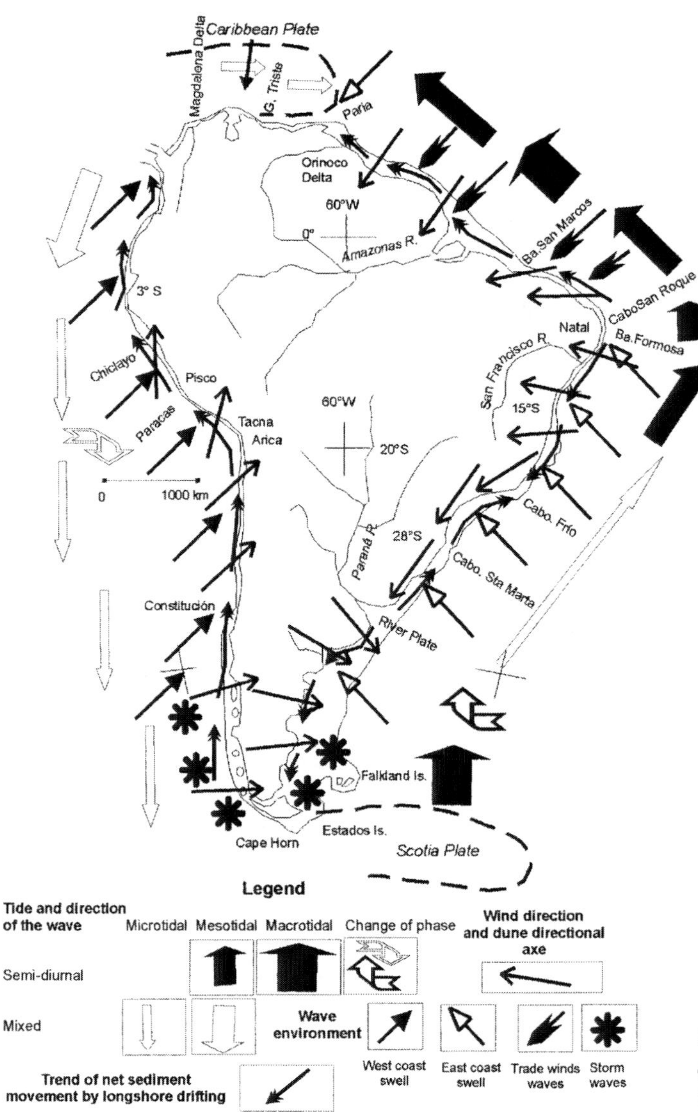

Figure 15.4 Oceanic factors influencing the South American coast. Wave environments and net sediment movement based partly on Davies (1977, 1982).

of the Atlantic and Pacific oceans. High storm waves interact with the regional westerly swells to create a very high-energy environment. The Caribbean coast is a *Protected Sea Environment* with almost continuous low energy. Protected from southerly swells by its north-facing aspect, and from trade-wind waves by the Caribbean islands, this coast also lies just beyond the normal tracks taken by tropical cyclones (hurricanes) passing westward through the Caribbean Sea.

These wave environments directly influence wave-induced longshore currents, and therefore net sediment transport and depositional features along the coast (fig. 15.4). On the Atlantic coast, from Bahía Formosa just south of Cabo São Roque to the Guianas, predominant littoral drift is toward the northwest, thereby allowing finer sediment brought down by the Amazon to supply mudflats off the Guianas. In contrast, from Bahía Formosa south to Cabo Frío, littoral drift is predominantly toward the southwest. Beyond Cabo Frío, the principal source of sediment is the Paraná-Paraguay river system entering the Río de la Plata estuary, and littoral drift is regulated by coastal orientation relative to SE swells. Thus, one portion of this sediment load moves northeast to supply the large barrier beaches of Uruguay and southernmost Brazil, while the other portion is directed to the southwest, providing sand to the beaches and dunes of the Buenos Aires-Bahía Blanca region. In contrast, along the Pacific coast, high-energy SW swells approach the coast obliquely, producing predominantly northward littoral drift. Littoral drift off Patagonia is more complex because of exposure to high-energy storm waves, frequent changes in coastal orientation, and in Chile deep tidewater passages close to shore.

Tides determine the level, and range of levels, through which waves and wave-induced and tidal currents function most effectively. Thus tidal direction, frequency, and magnitude are important factors, reflecting a large number of astronomic and oceanic variables, including shelf width (fig. 15.4). Tidal waves propagate southward along the Pacific coast, northward along the Atlantic coast, and eastward through the Caribbean. Tidal regimes are mixed with a semi-diurnal trend, semi-diurnal with a mixed trend, and mixed, respectively (Davies, 1982). In the Pacific, the range of spring tides is lower macrotidal (3.5–5.5 m) along the Columbian coast, declines southward to microtidal (<1 m) in Perú, and is lower mesotidal (1–2 m) farther south. On the west coast of Patagonia, ranges are again higher, becoming macrotidal in some fjords and generating strong tidal currents (Araya-Vergara, 1985a). Along the Atlantic coast of Patagonia and the eastern Strait of Magellan, ranges are upper macrotidal (>5.5 m) and reach 9 m in San Antonio Bay. Tidal ranges decline northward to microtidal in southern Brazil, then increase to upper macrotidal off the Amazon mouth (Fleming, 2005). The Caribbean coast is microtidal.

The onshore effects of locally dominant winds are well expressed in the directional axes of coastal dunes. These axes show that barchan dunes along the coast of Perú are shaped by south and southeast winds, while dunes in north-central Chile are shaped predominantly by southwest winds. Despite the prevalence of strong westerly winds, dune systems are rare in Patagonia, due partly to the paucity of suitable sand, partly to the ruggedness of its rocky coasts, and partly because winds blow offshore to the Atlantic. From just south of the Río de la Plata estuary northward to the Santos embayment, coastal-dune axes reflect locally dominant NNE winds blowing subparallel with the coast, but from Cape Frío to Bahía Formosa dominant winds move around from NE to SE. Beyond Bahía Formosa, dunes along the north coast of Brazil reflect the dominance of NE trade winds, for example in the dunes near the Parnaíba River delta.

15.3 Regional Coastal Groups

Despite many attempts, coastal classification schemes have had mixed success, owing to differences in analytical scales, systematic knowledge, and interpretive goals between investigators. This chapter seeks to clarify the essential regional characteristics of the South American coast by reference to hierarchies of landforms and tectonic classes, following Valentin (1969) and Inman and Nordstrom (1971). Based on the tectonic groups identified earlier, the scheme presented in table 15.4 recognizes four orders of coastal features in a nested hierarchy: the dominion, the principal morphogenetic feature of the coast; the regional type, commonly expressed by tectonic style; and two orders of key regional landforms. This scheme ignores marine terraces and continental shelves, reviewed earlier, and also ubiquitous features such as beaches and cliffs because they are not regional keys.

15.3.1 Pacific

Colombian-Ecuadorian Group The tectonic identity of this group is based on several criteria, such as the regional volcano-tectonic segmentation of the Andes (Hall and Wood, 1985), the kinematics of diffuse faulting (Marshall and Fisher, 2000), the geodynamics of slab subduction beneath the Andean margin (Gutscher et al., 2000), and the effects of intracontinental deformation (Taboada et al., 2000). The interaction between these mechanisms is expressed in the broad arcuate form of the coast, which also includes some transverse megashears that may explain certain offset features (Clapperton, 1993). Within this arcuate coastline are various rías and elliptical or zetaform embayments. In southern Colombia, this crenellated coastline gives way to a *cala* coast of small bays and inlets that reflect significant erosional processes. Mangrove swamps and coral reefs are nested within these features (fig. 15.5; Ayón and Jara, 1985; Schwartz, 1985a). In Ecuador, the mangrove swamps and salt marshes of Ancon de Sardinas Bay and the Guayas River estuary are notable, but owing to cool ocean waters, mangroves extend no farther south than Punta Malpelo (3°S) in northernmost Perú (Bird and Ramos, 1985).

Peruvian Group The coast of Perú is semi-arid to arid. Its tectonic dominion is largely explained by the geodynamics of flat-slab subduction of the Nazca plate beneath the Andean margin (see chapter 2; Gutscher et al., 2000). Arc tectonics seem to have produced a transverse megashear that divides the coast into two sectors, north and south of the Paracas Peninsula (Clapperton, 1993), and may explain the combination of longitudinal, arcuate, and offset tectonic styles along this coast. North of Paracas, the predominantly rectilinear coastline seems to be of tectonic origin, and the open embayments here have strong structural controls. The coast straightens where it has formed in the unconsolidated material of alluvial fans and bajadas of the Andean piedmont, often forming high cliffs such as those near Lima (fig. 15.5; Le Roux et al., 2000). Coastal erosion here is thought to reflect the combined action of Pleistocene eustatic changes of sea level and strong tectonic uplift. These effects are more common south of Paracas, where the relatively straight coast reflects a coast range in the north and the straightening and truncation of erodible Andean bajadas and pediments in the south. Some zetaform bays and pocket beaches favor coastal dune development, particularly where fluvial processes have delivered Andean sands to the shore (Gay, 1999).

North-Central Chilean Group The orientation of the Chilean coastline is strongly influenced by the north-south

Table 15.4 Coastal dominions, regional types, and key regional landforms

Tectonic group	Dominion (Order 1)	Regional type (Order 2)	Key regional landforms (Order 3)	(Order 4)
Pacific				
Colombian-Ecuadorean	Tectonic-erosional	Arcuate, offset	Rías, calas, zetaform bays	mangrove swamps coral reefs
Peruvian	Tectonic-abrasional	Longitudinal, arcuate, offset, coast range	Semi-structural bights, piedmont	Pocket beaches, dunes, ramps
North-central Chile	Tectonic, abrasional-accumulative	Longitudinal, aligned, offset	Zetaform and elliptical bights, rías	Estuarine deltas, dunes
Patagonian	Tectonic-glacial	Reticulate, arcuate	Glacial troughs, fjords, piedmont lobes	Calving glaciers, fjord deltas, glacial drift cliffs
Atlantic				
Argentinean	Erosional-abrasional	Lobate	Rías	Beach ridges salt marshes
Uruguayan	Tectonic-accumulative	Littoral-drift aligned	Segmented barrier-lagoons	Zetaform bays, marshes, dunes
Santos embayment	Accumulative-abrasional	Tectonic-offset	Zetaform and elliptical bights, barrier-lagoons	beach ridges, barriers, spits, dunes, mangroves
Bahía	Accumulative-abrasional	Tectonic-offset, littoral-drift aligned	Zetaform bights deltas, cuspate forelands	Elliptical bays, dunes, mangroves
Pernambuco-Paraíba	abrasional	Offset	Zetaform and elliptical bights,	Coral reefs, soft cliffs
Northeastern	Abrasional-accumulative	Offset	Zetaform bights, transgressive dunes	Coral reefs, mangrove swamps
Amazon-Guianas-Orinoco	Erosional-abrasional-accumulative	Offset	Rías, estuarine and lobate deltas	Calas, mudflats, beach ridges, mangrove swamps
Caribbean				
Venezuelan-Colombian	Tectonic-erosional	Offset, arcuate lobate	elliptical bights arcuate deltas aligned cliffs	Coral reefs, mangrove swamps, beach ridges

alignment of Andean orogenic structures (Araya-Vergara, 1976, 1985a; Paskoff, 1989). A coastal cordillera occurs throughout much of this region, its presence and offset forms linked with fore-arc tectonics. Key regional landforms, such as elliptical and zetaform embayments and mega-cliffs, are explained by tectonic segmentation. Abrasional features include extensive cliffs cut in relatively soft Cenozoic sediments. Accumulative features become increasingly common southward, as the climate becomes more humid, and fluvial processes and sediment delivery more assured. These changes are reflected in river outlets: prograded rías along the semi-arid coasts in the north, estuarine deltas in central Chile, and rías along the humid temperate coast farther south (fig. 15.5). The lack of marine deltas, or their stunted occurrence as blunt forelands, bear testimony to the high-energy conditions of this coast (Araya-Vergara, 1985b). Dune systems are common along the coast, from the semi-arid north to the humid south, where a large active dune field rises behind the zetaform embayment south of the Arauco Peninsula (Araya-Vergara, 1987, 1996). In this and similar embayments, the largest masses of dune sand, of both Pleistocene and Holocene age, occur to leeward of the prevailing southwesterly winds.

Patagonian Group Reticulate and arcuate tectonic structures and the legacies of Pleistocene glaciation are the principal influences on the west coast of Patagonia. Fjords, glacial troughs, and piedmont glacier lobes are the key

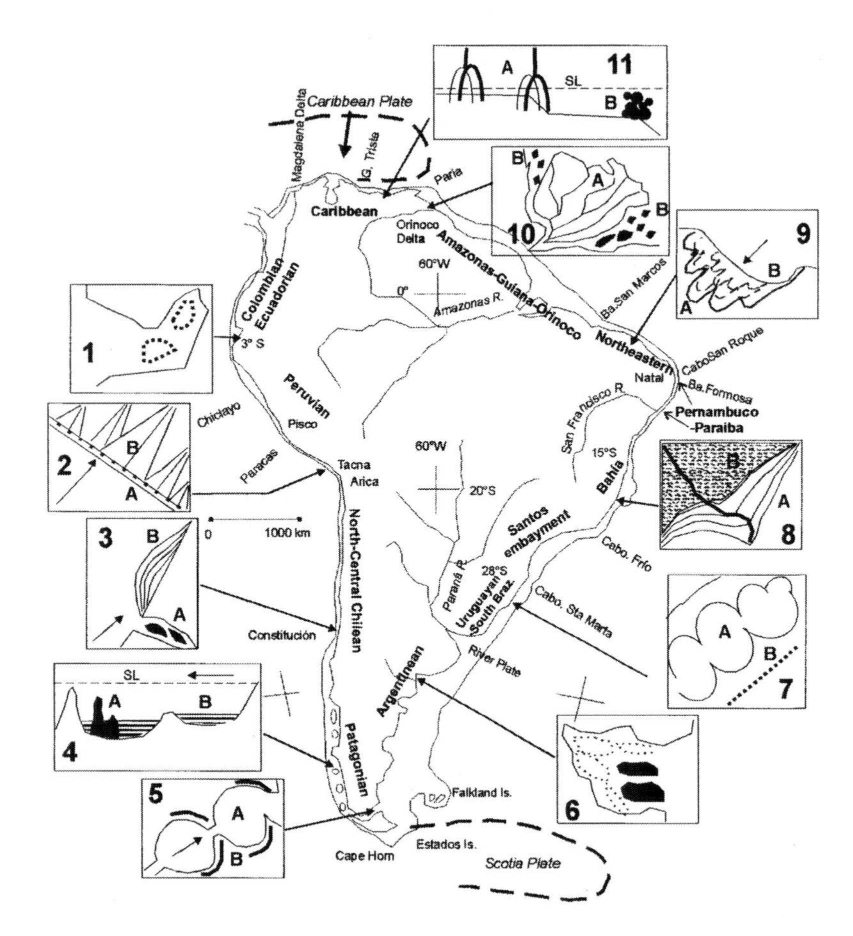

Figure 15.5 Regional coastal groups and key features

Key

1. Ría and mangrove swamp
2. A: cliff and ramp, B: piedmont
3. A: estuarine delta, B: rounded foreland
4. Fjord. A: moraine bank, B: ponded esplanade
5. Glacial piedmont. A: glacier lobe, B: barrier, spit
6. Ría and estuarine delta
7. Segmented lagoon A: lagoon, B: barrier, spit
8. Cuspate foreland. A: beach ridges, B: delta
9. A: transgressive dune, B: zetaform bight
10. Delta form. A: convex marine, B: estuarine
11. A: mangrove swamp, B: coral reef

regional landforms. Three types of fjord bottom have been identified from submarine profiles (Araya-Vergara, 1999). In the north, fjords contain ponding esplanades composed of laminated sedimentary beds, alternating occasionally with morainic banks or ridges (fig. 15.5). In the center, fjords backed by calving outlet glaciers draining from the Patagonian Ice Field have regular ponding esplanades and morainic ridges toward their distal ends. In the south (Magallanes), fjord bottoms have only embryonic development, lacking true ponding esplanades and morainic banks. Two types of submarine piedmont lobes also occur (Araya-Vergara, 2000). In the north, irregular bottoms are formed by dissected lobes inherited from the Last Glaciation (fig. 15.5). In Magallanes, piedmont lobes are smooth and usually shallow, corresponding to glacier grounding lines, and there is a close fit between coastal landforms and glacial and glacio-marine features produced mostly during late Pleistocene and Holocene glaciations (fig. 15.5).

5.3.2 Atlantic

Argentinean Group This coast is dominated by erodible sedimentary rocks fronting Patagonian plateaus and Pampas plains, and exhibits a wide array of erosional and abrasional features. The broadly scalloped outline of the coast reflects tectonic flexuring and basin formation across the continental margins of the South American plate, for example in the San Jorge Basin between the Somun Cura and Deseado massifs in Patagonia (see chapter 1; Urien and Martins, 1999). These structural styles produce mega-lobes, such as the Pampas coast south of Buenos Aires, and large gulfs such as San Jorge Gulf farther south (Psuty and Mizobe, 2005; Schnack, 1985). Key regional landforms within the gulfs include rías and, within more sheltered areas, estuarine deltas and marshes such as those of the Colorado River south of Bahía Blanca and the combined Paraná-Uruguay system entering the Río de la Plata. Along more exposed coasts, erodible cliffs and veneer ramps are common, and low rock strength ensures that cliff retreat is often rapid and the coast is smooth. Beach ridges are locally prominent, especially in Patagonia where Pleistocene glaciers and glaciofluvial outwash gravels reached the coast, and sequences of such ridges reflect post-glacial isostatic uplift (Schnack, 1985).

Uruguayan-South Brazilian Group Compared to its relatively smooth southern shore, the north shore of the Río de la Plata estuary is more rugged, especially where metamorphic and igneous rocks of the Río de la Plata craton and adjacent orogenic belts reach the shore (see chapter 1; Jackson, 1985). In regional terms, this coast is dominated by northward littoral drift and large zetaform bights, and accumulation is favored by the abundant sediment delivered to the Río de la Plata estuary. Within this context, the open coast as far north as Cabo Santa Marta is dominated by large barrier-lagoon systems, notably Lagoa Mirim and Lagoa dos Patos, associated with extensive dunes, salt marshes, and lagoon segmentation (fig. 15.5; Cruz et al., 1985).

Santos Embayment Group The coastline from Cabo Santa Marta to Cabo Frío forms a large bight dominated by a mix of accumulative and abrasional features, framed by tectonic offsets. Key regional features include elliptical and zetaform bays, and barrier-lagoon systems whose segmentation is associated with the formation of internal spits and barrier beaches (Cruz et al., 1985). The broad features of the open coast have been explained in terms of interactions between shore dynamics and the inner continental shelf, with barrier sands derived mostly from the adjacent sea floor rather than from rivers whose sediment is trapped inside coastal lagoons (Muehe and Valentini, 1998). The principal barriers also carry extensive dune systems. Geochronologic data from some of these barriers confirm a relative fall of sea level for later Holocene time.

Bahía Group Between Cabo Frío and the São Francisco River, the coastline is a spatial and temporal mix of accumulative and abrasional features, offset by regional tectonic structures. Accumulation is clearly represented by deltas and barrier beaches, but abrasion is deduced from the truncation of deltas and cuspate forelands (Cruz et al., 1985). Within these deltas and forelands, geochronologic data indicate that beach ridges have formed in response to estuarine processes and relative sea-level changes throughout later Pleistocene and Holocene time, and are still being formed (Martin et al., 1993). Relict Pleistocene deltas and Holocene delta-ridge complexes are notable at the mouths of the Paraíba do Sul, Doce, Jequitinhonha, and São Francisco rivers, while cuspate forelands are most prominent at Ponta do Baleia and Ponta do Corumbaú (fig. 15.5; Cruz et al, 1985).

Pernambuco-Paraíba Group From the São Francisco delta north to Bahía Formosa, the coastline is smooth in detail because it is backed by sandstone cliffs, including those developed in Pliocene sedimentary formations of the Barreiras Group, giving it an abrasional and aligned character. The open elliptical and zetaform embayments are explained by offset tectonics. South-east trade winds and more southerly swells move high waves onshore, forming massive beach ridges, but sheltered rías and estuaries are widely fringed by mangrove swamps (Cruz et al., 1985). Coral reefs extend for about 500 km along the coast, which also features extensive shore-parallel beach calcarenites and aeolianites, both onshore and submerged offshore, the dating of which helps to define relative sea-level changes during Holocene time (Psuty and Mizobe, 2005).

Northeastern Group From Bahía Formosa to Bahía de San Marcos, abrasional and offset tectonic features of the coast are expressed in rounded headlands located at the updrift end of well-developed zetaform embayments (fig. 15.5). Net accumulation is indicated by widespread dune systems whose transgressive nature becomes more evident toward the downwind end of the zetaform embayments. Beach ridges and dunes are more common where rivers introduce sand suitable for reworking by powerful onshore waves and winds driven by the north-east trade winds. Mangrove swamps are extensive in more sheltered locations (Psuty and Mizobe, 2005; Cruz et al., 1985). Deltas, such as that of the Paraíba do Norte River, offer useful data for reconstructing patterns of deposition during Pleistocene and Holocene times (Martin et al., 1993).

Amazonas-Guianas-Orinoco Group From Bahía de San Marcos to the Orinoco delta and the Paria Peninsula, the erosional nature of the coast is expressed by rías and smaller *calas*. The distribution of the inlets seems to be related to structural offsets. Some rías contain vast estuarine deltas, including those of the Amazon and Orinoco rivers, which give the coast its accumulative character (fig. 15.5; Cruz et al., 1985; Psuty, 1985; Psuty and Mizobe, 2005; Schwartz, 1985b; Turenne, 1985). The abundance of Amazon sediment and powerful littoral drift toward the west provide this coast with extensive mudflats, mangrove swamps, and beach-ridge plains. Sedimentation among the many islands and distributaries in the mouths of the Amazon and Pará (Tocantins) is subject to rapid change, resulting from the interaction of waves, strong tidal currents, and massive water and sediment discharge from these rivers. Suspended sediment, readily identified far beyond the continental shelf, eventually contributes to the vast Amazon submarine fan beyond the continental slope.

15.3.3 Caribbean

From the Paria Peninsula westward to the Isthmus of Panamá, the Caribbean coasts of Venezuela and Colombia reflect the arcuate and digitate structures of the Andean orogenic system and their subsequent erosion by rivers (Schubert, 1981; Kellog, 1984). Two different sectors are separated by the Triste Gulf. To the east, the coast is aligned parallel with the dominant east-west structures of the Coastal Cordillera, and coastal cliffs occur on the Paria Peninsula and around Caracas. Between these cliff segments, a breach in the coastal range harbors a large elliptical embayment containing the principal depositional landforms of this sector, barrier-lagoon systems showing much recent sedimentation (Ellenberg, 1985). Mangrove fringes also occur, for example around Laguna de Unare, and fringing reefs appear under favorable conditions (fig. 15.5).

To the west of Triste Gulf, the coast features lobate peninsulas and gulfs that reflect tectonic offsets within and between the several fingers of the Northern Andes and the Maracaibo lowlands and valleys of the north-flowing Magdalena, Cauca, and Atrato rivers. Continuing tectonic activity in the rising Guajira Peninsula and the Coastal Cordillera farther west, and in subsidence elsewhere, adds to coastal instability. Barrier beaches, mangrove swamps, and fringing coral reefs are common (Ellenberg, 1985). *Cayos*, small islands offshore, are similar to the Florida Keys. Interactions between fluvial and marine processes are most evident in the Magdalena delta, whose distributaries provide abundant Andean sediment for reworking into beach ridges by waves and currents. Truncation of these features reflects changes in accretion and erosion patterns during Holocene time (Psuty and Mizobe, 2005).

15.4 Conclusion

Compared to the coasts of Europe and parts of North America and Asia, the coast of South America has as yet been little modified by human activities. There are, however, notable exceptions, for example in the Río de la Plata estuary and the major port cities of central Chile, the Santos embayment, and the Caribbean shore, where engineering structures interfere with natural processes and pollution is a growing problem. More significant in the long term are the changes occurring in drainage basins that contribute sediment to the coast, notably in the Amazon basin and various Andean Pacific basins where increased sediment yields resulting from land-use changes are likely to impact coastal sediment budgets and lead to modifications at the shore.

References

Araya-Vergara, J.F., 1976. Reconocimiento de tipos e individuos geomorfológicos regionales en la costa de Chile. *Informaciones Geográficas de Chile*, 23, 9–30.

Araya-Vergara, J.F., 1985a. Chile. In: E.C.F. Bird and M.L. Schwartz (Editors), *The World's Coastline.* Van Nostrand Reinhold, New York, 57–67.

Araya-Vergara, J.F., 1985b. Sediment supply and morphogenetic response on a high wave energy west coast. *Zeitschrift für Geomorpholgie, Supplementband,* 57, 67–79.

Araya-Vergara, J.F., 1987. The evolution of modern coastal dune systems in central Chile. In: V. Gardiner (Editor), *International Geomorphology II.* John Wiley and Sons, Chichester, 1231–1239.

Araya-Vergara, J.F., 1996. Sistema de interacción oleaje-playa frente a los *ergs* de Chanco y Arauco, Chile. *Gayana Oceanología*, 4(2), 159–167.

Araya-Vergara, J.F., 1999. Perfiles longitudinales de fiordos de Patagonia central. *Ciencia y Technología del Mar*, 22, 3–29.

Araya-Vergara, J.F., 2000. Perfiles submarinos por los piedmonts del Estrecho de Magallanes y Bahía Nassau, Chile Austral. *Anales del Instituto de la Patagonia*, 28, 23–40.

Auer, V., 1970. *The Pleistocene of Fuego-Patagonia. Part V: Quaternary Problems of Southern South America.* Suomalainen Tiedeakademia, Helsinki.

Ayón, H., and W. Jara, 1985. Ecuador. In: E.C.F. Bird, and M.L. Schwartz (Editors), *The World's Coastline.* Van Nostrand Reinhold, New York, 49–52.

Barker, P.E., and J. Burrell, 1977. The opening of Drake Passage. *Marine Geology*, 25, 15–34.

Bird, E.C.F., 1985. *Coastline Changes: A Global Review.* John Wiley and Sons, Chichester.

Bird, E.C.F., and K. Koike, 1981. *Coastal Dynamics and Scientific Sites.* Commission on Coastal Environments, International Geographical Union, Department of Geography, Komazawa University, Tokyo.

Bird, E.C.F., and V.T. Ramos, 1985. Perú. In: E.C.F. Bird and M.L. Schwartz (Editors), *The World's Coastline.* Van Nostrand Reinhold, New York, 53–56.

Bird, E.C.F. and M.L. Schwartz (Editors), 1985. *The World's Coastline.* Van Nostrand Reinhold, New York.

Bullard, E.C., J.E. Everett, and A.G. Smith, 1965. The fit of the continents around the Atlantic. *Philosophical Transactions of the Royal Society, London*, 258A, 41–51.

Chappell, J., and N.J. Shackleton, 1986. Oxygen isotopes and sea level. *Nature*, 324, 137–140.

Clapperton, C.M., 1993. *Quaternary Geology and Geomorphology of South America.* Elsevier, Amsterdam.

Clark, J.A., and A.L. Bloom, 1979. Hydro-isostasy and Holocene emergence of South America. In: Suguio, K., T.R. Fairchild, L. Martin, and J.M. Flexor (Editors), *Proceedings of the 1978 International Symposium on Coastal Evolution in the Quaternary.* IUGS, IGCP, UNESCO, Sao Paulo, 41–60.

Cruz, O., P.N. Coutinho, G.M. Duarte, A. Gomes, and D. Muehe, 1985. Brazil. In: Bird, E.C.F. and M.L. Schwartz (Editors), *The World's Coastline.* Van Nostrand Reinhold, New York, 85–91.

Davies, J.L., 1977. *Geographical Variation in Coastal Development.* Longman, London.

Davies, J.L., 1982. World net sediment transport. In: Schwartz, M.L. (Editor), *Encyclopedia of Beaches and Coastal Environments.* Hutchinson Ross, Stroudsburg, 881–882.

De Wit, M.J., 1977. The evolution of the Scotia arc as a key to the reconstruction of Gondwanaland. *Tectonophysics*, 30, 53–81.

Ellenberg, L., 1985. Venezuela. In: E.C.F. Bird and M.L. Schwartz (Editors), *The World's Coastline*, Van Nostrand Reinhold, New York, 105–113.

Fleming, B.W., 2005. Tidal environments. In: M.L. Schwartz (Editor), *Encyclopedia of Coastal Science*, Springer, Dordrecht, 954–958.

Gay, S.P., Jr., 1999. Observations regarding the movement of barchan sand dunes in the Nazca to Tanaca area of southern Perú. *Geomorphology*, 17, 279–293.

Gutschner, M.A., W. Spakman, H. Bijwaard, and E.R. Engdahl, 2000. Geodynamics of flat subduction: Seismicity and tomographic constraints from the Andean margin. *Tectonics*, 19, 814–833.

Hall, M., and C.A. Wood, 1985. Volcano-tectonic segmentation of the Northern Andes. *Geology*, 3, 203–207.

Inman, D.L., and C.E. Nordstrom, 1971. On the tectonic and morphologic classification of coasts. *Journal of Geology*, 79, 1–21.

Jackson, J.M., 1985. Uruguay. In: Bird, E.C.F. and M.L. Schwartz (Editors), *The World's Coastline.* Van Nostrand Reinhold, New York, 79–84.
Kellog, J.N., 1984. Cenozoic tectonic history of the Sierra Perija, Venezuela-Colombia, and adjacent basins. *Geological Society of America Memoir*, 162, 239–261.
Le Roux, J.P., C. Tavares, and F. Alayza, 2000. Possible influence of the Nazca Ridge on the development of the Rimac-Chillón alluvial fan. *Actas, IX Congreso Geológica Chileno,* Puerto Varas, Chile, 1, 792–795.
Marshall, J.S., and D.M. Fisher, 2000. Central Costa Rica deformed belt: Kinematics of diffuse faulting across the Western Panama block. *Tectonics*, 19, 468–492.
Martin, L., K. Suguio, and J.M. Flexor, 1993. As fluctuações de nível do mar durante o Quaternário superior e a evolução geológica de "deltas" brasileiros. *Boletin IGUSP,* 15, São Paulo.
Muehe, D., and E. Valentini, 1998. *O Litoral do Estado do Rio de Janeiro, Uma Caracterização Físico Ambiental.* Fundação de Estudos do Mar, Rio de Janeiro.
Orme, A.R., 2005. Africa: Coastal Geomorphology. In: M.L. Schwartz (Editor), *Encyclopedia of Coastal Science*, Springer, Dordrecht, 9–21.
Naumann, T., and D. Geist, 2000. Physical volcanology and structural development of Cerro Azul Volcano, Isabela Island, Galapagos. *Bulletin of Volcanology*, 61, 497—514.
Parra, J.C., O. González-Ferrán, and J. Bannister, 1984. Aeromagnetic survey over the South Shetland Islands, Bransfield Strait and part of the Antarctic Peninsula. *Revista Geológica de Chile*, 23, 3–20.
Paskoff, R., 1989. Zonality and main geomorphic features of the Chilean coast. *Essener Geographische Arbeiten*, 18, 237–267.
Potter, P.E., 1994. Modern sands of South America: Composition, provenance and global significance. *Geologische Rundschau*, 83, 212–232.
Psuty, N.P., 1985. Surinam. In: E.C.F. Bird, and M.L. Schwartz (Editors), *The World's Coastline.* Van Nostrand Reinhold, New York, 99–101.
Psuty, N.P., and C. Mizobe, 2005. South America, coastal geomorphology. In: M.L. Schwartz (Editor), *Encyclopedia of Coastal Science.* Springer, Dordrecht, 905–909.
Radtke, U., 1987. Marine terraces in Chile (22°–32°S)—Geomorphology, chronostratigraphy and neotectonics: Preliminary results II. In: *Quaternary of South America and Antarctic Peninsula.* Balkena, Rotterdam, 5, 239–256.
Schnack, E.J., 1985. Argentina. In: E.C.F. Bird and M.L. Schwartz (Editors), *The World's Coastline.* Van Nostrand Reinhold, New York, 69–78.
Schubert, C., 1981. Are the Venezuelan fault systems part of the Caribbean plate boundary? *Geologishe Rundschau*, 70, 542–551.
Schwartz, M.L., 1985a. Pacific Colombia. In: E.C.F Bird, E.C.F. and M.L. Schwartz (Editors), *The World's Coastline.* Van Nostrand Reinhold, New York, 45–47.
Schwartz, M.L., 1985b. Guyana. In: E.C.F. Bird and M.L. Schwartz (Editors), *The World's Coastline.* Van Nostrand Reinhold, New York, 103–104.
Schwartz, M.L., 2005. *Encyclopedia of Coastal Science*, Springer, Dordrecht.
Stern, C.R., K. Futa, S. Saul, and M.A. Skewes, 1986. Nature and evolution of the subcontinental mantle lithosphere below southern South America and implications for Andean magma genesis. *Revista Geológica de Chile*, 27, 41–53.
Suess, E., 1900. *La Face de la Terre* (*Das Antlitz der Erde*). Volume 2, Colin, Paris.
Taboada, A., L.A. Rivera, A. Fuenzalida, A. Cisternas, P. Hervé, H. Bijwaard, J. Olaya, and C. Rivera, 2000. Geodynamics of the northern Andes: Subduction and intracontinental deformation (Colombia). *Tectonics*, 19, 787–813.
Turenne, J.F., 1985. French Guiana. In: E.C.F. Bird and M.L. Schwartz (Editors), *The World's Coastline.* Van Nostrand Reinhold, New York, 93–97.
Urien, C.M., and L.R. Martins, 1999. The southern South America continental terrace basins. In: R.L. Martins and C.I. Santana (Editors), *Non-living Resources of Southern Brazilian Coastal Zone and Continental Margin.* IOC-Unesco, Porto Alegre, 92–109.
Valentin, H., 1969. Principles of a handbook on regional coastal geomorphology of the world. *Zeitschrift für Geomorphologie*, 13, 124–129.

III

NATURE IN THE HUMAN CONTEXT

16

Pre-European Human Impacts on Tropical Lowland Environments

William M. Denevan

An ecology without Man . . . is true only
for an environment without Man.
Sauer (1958: 107)

The topic of early human impacts on New World environments, including Amazonia, is controversial as to degree and extent (Balée, 1989; Cleary, 2001; Denevan, 1992b; Gómez-Pompa and Kaus, 1992; Hayashida, 2005; Krech, 1999; Stahl, 1996; Vale, 2002). Certainly, whenever and wherever people were present, even sparse populations, there was some change in the landscape. People cannot live on the land and use plants and animals for subsistence and other needs without changing that land and the plants and animals present. Human-induced changes may involve equilibrium in which the natural ecosystem basically recovers, even though composition is changed, or in which the human ecosystem is sustainable; or the changes may involve disequilibrium in which the regenerative capacities (i.e., biological diversity and productivity) of either system are retarded or destroyed (i.e., degradation) (Sponsel, 1992). These changes may have been intentional or not, ephemeral or long lasting, localized or regional, one-time events or cumulative, highly visible or not readily apparent.

Vegetation is the most widespread focus of change; other impacts have been on wildlife, soil, river patterns, microclimate, and the land surface itself. The forces of change are settlement, cultivation, grazing, hunting and gathering (foraging), burning, and various earthworks and river works. All of these impacts and forces were present in pre-European South America. Furthermore, native people "recognize that human beings, past and present . . . have affected the distribution of the biota and the formation of the landscape" (Balée, 2003: 285–286). "Far from being a wild world . . . the forest is perceived as a superhuman garden" by the Achuar in Ecuador (Descola, 1994: 324).

Lacking written records, it is difficult to reconstruct early indigenous impacts, and any attempt to do so involves speculation and inference. Some alterations have persisted to the present, but most have been masked by either landscape recovery or by human destruction. Furthermore, nature and culture merge, the dichotomy being artificial, conceptual (Descola, 1994: 1–6). And some human disturbances that seem old are actually recent, but the reverse is also true. The distinction may be difficult to ascertain.

Archaeology and landscape analysis can be informative, but they are nevertheless limited in spatial coverage. Ethnohistorical and ethnographical information can be projected backward and this is done here; however, we do not know for certain whether such information is representative of pre-1492 people (Erickson, 1995; Roosevelt,

1989). There is paleoecological evidence of human impacts: soils, pollen, charcoal, and plant and animal remains. Finally, reconstruction of past human impacts is complicated by the fact that natural environmental change (especially climate) has occurred along with anthropogenic change. As result, separation of human from natural factors in explaining past change is difficult. Also, fire is a major cause of vegetation change, but often it is not clear whether the source of ignition is natural (lightning) or human, even today. Geographers have been particularly concerned with sorting out the role of humans in environmental change; however, the topic is interdisciplinary.

There is considerable awareness now of prehistoric human impacts on the Neotropical environment, but there are earlier antecedents. The pioneer botanist, Carl F.P. von Martius, who was in Brazil in 1817–1820, "noted that in the Amazon nature had already been transformed by humans for several millennia" (in Barreto and Machado, 2001: 243). And Alexander von Humboldt (1869: 193) in 1808 said that it was a "misconception" that tropical forests were "primeval."

This chapter focuses on the Amazon basin, the area I know best, and on the Orinoco region, the Guyanas, northern parts of Colombia and Venezuela, Panama, the humid Pacific coast of Ecuador, and the Brazilian coastal region.

Amazonia, scene of great human activity today, is generally perceived to have had sparse pre-European populations, with widely dispersed, small, temporary villages, with extensive forms of subsistence and with a simple stone-tool technology. This image has undergone a radical transformation in recent years (Denevan, 2001: 130–131; Heckenberger, 2005; Roosevelt, 1999a; Stahl, 2002; Viveiros de Castro, 1996).

Early populations of course were small and scattered. The oldest known human presence in Amazonia, around 11,200 B.P. (before present), is at Pedra Pintada in the upland near Monte Alegre just north of the lower Amazon River (Roosevelt, 1999b). Here, subsistence was initially based on the collection of fruits, fish, shellfish, and small game. Trees with edible fruits were undoubtedly spread by Paleo-Indians (pre-agricultural). By 8,000 B.P., these foragers had become specialized fishers and shellfish collectors, leaving enormous shell mounds along the rivers. In the uplands, subsistence was broad-spectrum foraging for wild plants, fish, birds, and small animals. Projectile points were fine and rare, probably used for large fish and aquatic mammals. Big game was of minor importance, as confirmed by the bone assemblages. There is no indication of overhunting or severe depletion of fish or shellfish by Paleo-Indians or subsequent hunters and gatherers.

Hunting and gathering people were fully capable of having an impact on the distribution of useful plants and animals, as evidenced by groups that still survive, such as the Nukak: "The foragers of the past were without doubt builders of their environment" (Politis, 1999, 2001: 48).

In Panama and in Colombian Amazonia, several domesticated crops were probably being cultivated by 7,000–10,000 B,P., including yams (*Dioscorea* spp.), lerén (*Calathea allouia*), arrowroot (*Maranta arundinacea*), bottle gourd (*Lagenaria siceraria*), squash (*Cucurbita* spp.), and sweet potato (*Ipomoea batatas*). In Panama, manioc (*Manihot esculenta*) was present at 7,000 B.P., and maize (*Zea mays*) between 5,500 and 7,000 B.P. (Piperno and Pearsall, 1998: 203–227; Piperno et al., 2000). By 4,000 B.P. there is evidence of major forest clearance in several regions, apparently for maize agriculture (Piperno and Pearsall, 1998: 312–313).

The population of Greater Amazonia in 1492 is now believed to have numbered at least five and one-half million (Denevan, 1992a). There were large permanent villages stretching for kilometers along the bluffs of the Amazon River and major tributaries, with some settlements probably totaling 5,000 or even more people (Denevan, 1996). In the interior interfluves or uplands (*terra firme*), there were both small villages and scattered, large semi-permanent settlements, in contrast to historical villages, which are invariably small and shifting. In some of the savannas, there were also substantial populations (Denevan, 1966; Roosevelt, 1991: 38–39). In the floodplains, cultivation was seasonal and very productive. In the uplands, where shifting cultivation dominates today, it is apparent that intensive, semipermanent forms of cultivation were practiced in the past in some areas (Denevan, 2001: 115–127). Given such populations, settlements, and intensive cultivation, an argument can be made that in 1492 much of Amazonia was not pristine but had been modified in various ways, ranging from insignificant and subtle to complete transformation. Even sparse hunting, gathering, and fishing people had some effect.

16.1 Forest Modification

"The indigenous populations of Central and South America have been using, manipulating, and managing tropical forests for several thousand years . . . [but] to the untrained eye, the managed and the pristine can easily merge into one" (Peters, 2000: 203–204). Amazonian forests occupied by native people today are anthropogenic in various ways, but even more so in pre-European times when rural populations were in some places greater than they are now. Current forms of biotic modification are indicators of what may have taken place in the past—change in species presence, distribution, and density, as well as in forest biomass, age, structure, and other characteristics (Balée, 1994: 116–165; Peters, 2000; Rival, 2002: 68–93; Roosevelt, 2000; Smith, 1995). Some of these changes can persist in some form for hundreds of years. In abandoned fields, full recovery of forest biomass can take 190 years in the upper Río Negro region; recovery of species diversity can take up to 80 years (Saldarriaga et al., 1988; Brown and Lugo, 1990).

Fallows as successional vegetation may actually have a species diversity comparable to or greater than that of a mature forest, including a number of rare species, although a different kind of diversity (Balée, 1993, 1994: 136–137; Brown and Lugo, 1990). Fallows may be actively managed and utilized for several decades by maintaining a high proportion of useful plants (Denevan and Padoch, 1988; Peters, 2000). After a plot is last harvested, residual crops may be protected, fruit and other useful trees may be planted, and certain secondary species may be favored. Examples from the Bora fallows in Perú are uvilla (*Pourouma cecropiaefolia*), peach palm (*Bactris gasipaes*), guaba (*Inga* spp.), caimito (*Pouteria caimito*), umarí (*Poraqueiba sericea*), and cashew (*Anacardium occidentale*) (Denevan and Padoch, 1988: 41).

Mature fields are also manipulated. Unwanted saplings are replaced by weeding or felling, and new useful trees are introduced by planting or by protection of desirable volunteers that otherwise might be shaded out. Useful plants may be established at tree-fall openings, which are more frequent than realized (Denevan, 2001: 125–126). The result is to "increase the density of desirable species by decreasing the density of undesirable ones" (Peters, 2000: 211). The early successional vegetation in clearings and at burns tends to have a higher concentration of edible plants than does mature forest (Piperno et al., 1992).

Some abandoned plots enriched by one or a few species may be able to maintain that dominance for centuries (Smith, 1995) or indefinitely (Peters, 2000). Old Ka´apor village sites often contain dense stands of tucumá palms (*Astrocaryum vulgare*) and babassu palms (*Orbignya phalerata*), which may also dominate burned clearings (Balée, 1992: 47–48). The Kalapalo manage groves of piqui (*Caryocar brasilensis*), which extend for miles around abandoned villages (Basso, 1973: 35). Various other types of dense stands of palm and other fruit trees occur on archaeological sites (Balée, 1988; Peters, 2000). Numerous forest trees (wild orchards) develop from discarded seeds around settlements, camps, and activity areas (Politis, 1999, 2001). Other examples of trees providing prehistoric cultural signatures, including the Brazil nut (*Bertholletia excelsa*), are given by Smith (1995).

In addition, there may be intentional planting of cultivated seeds and tubers carried from fields and gardens to forest areas adjacent to villages, camps, trails, and fishing sites. Aside from settlements, there are thousands of kilometers of indigenous trails today in central Brazil and undoubtedly there were many more in the past. Posey (1985) along just 3 km of a Kayapó trail found 185 planted trees, about 1,500 medicinal plants, and about 5,500 food plants. Also, there are small hidden patches in the forest consisting of both cultivated and semi-wild plants that serve for emergency food supplies. These may occur at tree falls or where a tree has been felled to get at honey or fruits.

Foraging for useful wild plants can affect species presence and distribution. Undesired plants are removed, but other selected plants may be protected and even planted. "More than 76 per cent of the species used by the Kayapó that are not domesticates are nonetheless systematically selected for desirable traits [genetic manipulation] and [are] propagated" (Posey, 1992: 28–29).

Hunting is another factor. Some hunted species disperse plant seeds, so that plant patterns are affected by game depletion. Various mammals, especially monkeys and rodents such as agoutis (*Dasyprocta* spp.), are dispersers of fruit and nut seeds. Likewise, bird populations, important for pollination as well for dispersing seeds, are affected by hunters. In addition, there is an abnormal seed dispersal by animals attracted to human settlements and fields, thus influencing fallow compositions (Balée, 1993).

A good example of the creation of a partly humanized rainforest comes from the Huaorani (Jivaro) of the Ecuadorian Amazon (Rival, 2002: 68, 70, 80, 83, 91–92):

> [M]en, women, and children spend hours "cruising" in the forest . . . collecting food within a radius of 5 kilometers. . . . They explored the forest systematically, looking for . . . evidence of previous occupation [where useful plants are concentrated] . . . [the forest] is conceptualized as a patchwork of successional fallows . . . numerous plant species are encouraged to grow outside cultivated areas. . . . [They] exploit plants where they find them in the forest. As a result, they actively manage the forest . . . by altering the natural distribution of plant and animal species. . . . Past people are thought to [have] provide[d] in abundance for their descendants . . . the forest, far from being a pristine environment, is the product [or "historical record"] of the life activities of past generations that have transformed it into an environment rich in resources.

Other excellent studies of the impact on tropical forests by foraging activity are for the Hotï in the Guayana region of Venezuela (Zent and Zent, 2004) and for the Nukak in Colombian Amazonia (Politis, 1999, 2001).

Paleoecological data from the rainforests of Darién in southern Panama provide a 4,000-year record of human disturbance. The forest and wildlife of today have recovered from that disturbance, following the Spanish conquest and the subsequent demise of the native population. However, early disturbance "may have had a profound effect upon the modern forest associations" (Bush and Colinvaux, 1994: 1761). In central Panama, major forest disturbance and deforestation had taken place by 3,000 B.P., if not earlier. "That prehistoric hunter/gatherers and shifting cultivators using 'primitive' technologies exerted considerable influences on the structure and species composition of tropical ecosystems [in Panama] no longer seems in doubt" (Piperno et al., 1992: 123). Although separating human impacts from climate-induced change is difficult, pollen studies indicate that human agency in rain-forest disturbance in western Columbia probably has been "significant from the [early] Holocene onwards" (Marchant et al., 2004: 833).

The overall impacts of these types of activities may seem to be minor, but extent and cumulative effect can be significant. Fallows are managed, not natural, regrowth. In a study of young Bora fallows, 645 plants were collected along transects, and of these 207 (32%) were identified as being useful; there was a total of 118 useful species in these fallows (Denevan and Padoch, 1988: 52). Peters (2000) reports small tracts of forest with as many as 300 useful species present. Cultural surveys of forest trees in Amazonia indicate proportions of species that are useful for several indigenous groups: 82% for the Chácobo, 77% for the Ka´apor, 61% for the Tembé, and 49% for the Panare (Balick and Cox, 1996: 183). Use of any kind means a degree of human impact on the plants involved. And human presence increases the density of useful plants. In Yucatán, botanist David Campbell (personal communication, 2000) found that in a 10-km radius of Mayan villages today, 70–90% of the trees present have utility, compared to only 50% or less in more distant areas. Also, traditional pharmacopoeias (medicinal plants) in the Neotropics mostly occur in disturbed forest, not in primary forest (Voeks, 2004).

Balée (1989) estimates that at least 12% of the current Amazonian forest in Brazil is mostly anthropogenic, consisting of babassu forest, bamboo forest, liana forest, and Brazil-nut forest. This does not include smaller patches of artificial forest or extensive areas of manipulated forest. These anthropogenic forest modifications involve recent impacts by indigenous people; however, they are indicative of human impacts that undoubtedly occurred in pre-European times.

Amazonian forests, even where semi-deciduous with a long dry season, or during occasional severe droughts, are infrequently ignited from natural causes, major fires "perhaps occurring only once or twice per millennium on average" (Cochrane and Laurance, 2002: 321). This is because lightning without heavy rain is rare during dry periods. However, early Indians burned their clearings and burned for other reasons. During droughts, these fires could escape, causing forest fires that could be extensive, as we know from recent conflagrations. Forest recovery of biodiversity from hot forest fires can take many years (Fearnside, 1990), varying with fire frequency and extent (Cochrane and Schulze, 1999).

16.2 Savanna from Forest

Savanna (tropical grassland with or without scattered or clustered trees) occupies about 30% of Greater Amazonia. The largest savannas are the scrub grasslands (*cerrados*) of central Brazil, the Pantanal of Mato Grosso, the Orinoco Llanos, and the Llanos de Mojos of northeastern Bolivia. Some are well drained (*cerrados*) and others are subject to seasonal flooding (Mojos, Pantanal), while the Orinoco region has sectors of both.

Grasslands are explained in the wet tropics as being the result of extremely weathered (senile) soils (*cerrado* landscapes), alternating flooding and desiccation (Mojos, Pantanal, Orinoco), clearing and burning, or a combination of these in conjunction with long dry seasons. There has been a long and continuing debate over the extent to which the South American savannas are natural versus anthropogenic, with the consensus now being that most are natural (Sarmiento, 1984). They have clearly been expanded, however, by forest clearing in association with savanna burning, and some small savannas are primarily the result thereof.

Today, all the South American savannas are burned frequently, even annually, by both natives and other people, and there are historical reports of past indigenous burning. It has been assumed that such fires will gradually push back a forest-savanna boundary at the expense of the forest. However, observations by myself in Mojos and by others elsewhere (Eden, 1986) suggest that boundaries may be fairly stable and that in some places the forest is even advancing. This is because the forest is moist and relatively fire resistant, so that grass fires do not penetrate. The forest/grassland boundary in Mojos can contain fire-tolerant plants, such as Bromeliaceae, *Heliconia* spp., certain palms such as motacú (*Attalea princeps*), and species of *Acacia*, *Mimosa*, *Cassia*, and *Curatella*. Where clearings occur at a forest edge, however, savanna fires can move into abandoned fields, thus expanding the savanna. The extent of such expansion over thousands of years is difficult to estimate. On the other hand, recent research in eastern Brazilian Amazonia indicates that anthropogenic fire frequencies of less than 20 years, common in savannas, can by themselves eradicate rain-forest trees for a distance of 500–2,500 m from the forest edge during a short period of time (Cochrane and Laurance, 2002).

Regardless of how much savanna fires can cause forest retreat, fires do have an effect on the woody plants and grasses within savanna, favoring fire-tolerant species and affecting biodiversity. And while natural fires do occur, their frequency is far less than human burning, which, as result, has a different type of impact on vegetation.

There are examples of savannas created by indigenous people at various times. Figure 16.1 shows forest clearing from 2,200 to 9,000 BP, based on a long-term study of sediment cores in central and western Panama (Ranere, 1992; Piperno and Pearsall, 1998: 290–297). Over half this region was cleared of primary forest by 2,200 B.P. In eastern Panama, pollen data for the past 4,000 years show a surge

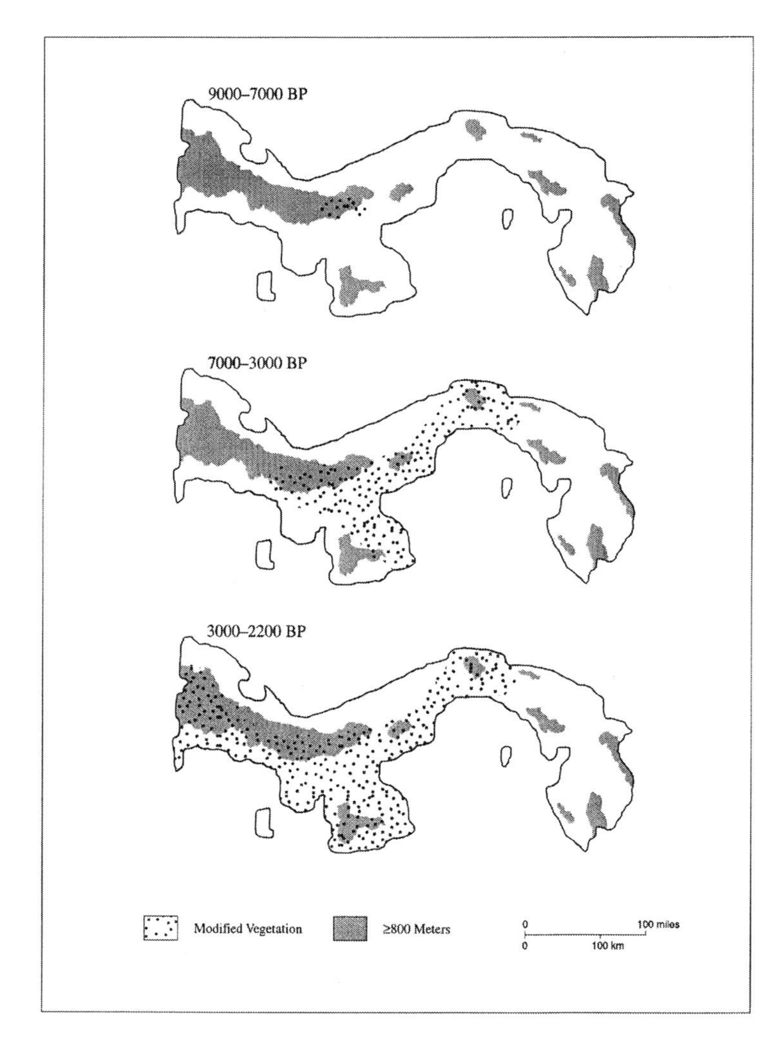

Figure 16.1 The known distribution of disturbed vegetation in Panama, including deforestation and conversion to savanna, for three periods prior to 2,200 B.P. based on pollen, phytoliths, and other indicators (from Ranere, 1992: 41)

of Gramineae (Poaceae), indicating grassland, and disturbed forest taxa, with charcoal influxes indicating burning. The forest recovered after A.D. 1600, which would be after native depopulation (Bush and Colinvaux, 1994). By about 3,300 B.P., "significant forest modification was certainly underway" on the north coast of Ecuador (Stahl, 2000: 254). In the sixteenth century, Spaniards described open, unforested sectors, including in Darién at the time of Balboa and in adjacent Dabeiba in present Colombia, as well as in western Panama. Soon after, much of this reverted to forest, except on cattle ranches (Sauer, 1966: 285–288; Bush and Colinvaux, 1994).

In the northern Colombian lowlands, west of the Río Cauca, a large sector was in anthropogenic savanna in the sixteenth century (fig. 16.2). With subsequent native depopulation and land abandonment, this region had recovered to forest by the eighteenth century, while much of the area north to Cartagena had been cleared for pasture. By the mid-twentieth century, most of the remaining northern forest had been cleared and the southern forest had once more reverted to savanna (Gordon, 1957: 57–78).

In the central-eastern Peruvian foothills, there is a low-lying plateau called the Gran Pajonal or Great Grassland. Actually it consists of dozens of small *pajonales* on the hillsides. Study of vegetation, soils, and climate indicates that these savannas originated as Campa (Ashánika) Indian swiddens, which were continuously burned after crop abandonment, with coalescence into larger patches. The anthropogenic origin of specific *pajonales* is confirmed over a period of 24 years by field observations, interviews with the local Campa, and aerial photographs. Historical records show that savannas were in existence in this remote region in the early eighteenth century, when Franciscan missionaries first arrived and gave the Gran Pajonal its name (Scott, 1978).

The open, white sand *campinas* (scrub savannas) of the lower Rio Negro in Brazil are believed to have once been closed *campina* or *caatinga* (scrub forest) that had been cleared and cultivated by Indians (Prance and Schubart, 1978). Ceramics and radiocarbon ages of charcoal indicate that these places were occupied about A.D. 800–1200. Subsequently, limited succession back to closed *campina* has taken place, and only very slowly due to the extremely poor

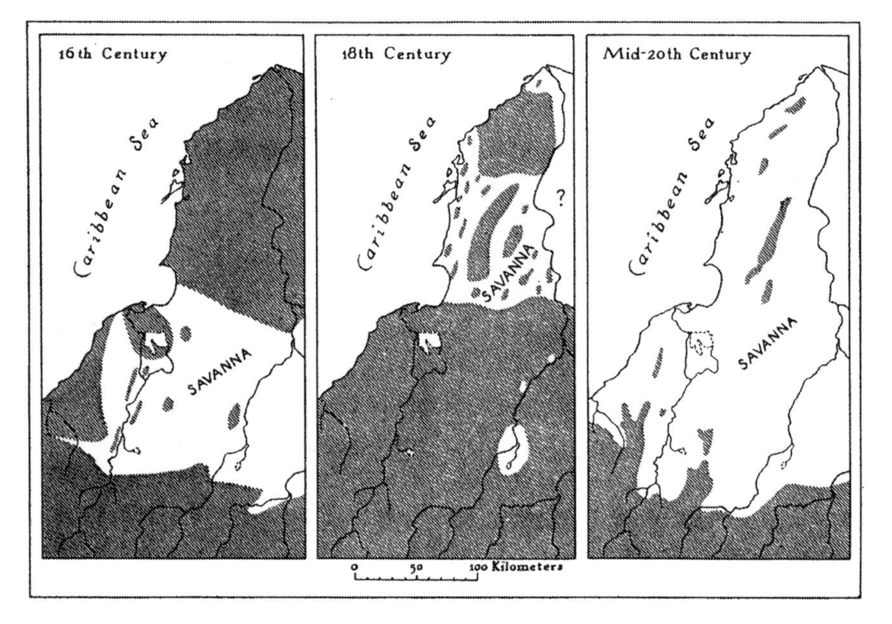

Figure 16.2 Changing extent of savanna vegetation in northwest Colombia: anthropogenic (indigenous) savanna in the south in the sixteenth century; forest recovery after depopulation in the south and deforestation (Spanish) in the north in the eighteenth century; and deforestation for pasture throughout in the mid-twentieth century (based on historical records, from Gordon, 1957: 69).

soil, except on patches of fertile *terra preta* soil (see below) that now contain closed *campina* forest.

Where landscapes were deforested and converted to savanna, whether gradually or rapidly, there would have been significant local and regional changes in water runoff, soils, microclimate, and wildlife.

16.3 Domesticated Plants

Indigenous management of tropical vegetation not only affected forest and savanna structure and composition, but also led to genetic alteration and creation of new species in the form of domesticated crops.

The most important food crops domesticated in Amazonia, northern Colombia-Venezuela, and southern Central America, are manioc and sweet potato. Others include squash, peanut (*Arachis hypogaea*), arrowroot, chili (*Capsicum* spp.), jack bean (*Canavalia* spp.), cocoyam (*Xanthosoma sagittifolium*), lerén, and yampee (*Dioscorea trifida*) (Piperno and Pearsall, 1998: 164–165).

Tree crops are a large and somewhat unclear category. The one true Amazonian tree domesticate is the peach palm or pejibaye (*Bactris gasipaes*). However, there are dozens of others, many of them palms, used for fruit, nuts, fibers, and medicinals, which can be categorized as semi-domesticates. These were modified to varying degrees by pre-European people and are now cared for and may be planted. For Amazonia and lowland northern South America, Clement (1999) lists 138 cultivated or managed plant species at European contact, of which 52 are domesticates, 41 are semi-domesticates, and 45 are incipient domesticates; 68% are trees. For discussion, lists, and identifications of domesticated, semi-domesticated, and other cultivated plants in South America, see Piperno and Pearsall (1998) and Denevan (2001: 307–323).

16.4 Animals and People

The distributions and densities of animals in the Neotropics were affected in the past by hunting, vegetation modification, and transfers from one place to another (Grayson, 2001). The game animals hunted by Amazonian Indians for both food and other products include large mammals such as monkeys, tapir (*Tapirus terrestris*), peccary (*Tayassu* spp.), deer (*Mazama* spp.), and smaller animals including paca (*Agouti paca*), agoutis, armadillo (*Dasypus* spp.), and capybara (*Hydrochaeris hydrochaeris*), as well as numerous birds, aquatic mammals, and fish.

Most hunting is done today within a 3–10 km radius of a village, with some sectors favored over others and with more infrequent distant hunts that last for several days. For Ache hunters in Paraguay, most hunting is within 8.5 km of their village, with normal game densities beyond (Hill et al., 1997). In the upper Río Napo region of Ecuador, 81% of the hunting days occurred in a territory of 590 km^2 (one day's travel or less) around a Siona-Secoya village, but intermittent hunting occurred over an additional 1,910 km^2 (Vickers, 1991). Thus game patterns can be influenced over large zones by a single small village or camp. Particularly within the closer zones, large game can be depleted rapidly, followed by a greater concentration on small game. Around permanent villages, depletion is accordingly severe (Hames, 1980). As foraging efficiency declines, either travel time increases or villages are relocated. Also, there is species selectivity in hunting, which varies culturally.

Some idea of the total game killed per year by indigenous people can be derived from data for the Waorani in Ecuadorian Amazonia (Yost and Kelley, 1983). A population of 230 people in four villages killed 3,165 mammals and birds in 11 months. This is equivalent to 3,453 in 12 months, which equals only 15 per person and is likely conservative. However, if this is extrapolated to Greater Amazonia for the year 1492, given an estimated population then of 5,664,000 people (Denevan, 1992a), the annual game kill would have been 85 million. Of course this is based on only one sample and on a crude estimate of the 1492 population. Redford (1992) estimates that for Brazilian Amazonia today, the annual kill of mammals is 14 million per year.

Game recovers where villages are abandoned or shift location. A zone farther distant may be hunted, allowing a zone closer in to recover. Thus we can envision a patch pattern of stable, depleted, and recovering zones of wildlife density related to population density and to village size and permanence or shifting (Hames, 1980; Robinson and Bennett, 2000; Vickers, 1991). Some species are able to "maintain population levels and genetic diversity in a patchwork landscape and across a heterogeneous region through metapopulation processes"; however, many other species are threatened by habitat fragmentation (Young, 1998).

Game depletion and densities are also affected by habitat modification, such as conversion of forest to savanna or open woodland, or of mature forest to secondary forest. It has been suggested that the dispersal and numbers of white-tailed deer (*Odocoileus virginianus*) and cottontail rabbit (*Sylvilagus floridanus*) in Panama and adjacent Colombia were influenced by prehistoric forest disturbance and savanna expansion (Bennett, 1968:40–41). Indigenous activities that increase the frequency of fruit-bearing trees increase the frequency of animals feeding on those fruits, including tapir, peccary, deer, rodents, monkeys, and birds (Piperno and Pearsall, 1998: 62).

Another aspect of the alteration of natural wildlife patterns is the attraction of animals to the crops in gardens or swiddens, those animals then being hunted by the farmers (Linares, 1976). Data from Tambopata in southeastern Perú indicate that large ungulates (tapir, peccary), primates, and felids are rare in swiddens or nearby, whereas small game species such as rodents and armadillos thrive and provide important game meat without being depleted (Naughton-Treves, 2002).

While zones may be depleted of certain species because of human presence, there is no evidence, apparently, of any definite, pre-European animal extinctions in Amazonia. Bennett (1968: 28, 49–50) has suggested that in lower Central America-Colombia, pre-agricultural people may have contributed to regional extinctions of large game (bison, horse, mastodon, mammoth), and that by A.D. 1500, brocket deer and tapir were absent in densely populated western Panama. Regional depletions and extinction of game result in those animals no longer being able to form critical ecological functions such as seed dispersal and interaction with other animals (Redford, 1992). The only animal domesticated in the tropical lowlands of South America is the Muscovy duck (*Cairina moschata*), possibly in the Caribbean region of Colombia-Venezuela (Donkin, 1989: 70–71).

16.5 Anthropogenic Soils: *Terra Preta*

Wherever cultivation and settlement activity take place, soils to some degree are altered physically and chemically, and erosion may occur. In Amazonia, there is a soil that was completely altered by human activity, hundreds, even thousands of years ago, and is still present and fertile today. This is *terra preta* or *terra preta do índio* (Indian black earth or dark earth) (Lehmann et al., 2003; Glaser and Woods, 2004). These soils occur in long strips along the river bluffs and in patches in the interior uplands, ranging in extent from less than a hectare to several hundred hectares, and in depth up to 2 m. Dates of ceramics in *terra preta* are as old as 2,360–2,450 B.P. (Peterson et al., 2001). Because of their high fertility, these soils are sought out today by local farmers, as undoubtedly they were in the past.

Terra preta, compared to surrounding reddish Ferralsols and Acrisols, has higher levels of pH, potassium, phosphorus, calcium, organic matter, and nutrient- and moisture-holding capacity. This has been attributed to high concentrations of charcoal (black carbon residues of incomplete combustion), along with intense, associated microbiological activity (Glaser et al., 2001).

While the blacker form (*terra preta*) of dark-earth soil seems to have been formed from pre-European village kitchen fires and middens containing large quantities of bones, ceramics, ash, and charcoal, there is a lighter, brownish form (*terra mulata*), which is much more extensive, usually surrounding the black patches. This brown soil is believed to have resulted from intensive cultivation, involving frequent in-field burning of logs, branches, and leaf litter within or brought to a field, along with crop remains and weeds, plus the incorporation of organic material through mulching and composting (Woods and McCann, 1999). Thus, if this thesis is correct, intensive cultivation in pre-European times, rather than depleting soil fertility, actually created a persisting fertile soil. Such soils are seldom formed today, given the frequent shifting of villages and fields.

Dark-earth soils tend to have a greater frequency of certain plant species than adjacent soils, including economic plants such as babussu palm, Brazil nut, cacao (*Theobroma cacao*), hog plum (*Spondias mombin*), papaya (*Carica candicans*), and guava (*Psidium guajava*), as well as more vines and thorny plants. After clearing, biomass accumulation is more rapid on dark earths, and weed growth in general is more rapid; some animals and birds are more common in dark-earth vegetation (German, 2003; McCann, 2004: 116–144).

16.6 Cultivated Fields

There is evidence for the cultivation of both tubers and maize in Panama by 6,000–7,000 B.P. (Piperno et al., 2000), and for maize in Ecuadorian Amazonia by 5,300 B.P. with dated charcoal indicating forest burning for fields as early as 7,000 B.P. (Piperno and Pearsall, 1998: 258–261). The practice of agriculture, of course, completely modifies the natural environments where fields and associated features are located. Cultivated fields may be short lived, as in shifting cultivation, with their impacts being ephemeral with relatively rapid habitat recovery. More permanent cultivation has long-term impacts on soil, and full vegetation recovery can take centuries. In addition, cultivation can include significant and highly varied landform-altering features such as terraces, canals, raised surfaces, excavations, and water control embankments and ditches (Denevan, 2001). These features can survive hundreds and even thousands of years following field abandonment, thus continuing to exert an environmental impact.

In Amazonia today, native cultivation invariably consists of short-cropping/long-fallowing shifting cultivation (Meggers, 1996: 19–23), and the common assumption is that the same was true in the past, even though there are few reports of shifting cultivation in the early colonial records. An argument can be made, however, that because of the inefficiency of stone axes, compared with metal axes introduced after 1492, a field once established (especially at tree falls, forest burns, and tree blow-downs) would be farmed more or less continuously, utilizing soil-enhancement techniques (Denevan, 2001: 116–119). Such fields can have long-lasting effects. The anthropogenic *terra mulata* soils discussed previously are a good example of this. In the interior, permanent villages must have been surrounded by large zones of disturbed vegetation consisting of rotations of fields with house gardens, orchards, and managed secondary forest. Heckenberger (2005: 96, 98, 118, 122) shows aerial photographs of pre-European village sites in central Brazil, each surrounded by secondary forest of unknown age. Along the main rivers, most settlement was on the edges of bluff tops, not in the floodplains where annual flooding could be destructive. *Terra preta* soils and cultural middens extend, in places, for several kilometers along the bluffs, evidence of intensive disturbance in the past (Denevan, 1996).

Two archaeological studies confirm ancient permanent settlement and cultivation. The first is for Araracuara on a bluff overlooking the Río Caquetá in Colombian Amazonia (Herrera et al., 1992). Here, at two of the sites excavated, settlement and cropping were more or less continuous from A.D. 385–1175 and from A.D. 1–1800. Agriculture consisted of permanent fields and agroforestry systems with fruit trees. Soil fertility was apparently maintained with additives of river silt and algae, and terrestrial organic material. The two sites total 20.5 ha of black and brown anthropogenic soils.

Pre-European village sites have been examined in the uplands of the upper Rio Xingu basin in southern Amazonia (Heckenberger, 1998, 2005). Two of these cover 40 and 50 ha, and each may have contained over 1,500 people (villages nearby today may only cover 2–5 ha, with a few hundred people or less). These sites were occupied from A.D. 800–900 to at least A.D. 1590. The central plazas of these villages are surrounded by multiple defensive ditches or moats. Between the plazas and the outer ditches there are anthropogenic soils, apparently created by both village refuse and intensive cultivation over many years.

In the dry savannas of South America (*campo cerrado* of Brazil and the higher-lying Orinoco Llanos), soil fertility is extremely poor, and industrial cultivation today uses heavy inputs of chemical fertilizers. Native cultivation, past and present, is rare except in gallery forests and forest patches (*islas*). However, the Kayapó create small (<4 ha) artificial scrub and forest patches (*apêtês*) within savanna and manage them (Anderson and Posey, 1989; but see Parker, 1992 for a dissent). These are initiated from compost heaps in nearby forest, consisting of sticks and leaves plus soil from termite and ant nests. The rotting material is used to make small mounds in the savanna, which are enlarged over time by natural and human processes. Most of the species present can be planted or managed. Whether completely artificial or not, the vegetation of *apêtês* is clearly human influenced and, in some instances, may be of considerable antiquity.

16.7 Raised Fields

In the seasonally flooded savannas of South America, pre-European people made cultivation possible by digging drainage ditches and constructing raised platforms, ridges, and mounds. Some measure over 350 m in length, up to 20 m in width, and a meter or more in height. Not only did such features alter drainage, vegetation, soil, and wildlife when constructed and used, but hundreds of thousands of them have also survived to the present, thus maintaining anthropogenic landscapes in now sparsely inhabited regions.

Water both accumulates in and is drained from the ditches between the fields. Aquatic vegetation and wildlife are concentrated in the ditches. Muck from the ditches was transferred to the raised surfaces to improve fertility. Today the vegetation, wildlife, and soils of the ditches and fields continue to be different from those of surrounding areas.

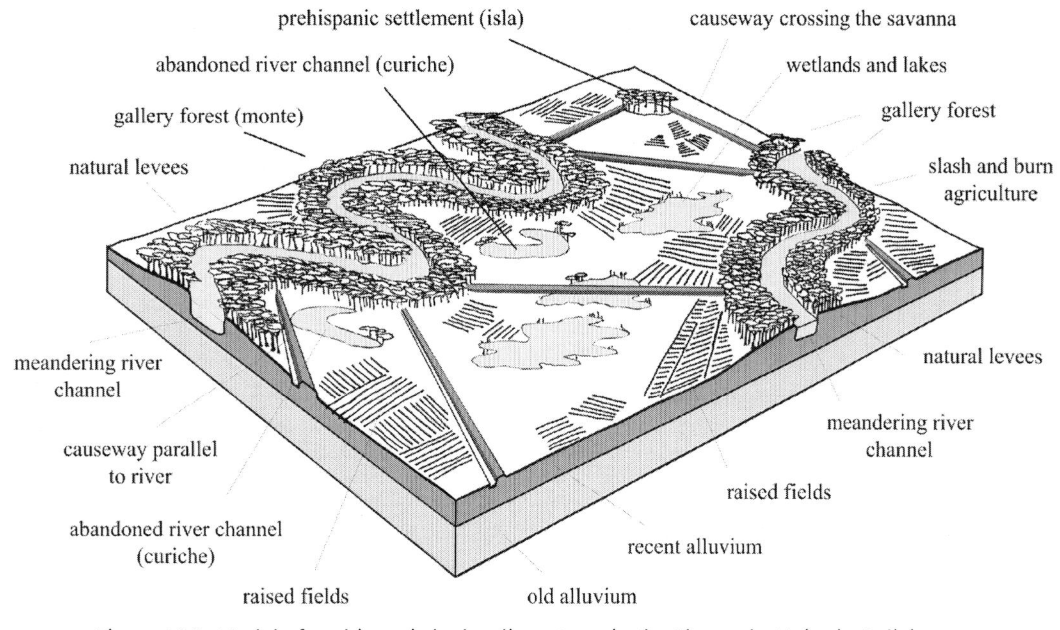

Figure 16.3 Model of prehispanic hydraulic systems in the Llanos de Mojos in Bolivian Amazonia, showing raised fields, an artificial mound (*isla*), and causeways and water-control walls or dikes with canals adjacent (from Erickson, 2001: 28)

Raised-field landscapes are not rare, localized, or minor in extent. They occur in the Llanos de Mojos of Bolivia (fig. 16.3), the San Jorge savannas of northern Colombia (fig. 16.4), the coastal savannas of the Guianas, the Orinoco Llanos of Venezuela, and the Guayas coast savannas of Ecuador, as well as in the Andes (Denevan, 2001: 215–277). Remains of pre-European raised fields cover at least 150,000 ha (1,500 km^2) in the tropical lowlands. This is only a portion of what once existed, the greater part having been buried under sediment or destroyed by erosion or by cattle and other modern land use.

16.8 Other Earthworks

Throughout the tropical lowlands of South America, there are many pre-European earthworks of various types that have survived to the present—mounds, embankments,

Figure 16.4 Prehispanic raised fields in the Río San Jorge floodplain, northern Colombia (from Parsons and Bowen, 1966: 331)

causeways, and ditches, in addition to the raised agricultural fields. Only a portion of these features have been discovered, mostly recently and mostly in savannas where they are visible. Others have been destroyed. They alter the land surface, and have a lasting effect on drainage, soils, vegetation, and wildlife.

All of these forms of early earthworks are present in great numbers, readily discerned from the air, in the Llanos de Mojos in Bolivian Amazonia. Thus much of that landscape today is a relic pre-European landscape (fig. 16.3). Mounds (settlement, burial, and ceremonial) are particularly prominent (Erickson, 2000b). There are 200–300 large mounds, each covering several hectares and 3–18 m in height; several thousand medium-sized mounds, 1–2 ha in size and 1–3 m high; and many thousands of small mounds, less than 1 ha in size and less than 1 m high. Many occur as circular patches of forest (*islas*) in open, seasonally flooded savanna. The origin of mounds in Mojos is complicated, however, by the fact that many of them are natural (clustered termite and ant mounds, remnants of river levees) (Langstroth, 1996). It is not easy to differentiate these from artificial mounds without careful examination; both may contain human bones and artifacts.

Causeways are elevated earthen roads. In Mojos, they measure up to 10–12 km in length, 4–6 m in width, and 0.5 to 1.5 m in height (fig. 16.3). Aerial photographs of the region reveal causeways at least 1,500 km in total length (Denevan, 1991), but this may be too low (Erickson, 2000c). Most are in poorly drained savanna and connect artificial mounds, but some are within or cross through forest. One mound has at least 40 causeways radiating from it, or nearby (Erickson, 2000c). The causeways served for foot travel during floods. Adjacent and parallel to many are the ditches from which the excavated earth was removed and that served as canals for canoe travel. Some causeways may have served as field boundaries and as dikes to control floodwaters in zones of raised fields, either to keep water out for drainage or to keep water in for irrigation, or both; there also may have been political and ritual functions (Erickson, 1995, 2000c, 2001).

A curious form of zigzag embankment exists in northeastern Mojos (Baures), which is neither causeway nor dike (Erickson, 2000a). Some of these are up to 3.5 km long, and a sampling from aerial photographs indicates that they total an estimated 1,515 km in length. They are 1–2 m wide and about 0.5 m high. They change directions frequently, and at the angles there are funnel-like openings that could be easily opened or closed, with small artificial ponds adjacent. There is no evidence of associated crop furrows or raised fields. Erickson believes these features are fish weirs and that the ponds served for fish storage. Also, enormous quantities of edible snail shells (*Pomacea gigas*) are found nearby. The zigzag walls probably also diverted water and affected water depth and duration, whether intentional or not.

Moats circling former village sites, often with mounds, are another earthwork feature in northeastern Mojos. In some places there is only one, elsewhere two or three around a site. They contain areas of up to 3–4 ha and are 1–3 m deep. They presumably had a defensive function and some are associated with low walls and perhaps palisades (Denevan, 1966: pl. 7; Erickson et al., 1997: 9–14, figs. 3–33).

Although the greatest concentrations of earthworks are in Mojos, they are also found elsewhere. Artificial mounds occur in other wet savannas, including the Orinoco Llanos, the coastal Guyanas, coastal Ecuador, the San Jorge region of northern Colombia, and they are scattered in the forests of the Amazon basin. Some of the largest are in the savannas of eastern Marajó Island at the mouth of the Amazon River. Here there are hundreds of mounds ranging in height from 3 to 20 m and in area up to 90 ha; the largest had villages on top, containing possibly a thousand or more people (Roosevelt, 1991: 30–31, 38–39).

Causeways occur in the Orinoco Llanos, and a few survive in other regions but nowhere as numerous as in Mojos (Denevan, 1991). Sunken trails or roads (from foot traffic and erosion) of possible pre-European origin have been reported in the Amazonian forest. Carvajal, on the first European voyage down the Amazon in 1541, observed roads leading from the interior to the river (Denevan, 1996). Nimuendajú (2004: 123) in 1939 reported old sunken roads near the Rio Tapajós that were 1–1.5 m wide and 30 cm deep, "running almost as straight as arrows from one *terra preta* to the next." Today, the numerous indigenous trails in Amazonia usually become rapidly overgrown with vegetation after abandonment, and we do not know how many of them may be pre-European. There was a major Inca road through the Guayas Basin of coastal Ecuador, and stone-paved Inca roads reached the eastern foothills of the Andes. Roads and trails are corridors for adjacent plant and animal disturbance.

Old moats (circular, semi-circular, rectangular) are also found in parts of Amazonia other than Mojos. They have been described in particular in the upper reaches of the Rio Xingu (Heckenberger, 2005: 75–112). Associated are circular ring mounds with and without moats, mounds, linear walls and ditches, and roads. Some of the moats enclose as many as 20 ha. Moats and other earthworks have also been described and photographed recently in deforested areas in Acre in western Brazil (Pärssinen and Korpisaari, 2003: 91–172).

16.9 Waterway Management

Less known in Amazonia than artificial earthworks are various fluvial modifications. A 1916 paper by Erland Nordenskiöld discussed such features in Mojos and elsewhere in Amazonia. He even suggested that the beginning section of the Casiquiare "Canal," which connects the Río

Orinoco with the Río Negro and the Amazon, may not be natural but rather was cut by pre-European people.

The Casiquiare thesis may not be valid, but there are documented instances of such artificial canals between river branches. Again the best examples come from the Llanos de Mojos in the Bolivian Amazon, and these still show up on aerial photographs (Denevan, 1966: 74–77, pl. 10). None have been dated but most seem quite old. Around 1710, Padre Diego Altamirano said that "canals made by hand" could be used for small boats (Denevan, 1966: 74). The Jesuit missionaries maintained and used these canals, and they may have constructed some of them. Pinto Parada (1987: 233, 269) mapped several canals, including one complex 120 km in total length, connecting lakes and streams. The large canals are up to 7 m wide and several meters deep. In addition there are hundreds of smaller canals crossing the savanna from one forest *isla* to another, some adjacent to causeways (Erickson et al., 1997: 15–17). All of these canals would have easily served for canoe transport. Some are simply paths cut through the grass for use during floods (Denevan, 1966: pl. 9b).

Canals may also have served to divert water away from or drain water from raised-field complexes. Causeways, whether intended or not, may block the movement of floodwater, thus creating deep water on one side and little or no water on the other, as can be seen today.

Another feature is the *corte* or river short-cut. These are dug out where river meander necks are narrow and would eventually break through naturally. They may be initiated by paths being cleared across necks for canoe portages to reduce travel time. They are gradually deepened, allowing water to fill them. They are cut today in Mojos, ranging in length from tens of meters to several kilometers (Denevan, 1966: 76–77; Erickson et al., 1997: 14, fig. 34). *Cortes* of unknown age have been reported throughout Amazonia (Raffles and WinklerPrins, 2003).

In the Amazon estuary, on Ituqui Island near Santarém, at Careiro Island near Manaus, and in the Orinoco floodplain, some waterways thought to be natural are actually anthropogenic. They were created to provide better access to fields and forest products and to concentrate sedimentation in back swamps for cultivation (Raffles and WinklerPrins, 2003; Sternberg, 1998: 98). These canals (*cavados*) are up to several kilometers long and a few meters deep. Over time, they are enlarged by natural processes that erase evidence of their origins. All of these canals were made recently; although native people likely dug similar ones in the past.

16.10 Ponds and Wells

Ponds or reservoirs dug for fish and for storing water, and also wells, occur in several parts of Amazonia. In northeastern Mojos, shallow, circular depressions (*pozas*), up to 50 m in diameter and 1.5 m deep, are associated with pre-European causeways and zigzag walls and artificial forest islands. The depressions may have been the sources of earth for these features. When they fill with rainwater, they contain fish and attract aquatic birds, deer, and other game (Erickson et al., 1997: 19, fig. 55).

Wells are smaller excavations that fill with water from shallow water tables during the dry season and provide a source of drinking water. In northeastern Mojos, they are concentrated within 100 m of pre-European settlement mounds and thus seem to be associated with them (Erickson et al., 1997: 19). Old wells, about 2 m in diameter and 2 m deep, occur within and near *terra preta* sites in the uplands east and west of the lower Rio Tapajós in Brazil. These sites are distant from flowing water (Nimuendajú, 2004: 123).

16.11 Conclusion

La Selva Humanizada

Correa (1993)

Human impacts on South American environments have been continuous, but highly variable in form and in intensity, from pre-agricultural hunters and gatherers to the present. Europeans entered the lowland tropics with the discovery of the Amazon River by Cabral in A.D. 1500. From slavery and epidemics, the native population was rapidly reduced. Fields and villages were abandoned and hunting declined. As result, the environment recovered in many respects. Fields were replaced by second growth, villages were grown over, and anthropogenic savannas reverted to forest, as we have seen for Panama and western Colombia. However, by the late nineteenth century, with the rubber boom, another cycle of destruction had begun, diminished after 1915, and then intensified after 1960.

There has been considerable debate over whether prehistoric indigenous land use conserved or depleted resources. The best response is that sometimes there was intentional conservation; other times there was unintentional conservation; and at other times practices were destructive and not sustainable of resources and habitat.

Today, with deforestation for agriculture and pasture, logging, settlement, roads, and mining, human disturbance is massive, obvious, and often irreversible. Much of this is by corporate entities and large estates, as well as by new settler farmers with little experience with rain-forest living. However, surviving Indians and long-time residents of mixed background are also present. Although details differ, land management and effect on the environment by these people are often comparable to those of pre-Europeans—use with recovery, sustainability within change, and major impacts mostly being concentrated and lesser impacts being dispersed.

Traditional societies in the past were not few and environmentally insignificant, but rather comprised vital components of the landscape.

References

Anderson, A.B., and D.A. Posey, 1989. Management of a tropical scrub savanna by the Gorotire Kayapó of Brazil. *Advances in Economic Botany*, 7, 159–173.

Balée, W., 1988. Indigenous adaptation to Amazonian palm forests. *Principes*, 32, 47–54.

———, 1989. The culture of Amazonian forests. *Advances in Economic Botany*, 7, 1–21.

———, 1992. People of the fallow: A historical ecology of foraging in lowland South America. In: K.H. Redford and C. Padoch (Editors), *Conservation of Neotropical Forests: Working from Traditional Resource Use*. Columbia University Press, New York, 35–57.

———, 1993. Indigenous transformation of Amazonian forests: An example from Maranhão, Brazil. *L'Homme: Revue Française d'Anthropologie*, 126–128, 231–254.

———, 1994. *Footprints of the Forest, Ka'apor Ethnobotany: The Historical Ecology of Plant Utilization by an Amazonian People*. Columbia University Press, New York.

———, 2003. Native views of the environment in Amazonia. In: H. Selin (Editor), *Nature across Cultures: Views of Nature and the Environment in Non-Western Cultures*. Kluwer, Dordrecht, 277–288.

Balick, M.J., and P.A. Cox, 1996. *Plants, People, and Culture: The Science of Ethnobotany*. Scientific American Library, New York.

Barreto, C., and J. Machado, 2001. Exploring the Amazon, explaining the unknown: Views from the past. In: C. McEwan, C. Barreto, and E. Neves (Editors), *Unknown Amazon: Culture in Nature in Ancient Brazil*. British Museum Press, London, 232–251.

Basso, E. B., 1973. *The Kalapalo Indians of Central Brazil*. Holt, Rinehart, and Winston, New York.

Bennett, C. F., 1968. *Human Influences on the Zoogeography of Panama*. Ibero-Americana 51. University of California Press, Berkeley.

Brown, S., and A.E. Lugo, 1990. Tropical secondary forests. *Journal of Tropical Ecology*, 6, 1–32.

Bush, M.B., and P.A. Colinvaux, 1994. Tropical forest disturbance: Paleoecological records from Darien, Panama. *Ecology*, 75, 1761–1768.

Cleary, D., 2001. Towards an environmental history of the Amazon: From prehistory to the nineteenth century. *Latin American Research Review*, 36, 65–96.

Clement, C.R., 1999. 1492 and the loss of Amazonian crop genetic resources. *Economic Botany*, 53, 188–216.

Cochrane, M.A., and W.L. Laurance, 2002. Fire as a large-scale edge effect in Amazonian forests. *Journal of Tropical Ecology*, 18, 311–325.

———, and M.D. Schulze, 1999. Fire as a recurrent event in tropical forests of the eastern Amazon: Effects on forest structure, biomass, and species composition. *Biotropica*, 31, 2–16.

Correa, F. (Editor), 1993. *La Selva Humanizada: Ecología Alternativa en el Trópico Húmedo Colombiano*. Instituto Colombiano de Antropología, Bogotá.

Denevan, W.M., 1966. *The Aboriginal Cultural Geography of the Llanos de Mojos of Bolivia*. Ibero-Americana, 48. University of California Press, Berkeley.

———, 1991. Prehistoric roads and causeways of lowland tropical America. In: C. D. Trombold (Editor), *Ancient Road Networks and Settlement Hierarchies in the New World*. Cambridge University Press, Cambridge, 230–242.

———, (Editor), 1992a [1976]. *The Native Population of the Americas in 1492*. Second edition. University of Wisconsin Press, Madison.

———, 1992b. The pristine myth: The landscape of the Americas in 1492. *Annals of the Association of American Geographers*, 82, 369–385.

———, 1996. A bluff model of riverine settlement in prehistoric Amazonia. *Annals of the Association of American Geographers*, 86, 654–681.

———, 2001. *Cultivated Landscapes of Native Amazonia and the Andes*. Oxford University Press, Oxford.

———, and C. Padoch (Editors). 1988. Swidden-fallow agroforestry in the Peruvian Amazon. *Advances in Economic Botany*, 5.

Descola, P., 1994. *In the Society of Nature: A Native Ecology of Amazonia*. Cambridge University Press, Cambridge.

Donkin, R.A., 1989. *The Muscovy Duck, Cairina moshata domestica: Origins, Dispersal, and Associated Aspects of the Geography of Domestication*. A. A. Balkema, Rotterdam.

Eden, M. J., 1986. Monitoring indigenous shifting cultivation in forest areas of southwest Guyana using aerial photography and Landsat. In: M. J. Eden and J. T. Parry (Editors), *Remote Sensing and Tropical Land Management*. John Wiley, Chichester, 255–277.

Erickson, C. L. 1995. Archaeological methods for the study of ancient landscapes of the Llanos de Mojos in the Bolivian Amazon. In: P. W. Stahl (Editor), *Archaeology in the Lowland American Tropics: Current Analytical Methods and Applications*. Cambridge University Press, Cambridge, 66–95.

———, 2000a. An artificial landscape-scale fishery in the Bolivian Amazon. Nature, 408, 190–193.

———, 2000b. Lomas de ocupación en los Llanos de Moxos. In: A.D. Coirolo and R.B. Boksar (Editors), *Arqueología de las Tierras Bajas*. Ministerio de Educación y Cultura, Montevideo, 207–226.

———, 2000c. Los caminos prehispánicos de la Amazonía boliviana. In: L. F. Herrera and M. Cardale de Schrimpff (Editors), *Caminos Precolombinos: Las Vías, los Ingenieros y los Viajeros*. Instituto Colombiano de Antropología y Historia, Bogotá, 15–42.

———, 2001. Pre-Columbian Roads of the Amazon. *Expedition*, 43(2), 21–30.

———, W. Winkler, and K. Candler. 1997. *Las Investigaciones Arqueológicas en la Región de Baures en 1996*. Projecto Agro-Arqueológico del Beni. Department of Anthropology, University of Pennsylvania, Philadelphia.

Fearnside, P.M., 1990. Fire in the tropical rain forest of the Amazon Basin. In: J.G. Goldammer (Editor), *Fire in the Tropical Biota: Ecosystem Processes and Global Challenges*. Springer-Verlag, Berlin, 106–116.

German, L. 2003. Ethnoscientific understandings of Amazonian dark earths. In: J. Lehmann, D.C. Kern, B. Glaser, and W.I. Woods (Editors), *Amazonian Dark Earths: Origin, Properties, Management*. Kluwer, Dordrecht, 179–201.

Glaser, B., L. Haumaier, G. Guggenberger, and W. Zech, 2001. The 'terra preta' phenomenon: A model for sustainable

agriculture in the humid tropics. *Naturwissenschaften*, 88, 37–41.
———, and W. I. Woods (Editors). 2004. *Amazonian Dark Earths: Explorations in Space and Time.* Springer-Verlag, Berlin.
Gómez-Pompa, A., and A. Kaus. 1992. Taming the wilderness myth. *BioScience*, 42, 271–279, 579–581.
Gordon, B. L., 1957. *Human Geography and Ecology in the Sinú Country of Colombia.* Ibero-Americana, 39. University of California Press, Berkeley.
Grayson, D.K., 2001. The archaeological record of human impacts on animal populations. *Journal of World Prehistory*, 15, 1–68.
Hames, R.B., 1980. Game depletion and hunting zone rotation among the Ye'kwana and Yanomamö of Amazonas, Venezuela. In: *Working Papers on South American Indians* 2. Bennington College, Bennington, Vermont, 31–36.
Hayashida, F.M., 2005. Archaeology, ecological history, and conservation. *Annual Review of Anthropology*, 34, 43–65.
Heckenberger, M.J., 1998. Manioc agriculture and sedentism in Amazonia: The Upper Xingu example. *Antiquity*, 72, 633–648.
———, 2005. *The Ecology of Power: Culture, Place, and Personhood in the Southern Amazon, A.D. 1000–2000.* Routledge, New York.
Herrera, L.F., I. Cavelier, C. Rodríguez, and S. Mora, 1992. The technical transformation of an agricultural system in the Colombian Amazon. *World Archaeology*, 24, 98–113.
Hill, K., et al., 1997. Impact of hunting on large vertebrates in the Mbaracayu Reserve, Paraguay. *Conservation Biology*, 11, 1339–1353.
Humboldt, A. von, 1869 [1808]. *Views of Nature: or, Contemplations on the Sublime Phenomena of Creation.* Bell and Daldy, London.
Krech III, S., 1999. *The Ecological Indian: Myth and History.* W.W. Norton, New York.
Langstroth, R. P. 1996. Forest Islands in an Amazonian Savanna of Northeastern Bolivia. Ph.D. dissertation. University of Wisconsin, Madison.
Lehmann, J., D.C. Kern, B. Glaser, and W.I. Woods (Editors), 2003. *Amazonian Dark Earths: Origin, Properties, Management.* Kluwer, Dordrecht.
Linares, O. F., 1976. Garden hunting in the American tropics. *Human Ecology*, 4, 331–349.
Marchant, R., et al., 2004. Vegetation disturbance and human population in Colombia—A regional reconstruction. *Antiquity*, 78, 828–838.
McCann, J.M., 2004. *Subsidy from Culture: Anthropogenic Soils and Vegetation in Tapajônia, Brazilian Amazonia.* Ph.D. dissertation. University of Wisconsin, Madison.
Meggers, B J., 1996 [1971]. *Amazonia: Man and Culture in a Counterfeit Paradise.* Revised edition. Smithsonian Institution Press, Washington, DC.
Naughton-Treves, L., 2002. Wild animals in the garden: Conserving wildlife in Amazonian agroecosystems. *Annals of the Association of American Geographers*, 92, 488–506.
Nimuendajú, C., 2004. *In Pursuit of a Past Amazon: Archaeological Researches in the Brazilian Guyana and in the Amazon Region.* Per Stenborg (Editor), Ethnological Studies 45. Elanders Infologistik Väst AB, Göteborg.
Nordenskiöld, E., 1916. Die anpassung der indianer an die verhältnisse in den überschwemmungsgebieten in Südamerika. *Ymer* (Stockholm), 36, 138–155.
Parker, E.P., 1992. Forest islands and Kayapó resource management in Amazonia: A reappraisal of the *apêtê*. *American Anthropologist*, 94, 406–428.
Parsons, J.J., and W.A. Bowen, 1966. Ancient ridged fields of the San Jorge River floodplain, Colombia. *Geographical Review*, 56, 317–343.
Pärssinen, M., and A. Korpisaari (Editors), 2003. *Western Amazonia: Multidisciplinary Studies on Ancient Expansionistic Movements, Fortifications and Sedentary Life.* Renvall Institute Publications 14. University of Helsinki, Helsinki.
Peters, C.M., 2000. Precolumbian silviculture and indigenous management of Neotropical forests. In: D. L. Lentz (Editor), *Imperfect Balance: Landscape Transformations in the Precolumbian Americas.* Columbia University Press, New York, 203–223.
Petersen, J.B., E. Neves, and M.J. Heckenberger, 2001. Gift from the past: *Terra preta* and prehistoric Amerindian occupation in Amazonia. In: C. McEwan, C. Barreto, and E. Neves, editors, *Unknown Amazon: Culture in Nature in Ancient Brazil.* British Museum Press, London, 86–105.
Pinto Parada, R., 1987. *Pueblo de leyenda.* Editorial Tiempo del Beni, Trinidad, Bolivia.
Piperno, D.R., M.B. Bush, and P.A. Colinvaux, 1992. Patterns of articulation of culture and the plant world in prehistoric Panama: 11,500 B.P.–3000 B.P. In: O. R. Ortiz-Troncoso and T. van der Hammen (Editors), *Archaeology and Environment in Latin America.* Universiteit van Amsterdam, Amsterdam, 109–127.
———, and D. M. Pearsall, 1998. *The Origins of Agriculture in the Lowland Neotropics.* Academic Press, San Diego.
———, A.J. Ranere, P. Hansell, and I. Hoist, 2000. Starch grains reveal early root crop horticulture in the Panamanian tropical forest. *Nature*, 407, 894–897.
Politis, G.G., 1999. Plant exploitation among the Nukak hunter-gatherers of Amazonia: Between ecology and ideology. In: C. Gosden and J. Hather (Editors), *The Pre-History of Food: Appetites for Change.* Routledge, New York, 99–125.
———, 2001. Foragers of the Amazon: The last survivors or the first to succeed? In: C. McEwan, C. Barreto, and E. Neves (Editors), *Unknown Amazon: Culture in Nature in Ancient Brazil.* British Museum Press, London, 26–49.
Posey, D.A., 1985. Indigenous management of tropical forest ecosystems: The case of the Kayapó Indians of the Brazilian Amazon. *Agroforestry Systems*, 3, 139–158.
———, 1992. Interpreting and applying the 'reality' of indigenous concepts. In: K.H. Redford and C. Padoch (Editors), *Conservation of Neotropical Forests: Working from Traditional Resource Use.* Columbia University Press, New York, 21–34.
Prance, G.T., and H.O.R. Schubart, 1978. Notes on the vegetation of Amazonia I. A preliminary note on the origin of the open white sand campinas of the lower Rio Negro. *Brittonia*, 30, 60–63.
Raffles, H., and A.M.G.A. WinklerPrins, 2003. Further reflections on Amazonian environmental history: Transformations of rivers and streams. *Latin American Research Review*, 38, 165–187.
Ranere, A.J., 1992. Implements of change in the Holocene

environments of Panama. In: O.R. Ortiz-Troncoso and T. van der Hammen (Editors), *Archaeology and Environment in Latin America*. Universiteit van Amsterdam, Amsterdam, 25–44.

Redford, K.H., 1992. The empty forest. *BioScience*, 42, 412–422.

Rival, L.M., 2002. *Trekking Through History: The Huaorani of Amazonian Ecuador*. Columbia University Press, New York.

Robinson, J.G., and E.L. Bennett (Editors), 2000. *Hunting for Sustainability in Tropical Forests*. Columbia University Press, New York.

Roosevelt, A.C., 1989. Resource management in Amazonia before the conquest: Beyond ethnographic projection. *Advances in Economic Botany*, 7, 30–62.

———, 1991. *Moundbuilders of the Amazon: Geophysical Archaeology on Marajó Island, Brazil*. Academic Press, San Diego.

———, 1999a. The development of prehistoric complex societies: Amazonia, a tropical forest. In: E.A. Bacus and L.J. Lucero (Editors), *Complex Politics in the Ancient Tropical World*. Archeological Papers of the American Anthropological Association, 9, Arlington, 13–33.

———, 1999b. Twelve thousand years of human-environment interaction in the Amazon floodplain. In: C. Padoch et al. (Editors), *Várzea: Diversity, Development, and Conservation of Amazonia's Whitewater Floodplains*. Advances in Economic Botany, 13, 371–392.

____, 2000. The lower Amazon: A dynamic human habitat. In: D. L. Lentz (Editor), *Imperfect Balance: Landscape Transformations in the Precolumbian Americas*. Columbia University Press, New York, 455–491.

Saldarriaga, J.G., D.C. West, M.L. Tharp, and C. Uhl, 1988. Long-term chronosequence of forest succession in the upper Río Negro of Colombia and Venezuela. *Journal of Ecology*, 76, 938–958.

Sarmiento, G., 1984. *The Ecology of Neotropical Savannas*. Harvard University Press, Cambridge.

Sauer, C.O., 1958. Man in the ecology of tropical America. *Proceedings of the Ninth Pacific Science Congress*, Bangkok, 20, 104–110.

———, 1966. *The Early Spanish Main*. University of California Press, Berkeley.

Scott, G.A.J., 1978. *Grassland Development in the Gran Pajonal of Eastern Peru: A Study of Soil-Vegetation Nutrient Systems*. Hawaii Monographs in Geography 1. University of Hawaii, Honolulu.

Smith, N.J.H., 1995. Human-induced landscape changes in Amazonia and implications for development. In: B. L. Turner II et al. (Editors), *Global Land-Use Change: A Perspective from the Columbian Encounter*. Consejo Superior de Investigaciones Científicas, Madrid, 221–251.

Sponsel, L.E. 1992. The environmental history of Amazonia: Natural and human disturbances, and the ecological transition. In: H.K. Steen and R.P. Tucker (Editors), *Changing Tropical Forests: Historical Perspectives on Today's Challenges in Central and South America*. The Forest History Society, Durham, 233–251.

Stahl, P.W., 1996. Holocene biodiversity: An archaeological perspective from the Americas. *Annual Review of Anthropology*, 25, 105–126.

———, 2000. Archaeofaunal accumulation, fragmented forests, and anthropogenic landscape mosaics in the tropical lowlands of prehispanic Ecuador. *Latin American Antiquity*, 11, 241–257.

———, 2002. Paradigms in Paradise: Revising standard Amazonian prehistory. *The Review of Archaeology*, 23, 39–51.

Sternberg, H.O'R, 1998 [1956]. *A Água e o Homen na Várzea do Careiro*. Second edition. Museu Paraense Emilio Goeldi, Belém.

Vale, T.R., 2002. The pre-European landscape of the United States: pristine or humanized? In: T.R. Vale (Editor), *Fire, Native Peoples, and the Natural Landscape*. Island Press, Washington, DC, 1–39.

Vickers, W T., 1991. Hunting yields and game composition over ten years in an Amazon Indian territory. In: J.G. Robinson and K.H. Redford (Editors), *Neotropical Wildlife Use and Conservation*. University of Chicago Press, Chicago, 53–81.

Viveiros de Castro, E., 1996. Images of Nature and Society in Amazonian Ethnology. *Annual Review of Anthropology*, 25, 179–200.

Voeks, R.A., 2004. Disturbance pharmacopoeias: Medicine and myth from the humid tropics. *Annals of the Association of American Geographers*, 94, 868–888.

Woods, W.I., and J.M McCann, 1999. The anthropogenic origin and persistence of Amazonian dark earths. *Yearbook, Conference of Latin Americanist Geographers*, 25, 7–14.

Yost, J.A., and P.M. Kelley, 1983. Shotguns, blowguns, and spears: The analysis of technological efficiency. In: R.B. Hames and W.T. Vickers (Editors), *Adaptive Responses of Native Amazonians*. Academic Press, New York, 189–224.

Young, K.R., 1998. Deforestation in landscapes with humid forests in the central Andes: Patterns and processes. In: K.S. Zimmerer and K.R. Young (Editors), *Nature's Geography: New Lessons for Conservation in Developing Countries*. University of Wisconsin Press, Madison, 75–99.

Zent, E.L., and S. Zent, 2004. Amazonian Indians as ecological disturbance agents: The Hotï of the Sierra de Maigualida, Venezuelan Guayana. *Advances in Economic Botany*, 15, 79–112.

17

The Legacy of European Colonialism

Gregory Knapp

South America was first "encountered" by Europeans during Columbus' third voyage in 1498. This marked the end of the pre-Columbian period of the continent, and the beginning of the colonial period that lasted until the end of the wars of independence in the early nineteenth century. Total liberation of the continent from Spain was finally achieved at the Battle of Ayacucho in 1824. Brazilian independence from Portugal was achieved more peacefully in 1822, when Dom Pedro became constitutional emperor. The Guianas remained colonies far longer; indeed Guyane (French Guiana) is still an overseas department of France, while Suriname (Dutch Guiana) became independent in 1975, and Guyana (originally a Dutch colony, later British) became independent in 1966. It could be suggested that dependency remained after the end of formal colonial rule, owing to the continued influence of global economic powers on the continent. However, for the purposes of this chapter, the colonial period can be considered as lasting for 326 years from 1498 to 1824.

If recent research has tended to enhance our appreciation of the impact of pre-Columbian peoples on the South American environment, it has also corrected some stereotypes concerning European colonial impacts. Europeans were not the first to substantially impact the South American environment. The colonial period was generally marked by depopulation and agricultural disintensification, with the result that many environments were more "pristine" at the end of the eighteenth century than at the end of the fifteenth century. Migrations, cultural hybridities, and new local, regional, and global economic linkages led to changes in demands on agriculture and resource extraction. New technologies, crops, and social structures also had an impact. These impacts were not always as negative as sometimes portrayed, and local populations often had a substantial say in the outcome. Many of the most noticeable impacts resulting from the encounter with Europeans did not become widespread until after independence (McAlister, 1984; Bethell, 1987; Hoberman, 1996; Hoberman et al., 1996; Mörner, 1985; Newson, 1995; Robinson, 1990; Butzer and Butzer, 1995).

In contrast to the case of pre-Columbian South America, we have substantial archival materials for the colonial period, and written accounts by official visitors and other travelers, some of whom were erudite and reliable (Jimenez de la Espada, 1965; Caillavet, 2000; Cieza de León, 1962; Cobo, 1956; Guamán Poma de Ayala, 1980). Only a small amount of research, however, has focused on the environmental dimension of the colonial South American world, in comparison with the larger amount of writing on political events, socioeconomic formations, and cultural characteristics of the Spanish, French, Dutch, British, and Portuguese empires. Many of the recent debates on the environmental dimensions of colonialism have focused on Mexico (e.g., Butzer and Butzer, 1995), but we need to be cautious in applying insights from Middle America to South America.

This chapter will focus on the Andes, the main center of activity during most of the colonial period. Coastal Brazil was another arena of action, but will not be discussed in detail here. Much of the interior and the far southern portion of South America remained under the control of indigenous peoples, pursuing modified versions of land management as described in chapter 16.

In this chapter, for the sake of clarity, sites will be located in terms of the boundaries of modern nations, but of course these nations did not exist in colonial times. Instead, the continent was divided into administrative and juridical districts under the control of the Spanish, Portuguese, French, Dutch, and British empires (figure 17.1). The boundaries of these districts were often vague or contested, and considerable areas were not under effective imperial control; furthermore, districts and their nomenclature changed throughout the colonial period. The map indicates the general pattern and nomenclature of administrative districts toward the end of colonial times, around 1790. Most areas were under the overall supervision of viceroys, located in the Spanish viceregal capitals of Bogotá, Lima, and Buenos Aires, and the Portuguese viceregal capital of Rio de Janeiro. Other subdivisions included captaincies and *audiencias;* significant administrative and juridical centers of the Spanish part of South America included Caracas, Quito, Cuzco, Santiago, La Paz, and Chuquisaca (Sucre). These cities were not only centers of imperial control but also centers of European settlement and the introduction of European ideas, technologies, plants, and animals.

17.1 Late Pre-Columbian Situation

In late pre-Columbian times, most of South America was occupied by agriculturalists. The major exceptions were far southern Chile, Patagonia, and the Pampas, which were occupied by hunters and gatherers, and the Andes above 4,000 meters, which were used for hunting and camelid (llama, alpaca) grazing. For the most part, settlement was dispersed in isolated households or small hamlets or villages. Even in areas where the population density was high,

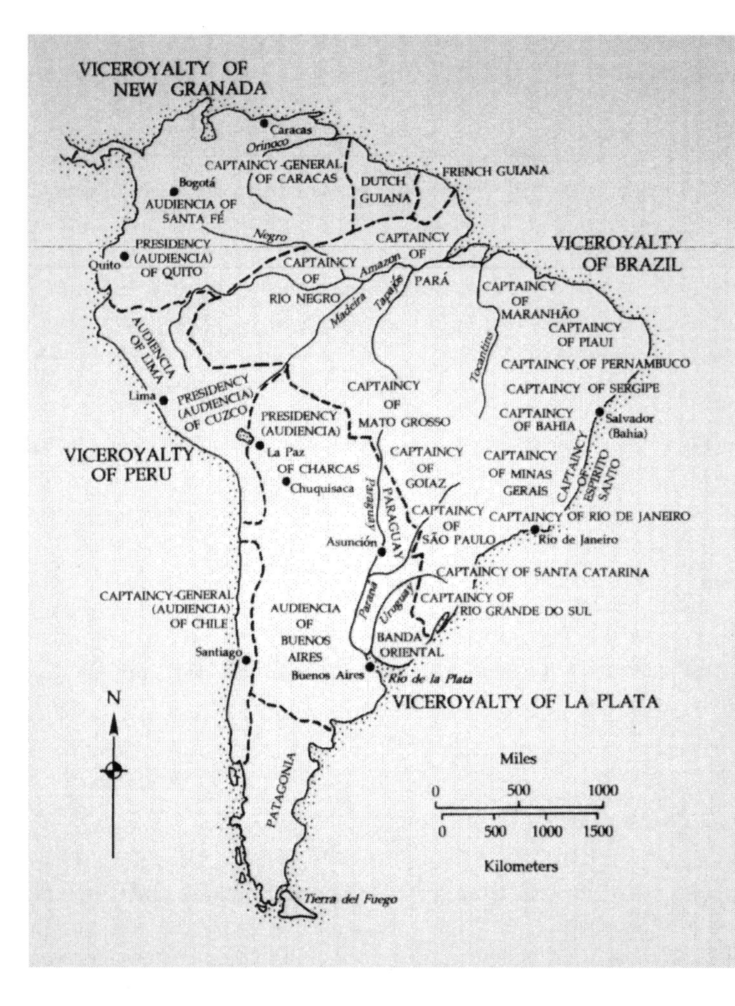

Figure 17.1 Colonial administrative districts and selected cities in South America, about 1790. Boundaries are generalized.

as in northern highland Ecuador, quite commonly most people lived close to their fields, with the major architectural and ceremonial centers being occupied only by chiefs, subsidiary ethnic lords, and their families and servants (Knapp, 1991). Large towns and cities did exist however in coastal Peru, and in parts of the highlands of Peru and Bolivia; impressively large villages also existed along the Amazon River, according to Carvajal's account of the first expedition down the river in 1541–1542 (Medina, 1988).

In late pre-Columbian times, the northern and central Andes (and adjacent coastal lowlands) were the main locus of intensive agriculture and landscape impacts. The highest arable slopes were used for potato cultivation that employed long fallows and rotational systems. Below about 3,000 meters, maize cultivation had led to massive landscape transformation. Irrigation canals snaked through the landscape, enabling farmers to produce reliable harvests even in areas with uncertain rainfall (Knapp, 1992b). In the south-central Andes, from what is now central Perú to northern Chile, true agricultural terracing (bench terracing) was widespread (Donkin, 1979). Raised fields were deployed in highland plains (*altiplanos*) from Colombia to Bolivia. Desert coastal valleys were irrigated for a variety of crops; and canals, water-table farming, raised fields, embanked fields, and sunken fields were deployed (Knapp, 1978, 1982). Humid coastal and Amazon lowland areas were also sometimes cultivated by using raised fields. Denevan (2001) has provided considerable evidence for intensive agriculture in late pre-Columbian South America.

Prior to European contact, the population of South America has been estimated at roughly about 24 million, with about half of the total in the central Andes (Denevan, 1992a, 1992b). Estimates such as this rely on collating and comparing evidence from archaeology (settlements, abandoned agricultural features), early accounts, carrying capacity calculations, as well as analogy with areas outside South America and extrapolation and interpolation to fill in gaps in knowledge. Even the best local estimates have a high margin of error. For example, in the Cara subregion of highland Ecuador north of Quito, 22 pyramid (ramp *tola*) sites are known; if all were occupied prior to the Inca conquest, and each corresponded to a chiefdom of 3,000 persons, this would have resulted in a population of 66,000 (Athens, 1978). The same region shows an abundance of abandoned agricultural landforms, including raised fields. Since raised fields are an inefficient way to produce food (in labor terms), their use suggests that alternatives niches (slopes, irrigated zones) had been subjected to intensification to the point of comparable labor inefficiency. Given assumptions about technology, crops, and available fertilizers (mostly night soil and guinea pig dung), this in turn suggests a regional population between 75,000 and 170,000 persons (Knapp, 1991). These figures can be seen as broadly consistent with a statement of Hernando de Santillán, who lived in northern Ecuador between 1563 and 1568. He claimed there had been a 4:1 depopulation since Inca times. Tribute figures suggest over 11,000 lived in the Otavalo encomienda alone, so Santillán's statement would indicate a population much greater than 44,000 for the Cara northern highlands (Knapp, 1991). Taken together, multiple kinds of evidence begin to produce more confidence that many tens of thousands lived in this region. In theory, these population densities could be extrapolated to other areas with less archaeological or ethnohistorical evidence.

A key problem with all population reconstructions based on agricultural landforms or settlement features is the question of simultaneous occupation. Were all *ramp tola* sites, or all raised field sites, used at the same time, or were they in use at different periods? Although the Cara region described above was likely to have been at a population peak prior to the Inca conquest, it is clear that substantial parts of it were depopulated by the Inca conquest prior to the Spanish arrival (Knapp, 1991). In other parts of the Andean highlands, it is clear that some areas were depopulated by volcanic eruptions (Knapp, 1999) or perhaps climate change (Kolata et al., 1999) well before the rise of the Inca state.

Similar questions have arisen on the Peruvian desert coast. Were all of the impressive prehistoric irrigation systems in full use at the time of the Spanish arrival, or were some abandoned (Schaedel, 1992)? In the savanna grasslands of the interior, were the raised fields in use at the time of the Spanish conquest or abandoned earlier (Denevan, 2001)? Further archaeological work is needed to answer these questions, but there is no doubt from Spanish accounts that substantial areas of the tropical Andean highlands and adjacent lowlands were under intensive cultivation at the end of the fifteenth century.

17.2 Colonial Impacts

Although there has been continued speculation about earlier contacts, particularly by voyagers from Africa, Polynesia, and Japan, the north coast of South America was certainly contacted by Columbus during his third voyage in late 1498, followed in swift succession by other navigators and leaders of military expeditions. The conquest of the Inca Empire may have been facilitated by an early smallpox epidemic in the 1520s that propagated through indigenous populations in advance of the Spanish military (Crosby, 1973; Dobyns, 1963; Cook, 1998). Caviedes (2001) has presented evidence that this conquest was also facilitated by an El Niño event in 1532. Expeditions included extensive forays into the interior, including Orellana's expedition down the Amazon (Medina, 1988). These activities led to the establishment of European mercantile empires and the beginning of the "Columbian Exchange" of organisms

(Crosby, 1973; Crosby, 1986; Viola and Margolis, 1991), tools, technologies, institutions, and peoples—indeed the beginnings of globalization in the modern sense of the term. The environmental results have, in the long term, been massive. Linkages involved not only Europe, the home of the Spanish and Portuguese empires, but also Africa and Asia. African slaves were present on the earliest expeditions, and African plants (bananas, plantains, and a variety of rice) were among the early introductions (Carney, 1998). Migration throughout the colonial period contributed to the process of diffusion (Robinson, 1990).

Butzer has pointed out lack of concrete empirical data to prove the point of a hypothesized "devastated colonial landscape" for New Spain. He also has discussed the evidence that European and Mediterranean land use was "overwhelmingly conservationist since prehistoric times" (Butzer, 1992). Unlike New Spain, there have been no scholarly claims for widespread colonial devastation in South America, and indeed there is little empirical evidence for such claims. However, local impacts were sometimes substantial. South America is a diverse continent, and colonial impacts varied from place to place. In general, one can distinguish between (1) zones of conquest of indigenous labor, (2) zones of relative colonial neglect, and (3) zones of indigenous suppression and creation of new economies and peoples.

The central Andes (roughly, modern Ecuador, Perú, and Bolivia) were the center of the Inca Empire and remained the region of greatest indigenous population density. Spaniards reorganized the economy around the needs of the silver export sector, including transportation infrastructure (oxcarts, mules), labor supply provisions, and the creation of food and textile supply networks for mine and plantation workers, and urban dwellers. This has been considered an example of an enclave economy (Glade, 1969), segregated from but dependent on the surrounding subsistence agricultural economy of surviving indigenous farmers and, to some extent, large agricultural estates. At the same time, there were many interchanges, transactions, and hybridities generated by the close proximity of European, indigenous, and other *castas* (Hoberman and Socolow, 1996).

Other areas of South America were relatively neglected during colonial times, either through fierce local resistance to colonial rule or because of a lack of resources attractive to Europeans. These include what are now southern Chile, Argentine Patagonia, and much of the Amazon basin.

Finally, some areas of South America were sites of long-term extermination or absorption of indigenous cultures in the context of an inflow of European and especially African populations, and the absence of any successful policy for protecting and sustaining indigenous communities with a chiefdom or villager form of social organization. These regions included much of what are now Colombia and Venezuela, eastern Brazil, and Uruguay, as well as northern Chile and Argentina. The censuses of the late colonial period show the demise of indigenous populations and the rise of African-American populations in these regions (Caviedes and Knapp, 1996; Schaedel, 1992).

17.3 The Central Andes

Even in the Central Andes, the conquest began a long period of depopulation, as European and African diseases such as smallpox, measles, and malaria spread among local populations (Dobyns, 1963; Cook, 1998; Newson, 1995). The drafting of local peoples into plantations and mining activities as labor and slaves may also have played a part in population decline. Populations were relocated or "reduced" to nucleated planned settlements for the purposes of improved health, labor supply, taxation, religious indoctrination, and governability. Spanish towns were established according to then understood principles of town planning. Thus, Trujillo, Lima, and other coastal Peruvian cities were sited next to rivers, placing them at risk for flooding during El Niño events (Schaedel, 1992; Caviedes, 2001). Under Viceroy Francisco de Toledo (1569–1585), the settlement hierarchy of the central Andes was worked out, so that many municipal capitals today are old *pueblos de indios* founded in his time, although quite a few were forced to relocate due to disasters such as floods or earthquakes (Schaedel, 1992), and their character became mestizo rather than Indian (Gade and Escobar, 1982). The separation of Spanish and Indian towns, and residential segregation in cities, was as much to preserve a measure of Indian autonomy (Butzer, 1992) as to exclude indigenous people from Spanish society.

Spain attempted to settle colonists in its new territories, but despite the experience with the *reconquista* in Iberia, had no experience with distant overseas colonization; as a result it experimented, using the institution of the land grant (*merced*), which tended in the New World to work against a small freehold tradition (Butzer, 1992). Although land grants could not include lands under cultivation by indigenous people, and *repartimientos* of indigenous labor precluded land rights, Borchert de Moreno (1981) has provided archival details from highland Ecuador on how in practice Spaniards were able to amass large estates, in some cases through strategic marriages with wealthy indigenous women who bequeathed their inheritances to their husbands. Indigenous depopulation and resettlement created further opportunities for the expansion of estates. The estates depended on indigenous labor, so the landscape in more densely populated regions tended toward a pattern of latifundia surrounded by indigenous minifundia. In areas like Colquepata in highland Perú, indigenous people were assigned subsistence plots that ignored previous management plans involving sectoral fallows (Zimmerer, 1996; Gade and Escobar, 1982; Orlove and Godoy, 1986). Although intended to provide for indigenous survival, such plans were disruptive and often resisted by local Indians, who attempted, often successfully, to re-establish dispersed settlement and old ways of territorial control.

The Spanish were familiar with irrigation and admired the sophistication of indigenous irrigation systems. Although some irrigated areas may have been abandoned on the Peruvian coast due to the greater water requirements of introduced crops (Schaedel, 1992), and there is evidence of abandoned irrigated terraces in southern Perú (Sherbondy and Villanueva, 1979), detailed study of the irrigation systems in northern highland Ecuador shows that irrigated areas were usually highly prized and expanded during colonial times (Knapp, 1992b). Irrigation is labor efficient; the effort of building and maintaining canals is more than repaid by the increased productivity due to improved soil fertility and optimization of water provision throughout the growing season. Also, irrigation permitted cultivation of lower, drier parts of intermontane valleys suitable for the production of long-cycle crops. Thus, for example, after Native Americans were relocated from the Chota Valley in Ecuador, lands previously devoted to coca production were put into sugar cane production with the use of imported slave labor. Elsewhere in Ecuador, new canals were installed upslope and downslope of pre-Columbian canals, providing for a net expansion of irrigated land (Knapp, 1992b).

Spaniards were familiar with both personal alienable water rights and with the concept of inalienable rights of water attached to pieces of land. The latter was the indigenous understanding of water, and was readily accepted by the colonial power as water rights were codified and litigation introduced. Many of the local structures and calendars of irrigation management were respected and retained, but water was managed by the elected officials of the newly established towns rather than by traditional chiefs and water officials (Guillet, 1992; Sherbondy, 1987; Mitchell and Guillet, 1994; Mitchell, 1991). Spaniards introduced the *óvalo* (water outlet of precise diameter to provide a measured flow) to help allocate water to different branch canals, and occasionally deployed *qanats* and water lifting devices, but otherwise their contribution to irrigation technology in South America was less important than the case in Mexico, where they deployed a variety of aqueducts and dams, especially in late colonial times (Knapp, 1992b). Much research remains to be done to understand better the patterns and processes of change in water management during the colonial period.

Many valley sides of the central Andes were terraced at the time of the conquest (Donkin, 1979). After the conquest, many of these terraces were abandoned, probably in part due to the declining subsistence needs of a shrinking population. Such was the case, for example, in the Colca Valley, the site of the largest single complex of pre-Colombian terraces. Here, farmlands remained under the control of maize-growing indigenous farmers, but most of the less accessible terraces seem to have been abandoned after the conquest (Treacy, 1994). Spaniards were familiar with terracing, but there is little evidence for Spanish construction of terraces or other slope management features in colonial times.

As for the raised fields that occupied large areas in highland flats as well as adjacent coastal lowlands and the Llanos de Mojos, most seem to have fallen into abandonment by colonial times, and in some cases they were abandoned much earlier (Denevan, 2001; Knapp, 1999; Kolata et al., 1999). In highland Ecuador, for example, former areas of raised fields became wetlands or areas of unmanaged pasturage, habitat for migratory ducks (Knapp, 1991).

Large areas of the Andes were being cultivated without either terracing or irrigation at the time of the conquest. In highland Ecuador, for example, the higher elevations were used for medium-fallow potato cropping and lower elevations for short-fallow maize cropping. Many other crops were also grown, including a variety of tubers, tarwi, quinoa, chili pepper, and beans. We know little of the exact nature of intercropping or rotational practices, but it is likely that in some areas a community-wide sectoral fallow system was followed (Zimmerer, 1996; Orlove and Godoy, 1986), and there were strategies for taking advantage of different ecological niches (Murra, 1972; Masuda et al., 1985; Caillavet, 2000). At higher elevations, and at locations with excessive or deficient rainfall, fallows were long enough to allow for the regrowth of brushland and woodland habitat for foxes (*Canis azarae*), rabbits, deer, and gallinaceous fowl. A wide range of fruit trees were grown, especially at somewhat lower elevations, including avocado, papaya, lúcuma, cherimoya, pacay, and tree tomato (Knapp, 1991); willow (*Salix humboldtiana*) and pepper tree (*Schinus molle*) may also have been diffused by humans in pre-Columbian times (Gade, 1999). Native trees were also planted on field margins, and quishuar species *(Buddleja incana, Buddleja coriacea)* were widely cultivated for wooden implements (Gade, 1999). Large quantities of firewood were harvested, but much of this was managed by the Inca empire; there were state-managed storehouses and groves (Gade, 1999). At higher elevations grasslands were maintained by clearing, grazing, and burning (Gade, 1999).

Zimmerer (1996) has pointed out that the Inca state promoted only a few land races of maize and potatoes in state-managed agricultural fields, leaving the bulk of land-race biodiversity in the hands of commoners. He attributed high crop diversity among commoners primarily to culturally given livelihood norms, but attributed land-race diversity to other (rather accidental) factors. It may be argued, however, that the growing of maize and potatoes in separate zones can be explained parsimoniously in terms of maturation times and labor inputs, given a household-level orientation to labor efficiency in achieving subsistence goals (Knapp, 1991). Potatoes require more labor to harvest per calorie than maize; thus they normally will not be grown in niches suitable for maize, and where they are grown, fallow cycles and fertilization will be sufficient to make their yields per unit of land so high that their planting and weeding costs will be low relative to short-cycle grain crops elsewhere (Knapp, 1991). While cultural preferences clearly lead farmers to find ways to provide for habitual culinary needs, it is argued that environmental factors also have long-term significance. This divergence of opinion calls for further research on the decision-making processes of the farmers. Zimmerer (1996) has suggested that land-race diversity in the Andes is in part an unanticipated consequence of localized seed-cycling processes, and has denied that this diversity has an ecological function. This fact is important for future crop biodiversity conservation policies.

After the Spanish conquest, these patterns of land use continued, albeit with some new crops and changes to tools. The *relaciones geográficas* (Jimenez de la Espada, 1965) and other sources show that a wide range of European crops were introduced into the Andes early in the colonial period. Gade (1992) suggested that, given the rigors of crossing the Panama isthmus, the preferred route of transfer of European crops and animals was from Hispaniola to Mexico and Nicaragua, and thence to the Rimac Valley near Lima. Here seeds, cuttings, and animals were picked up for diffusion along the Inca trail system through the Andes, where they had become part of a hybrid pan-Andean complex by the 1590s. This involved fewer than fifteen thousand Spaniards in a region of one million surviving Indians, so that clerics, indigenous leaders, and colonists took on special importance (Gade, 1992). The high elevations in the Andes provided a welcome habitat for European crops adapted to cooler weather. Some European crops were eventually incorporated into local rotation or intercropping, especially barley, wheat, and broad beans. Wheat and barley were grown for tribute or sale, as well as an element in local soups and stews; straw and stubble were also a livestock feed (Gade, 1992). These small grains tended to displace quinoa in the same way broad beans displaced tarwi (Gade, 1992). Wine grapes were introduced with some success in coastal oases and highland valleys, but never became widespread. Sugar cane was introduced successfully from the coast up to the warmer inter-montane valleys (Knapp, 1992b), but its use for producing sugar and alcohol remained a Spanish monopoly (Gade, 1992). Successful fruit-tree introductions included the orange, apple, pear, plum, peach, and capulin cherry (from Mexico) (Gade, 1992). Under Spanish rule, commoner fields remained the home of most of the crop and land-race diversity; only a few crops and land-races entered into markets or tribute (Zimmerer, 1996).

Much remains to be done to reconstruct the history of livestock introductions. The Europeans understood local concepts of commons, and deployed privately owned herds in common areas such as *ejidos* near towns and cities. Donkeys were widely adopted by Indians as pack animals, and mules became the main beast of burden for longer trips (Gade, 1992). Burning of high grasslands continued, sometimes under the aegis of the Feast of St. John the Baptist, a traditional occasion for bonfires (Gade, 1999). In general, sheep seem to have replaced llamas and alpacas at high elevations below 3,500 meters, but the pace and timing remains to be determined. There does not seem to be any equivalent to the massive transhumant sheep economy introduced into Mexico, and no crisis of overgrazing or erosion has been documented (Schaedel, 1992: 233; Gade, 1992). Pigs and chickens were also widely adopted by indigenous farmers, and a few also began raising cattle. One might speculate that it was in colonial times that farmers began to develop new manuring systems, including movable corrals in fields. Coupled with population decline, this new supply of dung (Winterhalder et al., 1974) probably reduced the labor input required to feed a family, and greatly reduced production pressures on more marginal lands. Although ox-drawn plows were an early introduction, their use remained localized in the colonial Andes (Schaedel, 1992; Gade, 1992: 468). Despite the field evidence for considerable gullying of unknown age and cause on many slopes, the influence of livestock on sheet or gully erosion remains to be documented in the Andes for the colonial period (but see Gade, 1999). The black rat (*Rattus rattus*) was introduced to South America in the 1540s, and other rats, mice, and vermin also arrived in these early decades of colonialism (Gade, 1999).

The broad outlines of Spanish impact can be seen in the Chilca Valley, just south of Lima. At the time of the conquest, up to 10,000 people were living in this valley, growing maize, pacay, lúcuma, cotton, peanuts, gourds, squash, and other crops, using a complex system of embankments to manage flood water farming, as well as sunken fields to enable direct use of water from the lens of fresh water under the beach sand. By 1539, Chilca was part of a *repartimiento*, and was still being largely cultivated for maize, tubers, and fruit trees. However by the close of the sixteenth century, epidemics and other factors had reduced the population of the valley to less than a thousand; grapes, figs, pomegranates, quince, and melons were being grown, alongside traditional maize cultivation and fishing activities. By 1653 most of the fields were abandoned (Cobo, 1956), and by the late 1700s the region was producing only salt and fish products, with very little farming, at least as visible to visitors (Knapp, 1978, 1982). Rostworowski de Diez Canseco (1981) has pointed out the adverse impacts of grazing, firewood collecting, and hunting on the dry woodlands and *lomas* of coastal Perú, alongside the maintenance of fishing and salt gathering activities. This case example is interesting in that the valley seems always to have been under the substantial control of an indigenous population resistant to outsiders; no haciendas were established here. The overall trajectory of depopulation and disintensification reaching a nadir in the 1600s and 1700s was characteristic of the entire Andean region.

The Spanish introduced iron technology, the sickle, tanning, tallow-processing (for lighting), soap making, wine making, sugar milling and refining, distilling, bread ovens and water-driven grist milling, among many other technologies (Gade, 1992; Schaedel, 1992). Spanish use of wood for bread ovens, brick factories, tile works, roof beams, doors, floors, coffins, furniture, window sashes, and charcoal, combined with browsing from sheep, provided stress on remaining woodlands in high Andean and coastal Peruvian valleys near towns and cities, even given the countervailing reduction of overall population and attempts at forest regulation through viceregal decrees. The result was most likely a reduction of woodland from the already sparse pre-Columbian situation (Gade, 1999; Rostworowski de Diez Canseco, 1981). Particularly devastated were *cedro* (*Cedrella spp.*) and *nogal (Juglans neotropica),* both of which were valued for furniture and other craft applications. Although not driven to extinction, both were largely removed from more accessible locations (Gade, 1999).

In the high Andes, the main objects of long-distance trade were silver and gold. Mercury mines discovered in Huancavelica, Perú, helped with the patio process of extracting silver but had long-term health consequences for those working with this substance. Logging for smelting and mine timbers helped deforest slopes near mining centers such as Potosí and Huancavelica (Gade, 1999). Mining pressures affected forests as far away as Chuquisaca (Sucre), 130 km away (Gade, 1999). The mining economy required provisioning of textiles and mules from distant areas in the Andes.

As a result of these impacts, the concept of "*lo Andino*" (characteristically Andean) as a timeless constant can be called into question, especially insofar as some of its advocates seem to imply that the Andes are a marginal, unproductive environment (Salman and Zoomies, 2003).

17.4 Other Regions

Coastal Brazil had numerous people and open, garden-like landscapes prior to European settlement (Parsons, 1989), but after the conquest forests expanded due to indigenous depopulation. Sugar plantations near the coast required clearings, and sugar mills and boiling houses required firewood and construction material (Sternberg, 1968). Much of the forest was worked by shifting cultivation during the colonial period, in some cases leading to a loss of soil fertility and forest degradation. The gold rush of late colonial times created one of the most serious episodes of environmental impact; over a million migrants were attracted to Brazil, and mining processes included the use of hydraulic methods, designed to reveal placer gold, which commonly caused hillsides to be washed away (Dean, 1995). Forays into the Brazilian interior to exploit various forest products had more localized impacts, including the removal of dyewoods such as various species of *Caesalpina* (Parsons, 1989; Dean, 1995).

As in the case of coastal Brazil, many of the indigenous groups of northwestern South America were eliminated or displaced by the growth of new societies of mixed heritage. The Antioqueños of Colombia, for example, expanded through much of the highlands northwest of Bogotá, initially in search of gold. Miners and charcoal burners began the process of removing woodlands and forests, although much of the environmental impact would occur later with the development of the coffee economy (Parsons, 1968).

The introduction of cattle, horses, sheep, goats, and other large domesticated animals resulted in a greater economic value for tropical grasslands and savannas. In tandem with the growth of herds, fodder crops such as alfalfa were introduced, and a long process began of importing new grass species, many from Africa, suitable for South American conditions. The resulting "Africanization" of South American grasslands has been one of the most important environmental processes over the last five centuries. Guinea Grass (*Panicum maximum*), Pará Grass (*Brachiaria mutica*), Jaraguá (*Hyparrhenia rufa*) and Molasses Grass (*Melinis minutifloria*) were apparently all present in Brazil by the end of colonial times; they also spread into Venezuela, Colombia, and other tropical grasslands (Parsons, 1970). Cattle were introduced to the Venezuelan Llanos by the 1540s, and by 1800 there may have been about a million cattle on these grassy plains (Parsons, 1989). Horse nomadism began in the sixteenth century in south-central Chile and Argentina, where it helped provide mobility for Native American resistance to European occupance until the nineteenth century (Schaedel, 1992). Cattle herding and ranching in various forms also expanded throughout the continent, providing meat and leather with a minimum of effort where land was abundant.

Metal tools (axes and ploughs) and animal power also meant that forest and grassland sod could be removed with much less effort than previously was the case. Thus, as Denevan (1992c) has argued, swidden became more efficient and forest agriculture more extensive: it was more labor efficient to open new fields than continue to recultivate the same field. This may help to explain the abandonment of dark earth sites in Amazonia during colonial and later times. These are the prehistoric anthropogenic black and brown soils discussed in chapter 16.

17.5 Legacy of Colonialism

In general, the colonial period in South America was one of recovery for forests and some wetlands, as population decline and relocation led to reduced human impacts. The high cost of transportation away from the coasts, and restricted trade with Europe and Asia, further reduced environmental impacts. Locally severe impacts were, however, associated with particular high-value export activities, including silver mining in the central Andes, gold mining in Brazil and Colombia, and sugar cultivation in coastal Brazil. In particular, the demand for wood led to stress on woodlands and forests, especially in the Andes and coastal Brazil.

There were however four major colonial legacies that would become even more important after 1800: (1) the legacy of insertion into a global system of trade and transfer, (2) the legacy of a coastal focus of development, (3) the legacy of cultural and caste diversity and divisions, and (4) a legacy of ethics, aesthetics, and attitudes toward nature.

Colonial regimes organized regional economies and infrastructure to serve the needs of towns, cities, and Iberia. In this early global division of labor, the emphasis was on high-value goods that could withstand high transportation and transaction costs. These included precious metals and sugar. At a more local scale, trade was mobilized in grains, mules, leather, woolen and cotton textiles, and other goods. After independence, South America continued to participate in global trade, developing new metallic and plantation resources. This would result in major environmental impacts in areas devoted to the development of exports, for example of coffee, rubber, wheat, cattle products, and soybeans. Furthermore, these global linkages continued to facilitate the exchange of plants and animals, involving many unpredictable impacts that have continued to the present day.

The spread of tropical lowland diseases and the focus of administrative, cultural, and economic centers near the coasts helped establish a "hollow continent," still visible in the distribution of population and intensive land use today. Although areas in the interior were never truly isolated from global processes, the extensive habitats still remaining are a colonial legacy. The major colonial administrative centers became the capitals of nations, and in some cases came overwhelmingly to concentrate political, economic, and cultural functions, and become primate cities. South America's relatively high level and concentration of urbanization continues to create its own special environmental impacts and challenges for conservation (see chapter 20).

The creation of a segmented labor system with African slave, native American, mestizo, and Spanish castes, with corresponding rights and legal frameworks, has been augmented more recently with other migrations, for example by Japanese, Lebanese, and German immigrants, with corresponding impacts on the ethnic division of labor and the structure of consumer demand. In some cases this has made it more difficult to create a policy consensus, necessary for certain types of conservation.

Finally, the legacy of an Iberian ethic, aesthetic, and perception of nature continues to affect the politics and science of environment in South America. Only in the last few years has wilderness travel and ecotourism become popular within the region (as opposed to with tourists from other regions). Similarly, the development of environmental science in South America has been equally slow. Proposals for environmental protection will benefit from the long Iberian tradition of interest in agronomy and humanist relations with nature but will continue to struggle with the lack of a wilderness ideal.

References

Athens, J.S., 1978. *Evolutionary Process in Complex Societies and the Late Period—Cara Occupation of Northern Highland Ecuador*. Ph.D. Dissertation, University of New Mexico, Albuquerque.

Bethell, L. (Editor), 1987. *Colonial Spanish America*. Cambridge University Press, Cambridge.

Borchert de Moreno, C., 1981. El Período Colonial. In: S.M. Yánez (Editor), *Pichincha: Monografía Histórica de la Region Nuclear Ecuatoriana*, Consejo Provincial de Pichincha, Quito, 195–276.

Butzer, K.W., 1992. The Americas before and after 1492: An introduction to current geographical research. *Annals of the Association of American Geographers,* 82/3, 543–556.

Butzer, K.W., and E.K. Butzer. 1995. Transfer of the Mediterranean livestock economy to New Spain: Adaptation and ecological consequences. In: Turner, B.L., II, et al. (Editors), *Global Land-Use Change: A Perspective from the Columbian Encounter,* Consejo Superior de Investigaciones Científicas, Madrid, 151–193.

Caillavet, C., 2000. *Etnias del Norte: Etnohistoria e Historia del Ecuador*. Abya Yala, Quito.

Carney, J., 1998. The role of African rice and slaves in the history of rice cultivation in the Americas. *Human Ecology,* 26, 525–545.

Caviedes, C.N., 2001. *El Niño in History: Storming Through the Ages*. University Press of Florida, Gainesville.

Caviedes, C.N., and G. Knapp, 1995. *South America*. Prentice Hall, Englewood Cliffs.

Cieza de León, P., 1962 [1553]. *La Crónica del Perú*. Espasa-Calpe, Madrid.

Cobo, B., 1956 [1653]. *Historia del Nuevo Mundo*. Biblioteca de Autores Españoles, Volumes 91 and 92, Madrid.

Cook, N.D., 1998. *Born to Die: Disease and New World Conquest, 1492–1650*. Cambridge University Press, Cambridge.

Crosby, A.W., Jr., 1973. *The Columbian Exchange: Biological and Cultural Consequences of 1492*. Greenwood Press, Westport.

———, 1986. *Ecological Imperialism: The Biological Expansion of Europe, 900–1900*. Cambridge University Press, Cambridge.

Dean, W., 1995. *With Broadax and Firebrand: The Destruc-*

tion of the Brazilian Atlantic Forest. University of California Press, Berkeley.
Denevan, W.M., 1992a. Native American populations in 1492: Recent research and a revised hemispheric estimate. In *The Native Population of the Americas in 1492*, edited by W.M. Denevan, 2nd Edition, pp. xvii–xxix. University of Wisconsin Press, Madison.
———, 1992b. The Pristine Myth: The landscape of the Americas in 1492. *Annals of the Association of American Geographers,* 82/83, 369–385.
———, 1992c. Stone vs. metal axes: The ambiguity of shifting cultivation in prehistoric Amazonia. *Journal of the Steward Anthropological Society,* 20, 153–165.
———, 2001. *Cultivated Landscapes of Native Amazonia and the Andes.* Oxford University Press, New York.
Dobyns, H.F., 1963. An outline of Andean epidemic history to 1720. *Bulletin of the History of Medicine,* 37, 493–515.
Donkin, R.A., 1979. *Agricultural Terracing in the Aboriginal New World.* Viking Fund Publications in Anthropology Number 56. University of Arizona Press, Tucson.
Gade, D.W., 1992. Landscape, system, and identity in the Post-Conquest Andes. *Annals of the Association of American Geographers*, 82, 460–477.
———, 1999. *Nature and Culture in the Andes.* University of Wisconsin Press, Madison.
Gade, D.W., and M. Escobar, 1982. Village settlement and the colonial legacy in southern Peru. *Geographical Review,* 72, 430–449.
Glade, W.P., 1969. *The Latin American Economies: A Study of Their Institutional Evolution.* Van Nostrand, New York.
Guamán Poma de Ayala, F., 1980 [1613]. *El Primer Nueva Crónica y Buen Gobierno.* Edited J.V. Murra and R. Adorno, translated J.L. Urioste. Siglo XXI, Mexico City.
Guillet, D.W., 1992. *Covering Ground: Communal Water Management and the State in the Peruvian Highlands.* University of Michigan Press, Ann Arbor.
Hoberman, L.S., 1996. Interpretations of the colonial countryside. In: L.S. Hoberman and S. M. Socolow (Editors), *The Countryside in Colonial Latin America*, University of New Mexico Press, Albuquerque, 235–251.
Hoberman, L.S., and S.M. Socolow (Editors), 1996. *The Countryside in Colonial Latin America.* University of New Mexico Press, Albuquerque.
Jimenez de la Espada, M. (Editor), 1965 [1881–1897]. *Relaciones geográficas de Indias. Perú.* Three volumes. Ediciones Atlas, Madrid.
Knapp, G., 1978. *The Sunken Fields of Chilca: Horticulture, Microenvironment, and History in the Peruvian Coastal Desert.* M.S. thesis, University of Wisconsin, Madison.
———, 1982. Prehistoric flood management on the Peruvian coast: Reinterpreting the "Sunken Fields" of Chilca. *American Antiquity*, 47, 144–154.
———, 1991. *Andean Ecology: Adaptive Dynamics in Ecuador.* Westview Press, Boulder.
———, 1992a. Cultural and historical geography of the Andes. *Yearbook of the Conference of Latin Americanist Geographers*, 17/18, 165–175.
———, 1992b. *Riego precolonial y tradicional en la Sierra Norte del Ecuador.* Hombre y Ambiente, 22, Abya Yala, Quito.
———, 1999. Quilotoa ash and human settlements in the Equatorial Andes. In: P. Mothes (Editor), *Actividad Volcánica y Pueblos Precolombinos en el Ecuador.* Ediciones Abya Yala, Quito, 139–155.
Kolata, A., et al., 1999. *Tiwanaku and Its Hinterland: Archaeology and Paleoecology of an Andean Civilization Volume 1: Agroecology.* Smithsonian Institution Press, Washington D.C.
Masuda, S., I. Shimada, and C. Morris (Editors), 1985. *Andean Ecology and Civilization: An Interdisciplinary Perspective on Andean Ecological Complementarity.* University of Tokyo Press, Tokyo.
McAlister, L. N. 1984. *Spain and Portugal in the New World, 1492–1700.* University of Minnesota Press, Minneapolis.
Medina, J.T. (Editor), 1988. *The Discovery of the Amazon.* Dover Publications, New York.
Mitchell, W.P., 1991. *Peasants on the Edge: The Struggle for Survival and the Transformation of Social and Religious Organization in the Andes.* University of Texas Press, Austin.
Mitchell, W.P., and D. Guillet (Editors), 1994. *Irrigation at High Altitudes: The Social Organization of Water Control Systems in the Andes.* Society for Latin American Anthropology Monograph Number 12. American Anthropological Association, Washington D.C.
Mörner, M., 1985. *The Andean Past: Land, Societies, and Conflicts.* Columbia University Press, New York.
Murra, J., 1972. El 'Control Vertical' de un máximo de pisos ecológicos en la economía de las sociedades andinas. In: J. Murra (Editor), *Visita de la Provinicia de Leon de Huánuco en 1562*, Volume 2. Universidad Nacional Hermilio Valdizán, Huánuco, Peru.
Newson, L., 1995. *Life and Death in Early Colonial Ecuador.* University of Oklahoma Press, Norman.
Orlove, B.S., and R. Godoy, 1986. Sectoral fallowing systems in the Central Andes. *Journal of Ethnobiology* 6, 169–204.
Parsons, J. J., 1968. *Antioqueño Colonization in Western Colombia.* Revised Edition. University of California Press, Berkeley.
———, 1970. The Africanization of the New World tropical grasslands. *Tübinger Geographische Studien,* 34, 141–153.
———, 1989. *Hispanic Lands and Peoples: Selected Writings of James J. Parsons.* Edited by W.M. Denevan, Westview Press, Boulder.
Ravines, R. (Editor), 1978. *Tecnología Andina.* Instituto de Estudios Peruanos, Lima.
Robinson, D.J. (Editor), 1990. *Migration in Colonial Latin America.* Cambridge University Press, Cambridge.
Rostworowski de Diez Canseco, M., 1981. *Recursos Naturales Renovables y Pesca, Siglos XVI y XVII.* Instituto de Estudios Peruanos, Lima.
Salman, T., and A. Zoomies (Editors), 2003. *Imaging the Andes: Shifting Margins of a Marginal World.* Aksant (CEDLA Latin American Studies 91) Amsterdam.
Schaedel, R.P., 1992. The archaeology of the Spanish colonial experience in South America. *Antiquity,* 66, 217–242.
Sherbondy, J.E., 1987. Organización hidráulica y poder en el Cuzco de los Incas. *Revista Española de Antropología Americana,* 17, 117–153.
Sherbondy, J.E., and H. Villanueva U., 1979. *Aguas y Poder.* Centro de Estudios Rurales Andinos Bartolomé de las Casas, Cuzco.

Sternberg, H. O'R., 1968. Man and environmental change in South America. In: F.J. Fittkau et al. (Editors), *Biogeography and Ecology in South America.* N.V. Junk, The Hague, 413–445.

Treacy, J.M., 1994. *Las Chacras de Coporaque: Andenería y Riego en el Valle del Colca.* Instituto de Estudios Peruanos, Lima.

Viola, H.J., and C. Margolis (Editors), 1991. *Seeds of Change: Five Hundred Years since Columbus.* Smithsonian Institution Press, Washington D.C.

Winterhalder, B., R. Larsen, and R.B. Thomas, 1974. Dung as an essential resource in a highland Peruvian community. *Human Ecology,* 12, 115–133.

Zimmerer, K.S., 1996. *Changing Fortunes: Biodiversity and Peasant Livelihood in the Peruvian Andes.* University of California Press, Berkeley.

Agriculture and Soil Erosion

Carol P. Harden
Glenn G. Hyman

People have manipulated the natural environments of South America for agricultural purposes for several millennia. While agriculture is strongly affected by the physical attributes of a place—soil, water, climate, biota, and topography—agriculture changes a landscape's physical and biological characteristics and processes. Agriculture may involve short- and long-term conversion of forest to cropland and pasture, modification of topography and drainage, and the introduction and propagation of exotic species. Soil erosion, much of which is caused by agriculture, is a major concern in South America. This chapter introduces the patterns of agriculture in South America and examines agricultural trends. It then reviews the causes and consequences of soil erosion at continental to local scales, providing examples from research conducted across the continent. As population grows and demand for agricultural production increases, knowledge of the physical geography of soil erosion will be even more critical for the sustainability of agriculture in South America.

18.1 Agriculture

18.1.1 Spatial Distribution of Agriculture in Contemporary South America

Agriculture is broadly defined here to encompass annual and permanent crops, tree crops, and livestock. Agricultural patterns of South America today reflect great differences in the continent's natural environments. They also reflect the influence of international and global markets, the impacts of national policies, and the imprints of pre- and post-colonial settlement patterns, preferred species, and cultural preferences. The wide range of climates in South America allows a great variety of temperate and tropical fruits, vegetables, and grains to flourish. Historically, the diverse agricultural capabilities of different parts of the continent have been fundamental influences in the development of pre- and post-colonial human habitation and economic patterns (U.S. Agency for International Development, 1993; see chapters 16 and 17).

At the continental scale, agriculture occurs across almost all regions of South America. It is notably absent only in the Gran Chaco, rugged portions of the high Andes, and desert landscapes along the Pacific coast of northern Chile and southern Perú (figs. 18.1, 18.2). In practice, there is little cropland in sparsely populated regions, especially in the Amazon basin, and in densely populated urban areas, even where the lands and climates of those places are capable of supporting agriculture. The contemporary pattern of agriculture in South America still generally resembles that documented by Whittlesey in 1936 (fig. 18.2). One difference, which can be seen by comparing figures 18.1 and 18.2, is that agricultural activity has increased in the Amazon basin, especially around its southeastern rim.

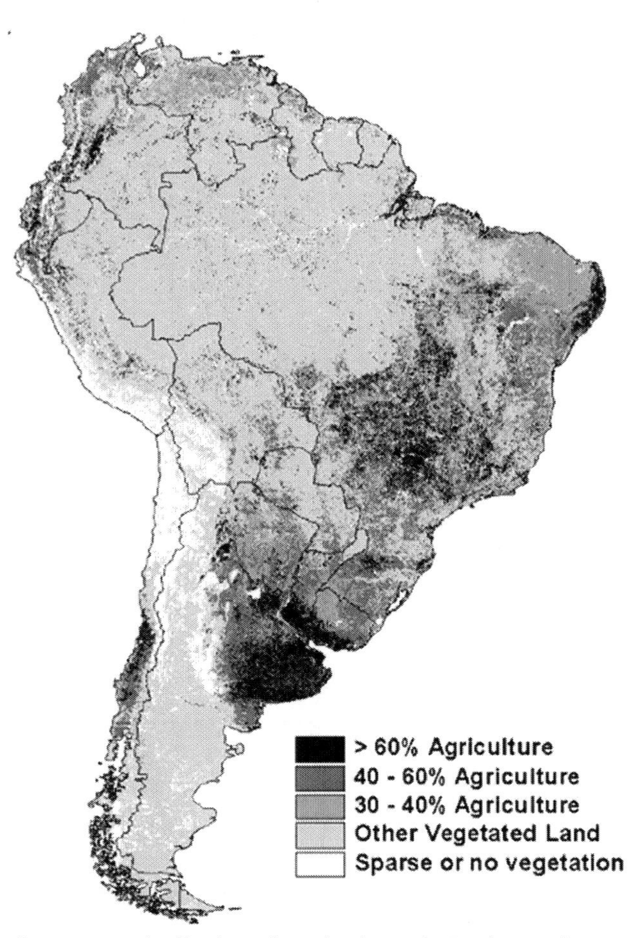

Figure 18.1 Distribution of cropland area in South America, based on satellite imagery from April 1992 to March 1993 (from Wood et al., 2000)

Regional patterns provide a generalized view of agriculture in South America. The Pampas region of Argentina supports intensive commercial agriculture, particularly for grains and oil seeds. Tropical plantation agriculture fringes the Pacific, Atlantic, and Caribbean coasts north of the Tropic of Capricorn, and also certain portions of the Amazon River in central Brazil. Plantations produce sugar cane (Brazil, Colombia, Ecuador, Perú, Venezuela), bananas (Colombia, Ecuador, French Guiana, Suriname, Venezuela), and cacao (Brazil, Ecuador, French Guiana) for export. Coffee is grown on plantations primarily in Colombia and Brazil, but also in Ecuador, Perú, Bolivia, and Venezuela. A narrow zone of Mediterranean-type climate in Chile favors the growth of crops from Mediterranean regions, notably grapes. Beyond the Amazon basin, the predominant agricultural activity in most of the interior of the continent is extensive livestock ranching. More than half of the land in Argentina, Paraguay, and Uruguay, for example, is in permanent pasture [Food and Agriculture Organization (FAO), 2000a].

Crops grown in South America today reflect the botanical history of the continent and the strength of local and international markets. All of the major export crops are introduced species. Crops native to the Andean region include quinoa (*Chenopodium quinoa*), amaranth (*Amaranthus caudatus*), potatoes (*Solanum* spp.), and numerous other tubers. Common beans (*Phaseolus vulgaris*), lima beans (*Phaseolus lunatus*), and peanuts (*Arachis hypogaea*) are considered native, as are several squashes (*Cucurbita* spp.), chili peppers (*Capsicum* spp.), coca (*Erythroxylon coca*), and various tropical fruits: chirimoya (*Annona cherimolia*), guanabana (*Annona muricata*), passion fruit (*Passiflora ligularis*), and guava (*Psidium guajava*) (Brush, 1982). Also South American in origin are tomatoes (*Lycopersicon* spp.), cacao (*Theobroma* spp.), pineapple (*Ananas* spp.), cassava (*Maniot esculenta*), papaya (*Carica papaya*), cashews (*Anacardium occidentale*), rubber tree (*Hevea brasiliensis*), cotton (*Gossypium* spp.), and the American oil palm (*Elaeis oleifera*) (Oldfield, 1984). Maize (*Zea mays*), which was domesticated in Mesoamerica, attained further genetic modification and diver-

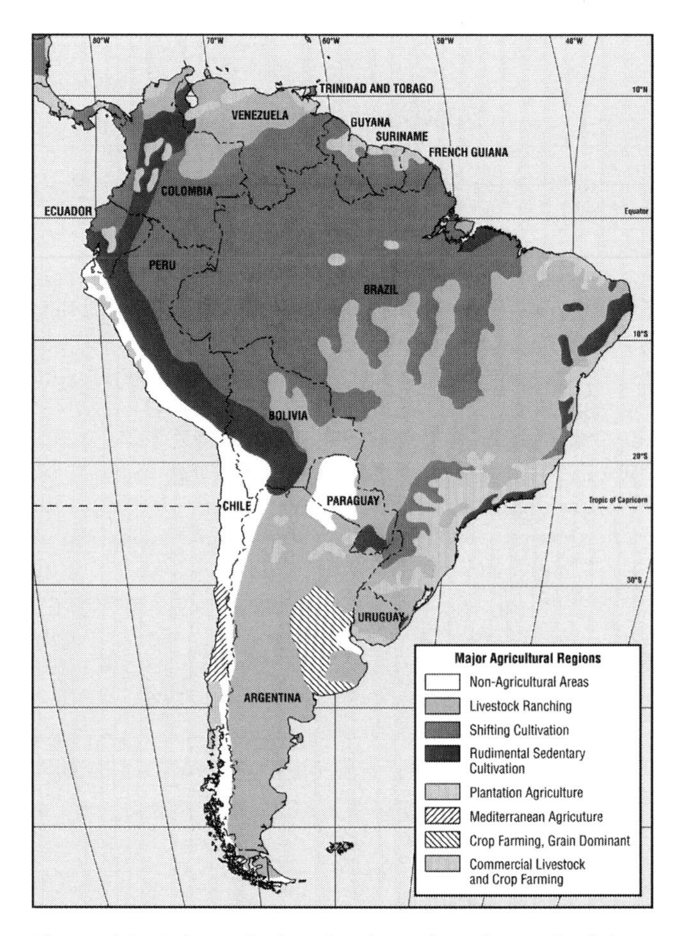

Figure 18.2 Major agricultural regions of South America (after Whittlesey, 1936)

sity in the central Andes, where it was an important part of the culture by 500 A.D. (Museo Arqueológico, 1997).

Agriculture is an important component of South American economies, cultures, and landscapes. Brazil, the largest country, has the most arable land (table 18.1), although Uruguay has the highest percentage (77%, 14 × 10^6 ha) of potentially arable land (FAO, 2000a). Over 75% of the total land area of South America is in forest, woodland, or permanent pasture. By far the most extensive use of agricultural land is as permanent pasture (29% of total land area, 81% of agricultural land) (FAO, 2000a).

Agricultural practices in South America strongly reflect the influence of the European conquest. The Spanish brought the plow and the wheel, both of which are essential to South American agriculture today. Since the conquest, there have been two primary types of agricultural systems in Latin America: (1) large, monocultural plantations and ranches, and (2) small, heterogeneous farms. Large plantations and ranches (*haciendas* and *latifundia*) have traditionally occupied the major proportion of the arable land and provided a major source of foreign exchange from export crops such as coffee, bananas, cotton, sugar cane, and beef. In spite of land-reform efforts, land-concentration indices have remained almost the same since 1950 (Winograd and Farrow, 1999). *Minifundia* (small farms) are scattered throughout the central and northern Andean region, especially on the interior slopes and valleys of the Andean ranges. A 1975 study by the Ecuadorian Ministry of Agriculture found 37.7% of all farms in the central Ecuadorian *Sierra* (Andean sector), or 40,948 farms, had less than 1 ha, while 46% had less than 5 ha, and only 0.10% had more than 1,000 ha (Cañadas and Salvador, 1982).

Minifundistas (small farmers) have traditionally relied upon fallow periods to regenerate soil fertility (Quinton and Rodriguez, 1999). Where population pressures on the land have increased, fallow periods have been reduced, and soil fertility has decreased. Lower crop productivity is one factor that motivates rural to urban migration. Cities and their suburbs now occupy many lands formerly used for agriculture.

18.1.2 Factors Controlling Present-day Distribution of Agriculture

The geography of agriculture is controlled by environmental opportunities and constraints, and also by economic and cultural factors. Much of the land area of South America lies within the tropics, but, owing to its longitudinal extent and altitudinal range, the continent contains many different environmental zones that support a great diversity of agricultural products. The major agricultural products of each of the 13 South American countries, listed in table 18.2, illustrate this variety.

Environmental Factors In the Andean regions, different environmental zones are juxtaposed by altitude. A transect of agricultural products through the Ecuadorian Andes, for example, ranges from tropical (e.g., cacao, mangos, bananas) to a wide range of tropical to temperate (e.g., tomatoes, broccoli, onions), and finally to the most cold-tolerant temperate crops, such as barley and potatoes, before the upper limit of agriculture is reached between 3,000 m and 4,200 m (Brush, 1982). Inhabitants of the Andean region have a long history of exchange between altitudinal zones. Before the arrival of the Spanish and the introduction of sheep to South America in the early sixteenth cen-

Table 18.1 Agricultural land use in South America

Country	Land area (1000 Ha)	Agricultural area[b] (1000 Ha)	Permanent[c] crops (1000 Ha)	Permanent pasture (1000 Ha)	Arable land (1000 Ha)	Irrigated cropland (1000 Ha)
Argentina	273,669	169,200	2,200	142,000	25,000	1,561
Bolivia	108,438	36,034	229	33,831	1,974	128
Brazil	845,651	250,200	12,000	185,000	53,200	2,656
Chile	74,880	15,219	315	12,925	1,979	1,800
Colombia	103,870	45,281	2,036	41,166	2,079	850
Ecuador	27,684	8,108	1,427	5,107	1,574	865
Paraguay	39,730	23,985	85	21,700	2,200	67
Peru	128,000	31,270	500	27,100	3,670	1,195
Uruguay	17,481	14,827	47	13,520	1,260	180
Venezuela	88,205	21,730	850	18,240	2,640	540
South America[a]	1,752,925	618,822	19,718	502,981	96,123	10,043

[a]Total for South America also includes Malvinas (Falkland) Islands, French Guiana, Guyana, South Georgia, and Suriname.
[b]Agricultural area is sum of permanent crops, permanent pasture, and arable land.
[c]Permanent crops include cocoa, coffee, rubber, fruit trees, nut trees, and vines.

Source: FAO (2000a).

tury, Andean residents wore clothing made of cotton, for which they depended on travel and exchange to lower, warmer elevations (Salomon, 1980). Today, driving times of only a few hours link highland and lowland environments, and Andean region markets typically contain a wide range of tropical to temperate fruits and vegetables.

Environmental requirements for commercial crops are well known. Table 18.3, which gives climatic ranges for major commercial crops grown in South America, shows that some (e.g., coffee *robusta*, paddy rice) are more climate-sensitive than others. Soil fertility is an important environmental factor, with the most fertile soils developed in volcanic, alluvial, or grassland environments. Most of Argentina's commercial agriculture, for example, occurs in the rich Mollisols and temperate climate of the Pampas (see chapter 14). Today, irrigation, fertilization, and new crop varieties allow agriculture to be successful in areas that would not previously have been considered environmentally optimal.

The future of this trend, however, and especially the extent to which it applies to the Amazon basin and other regions of Brazil, continue to be debated by experts. Sanchez et al. (1982) argue that agricultural researchers who understand the differences in clay mineralogy between Oxisols and temperate soils can develop strategies to alleviate the specific conditions, including aluminum toxicity, subsoil acidity, and phosphorus deficiency, that lower crop productivity. Sanchez (2000) cites the very successful growth of agricultural production in the Brazilian Cerrados from 1970 to 1990 as an example of the excellent potential for continuous cultivation in the Amazon region. Fearnside (1987), on the other hand, questions the sustainability of continuous cultivation in the Amazon, arguing that it occurs at carefully controlled sites in research stations, but not under normal field conditions managed by farmers with ordinary resources. Smith et al. (1999) report that large areas cleared for agriculture in the Peruvian Amazon have been abandoned to secondary forest because of the decline in fertility caused by the local practice of slash and burn agriculture.

Cultural preferences affect the distribution of agricultural products. For example, maize is grown extensively on hillslopes throughout southern Andean Ecuador, not because environmental factors limit other crops, but because maize is central to the diet of the region. All South American countries essentially feed themselves, but richer regions have more choice of foodstuffs, and poorer regions have fewer calories per person and greater incidence of malnutrition.

Market Factors and Governmental Intervention Within a given environmental zone, local and international market forces profoundly influence the choice of crops. All South American countries export agricultural products. Major export products are wheat (Argentina), soybeans and vegetable oils (Argentina, Bolivia, Brazil, Paraguay), coffee (Brazil, Colombia), bananas (Ecuador, Colombia), sugar (Guyana), meat (Paraguay, Uruguay), and cut flowers (Colombia, Ecuador) (table 18.2). Crop lands change as markets change. Since 1985, cotton-growing areas have almost

Table 18.2 Agriculture in South American countries

Country	% GDP in agriculture	% Labor force in agriculture	Agricultural products
Argentina	7	12	wheat, corn, livestock, soybeans, sunflower seeds, lemons, grapes, lemons, tea, tobacco, peanuts
Bolivia	17	47	soybeans, coca, cotton, corn, sugarcane, rice, potatoes
Brazil	14	23	soybeans, coffee, wheat, rice, corn, sugarcane, cocoa, citrus, beef
Chile	6	19	fruit, sugar beets, potatoes, beef, poultry, wool
Colombia	19	27	coffee, bananas, rice, corn, sugarcane, cocoa, oilseed, vegetables, cut flowers
Ecuador	14	33	bananas, coffee, cocoa, rice, potatoes, manioc, sugarcane, cocoa, beef, pork, plantains
Fr. Guinea	n/a	n/a	rice, manioc, sugar, cocoa, vegetables, bananas, livestock
Guyana	35	22	sugar, rice, molasses, vegetable oils, livestock
Paraguay	28	39	soybeans, cotton, sugarcane, corn, wheat, tobacco, fruits, vegetables, livestock
Peru	13	36	coffee, cotton, sugarcane, rice, wheat, potatoes, plantains, coca, livestock, wool
Suriname	13	21	paddy rice, bananas, palm kernels, coconuts, plantains, peanuts, beef, chickens
Uruguay	10	14	livestock, wheat, rice, barley, corn, sorghum
Venezuela	4	12	corn, sorghum, sugarcane, rice, bananas, vegetables, livestock

Source: Data from FAO (2000a) and World Bank (1999).

Table 18.3 Climatic ranges of important South American crops

Crop and Scientific name	Koppen climate classes	Minimum growth cycle (days)	Minimum suitable[a] temp (°C)	Maximum suitable[a] temp (°C)	Minimum optimal[b] rain (mm/yr)	Maximum optimal[b] rain (mm/yr)
Banana						
Musa acuminata	A C Bs	180	23	33	1200	3600
Barley						
Hordeum vulgare L.	Bw Bs C D A	90	15	20	500	1000
Bean, Common						
Phaseolus vulgaris L.	Aw C D Bs	50	16	25	500	2000
Cassava						
Manihot esculenta	A Bs	180	20	29	1000	1500
Coffee						
Coffea arabica L.	Aw Cf Cw	210	14	28	1400	2300
Coffea canephora	Aw	270	20	30	1700	3000
Cotton, American						
Gossypium hirsutum L.	A C D Bs	150	22	36	750	1200
Maize						
Zea mays L. s. mays	Aw C D Bs	65	18	33	600	1200
Potato						
Solanum tuberosum L.	Aw Bs C D	90	15	25	500	800
Rice						
Oryza sativa L. s. in	A	80	25	35	1500	2000
Oryza sativa L. s. ja	C D	80	20	30	1500	2000
Soybean						
Glycine max (L.) Merr	Aw Cs Bs	75	20	33	600	1500
Sugarcane						
Saccharum robustum Br	A Cs Cf	300	24	30	1400	2500
Wheat						
Triticum aestivum L.	A C D Bs Bw	90	15	23	750	900

[a]The minimum or maximum most suitable tempreature requirement, in degrees Celsius, for practical production and growth under average conditions.
[b]The minimum or maximum optimal annual rainfall for optimal growth and yield under average growing conditions.

Data source: FAO (1998).

disappeared in Latin America, but banana areas and banana production have doubled (Winograd and Farrow 1999). South America's location, mostly in the Southern Hemisphere, creates significant market opportunities for counter-seasonal produce for Northern Hemisphere markets. Chile, in particular, markets a variety of temperate fruits (e.g., grapes, pears, apples, peaches, lemons) to the Northern Hemisphere during the latter's winter.

Coca (*Erythroxylon coca*) provides an unusual example of a native South American tree crop simultaneously encouraged by market forces and deterred by governments. Grown in the eastern foothills of the Andes, coca is important within the Andean region. Coca leaves are used for medicinal purposes and for tea. Also, local people chew the leaves as a stimulant, to suppress appetite, and to combat altitude sickness. The motivation for eradication and much of the power of the coca market come from the illegal, international cocaine trade. Aggressive programs to eliminate coca-leaf production, along with decreased demand and lower prices following the capture of drug traffickers, caused coca leaf production in Perú to decrease from 115,300 ha in 1995 to 38,700 ha in 1999. In the same period, the areal extent of coca-leaf production was cut by half in Bolivia (48,600 ha to 21,800 ha), but more than doubled, from 50,900 ha to 122,500 ha, in Colombia (USAID, 2000). New efforts to eradicate coca-leaf production in Colombia still face the countering force of strong markets for cocaine.

18.1.3 Agricultural Trends in South America

Several major trends emerge from agricultural data. Overall, South American agriculture has become more intensive and has responded to changing population patterns over the past half-century. However, South America's genetic and soil resources continue to decline.

Agricultural Intensification From a global perspective, agriculture in South America has been very successful in recent decades (World Resources Institute, 1996–1997). Since 1961, the total land area in agriculture has increased only slowly, except in Brazil, where it increased by 66%, and in Uruguay, where it decreased by 3% (fig. 18.3). Brazil opened up nearly 70 million ha between 1960 and 1980 and added another 30 million ha of agricultural land between 1980 and 1998 (FAO, 2000a). The slow increase in

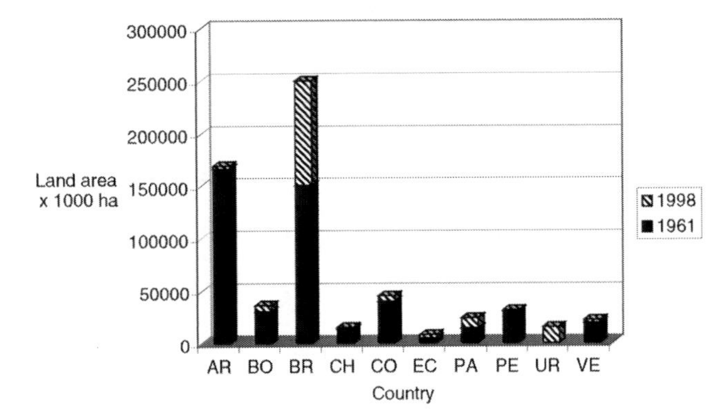

Figure 18.3 Agricultural land in ten South American countries, 1961 and 1998. Argentina (AR), Bolivia (BO), Brazil (BR), Chile (CH), Colombia (CO), Ecuador (EC), Paraguay (PA), Peru (PE), Uruguay (UR), and Venezuela (VE) (data: FAO, 2000a).

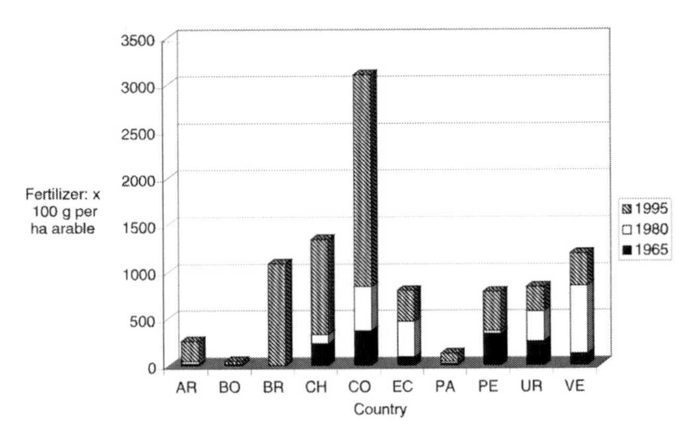

Figure 18.5 Fertilizer use, 1965–1995 for the ten largest South American countries (data: World Bank, 1999)

agricultural land area across most of the continent has been greatly exceeded by increased agricultural production, especially for export crops (fig. 18.4; World Bank, 1999). Increased production is also driven by the need to feed a growing population. The population of South America more than doubled between 1961 (152.5 million) and 1999 (340.7 million) (FAO, 2000a).

Agricultural intensification during the 1960s and 1970s involved a shift to higher value crops, which were in increasing demand by the growing number of urban consumers. The Green Revolution played a key role in this shift, with high-yield varieties of wheat and more fertilizers and pesticides coming into use (Kay, 1999). Figure 18.5 shows the proportional increase in fertilizer use over a 30-year period in the 10 larger South American countries. In all cases, fertilizer use has increased, and, in most cases, the increase has been very pronounced. In 1965, the most intensive fertilizer use was 36,700 g ha^{-1} in Colombia. Colombia's fertilizer use, centered in the broad Interandean valleys of the Cauca River and the Sabana de Bogotá, increased to 270,000 g ha^{-1} by 1995. Only Argentina, Bolivia, and Paraguay still used fertilizer at rates less than 30,000 g ha^{-1} (World Bank, 1999). United Nations data show that total fertilizer use in South America is now about nine times that of 1965 (FAO, 2000a). Brazil, the largest country, uses most (5.7×10^6 Mt in 1998 (Mt = 1,000 tonnes); FAO, 2000a).

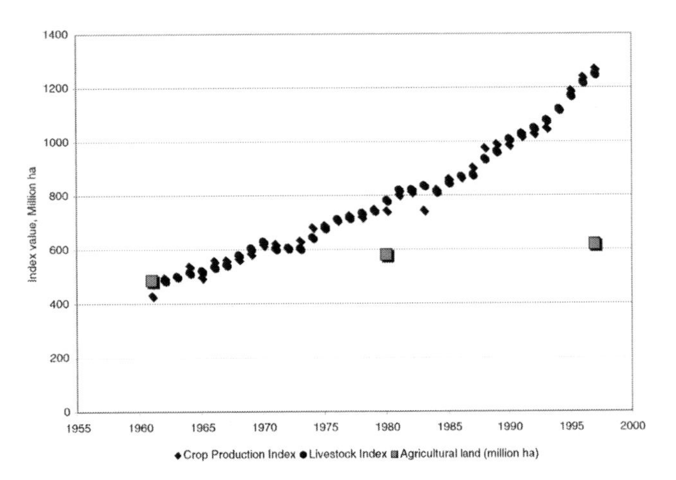

Figure 18.4 Indices for livestock and crop production for South America compared to the amount of agricultural land, 1961–1997 (data: World Bank, 1999)

As agriculture has intensified and modernized, it has become more mechanized. The total number of tractors used in South America increased by a factor of 43 in the 38 years from 1961 to 1999 (FAO, 2000a) (fig. 18.6). Agricultural intensification has also involved increased irrigation. Brazil, followed by Argentina, Perú, Chile, and Colombia, irrigates the most land (World Bank, 1999).

Improvements in communication and transportation in recent decades have strengthened the links between South America and other countries, and enabled South American countries to participate in the increasing globalization of international commerce. During the period of colonization, agricultural products were exported to Europe. Bananas and sugar cane, grown on plantations, were the leading export crops (Murray, 1999). In the late eighteenth century, wheat and beef also became important exports and, in the early nineteenth century, coffee from Brazil and other cash crops were added to the export system (Murray, 1999). Agriculture continues to provide a major share of Latin American foreign exchange, although its contribution declined in the 1970s and 1980s (Kay, 1999).

Today, agriculture in South America is becoming increasingly agribusiness, notably in Brazil, Argentina, Colombia, and Chile. Agricultural products are exported to other South American countries, as well as to other continents. The most important agricultural exports are coffee,

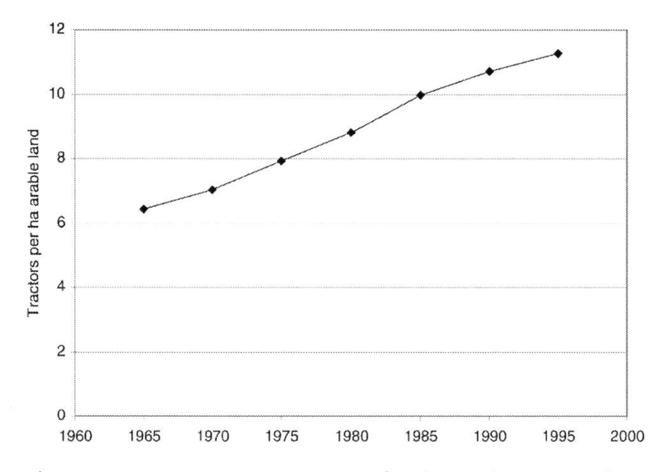

Figure 18.6 Tractor use, 1965–1995 for the ten largest South American countries (data: World Bank, 1999)

bananas, sugar, meat, cotton, flowers, and vegetable oils. One new trend has been the growth in production of oil crops. Soybean production in the three countries of Argentina, Brazil, and Paraguay grew from 275,000 Mt in 1961 to 53,651,100 Mt in 2000 (FAO, 2000a). Brazil, which strongly leads soybean production in South America, has doubled its share of the global soybean export market since 1995. Much of Brazil's production of soybeans and cotton occurs in the Cerrados, flat, scrubby lands in the state of Mato Grosso (Rich, 2001).

Another market-driven trend is the recent growth in counter-seasonal, nontraditional export items to the Northern Hemisphere. Beginning in the mid-1970s, Chilean farmers and exporters began new programs in grapes and apples (Arnade and Sparks, 1993). In the 1980s, Chileans added peaches and pears as major nontraditional export crops. In Bolivia, Perú, and Ecuador, export earnings from nontraditional crops grew from US $30 million in 1982 to US $100 million in 1992 (USAID, 1994). In Colombia alone, the export earnings from cut flowers grew from US $20,000 in 1965 to US $580 million in 2000. Flowers, principally roses, secondarily carnations, are cut and flown to markets in North America (84%), Europe, and Asia on the same day (Colombian Association of Flower Exporters, 2001). Despite successes in the development of nontraditional agricultural export crops in Colombia, Chile, and other South American countries, it remains to be seen whether development of these crops will be part of a broader trend in South American agriculture. So far, such efforts have dramatically succeeded in a few places, failed in others, and remain undeveloped in most countries (Carter et al., 1993).

Responses to Changing Population Patterns In 1960, half of the South American population was rural; in 1999, only one-fourth of the population was rural (Kay, 1999). Nonetheless, the rural population has increased in some locations and, overall, has remained relatively constant compared to the strong population increase in urban areas (figure 18.7). Locally, rural population growth increases the human pressure on the landscape. This is particularly notable in highland Bolivia, where increasing population pressure has caused fallow periods to decrease (Herve and Silvia de Cary, 1997). In other places, however, emigration from rural areas has reduced the rural labor force and agricultural production has declined.

Urban growth throughout South America has shifted consumers from local markets toward urban supermarkets, thereby changing domestic food distribution patterns and influencing consumer preferences. The extent of urban and periurban agriculture in South America has not yet been quantified, but concern for food security and nutrition in growing cities and megacities has focused international attention on strengthening this agricultural component (see chapter 20; Nugent, 2000). An international electronic conference on urban and periurban agriculture in 2000 attracted 720 participants, including South Americans (Drescher et al., 2000). The growth of cities in the Amazon region has been dramatic in the latter part of the twentieth century (Browder and Godfrey, 1997), and has affected the surrounding rural areas, where farmers and others serve urban markets and growing urban demand for agricultural commodities and forest products has put increasing pressure on the periurban resource base. For example, about 50,000 people live in a rural area near the city of Pucallpa (population 250,000) in the Central Peruvian Amazon, and about half of the local rural farm production goes to this urban market (INEI, 1995a, 1995b).

Biodiversity Loss and Land Degradation The growing commercialization of agriculture in South America has led to increasing monoculture (Clawson and Crist, 1982) and

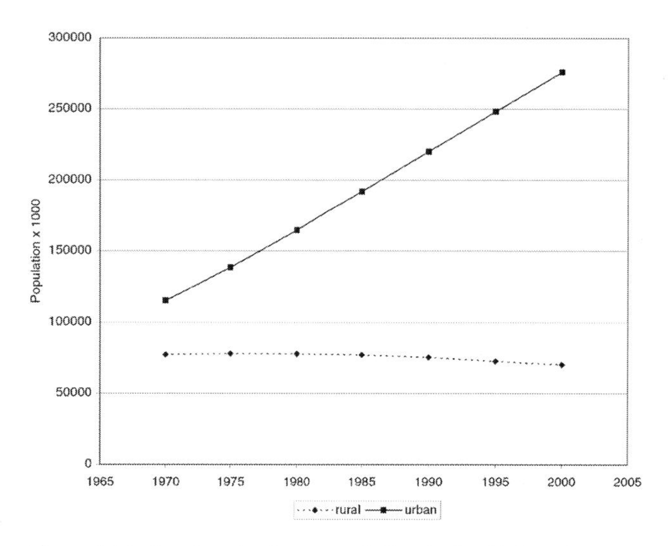

Figure 18.7 Rural and urban population in South America (13 countries), 1970–2000 (data: FAO, 2000a)

less diversity in the marketplace. Brush (1982) reported that individual farmers in the central Andes might cultivate as many as 50 distinct varieties of potatoes and that it was possible to find up to 100 locally named varieties in a single village. Today, almost 90% of agricultural production in Latin America is based on 15 cultivated crop species, most of which involve fairly homogeneous genotypes (Winograd and Farrow, 1999). Many varieties of native domesticates, especially on hillside agricultural lands, are being displaced by high-yield crops (Winograd and Farrow, 1999). Monocultures may yield well in the short-term, but they are more vulnerable to severe outbreaks of disease or pest. Emphasis on monoculture leads to loss of genetic diversity, thus limiting the genetic resources available for improving crops and increasing agricultural risk in the future. South America is not alone with these problems. Globally, only three species—maize, wheat, and rice—supply two-thirds of the world's grain crop (Oldfield, 1984).

International debate about the magnitude of genetic erosion resembles the debate about climate change in that logical arguments and theory provide reason for serious concern, but much research remains to be done to prove conclusively the human role in these changes. In both issues, however, there is substantial risk of greater, even irreversible, human impact if no counter measures are taken before the research is completed. Brush (1999), advocating further research on genetic erosion, points out that the situation becomes far more complex, and perhaps less grim, as the practices of traditional farmers of small holdings are better understood.

Increased pressure to achieve greater crop production puts land and soil resources at increased risk of degradation. Symptoms of land degradation include lower yields, greater need for commercial fertilizer, and erosional losses of the land itself. In the Cerrados of central Brazil, where 60% of Brazilian beef is produced on 100 million ha of pasture, pasture degradation, including loss of vigor, productivity, natural capacity for recovery, and ability to overcome the detrimental effects of insects, diseases, and weeds, is a significant problem. Where beef production in the Cerrados is replaced by soybean production, plowing causes soils to erode more rapidly (Macedo and Zimmer, 1993). Because degraded lands store less carbon, widespread degradation of pastures and soils is of global concern. Land degradation has been called the "silent crisis" of Latin America (Benites et al., 1993). Correcting for the negative physical, chemical, and biological changes brought about by agriculture remains a major challenge.

18.2 Soil Erosion

Soil erosion has been cited as a major environmental concern in South America, especially in regard to the sustainability of agriculture. Within a given country, soil erosion processes and rates vary with terrain, climate, and land use. In Argentina, for example, soil erosion occurs throughout the country, with rainfall erosion predominant in the eastern half and wind erosion in the arid and semi-arid zones. About 25% of Argentina's soils are considered eroded (Morrás, 1999).

Soil erosion, the displacement of soil particles, is a natural function of the land-surface system; however, human activities accelerate soil-erosion rates. The key components of soil erosion are well known to be bare soil and exposure to water or air with sufficient energy to dislodge and move soil particles (e.g., Bennett, 1939; Morgan, 1986). Across the South American continent, national and foreign scientists continue to determine local causes, extents, and rates of soil erosion. Soil erosion is a subtle and locally occurring chain of processes, which cannot be measured precisely at the continental scale. Nonetheless, it is possible to derive a geographic perspective of soil erosion in contemporary South America by integrating empirical evidence with an understanding of the relationships between soil erosion and its causal factors.

In natural settings throughout the world, the highest soil-erosion rates occur in semi-arid environments, where vegetative cover is discontinuous, and at sites where natural disturbances, including fire, landslide, wind-throw, and animal activity, expose soil to the forces of rain and wind (Morgan, 1986). Erosion by water is greatest where high-intensity rainfall, low soil-infiltration capacities, and steep slopes increase the quantity and energy level of overland flow. High relief in the Andes (Robles, 1987) and the high kinetic energy of rainfall in parts of the lowland tropics (Castro et al., 1999) contribute to greater risks of soil erosion in some South American regions. Natural processes of soil erosion operate at rates on the order of 1–100 mm per 1,000 years (Selby, 1985). They change the landscape very slowly, but their cumulative effects over millions of years wear away mountains and redistribute sediment particles in lowlands and eventually to the coast (see chapters 2 and 5).

18.2.1 Soil Erosion and Human Activity

In South America, as on other continents, human activities accelerate soil-erosion rates directly, for example by exposing the soil surface, or indirectly, for example by increasing the flow of rainfall runoff. Land clearing for agriculture, highway construction, or urbanization exposes soil to the erosional forces of rainwater and wind. The main soil-erosion problems in eastern Perú, where the Amazon lowland meets the eastern flank of the Andes, occur when vegetation is cleared (Robles, 1987). Agricultural fields are especially vulnerable to soil erosion at times of planting, before the new crop provides protective cover, and after the harvest, when the soil surface is bare. In some parts of South America, the market-driven opportunity to produce cash crops year-round has led to intensive use of erodible soils. One example of this occurs in the coffee and sugarcane producing region of the Andes in northern Colombia (Stini, 1982).

Some agricultural practices minimize erosional losses by providing vegetative cover and maintaining and improving soil permeability. These include fallow periods, minimum tillage, avoidance of steep slopes, mulching, and contour hedging (Gil, 1977; Alegre and Rao, 1996; Castro et al., 1999). Hillside terraces of the Inca empire in Perú increased rainfall infiltration by decreasing the effective slope and providing barriers to overland flow. By improving soil moisture, terraces promoted the better growth of plants, which then protected the soil against soil erosion until they were harvested.

Agricultural practices that increase rainfall runoff, or reduce the soil's ability to support plants, accelerate erosion processes. Steeper slopes are more erosion-prone because rainfall runoff can flow down them at higher energy levels. In Bolivia and Ecuador, cultivated slopes are commonly steeper than 25% and occasionally exceed 100% (Quinton and Rodriguez, 1999; fig. 18.8). A trend that increases the risk of soil erosion on sloping lands is the increased use of agricultural machinery. Compaction of soil by heavy equipment, as well as by heavy animals or repeated human walking, decreases permeability and increases rainfall runoff. In the rolling hills of southern Brazil (Rio Grande do Sul), forest clearing and mechanical tillage for wheat, soybeans, and corn were found to decrease infiltration rates from 135 mm hr^{-1} to 0.2 mm hr^{-1} and lead to severe erosion (Busscher et al., 1996). Trujillo, de Noni, and Viennot (1989) reported tractor use on steep slopes (up to 66% slope) in the Ecuadorian Andes, observing that, because tractors must go from top to bottom on such steep slopes, their tracks promote erosion by breaking down soil structure and channeling rainfall runoff directly downhill.

Throughout South America, soil erosion by water and wind are linked to overgrazing. Baldwin (1954) specified overgrazing as a cause of severe erosion in the Venezuelan lowlands north of Lake Maracaibo, in northeastern Brazil, in the Patagonian region of Chile and Argentina, and in mountain valleys in Colombia, Perú, and Bolivia. Overgrazing is frequently cited as a cause of erosion in the Andean region (Millones, 1982) and in Patagonia (Morrás, 1999). In northwestern Perú, Cotler (1998) reported that soils become very hard during the dry season due to the combined effects of cattle trampling and desiccation. He observed that the first rain of the rainy season on this nearly impermeable surface creates sheets of runoff, preferentially on cattle paths. The sharp hooves of horses, cattle, and sheep, all introduced by the Spanish, damage vegetation and expose soil to erosion. The soft footpads of the native camelids, on the other hand, are far less damaging to plants. Pilot studies in Ecuador have replaced sheep with alpacas, and cattle with llamas as a means of maintaining and increasing wool and meat production with less erosional impact (White and Maldonado, 1991).

Agriculture is not the only human activity that leads to higher rates of soil erosion. Construction, mining, forest clearance, and fire expose soil to the forces of wind and water. Impervious surfaces (i.e., roofs, roads) decrease soil absorbency and increase rainfall runoff. Roads and trails, for example, have higher rates of rainfall runoff (lower infiltration capacities) than their uncompacted surround-

Figure 18.8 Unterraced hillsides, such as this slope near Ambato, Ecuador, are more typical of Andean agriculture than the extensive terraces of the Incas, most of which are abandoned and in disrepair (photo: C. Harden, 1985).

ings, leading to accelerated erosion on the compacted and downslope surfaces that receive the runoff (Harden, 1992).

Unintentional acceleration of soil erosion by human activity has both on- and off-site consequences. On-site, soil erosion can significantly reduce the productivity and long-term sustainability of agricultural lands. Research is in progress to explain better the causative basis of this relationship in tropical soils (Tengberg et al., 1998; Lal, 1999). Soil erosion may influence productivity through the loss of effective rooting depth, decrease in plant-available water capacity, reduction in soil fertility, or any combination of these (Lal, 1999). Where rainfall is particularly erosive and overland flow combines with subsurface piping, gullies form, thereby accelerating rates of landscape change.

Erosion reduces productivity more dramatically when the eroded soil is topsoil rather than a subsoil layer. Recent research from erosion plots in tropical soils at three locations in Brazil (Chapeco, Itapiranga, and Campinas) demonstrates that productivity decreases greatly with the first 5 cm loss of topsoil, but that further erosion leads to less change (Tengberg et al., 1998). In a multinational initiative promoted by FAO, eight research groups in six South American countries (Brazil, Bolivia, Argentina, Paraguay, Chile, and Colombia) began parallel studies between 1989 and 1995 on the effects of soil erosion on crop productivity. Each group cleared experimental plots, allowed them to erode for two years, and removed surface soil from one subgroup of plots. Then, in the last two years of the experiment, each experimental plot was planted with a crop. Chemical, physical, and biological parameters were monitored throughout four-year study cycles. Of the sites tested, a Ferralsol in Chapecó, Brazil, had the highest erosion rate, nearly 300 t ha^{-1} yr^{-1}. These experiments drew attention to the chemical and physical roles of organic matter in soil and the decline in productivity when organic material is eroded from the soil. In eroded, iron-rich soils, decreased productivity is linked to loss of organic matter, pH, P, and K, and an increase in free Al (Tengberg et al., 1998).

Major off-site consequences of soil erosion are sediment deposition in river channels and reservoirs and increased turbidity of rivers. Turbidity alters aquatic habitats and damages hydroelectric turbines. Deposition changes habitats and reduces navigation and reservoir storage. For every tonne of cereal exported from the port of Buenos Aires, one tonne of silt (20×10^6 Mg yr^{-1}) is dredged from the channel (Molina Buck, 1993). Although this numerical correspondence neither requires nor proves a cause-and-effect relationship, agricultural lands in the Argentine Pampas are widely recognized as being especially prone to soil erosion (Molina Buck, 1993). Further examples of river and reservoir siltation can be found throughout South America. In eastern Perú, a set of Amazon-draining rivers became less navigable after the high jungle region was colonized in the mid-1900s to expand commercial agriculture (Llenera, 1987). Increased reservoir siltation has followed the rapid growth of soybean cultivation in southern Brazil in the 1970s (Castro et al., 1999), and sediment yields of the upper Orinoquia and Amazon basins in Colombia exceed 2000 t km^{-2} yr^{-1} (Malagon et al., 1995; see also chapter 5).

Siltation of reservoirs by eroded particles has important economic consequences. In 1957, less than two years after the Anchicaya Dam in Colombia had been completed, nearly one-fourth of its reservoir was filled with sediment. After seven years, the reservoir became functionally full of sediment, and dredging was required (Eckholm, 1976). A similar problem occurred in southern Andean Ecuador, where the reservoir of the hydroelectric power plant on the Paute River, completed in 1983, was so full of sediment after six years that Ecuador obtained a $14M international loan for dredging and watershed conservation efforts (El Comercio, 1991). Dredging just managed to keep up with an ongoing sediment influx of about 3×10^6 m^3 yr^{-1} (Carpio, 1991; figure 18.9). The Paute hydroelectric project supplies over half of Ecuador's electric power.

Great differences in elevation (up to 6,000 m) between the high Andes and surrounding lowlands create excellent opportunities for hydroelectric power generation. The high potential energy along the Andean cordillera drives natural denudation processes too, which contribute to the sediment loads of Andean-source rivers (see chapter 5). Therefore, it is often difficult to separate natural from anthropogenic causes of channel and reservoir sedimentation. Much of the sediment behind the Anchicaya (Colombia) Dam is thought to have become eroded and transported into the river as a result of the opening of new farms and roads in the vicinity of the dam (Eckholm, 1976). In the Paute River (Ecuador) a major landslide about 50 km upstream from the dam filled one kilometer of the river channel with small rock fragments in 1993, damming the river (Plaza and Zevallos, 1994; Harden, 2001). The landslide dam was artificially released 33 days later, creating a flood wave that carried landslide debris downstream and mobilized much more sediment by scouring the river channel and nearby roads and fields. The hydroelectric power dam held, but the river channel now contains more sediment, working its way downstream toward the reservoir.

18.2.2 Scales and Patterns of Soil Erosion

Soil erosion can be very site-specific (figure 18.10). Some soils, when exposed to wind and rain, are more erodible than others. However, local biophysical and anthropogenic factors can be of greater importance than soil type in determining specific locations of soil erosion. Local controls include slope (for water), exposure (for wind), fetch (for wind), drainage, land cover, and proximity to an impervious surface.

Figure 18.9 A dredge pumps sediment from the reservoir of the Paute River (Ecuador) and sends it through a floated tube to be discharged downstream of the dam (photo: C. Harden, 2002).

Harden (1991), in highland Ecuador, and Dunne (1979), in Kenya, found land use to be stronger than soil classification as a predictor of soil erodibility. Such studies illustrate the multiplicity of interactive, site-specific, biophysical and management factors that promote or deter soil erosion.

Recent research in South America highlights the erosional importance of land management at local scales. Zimmerer (1993) emphasized the local scale of conditions responsible for erosion and identified farm labor shortages, created by employment in nonfarm work by rural peasants, as primary factors controlling the location and intensity of soil erosion in Andean Bolivia. In a case study of the Calicanto watershed of the Bolivian Andes, he reported that local knowledge of soil management had not been lost, but that soil erosion practices had been reduced due to labor shortages, cropping areas had expanded, tillage with rented tractors had increased, fallow periods had been shortened, and increased grazing had put additional pressure on the land. In the Ecuadorian Andes, Harden (1996) documented higher rates of rainfall runoff and soil erosion on abandoned lands, compared to cropland, pasture, or forest. This difference stems from characteristically lower productivity and from unmanaged, informal grazing practices on abandoned land.

Figure 18.10 Rills in some barley fields, but not all adjacent fields, on this hillslope in the Ecuadorian Andes illustrate the local scale and spatial heterogeneity of soil erosion (photo: C. Harden, 1985).

At the global scale, various attempts have been made to map and compare erosion rates between continents. Walling (1988) identified the Andes as having some of the higher rates (>1,000 t km^{-2} yr^{-1}) of suspended sediment yield in the world. His extrapolation from drainage basins to continent, based on climatic, geologic, topographic, and other physiographic controls, showed generalized sediment delivery rates for rivers in the nonmountainous portions of South America to be less than 250 t km^{-2} yr^{-1}. Lal (1994) used the suspended sediment yield estimates of L'Vovich et al. (1990) with estimates of soil bulk density to calculate denudation rates for selected regions of the world (see also chapter 2). He estimated a denudation rate of 0.03–0.13 mm yr^{-1} for the Amazon basin and 0.33–0.67 mm yr^{-1} for the Andean region.

Denudation rates derived from suspended sediment measurements help us understand patterns of rainfall erosion, but do not advance understanding of wind erosion and desertification. Desertification is a threat in the Argentine states of La Rioja, San Luis, and La Pampa; and southward encroachment by the Atacama Desert poses a problem for northern Chile (Eckholm, 1976). During a decade-long drought in Chile in the 1960s, desert conditions advanced across an 80–160-km front at about 1.5 to 3 km yr^{-1} (Eckholm, 1976). The Bolivian government recently determined that desertification is occurring in the three subregions of Chaco, Altiplano, and Valle, and identified 275,544 km^2 as at risk of desertification (Bolivia, 1996).

The first continental-scale assessment of soil erosion by water and wind in South America was conducted by the Conservation Foundation and FAO of the United Nations (Baldwin, 1954). It employed air photo interpretation and field work to generate a detailed map and country-by-country summary of the extent and probable causes of soil erosion. The most severe erosion mapped was in mountain valleys of Perú and Colombia and in small patches of southeastern Brazil. During the past half century, major changes have occurred in the Amazon basin, where considerable clearing has taken place in both the lowlands and highlands. Much of the Peruvian land that was cleared in the mid-twentieth century, and later eroded, is located in the high jungle of the upper Amazon basin where land was colonized to expand commercial crops such as coffee, rice, fruits, coca, cacao, and tea (Llenera, 1987). Evidence of erosion in the interandean region is still very apparent. Zimmerer (1993) reported that moderate and severe soil erosion occurs in most of the mountainous terrain in the major Andean countries.

In 1998, researchers at the U.S. Natural Resources Conservation Service generated digital maps showing the vulnerability to wind and water erosion based on global coverages of climate and soils. The maps have a minimum scale of 1:5,000,000 and are rasterized on a 2-minute grid cell (USDA-NRCS, 1998a, 1998b). Figure 18.11, derived from the USDA-NRCS maps, provides a generalized view of water and wind erosion vulnerability on the South American continent. The Andes, the Argentine Pampas, and higher relief areas of Brazil and Venezuela are especially vulnerable to such erosion.

More recently, FAO (2000b) developed an international database that quantifies the extent of land degradation. Table 18.4, which reports the FAO data for countries in South America, shows the degree of contrast between countries with severe and extensive soil degradation (the Andean countries, Brazil, Venezuela) and those with little soil degradation (French Guiana, Guyana, Suriname, and Uruguay). Quantification of soil erosion at the scale of countries to continents is typically achieved by using models, most often forms of the Universal Soil Loss Equation (USLE), adjusted with data from in-country plots (Espinoza et al., 1993; Moriya and Alfonso, 1994; Ruiz et al., 1994).

18.2.3 Challenges of Accelerated Soil Erosion

The geography of soil erosion at the scale of the South American continent is primarily determined by topography, climate, and land use practices, the last-mentioned strongly affected by socioeconomic and historical factors. Direct and indirect consequences of accelerated soil erosion—decreased land productivity, flashy runoff, sedimentation, and desertification—are not always apparent in the short term, and they may be hidden, at least temporarily, by adding fertilizer, dredging, or irrigating. Because the countries of South America are engaged in economic development, immediate concerns, such as opening up new land or constructing new highways, often obscure or override longer range concern for the health of the soil resource.

Efforts to compare the rates and effects of soil erosion between countries or at the continental scale are all subject to problems of data availability, scale issues, and the difficulty of calibrating and comparing nonstandardized estimates of erosion from place to place (Lal, 1994). In some cases, good quality data on soil erosion exist for part of a country, but not for the entire country. International agencies, principally the Food and Agriculture Organization of the United Nations, have programs to coordinate research methodologies and establish global data bases (FAO, 2000b). Nonetheless, the spatial and temporal variability of soil erosion will continue to challenge those who develop continental-scale databases.

Soil-erosion specialists repeatedly discover that soil conservation is as much a problem for social scientists as natural scientists (e.g., Collins, 1986; Heath and Binswanger, 1996). Stadel (1989) encountered a degree of fatalism about soil loss among small farmers in the Ecuadorian Andes and recognized that poor people have other urgent priorities besides soil conservation. In a similar vein, spatial analysis of 1994 agricultural census data for Perú (INEI, 1995b) shows that terrorism, labor shortages,

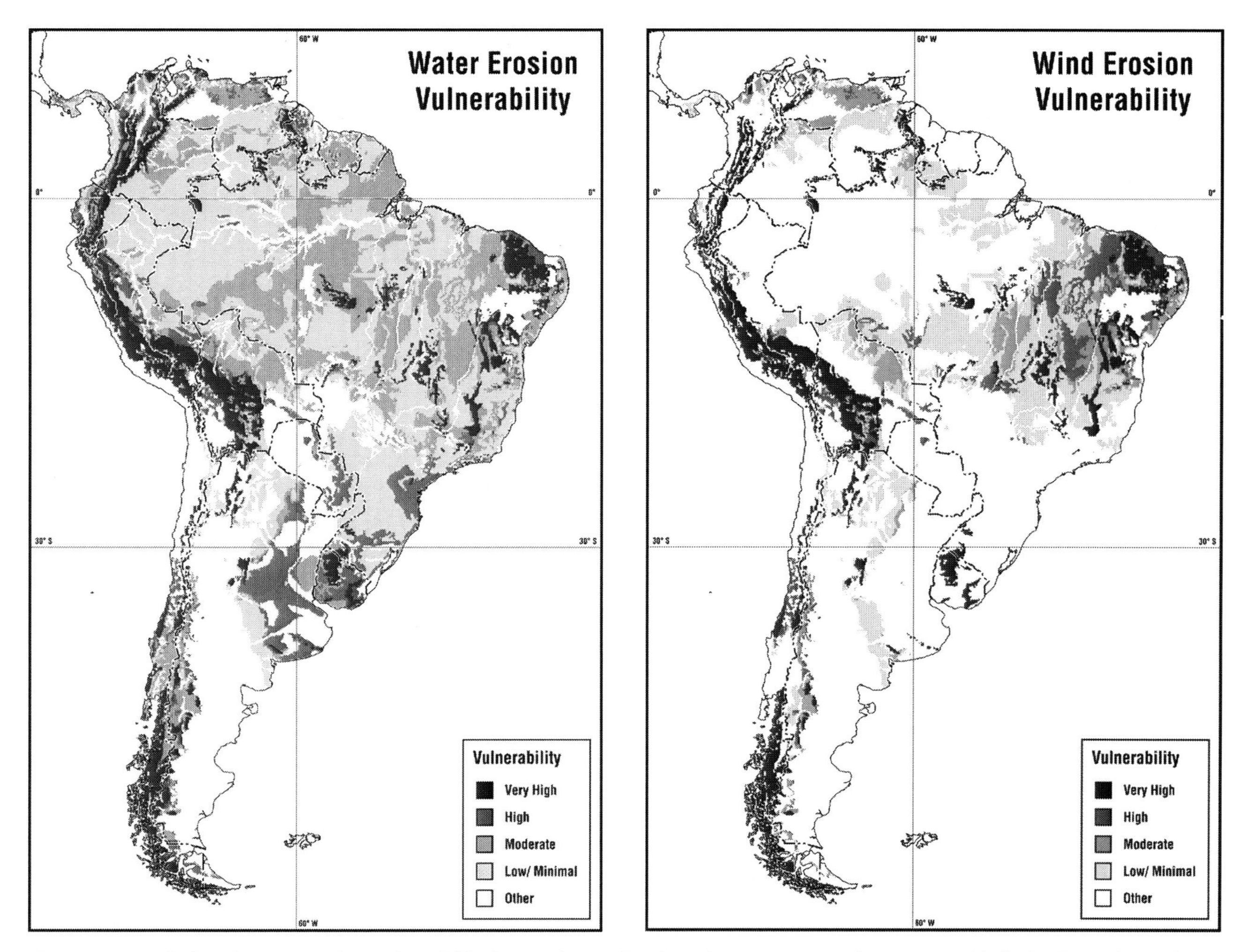

Figure 18.11 Wind and water erosion vulnerability in South America, based on USDA-NRCS (1998a, 1998b). "Other" consists of ice-covered, very cold, or very dry lands. For wind erosion, humid lands were combined with the "Low/Minimal" class; for water erosion, depositional areas were combined with the "Low/Minimal" class.

Table 18.4 Land degradation and soil erosion in South American countries

Country	Total area (x1000 km^2)	None to light Degradation % of area	Severe Degradation % of area	Very severe Degradation % of area	Cause[a]	Type[b]	% Area at risk for erosion
Argentina	2772	37	11	0	A,O,D	W, N, C	10
Bolivia	1096	63	23	6	O,D	W, N, C	23
Brazil	8479	48	24	4	D,A	W, C	16
Chile	749	52	14	5	D	W, N	33
Colombia	1136	63	18	0	D,O	W, C	20
Ecuador	283	71	2	4	D	W, C	29
Fr. Guinea	91	88	0	0	A	P	1
Guyana	215	86	13	0	A,O	W	18
Paraguay	407	70	19	8	A,D	W, C	4
Peru	1281	46	32	1	D,O	W, C	30
Suriname	164	81	0	0	A	P	7
Uruguay	186	87	12	0	A,O	W	6
Venezuela	910	61	21	0	D,O	W, C	30

[a]Causes: A = agriculture, O = overgrazing, D = deforestation
[b]Types: W = water erosion, N = wind erosion, C = chemical deterioration, P = physical deterioration

Source: FAO (2000b).

and lack of credit accounted for the lack of cultivation on far more hectares of potentially arable land than physical constraints, including soil degradation and lack of water.

Good land stewardship can play a critical role in minimizing accelerated soil erosion. Migration away from rural areas, on the other hand, has left many parcels of agricultural land abandoned. Especially in semi-arid conditions, as exist in much of the interandean region, abandoned lands have higher rates of rainfall runoff and soil erosion than well-managed agricultural lands (Harden, 1996). In Perú, lack of active management has led to the deterioration of old terraces (Llerena, 1987). Like other cultural properties, knowledge of land management can be lost. Eckholm (1976), among others, has written that the decimation of the Inca population and its social order by the arrival of the Spanish was accompanied by loss of their empire's conservation ethic and knowledge of agricultural engineering. In a more recent example, Llerena (1987) observed that logging companies and non-native agricultural colonists introduced and led indigenous people to adopt soil-eroding practices in the high jungle of eastern Perú.

Despite the widespread occurrence of soil erosion, much of South America remains rich with agricultural land and fresh water resources. Agricultural production continues to rise, and agriculture continues to be an important sector of national economies. In the Andes, where anthropogenically related soil-erosion rates are relatively high, natural rates of soil erosion, landsliding, and river incision are also high. Sustainable human activity in high energy Andean environments, then, requires special effort to avoid accelerating natural erosion processes. Except for lands recently opened to agriculture, such those in Brazil and eastern Perú, most land now considered eroded has a legacy of soil erosion that predated the arrival of Europeans and accelerated following their arrival (de Noni, 1986). Soil loss can be rapid, but soil formation can take millennia. Therefore, soil conservation today must remedy old as well as recent problems, and sustainable land uses require conscious stewardship of the land resources that support them.

References

Alegre, J., and M. Rao., 1996. Soil and water conservation by contour hedging in the humid tropics of Peru. *Agriculture, Ecosystem and Environment,* 57, 17–25.

Arnade, C., and A. Sparks, 1993. Chile's agricultural diversification. *Agricultural Economics,* 9, 1–13.

Baldwin, M., 1954. Soil erosion survey of Latin America, Part II: South America. *Journal of Soil and Water Conservation,* 9, 214–237.

Benites, J., D. Saintraint, and K. Morimoto, 1993. Degradación de suelos y producción agrícola en Argentina, Bolivia, Brasil, Chile y Paraguay. In: *Erosión de Suelos en America Latina.* Food and Agriculture Organization of the United National (FAO), Rome, Italy.

Bennett, H.H., 1939. *Soil Conservation.* McGraw-Hill, New York.

Bolivia, 1996. *Mapa preliminar de desertificación de tierras: región árida, semiárida, y subhumeda seca de Bolivia.* Ministerio de Desarrollo Sostenible y Medio Ambiente, Secretería Nacional de Recursos Naturales y Medio Ambiente, La Paz, Bolivia.

Browder, J., and B. Godgrey, 1997. *Rainforest cities, urbanization, development, and globalization of the Brazilian Amazon.* Columbia University Press, New York.

Brush, S., 1982. The natural and human environments of the central Andes. *Mountain Research and Development,* 2, 19–38.

Brush, S., 1999. Genetic erosion of crop populations in centers of diversity: a revision. *Proceedings of Technical Meeting on Methodology of World Information Systems and Early Warning Systems on Plant Genetic Resources,* Prague, 21–23 June, 1999.

Busscher, W., D. Reeves, R. Kochhann, P. Bauer, G. Mullins, W. Clapham, W. Kemper, and P. Galerani, 1996. Conservation farming in southern Brazil: using cover crops to decrease erosion and increase infiltration. *Journal of Soil and Water Conservation,* 51, 188–192.

Cañadas, L., and H. Salvador, 1982. Agrometeorological assessment models for rural development in central Sierra of Ecuador. Final report to National Program for Agrarian Regionalization and National Institute of Meteorology and Hydrology, Quito, Ecuador for A.I.D. NOAA/CEAS Models Branch, Columbia, Missouri.

Carpio, W., 1991. La Verdadera Situación del Proyecto Paute. *El Comercio,* Quito, Ecuador.

Carter, M., B. Barham, D. Mesbah, and D. Stanley, 1993. Agroexports and the rural resource poor in Latin America: policy options for achieving broadly-based growth. *Agricultural Economics Staff Paper Series,* No. 364. University of Wisconsin, Madison, Wisconsin.

Castro, N., A-V. Auzet, P. Chevallier, and J.C. Leprun, 1999. Land use change effects on runoff and erosion from plot to catchment scale on the basaltic plateau of Southern Brazil. *Hydrological Processes,* 13, 1621–1628.

Clawson, D, and R. Crist, 1982. Evolution of land-use patterns and agricultural systems. *Mountain Research and Development,* 2, 265–272.

Collins, J.L., 1986. Smallholder settlement of tropical South America: the social causes of ecological destruction. *Human Organization,* 45, 1–10.

Colombian Association of Flower Exporters, 2001. *Export Data.* Updated April, 2001.

Cotler, H., 1998. Effects of land tenure and farming systems on soil erosion in northwestern Peru. *Advances in GeoEcology,* 31, 1539–1543.

De Noni, G., 1986. Breve historia de la erosión en Ecuador. *Centro Ecuatoriano de Investigación Geográfica Documentos de Investigación,* 6, 15–23.

Drescher, A., R. Nugent, and H. de Zeeuw, 2000. Final report: Urban and peri-urban agriculture on the policy agenda. FAO/ETC Joint Electronic Conference, August 21–September 30, 2000.

Dunne, T., 1979. Sediment yield and land use in tropical catchments. *Journal of Hydrology,* 42, 281–300.

Eckholm, E., 1976. *Losing Ground.* W.W. Norton and Company, New York.

El Comercio, 1991. Al Rescate de Paute. May 1, 1991. *El Comercio,* Quito, Ecuador. A-11.

Espinoza, Q., M. Lagos, and R. Ortiz, 1993. Erosión de los

suelos en Chile. In: *Erosión de Suelos en America Latina*. Food and Agriculture Organization of the United Nations Regional Office, Santiago, Chile.
Fearnside, P.M., 1987. Rethinking continuous cultivation in Amazonia. *BioScience*, 37, 209–214.
Food and Agriculture Organization of the United Nations (FAO), 1998. ECOCROP: The crop environmental requirements database and the crop environment response database. *Land and Water Digital Media Series*, 4. (CD-ROM). Food and Agriculture Organization, Rome.
Food and Agriculture Organization of the United Nations (FAO), 2000a. *FAOSTAT Database*. Food and Agriculture Organization, Rome.
Food and Agriculture Organization of the United Nations (FAO), 2000b. Land degradation: severity of human-induced degradation. *TERRASTAT—Land Resource Potential and Constraints: Statistics at Country and Regional Level*. FAO/AGL, Food and Agriculture Organization, Rome.
Gil, N., 1977. The role of soil conservation in watershed management on agricultural lands. *FAO Soils Bulletin*, 33, 89–95.
Harden, C.P., 1991. Andean soil erosion: a comparison of soil erosion conditions in two Andean watersheds. *National Geographic Research & Exploration*, 7, 216–231.
Harden, C.P., 1992. Incorporating the effects of road and trail networks in watershed-scale hydrologic and soil erosion models. *Physical Geography*, 13, 368–385.
Harden, C.P., 1996. Interrelationships between land abandonment and land degradation: a case from the Ecuadorian Andes. *Mountain Research and Development*, 16, 274–280.
Harden, C.P., 2001. Sediment movement and catastrophic events: the 1993 rockslide at La Josefina, Ecuador. *Physical Geography*, 22, 305–320.
Heath, J., and H. Binswanger, 1996. Natural resource degradation effects of poverty and population growth are largely policy-induced: the case of Colombia. *Environment and Development Economics*, 1, 65–83.
Herve, D., and R. Silvia de Cary, 1997. Efecto de la duración del descanso sobre la capacidad de producir en las tierras altas de Bolivia. In: M. Lieberman and C. Baied, (Editors) *Desarollo Sostenible de Ecosistemas de Montaña: Manejo de Areas Frágiles en los Andes*. The United Nations University and Instituto Ecológico, Artes Gráficas Latina, La Paz, Bolivia, 189–199.
Instituto Nacional de Estadística e Informática (INEI), 1995a. *IX Censo Nacional de Población y IV de Vivienda* [National Census of Population and Households]. Instituto Nacional de Estadística e Informática, Lima, Perú.
Instituto Nacional de Estadística e Informática (INEI), 1995b. *III Censo Nacional Agropecuario*. Instituto Nacional de Estadística e Informática, Lima, Peru.
Kay, C., 1999. Rural development: from agrarian reform to neoliberalism and beyond. In: R. Gwynne and C. Kay (Editors), *Latin America Transformed: Globalization and Modernity*. Arnold, London, 272–304.
Lal, R., 1994. Global overview of soil erosion. In: R. Baker, G. Gee, and C. Rosenzweig (Editors), *Soil and Water Science: Key to Understanding Our Global Environment*. Soil Science Society of America, Madison, 39–51.
Lal, R., 1999. Erosion impact on soil quality in the tropics. In: R. Lal (Editor), *Soil Quality and Soil Erosion*. CRC Press, Boca Raton, Louisiana, 285–305.
Llenera, C., 1987. Erosion and sedimentation issues in Perú. In: R. Beschta, G. Grant, G. Ice, and F. Swanson (Editors), *Erosion and Sedimentation in the Pacific Rim*, IAHS Publication 165, 3–14.
L'Vovich, M., N. Bratseva, G. Ya Karasik, G. Medvedeva, and A. Meleshko, 1990. A map of contemporary erosion of the Earth's surface. *Mapping Science Remote Sensing*, 27, 61–67.
Macedo, M., and A. Zimmer, 1993. Sistema pasto-lavoura e seus efeitos na produtividade agropecuaria. In: V. Favoreto, L. de Rodrigues, R. Reis (Editors), *Simposio Sobre Ecosistema de Pastagens, 2,* FUNEP, Sao Paulo, 216–245.
Malagon, M., C. Pulido, R. Llina, and C. Chamorro, 1995. *Suelos de Colombia: Origin, Evolución, Clarification, Distribución, y Uso*. Instituto Geográfico Agustin Codazzi, Bogotá, Colombia.
Milliones, J., 1982. Patterns of land use and associated environmental problems of the central Andes: an integrated summary. *Mountain Research and Development*, 2, 49–61.
Molina Buck, J.S., 1993. Soil erosion and conservation in Argentina. In: D. Pimentel (Editor), *World Soil Erosion and Conservation*. Cambridge University Press, Cambridge, 171–192.
Morgan, R.P.C., 1986. *Soil Erosion and Conservation*. Longman Scientific & Technical, Harlow, Essex.
Moriyo, M., and L. Alfonso, 1994. Erosión actual y potencial del suelo en Paraguay. In: *Erosión de Suelos en America Latina*. Food and Agriculture Organization of the United Nations Regional Office, Santiago, Chile.
Morrás, H., 1999. La erosión de los suelos en la Argentina. *Ciencia Hoy*, 9, 54.
Museo Arqueológico Rafael Larco Herrera, 1997. *The Spirit of Ancient Perú: Treasures from the Museo Arqueológico Rafael Larco Herrera*. Thames and Hudson, New York.
Murray, W.E., 1999. Natural resources, the global economy and sustainability. In: R. Gwynne and C. Kay (Editors), *Latin America Transformed: Globalization and Modernity*. Arnold, London, 127–152.
Nugent, R., 2000. Urban and periurban agriculture, household food security and nutrition. Discussion paper for FAO-ETC/RUAF electronic conference "Urban and periurban agriculture on the policy agenda." August 21–September 30, 2000.
Oldfield, M., 1984. *The Value of Conserving Genetic Resources*. U.S. Department of the Interior, National Park Service, Washington, D.C.
Plaza, G., and Zevallos, O., 1994. The 1993 la Josefina rockslide and Río Paute landslide dam, Ecuador. *Landslide News*, 8, 4–6.
Quinton, J., and F. Rodriguez, 1999. Impact of live barriers on soil erosion in the Pairumani sub-catchment, Bolivia. *Mountain Research and Development*, 19, 292–299.
Rich, J., 2001. U.S. farmers look back and see soy growers in Brazil shadowing them. *New York Times*, Tuesday, July 21, 2001, C1–C2.
Robles, I., 1987. Processes diagnosis in Perú. In: I. Pla (Editor), *Soil Conservation and Productivity*. Proceedings IV International Conference on Soil Conservation, November 3–9, Maracay, Venezuela, 221–231.
Ruiz M., J., M. Ruiz B., and J. Karisen, 1994. Situación ambiental en relación con la erosión en Bolivia. In: Erosión de Suelos en America Latina. Food and Agri-

culture Organization of the United Nations Regional Office, Santiago, Chile.
Salomon, F., 1980. *Los Señores Etnicos de Quito en la Epoca de los Incas.* Instituto Otavaleño de Antropología, Otovalo, Ecuador.
Sanchez, P.A., D.E. Bandy, J.H. Vallachica, and J.J. Nicholaides III, 1982. Amazon basin soils: management for continuous crop production. *Science,* 216, 821–827.
Sanchez, P.A., 2000. Tropical soils, climate and agriculture: An ecological divide? Paper presented at the Harvard Conference on Raising Agricultural Productivity in the Tropics: Biophysical Challenges for Technology and Policy. October 19, 2000.
Selby, M.J., 1985. *Earth's Changing Surface.* Clarendon Press, Oxford.
Smith, J., P. van de Kop, K. Reategui, I. Lombardi, C. Sabogal, and A. Diaz, 1999. Dynamics of secondary forests in slash-and-burn farming: interactions among land use types in the Peruvian amazon. *Agriculture, Ecosystems and Environment* 76, 85–98.
Stadel, C., 1989. The perception of stress by campesinos:a profile from the Ecuadorian Andes. *Mountain Research and Development* 9, 35–49.
Stini, W., 1982. The interaction between environment and nutrition. *Mountain Research and Development* 2, 281–288.
Tengberg, A., M. Stocking, and S.C.F. Dechen, 1998. Soil erosion and crop productivity research in South American. In: H. -P. Blume, H. Eger, E. Flieischhauer, A. Hebel, C. Reij, and K.G. Stiner (Editors), *Towards Sustainable Land Use, Volume I.* Catena Verlag, Reiskirchen, Germany, 355–362.
Trujillo, G., G. de Noni, and M. Viennot, 1989. *Erosión y Conservación de Suelos: la Experimentación–Demonstración, una Herramienta en la Lucha Antierosiva.* Dirección Nacional Agrícola, Departamento de Suelos, Ministerio de Agricultura y Ganadería, Quito, Ecuador.
U.S. Agency for International Development (USAID), 1993. *Green Guidance for Latin America and the Caribbean: Integrating Environmental Concerns in A.I.D. Programming.* USAID and World Resources Institute, Washington, D.C.
U.S. Agency for International Development (USAID), 1994. *Harvest of Progress: A Quiet Revolution in Latin American and Caribbean Agriculture.* Office of Sustainable Development for Latin America and the Caribbean, USAID, Washington, D.C.
U.S. Agency for International Development (USAID), 2000. *Andean Narcotics Cultivation and Production Estimates 1999.* DI Design Center, Washington, D.C.
U.S. Department of Agriculture and National Resources Conservation Service (USDA-NRCS), 1998a. *Vulnerability for Water Erosion.* USDA-NRCS, Soil Survey Division, World Soil Resources, Washington, D.C.
U.S. Department of Agriculture and National Resources Conservation Service (USDA-NRCS), 1998b. *Vulnerability for Wind Erosion.* USDA-NRCS, Soil Survey Division, World Soil Resources, Washington, D.C.
Walling, D., 1988. Measuring sediment yield from river basins. In: R. Lal (Editor), *Soil Erosion Research Methods.* Soil and Water Conservation Society, Ankeny, Iowa, 39–73.
White, S., and F. Maldonado, 1991. The use and conservation of natural resources in the Andes of southern Ecuador. *Mountain Research and Development,* 11, 37–55.
Whittlesey, D., 1936. Major agricultural regions of the Earth. *Annals of the Association of American Geographers.* 26, 199–241.
Winograd, M., and A. Farrow, 1999. *Agroecosystem Assessment for Latin America: Agriculture Extent, Production Systems and Agrobiodiversity.* Center for Tropical Agriculture, Cali, Colombia.
Wood, S., K. Sebastian, and S.J. Scherr, 2000. *Pilot Analysis of Global Ecosystems: Agroecosystems.* International Food Policy Research and World Resources Institute, Washington, D.C.
World Bank, 1999. *World Development Indicators on CD-ROM.* The World Bank, Washington, D.C.
World Resources Institute, 1996–97. *World Resources 1996–97.* World Resources Institute, Washington, D.C.
Zimmerer, K. S., 1993. Soil erosion and labor shortages in the Andes with special reference to Bolivia, 1953–91: Implications for "conservation-with-development." *World Development,* 21, 1659–1675.

19

Impacts of El Niño-Southern Oscillation on Natural and Human Systems

César N. Caviedes

Off the coasts of northern Perú and southern Ecuador, warm equatorial waters meet the cold Humboldt Current. Variations in sea temperatures and associated fauna have been known to fishing folk since colonial times. They noticed that toward the end of every year tepid waters appeared between the Gulf of Guayaquil (Ecuador) and Point Pariñas (Perú) and persisted until late February, causing tropical species to be added to the fish they commonly caught. Coupled with the arrival of warm waters was a surge in air humidity and an increase in summer showers. Since this environmental phenomenon occurred around Christmas, the local fishermen called it *El Niño*, or Child Jesus.

Early scientific observations on the nature and extent of these phenomena revealed that they were not regionally restricted to coastal Perú and Ecuador, but extended over the whole tropical Pacific, involving pressure fields and wind flows across the basin. Thus, when referring to this coupled ocean-atmospheric system, both variations of sea temperature across the tropical Pacific and changes of the atmosphere in contact with the ocean must be considered (Neelin et al., 1998).

19.1 Oceanic Aspects of El Niño

Normally, the tropical Pacific Ocean, from the coast of Ecuador and Perú to longitude 120°W, is dominated by westward-flowing cold waters, which are the prolongation of the Humboldt Current. Near longitude 120°W, sea surface temperatures approach normal equatorial values of ~28°C. When the flow reaches the western Pacific, it creates a sea-level rise of nearly 40 cm, which is maintained by the wind shear of the equatorial easterlies. The *thermocline*, which marks the lower boundary of the sun-heated water layer, runs at a depth of 40 m between Perú and the Galápagos Islands, but on the Asian side of the Pacific it dips to 120 m, revealing a marked asymmetry in the thickness of the sun-heated layer across the Pacific (fig. 19.1).

During El Niño years, the westward flow of cooler waters is weak because there is less wind shear from the easterly winds, and the thermocline plunges to 80 m in the eastern equatorial Pacific. The horizontal movement of warm water from west to east is effected by Kelvin waves that propagate within the upper 100-m layer of the tropical Pacific. This eastward advection was originally believed to be a massive outflow and, until the 1950s, prompted the use of the term "El Niño current," but today the process is better envisaged as the spread of an oil slick, with patches constantly changing shape and direction under the influence of variable winds.

As these Kelvin waves travel eastward across the Pacific they cause rises in sea level from 30 to 80 cm, depending on the intensity of the advection. Indications of such a condition appear toward the peak of the southern spring

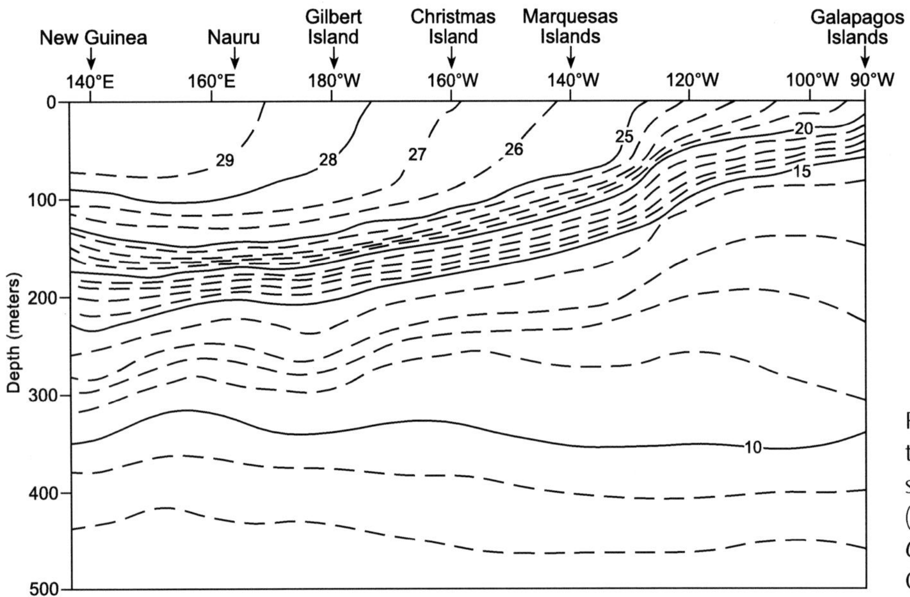

Figure 19.1 The thermocline across the equatorial Pacific. Isolines are sea temperatures in degrees Celsius (compiled with data from the *Climate Diagnostics Bulletin*, Climate Analysis Center, NOAA).

(October-November). At the onset of summer, the first Kelvin waves arrive in the triangle between the Galápagos Islands, Ecuador, and northern Perú. Successive waves in the course of the southern summer reach the west coast of South America and are deflected westward in the form of Rossby waves, contributing to the extreme water mixing in the tropical Pacific during El Niño episodes. Since the oceanic El Niño peaks during the southern summer, its effects are felt in North and Central America during the northern winter. As the South Pacific enters the phase of winter cooling in the southern fall, the warm water waves subside. Nevertheless, during some pronounced El Niño events, the cooling fails and the warm sea temperatures of the preceding summer linger through the winter. In the following spring, when the warming is reactivated, a prolonged El Niño event that can last up to 18 or even 24 months ensues.

The return to normal conditions occurs briskly and frequently through a shift to the opposite phase of the Southern Oscillation, or La Niña, which is characterized by cold waters in the eastern tropical Pacific, a shallow thermocline in the equatorial Pacific, and a greater "piling up" of warm waters in the western Pacific. Little evaporation from the cold surface waters and scarce heat transfers to the atmosphere cause droughts in many areas conterminous to the Pacific Ocean. In the eastern hemisphere of the Southern Oscillation (see next section), La Niña displays different traits: high sea surface temperatures spread across the tropical Atlantic, a condition favorable for the generation of tropical depressions and hurricanes.

Originally, it was believed that the cold episodes in the Pacific Ocean occurred midway between El Niño events, but recent evidence has shown that the cold episodes tend closely to *follow* major El Niños. In a few instances they have even *preceded* warm ocean events in the Pacific, suggesting that in some cases the sequence of the two phases of the Southern Oscillation may be reversed (Caviedes and Waylen, 1991).

19.2 Meteorological Aspects of El Niño

During El Niño years, the centers of action and the wind fields in the Pacific basin and over South America are altered. The South Pacific high pressure cell, centered near Easter Island (27°S, 110°W), and its counterpart in the North Pacific are weakened. When the South Pacific anticyclone loses intensity, the Trade Winds slacken, the thermocline thickens in the eastern tropical Pacific, and westerly winds replace the usual equatorial easterlies. Gliding atop the eastward-flowing warm waters, humid air masses progress to the east and release torrential rains over equatorial islands, the Galápagos Islands, and the arid coasts of Ecuador and northern Perú. During La Niña episodes, the Trade Winds, the equatorial easterlies, coastal upwelling, and westward surface-water transport are abnormally strong, the waters of the eastern Pacific are cold, and the thermocline runs nearer the ocean surface. Dryness prevails over most of the eastern half of the Pacific basin.

Non-annual fluctuations of the atmosphere in the tropical Pacific—such as El Niño and La Niña—are ruled by the *Southern Oscillation*, a counterbalance between the South Pacific high pressure cell and the low pressure center over Indonesia. The Southern Oscillation is expressed as the standardized pressure difference between both centers (Wright, 1984, 1989). A high Southern Oscillation Index

(SOI) denotes strong easterly winds across the tropical Pacific, air dryness, and cold sea surface temperatures; a low SOI means dominance of westerly winds, weakness of equatorial easterlies, warm ocean conditions, and high humidity. The Southern Oscillation Index has become the established yardstick against which weather variations in the tropics and middle latitudes are gauged. During major El Niño events, the area influenced by the Southern Oscillation can be divided into two hemispheres (fig. 19.2). A western hemisphere, comprising the Pacific basin, western South America, Central America, and the western half of North America, is characterized by abnormal low air pressures, warm coastal waters, and heavy rains. The eastern hemisphere, centered over Indonesia but extending from Australia across Africa to the eastern front of South America, is dominated by high air pressures, cooler ocean waters, and devastating droughts (fig. 19.2).

19.3 Rainfall and Thermal Anomalies

The west coast of South America, under the influence of the cold Humboldt Current, is noted for its clear skies and dry air, although low level fog (*garúa*) is common in winter. Rain occurs on islands of the equatorial Pacific, the Galápagos Islands, and the coasts of Perú, Ecuador, and northern Chile only during unusual humid air onslaughts. These rainfall anomalies are closely associated with the phases of ENSO and the surface sea temperatures in the tropical Pacific.

Oceanographers distinguish four major quadrants in the tropical Pacific (fig. 19.3). El Niño 1 region comprises the coastal waters off Perú, which heat up only during major oceanic warming events. High temperatures in this quadrant correlate with simultaneous torrential rains in the conterminous countries, except in the Peruvian Andes and Bolivian Altiplano, where severe droughts are experienced. Across the continent, in northeastern Brazil, conspicuous El Niño episodes in this first quadrant are associated with devastating droughts. Warm waters in El Niño 2 region, between the Galápagos Islands and the coast of Ecuador, correlate with rains in the Pacific lowlands and the Andes of Ecuador, and with lower seasonal precipitation in the interior Amazon basin. When the ocean in El Niño 3 region warms, high precipitation falls, with a two or three month lag, in central Chile and the Rio de la Plata basin east of the Andes. The coast of California is also affected while, in most of western Mexico and on the Pacific coast of Central America, deficit rains occur in phase with anomalies in the tropical eastern Pacific. Oceanic warming that is confined to El Niño 4 region, without spreading into the coastal waters off Perú or Ecuador, tends to have a mellowing effect on the winters of northern California, Oregon, Washington, British Columbia, as well as the American and Canadian prairies, and to reduce precipitation in most of the Antilles and along the Caribbean coasts of Colombia and Venezuela.

19.4 Hydrologic Responses

While instrumentally measured precipitation primarily denotes conditions at a meteorological station, runoff data reflect the collective response of a watershed to precipitation and the hydric balance. Therefore, since El Niño causes abnormally high discharges in certain regions and severe lows in others, the runoff of South American rivers reflects ENSO phases so well that a regionalization of runoff responses to these phases can be proposed (Caviedes, 1992, 1998).

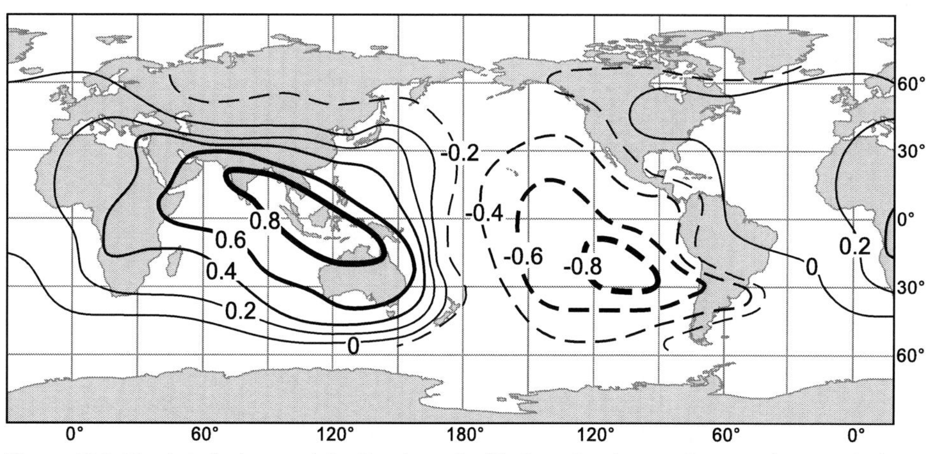

Figure 19.2 The hemispheres of the Southern Oscillation. Continuous lines are iso-correlations of sea-level pressures of west Pacific and Indian Ocean stations with Jakarta; broken lines are iso-correlations of east Pacific stations with Tahiti (from Caviedes, 2001).

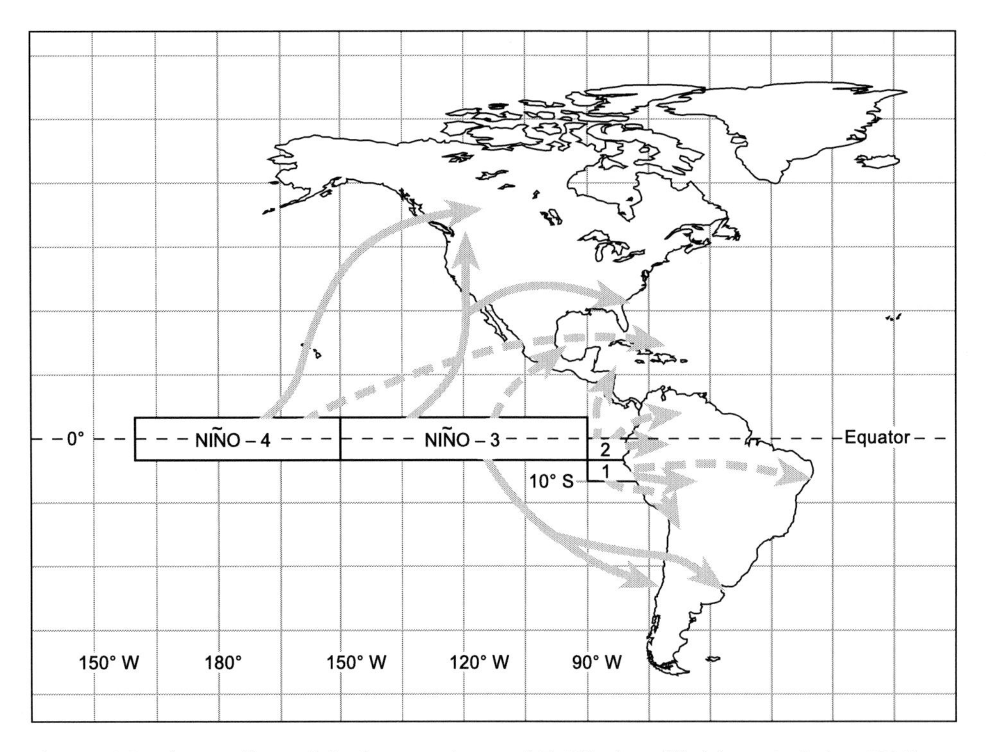

Figure 19.3 Distant effects of the four quadrants of El Niño (modified from Caviedes, 2001)

The rivers of northern Perú and coastal Ecuador display the most direct reactions to El Niño in the form of massive floods during wet summers (Waylen and Caviedes, 1986). With elevation, the ENSO signal fades in the watercourses of the Peruvian and Ecuadorean Andes and even shifts to deficit discharges in rivers of the Sierra, the Altiplano, and the eastern slopes of the Andes. These findings reveal that the flood impacts of El Niño are restricted to the Pacific coast and that the Andes play an important role in reducing precipitation to leeward. In temperate central Chile, where the annual precipitation maxima happen in winter, high discharges usually occur in the winter *following* an El Niño event or during the spring snowmelt. In some El Niño years, however, excessive rains fall during the winter *before* a warm ENSO event, suggesting that a warming in the Pacific had begun in the winter preceding that particular event (Waylen and Caviedes, 1990). Rivers to leeward of the temperate Andes also react in phase with Chilean rivers (Compagnucci et al., 1996; Berri and Flamenco, 1999). Across the lowlands of the River Plate, from eastern Bolivia to southern Brazil, high waters and floods tend to occur in the late summer and autumn of El Niño years (Berbery and Barros, 2002; Mechoso and Perez-Irribarren, 1992). They reveal the impact of the altered atmospheric circulation and humidity exports from the tropical Pacific into this region caused by the southward bending of a low-level jet flowing across the mountains and the development of a trough over Argentina (Marengo et al., 2004).

In the eastern tropical lowlands of the continent the responses to ENSO stimulations are negative (Foley et al., 2002). At the eastern foothills of the Andes, the high and low waters of rivers such as the Mamoré, in eastern Bolivia, occur in conjunction with sea surface temperatures in the Atlantic Ocean and are not modulated by El Niño events in the tropical Pacific (Ronchail et al., 2005). Rivers in the lowlands north of the Brazilian Highlands, including northeast Brazil, display low water levels at the height of El Niño episodes, but brisk highs afterward, as exemplified by the runoff of streams in the central Amazon basin (Richey et al., 1989; Marengo, 1995), and by sedimentation rates in the Amazon River itself (Aalto et al., 2003). This pattern intensifies in northern South America, as revealed by the low waters of Guyanese and Venezuelan rivers (Hastenrath, 1990). In Colombia, where there is a convergence of influences from Caribbean flows, from the equatorial Pacific, from the seasonal shifts of the ITCZ, and from the easterly running Chocó jet (Poveda and Mesa, 2000), the reactions are more complex, reflecting the seasonal variabilities of precipitation to the warm and cold phases of ENSO (Spence et al., 2004). While the rivers emptying into the Caribbean Sea undergo severe lows during El Niño years, those flowing into the Orinoco display normal discharges (Quesada and Caviedes, 1992). The rivers more open to Pacific influences tend to yield high discharges during warm ENSO events (Riehl, 1984; Blanco et al., 2003). Equally mixed are the responses in Ecuador. While rivers flowing into the Pacific display pronounced highs during El Niño events, those to leeward of the Andes or close to the Colombian border tend to experience low discharges. Other rivers in the Ecuadorean Andes have an even more complex regime in that they may react with high flows to strong warm ENSO events in the Pacific, but exhibit normal to low flows during moderate El Niño events (Rossel, 1997). Obviously, the behavior of Ecuador's rivers is testimony to that country's location at the crossroads of Pacific, Amazonian, and Caribbean atmospheric circulation influences.

Not all regions of the continent display responses coherent with the two extreme phases of ENSO. In Patagonia, including the southern portion of the Pampa plains, discharge records reveal a runoff regime without drastic interannual variations associated with either of the two phases of the Southern Oscillation (Caviedes, 1998), probably because the westerly circulation, unaffected by either phase of ENSO, remains consistently steady year after year (Minetti and Sierra, 1989). On the other hand, investigations of seasonal temperature and precipitation variability in the Patagonian-Andean region (Daniels and Veblen, 2000) suggest distant influences from ENSO phases in the Pacific, an indication that there exists some muddling of the teleconnections due to the complex interlacing of nival-glacial processes, zonal wind flows, and snowmelt periods. The coastal fringe of central Brazil reveals a precipitation regime totally off-phase with that of the rest of the continent, since it depends chiefly on the strength of the South Atlantic anticyclone, the temperature variations of the Brazil Current, and topographic controls along the oceanic façade of its regional ranges (Ronchail et al., 2002).

Among the lakes of the continent, Lake Titicaca exhibits the most sensitive response to ENSO (Keller, 1990), displaying low water levels during El Niño, when the Altiplano undergoes precipitation deficits, and rising water levels during La Niña due to heightened airflows from the humid lowlands to the east of the Bolivian cordillera (Garreaud et al., 2003). These variations have been documented as far back as the first millennium B.C. in the Altiplano of Bolivia (Binford et al., 1997), opening new perspectives on the ecological effects of ancient El Niños. Studies of smaller Andean lakes have revealed distinct sediment laminations in accordance with wet and dry years modulated by ENSO (Rodbell et al., 1999).

19.5 Responses of Marine Ecosystems

El Niño-Southern Oscillation events trigger significant changes in marine ecosystems through their impact on ocean surface and near-surface waters, and on the ocean-atmosphere interface, which in turn affects terrestrial ecosystems (fig. 19.4). Ocean warming alters the chemical balance of its waters, causing variations in primary production that affect the food chain of fish populations and lead to crises among the mammals and seabirds that feed on them. High mortality besets other marine organisms such as corals, while massive migrations occur involving species in both pelagic and littoral habitats. Inputs of heat and humidity to the atmosphere result in abundant precipitation in certain regions and insufficient rains in others, both

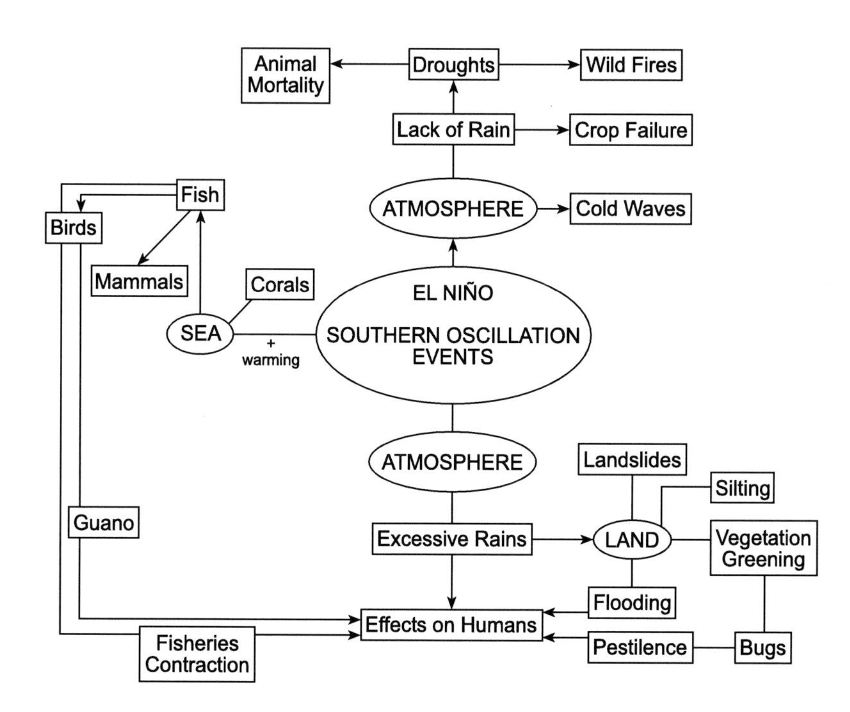

Figure 19.4 Natural systems of El Niño

at sea and over land. These impacts in turn have dire consequences for the affected human populations and their activities.

Considering the preeminent oceanic character of ENSO events, a review of their consequences begins, appropriately, with a treatment of marine responses. Carbon and oxygen availability are essential for the maintenance of the Humboldt Current biomass, estimated at 30 million metric tons per year (Muck, 1989). Primary productivity is calculated to be 3.18 GCM day^{-1} and depends on the upwelling of nitrate consumed by phytoplankton (Chavez et al., 1989). Anchoveta and sardine larvae feed on phytoplankton, although in their further development the feeding habits shift to zooplankton. Primary productivity is optimal at 14°C and around a depth of 60 m, precisely the realm where El Niño-induced anomalies are at their strongest, thus explaining why El Niño has such an impact on primary productivity and phytoplankton production, and ultimately on fish stocks.

19.5.1 *Effects on Fish Populations and Fish Migration*

Perú is still one of the top fish-landing nations of the world, its catches consisting mostly of anchovies (*Engraulis ringens*), a clupeoid that lives in voluminous schools in the plankton-rich waters of the Humboldt Current. In 1970, Perú harvested close to 10 million metric tons, which led marine biologists to issue a serious warning concerning the vulnerability of the ecosystem. Then, El Niño struck in 1972–1973, and the anchovy catches that in 1971 had soared to 12 million metric tons dropped to 2.3 million. Subsequent studies related the depression of anchovy populations to excessive recruitment of young individuals, southward migration of the stocks, and high predation or cannibalism on anchovy eggs during times of stress (Pauly and Soriano, 1989; Muck, 1989). Overfishing and recruitment failures proved deleterious for the entire stock in subsequent years (Caviedes and Fik, 1993).

The 1973 drop in catches indicates the magnitude of the anchovies' collapse. That year, Perú landed only 2.3 million metric tons and Chile as little as 668 thousand metric tons (fig. 19.5). Thereafter and up to 1983 (another El Niño year), Perú endured an unstable fishing period with annual catches hovering around 2 million metric tons. After the 1972–1973 low, Chile's fishing efforts concentrated on jack mackerel (*Trachurus murphii*), a large subtropical fish, and on sardines (*Sardinops sagax*), which thrive in temperate waters. With this move Chile jumped ahead of Perú between 1979 and 1985. Only after 1985 did Perú begin the road to recovery, initially also by switching to sardines and jack mackerel. Since the early 1990s, a recovery of the anchovy stock has allowed the industrial fisheries of Perú and Chile to concentrate again on that species.

During the 1990s, world fisheries underwent some fundamental changes for which El Niño is to be considered a catalyst. After oceanic warming in 1982–1983 and 1986 again led to contractions in fish catches, the combined an-

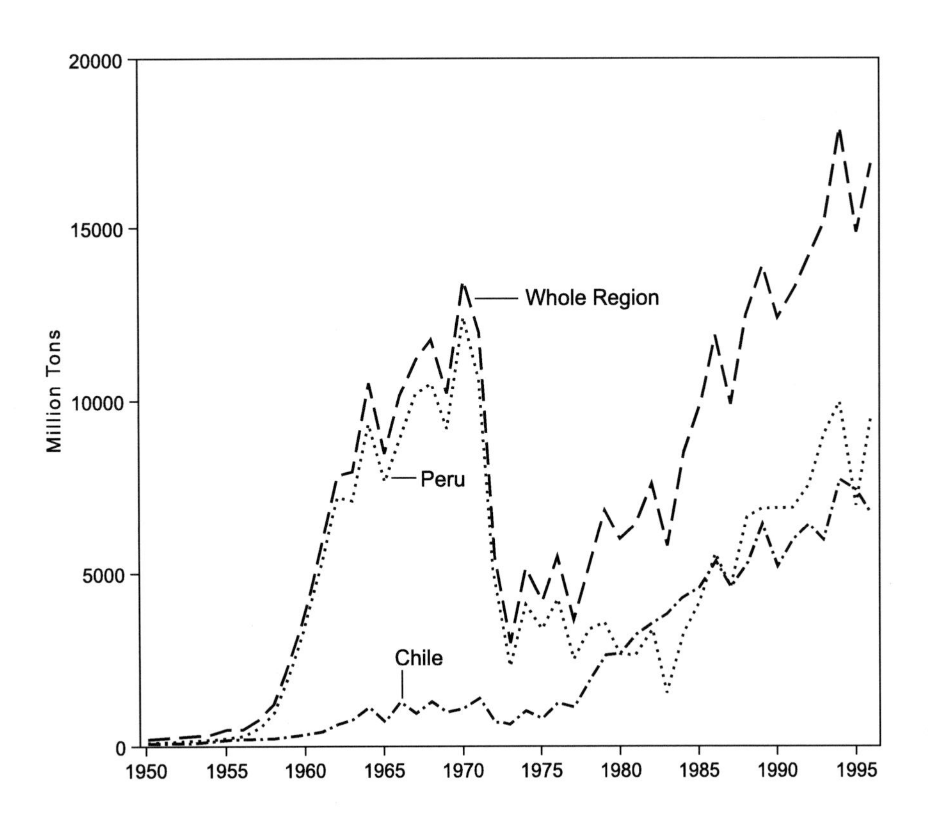

Figure 19.5 Fish landings of western South America, 1952–1997 (source: FAO *Fisheries Statistics*)

nual landings of Perú, Chile, and Ecuador soared to a total of 14 million metric tons in 1990, a volume higher than in any year prior to the 1972–1973 El Niño. Sardine stocks increased with repeated sea warming, and the anchovies made a comeback during the ensuing cooler periods. In 1994, landings from the cold Humboldt Current totaled 18 million metric tons, a volume thought to be unattainable twenty years before and a demonstration of the resilience of the Humboldt Current ecosystem. Nevertheless, the system's vulnerability was evinced again in the wake of the 1997–1998 El Niño, when total regional landings dropped to 7.9 million metric tons (fig. 19.5). This notwithstanding, Perú has regained the top position among the world's fishing nations that it enjoyed in the 1960s, and Chile has secured third place. In recent years, China has taken the second place formerly occupied by Japan, and the latter has slipped to fourth. However, while Perú and Chile restrict their activities to the adjacent coastal waters, Chinese and Japanese fishing vessels harvest resources from waters near and far.

Warm water invasions bring unexpected visitors to the South American coast bathed by the Humboldt Current. Sharks, tuna, albacore, bonito, and swordfish from equatorial waters increase noticeably during El Niño years. Among the exotic species caught in coastal waters are the prickly globe-fish (*Diodon hystrix*), the big-scale pomfret (*Taractichthys steindachneri*) common in the warm Kuro Shio off Japan, and the humped fish *jorobado* (*Selene sp.*). Particularly welcome by Peru's artisan fishermen are the "dorado" (*Coryphaena hippurus*), which, like the dolphin or mahe-mahe of Northern Hemispheric tropical waters, is coveted for its exquisite firm white flesh, and the "pez guitarra" (*Rhinobatus planicep)*, an edible ray that is often sun-dried. Other fish normally occurring only at the interface between warm tropical waters and the northern edge of the cold Humboldt Current, such as "mero" (tropical perch) and skipjack *(Katsuwonus pelamis*), move to cooler waters farther south during El Niño events.

19.5.2 Effects on Mollusks and Corals

Massive southward migrations of coastal mollusks and crustaceans also occur during warm-water episodes. The Pacific scallop (*ostión*), a tasty shellfish normally found in protected bays, experiences explosive population increases during El Niño years (Arntz and Fahrbach, 1991). On the swampy Pacific shores of Ecuador and Colombia, crabs and shrimp, otherwise confined to propitious coastal locations, multiply at increased rates during El Niño events. Following El Niños in 1972–1973 and 1982–1983, shrimp nurseries were established in huge ponds in the shallow bays of Ecuador and the products exported to North America and Europe. Competition with small artisan shrimpers and the pollution of coastal waters by artificial feed caused such concern among environmentalists that today this industry operates in a low-key manner.

A particularly damaging El Niño effect besets the coral populations in the Pacific basin. Corals are very sensitive to environmental alterations, which they reflect in their biological development. Colonies of these invertebrates need clear water, good circulation, and temperatures within a narrow range. A slight change in these conditions stunts their development and larger changes cause death. By means of carbonate secretion, the bodies of individual corals grow radially, indicating annual growth phases in the latticed structures of their skeleta. Aragonite, a mineral that incorporates into calcium trace metals such as cadmium, barium, manganese, as well as oxygen isotopes, is the carbonate that makes the coral skeleton. Cadmium and barium are regarded as indicators of cool upwelling waters. Manganese, an indicator of fine sediment of continental provenance, decreases during warm ENSO episodes and is replaced by coarse suspended sediment. The O^{18} oxygen component of the skeletal aragonite is precipitated from sea water and becomes diluted as water temperature and rainfall increase, two major concomitants of El Niño conditions (Cole et al., 1992). By analyzing thin coral sections, marine biologists and geologists have correlated deficits in O^{18}, calcium/cadmium, barium/calcium, and manganese/calcium with warm ENSO phases (Evans et al., 1998) and are now able to identify El Niño episodes in the past far beyond the scope of contemporary chroniclers and environmental historians.

Apart from ENSO impacts on the chemistry of corals, another effect of ocean warming is coral bleaching, a whitening of coral surfaces due to the loss of zooxanthellae—microorganisms that entertain a symbiotic relationship with growing corals. The absence of the temperature-sensitive zooxanthellae precipitates coral mortality and hampers the rapid reconstruction of young colonies. In addition, unusual warming of the water leads to the disappearance of large and small crustaceans that feed on sea snails and sea urchins, which in turn favor the growing coral sprouts. Finally, coral mortality is accelerated by the increased turbidity of coastal waters that results from the unusually high discharge of river sediment during El Niño floods. Glyn and Colgan (1992) estimated that the ENSO event of 1982–1983 caused the extermination of nearly 95% of the coral colonies of the Galápagos Islands and resulted in a 70–95% reduction in the corals on the coasts of Costa Rica, Panama, Colombia, and Ecuador. Fortunately, corals have a remarkable regenerative ability by recruiting coral larvae from nearby colonies.

19.5.3 Effects on Coastal Fauna

El Niño events disrupt the thermal and biological balance of ocean waters, raising sea temperatures and reducing oxygen and carbon levels. This produces drastic contractions in microorganisms that in turn trigger breaks in the ocean food chain. Fish die or are so weakened that they fail to spawn. Consequently, coastal communities who depend on these resources for their livelihood go into periods of depression and impoverishment. Collapses in the ocean's food chain also affect sea mammals, such as seals, several species of whales, and otters that thrive in the cool waters of the eastern Pacific. These mammals may starve when fish, squid, or mollusks become scarce during warm water episodes. Their corpses, littering the beaches of western South America and the Galápagos Islands, are grim testimonies to the ravages caused by oceanic El Niño events.

The fish stocks of the eastern Pacific feed large populations of sea birds that nest on the cliffs along the coast or on offshore islets. The accumulated bird droppings, *guano*, supported a lucrative mining industry before artificial fertilizers came on the market. Even today, guano remains the preferred fertilizer for decorative plants. High mortality among sea birds causes a parallel contraction in guano production and export. Of the 27.5 million sea birds estimated in 1970, only 1.8 million survived the El Niño crisis of 1972–1973. By 1982, numbers were up to 8 million but, in the course of the 1990s, never rose above the 10 million mark. Among these birds, the *guanay* (cormorant) was the most severely affected. The torrential rains of El Niño events in 1972–1973 and 1982–1983 also destroyed the nests of sooty terns, frigate birds, the red-footed booby, and seagulls on the equatorial Pacific islands of Malden and Christmas, thereby reducing to a dangerous minimum the marine bird populations of these islands.

Another development prompted by disturbances to the ocean's ecological balance by El Niño in 1972–1973 has been the transition from fishmeal to soybean meal in the production of animal feed. Until the early 1970s, the supply of Peruvian fishmeal had satisfied the international demand for inexpensive animal feed in North America and western Europe, but the slump in this supply following the 1972–1973 event forced major international feeder companies to seek a replacement; they turned to soybeans. Today, this grain has become (behind wheat, rice, and maize) one of the most largely cultivated commodities in the world, but the Peruvian fishmeal industry has collapsed.

19.6 Responses of Terrestrial Ecosystems

The impact of El Niño on terrestrial ecosystems is most evident through excessive raininess (Holmgren et al., 2001). Subsequent effects are flooding, erosion or silting of riverine areas, greening of dormant shrubs and grasses, and proliferation of noxious insects and rodents.

19.6.1 Effects on Terrestrial Fauna

In northern Perú, the wet and warm summers of El Niño cause unusual alterations to the fauna. From the forests of the Peruvian *montaña* and from the Gulf of Guayaquil, birds and insects descend onto the greening river oases. Egrets temporarily settle in ephemeral ponds and swamps to feed on an overabundance of toads and frogs. At the height of El Niño in 1982–1983, this author observed the mud-covered streets and parks of northern Peruvian towns such as Piura, Sullana, and Chulucanas littered with bull frogs. In pre-Andean villages the explosion of a centipede that inflicts painful skin lacerations (known therefore as *el latigazo*) caused as much fear as the outbreaks of Leishmaniasis, typhoid fever, cholera, and malaria (Caviedes, 1984).

Flooding in the Amazon basin, whose rivers overflow during normal rainy seasons (March–August), forces rodents, reptiles, and insects to move to higher ground. Zoologist Arim Addis noticed that spiders and beetles sense the increased humidity and start crawling up tree trunks at the onset of the rainy season, moving higher as the rains and flooding progress. However, El Niño coincides with reduced rains in the Amazon basin and the rainy season is shortened to April, May, and June (Caviedes, 1982). Addis noted that during these low-precipitation years, the upward crawling started later than usual and remained at lower levels, while during abnormally rainy years the movement began earlier and reached higher levels (Addis and Latif, 1996).

Little known, but no less significant, are the effects of El Niño on coastal and freshwater crustaceans along the west coast of South America. In the ephemeral rivers and creeks of northern and central Chile, the excessive rains caused by El Niño produce an unusual multiplication of crayfish and river shrimp that during dry years are reduced to minimum sustainable populations (Báez, 1985). In a reversal, along the Pacific coast of Colombia, unprecedented reductions in the shrimp populations of *Xiphopeneaus iverti, Peneaus occidentalis,* and *Thachypeneaus byrdi* were caused by the ocean warming of El Niño in 1982–1983 (ERFEN, 1983).

19.6.2 Effects on Vegetation

Many and varied are the responses of terrestrial plants across South America to El Niño events. While along the west coast the intense summer or winter rains during warm-humid events cause unusual greening of vegetation and increased plant productivity, across most eastern parts of the continent, in the Andes, and along the Caribbean front, precipitation deficits result in reduced plant productivity with severe ecological implications.

The most dramatic vegetation changes due to El Niño occur along the coastal desert and in the semiarid belts of southern Ecuador, Perú, and northern Chile. On elevated marine terraces or structural plains, called *tablazos* in coastal Perú, the monotony of the desert is broken by the greening of a variety of grasses, dormant bulbs, and seasonally green shrubs (Koepcke, 1961). These plant communities, known in Perú as *lomas*, are particularly visible in the Tablazo de Paita, in the Desert of Sechura, and in the Lomas de Lachay. Arntz and Fahrbach (1991) report that in the vicinity of Lima the rains of El Niño in 1982–1983 caused an 89% expansion of the lomas, while in normal years the surface covered by lomas is no more than 5% of the coastal tablelands. These numbers alone offer a measure of the substantial ecological alterations that El Niño causes in the Peruvian desert. In the coastal thorn savanna of the Santa Elena peninsula, in semi-arid southern Ecuador, where grasses are more dense and thorny shrubs more abundant, an extraordinary greening of thorny species occurred in the wake of the copious summer rains of El Niños in 1972–1973 and 1982–1983.

A few thousand kilometers to the south, in the subtropical shrubs and the *matorral* of north-central Chile, the greening effect is similar, except that, owing to the seasonal timing of incoming cyclonic storms, the rains in these regions occur during the winters preceding or following El Niño episodes. At the southern margin of the Chilean desert the greening of the dormant vegetation results in the *desierto florido* (blooming desert), while in semiarid Chile the local matorral explodes into a verdant cover that encourages the multiplication of rodents. Also in temperate middle Chile, the raininess of El Niño years results in an explosion of grasses, cacti, shrubs, and low trees not seen in normal years. Farther south, in Chile's temperate rain forest domain and in adjacent forests on the Argentine side of the Andes, winter-spring precipitation and summer temperatures tend to be above average during El Niño events, although the weight of these relationships is substantially less than for most other regions of South America influenced by ENSO variabilities (Daniels and Veblen, 2000). Since the precipitation maximum of this region occurs in winter, variations in the summer onset of El Niño events have a major influence on the ecology. Usually, the cool and dry La Niña conditions in the Pacific realm promote wildfires through reduced winter and spring precipitation, but late El Niño episodes can also result in increased fire due to extremely warm summers (Veblen et al., 1999). Nevertheless, the fact that this southern region receives echoes from El Niño developments farther north can be recognized in the signatures left on tree rings, whose chronology permits hints of major El Niño events to be traced back to 1500 B.C. (Lara and Villalba, 1993).

In northeast Brazil, today as in the past, El Niño episodes in the Pacific correlate with severe droughts or *sêcas* (Caviedes 1973). The cause of these summer droughts (the winters here are intrinsically dry) is the failure of the Intertropical Convergence Zone over the equatorial Atlantic to shift southward during the austral summer due to the intensification of the South Atlantic anticyclone. Cooler sea temperatures in the tropical Atlantic help stabilize the air near the surface, adding to the dryness. During these years, the *caatinga*, the dominant vegetation of arid northeastern Brazil, does not green up for lack of summer rains and prolongs its dormancy through the subsequent rainless months. Brazilians refer to this arid landscape as *sertão*, and the different degrees of dryness evinced by the natural vegetation and surface hydrology allow them to differentiate between catastrophic droughts (*grandes sêcas*), usually stretching over several years, droughts (*sêcas*), when one dry summer is caught between two arid winters, and green droughts (*sêcas verdes*), when the rains are sufficient to awaken the caatinga vegetation, but insufficient to grow summer crops.

Information as to the effects of El Niño on the tropical forests of the continent is fragmentary. An indication of what happens with the forest canopies during warm ENSO episodes is inferred from studies conducted in the wet-dry forests of the Isthmus of Panama. During El Niño events, the steady flow of northeast trade winds and dominant clear skies favor high levels of carbon fixation so that flowers and fruits in the high canopy become abundant immediately after El Niño episodes, as does the animal life that they support. Inversely, during non–El Niño years (and more so during La Niña), cloudiness and shortened sunshine hours cause a reduction in canopy growth and fruit production (Wright et al., 1999).

19.6.3 ENSO and Carbon Sinks in the Rain Forests of South America

A subject that is attracting the attention of current environmentalists and ecologists is the relationships between the warm or cold phases of ENSO, the primary productivity of the Amazon rain forest, and the variations of carbon dioxide in natural sinks.

The occurrence of some of the strongest ENSO episodes of the twentieth century in 1972–1973, 1982–1983, and 1997–1998, that is during the past three decades, coincided with marked increases in global temperatures and in atmospheric carbon dioxide. These trends were coupled with heightened net primary production due to stimulated plant growth and with accelerated gas exchanges between biosphere and atmosphere. Most of the increase in net plant production between 1982 and 1999 occurred in the tropics, and this tendency was especially strong in South America as the result of higher solar radiation due to a decline in cloud cover. According to Nemani et al. (2003), the contribution of the immense Amazon forests to the global increase in net primary production in those years was as much as 42%.

Studies by Mahli and Wright (2004) point out that the humidity increases during and after El Niño events have impacted the carbon balance in the Amazon Basin ecosystem. When analyzing two-hundred-year-old live hardwoods (*Euphorbiaceae*) in the seasonally flooded plains of the Amazon River, Schöngart et al. (2004) detected that the vegetative growth during nonflooding periods (El Niño years) resulted in wider radial growth rings. Thus, tree-ring records serve as accurate indicators of climate conditions dominated by the ENSO events in the tropical Pacific, and allow the use of proxies in order to prolong the existing instrumental rainfall time series for that region, which do not go beyond one hundred years. Furthermore, this dendroclimatic indicator offers sound evidence that the effects of El Niño events on tree growth in the Amazon basin have intensified during the last two hundred years as demonstrated by more frequent and wider growth bands.

As to whether the Amazon forests represent significant sinks or pools for global atmospheric carbon dioxide, Rice at al. (2004) point out that, notwithstanding the carbon gains in biomass, there are also considerable losses due to respiration emissions from coarse woody debris, which seem to increase during periods of high tree mortality, such as those that resulted from the drying episodes associated with the strong ENSO events of the 1990s. With increased coarse woody debris, the carbon transfer from live to dead biomass pools appears to have led to substantial amounts of net carbon release. Primary productivity, tree mortality, and carbon storage all appear to be on the rise in the South American rain forests. Even forest composition and plant community dynamics have been affected, as fast-growing trees have achieved wider spatial dominance and the numbers of slower-growing trees have declined. The changes seem to be the result of an upset equilibrium caused by past ecological disturbances, of differential tree vulnerability to El Niño–related drought, and of multidecadal changes in rainfall amounts (Laurance et al., 2004). Thus, climate variability lies at the root of many of the biotic developments in the largest of all South American forested regions.

In an examination of the ecosystem carbon balance and river responses to the different phases of the ENSO, Foley et al. (2002) suggested that, during El Niño occurrences, there is a diminished carbon dioxide source from terrestrial ecosystems due to reduced net primary production, particularly in the northern segment of the Amazon basin. There is also a decline in discharge along many of the rivers there, which reduces the flooded areas along the main channel of the Amazon River. During La Niña phases, an increased incorporation of carbon dioxide *into* terrestrial ecosystems occurs, largely due to activation of the net primary production in the northern basin. In addition, river discharge in the Amazon basin increases, especially through the contributions of northern and western tributaries, and with it a corresponding increase in flooded area, again largely along the northern rivers. Thus, ENSO seems to be a complex modulator of environmental changes and carbon flux across the whole basin.

Augmentation of plant biomass in the Amazon basin during dryness linked with El Niño seems incompatible with the belief that scant precipitation hinders healthy forest development. However, Williamson et al. (2000) noticed that during El Niño of 1997–1998, a value of 1.91% for tree mortality in the Amazon basin was in fact surprisingly low compared with the 1.23% value measured during the normal years that followed. Nor did the dead trees differ significantly in size or species composition from those that had died previously, and neither did soil texture—when affected by the drying period—have a detectable effect on the mortality rates. A different view is expressed by Barlow et al. (2003), who believe that the fires in the Amazon forests, which are more frequent during El Niño years, contribute as much as 5% more of the annual carbon emissions than anthropogenic sources, and that this flux comes mostly from young trees. The widely spread bigger and thicker-barked trees appear initially quite fire resistant, but undergo high mortality between one and three years after the event, a finding which would more than double current estimates of biomass losses and committed carbon emissions from low intensity fires in tropical forests.

All this indicates that even when some undisturbed rain forests in the Amazon basin are rather resilient to severe drought events, forest fires during El Niño years have serious consequences for ecological symbiotic relations. They kill seedlings, saplings, and small trees and reduce forage and prey for understory vertebrates, and destroy fruit in the overstory and damage domestic fruit trees and food crops of native people. Wildlife becomes more vulnerable to subsistence hunting as shown by a reduction in the number of birds, but not in bird species diversity (Barlow and Peres, 2004). After the fires of the 1997–1998 El Niño, only 22 out of the former 35 understory vertebrate species were left in hunted and burned forests, and the populations of seven of those species had declined severely. This disappearance of entire species—while also caused by greater hunting activity—was primarily the result of dramatic alterations in the amount of light reaching the understory after the fires had destroyed more than a third of the large trees. As more light was streaming through the thinned canopy cover, the understory plant composition evolved into a secondary forest habitat of impenetrable plant growth, in which some vertebrates thrived, while others perished (Peres et al., 2003). The same authors believe that canopy reductions through an increase in wild fires in connection with mounting ENSO events may eventually lead to an irreversible transition from tropical forest to scrub savanna.

19.7 Impacts on People

The effects of El Niño on the human population of South America are varied. In some regions El Niño means raised temperatures, high humidities, and excessive rainfall, while in others it is a harbinger of dryness. Heavy rains have devastating sequels because they often occur in areas that are chronically dry, and because it is not the total amount of precipitation but its intensity that wreaks havoc. Such rains produce three major environmental consequences—widespread flooding, increased erosion and sediment yields, and mass movement—each of which impacts human life and livelihood.

El Niño rains cause rivers to rise and flood far beyond their normal channels, notably in coastal Perú and Ecuador during the southern summer, in the lowlands of the Rio de la Plata basin and the rivers of southern Brazil during autumn, and in central Chile and the pre-Andes of western Argentina in winter. These floods cause sizeable losses in property and damage to public works such as roads, bridges, and water management facilities, whose costs have never been properly quantified. States, provinces, and municipalities prone to such hazards expend more in relief funds trying to mitigate the effects of these disasters than to prevent them. In basins that normally see scant runoff, high discharges induce flooding of sewage systems, thus severely contaminating surface and ground water.

Increased soil erosion and sediment yields attributable to El Niño events are recurrent features in Perú where, following heavy rains, fertile floodplains under cultivation for sugar cane, cotton, or maize become blanketed by thick layers of sediment that may prevent agricultural use for several years. Resulting crop losses and damage to agricultural infrastructure may cause lasting impacts to local and regional economies and the people that depend on them. However, since the 1972–1973 event, certain agricultural production indicators have shown some positive effects of El Niño, including productivity gains in grains, soybeans, peanuts, and tobacco (Handler and Handler, 1983). Grain production in the Rio de la Plata lowlands tends to rise following the humid summer of an El Niño in the Pacific. This tendency is more evident in wheat than in maize, which also experiences rises in certain districts of the Argentine *pampa* following the summer of a La Niña event (Podestá et al., 1999). Maize, peanuts, and soybeans have yielded bumper crops in both North and South America during El Niño years when compared with normal or La Niña years (Hansen et al., 1998, 1999), thus illustrating the potentially beneficial effects of warm ENSO events on agrarian economies generally.

Of more local significance are the damaging landslides and debris flows that result from loose masses of debris heavily saturated with water from the torrential rains of El Niño. Known as *huaicos* in Perú, these disasters add to the general misery in that they frequently block or destroy rural roads needed to get relief to isolated communities in distress. During El Niño in 1997–1998, roads from the lowlands of western Ecuador into the densely populated areas of the Andean interior were blocked, sometimes for weeks, so that communication between damaged coastal communities and the Ecuador heartland had to be reestablished via long and costly detours. In northern and central Perú, *huaicos* cut vital links between the coast and the interior, thus making it even harder to provide emergency help to rural areas devastated by the rains.

One palliative for the evils caused by excessively rainy El Niño years in arid regions is that the abundance of water has some beneficial effects for agriculture. Most of the river oases of Perú and northern Chile draw their water from limited aquifers fed by scant summer rains and by winter snowfall in the high Andes. El Niño rains replenish these aquifers, thereby assuring water availability for the lean years to follow. The rains are also a bonus for Peruvian cattle ranchers and dairy farmers struggling in otherwise arid coastal environments, because they generate, at least temporarily, additional grazing land to augment the irrigated pastures.

19.8 El Niño and Disease

Excessive humidity, the multiplication of insects, flooding, and the clogging of drinking-water and waste-water facilities from excessive sedimentation in water courses, all resulting from El Niño events, create conditions favorable for the spread of vector-transmitted epidemics.

Notwithstanding earlier recognition by geographers concerning the sequels of El Niño events in 1972–1973 and 1982–1983 in South America, including epidemics of typhoid fever, dysentery, hepatitis, intestinal ailments, and skin infections (Caviedes, 1974, 1984), these problems only came to the attention of international epidemiologists at the time of the 1997–1998 El Niño episode, when bibliographies worldwide became flooded with reports about health deteriorations caused by El Niño. Following that event, Epstein (1999) showed that outbreaks of epidemics throughout the world's tropical belt during the 1997–1998 ENSO were most pronounced in South America, and most particularly in Perú, Ecuador, and the countries of the Río le la Plata basin (fig. 19.6). This is precisely where flooding caused waste water to spill into rivers and wells, and where pooled surface waters provided favorable grounds for the breeding of obnoxious insects.

In Perú, vector-borne diseases such as cholera, malaria, encephalitis, and dengue have surged significantly in the wake of the last two El Niño events (Zhou et al., 2002). Cholera affected many backward villages in northern Perú and along the coast (Gil et al., 2004), while diarrhea was responsible for large numbers of infant deaths in native villages (Lama et al., 2004). In Ecuador, the contamination of water by fecal matter became so widespread from 1997 to 1999 that a national emergency was decreed; but even the most stringent measures for isolating existing pockets and preventing further spread of the ensuing diseases were in many cases unsuccessful (Gabastou et al., 2002).

In northern Chile, a different sequel was observed, one associated to the oceanic character of El Niño. In November

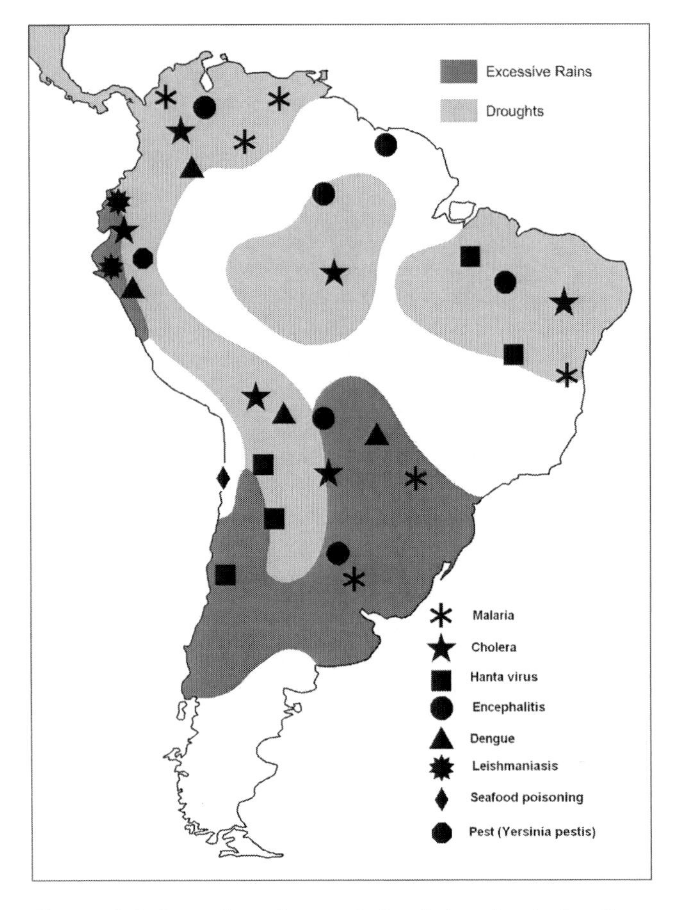

Figure 19.6 Contagious diseases in South America during El Niño years (from an idea of P.R. Epstein, 1999)

1997 and April 1998, several human gastroenteritis cases surfaced in the northern city of Antofagasta, with the offending agent being identified as the bacterium *Vibrio parahaemolyticus*. This bacterium enters shellfish when warming of coastal waters leads to a bacterial bloom and is contracted by humans through consumption of the shellfish (Córdova et al., 2002).

Malaria outbreaks were mostly concentrated in the northern tropical countries of the continent—Perú, Ecuador, Colombia, Venezuela, Guyana, Surinam, and French Guiana. Interestingly, these outbreaks occurred not only under extremely wet conditions—which foster the incubation of mosquitoes—but also under drought and receding water conditions that exposed low-lying wetlands, the favorite breeding grounds of mosquitoes (Barrera et al., 1999). The relationship between malaria outbreaks in Colombia and dry conditions in that country during major El Niño events had been noticed earlier, but recently Bouma et al. (1997) and Poveda et al. (2001) investigated the correlations between these outbreaks and the humid-warm phases of ENSO. In Venezuela, there was a 36% increase of malaria cases in the wake of an El Niño (Bouma and Dye, 1997). Bouma et al. (1997) sought to predict outbreaks of malaria in Colombia by monitoring water temperatures in the tropical Pacific.

Dengue, an infectious disease caused by the genus *Flavivirus* that is transmitted by *Aedes aegypti*—a domestic, day-biting mosquito that likes to feed on humans—has generally been on the increase in South America, more particularly during recent ENSO episodes. Researchers have estimated with a 95% degree of confidence that there are dengue epidemics in French Guiana during years with El Niño in the Pacific; in Colombia and Surinam the relationship stands at a 90% confidence level. Dengue manifests itself with a spectrum of clinical illnesses ranging from nonspecific viral syndrome to severe and fatal hemorrhagic disease. The primarily tropical virus propagates in a cycle that involves humans and insects. Since the 1970s, there has been a resurgence of the *Aedes* mosquito, and dengue has spread through tropical South America and into Central America and the Caribbean, all regions experiencing, during El Niño years, the statistically warmer air temperatures and reduced rainfall that are conducive to the outbreak of the epidemic. Dengue was also introduced to the United States by unsuspecting migrants.

Encephalitis has surfaced with great virulence in the swamps of northern Venezuela and Colombia, as well as in French Guyana, Brazil, and northern Argentina since the early 1970s. This often fatal infection, which is also transmitted to humans directly by mosquitoes or through rodents or domestic animals, has now become a common sequel in the wet periods following El Niño events (Weaver et al., 1996). A variety called "Venezuelan equine encephalitis," which can be transmitted to humans, is also on the upsurge and has spread into Central America, Mexico, and the southern United States.

In the semi-arid environments at the margins of the South American tropics, where year-to-year precipitation varies widely, El Niño rains provoke an unusual greening of dormant vegetation that increases the feeding capabilities of these environments. In Chile's *Norte Chico*, there have been sizeable explosions of rodent populations after El Niño events (Fuentes and Campusano, 1983; Meserve et al., 1995). This shows a truly fascinating connection between rodent proliferation due to El Niño and the spread of epidemics. Outside the continent, in the Four Corners region of the United States (the border area between Utah, Colorado, New Mexico, and Colorado), a similar incident made headlines in 1993 when a viral illness broke out in the Hanta Navajo reservation there. This virus, pertinently called the Hanta virus (Engelthaler et al., 1999), resides in the desiccated feces of the deer mouse and attacks the human respiratory system. Along with other rodents that feed on grasses and succulents, the deer mouse population there had exploded following El Niño rains of 1992–1993. After that, the fatal Hanta virus surfaced in other parts of the North America, and then in South America in the aftermath of the 1997–1998 El Niño. Hanta pulmonary deaths were recorded in northern Chile, the highlands of southern Bolivia, and northwestern Argentina, all of them arid regions like the Four Corners, where rodents may multiply twenty-fold during rainy El Niño years.

Interestingly, the relationship between rodent explosions and outbreaks of contagious diseases in which fleas carry the virus from animals to humans seems to find confirmation in the fact that, in the wake of a recent El Niño, there were five cases of bubonic plague in the Andean native community of Jacocha-Huancabamba in February 1999, one of them fatal. The presence of *Yersinia pestis* was confirmed when antibodies were discovered in neighboring dogs and confirmed by serology (Dávalos et al., 2001). In *El Niño in History*, I remark on the interesting biological "convergences" in the outbreak of these kinds of epidemics in the Old and New Worlds that seem to have been triggered by climatic anomalies of El Niño type (Caviedes, 2001).

This observation leads to a related aspect that until recently was overlooked by many scientists interested in relationships between climate variabilities and human well-being. As posited by Githeko et al. (2000), the climate fluctuations of recent decades have caused environmental temperature increases in the order of 1.0°C to 3.5°C, and this has raised the upper range for the occurrence of vector-transmitted diseases from 35°C to 40°C. However, it is not only the increase in temperature but also in humidity associated with El Niño in many regions that constitutes an important concomitant for the spread of these epidemics. Pivonia et al. (2004) have suggested that the accrued air humidity during major El Niño episodes may been conducive to the spread of disease by air; plant pathologists noticed that stripe rust *(Puccinia striiformis f. sp. hordei)* and sorghum ergot (*Claviceps africana*) were swept from the South American tropics into North America during El Niño events in 1986–1987 and 1997–1998, when southerly winds advanced abnormally far into the Northern Hemisphere. Last but not least, French physicians have suggested that influenza epidemics in Europe between 1972 and 2002 correlate very well with the lower phases of ENSO (El Niño episodes) in the distant tropical Pacific (Viboud et al., 2004).

19.9 Paleoclimatic Records of El Niño

In view of the varied effects of contemporary ENSO events on nature and people, as well as the areas involved, the question arises as to whether such oscillations also occurred in the undocumented past and, if so, what were their implications? South America, the continent most directly affected by El Niño, has been the object of diligent research to unearth traces of past ENSOs. The fact that ENSO-related phenomena have increased in frequency and intensity since the 1970s suggests that recent global warming has a bearing on their inception (Zwier and Weaver, 2000). Historical accounts of catastrophic events caused by conditions similar to those of recent El Niño events indicate that such climatic fluctuations did indeed occur in past centuries, but not with the frequency of recent decades. This trend toward more frequent El Niño events has accentuated since the end of the Little Ice Age (LIA), around the mid-nineteenth century, when increased carbon dioxide emissions to the atmosphere from industrializing nations may have inaugurated more drastic climate changes. Before that, between the fifteenth and nineteenth centuries, the relative cool of the LIA is thought to have favored a quieter global circulation that discouraged frequent major El Niño outbursts. For this reason, whenever such events did occur in South America during colonial times, they did not escape the attention of chroniclers. Before then, a short-lived warming period in the early 1500s, coinciding with European discovery of the Americas, explains why many climatic events with indicators similar to contemporary ENSOs were recorded in the discoverers' chronicles (Caviedes, 2001). During the fourteenth and fifteenth centuries, another cooling episode had put an end to the Medieval Warm Period (MWP), 1000–1300 A.D., which had permitted the expansion of settlements in northernmost Europe and into Iceland and Greenland (Lamb, 1995).

In South America, archeological, anthropological, and geological evidence suggests that the MWP may have had profound human and environmental consequences. Nials et al. (1979) were the first to advance the idea of earlier El Niño events when they observed massive alluvial deposits in coastal northern Perú that pointed to mighty rains and devastating floods in pre-Hispanic times. Similar evidence has been used to date cultural crises and relocations of political centers in the coastal oases of northern Perú (Moseley, 1983, 1997), and to provide a chronology of catastrophic El Niño events (Sandweiss et al., 1997; De Vries, 1987; Wells, 1990). From this evidence arose the concept of *mega-Niño* to designate prolonged El Niño episodes during the MWP and earlier. Several cultural changes in South American prehistory have been related to such episodes. The flourishing periods and ultimate decay of the Tiwanaku culture around Lake Titicaca seem to have been connected with humid intervals and prolonged dry periods, respectively (Abbott et al., 1997; Binford et al., 1997). Ice cores extracted from the Quelccaya, Huascarán, and Sajama ice caps (Thompson et al., 1992, 1995, 1998), reveal that alternating periods of humidity and dryness very similar to those prevailing during contemporary El Niño episodes (wetness on the Pacific slope of the Andes and dryness in the Peruvian Sierra, the Altiplano, and most of the Amazon basin) also occurred in pre-Hispanic times. Particularly strong El Niño events occurred around 1100, 750, 602–635, and 500 A.D., causing the abandonment of human settlements in north Peruvian oases due to destructive alluviation (Moseley, 1992). The Peruvian Andes experienced severe droughts in phase with these events (Shimada et al., 1991).

In the Amazon basin, Meggers (1994) notes remarkable cultural revolutions, evinced in changes of ceramic styles, intrusions of alien linguistic elements, and site relocations that were caused by prolonged aridity during the prehistoric El Niño events mentioned above. She also mentions traces of sites abandoned due to climatic crises in northern Colombia, coastal Venezuela, and the Orinoco basin (Meggers, 1996) that she considers plausible proof of early periods of dryness in the tropical lowlands east of the Andes prompted by ENSO events.

Recent research on South American climate history has permitted the construction of time series that go back some 25,000 years (Baker et al., 2001). Within these series, peculiar fluctuations in deposition styles, pollen assemblages, and microfaunal facies can be attributed to warm humid episodes related to possible El Niño events, although caution must be exercised in attributing all such changes to ENSO when longer-term oscillations related to Earth's orbital relations with the Sun may also be involved. Relying on dust particles and oxygen isotopes from ice cores, Thompson et al. (2000) suggest the occurrence of warming episodes around 2,000, 4,500, 5,000, 6,000, 7,000, and 9,000 years B.P. (before present, 1950). A mighty warming period around 15,000 years B.P. marks the end of the Late Glacial Stage in the subtropical Andes (see chapter 4). Since interannual rainfall variations in the central Andes are modulated by the Southern Oscillation, it can be argued that ENSO-like episodes may have set in at that time and that air-ocean circulation patterns in the tropical eastern Pacific similar to those of the present were also established then. Of relevance, Lea et al. (2000) have recreated sea surface temperatures in the equatorial Pacific for the last 450,000 years based on foraminifera extracted from sediment overlying the Cocos Ridge. These suggest that, during the cold stages of the Quaternary, sea surface temperatures were as low as 22°C in the eastern equatorial Pacific, while during the interglacials they rose to 27–28°C. A comparison of these values with contemporary sea surface temperatures, of 23°C off Perú and Ecuador during La Niña episodes and of 29°C during El Niño events, suggests that the effects on sea surface temperatures of contemporary El Niños and La Niñas are not much different. This implies that the atmospheric circulation characteristics of today's extreme phases of the Southern Oscillation are very similar to those which dominated over the tropical Pacific during glacial and interglacial stages.

Based on sediment deposited in a mountain lake in Ecuador, Rodbell et al. (1999) propose that, if ENSOs did occur between 15,000 and 7,000 years B.P., they were weak, and that events comparable to those of today are recognizable in lacustrine beds only after 5,000 B.P.. ENSO events may be identifiable in coastal deposits as far back as 40,000 years B.P. (Wells and Noller, 1997), but Sandweiss et al. (1997) argue that there is no sedimentary evidence for ENSOs between 8,900 and 5,800 B.P. because the eastern Pacific was warmer than what is normal for eastern boundary currents and that only after 5,800 B.P. were phenomena of this kind likely to occur. Analyzing fossil rodent middens and wetland deposits, Betancourt et al. (2000) concluded that humidity advection from the east (La Niña) and dry westerlies from the Pacific (El Niño) seem to have occurred in the Atacama Desert since sometime between 11,800 and 10,500 years ago.

Discussions about the time when the first El Niño events appeared will probably continue until a reliable chronological series is produced for the tropical and extra-tropical regions of the continent. Independent from this, the environmental implications and human relevance of these oscillations for South America will continue to underline the enormous bearing of El Niño as a main modulator of variabilities in contemporary and past climates.

References

Aalto, R., L. Maurice-Bourgoin, T. Dunne, D.R. Montgomery, C.A. Nittrouer, and J. Guyot, 2003. Episodic sediment accumulation on Amazonian flood plains influenced by El Niño/Southern Oscillation. *Nature,* 425, 493–497.

Abbott, M.B., M.W. Binford, M. Brenner, and K.R. Kelts, 1997. A 3500 ^{14}C yr. high-resolution record of water level changes in Lake Titicaca, Bolivia/Peru. *Quaternary Research,* 47, 169–180.

Addis, J., and M. Latif, 1996. Amazonian arthropods respond to El Niño. *Biotropica,* 28, 403–408.

Arntz, W.E., and E. Fahrbach 1991. *El Niño: Klimaexperiment der Natur.* Birkhäuser Verlag, Basel.

Báez, P., 1985. Fenómeno El Niño, elemento importante en la evolución del camarón de río (*Cryphiops caementarius*). *Investigación Pesquera,* 32, 235–242.

Baker, P.A., G.O. Seltzer, S.C. Fritz, R. Dunbar, M. Grove, P. Tapia, S. Cross, H. Rowe, and J. Broda, 2001. The history of South American tropical precipitation for the past 25,000 years. *Science*, 291, 640–643.

Barlow, J., and C.A. Peres, 2004. Avifaunal responses to single and recurrent wildfires in Amazonian forests. *Ecological Applications,* 14, 1358–1373.

Barlow, J., C.A. Peres, B.O. Lagan, and T. Haugaasen, 2003. Large tree mortality and the decline of forest biomass following Amazonian wildfires. *Ecology Letters*, 6, 6–8.

Barrera, R., M.E. Grillet, Y. Rangel, J. Berti, and A. Ache, 1999. Temporal and spatial patterns of malaria reinfection in northeastern Venezuela. *American Journal of Tropical Medicine and Hygiene*, 61, 784–790.

Berbery, E.H., and R. Barros, 2002. The hydrologic cycle of the La Plata Basin in South America. *Journal of Hydrometeorology*, 3, 630–645.

Berri, G.J., and E.A. Flamenco, 1999. Seasonal volume forecast of the Diamante River, Argentina, based on El Niño observations and predictions. *Water Resources Research*, 35, 3803–3810.

Betancourt, J.L., C. Latorre, J.A. Rech, J. Quade, and K.A. Rylander, 2000. A 22,000-year record of monsoonal

precipitation from northern Chile's Atacama Desert. *Science*, 289, 1542–1546.
Binford, M.W., A. Kolata, M. Brenner, J.W. Janusek, M.T. Seddon, M. Abbott, and J.H. Curtis 1997. Climate variation and the rise and fall of an Andean civilization. *Quaternary Research*, 47, 235–248.
Blanco, J.F., G.L. Vásquez, J.C. Ramírez, and A.M. Navarrete, 2003. Variación de algunos parametros fisicoquímicos en el Río Pescador, Valle del Cauca, durante el ciclo El Niño 1997/1998—La Niña 1998/1999. *Actualidades Biológicas*, 25, 59–69.
Bouma, M.J., and C. Dye, 1997. Cycles of malaria associated with El Niño in Venezuela. *Journal of the American Medical Association*, 278, 1772–1774.
Bouma, M.J., G. Poveda, W. Rojas, D. Chavasse, M. Quiñones, J. Cox, and J. Patz, 1997. Predicting high-risk years for malaria in Colombia using parameters of El Niño Southern Oscillation. *Tropical Medicine and International Health*, 2, 1122–1128.
Caviedes, C.N., 1973. Sêcas and El Niño: Two simultaneous climatological hazards in South America. *Proceedings of the Association of American Geographers*, 5, 44–49.
———, 1982. On the genetic linkages of precipitation in South America. *Freiburger Geographische Hefte* 18. Fortschritte Landschaftökologischer und Klimatologischer Forschungen in den Tropen. Festschrift zum 60. Geburtstag von Professor Wolfgang Weischet, 55–77.
———, 1984. El Niño 1982–83. *Geographical Review*, 74, 267–290.
———, 1992. South American hydrology and its association with El Niño/Southern Oscillation. Proceedings of the International Meeting of NOAA and Instituto de Pesquisas Espaciais on Global Climate Predictions, Fortaleza, Ceará.
———, 1998. Influencia de ENOS sobre las variaciones interanuales de ciertos ríos en América del Sur. *Bulletin de l'Institut Français d'Études Andines*, 27, 627–641.
———, 2001. *El Niño in History*. University Presses of Florida, Gainesville.
Caviedes, C.N., and T.L. Fik, 1993. Modeling change in the Chilean/Peruvian Eastern Pacific fisheries. *GeoJournal*, 30, 369–380.
Caviedes, C.N., and P.R. Waylen, 1991. Chapters for a climatic history of South America. In: W. Endlicher and H. Gossmann (Editors), *Beiträge zur Regionalen und Angewandte Klimatologie. Freiburger Geographische Hefte*, 32, 149–180.
Chavez, F.P., R.T. Barber, and M. Sanderson, 1989. The potential primary production of the Peruvian upwelling ecosystem, 1953–1984. In: D. Pauly, P. Muck, J. Mendo, and I. Tsukayama (Editors), *The Peruvian Upwelling Ecosystem: Dynamics and Interactions* IMARPE/GTZ/ICLARM, Manila.
Cole, J.E., G.T. Shen, R.G. Fairbanks, and M. Moore, 1992. Coral monitors of El Niño/Southern Oscillation dynamics across the equatorial Pacific. In: H.F. Diaz and V. Markgraf (Editors), *El Niño. Historical and Paleoclimatic Aspects of the Southern Oscillation*. Cambridge University Press, New York.
Compagnucci, R., S. Blanco, A. Costa, P. Jacobski, and O. Rosso, 1996. Analysis of Argentinean river streamflows by nonlinear dynamic metric tools. *Abstracts of the 13th Conference on Probability in the Atmospheric Sciences*, San Francisco, 21–23.
Córdova, J.L., J. Astorga, W. Silva, and C. Riquelme, 2002. Vibrio parahaemolyticus isolates collected during the 1997–1998 Chilean outbreak. *Biological Research*, 35, 433–440.
Daniels, L.D., and T.T. Veblen, 2000. ENSO effects on temperature and precipitation of the Patagonian-Andean region: Implications for biogeography. *Physical Geography*, 21, 223–243.
Dávalos, V.A., M.A. Torres, C.O. Mauricci, V. Laguna-Torres, and M. Chinarro, 2001. Surto de peste bubonica na localidade de Jacocha, Huancabamba, Peru. *Revista da Sociedade Brasileira de Medicina Tropical*, 34, 87–90.
De Vries, T.J., 1987. A review of geological evidence for ancient El Niño activity in Peru. *Journal of Geophysical Research*, 92, 14471–14479.
Engelthaler, D.M., D.G. Mosley, J.E. Cheek, C.E. Levy, K.K. Komatsu, P. Ettestad, T. Davis, D.T. Tanda, L. Miller, J.W. Frampton, R. Porter, and R.T. Bryan, 1999. Climatic and environmental patterns associated with Hantavirus pulmonary syndrome, Four Corners Region, United States. *Emerging Infectious Diseases*, 5, n.p.
Epstein, P.R., 1999. Climate and health. *Science*, 285, 347–348.
Estudio Regional del Fenómeno El Niño (ERFEN), 1983. *Informe de la Tercera Reunión del Comité Científico del ERFEN. Cali, Colombia*, ERFEN, Cali.
Evans, M.N., R.J. Fairbanks, and J.L. Rubenstone, 1998. A proxy of ENSO teleconnections. *Nature*, 394, 732–734.
Foley, J.A., A. Botta, M.T. Coe, and M.H. Costa, 2002. El Niño-Southern Oscillation and the climate, ecosystems and rivers of Amazonia. *Global Biogeochemical Cycles*, 16, 35–44.
Fuentes, E.R., and C. Campusano, 1983. Pest outbreaks and rainfall in the semiarid region of Chile. *Journal of Arid Environments*, 81, 67–72.
Gabastou, J.M., C. Pesantes, S. Escalante, Y. Nervier, E. Vela, L. García, D. Zabala, and Z.E. Yadon, 2002. Características de la epidemia de cólera de 1998 en Ecuador, durante el fenómeno de El Niño. *Revista Panamericana de Salud Pública*, 12, 157–164.
Garreaud, R., M. Vuille, and A.C. Clement, 2003. The climate of the Altiplano: Observed current conditions and mechanisms of past changes. *Palaeogeography, Palaeoclimatology, Palaeoecology*, 194, 5–22.
Gil, A.I., V.R. Louis, I.N. Rivera, E. Lipp, A. Huq, C.F. Lanata, D.N. Taylor, E. Russek-Cohen, N. Choopun, R.B. Sack, and N. Colwell, 2004. Occurrence and distribution of Vibrio cholerae in the coastal environment of Peru. *Environmental Microbiology*, 6, 699–706.
Githeko, A.K., S.W. Lindsay, U.E. Confalonieri, and J.A. Patz, 2000. Climate change and vector-borne diseases: A regional analysis. Bulletin of the World Health Organization, 78, 1136–1147.
Glyn, P.W., and M.W. Colgan, 1992. Sporadic disturbances in fluctuating coral reef environments: El Niño and coral reef development in the Eastern Pacific. *American Zoologist*, 32, 707–719.
Handler, P., and E. Handler, 1983. Climatic anomalies in the tropical Pacific Ocean and corn yields in the United States. *Science*, 22, 1155–1157.
Hansen, J.W., A. Hodges, and J.W. Jones, 1998. ENSO influences on agriculture in the southeastern United States" *Journal of Climate*, 11, 404–411.
Hansen, J.W., J.W. Jones, C.F. Kiker, and A. Hodges, 1999.

El Niño-Southern Oscillation impacts on winter vegetable production in Florida. *Journal of Climate*, 12, 92–102.
Hastenrath, S., 1990. Diagnostics and prediction of anomalous river discharge in northern South America. *Journal of Climate*, 3, 1080–1096.
Holmgren, M., Scheffer, M., Ezcurra, E., Gutiérrez, J.R., and G.M. Mohren, 2001. El Niño effects on the dynamics of terrestrial ecosystems. *Trends in Ecology and Evolution*, 16, 89–94.
Keller, A., 1990. Das El Niño-Phänomen und der Titicacaseespiegel. Festschrift für Wendelin Klaer 65. Geburtstag. *Mainzer Geographische Studien*, 34, 91–100.
Koepcke, H-W., 1961. Synökologische Studien an der Westseite der peruanischen Anden. *Bonner Geographische Abhandlungen*, 29, 1–320.
Lama, J.R., C.R. Seas, R. Leon-Barua, E. Gotuzzo, and R.B. Sack, 2004. Environmental temperature, cholera, and acute diarrhea in adults in Lima, Peru. *Journal of Health, Population, and Nutrition*, 22, 399–403.
Lamb, H.H., 1995. *Climate, History, and the Modern World*. Methuen, London.
Lara, A., and R. Villalba, 1993. A 3620-year temperature record from *Fitzroya cupressoides* tree rings in southern South America. *Science*, 260, 1104–1106.
Laurance, W.F., A.A. Oliveira, A.A., S.G. Laurance, R. Condit, and F.M. Nascimento, 2004: Pervasive alteration of tree communities in undisturbed Amazonian forests. *Nature*, 428,171–175.
Lea, D.W., D.K. Pak, and H. Spero, 2000. Climate impact of late Quaternary equatorial Pacific sea surface temperature variations. *Science*, 289, 1719–1724.
Mahli, Y., and S.J. Wright, 2004: Spatial patterns and recent trends in the climate of tropical rainforest regions. *Philosophical Transactions of the Royal Society of London. Series B, Biological Sciences*, 359, 311–329.
Marengo, J.A., 1995. Variations and change in South American streamflow. *Climate Change*, 5, 99–117.
Marengo, J.A., W.R. Soares, C. Saulo, and M. Nicolini, 2004. Climatology of the low-level jet east of the Andes as derived from the NCEP-NCAR reanalyses: Characteristics and temporal variability. *Journal of Climate*, 17, 2261–2280.
Mechoso, C.R., and G. Perez-Irribarren, 1992. Streamflow in southeastern South America and the Southern Oscillation. *Journal of Climate*, 6, 1535–1539.
Meggers, B.J., 1994. Archeological evidence for the impact of mega-Niño events on Amazonia during the past two millennia. *Climatic Change*, 28, 321–328.
———, 1996. Possible impact of the mega-Niño events on Precolumbian populations in the Caribbean. In: M.V. Maggiolo and A.C. Fuentes (Editors), *Ponencias del Primer Seminario de Arqueología del Caribe*, Altos de Chavón: Museo Arqueológico Regional and Organización de Estados Americanos.
Meserve, P., J. Yunger, J. Gutiérrez, L. Contreras, B. Milstead, K. Cramer, S. Herrera, V. Lagos, S. Silva, E. Tabilo, M. Torrealba, and F. Jaksic, 1995. Heterogeneous responses of small mammals to an El Niño-Southern Oscillation event in north-central semiarid Chile. *Journal of Mammology*, 76, 580–596.
Minetti, J.L., and E.M. Sierra, 1989. The influence of general circulation patterns on humid and dry years in the Cuyo Andean region of Argentina. *International Journal of Climatology*, 9, 55–68.
Moseley, M.E., 1983. The good old days were better: Agrarian collapse and tectonics. *American Anthropologist*, 85, 773–799.
———, 1992. *The Inca and Their Ancestors*. Thames and Hudson, London.
———, 1997. Climate, culture, and punctuated change: New data, new challenges. *The Review of Archaeology*, 17, 19–27.
Muck, P., 1989. Relationships between anchoveta spawning strategies and the spatial variability of sea surface temperature off Peru. In: D. Pauly, P. Muck, J .Mendo, and I. Tsukayama (Editors), *The Peruvian Upwelling Ecosystem: Dynamics and Interactions*, IMARPE/GTZ/ICLARM, Manila.
Neelin, J.D., D.S. Battisti, A.C. Hirst, F-F. Jin, Y. Wakata, T. Yamagata, and S.E. Zebiak, 1998. ENSO theory. *Journal of Geophysical Research*, 103, 14261–14290.
Nemani, R.R., C.D. Keeling, H. Hashimoto, and W.M. Jolly, 2003. Climate-driven increases in global terrestrial net primary production from 1982 to 1999. *Science*, 300, 1560–1563.
Nials, F.L., E.E. Deeds, M.E. Moseley, S.G. Pozorski, T.G. Pozorski, and R.A. Feldman, 1979. El Niño: The catastrophic flooding of coastal Peru. *Field Museum of Natural History, Bulletin*, 50(7), 4–14; 50(8), 4–10.
Pauly D., and M. Soriano, 1989. Production and mortality of anchoveta (*Engraulis ringens*) eggs off Peru. In: D. Pauly, P. Muck, J. Mendo, and I.Tsukayama (Editors), *The Peruvian Upwelling Ecosystem: Dynamics and Interactions*. IMARPE/GTZ/ICLARM, Manila.
Peres, C., J. Barlow, and T. Haugassen, 2003. Vertebrate responses to surface wildfires in a Central Amazonian forest. *Oryx*, 37, 97–109.
Pivonia, S., Z. PAN, and X.B. Yang, 2004. Possible link between El Niño events and introductions of airborne fungal pathogens from South America to North America. *Phytopathology*, 94, 21–34.
Podestá, G.P., C. Messina, M.O. Grondona, and G. Magrin, 1999. Associations between grain crop yields in central-eastern Argentina and El Niño-Southern Oscillation. *Journal of Applied Meteorology*, 38, 1488–1497.
Poveda, G., and O.J. Mesa, 2000. On the existence of Lloró (the rainiest locality on Earth): Enhanced ocean-atmosphere-land interaction by a low-level-jet. *Geophysical Research Letters*, 27, 1675–1678.
Poveda, G., W. Rojas, M.L. Quiñones, I.D. Vélez, R.I. Mantilla, D. Ruiz, J.S. Zuluaga, and G.L. Rua, 2001. Coupling between annual and ENSO timescales in the malaria-climate association in Colombia. *Environmental Health Perspectives*, 109, 489–493.
Quesada, M., and C.N. Caviedes, 1992. Características estadísticas de algunos ríos de Colombia. *Revista Geográfica I.P.G.H.*, 116, 53–66.
Rice, A.H., E.H. Pyle, S.R. Saleska, L. Hutyra, M. Palace, M. Keller, P.B. de Camargo, K. Portilho, D.F. Marques, and S.F. Wofsy, 2004. Carbon balance and vegetation dynamics in an old growth Amazonian forest. *Ecological Applications*, 14, 55–71.
Richey, J.E., C. Nobre, and C. Deser, 1989. Amazon River discharge and climate variability: 1903 to 1985. *Science*, 246, 101–103.
Riehl, H., 1984. El Niño north of the Equator in South America. *Tropical Ocean-Atmosphere Newsletter*, 4, 2–3.
Rodbell, D.T., G. Seltzer, D.M. Anderson, M.B. Abbott, D. Enfield, and J.H. Newman, 1999. An ~15,000-year

record of El Niño-driven alluviation in southwestern Ecuador. *Science*, 283, 516–520.
Ronchail, J., G. Cochonneau, M. Molinier, J.-L. Guyot, A.G. De Miranda Chaves, V. Guimaraes, and E. De Oliveira, 2002. Interannual rainfall variability in the Amazon Basin and sea surface temperatures in the Equatorial Pacific and the Tropical Atlantic Oceans. *International Journal of Climatology*, 22, 1663–1686.
Ronchail, J., L. Burrel, G. Cochoneau, P. Vauchel, L. Phillips, A. Castro, J.-L. Guyot, and E. De Oliveira, 2005. Inundations in the Mamoré basin (south-western Amazon, Bolivia) and sea surface temperature in the Pacific and Atlantic Oceans. *Journal of Hydrology*, 302, 223–238.
Rossel, F., 1997. *Influencia del El Niño sobre los Regímenes Hidro-Pluviométricos del Ecuador*. Quito. Instituto Nacional de Meteorología e Hidrología y Institut Français de Recherche Scientifique et Coopération (ORSTOM).
Sandweiss, D., J. Richardson, E. Reitz, H. Rollins, and K. Maasch, 1997. Determining the early history of El Niño. A Response. *Science*, 276, 966–967.
Schöngart, T, J., W.J. Junk, M.T. Piedale, J.M. Ayres, A. Hüttermann, and M. Worbes, 2004. Teleconnection between tree growth in the Amazonian floodplains and the El Niño-Southern Oscillation effect. *Global Change Biology,* 10, 683–692.
Shimada, I., C.B. Schaaf, L.G. Thompson, and E. Mosley-Thompson, 1991. Cultural impacts of severe droughts in the prehistoric Andes: Application of a 1500-year ice-core precipitation record. *Archaeology and Arid Environments*, 22, 247–270.
Spence, J.M., Taylor, M.A. and A. Chen, 2004. The effect of concurrent sea-surface temperature anomalies in the tropical Pacific and Atlantic on Caribbean rainfall. *International Journal of Climatology*, 24, 1531–1541.
Thompson, L.G., and E. Mosley-Thompson, 1992. Reconstructing interannual climate variability from tropical and subtropical ice-core records. In: H.F. Diaz and V. Markgraf (Editors), *El Niño. Historical and Paleoclimatic Aspects of the Southern Oscillation*, Cambridge University Press, New York.
Thompson, L.G., E. Mosley-Thompson, M. Davis, P-N. Lin, K. Henderson, J. Cole-Dai, J. Bolzan, and K-B. Liu, 1995. Late Glacial stage and Holocene ice-core records from Huascarán, Peru. *Science*, 269, 46–50.
Thompson, L.G., M.E. Davis, E. Mosley-Thompson, T.A. Sowers, K.A. Henderson, V.S. Zagorodnov, P-N. Lin, V.N. Mikhalenko, R.K. Campen, J.F. Bolzan, J. Cole-Dai, and B. Francou, 1998. A 25,000-year tropical climate history from Bolivian ice cores. *Science*, 282, 1858–1861.
Thompson, L.G., M.E. Davis, E. Mosley-Thompson, and K.A. Henderson, 2000. Ice-core palaeoclimate records in tropical South America since the Last Glacial Maximum. *Journal of Quaternary Science*, 15, 377–394.
Veblen, T.T., T. Kitzberger, R. Villalba, and J. Donnegan, 1999. Fire history in northern Patagonia: The roles of humans and climatic variation. *Ecological Monographs*, 47–67.
Viboud, C., K. Adman, P.Y. Boelle, M.L. Wilson, M.F. Myers, A.J. Valleron, and A. Flahault, 2004. Association of influenza epidemics with global climate variability. *European Journal of Epidemiology*, 19, 1055–1059.
Waylen, P.R., and C.N. Caviedes, 1986. El Niño and flooding on the north Peruvian littoral. *Journal of Hydrology*, 89, 141–156.
———, 1990. Annual and seasonal fluctuations of precipitation and streamflow in the Aconcagua River basin. *Journal of Hydrology*, 120, 79–102.
Weaver, S.C., R. Salas, R. Rico-Hesse, G.V. Ludwig, M.S. Oberste, J. Boshell, and R.B. Tesh, 1996. Re-emergence of epidemic Venezuelan equine encephalomyelitis in South America. *Lancet*, 348, 436–440.
Wells, L.E., 1990. Holocene history of El Niño phenomena as recorded in flood sediments of northern coastal Peru. *Geology*, 18, 1134–1137.
———, and J.S. Noller, 1997. Determining the early history of El Niño. *Science,* 276, 966–967.
Williamson, G.B., W.F. Laurance, A.A. Oliveira, P. Delamonica, C. Gascon, T.E. Lovejoy, and L. Pohl, 2000. Amazonian tree mortality during the 1997 El Niño drought. *Conservation Biology*, 14, 1538–1542.
Wright, P.B., 1984. Relationships between indices of the Southern Oscillation. *Monthly Weather Review*, 112, 1913–1919.
———, 1989. Homogenized long-period Southern Oscillation indices. *International Journal of Climatology*, 9, 1872–1985.
Wright, S.J., C. Carrasco, O. Calderón, and S. Patton 1999. The El Niño-Southern Oscillation, variable fruit production and famine in a tropical forest. *Ecology*, 80, 1632–1647.
Zhou, J., W-M. Lau, P.M. Masuoka, R.G. Andre, J. Chamberlin, P. Lawyer, and L.W. Laughlin, 2002. El Niño helps spread Bartonellosis epidemics in Peru. *EOS, Transactions, American Geophysical Union*, 83, 160–161.
Zwiers, F.W., and A.J. Weaver, 2000. The causes of 20th century warming. *Science*, 290, 2081–2083.

Environmental Impacts of Urbanism

Jorgelina Hardoy
David Satterthwaite

This chapter describes the environmental impacts of urbanization in South America, and the difficulties that governments have had in managing them. The discussion focuses initially on the rapid urbanization of the continent and its environmental implications and then reviews the quality of the urban environment within the homes and neighbourhoods in which the urban population lives, in the workplace, and in the wider city (the ambient environment). The environmental impacts of these urban areas on their surroundings are then described and their wider and more diffuse impacts considered, including an evaluation of global climate change. Lastly, some of the new directions taken by governments in the region toward addressing these problems are noted. Table 20.1 provides a summary of the main city-related environmental problems in terms of their spatial context and the nature of the hazard or problem.

The urban environment is taken to mean the physical environment in urban areas, with its complex mix of natural elements (including air, water, land, climate, flora, and fauna) and the built environment, in other words a physical environment constructed or modified for human habitation and activity encompassing buildings, infrastructure, and urban open spaces (Haughton and Hunter, 1994; OECD, 1990). Its quality is much influenced by: (1) its geographical setting; (2) the scale and nature of human activities and structures within it; (3) the wastes and emissions these activities create and their environmental impacts; and (4) the competence and accountability of the institutions elected, appointed, or delegated to manage it.

In summarizing the environmental impacts of urbanization, this chapter concentrates on some of the region's most serious urban problems. However, it should always be remembered that this is also a region with rich and varied urban cultures. South America has some of the world's finest historic cities—for instance the historic centers of Cusco, Quito, and Salvador de Bahía. The urban cultures have evolved from a long history, including a rich pre-Colombian urban history in many places (Hardoy, 2000). The cities are widely known outside South America through the literature they have inspired—for instance, for the English-speaking world, the works of Garcia Marquez, Amado, and Vargas Llosa. Its cities are also known for the art, music, and dance that they incubated and inspired.

20.1 Background

20.1.1 Urban Change, 1950–2000

Urban areas are now home to four out of five South Americans and contain most of the continent's economic activities.

Table 20.1 A summary of the main city-related environmental problems

Context	Nature of hazard or problem	Some examples
Within house and its plot	Biological pathogens	Water-borne pathogens and those classified as water-washed (or related to water-scarcity). Airborne, food-borne and vector-borne pathogens (including some water-related vectors). Insufficient quantities of water may have as serious a health impact as poor water quality. The quality of provision for sanitation is also very important. Overcrowding/poor ventilation will aid transmission of many infectious diseases.
	Chemical pollutants	Indoor air pollution from fires, stoves or heaters. Accidental poisoning from household chemicals. Occupational exposure for home workers.
	Physical hazards	Household accidents—burns and scalds, cuts, falls. Physical hazards from home-based economic activities. In poor-quality housing, inadequate protection from rain, extreme temperatures, and other extreme weather conditions.
Neighborhood	Biological pathogens	Pathogens in waste water, solid waste (if not removed from the site) and local water bodies. Disease vectors often breeding in standing water, drains or garbage. If sanitation is inadequate, many people will defecate on open sites, bringing widespread fecal contamination. If a settlement is served by communal standpipes, latrines, and/or solid-waste collection points, these need intensive maintenance to keep them clean and functioning well.
	Chemical pollutants	Ambient air pollution from fires or stoves; also often from burning garbage if there is no regular collection service. Air and water pollution and wastes from cottage industries; air pollution from motor vehicles.
	Physical hazards	Site-related hazards, e.g. housing on slopes with risks of landslides; sites regularly flooded, sites at risk from earthquakes. Traffic hazards. Noise.
Workplace	Biological pathogens	Overcrowding/poor ventilation aiding transmission of infectious diseases.
	Chemical pollutants	Toxic chemicals, dust.
	Physical hazards	Dangerous machinery, noise.
City (or municipality within larger city)	Biological pathogens	The quality and extent of provision for piped water, sanitation, drainage, solid waste collection, disease control, and health care at city or municipal level are a critical influence on extent of the problems. Habitats for disease vectors on neglected or abandoned land and poorly drained areas.
	Chemical pollutants	Ambient air pollution (mostly from industry and motor vehicles; motor vehicles' role generally growing); water pollution; hazardous wastes mainly from industries. In many cities, problems of abandoned contaminated brownfield sites.
	Physical hazards	Traffic hazards. Natural disasters and their unnaturally large impact because of inadequate attention to prevention, mitigation, and preparedness.
	Citizen's access to land for housing	An important influence on housing quality directly and indirectly (e.g. through insecure tenure discouraging households investing in improved housing and discouraging water, electricity, and other utilities from serving them)
	Heat island effect and thermal inversions	Raised temperatures a health risk, especially for vulnerable groups (e.g. elderly, very young). Air pollutants may become trapped within cities, increasing their concentration and the length of people's exposure to them.
City-region (or urban periphery)	Resource degradation	Soil erosion from poor watershed management or land development or clearance; deforestation; water pollution; ecological damage from acid precipitation and ozone plumes arising from emissions within urban areas
	Land or water pollution from waste dumping	Pollution of land from dumping of conventional household, industrial, and commercial solid wastes, and toxic/hazardous wastes. Leaching of toxic chemicals from waste dumps into water. Contaminated industrial sites. Pollution of surface water and possibly groundwater from sewage and storm/surface runoff.
	Pre-emption or loss of resources	Freshwater for city pre-empting its use for agriculture; expansion of paved areas over good quality agricultural land. Agricultural land often lost through extraction of gravels, brick clays and building stone.
Links between city and global issues	Non-renewable resource use	Use of fossil fuels and other mineral resources; loss of biodiversity; loss of non-renewable resources in urban waste streams.
	Non-renewable sink use	Persistent chemicals in urban waste streams; greenhouse gas emissions (mostly carbon dioxide but also chlorofluorocarbons, nitrous oxide and methane); release of stratospheric ozone-depleting chemicals.
	Overuse of finite renewable resources	A scale of consumption (primarily by middle and upper income groups) that is incompatible with global limits for soil, forests, and freshwater.

Over the last five decades, the urban population grew more than fivefold to reach 277 million in 2000 (table 20.2). In this year, the rural population of 71 million was barely more than it had been in 1950. Projections by the United Nations (UN) now suggest that the region's rural population is declining (United Nations, 2004). In general, the nations with the largest expansion in their economies since 1950 have urbanized the most (UNCHS, 1996).

South America has the highest concentration of population in large cities (with 5 to 10 million inhabitants) and very large cities or mega-cities (with more than 10 million inhabitants) of any of the world's continents. By 2000, the region had 34 cities with more than a million inhabitants, including 15 million in Brazil, 4 million in Colombia and Venezuela, and 3 million in Argentina (table 20.3; United Nations, 2004). In general, the region's largest cities are concentrated in its largest economies—as is the case in all other regions of the world (Satterthwaite, 2005). South America also had two of the world's 10 largest cities in 2000 (São Paulo with 17.1 million inhabitants, Buenos Aires with 12.6 million inhabitants) and three more among the world's 30 largest cities (Rio de Janeiro with 10.8 million, Lima with 7.5 million, and Bogota with 6.7 million). However, half of South America's urban population live in urban centers with fewer than 500,000 inhabitants, including a significant proportion living in urban centers with fewer than 50,000 people.

The slower population growth rates evident in most of the region's largest cities in the most recent intercensus periods for which data are available, and the more decentralized patterns of urban development evident in many nations, suggest that the trend toward increasing proportions of the population in very large cities may slow. The continent's three largest cities have growth rates that have slowed down. Each had more people moving out than moving in during the 1980s and the 1990s. Each had significantly smaller populations by the year 2000 than had been predicted 20 years earlier. For example, Rio de Janeiro had around 11 million in 2000, rather than the 19 million predicted by the United Nations when projections were made 20 years ago (United Nations, 1980).

One reason for the slower population growth rates among most cities was slow economic growth (or economic decline), so fewer people moved there. A second reason was declining fertility rates, which reduced natural increase. A third, particularly evident in Brazil (and, north of the continent in Mexico), was the capacity of cities outside the very large metropolitan areas to attract a significant proportion of new investment. Not all large cities grew slowly—for instance Bogota's population grew more rapidly than expected during the 1985–1993 census period.

20.1.2 The Environmental Impacts of Urbanization

All urban development imposes a new built environment on natural landscapes, resource flows, and ecosystems. Land is cleared, hillsides are often cut or bulldozed into new shapes, valleys and swamps are filled with rocks and waste materials, water is generally extracted from beneath the city, and regimes of soil and groundwater are modified in many ways (Douglas, 1983, 1989). This in turn brings changes in total runoff, alterations in peak-flow characteristics (which often result in serious flooding) and a decline

Table 20.2 Urban Change in South America

Country	Urban population (thousands)		Percentage of total population living in urban areas		Latest census on which these figures drew
	1950	2000	1950	2000	
Argentina	11206	33181	65.3	89.5	1991
Bolivia	919	5149	33.8	61.9	2001
Brazil	19406	139403	36.0	81.1	2000
Chile	3553	13084	58.4	85.9	2002
Colombia	5292	31553	42.1	74.9	1993
Ecuador	958	7489	28.3	60.3	2001
French Guiana	14	123	53.7	75.1	1999
Guyana	119	275	28.0	36.3	1980
Falklands/Malvinas	1	2	44.3	78.8	2001
Paraguay	514	3027	34.6	55.3	2002
Peru	3129	18885	41.0	72.8	1993
Suriname	101	315	46.9	74.1	1971
Uruguay	1744	3071	77.9	91.9	1996
Venezuela	2384	21103	46.8	86.9	1990
South America	49340	276661	43.7	79.7	

Source: United Nations, 2004. Note that some nations have had no recent census and for others, the data from their most recent census have not been incorporated into this dataset.

Table 20.3 Populations of the largest cities in South America, 1800–2000

Urban Center	Country	Population (thousands)									Year of latest census
		c 1800	c 1850	c 1900	1950	1960	1970	1980	1990	2000	
São Paulo	Brazil		22	240	2,313	3,969	7,620	12,089	14,776	17,099	2000
Buenos Aires	Argentina	43		813	5,041	6,771	8,417	9,920	11,180	12,583	1991
Rio de Janeiro	Brazil	43	266	967	2,930	4,373	6,637	8,583	9,595	10,803	2000
Lima	Peru	53	94	130	973	1,688	2,927	4,401	5,825	7,454	1993
Santa Fé de Bogota	Colombia	24	30	100	676	1,303	2,391	3,664	4,970	6,771	1993
Santiago	Chile	21	70	288	1,330	2,035	2,807	3,725	4,571	5,266	1992
Belo Horizonte	Brazil	0	8	13	407	782	1,485	2,441	3,548	4,659	2000
Porto Alegre	Brazil	4	44	74	483	881	1,398	2,133	2,934	3,505	2000
Recife	Brazil	25	74	217	655	1,073	1,638	2,122	2,690	3,230	2000
Caracas	Venezuela	31	43	98	676	1,282	2,053	2,575	2,867	3,153	1990
Salvador	Brazil	100	112	206	400	671	1,069	1,683	2,331	2,968	2000
Fortaleza	Brazil	0		89	261	500	867	1,488	2,226	2,875	2000
Medellin	Colombia	6	14	55	376	727	1,260	1,744	2,147	2,866	1993
Brasilia	Brazil				36	137	525	1,293	1,863	2,746	2000
Curitiba	Brazil	0		96	155	378	651	1,310	1,829	2,494	2000
Campinas	Brazil	7	31	68	150	293	540	1,109	1,693	2,264	2000
Cali	Colombia	6	12	31	231	474	851	1,232	1,591	2,233	1993
Guayaquil	Ecuador	14	28	120	258	456	719	1,120	1,572	2,077	2001
Maracaibo	Venezuela	22	22	29	260	459	697	964	1,351	1,901	1990
Valencia	Venezuela	7		28	108	193	390	673	1,129	1,893	1990
Belem	Brazil	12		97	240	378	601	827	1,214	1,749	2000
Barranquilla	Colombia		6	40	294	455	691	977	1,244	1,683	1993
Goiania	Brazil				52	148	375	737	1,132	1,609	2000
Baixada Santista (Santos)	Brazil	6	9	50	244	395	625	949	1,184	1,468	2000
Asuncion	Paraguay	7	10	52	223	309	452	671	928	1,457	2002
Cordoba	Argentina	11	29	66	416	588	787	977	1,188	1,444	1991
Grande Vitoria	Brazil			12	84	171	339	716	1,052	1,398	2000
La Paz	Bolivia	21	68	53	319	438	600	809	1,062	1,394	1992
Manaus	Brazil	0	29	65	89	151	281	604	955	1,392	2000
Quito	Ecuador	28	36	52	206	319	501	780	1,088	1,357	2001
Montevideo	Uruguay	14	34	268	1,140	1,155	1,170	1,213	1,274	1,324	1996
Rosario	Argentina	5	10	123	554	671	816	953	1,084	1,231	1991
Santa Cruz	Bolivia	6		16	42	84	166	324	616	1,061	2001
Maracay	Venezuela			4	89	168	326	560	766	1,015	1990
Sao Jose dos Campos	Brazil	0		0	58	112	216	412	633	972	2000
Mendoza	Argentina	5		62	248	335	474	601	758	955	1991
Maceio	Brazil				123	182	278	420	660	952	2000
Bucaramanga	Colombia				110	190	325	471	648	937	1993
Barquisimeto	Venezuela				127	236	384	583	742	923	1990
Natal	Brazil				107	176	288	467	692	909	2000
Grande Sao Luiz	Brazil	12		29	119	158	263	445	672	876	2000
Cartagena	Colombia				107	177	274	416	576	845	1992
João Pessoa	Brazil				117	184	289	453	652	827	2000
Campo Grande	Brazil				33	66	137	299	495	821	2000
Norte/Nordest e Catarinense	Brazil				64	116	205	378	603	815	2000
Ciudad Guyana	Venezuela				5	29	126	294	494	799	1990
Teresina	Brazil				54	104	195	390	614	789	2000
Cucuta	Colombia				70	122	206	333	520	772	1993
San Miguel de Tucuman	Argentina				221	295	364	494	611	754	1991

Source: United Nations (2004) for data for 1950–2000; data for 1800–1950 drawn from many sources, including early censuses and from Chandler (1987).

in water quality, as well as greatly affecting the processes of erosion and sedimentation. To these changes must be added the network of pipes and channels for water collection, treatment, transmission, regulation, and distribution—and the culverts, gutters, drains, pipes, sewers, and channels of urban waste-water disposal and stormwater drainage systems (Douglas, 1983).

Urban development transforms not only the areas that become urbanized (which cover a very small proportion of South America's total area) but also much larger areas, as can be seen in the changes in the rural landscape and ecology driven by productive activities that respond to urban-based demand for inputs (water and raw materials), food, and other goods and services. Very large demands are made on the regions around cities for building materials and landfill as a result of the construction of buildings, roads, car-parks, industries, and other components of the urban fabric. Many of the urban-generated wastes impact the surrounding region—for instance as solid wastes are transported there, or through water pollution or acid precipitation. Many of the environmental impacts of urban development are felt off-site, down-valley, downstream, or downwind (Douglas, 1986).

A considerable part of the expansion in South America's urban population over the last five decades has occurred without the needed expansion in the infrastructure and services essential for a healthy urban environment and for managing solid and liquid waste flows. Much of it has taken place without a planning and regulatory framework that limits environmental costs, guides urban expansion away from unsuitable sites, and protects important natural resources. Few urban governments have adequately met their multiple responsibilities, including those for environmental management.

20.1.3 Diversities and Commonalities

It is difficult to summarize environmental problems for the thousands of urban centers on the continent. These urban centers range in size from the mega-cities of São Paulo and Buenos Aires to small urban centers with only a few thousand inhabitants. Colombia alone, with less than 50 million inhabitants, has over 1,000 urban centers with great diversity between them as each is shaped by its topography, the local and regional site characteristics and ecology, economic base, and links with other regions. In terms of environmental quality, at one extreme are cities such as Porto Alegre in Brazil with virtually all its population served with piped water and regular garbage collection and most with good provision for sanitation (Menegat, 2002). Porto Alegre is also well known for developing "participatory budgeting," which has strengthened local democracy and given citizens a more direct involvement in setting municipal priorities (Abers, 1998, Souza, 2001). Average life expectancy had reached 74.4 years in 1996 (similar to that in high-income nations) and infant mortality rates were below 20 per 1,000 live births (Menegat, 1998, 2002). This has been one of the region's most rapidly growing cities over the last 50 years, so it demonstrates that rapid growth does not necessarily mean serious environmental problems. Manizales in Colombia is another example of a well-managed urban environment, which shows that this is also possible in smaller, less wealthy cities (Velasquez, 1998).

At the other extreme are thousands of urban centers or urban districts within larger cities where only a small proportion of the population has piped water and adequate provision for sanitation, drainage, and solid waste management, and where the local authorities have little capacity to address environmental problems. Here it is common for a high proportion of the population to live in poor-quality housing—for instance whole households living in one or two small rooms in tenements, cheap boarding houses, or shelters built on illegally occupied or subdivided land. Here environment-related diseases and injuries cause or contribute much to disablement and premature deaths; they are often the leading cause of death and illness. Average life expectancies can be 20–30 years less than in Porto Alegre, while infant mortality rates can be five or more times higher. Large variations between well-managed and poorly managed urban settlements exist not only between countries but also within them. For instance, some major Brazilian cities have life expectancies 20 years lower than Porto Alegre (Mueller, 1995), while within cities, significant variations exist between the best and the worst-served districts (Arrossi, 1996; Hardoy et al, 2001).

Within this diversity, there are obvious characteristics that all urban centers share. They all combine concentrations of human populations (and their homes and neighborhoods) and a range of economic activities. They all have local and regional environmental impacts related to their roles as centers of production for goods and services and as centers of consumption for their inhabitants. The extent of the environmental problems within their boundaries is much influenced by the quality and capacity of their governments. All urban centers require some form of government to ensure an environment of adequate quality for their inhabitants and to ensure that needed infrastructure and services are in place (piped water supplies, provision for sanitation and drainage, waste collection and management, roads and paths, provision for electricity, schools, and health care). Ensuring good-quality environments becomes increasingly complex the larger the population and the greater the scale and range of the population's daily movements and the more industrial the production base. Local governments are also needed to manage the draw that city dwellers and enterprises make on natural resources and on natural sinks for their wastes—for instance to regulate land use, protect watersheds, and set limits on the generation of wastes and pollution and their disposal.

In all cities, environmental management is an intensely

political task, as different interests (including powerful interests) compete for the most advantageous locations, for the ownership or use of resources and waste sinks, and for publicly provided infrastructure and services. In the absence of good environmental management, many such interests contribute to the destruction or degradation of key resources. The political nature of urban management is widely recognized, as can be seen in the shift in the literature on urban development from *government* (where the concentration is on the role, responsibilities, and performance of government bodies) to *governance,* which also encompasses the relationship between government and civil society (McCarney, 1996; UNCHS, 1996).

20.2 The Urban Environment

Here, four environmental issues for city populations will be discussed in more detail: (1) the quality of the residential environment, with a particular focus on provision for water, sanitation, and solid waste collection, (2) the quality of the work environment, (3) the problem of hazardous urban sites and risks from disasters, and (4) air pollution. There are many other environmental problems but there is insufficient space to provide details. For instance, traffic accidents rank among the main causes of premature death and injury in many cities (one estimate for Latin America as a whole suggested that each year there are 100,000 traffic fatalities and 1.2 million serious injuries; della Porta, 2001). There are serious problems of noise pollution in many cities [see Rivas Roche (2000) for Buenos Aires and Medina and Uribe (1994) for Colombian cities], as well as high levels of violence, crime, and other psychosocial problems to which environmental conditions contribute (Blue, 1996).

20.2.1 Residential Environments

Virtually all South American cities include residential neighborhoods with very good quality environments. Around half the urban population live in houses or apartments with individual connections to piped water and sewers (WHO/UNICEF, 2000). The houses in such neighborhoods generally have two or more rooms per inhabitant, electricity, and regular house-to-house services to collect household waste. Virtually all cities also have many neighborhoods with very poor quality environments—for instance small shacks mostly fabricated with temporary materials with little or no provision for piped water, sanitation, drainage, or garbage collection. It is common for between 20 and 50% of the city population to live in such neighborhoods. The proportion is generally higher, the smaller the city and the lower the nation's per capita income—although there are exceptions, linked mostly to places with more competent and accountable local governments. If these residential neighborhoods have electricity and piped water, it is often through illegal connections. Much such housing is on land that is illegally occupied (squatter settlements) or in illegal subdivisions (where the inhabitants purchased the land from its legal owner but the land is used for housing without official permission, and without meeting official standards for lot sizes and infrastructure). Many major cities also have large inner city districts with overcrowded tenements and boarding houses. Here, there is generally some provision for urban infrastructure and services, but the amenities are poorly maintained and insufficient for the number of people, for example water taps and toilets have to be shared between many households (de los Rios Bernardini, 1997; Harms, 1997).

For provision for water and sanitation in cities in the region, the proportion of the population adequately served varies from more than 98% to under 30% (UN Habitat, 2003). For instance, in the wealthier cities in the south and southeast of Brazil, more than 90% of the population in 2000 were served by piped water supplies; over 95% in São Paulo and Porto Alegre were served (UN Habitat, 2003; Menegat, 1998). Provision for sanitation was less extensive, but by 2000, major cities in the south and southeast generally had more than three quarters of their population in housing with connections to sewer systems. Provision for water and sanitation was much less common in the poorer cities of north and northeast Brazil. For instance, in 2000, half of Belem's population and a quarter of Fortaleza's population were in accommodations that lacked piped water supplies, and most of the population of Belem, Recife, and Fortaleza were in accommodations that lacked connection to the city sewer system (UN Habitat, 2003). The level of provision in smaller urban centers is less well documented but in most such centers is likely to be much lower than in larger cities. For instance, a study of some Amazonian frontier towns showed that most of the population had inadequate provision for water and sanitation (Browder and Godfrey, 1997). However, in wealthier nations such as Chile or wealthier regions within nations such as Brazil, a high proportion of the population in many secondary cities do have good provision for water and sanitation (UN Habitat, 2003). In addition, the poorer and more peripheral municipalities in wealthy cities can still have the majority of their inhabitants lacking good provision, as in Moreno in Buenos Aires (Hardoy et al., 2005).

A combination of concentrated populations, high densities, and inadequate sanitation is particularly dangerous. Many diseases are associated with inadequate provision for water and sanitation, including not only those related to human contact with excreta, contaminated water, or insufficient water for personal hygiene but also those such as malaria and dengue fever where the disease is transmitted by insect vectors that breed in water. Inadequate provision for water, sanitation, and drainage causes or contributes to large health burdens, including high levels of infant and child mortality (Cairncross and Feachem, 1993). A study in Betim, Brazil, highlighted that it is not only the availability of water and sanitation infrastructure that influences health but also water quality, per capita consumption, regularity of supply, extent of indoor plumbing, and provision for drainage (Heller, 1999). The quantity of water available and the price paid for it can be as important to a family's health as its quality (Cairncross, 1990). Large sections of South America's urban population have to rely on water vendors for supplies (or to meet their water needs for drinking and cooking, as they use cheaper, poorer quality water for other tasks). In Guayaquil, Ecuador's largest city, 35% of the population of 1.6 million dwellers do not have access to adequate, and reliable water supplies and the whole city suffers from water shortages (Swyngedouw, 1995). Approximately 400 tankers service 35% of the population; these water merchants buy the water at a highly subsidized price and can charge up to 400 times the price per liter paid by consumers who receive water from the public water utility. The problem in Guayaquil is not a lack of water but a lack of the institutional capacity to install and maintain a good quality system.

There are also large differences between cities in the proportion of their inhabitants with regular solid-waste collection services. About 70% of the population in cities with one million plus inhabitants have waste collection services; in smaller cities, this coverage is estimated to range between 50 and 70% but can be under 30% (Sabaté, 1999). It is normally middle and higher income areas that enjoy regular service whereas low-income neighborhoods have erratic or no services. Many informal settlements are on land sites to which access by conventional garbage trucks is difficult or impossible. Most local authorities have made little attempt to develop garbage collection services more suited to such sites (e.g., Zevallos Moreno, 1996). In many capital cities, including Caracas, Lima, and Asuncion, waste collection coverage remains below 40% for low-income areas (Arroyo Moreno et al., 1999). By contrast, in some cities, virtually all the population have a regular collection service—for instance in Santiago (Dockemdorf et al., 2000) and Porto Alegre (Menegat, 2002).

Without a collection service, households generally discard their wastes on the nearest available empty site or in nearby ditches or along streets. Problems include odors, disease vectors, pests attracted by garbage (including rats, mosquitoes, and flies), and overflowing, waste-clogged drainage channels. Since provision for sanitation is also often deficient, many households dispose of toilet wastes into drains. When drains overflow, excreta contaminates the site. Uncollected garbage is a serious health hazard—especially for children playing in and around their homes (e.g., Suarez, 1999). Flies and cockroaches feeding on garbage can subsequently contaminate food (Cointreau, 1982). Leachate from decomposing garbage can also contaminate local water sources (UNCHS, 1988).

Few cities have efficient environmental management of their solid wastes in terms of promoting waste reduction, ensuring capture from waste streams of potentially recyclable materials, and managing solid-waste dumps. Most cities also have problems with illegal dumps. In Buenos Aires, for example, there are over 100 illegal open-air dumps, despite the existence of four sanitary landfill operations (Sabate, 1999).

A cluster of environmental factors influences the scale and nature of other environmental hazards in homes and residential areas, including level of overcrowding (which affects transmission of infectious diseases and frequency of household injury) and the nature of the equipment and fuel used for heating and cooking (which affects indoor air quality). For instance, overcrowding and poor ventilation can increase the transmission of tuberculosis (TB). South American cities, like cities all over the world, are facing an increase in the number of TB cases, many of them associated with human immunodeficiency virus (HIV) infection. High levels of overcrowding and poor-quality housing increase the risk of food contamination during food preparation or storage (McGranahan, 1991; Birley and Lock, 1998). Reliance on kerosene stoves and candles or kerosene lamps for light, along with overcrowding and the widespread use of flammable materials in house construction, produce high levels of risk for burns, scalds, and accidental fires (Hardoy et al., 2001).

20.2.2 The Workplace

Large sections of South America's working population face dangerous concentrations of toxic chemicals and dust, inadequate lighting, ventilation and space, and lack of protection from noise and machinery. Such hazards are evident in workplaces from large factories and commercial institutions down to small backstreet workshops and where people work from home. In 1985, a review of occupational health issues in Latin America commented: "Few health standards are applied to limit work-place exposures; in most of the region's countries, the standard-setting process is either just beginning or has not yet begun. In those nations where standards regulating work practices or toxic exposure do exist, the standards are often not enforced, either for political or economic reasons or because of a lack of trained inspectors" (Michaels et al., 1985, p. 538).

There has probably been some improvement in formal, registered workplaces in most nations more recently, in part as a result of the return to democratic rule (WHO, 1992). Rather less progress has occurred within informal enterprises (including home workers), since it is more difficult for government agencies to ensure that workplace standards are met. In addition, the proportion of the workforce working in informal, unregistered workplaces has increased in many cities (UNCHS, 1996).

20.2.3 Hazardous Urban Sites and Natural and Human-induced Disasters

Millions of urban households are located on hazardous land sites. Many of the settlements most at risk are easily visible—for instance, clusters of illegal housing on steep hillsides prone to landslides in Rio de Janeiro (Brazil), La Paz (Bolivia), and Caracas (Venezuela); housing on sandy deserts or hills, as in Lima (Perú); or settlements on land prone to flooding or tidal inundation, as in Guayaquil (Ecuador), Recife (Brazil), and Buenos Aires and Resistencia (Argentina).

In most cities, there is a complex interaction between natural hazards and human actions. Human actions can often greatly lessen or eliminate environmental hazards but they may also act to make them more frequent or increase the scale and severity of risk (Douglas, 1983). For instance, in Caracas, where close to 600,000 persons live on slopes with a high risk of landslide, most slope failures were associated with earthquakes until the 1960s (Jimenez Diaz, 1992). From the 1970s onward, they have been increasingly associated with rains and with areas where low-income settlements had been developed. The changes introduced to the slopes through development for housing increased the likelihood that rainfall could trigger slope failure. Many of 30,000 people who died and the 600,000 others who were severely affected by the catastrophic floods in December 1999 lived in low-income, urban households located on unstable hillslopes. Hundreds of low-income urban dwellers also lost their lives and thousands were made homeless in mudslides in Medellin in 1987, Rio de Janeiro in 1988, and Caracas in 1989 (Hardoy et al., 2001). In each of these disasters, large numbers of schools, health centers, roads, bridges, and other structures were destroyed or severely damaged, and many citizens lost their livelihoods. Smaller disasters in which fewer people are killed are much more common and are often seen as routine events, but few studies have evaluated their economic impact (Lavell, 1994).

Most people living on hazardous sites have low incomes. Rarely do they live there in ignorance of the dangers. Such sites are chosen because homes can be constructed with less fear of eviction, as the hazards make the sites unattractive for commercial development; and hazardous sites are often the only locations close to work opportunities that are available to low-income groups. To the hazards inherent in the site are added those linked to a lack of investment in infrastructure and services. On steep hillsides, the introduction of water supplies with no provision for drainage can add significantly to the risk of slope failure. There may be a lack of knowledge among the settlers as to how to reduce risks, for instance by minimizing the amount of vegetation cleared from a slope as it is developed for housing to reduce the risk of landslides and mudflows (Greenway, 1987). Or the settlers may have the knowledge but not the collective organization to permit its effective use. For instance, those who have settled on a slope may be powerless to prevent new housing developments or a new road at the toe of the slope that puts the whole hillside at risk (Douglas, 1986). It is also common for the risks to informal settlements to be increased by urban developments elsewhere—for instance the intensity of flooding may be increased for low-income settlements in a flood plain by flood embankments to protect middle- or upper-income residential areas and commercial developments upstream.

Much urban development has expanded onto sites at risk from disasters. Many cities are in locations where the original site was relatively free from natural hazards but urban growth brought expansion over unsafe sites. For instance, for most of Caracas's history, the population was easily accommodated in the valley surrounded by hills; only in recent decades, with the large growth in population and commercial land markets, were poorer groups pushed in increasing numbers into settlements on steep hillsides (Jimenez Diaz, 1992). In Quito, the population has expanded more than fourfold since 1965. Much of the growth has been accommodated on mountain slopes, producing soil erosion and landslides and causing floods linked to sediment and other wastes that has blocked drainage or reduced river flow (Zevallos Moreno, 1996). And many cities located beside rivers, lakes, or estuaries, although formerly confined to sites with little risk of flooding, have expanded onto unsafe areas. One reason that city populations have been allowed to increase on hazardous sites is because it is rarely the rich and powerful groups who are at risk.

Another factors is that an increasing proportion of South America's urban population is living in regions or zones where the ecological underpinnings of urban development are more fragile. A considerable proportion of the rapid urbanization in recent decades has taken place in areas that were sparsely settled in the 1970s. Many of the most rapidly growing urban centers in the 1980s and 1990s are those that have developed as administrative and service centers in areas of agricultural colonization or mining or logging in what were previously uninhabited or sparsely populated areas. For instance, only since the 1970s has urban development spread to the hot and humid regions in the interior of Brazil, Bolivia, Paraguay, and Venezuela, and into Patagonia. Among the reasons that most of these regions had remained sparsely populated were the greater environmental hazards or presence of soils less suited to sustained commercial exploitation (di Pace et al., 1992).

Certain disease epidemics may be associated with natural disasters—especially floods, which can contaminate all available water supplies and be associated with epidemics of dysentery or other water-borne diseases, as well as diseases associated with inadequate water supplies. Flooding can trigger cholera-harboring marine plankton blooms by providing extra nutrients to the coastal environment (Scott et al., 2001). Outbreaks of leptospirosis (usually caused by drinking water infected by rat urine) have been associated with floods in Rio de Janeiro and São Paulo. People living in poor-quality, overcrowded settlements at risk of flooding with inadequate provision for garbage collection (or living close to garbage dumps) are particularly at risk (WHO, 1992).

Many urban sites have become hazardous because of dumping of toxic or otherwise hazardous wastes. In most urban centers in Latin America, few measures have been taken to stop industries and other generators of such wastes from disposing of them without treatment (Hardoy et al., 2001). A broad range of industries produce hazardous wastes, and most cities with such industries have land sites and water bodies contaminated as a result. Most industrial centers have contaminated industrial sites, including those that have been abandoned. Most cities lack the specialist facilities needed to process safely or store hazardous wastes. Many cities have no officially licensed treatment plants, so hazardous wastes have long been disposed of in conventional sanitary fills or open dumps used for domestic wastes, or disposed into water bodies or onto illegal land sites, usually without treatment to render them less hazardous, or kept in temporary stores within industrial sites. Even in countries where there has been progress, as in Argentina over the last fifteen years with new national and provincial laws on hazardous wastes and new facilities for such wastes, there are still serious problems with the many illegal dumps that have hazardous wastes and with industrial firms who still dispose of such wastes illegally. In Buenos Aires, as in nearly all other cities, there are difficulties of getting sites for facilities to process or store hazardous waste (no neighbourhood wants this facility in their area) and on reaching agreement as to who should pay for the facilities. For industries accustomed to disposing of hazardous wastes illegally, meeting good practice in disposal or storage can seem very costly. Decontaminating industrial sites and dealing with illegal dumps are also expensive. Addressing such problems is often particularly difficult when economies are stagnant, industries that have contributed most to the problem are unable to contribute to costs (or have the power to avoid doing so), and governments lack the funds to act.

20.2.4 Air Pollution

In many major cities, the concentrations and mixes of air pollutants are already high enough to cause illness in more susceptible individuals and premature death among the elderly, especially those with respiratory problems (WHO, 1992). One estimate for all of Latin America suggests that over 2 million children suffer from chronic coughs as a result of urban air pollution and that air pollution means an excess of 24,300 deaths a year. This same source estimated that some 65 million person-days of workers' activities were lost to respiratory-related problems caused by air pollution. While these are rough estimates, they give an idea of the size of the problem (Romieu et al., 1990).

The extent of air pollution, the relative importance of different pollutants, and the relative contribution of different sources vary considerably from city to city and often from season to season within cities. Citywide air pollution problems are generally most serious in industrial cities with inadequate pollution control, in cities where solid fuels are widely used for cooking and heating in homes and small-scale industries, and in large cities with a high concentration of motor vehicles where local conditions inhibit the dispersal of air pollutants. More localized problems with air pollution occur in and around particular industries, roads, or "hot spots" with particular combinations of emissions and weather conditions. In many city centers, road transport has become the dominant source of many of the pollutants that affect the health of urban residents (Elsom, 1996). One difficulty for the larger cities is that wealthier groups move out to suburbs (in part to avoid higher air pollution levels), thereby requiring increased reliance on private automobile use, which then exacerbates air pollution levels in central areas (Escudero, 1994). Not surprising, high levels of air pollution in many of the region's larger cities are associated with high levels of private automobile ownership and use, older motor vehicles (with less provision for cutting polluting emissions), and serious problems of road traffic congestion.

In many cities, high levels of air pollution are more associated with industrial activity than with motor vehicles. For instance, in the Peruvian city of Ilo, the very serious air pollution problems are largely the result of emissions from one of the world's largest copper foundries close-by. In May 1995, peak levels for sulphur dioxide were 30 times the World Health Organization's 10 minute guideline. Although emissions have since been reduced, largely as a result of sustained community pressure, the foundry still emits 1,400 metric tonnes of sulphur dioxide into the atmosphere each day, as well as suspended particulates that include heavy metals (Follegatti, 1999; Boon et al., 2001).

The extent to which air pollution constitutes a threat to human health depends largely on the city site and weather conditions. For instance, the location of Buenos Aires on the coast of a flat plain means that the prevailing wind may disperse pollutants. But a combination of topography and particular climate and weather conditions may trap pollutants in or over a city, for example in Santiago de Chile, where weather conditions and surrounding mountains help trap pollutants for much of the year (Durán de la

Fuente, 1999). In Medellin and Cali, two of Colombia's largest cities, problems of air pollution are exacerbated by limited natural air circulation (Medina and Uribe, 1994).

In some cities, air-pollution levels can be sufficiently high to cause demonstrable health impairment for the population in general or for particular groups. In the industrial city of Cubatão (Brazil) during the 1980s, pollution levels were linked to reduced lung functions in children (Hofmaier, 1991). Air-pollution levels there were particularly high because of the concentration of industry and because the government at that time made little attempt to address the problem. Under the military government of the time, there had also been restrictions on all press reports on Cubatão, up to 1978. Only with the removal of media censorship and the return to free direct elections in Brazil did the scale and nature of the pollution in Cubatão (and in many other cities in Brazil) become widely known and action taken to address it (Lemos, 1998). In South America, increasing numbers of industrial firms reduced emissions during the 1990s, although progress was uneven within most sectors and there was less progress among smaller firms and more traditional industries (Jenkins, 2000).

20.3 Environmental Impacts on the Urban Periphery

This section concentrates on two environmental impacts that cities have on their periphery: water pollution and ecological changes. It is difficult to draw a clear distinction between environmental impacts that occur within and outside urban areas. First, the urban boundary is uncertain. Official urban boundaries rarely match the exact extent of the built-up area; they may be too small (because urban development has spilled over boundaries set many years ago) or too large (because urban boundaries have been defined that encompass large areas of agriculture, forest, and water). Second, there are many material flows across them. Rivers that become polluted flowing through a city may still be highly polluted many kilometers beyond its boundaries. City-generated solid wastes are often disposed of on land sites outside city boundaries. High levels of urban air pollution may also cause acid precipitation or ozone plumes that affect vegetation in parts of the surrounding area. Acid precipitation is causing concern in many such areas in South America, but its extent and impact on vegetation, forests, and water bodies is not well known (McCormick, 1997). Mining centers often cause serious problems of soil contamination and air pollution. For instance, there is likely to be considerable damage to agricultural production in the areas around the copper foundry in Ilo, so much so that the company that sought to avoid cutting its pollution and taking responsibility for the damage it causes has, nevertheless, been forced to pay some compensation to farmers for crop losses (Boon et al., 2001).

In addition, urban systems in the wealthier areas have become more spread out, supported by more extensive highway systems, better communications, and industrial structures that encourage more decentralized production systems. An aerial view of most major cities is no longer one of a concentrated urban center surrounded by countryside but of a complex spatial pattern of urbanized and nonurbanized areas with the former often stretching along major transport corridors for long distances beyond city boundaries. In addition, it is common for residential communities and industrial and commercial concentrations to develop close to major cities but separated from the center.

2.3.1 Water Pollution

Problems with water are often both a city problem (the limited or poor quality of supplies within the city) and a regional problem (water drawn from the region around a city affects its ecology or productive activities, or liquid wastes and storm and surface runoff from the city damage water bodies outside the city). This section discusses water pollution and water scarcity and highlights their interlinkages.

Most urban centers have serious problems with water pollution from sewage—in part because of inadequate sanitation, in part because of little or no treatment for the sewage from households that are served by sewers. Many cities also have serious water-pollution problems because of lax pollution control. For instance, in Buenos Aires, the Reconquista, Lujan, and Matanza rivers have high levels of pollution, arising largely from waste waters, industrial effluents, and leakage from solid waste dumps (di Pace and Crojethovich, 1999). In Bogotá, the rivers Bogotá and Cauca are extremely polluted, as are many ground water sources (Medina and Uribe, 1994). Water pollution not only damages water bodies (and aquatic life within them, including fisheries) but also contaminates fresh water sources, causing health problems for those who subsequently use them. Although no urban center can exist without reliable sources of fresh water, few urban authorities have paid sufficient attention to safeguarding these sources and preventing water pollution. The cheapest and most convenient freshwater sources have often been overused or contaminated. In many coastal cities, for instance, local aquifers have been overpumped, resulting in saltwater intrusion. This usually implies much higher costs if water supplies are to be improved, because more expensive, distant sources need to be tapped (Bartone et al., 1994).

Surface-water sources on which cities draw are often of poor quality. For instance, such sources may be saline because of return water from irrigation, contaminated with agricultural chemical and human and livestock wastes, or heavily polluted by industries or other users upstream (Prestes and Booth , 1996). Ground water resources are also often contaminated, especially if industries have been disposing of their wastes down deep wells.

The contamination of rivers, lakes, seashores, and coastal waters is an example not only of the impact of city-generated wastes on the wider region but also of governments' negligence in protecting open areas. Untreated waste water and sewage from households and commercial and industrial enterprises are usually discharged into nearby rivers or lakes. In cities on or close to coasts, untreated sewage and industrial effluents often flow into the sea. Many coastal areas have serious problems with dirty, contaminated beaches and water that are major health risks to bathers. It is usually the most accessible beaches that are most polluted. In many cities, these are among recreational areas most widely used by lower income groups. Richer households suffer much less; those with automobiles can reach more distant, less polluted beaches.

Most urban centers have serious nonpoint sources of water pollution because large sections of their population are not served by sewers, drains, or solid waste collection services. This means that much of the liquid waste from households, businesses, and often industries, and a considerable proportion of the solid waste, end up in the nearest streams, rivers, or lakes. Sediment washed from construction sites or other bare ground is often a major source of water pollution.

Many cities also face problems with water shortages, which add to the problem of disposing of liquid wastes from industries and sewage. Large volumes of water dilute wastes and can render them much less dangerous; in addition, bacteria and other micro-organisms in the water can break down organic wastes, if the biochemical oxygen demand of these wastes is not too high relative to the water's dissolved oxygen. Problems of water scarcity are particularly acute for the many urban centers in relatively arid areas. Many of the coastal cities in Perú, including Lima, and La Rioja and Catamarca in Argentina, are among the many cities with severe constraints on expanding freshwater supplies.

Liquid wastes from city activities have environmental impacts stretching beyond the immediate hinterland. It is common for fisheries to be damaged or destroyed by liquid effluents from city-based industries (e.g., Kreimer et al., 1993 for Rio de Janeiro). Heavy metals from mining operations, metal smelting, or other industrial processes pose serious problems to river and coastal environments in some areas (Hardoy et al., 2001.; Hinrichsen, 1998). In many coastal areas, there are also problems with oil pollution from illegal discharges by tankers and other ships, offshore oilrigs, and the many oil refineries near the shore (Hinrichsen, 1998).

20.3.2 Ecological Changes on the Urban Periphery

Figure 20.1 illustrates changes that generally occur on the urban periphery; this figure is drawn from a study of Buenos Aires, but the kinds of changes it illustrates are

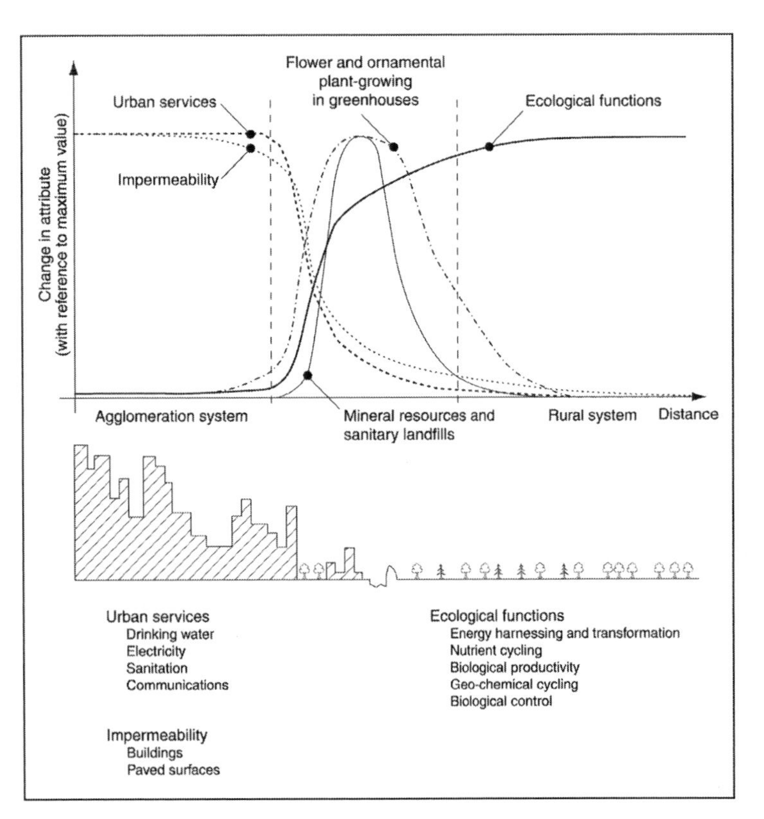

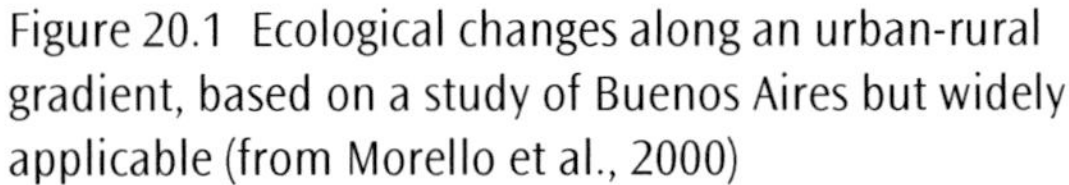

Figure 20.1 Ecological changes along an urban-rural gradient, based on a study of Buenos Aires but widely applicable (from Morello et al., 2000)

present in the surrounds of most cities in South America. In Buenos Aires, the changes included:

- land parcelling in city-block-sized lots;
- earth removal for building embankments, brickworks, and landfills for sites prone to flooding in order to raise the elevation;
- extraction of sections of turf;
- consumption of soil for plant nurseries;
- legal and illegal disposal of household refuse and industrial waste;
- urban wastelands;
- spontaneous settlements (including squatter settlements);
- country clubs, *barrios cerados* (closed communities), rustic clubs (*clubes campestres*);
- industries;
- heavy infrastructure equipment;
- storage or dumping grounds;
- water purification facilities;
- automobile salvage and junkyards;
- clandestine pig farms and farm abattoirs, greenhouse and aviary complexes;
- road construction yards; and
- garden cemeteries and neo-ecosystems—that is, semi-natural landscapes where the dominant or most frequently occurring species are not native but accompanying and subordinated species.

The expansion of the built-up area, including the construction of roads, water reservoirs, and drains, together with land clearance and deforestation, often changes the local ecology in ways that favor the emergence or multiplication of particular disease vectors (WHO, 1992). Natural foci for disease vectors may become entrapped within the suburban extension, and new ecological niches for the animal reservoirs may be created. Within urban conurbations, disease vectors may adapt to new habitats and introduce new infections to spread among the urban population. For instance, *Aedes aegypti,* the mosquito vector of dengue and urban yellow fever, proliferates in tropical urban settlements and frequently breeds in polluted water sources such as soak-away pits, septic tanks, and other breeding sites that have a high organic matter content.

The above discussion has outlined some of the environmental costs that city-based production and consumption impose on surrounding regions. However, a focus only on such problems can obscure the fact that many rural-urban linkages are positive. For instance, demand for rural produce from city-based enterprises and households often supports prosperous farms and rural settlements, where environmental capital is not being depleted—for instance as local producers invest in maintaining the quality of soils and water resources. Increasing agricultural production can support rising prosperity for rural populations and rapid urban development within or close to the main farming areas—the two supporting each other. Some of the most rapidly growing urban centers in South America in recent years have been those within or close to agricultural areas growing high-value-added crops such as fruit (including grapes for wine), tea, or coffee (e.g. Manzanal and Vapnarsky, 1986, Satterthwaite and Tacoli, 2003). There are many rural households for whom urban incomes are important components of total incomes. This often includes large numbers of low-income households who live in rural areas around cities and are considered rural inhabitants but who have one or more member working in the city or commuting there.

20.4 Environmental Impacts beyond the City-Region

Although much of the literature on the environmental impact of cities focuses on impacts in the peri-urban region, many of the demands that larger and wealthier cities concentrate for food, fuel, and raw materials are met by imports from more distant areas. If many consumers in (say) São Paulo, Buenos Aires, or Caracas are drawing their fruit, vegetables, cereals, meat, fish, and flowers from far beyond their hinterland, it is more difficult to establish the links between this consumption and the ecological consequences. In addition, in wealthy cities, many of the goods whose fabrication has large ecological impacts are imported—for instance goods produced with high levels of fossil-fuel use, water use, and other natural resource use, and from dirty industrial processes (including the generation of hazardous wastes) and hazardous conditions for the workforce. Wealthy cities do not have to live with many of the ecological consequences of the consumption they concentrate in their surrounds.

Other environmental cost transfers are into the future. Emissions of carbon dioxide (the main greenhouse gas) rise with economic growth. This is transferring costs to the future through the human and ecological costs of atmospheric warming. The generation of hazardous wastes, or wastes whose rising concentrations within the biosphere have worrying ecological implications (often referred to as persistent organic pollutants), is also transferring costs to the future. So, too, are current levels of urban consumption for the products of agriculture and forestry where the soils and forests are being destroyed or degraded, and biodiversity reduced. The work of William E. Rees on the "ecological footprint" of cities (and of particular households, enterprises, and nations) has made evident the large land area on whose production the inhabitants and businesses of any city depend for food, other renewable resources, and the absorption of carbon to compensate for the carbon dioxide emitted from fossil-fuel use (Rees; 1992, Wackernagel and Rees, 1995). Wackernagel (1998) demonstrates the scale of such an ecological footprint for Santiago de Chile.

In regard to global warming, the main contributor of city-based production and consumption to greenhouse gases is carbon dioxide emission from fossil-fuel combustion by industry, thermal power stations, motor vehicles, and domestic and commercial energy uses. City-based activities also contribute to emissions of other greenhouse gases—for instance of chlorofluorocarbons by certain industries or industrial products, of nitrous oxide by motor vehicle engines, and of methane by city-generated wastes. It is generally the wealthiest cities that have the highest levels of greenhouse-gas emissions per person. Certain cities with high concentrations of heavy industry or of thermal power stations generally have above-average levels per person. However, there are few figures on average greenhouse-gas emissions per person for cities, although in general, the average for urban areas in South America is likely to be much lower than those for North America or Europe.

20.5 Impacts of Global Warming on Cities

There is still uncertainty about the possible scale of global warming in the future and its direct and indirect effects. But there is a strong scientific consensus that if no action is taken to moderate the emission of greenhouse gases from human activities, it is likely to bring very serious problems. These problems can arise from the direct effects of rising temperatures (including heat waves) and the changes in weather that these bring, and the indirect effects induced by the changes in temperature and weather on (among other things) sea level, ecological systems, and air pollution.

With respect to cities, the most likely effects of climate warming include higher mean temperatures, rising sea levels, and changes in weather patterns. Changes in precipitation and temperature, including increased melting of snow and ice in the Andes, will also bring changes in river flow, leading to increased risks of flooding in many cities or particular city-districts. The effects of climate change will also involve changes in evapotranspiration rates and in the structure of ecosystems—for instance as a result of changes in plant growth rates and favored species, and in insect populations. There will also be changes in the frequency and severity of extreme weather conditions such as storms and sea surges (Scott, 1994).

The 2001 assessment of the Inter-Governmental Panel on Climate Change noted four types of urban settlement that were particularly vulnerable—those with coastal, river, or hillside locations; those in arid areas; those with particularly large population concentrations; and those whose economy depended on a resource base that was at risk from climate change (such as agriculture, forestry, fishing, or tourism) (Scott et al., 2001).

Sea-level rise will be most disruptive to low-lying coastal and estuarine settlements, and this is where a considerable proportion of the region's population lives. For historic reasons linked to South America's colonial past, most of the largest cities are ports. The high cost of land in the central areas of cities has often encouraged major commercial developments on land reclaimed from the sea or estuary, and these will be particularly vulnerable to storms and rising sea levels. So, too, will the many industries and thermal power stations concentrated at the coast because of their need for cooling water or because the sea becomes a convenient sink for their waste (Parry, 1992). Sea-level rise will also raise ground-water levels in coastal areas, which will in turn threaten ground-water resources, disrupt existing sewerage and drainage systems, and may undermine buildings. Coastal settlements are also most at risk from any increase in the severity and frequency of floods and storms induced by global warming. The combination of high winds, intense rainfall, higher sea levels, and storm-induced tidal surges will be particularly problematic (Smit, 1990; Turner et al., 1990; Scott et al., 2001). Coastal cities whose economies benefit from tourism may have considerable difficulty protecting tourist attractions such as beaches and nearby wetlands (Turner et al., 1990). There are already serious problems with coastal erosion in many regions—for instance in and close to Recife and Olinda (Prestes and Booth, 1996). Conversely, many cities will face shortfalls of freshwater because of reduced precipitation levels and the difficulties in protecting groundwater resources from contamination by seawater.

Changes in the frequency and magnitude of precipitation will also increase flooding, landslides, and mudflows in many places (di Pace et al., 1992; Scott et al., 2001). Many inland cities and smaller urban centers are alongside rivers and already face serious problems of flooding as a result of inadequate investment in drainage systems, flood control, and watershed management. Thousands of people in urban centers in the foothills of the Andes will face increased risks of flooding—including such major cities as Lima, Santiago, Quito, and Bogotá (di Pace et al., 1992). The floods in Venezuela in December 1999 demonstrated people's vulnerability to extreme weather events. There are also hazards related to possible increased frequency of El Niño phenomena, such as the deaths, injuries, and large economic losses that occurred during the 1997–1998 event (see chapter 19). Changes in rainfall regimes will reduce fresh water availability in many regions, which will pose particular problems for cities and smaller urban centers already facing serious water shortages.

Increased temperatures in cities may aggravate air-pollution problems, for instance by increasing concentrations of ground-level ozone. Warmer average temperatures permit an expansion in the area in which tropical diseases can occur by extending the range of the mosquito types that are the vectors for malaria, dengue fever, and filariasis (WHO, 1992). Rodent populations are likely to increase and with them increased risk of the many infectious diseases in which they are involved (McMichael et al., 1996). Climate-change induced reductions in water supply or increased floods (which often contaminate water supplies) could increase the incidence of waterborne and some other water-related diseases. Precise predictions are difficult because climate change will bring changes to so many factors that influence disease-causing agents and vectors—including temperature, humidity, precipitation, flora, fauna, predators, and competitors.

Global warming will also increase human exposure to exceptional heat waves. The elderly, the very young, and those with incapacitating diseases are most at risk (WHO, 1992). Low-income groups will generally be more vulnerable because their housing is less suited to moderating extreme temperatures and they lack air conditioning (McMichael et al., 1996). Those living in urban *heat islands*, where temperatures remain significantly above those of the surrounding regions, will be particularly at risk. High relative humidity will amplify heat stress (WHO, 1992).

One other aspect of global warming is difficult to predict—the capacity and readiness of societies to respond to the changes that warming and its associated effects will bring. There is the vast stock of buildings, roads, public transport systems, and basic urban infrastructure that has been built without making allowance for future global warming (Scott et al., 2001). Modifying or replacing these will be expensive and time consuming—and particularly difficult in low-income countries with weak and under-resourced city authorities, and where there are still large deficits in basic infrastructure.

20.6 Conclusions

20.6.1 Responses to Environmental Problems

The scale and nature of environmental problems in and around urban areas in South America often reflect the limitations of their governments. More attention is needed to ensure action in four areas: (1) enforcement of appropriate legislation, including that related to environmental health, occupational health, and pollution control; (2) adequate provision for water supply to, and solid and liquid waste collection from, all homes and neighbourhoods; (3) adequate provision of health care that not only treats environment-related illnesses but also implements preventive measures to limit their incidence and severity; and (4) integrated disaster prevention and preparedness in the form of specific plans and investment schedules. Most city governments have also failed to implement land-use policies that ensure sufficient land is available for housing developments for low-income groups and for the avoidance of hazardous sites.

As if good environmental management within city boundaries was not complex enough, "good urban governance" must extend to environmental management in the regions around the city and contribute to global goals such as minimizing greenhouse gas emissions and protecting biodiversity. All cities and smaller urban places need to develop a governance structure that better meets this multiplicity of needs. This has to provide the means by which compromises are reached between competing interests, and in which the needs of the least powerful groups receive adequate attention. It must ensure care and maintenance of the productive and protective functions of the ecosystems within which the city is located. It must also ensure no depletion of key natural (and other) assets, if the needs of future generations are to be safeguarded.

Throughout the continent, there are examples of innovation and better practice that give clues as to how current problems can be tackled. Most come from local developments, many of them related to more competent and democratic urban governments in nations where decentralization programs have given more power and resources to such governments. The following examples illustrate some of these improved practices.

Industrial pollution control has become more effective in many nations over the last twenty years. For instance, in Santiago, the control of industrial pollution moved from a relatively conflictual stance between government and industry during the late 1980s (as the government began to implement pollution control regulations and closed some industries) to a more cooperative phase in the late 1990s with government and industries working together to prepare and implement new regulations (Katz et al., 1999). Considerable reductions in industrial emissions were achieved in São Paulo during the 1980s, as the state government became more effective at pollution control (Wilheim, 1990). However, in both cities, there was less success in reducing pollution from motor vehicles (see Jacobi et al., 1999, for São Paulo; Escudero, 1996, for Santiago). Public pressure and a government more responsive to environmental issues and citizen concerns also produced sufficient improvement for the industrial city of Cubatão to be reported as a "success story" (World Bank, 1999), but perhaps it is more appropriately remembered as a cautionary story of what local and international companies can get away with, especially under the protection of a repressive political system (Lemos, 1998).

Some cities such as Curitiba and Porto Alegre in Brazil have high-quality living environments and innovative environmental policies. In Curitiba, a much admired public transport system was developed, based on express busways and feeder buses (Rabinovitch, 1992; Ceneiva, 1998) and this has encouraged comparable systems in other cities—for instance in Bogota (Montezuma, 2005). Many other cities have major programs to improve provision for water, sanitation, drainage, and garbage collection in lower income areas—for instance the *favela bairro* program and other programs in Rio de Janeiro (Fiori et al.,

2000; Magalhaes, 1998), the PREZEIS and ZEIS programs and participatory budgeting in Fortaleza (Melo et al., 2001), and the PROCASA program in El Alto in Bolivia (Bascon, 1998). Some cities have shown new institutional models for improving water and sanitation—for instance Santa Cruz's cooperative water company (Nickson, 1998) and the neighborhood level water and sanitation entrepreneurs in Asuncion (Solo, 1999). In many countries, the introduction of elected mayors and councillors over the last decade or so has helped make many city governments more accountable and responsive to their citizens. Participatory budgeting supported many environmental improvements in Porto Alegre, including a greater focus on environmental improvements in the poorer areas (Menegat, 2002).

Some cities have developed long-term environmental programs that combine attention to environmental health problems within the city with better environmental management of the city's (and the wider region's) natural resources—for instance the Bioplan developed in Manizales (Velasquez, 1998) and the environmental plans in Ilo (Follegatti, 1999; Boon et al., 2001). Manizales has also developed a much admired public information system—the "environmental traffic lights" through which environmental conditions and trends in all neighborhoods are measured and displayed (Velasquez, 1998). There are now new models of watershed management that seek to combine environmental goals and the needs of the inhabitants—for instance in the case study for Santo Andre, part of the watershed for São Paulo (van Horen, 2001).

20.6.2 Prospects for the Future

These examples of good practice are indications that progress is possible. However, among industries in South America, there is still a widespread perception that environmental protection is a cost to be minimized or avoided where possible (Jenkins, 2000). Among governments under pressure to cut public expenditures, provisions for many kinds of urban infrastructure and services have been or are being privatized; the hopes that privatization would ensure new investment that would extend provision of piped water and good quality sanitation to tens of millions of households have yet to be realized and many doubt that they will be (Hardoy and Schusterman, 2000; Loftus and McDonald, 2001; Budds and McGranahan, 2003). There are, as yet, few examples of the kind of intermunicipal cooperation and coordination needed for good environmental management within large urbanized regions, and developing these will present considerable political challenges (Cohen, 2001). The need then is to develop governance systems that encourage and support the multiple ways through which citizens and their organizations, non-government organizations, and businesses can improve urban environments while reducing the transfer of environmental costs to other people or other ecosystems, both now and in the future (Haughton, 1999). Such tasks are more easily stated and justified than actually implemented.

References

Abers, R., 1998. Learning democratic practice: distributing government resources through popular participation in Porto Alegre, Brazil. In: M. Douglass and J. Friedmann (Editors), *Cities for Citizens,* John Wiley and Sons, Chichester, 39–65.

Arrossi, S., 1996. Inequality and health in Metropolitan Buenos Aires. *Environment and Urbanization,* 8(2), 43–70.

Arroyo Moreno, J., F.R. Ríos, and I. Lardinois, 1999. *Solid Waste Management in Latin America: The Role of Micro-and Small Enterprises and Cooperatives.* IPES, ACEPESA and WASTE.

Bartone, C., J. Bernstein, J. Leitmann, and J. Eigen, 1994. *Towards Environmental Strategies for Cities: Policy Considerations for Urban Environmental Management in Developing Countries.* UNDP, UNCHS and World Bank Urban Management Program (18), World Bank, Washington, DC.

Bascon, R. 1998. Las familias de bajos ingresos acceden al financiamiento privado para la vivienda. In: E. Rojas and R. Daughters (Editors), *La Ciudad en el Siglo XXI. Experiencias Exitosas en Gestión del Desarrollo Urbano en América Latina,* BID, 317–325.

Birley, M.H., and K. Lock, 1998. Health and peri-urban natural resource production. *Environment and Urbanization,* 10(1), 89–106.

Blue, I., 1996. Urban inequalities in mental health: the case of Sao Paulo, Brazil. *Environment and Urbanization* 8(2), 91–99.

Boon, R.G.J., N. Alexaki, and H. Becerra, 2001. The Ilo Clean Air Project: a local response to industrial pollution control. *Environment and Urbanization,* 13(2), 215–232.

Browder, J.D., and B.J. Godfrey, 1997. *Rainforest Cities: Urbanization, Development and Globalization of the Brazilian Amazon.* Columbia University Press, New York.

Budds, J., and G. McGranahan, 2003. Are the debates on water privatization missing the point? Experiences from Africa, Asia and Latin America. Environment and Urbanization, 15(2), 87–114.

Cairncross, S., 1990. Water supply and the urban poor. In: J.E. Hardoy, S. Cairncross, and D Satterthwaite (Editors), *The Poor Die Young: Housing and Health in Third World Cities,* Earthscan Publications, London, 109–126.

Cairncross, S., and R.G. Feachem, 1993. *Environmental Health Engineering in the Tropics; an Introductory Text.* 2nd edition, John Wiley and Sons, Chichester.

Ceneiva, C., 1998. Curitiba y su red integrada de transporte. In: E. Rojas and R. Daughters (Editors), *La Ciudad en el Siglo XXI. Experiencias Exitosas en Gestión del Desarrollo Urbano en América Latina,* BID, 101–109.

Chandler, T., 1987. *Four Thousand Years of Urban Growth: An Historical Census.* Edwin Mellen Press, Lampeter, Wales.

Cohen, M., 2001. Urban assistance and the material world: learning by doing at the World Bank. *Environment and Urbanization,* 13(1), 37–60.

Cointreau, S., 1982. *Environmental Management of Urban Solid Waste in Developing Countries.* Urban Development Technical Paper No 5, The World Bank, Washington DC.

della Porta, A., 2001. Surviving the streets: An interview with Charles Wright, *IDB America,* November 26th, http://www.iadb.org/idbamerica/English/OCT01E/oct01e2.html.

de los Rios Bernardini, S., 1997. Improving the quality of life in low-income neighbourhoods occupied by tenants. *Environment and Urbanization,* 9(2), 81–99.

di Pace, M., S. Federovisky, J.E. Hardoy, J.E. Morello, and A. Stein, 1992. Latin America, In: R. Stren, R. White, and J. Whitney (Editors), *Sustainable Cities: Urbanization and the Environment in International Perspective.* Westview Press, Boulder, 205–227.

di Pace M., and A.D. Crojethovich, 1999. *La Sustentabilidad Ecológica en la Gestión de Residuos Sólidos Urbanos. Indicadores para la Región Metropolitana de Buenos Aires.* Informe de Investigación No. 3, Instituto del Conurbano, Universidad Nacional de General Sarmiento, San Miguel, Argentina.

Dockemdorff, E., A. Rodríguez, and L. Winchester, 2000. Santiago de Chile: metropolization, globalization and inequity. *Environment and Urbanization,* 12(1), 171–183.

Douglas, I., 1983. *The Urban Environment.* Edward Arnold, London.

Douglas, I., 1986. Urban geomorphology. In: P.G. Fookes and P.R. Vaughan (Editors), *A Handbook, of Engineering Geomorphology.* Surrey University Press, Blackie and Son, Glasgow, 270–283.

Douglas, I., 1989. The rain on the roof: a geography of the urban environment. In: D. Gregory and R. Walford (Editors), *Horizons in Human Geography.* Barnes and Noble, New Jersey, 217–238.

Durán de la Fuente, H., 1999. Propuesta para enfrentar la contaminación atmosférica en Santiago: descentralización y una nueva capital para Chile. *Ambiente y Desarrollo,* XV(3), 44–49.

Elsom, D., 1996. *Smog Alert: Managing Urban Air Quality.* Earthscan, London.

Escudero, J., 1994. Hasta donde aguanta Santiago? *Ambiente y Desarrollo,*10(1), 40–42.

Escudero, J. 1996. Situación global de la calidad del aire en la región metropolitana. Apuntes de ingenieria. *Ambiental,* 19(3), 7–20.

Fiori, J., L. Riley, and R. Ramirez, 2000. *Urban Poverty Alleviation through Environmental Upgrading in Rio de Janeiro: Favela Bairro.* Development Planning Unit, University College London, London.

Follegatti, J.L.L., 1998. Ilo: a city in transformation. *Environment and Urbanization,* 11(2), 181–202.

Greenway, D.R., 1987. Vegetation and slope instability. In: M.E. Anderson and K.S. Richards (Editors), *Slope Stability,* John Wiley and Sons, Chichester.

Hardoy, A., and R. Schusterman, 2000. New models for the privatization of water and sanitation for the urban poor. *Environment and Urbanization,* 12(2), 63–75.

Hardoy, A., J. Hardoy, G. Pandiella, and G. Urquiza, 2005. Governance for water and sanitation services in low-income settlements: experiences with partnership-based management in Moreno. Buenos Aires. *Environment and Urbanization,* 17(1), 183–200.

Hardoy, J.E, 2000. *Ciudades Precolombinas.* Ediciones Infinito, Buenos Aires.

Hardoy, J.E., D. Mitlin, and D. Satterthwaite, 2001. *Environmental Problems in an Urbanizing World: Finding Solutions for Cities in Africa, Asia, and Latin America.* Earthscan Publications, London, 470 pp.

Harms, H., 1997. To live in the city centre: Housing and tenants in the central neighborhoods of Latin American cities. *Environment and Urbanization,* 9(2), 191–212.

Haughton, G., 1999. Environmental justice and the sustainable city. *Journal of Planning Education and Research,* 18(3), 233–243.

Haughton, G., and C. Hunter, 1994. *Sustainable Cities.* Regional Policy and Development series, Jessica Kingsley, London.

Heller, L., 1999. Who are really benefited from environmental sanitation in cities: an intra-urban analysis in Betim, Brazil. *Environment and Urbanization,* 11(1), 133–144.

Hinrichsen, D., 1998. *Coastal Waters of the World: Trends, Threats, and Strategies.* Island Press, Washington, DC.

Hofmaier, V.A., 1991. *Efeitos de poluicao do ar sobre a funcao pulmonar: Un estudo de cohorte em criancas de Cubatão.* Doctoral Thesis, Sao Paulo School of Public Health.

Jacobi, P., D. Baena Segura, and M. Kjellén, 1999. Governmental responses to air pollution: summary of a study of the implementation of Rodízio in São Paulo. *Environment and Urbanization,* 11(1), 79–88.

Jenkins, R. (Editor), 2000. *Industry and Environment in Latin America.* Routledge, London.

Jimenez Diaz, V., 1992. Landslides in the squatter settlements of Caracas; towards a better understanding of causative factors. *Environment and Urbanization,* 4(2), 80–89.

Katz, R., G. Del Fávero, and L. Sierralta, 1999. Gestion ambiental en Chile: acción pública del sector privado en el área de la contaminación. *Ambiente y Desarrollo,* XV (3), 50–54.

Kreimer, A., T. Lobo, B. Menezes, M. Munasinghe, and R. Parker (Editors), 1993. *Towards a Sustainable Urban Environment: The Rio de Janeiro Study.* World Bank Discussion Papers No 195, World Bank, Washington DC.

Lavell, A., 1994. Prevention and mitigation of disasters in Central America: Vulnerability to disasters at the local level. In: A. Varley (Editor), *Disasters, Development and Environment.* John Wiley and Sons, Chichester, 50–63.

Lemos, M.C. de Mello, 1998. The politics of pollution control in Brazil: state actors and social movements cleaning up Cubatão. *World Development,* 26 (1), 75–87.

Loftus, A. J., and D.A. McDonald, 2001. Of Liquid Dreams: A Political Ecology of Water Privatization in Buenos Aires. *Environment and Urbanization,* 13(2), 179–199.

Magalhaes, S.F., 1998. Rio de Janeiro: una perspectiva actual de la ciudad. In: E. Rojas and R. Daughters (Editors), *La Ciudad en el Siglo XXI. Experiencias Exitosas en Gestión del Desarrollo Urbano en América Latina,* BID, 195–200.

Manzanal, M., and C. Vapnarsky, 1986. The Comahue Region, Argentina. In: J.E.Hardoy and D. Satterthwaite (Editors), *Small and Intermediate Urban Centres: Their Role in National and Regional Development in the Third World.* Hodder and Stoughton, London, 18–79.

McCarney, P. L., 1996. Considerations on the notion of 'governance'—new directions for cities in the developing world In: P.L. McCarney (Editor), *Cities and Governance: New Directions in Latin America, Asia and Africa.* Centre for Urban and Community Studies, University of Toronto, Toronto, 3–20.

McCormick, J., 1997. *Acid Earth: the Politics of Acid Pollution.* Earthscan, London.

McGranahan, G., 1991. *Environmental Problems and the Urban Household in Third World Countries.* The Stockholm Environment Institute, Stockholm.

McMichael, A.J., A. Haines, R. Sloof, and S. Kovats, 1996. *Climate Change and Human Health.* WHO, Geneva.

Medina, Y.G., and E.B. Uribe, 1994. La contaminación industrial en Colombia. *Planeacion y Desarrollo.* V(1), 59–72.

Melo, M., with F. Rezende and C. Lubambo, 2001. *Urban Governance, Accountability and Poverty: The Politics of Participatory Budgeting in Recife, Brazil.* Working Paper 27, Project on Urban Governance, Partnerships and Poverty, University of Birmingham.

Menegat, R. (main coordinator), 1998. *Atlas Ambiental de Porto Alegre,* Universidade Federal do Rio Grande do Sul, Prefeitura Municipal de Porto Alegre and Instituto Nacional de Pesquisas Espaciais, Porto Alegre.

Menegat, R., 2002. Participatory democracy and sustainable development: integrated urban environmental management in Porto Alegre, Brazil. *Environment and Urbanization,* 14(2), 181–206.

Michaels, D., C. Barrera, and M. Gacharna, 1985. Economic development and occupational health in Latin America: new directions for public health in less developed countries. *American Journal of Public Health,* 85(5), 536–542.

Montezuma, R., 2005. The transformation of Bogotá, Colombia, 1995–2000: Investing in citizenship and urban mobility. *Global Urban Development,* 1(1), 1–12.

Morello, J., G.D. Buzai, C.A. Baxendale, A.F. Rodríguez, S.D. Matteucci, R.E. Godagnone, and R.R. Casas, 2000. Urbanization and the consumption of fertile land and other ecological changes: the case of Buenos Aires. *Environment and Urbanization,* 12(2), 119–131.

Mueller, C.C., 1995. Environmental problems inherent to a development style: degradation and poverty in Brazil. *Environment and Urbanization,* 7(2), 67–84.

Nickson, A., 1998. Gobierno local. Una responsabilidad compartida. In: E. Rojas and R. Daughters (Editors), *La Ciudad en el Siglo XXI. Experiencias Exitosas en Gestión del Desarrollo Urbano en América Latina.* BID, 129–139.

OECD, 1990. *Urban Environmental Policies for the 1990s.* OECD, Paris.

Parry, M., 1992. The urban economy, presentation at *Cities and Climate Change*: a conference at the Royal Geographical Society, 31st March.

Prestes Barbosa M., and T. Booth, 1996. Lo Urbano, La Degradación Ambiental y Los Desastres: Cuestión Polémica. In: M.A. Fernandez (Editor), *Ciudades en Riesgo. Degradación Ambiental, Riesgos Urbanos y Desastres.* La Red, Peru, 141–150

Rabinovitch, J., 1992. Curitiba: towards sustainable urban development. *Environment and Urbanization,* 4(2), 62–77.

Rees, W.E., 1992. Ecological footprints and appropriated carrying capacity: What urban economics leaves out. *Environment and Urbanization,* 4(2), 121–130.

Rivas Roche, C., 2000. ¿Que Respiramos los Porteños? Primero, saber de qué estamos hablando. *Empresa y Medio Ambiente,* 7(50), 37–39.

Romieu, I., H. Weitzenfeld, and J. Finkelman, 1990. Urban air pollution in Latin America and the Caribbean: health perspectives. *World Health Statistics Quarterly,* 23(2), 153–167.

Sabaté, A.F., 1999. *El Circuito de los Residuos Sólidos Urbanos. Situación en la Región Metropolitana de Buenos Aires.* Informe de Investigación No. 5, Instituto del Conurbano, Universidad Nacional de General Sarmiento, San Miguel, Argentina.

Satterthwaite, D., 2005. *The Scale of Urban Change Worldwide 1950–2000 and Its Underpinnings.* IIED Working Paper, IIED, London.

Satterthwaite, D., and C. Tacoli, 2003. *The Urban Part of Rural Development: The Role of Small and Intermediate Urban Centres in Rural and Regional Development and Poverty Reduction.* Rural-Urban Working Papers Series, No 9, IIED, London.

Scott, M.J., 1994. Draft paper on Human settlements—impacts/adaptation. IPCC Working Group II, WMO and UNEP.

Scott, M., S. Gupta, E. Jáuregui, J. Nwafor, D. Satterthwaite, D.S. Wanasinghe, T. Wilbanks, and M. Yoshino, 2001. Human settlements, energy and industry. In: J.J. McCarthy, O.F. Canziani, N.A. Leary, D.J. Dokken and K.S. White (Editors), *Climate Change 2001: Impacts, Adaptation, and Vulnerability.* Contribution of Working Group II to the Third Assessment Report of the Intergovernmental Panel on Climate Change, Cambridge University Press, Cambridge, 381–416.

Smit, B., 1990. Planning in a climate of uncertainty. In: J. McCulloch (Editor), *Cities and Global Climate Change.* Climate Institute, Washington, DC, 1990, 3–19.

Solo, T.M., 1999. Small scale entrepreneurs in the urban water and sanitation market. *Environment and Urbanization,* 11(1), 117–131.

Souza, C., 2001. Participatory budgeting in Brazilian cities: limits and possibilities in building democratic institutions. *Environment and Urbanization,* 13(1), 159–184.

Suarez, F., 1999. Residuos. In: Di Pace, M. and E. Reese, 1999. *Diagnóstico Ambiental Preliminar del Municipio de Malvinas Argentinas. Manual de Gestión 2.* Institute del Conurbano Universidad Nacional de General Sarmiento, San Miguel. Argentina, 91–109

Swyngedouw, E.A., 1995. The contradictions of urban water provision a study of Guayaquil, Ecuador. *Third World Planning Review,* 17(4), 387–405.

Turner, R.K., P.M. Kelly, and R.C. Kay, 1990. *Cities at Risk,* BNA International, London.

UNCHS, 1988. *Refuse Collection Vehicles for Developing Countries.* HS/138/88E, UNCHS, Habitat, Nairobi, Kenya, 1988, 50 pp.

UNCHS, 1996. *An Urbanizing World: Global Report on Human Settlements, 1996.* Oxford University Press, Oxford and New York, 593 pp.

UN-Habitat, 2003. *Water and Sanitation in the World's Cities: Local Action for Global Goals.* Earthscan, London.

United Nations, 1980. *Urban, Rural and City Population, 1950–2000, as Assessed in 1978.* ESA/P/WP.66, June, New York.

United Nations, 2004. *World Urbanization Prospects: The 2003 Revision.* Population Division, Department for

Economic and Social Affairs, ESA/P/WP.190, New York.
van Horen, B., 2001. Developing community-based watershed management in Greater São Paulo; the case of Santo André. *Environment and Urbanization,* 13(1), 209–222.
Velasquez, L.S., 1998. Agenda 21: a form of joint environmental management in Manizales, Colombia. *Environment and Urbanization,* 10(2), 9–36.
Wackernagel, M., 1998. The ecological footprint of Santiago de Chile. *Local Environment,* 3(1), 7–25.
Wackernagel, M., and W.E. Rees, 1995. *Our Ecological Footprint: reducing Human Impact on the Earth.* New Society Publishers, Gabriola, Canada.
WHO, 1992. *Our Planet, Our Health.* Report of the Commission on Health and Environment, Geneva.
WHO/UNICEF Joint Monitoring Programme for Water Supply and Sanitation, 2000. *Global Water Supply and Sanitation Assessment 2000 Report.* World Health Organization and United Nations Children's Fund, Geneva.
Wilheim, Jorge, 1990. No hay buenos vientos para quien no sabe a donde quiere ir. *Ambiente y Desarrollo,* VI(1), 11—23.
World Bank, 1999. *Greening Industry: New Roles for Communities, Markets and Governments.* Oxford University Press, New York.
Zevallos Moreno, O., 1996. Occupation de laderas: incremento del Riesgo por Degradación Ambiental Urbana en Quito, Ecuador. In: M.A. Fernandez (Editor), *Ciudades en Riesgo. Degradación Ambiental, Riesgos Urbanos y Desastres.* La Red, Peru.

21

Future Environments of South America

Thomas T. Veblen
Kenneth R. Young
Antony R. Orme

An important goal of this book has been to provide a comprehensive understanding of the physical geography and landscape origins of South America as important background to assessing the probabilities and consequences of future environmental changes. Such background is essential to informed discussions of environmental management and the development of policy options designed to prepare local, national, and international societies for future changes. A unifying theme of this book has been the elucidation of how natural processes and human activities have interacted in the distant and recent past to create the modern landscapes of the continent. This retrospective appreciation of how the current landscapes have been shaped by nature and humans will guide our discussion of possible future trajectories of South American environments.

There is abundant evidence from all regions of South America, from Tierra del Fuego to the Isthmus of Panama, that environmental change, not stasis, has been the norm. Given that fact, the history, timing, and recurrence intervals of this dynamism are all crucial pieces of information. The antiquity and widespread distribution of changes associated with the indigenous population are now well established (see chapter 16). Rates and intensities of changes related to indigenous activities varied widely, but even in regions formerly believed to have experienced little or no pre-European impacts we now recognize the effects of early humans on features such as soils and vegetation. Colonization by Europeans mainly during the sixteenth century modified or in some cases replaced indigenous land-use practices and initiated changes that have continued to the present (see chapter 17). Complementing these broad historical treatments of human impacts, other chapters have examined in detail the environmental impacts of agriculture (chapter 18) and urbanism (chapter 20), and the disruptions associated with El Niño–Southern Oscillation events (chapter 19). The goal of this final synthesis is to identify the major drivers of change and to discuss briefly their likely impacts on South American environments and resources in the near and medium-term future. Our intent is not to make or defend predictions, but rather to identify broad causes and specific drivers of environmental change to inform discussions of policy options for mitigating undesirable changes and to facilitate potential societal adaptations to them.

The underlying causes of *global environmental change* can be grouped into two broad categories: (1) social, political, and economic changes that are often referred to in the aggregate as *globalization*, and (2) climate and atmospheric changes, including those caused by both anthropogenic and natural processes, which are often difficult to distinguish clearly because of interactions and feedbacks. Under the globalization rubric, we focus on two specific drivers of environmental change: changes in land use and biotic exchange (deliberate or accidental introduction of plants and animals to an ecosystem). Changes in land use, such as conversion of forest to pasture, are clearly important drivers. Biotic exchanges are less widely recognized because they often proceed rather slowly, with initially subtle effects on ecosystems, but in the longer term can result in widespread ecological and economic impacts (Mack et al., 2000; Mooney and Hobbs, 2000). Other sociological and political aspects of globalization are clearly important in shaping future land-use trends (e.g., Hecht and Cockburn, 1990; Painter and Durham, 1995; Wunder, 2001; Zimmerer and Bassett, 2003), but are largely beyond the scope of this book. In the second category of drivers of global environmental change, we discuss climate changes involving atmospheric CO_2 concentrations, nitrogen deposition, and acid rain.

We then probe current and probable future trends of environmental change in the context of the systematic and regional perspectives provided by earlier chapters, and assess their resource implications. Although synthesis at the continental scale requires broad generalizations, we stress the complexity and diversity of environmental changes within and among its major biomes.

21.1 Drivers of Environmental Change

21.1.1 Demographic and Socioeconomic Drivers

Current and continuing socioeconomic and technological changes operating at a global scale should be considered in a historical context, particularly in relation to their environmental consequences. In 1950, the population of South America was approximately 113 million; by 2000 it had reached 347 million (FAOSTAT data, August 2005). At a continental scale, public health advances following World War II have been the key factors driving rapid population growth. The most effective measures taken to reduce high rates of infant mortality have been improved sanitation (water and sewage facilities), mass vaccinations for diseases such as smallpox, public health education, and access to health care. During the early 1990s, the doubling time for the population of the continent was 37 years, but there is wide variation in population growth from country to country. In the 1990s, Bolivia, Ecuador, Paraguay, Perú, and Venezuela all had doubling times of less than 30 years, whereas Argentina and Uruguay had doubling times of 44 and 77 years, respectively (Caviedes and Knapp, 1995).

Variations in birth and death rates among South American countries fit a general theory of demographic transition widely applied as an explanation of changes in populations in relation to economic development and modernization of societies. In theory, countries with high birth and death rates such as Bolivia and Ecuador represent early stages, whereas countries with low birth and death rates such as Argentina and Uruguay represent a final, relatively stable stage similar to most western European countries. Research has shown that this demographic transition into a relatively stable population is associated with the transformation of a mainly rural population primarily engaged in agriculture and exploitation of primary resources into an urbanized society with greater access to health and educational facilities. Modernization and urbanization essentially remove the benefits of children as farm labor and other incentives to have high birth rates. Projections of size of the future population in South America assume that modernization and education will continue this demographic transition so that a stable population of about 560 million people may be attained before the end of the twenty-first century (Caviedes and Knapp, 1995). Recognizing the wide variations among South American countries in their stages of demographic transition, with countries such as Bolivia and Ecuador at one extreme and Uruguay and Argentina at the other, is crucial to any assessment of the effects of humans on environment and resources over the next few decades. In addition, such differences can also be found within individual countries, making subnational or regional data crucial for some kinds of predictions.

Although rapid population growth and related environmental changes during the second half of the twentieth century are unprecedented in their rates and magnitudes, South America has also experienced earlier major demographic changes with consequent environmental impacts. Early in the colonial period, for example, population declines and abandonment of cultivated areas led to significant changes in land-cover types (see chapter 16). Lethal Old World diseases introduced in the early phases of European conquest resulted in indigenous population declines of varying magnitudes and with variable environmental consequences. Thus, formerly large populations in riverine habitats of Amazonia nearly disappeared during the sixteenth century, triggering a large-scale change in the landscape from managed agricultural fields to forests that were perceived as pristine by early nineteenth century observers (see chapter 16). Analogously, introduced diseases and social chaos triggered by the *conquista* resulted in population declines and abandonment of intensively cultivated fields in the central Andes, as attested to centuries later by the presence of agricultural terraces beneath forests. On the other hand, remote areas of low population density such as Patagonia and Tierra del Fuego were initially less affected by the epidemics of the sixteenth century, and in many cases did not experience demographic declines until they were effectively colonized by Europeans in the mid-nineteenth century. Further, rapid population growth and deforestation are not entirely phenomena of the later twentieth century. For example, most of the lowland forests of the Central Depression of southern Chile were eliminated by logging and burning by European colonists during the later nineteenth century (see chapter 13). Likewise, the tropical Atlantic coastal forests of Brazil receded relatively early in the colonial period with settlement and the development of a sugarcane export economy. Some Andean deforestation probably began at least a millennium ago.

At a continental scale, however, the population growth of the second half of the twentieth century has occurred at an unprecedented rate and, along with socioeconomic effects associated with a globalized world economy, has resulted in massive deforestation in tropical lowlands. This population growth has fed agricultural expansion into lowland forests, which has been one, but not the sole, driver of deforestation in the tropical parts of South America. Often, agricultural expansion into these areas has been an explicit policy designed to satisfy demands for agricultural land among the landless poor where more accessible and more productive agricultural lands have been long occupied by politically powerful groups. National policies have thus often favored the expansion of agriculture into frontier zones rather than addressing land reform. This has not been an expansion into empty lands, as indigenous populations are already present, although typically with few legal or other resources with which to maintain their lands and land uses under such pressure.

In countries with rapidly growing populations, expansion of agriculture into tropical lowlands typically has been facilitated through policies aimed at providing access and infrastructure for the movement of people from densely settled regions into frontier areas of little or no settlement. Large national investments have been committed to road construction into tropical lowlands, mainly in Amazonia, and to the linkage of frontier agricultural areas to urban markets. Road construction and settlement in these frontier areas have also been motivated by geopolitical and security concerns along the borders of Amazonia. The result has been alarming rates of deforestation in Amazonia and other areas of lowland tropical forest (see chapter 9; Cardille and Foley, 2002; Viña et al., 2004)

The late twentieth and early twenty-first century period of explosive population growth has seen the rapid conversion of large areas of lowland forests to agricultural land use, analogous to demographic and land-use changes that occurred several centuries ago in Europe and eastern North America. Such deforestation, due to growth in the number of small land holdings practicing mostly subsistence agriculture, continues in many areas in South America and can be interpreted simplistically as a manifestation of national and local population pressures on resources. However, an increasing influence on the conversion of these lowland forests to other land uses has been world and national (often urban) demand for products such as meat and soybeans. For example, during the late twentieth century, cattle raising was the primary cause of deforestation in the Brazilian Amazon. At the beginning of the twenty-first century, soybean production, most of which is exported to China for animal feed, is the major cause of deforestation in seasonally dry forests of Brazil, Bolivia, Paraguay, and Argentina (Aide and Grau, 2004). This is a clear example of how globalization of the world economy is facilitating rapid changes in the landscapes of South America. These changes are now being amplified by increased demands for products by the urban populations in the tropical countries themselves.

Of course, earlier episodes of major vegetation change affected large parts of the continent due to world demand for products such as coffee, bananas, and sugar cane. However, in a globalized economy in which technology facilitates rapid communication and massive transportation projects, world demand for agricultural products such as meat and soybean is much more quickly transformed into major local and regional impacts on the environment. Since the 1970s, availability of relatively cheap, rapid transportation has facilitated major land-use changes for the production of cut flowers in Colombia and counter-seasonal fruit in central Chile for the Northern Hemisphere markets (see chapter 18). Likewise, in a globalized economy, world demand for forest, mineral, and energy resources is having an immediate and widespread impact on the landscapes of South America (see chapter 1; Caviedes and Knapp, 1995).

Consistent with the theory of demographic transition is the strong trend toward a higher ratio of urban to rural population during recent decades. In 1961, 48% of South America's population was rural whereas in 2003 only 19% was rural (FAOSTAT data, August 2005). More significantly, since 1980 the population whose livelihood directly depends on agriculture, fishing, hunting, or forestry has actually declined by about 20 million people (fig. 21.1). Given the rapid growth in the total population, this trend represents a major migration from rural to urban areas, as reflected in the five-fold increase in South America's urban population during the second half of the twentieth century (see chapter 20). The major factors driving this rural-urban migration include: (1) decreased demand for rural labor because of conversion from small farms to large-scale modern agriculture; (2) expansion of high-yield agriculture, resulting in price declines for products such as maize, grains, beef, and potatoes, which make it more difficult for small-scale farmers to compete; and (3) the cultural and economic attraction of urban life, in particular for young people (Aide and Grau, 2004). The net result of these factors is the abandonment of extensive areas of grazing and agricultural land, especially in habitats of marginal

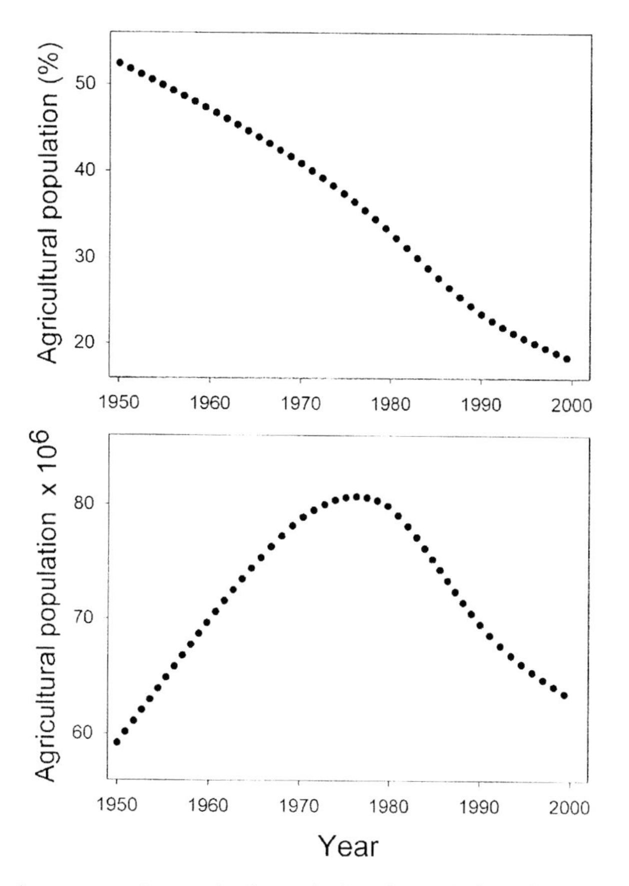

Figure 21.1 Changes in the agricultural population of South America, 1950–2000. The agricultural population is defined as individuals whose livelihood depends on agriculture, hunting, fishing, or forestry (source: FAOSTAT, August 2005).

productivity (Rudel, 2002; Aide and Grau, 2004). Thus, demographic transition and the rural to urban migration, in particular, have important implications for future environments and resources, but the rate of this transition will continue to vary considerably from country to country and from region to region. In addition, the distinction between what is urban and what is rural becomes complicated given increased transportation and communication in rural areas, and the heterogeneity of conditions in urbanized areas (Dufour and Piperata, 2004).

21.1.2 Atmospheric and Climate Changes

Elevated CO_2 Concentration in the Atmosphere Globally, enormous amounts of carbon are emitted into the atmosphere, due primarily to fossil-fuel combustion and deforestation. As a result, the concentration of CO_2 in the atmosphere has increased by 31% since A.D. 1750 (Houghton et al., 2001). Over the past 20 years, approximately 75% of anthropogenic emissions of CO_2 are the due to fossil-fuel burning, and the remainder is from land-use practices, primarily deforestation. In addition to its important role in modifying climate as a greenhouse gas, this increase in CO_2 will have, and probably already has had, major effects on the growth of plants and the structure of natural ecosystems (Bazzaz, 1990; Gifford, 1994; Körner, 2000). CO_2 has direct and specific effects on the physiology and growth of plants, which in turn can alter the structure and dominance of plant communities. The community-level outcomes of elevated CO_2 are complex but are likely to have the greatest impacts in biomes where plant growth is most limited by water availability and where there is a mix of species with different biochemical photosynthetic pathways (e.g., C_3 and C_4 species; Bazzaz, 1990). Because of known species differences in the effect of CO_2 on water efficiency, the greatest community-level impacts of elevated CO_2 are likely to be in semiarid and savanna environments with a mixture of woody and graminoid species (Sala et al., 2000).

Nitrogen Deposition and Acid Precipitation Human activities have also been increasing the concentration of nitrogen compounds in the atmosphere and their deposition in terrestrial and aquatic ecosystems (Melillo, 1996; Smil, 1997; Vitousek et al., 1997). The main sources of oxidized nitrogen (NO_x) are vehicles operated by fossil fuels and power plants that burn fossil fuels. Ammonia (NH_{3-}) emissions are largely from fertilization for agriculture. These sources, as well as some manufacturing processes (e.g. nylon production) produce elevated deposition of nitrate (NO_{3-}) in rain, and nitrogen dioxide (NO_2) and nitric acid (HNO_3) in dry deposition.

The emissions of nitrogen oxides (NO_x) and ammonia (NH_{3-}) deposition, and their transformation products, can cause a wide range of environmental effects at local to global scales, including soil acidification, eutrophication in aquatic ecosystems, formation of tropospheric ozone, and contributions to greenhouse gases. In terrestrial ecosystems, nitrogen deposition results in the acidification of soils and altered availability of nitrogen as a nutrient for plants. These changes can significantly affect plant growth rates and plant health, which in turn may result in changes in the structure and dominance of plant communities as well as altered ecosystem productivity and nutrient cycling. Emission of sulphur, primarily by coal-burning power plants, is also an important contributor to acid deposition and ecosystem change.

Climate Change As a driver of environmental change we consider climate change to be a consequence of both natural and anthropogenic factors. Thus, whereas the paleoclimatic record provides ample reason to expect that ecologically significant climatic variations will continue to occur at time scales relevant to humans (see chapters 4 and 19), anthropogenic changes are probably also significantly affecting climate trends at regional and global scales (Watson, 2001). In addition to CO_2, human activities have also increased the atmospheric concentrations of other greenhouse gases such as methane (CH_4), nitrous oxide (N_2O), and tropospheric ozone (O_3) through agricultural activities and burning of fossil fuels. For example, the atmospheric concentration of methane is believed to have increased by 151% since 1750 and continues to increase (Houghton et al., 2001). Slightly more than half of current CH_4 emissions are anthropogenic, originating from use of fossil fuels, cattle, rice cultivation, and landfills (Houghton et al., 2001). Similarly, the tropospheric concentration of ozone is estimated to have increased by 36% since 1750, but its concentration varies regionally. Overall, most atmospheric scientists believe that the global warming observed over the past 50 years is attributable to atmospheric changes associated with human activities (Houghton et al., 2001; Watson, 2001).

Annual and multidecadal scale climatic variation across South America has been strongly related to variations in large-scale drivers of climate such as changes in sea surface temperatures in the tropical Pacific (see chapters 3 and 19). Indisputably, the climate of South America will continue to vary significantly and affect natural processes as well as human activities, as exemplified by seasonality in Amazon forests, which alters carbon sequestration processes and balances (Saleska et al., 2003). Against this background of continuing natural variability, it is difficult to ascertain the extent to which increases in greenhouse gases have contributed to recent climate trends. Nevertheless, there is a strong consensus that global and regional climates already manifest the effects of increases in greenhouse gases (Houghton et al., 2001; Watson et al., 2001).

There is increasing evidence that large-scale human-induced changes in land cover, such as conversion of forest to cultivated land, burning, and irrigation of dry lands, have a significant impact on local and regional climate. Such changes may also be a significant factor altering the global climate (Pielke et al., 2002). Regional-scale simulations of the climatic impact of the Amazon basin show a reduction of about 20% in rainfall and a lengthening of the dry season when forest is replaced by grassland (Shukla et al., 1990). Further modeling suggests numerous species will be affected in terms of their distributions, abundances, and population persistence (Miles et al., 2004). Although large-scale deforestation of the Amazon would be likely to affect the climate of the rest of South America, the details of that impact are unknown.

Although the climatic effects of deforestation are likely to be significant, and perhaps already are, uncertainties about rates of deforestation and forest recovery make it difficult to create specific scenarios for future climates based on land-cover changes. Instead, such scenarios are typically developed from general circulation models in which CO_2 is the major variable. Using such a top-down approach, Labraga (1998) simulated climatic conditions for South America from five general circulation models under a doubled concentration of atmospheric CO_2. This comparison predicted five consistent climate scenarios with important implications for future ecosystems and human activities. (1) The models predict a southward shift of the summer continental low pressure zone and the Pacific anticyclone, the two principal drivers of the continent's atmospheric circulation. Among other effects, this would result in a southward displacement of the arid zone of central Chile and western Argentina, and increased temperatures in these areas. (2) Precipitation is predicted to increase in the tropical central and eastern part of the continent south of about 15°S and to decrease north of this latitude. This would probably result in substantial shifts in the boundaries of the various tropical forest and savanna vegetation types in the center of the continent. (3) Warming is predicted to be maximal in Paraguay and southern Brazil, which include large areas of seasonally arid lands. (4) Convective activity is predicted to increase in the Intertropical Convergence Zone along the Pacific coast north of the Equator. This is likely to affect the relatively sharp transition from arid conditions in northern Perú to wetter conditions in southern Ecuador. (5) At higher latitudes in southern Chile and Argentina both precipitation and temperature are predicted to increase. Thus, while recognizing uncertainty in these predictions, significant climatic impacts should be anticipated across South America in response to global warming linked to elevated atmospheric CO_2. The models also specify where change is most likely to occur: high elevations, high latitudes, coastal areas, and all the ecotones between different vegetation and landscape types.

While land-use and land-cover change on the one hand, and climate change on the other hand, are both major drivers of ecosystem change, they also interact in complex ways, making projections for specific areas difficult to assess. For example, deforestation certainly can affect local climates, and can potentially affect regional and global climates by changing the land-surface albedo so that less energy is absorbed and available to evaporate water back to the atmosphere. Deforested sites are also likely to increase surface runoff at the expense of infiltration and transpiration, leading to more variable stream flows, while CO_2 released by forest burning contributes to greenhouse warming at a global scale (Houghton et al 2001). Thus, although the top-down approach to predicting climate scenarios through general circulation models is a useful starting point, actual future regional climates are likely be affected in complex and uncertain ways by regional changes in land use as constrained by soil fertility (Moran et al., 2000) and socioeconomic capital (Walker et al., 2000).

21.2 Major Environmental Changes and Resource Implications

This section focuses on how drivers of global change may influence future environments of South America, organizing the discussion around the potential implications for terrestrial, freshwater, and coastal ecosystems, and for biodiversity. We also discuss some further implications for people to complement the evaluations on the environmental impacts of agriculture and urbanism found in chapters 18 and 20, respectively.

21.2.1 Terrestrial Ecosystems

Large areas of forest and grassland in South America are expected to be affected by projected climate changes (Watson et al., 1997). Transitional areas between different ecosystem types along precipitation gradients and in mountainous areas are likely to be most vulnerable to rapid climate-induced changes (see chapters 10, 11, and 12). Climate change is likely to increase the stress on tropical forests already affected by deforestation. For example, shifts toward drier conditions due to altered atmospheric controls on climate may augment trends toward reduced rainfall associated with deforestation. In an area as large as the Amazon Basin, the positive feedback of global climate change and deforestation could significantly reduce precipitation recycling through evapotranspiration and affect global carbon cycles, with multiple consequences for biodiversity (Watson et al., 1997; Ozanne et al., 2003).

Although not underestimating the probable dramatic effects of future climate change on terrestrial ecosystems, we stress that direct human impacts through land-use changes involving deforestation, overgrazing, burning, and agricultural expansion will continue to be severe and immediate problems in most of the continent. The areas most affected by deforestation are along the edges of the continent, where population is concentrated, and along major access routes. In terms of the proportion of original forest that has remained relatively intact since before European settlement, the areas most severely affected by deforestation are the tropical coastal forests of southeastern Brazil and Ecuador, and the temperate forests of central and south-central Chile. Montane tropical forests flanking the northern and central Andes have also been rapidly deforested by colonists in recent decades. The interior of the continent, especially its more seasonally dry environments, will continue to see the greatest absolute amount of deforestation with the most potential for affecting regional and global climate through feedbacks to the global carbon balance and the hydrological cycle. Rates of deforestation in Amazonia in the near future are likely to be driven primarily by world markets for tropical agricultural products but will also be significantly affected by national policies. In recent decades, for example, the greatest extent of deforestation has been along the southern margin of Amazonia where major new highways in Brazil have attracted settlers and ranchers (Caviedes and Knapp, 1995; Laurance et al., 2002). Policymakers in Amazonian countries clearly have options to facilitate transportation projects likely to exacerbate deforestation or to adopt strategies that will reduce incentives for large-scale deforestation (e.g., Soares-Filho et al., 2004).

Global policies aimed at mitigating the effects of greenhouse gases have the potential for severely affecting rangelands and native flora and fauna in large areas of South America. A policy of global trading of carbon credits is currently evolving in which it is likely that major polluting countries (or regions) will be allowed to continue to emit greenhouse gases at higher than desired rates based on the purchase of carbon credits derived from carbon sequestration activities in other countries or regions. Although reduction in greenhouse gases should be an urgent worldwide priority, the contribution of global carbon trading to such a reduction is of uncertain efficacy and raises serious concerns in the realm of environmental justice. Rich countries are likely to be allowed to continue to pollute while poor developing countries are expected to implement mitigation measures to absorb carbon dioxide emissions. Thus, wealthy polluting countries will be able essentially to export some of the environmental impacts of their production and consumption. Specifically, the trading of carbon credits on a global scale is an incentive to afforest large areas of nonforested ecosystems in South America, a policy that may have detrimental impacts on local resources (water and rangeland) and on the native flora and fauna. For example, large areas of the high Andean grasslands (*páramos*) in Ecuador are being planted with exotic trees, most typically Monterey pine (*Pinus radiata*), with a number of negative local consequences. Conversion of *páramos* to conifer plantations has been shown to reduce water quality and water-storage capacity in drainage basins that are often the primary sources of water for Ecuador's high elevation urban centers (Farley et al., 2004). Large-scale afforestation programs for the purpose of carbon sequestration have already begun, or will soon begin, in the grasslands and steppes of the temperate latitudes of Argentina and Uruguay (see chapters 13 and 14). These programs will result in loss of habitat for native fauna, reduction or disappearance of much of the native flora, and expansion of invasive alien plants, including noxious weeds capable of reducing productivity of nearby cropland and rangeland. In the lowland tropics, difficulties with plantations arising from diseases and environmental impacts suggest that allowing forest regrowth on abandoned lands would be a more effective strategy (Post and Kwon, 2000; Silver et al., 2000).

The terrestrial ecosystem changes expected under altered future climates clearly have implications for food and fiber production. Leaving aside the uncertainties of socioeconomic influences on agricultural and forest production (e.g., Swinton and Quiroz, 2003), consideration of just the climatic, atmospheric, and ecological determinants of future production is highly complex. For example, the effects of elevated atmospheric CO_2 on crops under controlled conditions lead to predictions of increased crop yields in the future. However, there is much uncertainty in such predictions for particular regions because of limitations imposed by the availability of water and nutrients, and the likelihood of increased damage from pests and diseases under warmer conditions. Simulation studies based solely on elevated CO_2 levels suggest that areas suitable as cropland will experience a small decrease in tropical South America, while higher latitudes of the Northern Hemisphere will be likely to see increases in cropland (Ramankutty et al., 2002). Under such scenarios and given the small area of South America at higher latitudes, the continent as whole will become a net loser in agricultural production relative to Northern Hemisphere temperate lands.

While not minimizing the severity of probable future disruptions due directly to climate change, the effects of socioeconomic and political influences on food and fiber production in South America are likely to be greater. This is because tropical South America, along with Africa, contains the most potential cropland as yet uncultivated, leading to the prediction that South American farmers will be feeding more of the world's expanding population (Ramankutty et al., 2002). Thus, future global demand for agricultural products is likely to have greater impact on biodiversity and forest habitat in tropical South America and Africa than in other regions.

Compared with other continents, the intensification of agricultural production in South America has been particularly great during the last few decades of the twentieth century, although the total land area in agriculture has grown slowly relative to the rapid growth in total production (see chapter 18). This successful intensification is due to a number of factors, including genetic improvements of crops (i.e., the green revolution), growth of mechanized farming, greater use of fertilizers, and the more efficient practices of large agribusiness, particularly for the production of export products. Intensification will undoubtedly continue to increase yields per land area in particular regions and for certain crops. However, there are also dangers associated with this intensification, such as increased dependence on monocultures susceptible to pests and pathogens, reliance on fossil fuels and fertilizers that are likely to continue to rise steeply in cost, and erosion and other forms of land degradation associated with intensive production (see chapter 18). Overall, while total cultivated area will expand and production will further intensify in response to increased world demand, the sustainability of crop production under environmental limitations, imposed for example by steep slopes, soil fertility, pests, and pathogens, will continue to be a major issue at local scales, and perhaps also at regional scales.

21.2.2 Freshwater Ecosystems

Freshwater ecosystems in South America are already severely affected by human activities because people live disproportionately near waterways and extensively modify riparian zones, wetlands, and lakes. Even in biomes that are otherwise sparsely populated, people alter these ecosystems through inputs of nutrients, sediment, and contaminants (see chapter 5). In addition, waterways are heavily used for their water resources and, less compatibly, for transportation and sewage disposal. Biotic exchange is more important for aquatic systems, especially lakes, than for terrestrial ecosystems because of intentional activities, such as stocking with fish, as well as unintentional introductions. Due primarily to land use, it is likely that change will occur more rapidly, and biodiversity will decline more severely, in freshwater ecosystems than in terrestrial ecosystems. But climate change will also affect these systems. In particular, biodiversity in rivers is likely to change more than in lakes because the former are more sensitive to climate-induced changes in runoff. In contrast, biotic exchanges are likely to impact lakes more than streams because the harsh environment of the latter reduces the success of introduced species.

The peoples of South of America have long been involved with managing water resources—from indigenous peoples using water for irrigation and travel, through European colonists seeking to secure water supplies for fledgling towns, to the demands of expanding commercial agriculture during the later nineteenth and twentieth centuries. The effects of these activities were, however, modest compared to the human impacts on water resources that began in the later twentieth century and which provide a portent of future problems and challenges.

South America is mostly a wet continent, albeit with seasonal variations, and truly arid environments are limited in extent and often near glacier, snowmelt, and stream discharges from the High Andes. It is not surprising, therefore, that recent decades have seen an escalation of activities designed to manage water resources on a grand scale for domestic supplies, agriculture, industry, hydropower, and navigation. The continent now contains some of the world's most capacious reservoirs, such as the Guri in the Orinoco Basin and the Sobradinho on the São Francisco River. In terms of megawatt production, its rivers host nine of the world's top 22 existing or planned hydropower facilities, and Paraguay provides virtually all its domestic electricity needs from water power, with a large surplus for export. The Paraná basin in Brazil has been particularly impacted by dam and reservoir construction, notably in the urban and industrial heartland of southern Minas Gerais and São Paulo states, where west-flowing tributaries such as the Paranaíba, Grande, Tieté, and Paranapanema harbor many reservoirs. Downstream, that portion of the Paraná mainstem shared between Brazil and Paraguay supports the Itaipu reservoir, presently the world's third largest producer of hydropower. Nor is the Amazon basin, with its seemingly unlimited water, exempt from impoundments such as the Balbina reservoir north of Manaus and the Tucurui reservoir on the Tocantins River south of Belém. Brazil alone has more than 2,000 reservoirs of varying size, most of the larger ones constructed during the hydropower frenzy of the 1970s and 1980s.

However, sequestering river water is not without cost—environmental, social, and political—and in this respect the future looks less certain. From a physical perspective, dams impound both water and sediment, and in basins with high sediment yields, such as those in the Andes foothills, the life of a reservoir is more limited than in clear water streams (see chapters 5 and 18). Impoundments also affect surface and groundwater hydrology, for example by waterlogging lands adjacent to and upstream from reservoirs, and by changing streamflow characteristics and increasing erosion potential below dams.

From a biological perspective, the construction of dams and reservoirs means loss of riparian and floodplain habitat, with serious impacts for dependent plants, animals, fish and other life forms. For example, completion of the 180–m high Barra Grande dam on the Pelotas River, separating Santa Catarina and Rio Grande do Sul states in southern Brazil, will submerge 20 km^2 of primary Atlantic Forest habitat, including stands of *Araucaria* pine, although most of the trees are being felled prior to filling the reservoir. Reservoirs also replace people. Despite some compensation, human lives and livelihoods are disrupted and, not surprising, displaced peoples and environmentalists have joined forces to resist such changes, for example in Brazil's Movement for Dam-Affected Peoples. Thus do water resources enter the geopolitical arena. Nevertheless, the impoundment of water resources also ensures a more dependable supply of water for irrigated agriculture, industry, electricity, and domestic use. Thus costs to the environment provide benefits for most of the people.

Climate change will also affect regional hydrological cycles, but the scope of these changes for specific regions is difficult to predict. Changes in the magnitude, intensity, and seasonal distribution of precipitation, such as those predicted by simulation studies (Labraga, 1998), will certainly affect surface runoff and groundwater recharge and have significant impacts on natural ecosystems and human activities. The melting of mountain glaciers in the tropical Andes, which began with the waning of the Little Ice Age (A.D. 1500–1850), has accelerated in recent decades (see chapter 4). The potential loss of these ice reservoirs will yield more variable stream flows, which in turn will affect hydropower generation, irrigated agriculture, and domestic water supplies, especially during the extended dry season. Farther south, in temperate Chile and Argentina where hydropower generation and water resources are already sensitive geopolitical issues, glacier retreat and changing water budgets could lead to future conflict. Semiarid areas such as the Monte of Argentina and *caatinga* of northeast Brazil, and seasonally dry areas such as the Chaco and Caribbean coastlands, are especially vulnerable to changes in water availability because much agricultural and other activity is already marginal.

Simulation studies that combine climate model outputs, water budgets, and socioeconomic information have been used to assess the vulnerability of global water resources to climate change and population growth (Vörösmarty et al., 2000). Total discharge for South America is predicted to decrease by 11% by 2025, relative to the baseline in 1985. This is the largest decrease in any of the six regions studied (Africa, Asia, Australia/Oceania, Europe, North America, and South America). At the same time, future water use under three scenarios—assuming only climate change, only population growth, or both—predicts increases in demand ranging from 12% to 121% (Vörösmarty et al., 2000). Again, South America leads all six regions in percentage increases in projected water demand for urban populations and intensive agriculture. Analogous to the findings for terrestrial ecosystems, rising water demands are expected to greatly outweigh greenhouse warming in defining the state of water systems in 2025.

21.2.3 Coastal Ecosystems

Assuming that global-warming scenarios persist into the immediate future, coastal ecosystems will be likely to respond both to rising ocean levels and to changing storm patterns and frequencies. The continent's present coastline was broadly outlined by the global marine transgression triggered by the melting of the world's last continental ice sheets, which began around 18–20 ka ago and had culminated by 5 ka ago (ka: thousand years). During the maximum phase of this transgression, the sea rose across the now submerged continental shelf at a mean rate of ~10 mm yr^{-1} (~10 m ka^{-1}). Over the past 5 ka, sea levels have continued to rise episodically but much more slowly relative to the land, at perhaps one-tenth of the former rate. Nevertheless, it is this recent rise that, linked to concerns over global warming, attracts most attention.

Recent estimates of global sea-level change based on tide-gauge records over the past 50–150 years suggest a relative sea-level rise of between 1.0 and 2.4 mm yr^{-1} (Douglas, 2001). This rise involves renewed melting of Earth's remaining continental ice bodies after the Little Ice Age, steric increases in ocean volume related to rising ocean temperatures, hydroisostatic adjustments to water loading on continental shelves, and other factors. Decomposing the time series revealed by tide-gauge data yields much uncertainty but suggests that melting of Antarctic and Greenland ice may be contributing ~0.8–1.2 mm yr^{-1} to rising sea level while steric changes contribute ~0.3–0.7 mm yr^{-1} and are highest in the tropics (Levitus et al., 2000; Church, 2001). Much of this rise may be due to natural warming after the Little Ice Age, but there is growing consensus that some of it is due to anthropogenic forcing of global climate during the industrial age of the past 150 years (Gehrels, 2005). Combining empirical data with modeling, and assuming a range of scenarios, it is likely that global sea level will rise at least 80 mm (and by some estimates up to 800 mm!) over the next 100 years (Watson et al., 2001).

What effect will this rise of sea level have on the South American coast? First, the predicted global (eustatic) changes will not be felt uniformly because steric, oceanographic, isostatic, and geoidal forcing factors are essentially regional. For example, the tide-gauge record for Buenos Aires reveals a mean rise of 1.1 mm yr^{-1} for the period 1905–1980, but a higher rate of 1.5 mm yr^{-1} if the record is extended to 1987. This latter blip reflects super-elevation of water levels in the 1980s by El Niño events, which caused excessive runoff from rivers draining into the Río de la Plata. (Douglas, 2001). In contrast, tide-gauge data from Quequén, 400 km south on the open Atlantic coast, shows no El Niño forcing and the rate of rise over 65 years is 0.8 mm yr^{-1}. Second, while recognizing regional variations, the predicted changes will mean little to steep coasts, such as those that often front the Pacific Ocean, or to the southern cone where storm waves predominate. Third, however, a modest sea-level rise will have a significant impact on soft targets, such as sandy beaches, and along low-lying shores, such as those extending from the Amazon to the Orinoco, where mangrove and estuarine ecosystems will be likely to migrate landward. Coral reef communities will also be affected where rates of reef growth fail to keep pace with rising sea-levels. Fourth, impacts on human infrastructure may be locally severe, from increased flooding of developed coastlands and saltwater intrusion into freshwater aquifers to impacts on port, industrial, and recreational facilities. Low-lying portions of the Caribbean coast and the Río de la Plata estuary will be particularly vulnerable. With a 1 m rise in sea level, Venezuela would lose ~5700 km^2 of land, Guyana ~2400 km^2, and Argentina ~3400 km^2 (IPCC, 1996).

Natural factors apart, coastal ecosystems of the future will also suffer human impacts commensurate with increasing globalization. These will involve the growth of coastal cities, industrial and port facilities, and recreational activities on a continent whose human focus is already predominantly coastal. In addition, rivers draining from expanding agricultural areas will flush more sediment, agrochemicals, and farm waste to the coast, with inevitable impacts on nutrient budgets and biotic responses. Plants and animals presently found in estuaries, beaches, dunes, cliffs, and coral reefs will all retreat under these impacts. In addition, there will be more pollution of coastal waters from urban and industrial waste, ship discharges, and oil spills, and incursions of marine and estuarine species foreign to South American waters. Some ecosystems will be especially impacted: coral reefs by increased sedimentation and ecotourism; and mangroves, which by their very nature coincide with shallow tidewaters suitable for mariculture and for industrial and urban reclamation. Thus, in nations with ocean frontage, governments at all levels should be alert to these probable changes and plan accordingly.

21.2.4 Biodiversity

Biodiversity losses in South America have been largely attributed to deforestation and habitat destruction. However, as noted above, there are also areas where the abandonment of agricultural land, due to rural to urban migration, poor soils, and withdrawal of subsidies, is leading to forest recovery. Further complexity is added to biodiversity issues by climate change and introduction and spread of introduced plants and animals. In face of such complexity, a useful approach is to create future scenarios based on changes in atmospheric CO_2, climate, vegetation, and land use, taking into account the known sensitivity of biodiversity to these changes. This approach was adopted by a group of scientists, including many from South America, in a comprehensive attempt to assess likely future changes in biodiversity globally (Sala et al., 2000). They identified and ranked the drivers of biodiversity change and used simulation models to create possible scenarios for different major terrestrial biomes and freshwater ecosystems. The biomes they analyzed that occur in South America include alpine, desert, Mediterranean, temperate grassland, tropical savanna, southern temperate forest, tropical forest, lakes, and streams. They used global models of climate, vegetation, and land use to estimate the change in magnitude of the drivers of biodiversity for each biome between the years 1900 and 2100. Application of these models for the creation of scenarios relied on current trends and expert opinions from a large body of literature (e.g., Houghton et al., 2001). Despite the uncertainties in this process, the consensus is instructive in our attempt to understand possible future trends for South American environments.

At a global scale, the model of land-use change used by Sala et al. (2000) projects that most change will continue to occur in tropical forests worldwide and in the temperate forests of South America. The extent of habitat modification is projected to be modest in desert environments and intermediate in savannas, grasslands, and Mediterranean-type ecosystems. Since CO_2 mixes globally, it was assumed that all biomes would experience similar change in CO_2

concentration. Nitrogen deposition was projected to be relatively high in large urban areas (but much less than in Northern Hemisphere urban areas) and smallest in remote southern temperate forests. Warming is expected to be least in tropical areas and most at polar latitudes. It was assumed that biotic exchanges would reflect patterns of human activity and would be least in remote areas.

When averaged across biomes, land-use change is the driver that is expected to have the largest global impact on biodiversity by the year 2100 (Sala et al., 2000). This is primarily due to its devastating effects on habitat availability and on species extinctions following deforestation. Climate change will be the second most important driver of biodiversity change, but will be more effective at high latitudes (especially in the Northern Hemisphere) than in the tropics. Thus, land-use change will be the primary driver of biodiversity change in South America because only a small area of the land mass extends poleward from 45°S and none is south of 56° S. In contrast, vast continental areas lie poleward of 45°N in the Northern Hemisphere and are likely to benefit from increased agricultural productivity on a warmer Earth. Although less important than land use and climate change, the other drivers (atmospheric CO_2, biotic exchange, and nitrogen deposition) will have substantial future effects but also show variability among biomes.

Biotic exchange is projected to have less effect in extreme environments (e.g., alpine and desert), where successful introductions are less likely (Sala et al., 2000). In the tropics, the impacts of introduced species on the diversity of intact ecosystems will be proportionally smaller because of the high initial diversity and apparent resistance to dominance by invaders. The greatest effect of biotic exchange is expected for Mediterranean-type ecosystems and the southern temperate zone because of their long isolation from similar ecosystems in the Northern Hemisphere and the pre-adaptation of invasive species by convergent evolution. Thus, for South America, high impacts are predicted for the temperate forests of southern Chile and Argentina and for temperate shrublands and woodlands living under a Mediterranean-type climate in central Chile. This is consistent with recent observations of widespread impacts of exotic plants in those biomes (see chapters 11 and 13). Atmospheric CO_2 showed the least variability among the drivers because it is well mixed in the atmosphere, and the other drivers showed intermediate levels of variability across biomes (Sala et al., 2000).

The tropical and southern temperate forest biomes showed the greatest impact from land-use change with relatively small impacts due to other drivers (Sala et al., 2000). In contrast, Mediterranean-type ecosystems, savannas, and grasslands are significantly affected by most drivers and, when synergistic interactions among drivers are included, these ecosystems are predicted to suffer the greatest impacts. In contrast, if diversity is determined only by the factor determining the greatest impact, then tropical and southern temperate forests will experience the greatest biodiversity changes (Sala et al., 2000). This analysis highlights the sensitivity of biodiversity changes to uncertain assumptions about interactions among the drivers of change.

Conceptually, it should now be possible to overlay such analyses on existing spatial patterns of biodiversity, and on places where land cover has been converted or altered by people. As examples, Davis et al. (1997) and Stattersfield et al. (1998) documented the richness and uniqueness, respectively, of the plants and birds of South America, identifying places with high numbers of species and high degrees of endemism. Dinerstein et al. (1995), Myers et al. (2000), and Brooks et al. (2002) used biodiversity data such as these, plus the degree of habitat conversion and fragmentation, to delineate suggested priority areas for conservation efforts. Distance to infrastructure and settlements can also be used as predictors of habitat conversion (e.g., Wilson et al., 2005). Other modeling approaches can suggest places likely to be affected by invasive species (Peterson and Vieglais, 2001) and shifts in disease-causing organisms (Rogers and Randolph, 2000). Eva et al. (2004) have prepared a land-cover map for South America, using a baseline for the year 2000, which will permit future modeling of change. Their map shows that the areas presently experiencing most dramatic change are in the southeastern Amazon basin, along the eastern base of the Andes, the pampas of Argentina, the Brazilian *cerrado* and *caatinga*, and the dry Chaco forests of southern Bolivia and northern Argentina. They estimate that more than 11% of South America is now converted to intensive agriculture, while 8.5% of the forests and 4.1% of the nonforest vegetation types have been degraded.

21.2.4 Implications for People

The scenarios of environmental changes described above have many implications for people everywhere in terms of loss of biodiversity and potential feedbacks of regional environmental changes into global climate. Ecosystem services such as watershed protection and wildlife habitat will be degraded in some environments, with potentially major impacts on human activities. Impacts of climate change on food and fiber production will vary widely within South America. Even after allowing for some increase in crop growth under elevated CO_2 scenarios, decreases in the production of major crops are projected for Brazil, Chile, Argentina, and Uruguay (Watson et al., 1997). In large areas where grasslands will experience decreased water availability, livestock production will decline. Livelihoods and cultural survival of traditional peoples, such as many tropical Andean communities, are particularly vulnerable to reductions in rangelands and climate impacts on production of traditional crops. Cultural survival of indigenous people, already threatened by accelerating contacts with the outside world, are likely to be exacerbated by increases in road construction and growing world demand for minerals and other nonrenewable resources (Caviedes and Knapp, 1995).

Socioeconomic impacts on the quality of urban environments are likely to become more severe in all the major urban areas of South America, due both to general population growth and to rural-to-urban migration (see chapter 20). Climate change will have both direct and indirect impacts on the quality of life in urban areas (Watson et al., 1997). Rising sea level will directly affect many people because the continent's population is largely located along the coast and many mega-cities, such as Rio de Janeiro and Buenos Aires, are ports for reasons of colonial history. Other direct effects include higher probabilities of more severe floods, landslides, and heat waves. Indirect effects include negative impacts on food and water supplies, and the deterioration of sanitation infrastructure. The people particularly vulnerable to these climate-related impacts are the populations of shanty towns surrounding all major urban areas, which will continue to grow from both natural reproduction and rural-to-urban migration. Typically, the growth of shanty towns has been on sites more susceptible to flooding and landslides than the sites of permanent urban settlement. Projected changes in climate could increase malnutrition and diseases in human populations. If temperature and precipitation increase, vector-borne diseases (e.g., malaria, dengue, Chagas' disease) and infectious diseases (e.g., cholera) are likely to expand upward in elevation and southward to higher latitudes (Watson et al., 1997; Epstein et al., 1998).

21.3 Conclusions

These predictions of global environmental change for the twenty-first century paint a rather bleak picture for South American environments and their inhabitants. The continent will experience increasing environmental deterioration (e.g. biodiversity changes, deforestation, cropland losses, and more variable water resources) due to both climate change and land-use changes. These changes will aggravate existing socioeconomic and health problems, and may deepen the potential for societal and political conflict. Despite this bleak outlook, there are also reasons for optimism.

The formulation of future scenarios for South American environments and resources through the application of various models consistently shows that socioeconomic drivers of change will have more severe impacts than climatic drivers. From water resources to biodiversity, modeling exercises show that land-use practices, demand for resources, and other socioeconomic factors will continue to dominate at the continental scale. Despite the potentially significant impact of climate change, the primacy of socioeconomic factors clearly increases need for societal flexibility in mitigating and adapting to future environmental changes. The effects of short- to medium-term trends in the socioeconomic causes of global environmental change are relatively well understood. Thus, the canard of uncertainty over the causation of climate change cannot be used as an excuse for lack of planning for and mitigation of environmental changes that are already under way and likely to accelerate.

Another reason for optimism about society's ability to cope with future environmental change is the increased understanding of environmental issues in South America. Improved knowledge and education in recent decades have led to profound changes in public attitudes toward the costs and benefits of environmental management, nature protection, and the relationship of public health to environmental quality. In the context of urban and peri-urban environments, where public concerns have the greatest chance of affecting the political process, there are numerous recent examples of success stories and good practices in environmental management (see chapter 20). New government policies and enforcement efforts since the 1980s have had measurable impacts on the amelioration of industrial pollution in some major cities such as Santiago, Chile, and São Paolo, Brazil. Many cities have invested wisely in improved water and sanitation systems, and issues of environmental health have entered the political discourse in many urban areas. The trend toward more elective government undoubtedly has contributed to the increase in governmental attention to environmental health and management issues. Although the environmental challenges of urban and peri-urban areas remain daunting, the existence of these success stories provides important clues to the types of political and educational strategies required for confronting and progressing against these challenges. At least in the context of urban and peri-urban environmental management, the most successful programs often have originated from local developments led by the more competent and democratic urban governments of the 1980s and 1990s.

Coping effectively with environmental problems in rural areas where large businesses and land scarcity often have disproportionately large influences on the political process will be an even greater challenge. There is a critical need to promote education and cooperation among local residents, nongovernmental organizations, and large businesses to deal effectively and equitably with environmental management in the hinterlands. This task is complicated by past and present injustices, and ethnic and linguistic differences, resulting in differential vulnerabilities. On the other hand, there is a wealth of traditional and modernized land-use practices that have proven sustainable and so could inform responsible development.

The history of changes repeatedly documented in the systematic and regional chapters of this book informs us that dynamic change, not stasis, has long been inherent in the physical environments and landscapes of South America. Throughout the continent there is a record of longer-term changes related to natural processes such as tectonic and volcanic activity, climate change, and sea-level fluctuations, and of shorter-term changes related to land-use practices extending from the pre-European period through the colonial era to modern times. While the magnitude and rates of environmental change measurable in the present and predicted for the near future are without precedent, largely owing to the addition of human factors to natural processes, the history of previous changes provides insights that can inform current environmental policies. Previous episodes of rapid climate change and recurrent El Niño events are invaluable for assessing the impacts of future climate change on nature and human societies. Prior experiences with indigenous agriculture and modern agriculture across the range of environments, from arid to humid and from lowland to highland, can inform our discussions of possible future agricultural productivity. Recolonization of areas devastated by debris flows and stream floods illustrates the resilience of nature and people but also warns of the need for better planning and preparedness in anticipation of future natural hazards, especially as flood-recurrence intervals change with changing climate. Recovery of some tropical lowlands that were intensively cultivated in the precolonial period to intact species-rich tropical forests today also demonstrates the resiliency of some environments. The long-lasting degradation of agricultural productivity associated with mechanized modern agriculture in some environments also teaches us about the sensitivity of different environments to different land uses. This type of knowledge, set out in detail in the chapters of this book, informs choices and enhances options in adapting to the future environmental consequences of both natural change and human globalization in South America.

References

Aide, T.M., and H.R. Grau, 2004. Globalization, migration, and Latin American ecosystems. *Science*, 305, 1915–1916.

Bazzaz, F.A., 1990. The response of natural ecosystems to the rising global CO_2 levels. *Annual Review of Ecology and Systematics*, 21, 167–196.

Brooks, T. M., R.E. Mittermeier, C.G. Mittermeier, G.A.B. da Fonseca, A.B. Rylands, W.R. Konstant, P. Flick, J. Pilgrim, S. Oldfield, G. Magin, and C. Hilton-Taylor, 2002. Habitat loss and extinction in the hotspots of biodiversity. *Conservation Biology*, 16, 909–923.

Cardille, J.A., and J.A. Foley, 2002. Characterizing patterns of agricultural land use in Amazonia by merging satellite classifications and census data. *Global Biogeochemical Cycles*, 16 (3), 10.1029/2000GB001386.

Caviedes, C., and G. Knapp (Editors), 1995. *South America*. Prentice Hall, Englewood Cliffs.

Church, J.A. (Editor), 2001. Changes in Sea Level. In: *Climate Change 2001: A Scientific Assessment*. Intergovernmental Panel on Climate Change.

Davis, S.D., V.H. Heywood, O. Herrera MacBryde, and A.C. Hamilton (Editors), 1997. *Centres of Plant Diversity: A Guide and Strategy for their Conservation. Volume 3: The Americas*. WWF and IUCN, London.

Dinerstein, E., D.M. Olson, et al., 1995. *A Conservation Assessment of the Terrestrial Ecoregions of Latin America and the Caribbean*. The World Bank and World Wildlife Fund, Washington, D.C.

Douglas, B.C., 2001. Sea-level change in the era of the recording tide gauge. In: B.C. Douglas, M.S. Kearney, and S.P. Leatherman, 2001. *Sea Level Rise: History and Consequences*. Academic Press, San Diego, 37–64.

Dufour, D.L., and B.A. Piperata, 2004. Rural-to-urban migration in Latin America: An update and thoughts on the model. *American Journal of Human Biology*, 16, 395–404.

Epstein, P.R., H.F. Diaz, S. Elias, G. Grabherr, N.E. Graham, W.J.M. Martens, E. Mosley-Thompson, and J. Susskind, 1998. Biological and physical signs of climate change: Focus on mosquito-borne diseases. *Bulletin of the American Meteorological Society*, 79, 409–417.

Eva, H.D., A.S. Belward, E.E. De Miranda, C.M. Di Bella, V. Gond, O. Huber, S. Jones, M. Sgrenzaroli, and S. Fritz, 2004. A land cover map of South America. *Global Change Biology*, 10, 731–744.

Farley, K.A., E.F. Kelly, and R.G.M. Hofstede, 2004. Soil organic carbon and water retention after conversion of grasslands to pine plantations in the Ecuadorian Andes. *Ecosystems* 7, 729–739.

Gehrels, W.R., 2005. Sea-level changes during the last millennium. In: M.L. Schwartz (Editor), *Encyclopedia of Coastal Science*. Springer, Dordrecht, 830–833.

Gifford, R.M., 1994. The global carbon cycle: A viewpoint on the missing sink. *Australian Journal of Plant Physiology*, 21, 1–15.

Hecht, S.B., and A. Cockburn, 1990. *Fate of the Forest*, 2nd Edition. Harper-Collins, New York.

Houghton, T., Y. Ding, D.J. Griggs, M. Noguer, P.J. van der Linden, and D. Xiaosu (Editors), 2001. *Climate Change 2001: The Scientific Basis*. Cambridge University Press, Cambridge.

IPCC, 1996. Intergovernmental Panel on Climate Change, *Global Climate Change and the Rising Challenge of the Sea*. Cambridge University Press, Cambridge.

Körner, C., 2000. Biosphere responses to CO_2 enrichment. *Ecological Applications*, 10, 1590–1619.

Labraga, J.C., 1998. The climate change in South America due to a doubling in the CO_2 concentration: Intercomparison of general circulation model experiments. International *Journal of Climatology*, 17, 377–398.

Laurance, W.F., A.K.M. Albemaz, G. Schroth, P.M. Fearnside, S. Bergen, E.M. Venticinque, and C. Da Costa, 2002. Predictors of deforestation in the Brazilian Amazon. *Journal of Biogeography*, 29, 737–748.

Levitus, S., J. Antonov, T. Boyer, and C. Stephens, 2000. Warming of the world ocean. *Science*, 287, 2225–2229.

Mack, R.N., D.Simberloff, W. M. Lonsdale, H. Evans, M. Clout, and F.A. Bazzaz, 2000. Biotic invasions: Causes, epidemiology, global consequences, and control. *Ecological Applications*, 10, 689–710.

Melillo, J.M., 1996. Carbon and nitrogen interactions in the terrestrial biosphere: Anthropogenic effects. In: B. Walker and W. Steffen (Editors), *Global Change and*

Terrestrial Ecosystems. Cambridge University Press, Cambridge, 431–450.

Miles, L., A. Grainger, and O. Phillips, 2004. The impact of global climate change on tropical forest biodiversity in Amazonia. *Global Ecology and. Biogeography,* 13, 553–565.

Mooney, H.A., and R.J. Hobbs (Editors), 2000. *Invasive Species in a Changing World.* Island Press, Washington, DC.

Moran, E.F., E.S. Brondizio, J.M. Tucker, M.C. da Silva-Forsberg, S. McCracken, and I. Falesi, 2000. Effects of soil fertility and land-use on forest succession in Amazonia. *Forest Ecology and Management,* 139, 93–108.

Myers, N., R.A. Mittermeier, C.G. da Fonseca, A.B. Gustavo, and J. Kent, 2000. Biodiversity hotspots for conservation priorities. *Nature,* 403, 853–858.

Ozanne, C.M.P., D. Anhuf, S.L. Boulter, M. Keller, R.L. Kitching, C. Körner, F.C. Meinser, A.W. Mitchell, T. Nakashizuka, P.L. Silva Dias, N.E. Stork, S.J. Wright, and M. Yoshimura, 2003. Biodiversity meets the atmosphere: A global view of forest canopies. *Science,* 301, 183–186.

Painter, M., and W. Durham (Editors), 1995. *The Social Causes of Environmental Destruction in Latin America.* University of Michigan Press, Ann Arbor.

Peterson, A.T., and D.A. Vieglais, 2001. Predicting species invasions using ecological niche modeling: New approaches from bioinformatics attack a pressing problem. *Bioscience,* 51, 363–371.

Pielke Sr., R.A., G. Marland, R. Betts, R.N. Chase, J.L. Eastman, J.O. Niles, D. Niyogi, S.W. Running, 2002. The influence of land-use change and landscape dynamics on the climate system: Relevance to climate change policy beyond the radiative effect of greenhouse gases. *Philosophical Transactions of the Royal Society of London (Series A),* 360, 1–15.

Post, W.M., and K.C. Kwon, 2000. Soil carbon sequestration and land-use change: Processes and potential. *Global Change Biology,* 6, 317–327.

Ramankutty, N., J.A. Foley, J. Norman, and K. McSweeney, 2002. The global distribution of cultivable lands: Current patterns and sensitivity to possible climate change. *Global Ecology and Biogeography,* 11, 377–392.

Rogers, D.J., and S.E. Randolph, 2000. The global spread of malaria in a future warmer world. *Science,* 289, 1763–1766.

Rudel, T.K., 2002. Paths of destruction and regeneration: Globalization and forests in the tropics. *Rural Sociology,* 67, 622–636.

Sala, O.E., F.S. Chapin, J.J. Armesto, E. Berlow, J. Bloomfield, R. Dirzo, E. Huber-Sanwald, L. F. Huenneke, R.B., Jackson, A. Kinzig, R. Leemans, D.M. Lodge, H.A. Mooney, M. Oesterheld, N.L. Poff, M.T. Sykes, B.H. Walker, M. Walker, and D.H. Wall, 2000. Global biodiversity scenarios for the year 2100. *Science,* 287, 1770–1774.

Saleska, S.R., S.D. Miller, D.M. Matross, M.L. Goulden, S.C. Wofsy, H.R. da Rocha, P.B. de Camargo, P. Crill, B.C. Daule, H.C. de Freitas, L. Hutyra, M. Keller, V. Kirchhoff, M. Menton, J.W. Munger, E.H. Pyle, A.H. Rice, and H. Silva, 2003. Carbon in Amazon forests: Unexpected seasonal fluxes and disturbance-induced losses. *Science,* 302, 1554–1557.

Shukla, J., C. Nobre, and P. Sellers, 1990. Amazon deforestation and climate change. *Science,* 247, 1322–1325.

Silver, W.L., R. Ostertag, and A.E. Lugo, 2000. The potential for carbon sequestration through reforestation of abandoned tropical agricultural and pasture lands. *Restoration Ecology,* 8, 394407.

Smil, V., 1997. *Cycles of Life: Civilization and the Biosphere.* Scientific American Library, New York.

Soares-Filho, B., A. Alencar, D. Nepstad, G. Cerqueira, M. del Carmen Vera Diaz, S. Rivero, L. Solórzanos, and E. Voll, 2004. Simulating the response of land-cover changes to road paving and governance along a major Amazon highway: The Santarém-Cuiabá corridor. *Global Change Biology,* 10, 745–764.

Stattersfield, A.J., M.J. Crosby, A.J. Long, and D.C. Wege, 1998. *Endemic Bird Areas of the World: Priorities for Conservation.* Birdlife International, Cambridge, U.K.

Swinton, S.M., and R. Quiroz, 2003. Is poverty to blame for soil, pasture and forest degradation in Peru's Altiplano? *World Development,* 31, 1903–1919.

Viña, A., F.R. Echavarria, and D.C. Rundquist, 2004. Satellite change detection analysis of deforestation rates and patterns along the Colombia-Ecuador border. *Ambio,* 33, 118–125.

Vitousek, P.M., J.D. Aver, R.W. Hobart, G.E. Likens, P.A. Matson, D.W. Schindler, W.H. Schlesinger, and D.G. Tilman, 1997. Human alteration of the global nitrogen cycle: Sources and consequences. *Ecological Applications,* 7, 737–750.

Vörösmarty, C.J., P. Green, J. Salisbury, and R.B. Lammers, 2000. Global water resources: Vulnerability from climate change and population growth. *Science,* 289, 284–288.

Walker, R., E. Moran, and L. Anselin, 2000. Deforestation and cattle ranching in the Brazilian Amazon: External capital and household processes. *World Development,* 28, 683–699.

Watson, R.T., M.C. Zinyowera, and R.H. Moss (Editors), 1997. *The Regional Impacts of Climate Change: An Assessment of Vulnerability.* Cambridge University Press, Cambridge.

Watson, R.T., and Core Writing Team (Editors), 2001. *Climate Change 2001: Synthesis Report.* Cambridge University Press, Cambridge.

Wilson, K., A. Newton, C. Echeverría, C. Weston, and M. Burgman, 2005. A vulnerability analysis of the temperate forests of south central Chile. *Biological Conservation,* 122, 9–21.

Wunder, S., 2001. Poverty alleviation and tropical forests—What scope for synergies? *World Development,* 29, 1817–1833.

Zimmerer, K.S., and T.J. Bassett, 2003. *Political Ecology: An Integrative Approach to Geography and Environment-Development Studies.* Guilford Press, New York.

Index

acid precipitation, 326, 331, 343
Aedes
  aegypti, 316, 333
  mosquito, 316
aeolian processes and landforms, 7, 9, 36, 170, 171, 177, 219, 236
  dune, 36
  sedimentation, 7
  volcanic-aeolian deposit, 219
afforestation, 243–245, 345
Africa, 3, 4, 112–121, 187, 201, 250, 253, 281, 282, 285, 307
African-American, 282
Africanization, 285
agouti, 267, 270
agriculture, 289–302
  intensification, 281, 293–295, 313, 346
  practice, 291, 297
  production, 108, 289, 315, 331, 346
agriculture and soils, 102–111
agroforestry, 150, 272
air mass, 26, 93, 96, 97, 98, 106, 135, 136, 160, 169, 185, 187, 202–204, 227, 233, 234, 306
alder, 205
alfisol, 106–110, 236
alien species. *See* exotic species
alpaca, 280, 284, 297
Altiplano, 14, 18, 26, 27, 32, 36, 38, 40, 67, 68, 70, 106, 125, 126, 160, 202, 300, 308
  climate, 47, 49, 52–54, 203, 307, 309, 317
  fauna, 118, 123
  flora, 40
  soil, 106
Altiplano Plateau, 14, 202
Amado, Jorge, 322
amaranth, 290
Amazon, 83, 124, 266, 292
  biogeography, 146–147
  campina, 144
  deforestation, 342, 344
  delta, 38, 80
  diversity, 93, 123, 124–125, 136, 144
  estuary, 76, 79–80, 84, 146, 252, 275
  fan, 35, 259
  fire, 314
  floodplain, 80, 86, 138, 313
  lowland, 15, 24, 40, 91, 93, 95, 98, 107, 108, 142, 281
  mouth, 256
  region, 295
  vegetation, 93, 94, 95, 115, 124, 125, 135–136, 142–146, 148, 268, 274, 313
Amazon Basin, 25, 26, 32, 37, 40, 45, 46, 47, 49, 51, 76–81, 83, 84, 88, 93, 95, 96, 98, 107, 108–109, 110, 112, 113, 118, 123, 124–125, 135–137, 138, 145, 146, 160, 202, 203, 254, 260, 266, 274, 282, 289, 292, 298, 300, 307, 308, 312, 313, 314, 344, 345, 347, 349
Amazon Cone, 80, 138
Amazon River, 7, 19, 32, 34, 35, 46, 76, 77–82, 84, 86, 95, 107,108, 109, 137–138, 142, 146, 254, 255, 259, 266, 274, 275, 281, 290, 296, 308, 314, 348
Amazonia, 7, 38, 40, 88, 125, 135–154, 202, 265–268, 270–272, 274, 275, 285, 345
  biome, 117
  climate, 27, 47, 49, 52, 136–138, 202–203, 309, 314
  environment, 136–142, 300, 341, 345
  fauna, 40, 41, 113–118, 122, 146
  flora, 141–142
  population, 327, 342
  western, 52, 118
Amazonian Shield, 7
amphibian, 40, 112, 114, 118, 121, 123, 124, 125, 167, 176, 211
amphisbaenid, 115
anchovy, 310
Andes, 4, 7, 9, 15, 17, 18, 19, 24, 27, 29, 32, 34, 36, 37, 38, 45, 47, 50, 52, 76, 84, 86, 102, 108, 109, 110, 111, 112, 123, 137, 150, 158, 159, 160, 161, 162, 163, 168, 170, 172, 185, 187, 190, 195, 217, 218, 219, 232, 234, 250, 253, 254, 273, 279, 280, 296, 300, 302, 308, 334, 346, 349
  central, 12, 13, 14, 15, 16, 18, 34, 36, 47, 116, 194, 281, 282–285, 291, 296, 341, 345
  climate, 47–49, 50, 52, 53, 219–220, 318
  fauna, 113, 114, 115, 116, 123–124
  flora, 26, 91–98, 194, 207, 208–209
  high Andes, 15, 26, 208–209, 289, 298, 315, 346
  northern, 9, 12, 13, 16, 35, 116, 281, 345
  orogeny, 12–16, 27

Andes (*continued*)
rain shadow effect, 26, 27, 36, 95, 97, 123, 187, 188, 192, 202, 206, 219, 221, 234
uplift, 12–16, 24, 26, 32, 34, 37, 39, 40, 77, 160, 187, 188, 194, 234, 236
soil, 204, 219
southern, 12, 15–16, 36, 217, 219, 226, 227
subtropical, 200–212
temperate forest, 217–230
tropical, 40, 60–73, 116, 125, 200–212, 347
vegetation, 91–98, 204–211, 217–230
wetland, 209–210
andisol, 106, 107, 110
andosol, 106, 107, 110, 219
angiosperm, 97, 220, 222, 230
Antarctica, 4, 10, 25, 38, 39, 112, 118, 120, 187, 222
Antarctic Circumpolar Current, 25, 26, 39, 187
Antarctic plate, 4, 15, 17, 31, 219, 250
anthropogenic influence,
on climate, 343–344, 348
on land use/land cover, 76, 191, 196, 212, 298
on landscape, 230, 266, 268, 269, 272, 275, 302
on mammalian extinction, 121
on natural processes, 340
on waterways, 275
Apuré Basin, 18, 86
aquept, 219
Araucaria, 149, 221, 222, 227, 228, 347
Arequipa craton, 9
Argentina, 4, 9, 12, 19, 39, 52, 81, 83, 97, 114, 118, 119, 121, 124, 125, 148, 158, 169, 187, 194, 225, 226, 228, 229, 232, 234, 237, 243, 244, 251, 254, 282, 285, 290, 292, 294, 295, 296, 297, 298, 309, 324, 329, 330, 341, 342, 344, 345, 347, 348, 349
Andes, 36, 108, 218
central, 47, 50, 52, 97, 110, 233
and climate change, 194, 349
fauna, 118, 124, 125
flora, 118, 194, 220
livestock, 243
northeast, 54, 55
northern, 15, 19, 96, 200, 202, 204, 316, 349
northwest, 34, 36, 194, 205, 317
soil, 110
southern, 82, 110, 202, 252
vegetation, 96–97, 114, 121, 125, 148, 169, 349
western, 110, 315, 344
Arica, 160, 163, 165, 167, 168
Arica Bend, 12, 15, 29, 162
aridisol, 107, 108, 109, 110, 171, 236
arrowroot, 266, 270
Atacama Desert, 19, 26, 27, 36, 96, 118, 158–169, 178, 184, 187, 217, 300, 318
protected area, 168–169
Atlantic Convergence Zone. *See* SACZ
Atlantic gyre, 27
Atlantic Ocean, 4, 7, 10, 25, 26, 34, 45, 52, 55, 69, 77, 84, 93, 96, 98, 138, 201, 202, 238, 240, 249, 308
Atlantic rain forest, 93, 94, 96, 117, 118, 125, 135, 146, 147–150, 342, 347
biogeography, 148–149
conservation, 149–150
diversity, 115, 116, 117, 125
See also Mata Atlântica
Atlantic Rift, 19
Atlantic Shelf, 252, 253
Atlantic Shield, 7
atmospheric circulation, 17, 26, 45–55, 91, 135, 202, 308, 309, 318, 344
Australia, 25, 39, 96, 97, 112, 113, 114, 117, 120, 187, 222, 307, 347
Australia Outback, 154
Austral temperate forest, 114, 115
Austrocedrus woodland, 228
avalanche, 34, 225, 226, 227
avifauna, 115–116, 124

babassu (palm), 144, 267, 268
forest, 268
Bactris, 267, 270
Baja California, 160
bamboo, 96, 145, 149, 205, 208, 210, 221, 225, 227, 228
forest, 145, 268
banana, 143, 282, 290, 291, 292, 293, 294, 295, 342
barley, 284, 291
barrier beach, 35, 254, 255, 259, 260
bat, 116, 117, 118,120, 144, 206
bean, 283, 290
broad, 284
jack, 270
Lima, 290
biodiversity, 40, 348–349
hotspot, 93, 125, 126, 147, 184
South America, 126
threat to, 168, 184, 230
biodiversity conservation, 177, 200, 284, 286, 349
Amazon, 125
Andes, 200
Chile, 184, 188, 196–197
coastal habitat, 168–169
Peru, 168
biogeography, 38–41, 93, 112–126, 166–168, 172–173, 221–224, 234–236
isolation, 201
links, 148–149, 151–152, 153
and tectonism, 23, 38
biotic exchange, 39, 341, 346, 349
bird, 115–116
black water, 140, 145
Bogota, 61, 280, 285, 324, 331, 334, 335
Bogota River, 331, 335
Bogota Sabana, 294
Bolivia, 18, 78, 83, 194, 211, 268, 273
and agriculture, 290, 292, 293, 294, 295, 297
Altiplano, 202, 203
Amazonian, 84, 95, 136, 145, 281
Andes, 15, 18, 19, 39, 64, 95, 123, 200
climate, 47, 52, 68, 70, 81, 136, 203
and coca, 293
dam, 88
deforestation, 342
desertification, 300
disturbance, 297, 298
diversity, 123, 125
and El Niño, 308, 317
fauna, 123, 125
flora, 39, 205, 209
glaciation, 36, 60, 65–66, 68, 72
indigenous population, 282
population, 72, 281, 295, 329, 341
soil, 297, 298, 299
urban area, 329, 336
vegetation, 96, 145, 349
volcanism, 29
Bolivian High, 48, 49, 53
bonytongue fish, 113
bottle gourd, 266
Bouguer, Pierre, 37
Brazil, 7, 35, 36, 82, 110, 194, 255, 256, 266, 315, 316
and agriculture, 272, 274, 275, 286, 290, 291, 292, 293, 294, 295, 297, 298, 302
Amazonian, 84, 93, 108, 135, 136, 290, 345
Atlantic coast, 12, 109, 118, 148, 254, 256, 280, 285, 342
and climate change, 344, 349
climate, 36, 46, 47, 48, 52, 54, 55, 136, 150, 233, 234, 309, 344
dam, 88, 346, 347
deforestation, 146, 152, 342, 274, 345
diversity, 117, 118, 125
and El Niño, 54, 55, 307, 308, 313
fauna, 117, 118, 125, 145
flora, 145, 271
fossil, 39
geology, 7, 10, 12, 15, 20, 39, 77, 86, 107, 138, 139, 250
indigenous population, 267, 272, 282, 285
introduced flora, 285
land degradation, 296, 297, 298, 300
population, 324, 326, 327
soil, 107, 108, 110
urban area, 88, 324, 326, 327, 329, 331, 335, 350
vegetation, 39, 40, 94, 96, 144, 146, 149, 150, 152, 153, 158, 232, 268, 269, 296, 313, 345
Brazil Current, 25, 27, 309
Brazilian Plateau, 12, 45, 52
Brazilian Shield, 4, 18, 19, 32, 35, 36, 37, 76, 77, 84, 86, 88, 91, 98, 107, 108, 109, 110, 138, 250, 252, 253
Brazil nut, 143, 267, 268, 271
bromeliad, 146, 152, 153, 164, 206
Buenos Aires, 9, 39, 110, 238, 255, 258, 280, 298, 324, 326–333, 348, 350
burning, 102, 105, 140, 147, 151, 196, 207, 211, 226, 228, 265, 268, 269, 271, 272, 283, 284, 342, 344, 345

caatinga, 36, 40, 93, 94, 117, 118, 135, 139, 144, 148, 150, 152–154, 158, 269, 313, 347, 349
disturbance, 154
Cabral, Pedro Alvares, 275
cactus, 95, 152, 153, 158, 159, 164, 165, 166, 167, 169, 190, 206, 313
caecilian, 114, 119
calcium, 104, 105, 106, 109, 140, 244, 271, 311
California, 96, 184, 188, 195, 210, 307
California current, 26
camanchaca, 160
Cambrian, 4, 9
camelid, 121, 123, 228, 280, 297
campina, 141, 144, 269, 270
campinarana, 144
campo cerrado. *See* cerrado
campo rupestre, 146
canid, 121
capybara, 121, 270
Caracas, 259, 280, 329, 330, 333
carbon, 41, 101, 102, 104, 146, 147, 154, 210, 230, 241, 243, 244, 296, 310, 311, 313–314, 333, 343, 345
credit, 345
sequestration, 230, 313–314, 344, 345
Carboniferous, 7, 19
Caribbean, 4, 19, 88, 94, 202, 254, 259–260, 309, 347, 348
fauna, 26, 39
forest, 94, 97
Caribbean Plate, 4, 13, 16, 17, 29, 39, 77, 120, 201, 249, 250
Caribbean Sea, 26, 39, 54, 249, 255, 308
cashew, 267, 290
Casiquiare, 274, 275
Catamarca, 170, 171, 172, 174, 175, 332
cattle, 41, 97, 107, 154, 176, 192, 228, 244, 273, 284, 285, 286, 297, 344
grazing, 107, 153, 176, 192, 195
production, 110, 147, 177, 244, 285, 342
Cecropia, 143, 144, 146
Cenozoic, 4, 9, 12, 16, 17, 40, 112, 123, 162, 221, 252
biogeography, 38–40, 120, 122, 194, 221–222, 223
climate, 18, 24–27, 40, 79, 172, 202, 236
glaciation, 60, 219
tectonism, 4, 7, 9, 14, 16, 17, 18, 19, 26, 32, 34, 40, 98, 112, 160, 161, 170, 187, 257
volcanism, 27
weathering, 19
Central American Isthmus, 13, 17, 24, 26, 39, 40, 121
Central Cordillera (Colombia), 13, 29
Central Depression, 15, 26, 96, 185, 187, 190, 191, 192, 195, 218, 219, 228, 342
Central Valley, 18, 160, 162, 163, 168
cerrado, 93, 94, 96, 107, 109, 110, 117, 118, 125, 141, 148, 153, 154, 268, 272, 349
cervid, 121, 228
Chaco, 15, 36, 117, 118, 126, 145, 150, 169, 172, 173, 177, 289, 300, 347, 349
climate, 27, 47, 51
diversity, 175, 176
vegetation, 94, 96, 98, 145, 150, 151, 152, 153, 349
Chaco Basin, 7, 76
Chile, 69, 124, 184–197, 200, 202, 327
and agriculture, 290, 293, 294, 295, 298, 342
biogeography, 40, 124, 159, 160, 168, 221–223
and biological conservation, 168, 196–197
central, 96
climate, 46, 47, 50, 51, 52, 53, 104, 158, 160, 185–188, 219, 223, 254, 307, 308, 310, 349
and climate change, 344, 347, 349
coast, 158, 217, 254, 255, 256, 260, 289
dam, 88
deforestation, 229–230, 342
desertification, 300
disturbance, 195, 196, 226–228, 229, 297
diversity, 118, 123, 124, 126, 168, 184
and El Niño, 55, 161, 186, 310, 312–313, 315, 317
exotic species, 228–229, 230
fauna, 115, 118, 167
fishery, 310–311
flora, 39, 159, 167, 168, 188–195, 190, 221, 225
geology, 4, 12, 13, 15, 16, 18, 19, 20, 26, 47, 161, 162, 163, 185, 187, 218–219, 250, 251, 252, 257
glacier, 36, 61, 79
indigenous population, 280, 281, 282, 285
population, 228
protected area, 168
soil, 110
urban area, 88, 330, 333, 350
vegetation, 39, 96, 97, 124, 158, 160, 168, 188–195, 208, 217, 220–221, 224–230, 232, 312, 313, 345, 349
volcanism, 204
Chilean Lake District, 225, 226, 229
chili pepper, 283, 290
Chiloé Island, 218, 219
Chimborazo, 29, 36, 69, 202
Chocó, 95, 115, 124, 125, 126, 135, 202
cholera, 312, 315, 330, 350
Chuquisaca, 280, 285
Chusquea, 221, 225–226, 227, 228
cichlid, 113, 119
clear water, 137, 145, 347
climate change (past), 24–27, 60, 64
and biodiversity, 67, 212
and glaciation, 62, 63, 71
Quaternary, 62
and tectonism, 23, 32, 35, 36, 60
climate change (future), 296, 322, 334, 335, 343–347
and biodiversity, 348–349
and people, 349–350
climatic variability, 45–55
climatology (South America), 45–55
coast, 249–260, 347–348
collision-edge, 250, 252
trailing-edge, 252
Coastal Cordillera (Chile), 218
coastal erosion, 256, 334
coastal fauna, 311–312
coca, 211, 283, 290, 293, 300
cocoa, 108
Cocos Plate, 13, 29, 201, 249, 250
Cocos Ridge, 40, 318
coffee, 107, 108, 110, 285, 286, 290, 291, 292, 294, 297, 300, 333, 342
cold surge, 52,
Colombia, 200
and agriculture, 270, 273, 274, 275, 281, 290, 292, 294, 295, 297, 298, 342
Amazonian, 95, 146
Andes, 13, 95, 202
biogeography, 40, 121, 123
Caribbean, 150, 158, 259, 307
climate, 202, 203, 254, 308
and coca, 293
dam, 298
and disease, 316
disturbance, 297, 298, 300
diversity, 125
and El Niño, 311, 316, 318
exotic species, 285
fauna, 115, 121, 123, 125
geology, 9, 13, 19, 26, 39, 78, 250
glaciation, 61
indigenous population, 271, 282, 285
Pacific, 256
population, 324, 326
soil, 109, 110
urban area, 326
vegetation, 94, 95, 158, 275
volcanism, 204
colonialism,
environmental dimensions, 279
European, 279–286
colonization, 94, 98, 152, 228, 229, 282, 294, 329, 340
biological, 180, 207, 208
Columbus, Christopher, 279, 281
conifer, 97, 124, 149, 194, 217, 220, 221, 223, 225, 230, 345
conservation,
and agriculture, 284
Atlantic rain forest, 149–150
Chile, 168, 169, 177, 196–197
fauna, 349
management, 93
Peru, 168
tropical Andes, 200
continental shelf, 4, 9, 12, 34, 35, 36, 80, 163, 252, 259, 347
coral, 39, 309, 311
bleaching, 311
reef, 253, 254, 256, 259, 260, 311, 348
cormorant, 312
Coropuna, 18, 29
Cotopaxi, 29, 36
cotton, 102, 284, 286, 290, 291, 292, 295, 315

Cretaceous,
and biogeography, 221, 222
climate, 24, 32
fossil, 38, 39, 119, 120
tectonics, 7, 9, 10, 12, 13, 17, 19, 25, 32, 113, 119, 120, 123, 161, 250, 253
crime, 327
crocodilian, 38, 39, 114, 115, 145
cryoturbation, 208
cultivation, 88, 111, 150, 210, 265, 266, 271, 272, 275, 281, 282, 283, 286, 292, 298, 302, 315, 344
shifting, 266, 272, 285
cut flower, 292, 295, 342
cutoff low, 50, 52
Cuzco, 280

dam, 76, 84, 87, 88, 186, 229, 283, 298, 347
Darwin, Charles, 40, 91, 93, 95, 97, 98, 236
deciduous forest, 94, 96, 107, 144, 150–152, 206
deer, 39, 40, 41, 121, 146, 167, 228, 244, 270, 271, 275, 283
deforestation, 88, 93, 96, 98, 125, 136, 137, 146, 149, 154, 177, 220, 229, 267, 275, 333, 342–345, 348–350
deglaciation, 38, 61, 65, 66, 68, 70, 71, 73, 223
Demerera Plateau, 10
demographic transition, 341–343
dengue, 315, 316, 327, 333, 334, 350
denudation, 27, 35, 36, 37–38, 202, 298, 300
depopulation, 269, 279, 281, 282, 285
desert, 7, 26, 34, 36, 47, 91, 93, 94, 95, 96, 108, 158–169, 177, 184, 208, 238, 254, 281, 312, 329, 348, 349
and agriculture, 289
biome, 40
fauna, 118, 167, 176
flora, 160, 166–167, 177
vegetation, 169, 313
desertification, 154, 172, 177, 234, 244, 245, 300
dinosaur, 38, 119, 120
disease, 103, 104, 121, 282, 286, 296, 315–317, 326, 327, 328, 330, 333, 334, 335, 341, 345, 346, 349, 350
disintensification, 279, 285
dispersal, 95, 112, 119, 120, 121, 122, 123, 124, 125, 144, 167, 218, 222, 271, 330
barrier, 125, 202, 211
long-distance, 38, 94, 95, 96, 144, 172, 222
route, 221
seed, 38, 144, 188, 192, 195, 206, 267, 271
waif, 120, 121
domestication, 41, 210–211, 270
Dom Pedro, 279
Doyle, Arthur Conan, 95
Drake Passage, 17, 25, 26, 187
drought, 32, 40, 41, 54, 55, 72, 150, 153, 154, 160, 168, 169, 184, 185, 186, 188, 195, 225, 226, 227, 268, 300, 306, 307, 313, 314, 316, 318
adaptation to, 104,144, 166
stress, 191
dry forest, 84, 94, 150, 150–153, 227, 313, 342
biogeography, 151–152, 153
tropical, 94, 95, 124, 150–153, 206
vegetation, 152, 153
earthquake, 3, 19, 20, 29, 32, 34, 41, 91, 110, 204, 226, 282, 329
East Coast Swell, 254
Easter Island, 306
ecological footprint, 333
ecotone, 84, 169, 173, 204, 227, 228, 230, 232, 344
Ecuador, 37, 125, 160, 168, 200
and agriculture, 273, 274, 283, 290, 292, 295, 297
Amazonian, 34
Andes, 95, 110, 202
climate, 48, 54
and climate change, 344
coast, 110, 254, 269, 273, 274
dam, 88, 298
deforestation, 146
and disease, 315–316
disturbance, 269, 298–299, 345
diversity, 125
and El Niño, 34, 203, 305–308, 311, 315, 318
exotic species, 345
fauna, 123, 311
fisheries, 310–311
glaciation, 36, 60, 69, 71, 72
indigenous population, 265, 270, 281, 282, 283
and La Niña, 318
population, 72, 341
soil, 109, 110
tectonics, 9, 12, 13, 17, 19, 32
urban area, 329
vegetation, 95, 135, 210, 256, 269, 312, 313, 345
volcanism, 27, 204
edge effect, 125, 150
ELA (Equilibrium line altitude), 60, 64, 65, 67, 68, 71
El Niño, 26, 32, 34, 46, 53, 54, 55, 64, 80, 96, 147, 158, 160, 161, 165, 166, 167, 186, 203, 254, 281, 282, 305–318, 334, 348, 351
and disease, 315–317
encephalitis, 315, 316
endangered species, 148
endemism, 94, 95, 114, 115, 116, 117, 124, 125, 126, 152, 166, 167, 168, 175, 177, 184, 188, 190, 197, 205, 211, 349
ENSO (El Niño Southern Oscillation), 45, 55, 71, 80, 160, 186, 203, 305–318, 340
environmental change, 63, 350, 351
biological, 38
cause, 341–343
and climate, 343
future, 212
natural, 265
environmental management, 326, 327, 328, 335, 336, 340, 350
Eocene, 10, 13, 24, 32, 120, 151
epiphyte, 95, 143, 148, 149, 152, 188, 204
equid, 121
erosion,
soil, 289–302
wind, 36, 209, 236, 296, 300
Espeletia, 209
Eucalyptus, 150, 195, 197, 244
evapotranspiration, 103, 104, 137, 153, 169, 175, 206, 234, 241, 243, 334, 345
evolution, 24, 38, 39, 40, 92, 93, 94, 95, 96, 119, 120, 125, 141, 172, 188, 202, 207, 210, 349
exotic species, 41, 163, 195, 197, 208, 210, 211, 228, 229, 230, 289, 290, 311, 346, 349
extinction, 24, 38, 41, 98, 118, 121, 125, 149, 271, 285

fallow, 267, 283, 284, 291, 295, 297, 299
sectoral, 283
Farallón plate, 13
faunistic region, 113–117
fern, 95, 145, 148, 149, 210
fertilizer, 102, 104, 105, 106, 107, 109, 110, 111, 163, 272, 281, 294, 296, 300, 312, 346
fire, 105, 125, 144, 147, 151, 197, 224, 229, 245
anthropogenic, 268
biological adaptations, 94, 143, 144, 150, 225, 227, 268, 314
and disturbance, 125, 190, 192, 195, 196, 208, 225, 226, 227, 229, 245, 265, 296, 297
frequency, 105, 195, 223, 228, 230, 245, 268, 313
natural, 187
regime, 227–228
firewood, 177, 196, 207, 283, 284, 285
fish, 40, 86, 88, 112, 113–114, 119, 121, 123, 124, 125, 145, 146, 266, 270, 274, 275, 284, 309–312, 333, 346, 347
fisheries, 39, 310, 331, 332, 342
Chile, 310
industrial, 310
Peru, 310
fishmeal, 312
Fitzroya cupressoides, 221, 223, 225, 229
forest, 227
fjord, 79, 218, 219, 254, 256, 257, 258
fluvial geomorphology, 84–87
fog, 27, 40, 159–161, 164–166, 167, 168, 187, 188, 203, 307
food production, 102, 105, 106, 110, 111
forest clearance, 34, 40, 147, 211, 229, 266, 297

fossil, 17, 26, 39, 40, 97, 119, 120, 121, 187, 190, 236
microfossil, 65, 190, 191, 223
rodent midden, 318
fossil fuel, 333, 334, 343, 344, 346
frog, 114, 115, 120, 123, 176, 312

Galapagos finch, 97
Galapagos Islands, 17, 29, 40, 97, 305, 306, 307, 311, 312
García Marquez, Gabriel, 322
garua, 160, 307
geomorphology, 27–38, 77–87, 138–139, 161–163, 169–172, 203–204
glacial stage, 68, 79, 123, 124, 125, 149, 188, 318
glaciation, 7, 23, 25, 26, 27, 32, 36, 60, 61, 62, 63, 64, 66, 67, 68, 69, 70, 71–72, 98, 236, 251, 252, 254
glacier, 26, 27, 36, 37, 54, 60–73, 106, 123, 203, 219, 227, 254, 258, 347
advance, 61, 67, 72, 223, 236
retreat, 66, 67, 69, 72, 209, 210, 223, 347
tropical, 64, 67, 72, 79
global environmental change, 340, 341, 350
globalization, 98, 282, 294, 340, 341, 342, 348, 351
global warming, 38, 60, 64, 251, 317, 334–335, 344, 347–348
gold, 19, 285, 286
gomphothere, 38, 41, 121
Gondwana, 4, 7–12, 13, 16, 17, 18, 32, 38, 40, 60, 97, 112, 118, 119, 120, 187, 221, 222, 249, 250, 253
governance, 327, 335, 336
Gran Chaco, 27, 117, 289
Gran Pajonal, 269
grape, 104, 284, 290, 293, 295, 333
grass, 93, 94, 97, 145, 152, 153, 175, 176, 191, 204, 207, 208, 209, 217, 238, 245, 312, 313, 316
C3, 194, 237, 243
C4, 194, 206, 237, 243, 245
exotic, 192, 208, 285
grassland, 41, 91, 93, 94, 95, 136, 141–142, 145, 174, 175, 200, 204, 206, 292, 344, 345, 348, 349
Andean, 200, 207–208, 210, 211, 283, 284, 345
temperate, 24, 217, 228, 232–246, 348
tropical, 145, 207–208, 268–270, 281, 285
grazing, 40, 95, 103, 107, 108, 109, 110, 125, 153, 165, 176, 191, 192, 195, 204, 207, 210, 229, 237, 238, 243–245, 265, 280, 283, 284, 299, 315, 342
and desertification, 177, 245
Great American Biotic Interchange, 39, 112, 120–123
Green Revolution, 294, 346
guanaco, 40, 166, 167, 168, 176, 195, 228, 244
guano, 312
island, 164
mine, 167
Guaporé, 84, 138, 145
Guaporé Shield, 4
guava, 271, 290
Guayaquil, 328, 329
Gulf of, 77, 305, 312
Guayas River, 256, 273, 274
Guianas Basin, 40
Guiana Shield, 4, 18, 32, 37, 40, 76, 84, 86, 91, 94, 95, 98, 107, 108, 109, 110, 138, 139, 145, 250, 252, 253
Guyana, 94, 107, 109, 125, 139, 254, 279, 292, 300, 316, 348
Guyana Shield. *See* Guiana Shield
gymnosperm, 97

halophyte, 174
Hanta virus, 226, 316
hazardous waste, 330, 333
Holocene, 15, 70, 79, 170, 223, 250, 251, 260, 268
climate, 36, 40, 60, 64, 73, 161, 191, 192, 193, 194, 223
and ENSO, 161
fire, 223
floodplain, 86, 139
glaciation, 26, 60, 64, 69, 71–72, 258
lahar, 29
sea-level, 79, 250–251, 259
volcanism, 34, 226
hotspot (biological), 93, 125, 126, 147, 184
house garden, 272
huaico, 34, 315
Huancabamba, 211
Huancabamba Bend, 12
Huaorani, 267
Huascarán, 14, 34, 68, 69, 70, 71, 202, 204, 317
Huaynaputina volcano, 29
human
habitation, 110, 161, 196, 289, 322
impact, 40, 125, 149, 164, 168, 177, 178, 192, 195–196, 210–211, 212, 227–229, 243, 265–276, 296, 326, 340, 345, 346
migration, 282, 291, 302, 342, 343, 348, 350
Humboldt, Alexander von, 91, 95, 98, 135, 266
Humboldt Current, 25, 26, 27, 159, 160, 161, 187, 305, 307, 310, 311
hunting, 121, 125, 126, 150, 228, 265, 266, 267, 270–271, 275, 280, 284, 314, 342
hydroelectric power, 64, 87, 186, 298
hyper-aridity, 26, 36, 96, 118, 119, 124, 160–161, 163, 178, 187, 234

ice core, 62, 63, 66, 68, 69, 70, 71, 317
igapó, 145
Iguaçu Falls, 36
impervious surface, 297, 298
inceptisol, 106, 107, 140, 236
infant mortality, 326, 341
interglacial, 27, 41, 223
cycle, 79
Last, 251
stage, 79, 123, 124, 188, 250, 318
introduced species. *See* exotic species
irrigation, 40, 64, 72, 93, 103, 104, 107, 108, 110, 111, 161, 177, 186, 210, 243, 244, 274, 281, 283, 292, 294, 331, 344
Isla Margarita, 94
Island Biogeographic Theory, 124, 149
island hopping, 39, 120, 129
isostasy, 24, 27, 37–38
Isthmus of Panama, 40, 259, 284, 313, 340
ITCZ (Intertropical Convergence Zone), 25, 26, 27, 46, 96, 136, 202, 233, 249, 313, 344

jet stream, 27, 47, 48, 50, 55, 308
Jurassic, 250
climate, 7
fossil, 119, 120
tectonics, 7, 9, 10, 12, 13, 18, 123, 185

Kayapó, 267, 272
Kelvin wave, 305, 306
Kerguelen Plateau, 25, 38
keystone species, 225–226

lahar, 27, 29, 31, 204
Lake Junin, 66, 70, 71
Lake Poopo, 14, 26
Lake Titicaca, 14, 26, 54, 63, 66, 68, 70, 71, 123, 211, 309, 317
land abandonment, 269, 272, 274, 283, 285, 317, 341, 342, 348
land degradation, 295–296, 300, 346
landslide, 19, 34, 41, 197, 204, 206, 224, 225, 226, 296, 298, 315, 329, 334, 350
land use,
change, 243–244, 260, 341, 342, 344, 345, 348, 349, 350
indigenous, 227, 275, 340
La Niña, 26, 53–55, 71, 186, 309, 313, 314, 315, 318
La Paz, 329
La Rioja, 169, 170, 172, 173, 300, 332
Larrea, 172, 173, 174, 175, 176
Last Interglacial, 251
Late Quaternary glaciation, 60–73
Laurasia, 38
Laurussia, 9, 119, 120
leishmaniasis, 312
LGM (Last Glacial Maximum), 37, 40, 63, 64–69, 86, 142
tropical Andes, 65–69
Lima, 161, 164, 168, 256, 280, 282, 284, 313, 324, 328, 329, 332, 334
limestone, 104–110, 150, 204
Little Ice Age, 72, 223, 317, 347, 348
livestock, 40, 95, 150, 177, 207, 208, 223, 228, 229, 230, 238, 243, 245, 284, 289, 290, 349

lizard, 115, 119
llama, 41, 280
Llanos, 19, 32, 94, 109, 118, 268, 272, 273, 274, 285
Llanos de Mojos, 268, 274, 275
loess, 36, 109, 110, 203, 234, 236
logging, 146, 147, 150, 275, 302
  impacts of, 145, 147, 190, 196, 225, 227, 229, 285, 329, 342
lomas, 158, 160, 163, 164–168, 284, 313
lungfish, 113, 119

Madeira River, 83, 86, 135, 137
Magdalena River, 18, 32, 35, 40, 54, 78, 80, 114, 260
mahogany, 144
maize, 95, 105, 266, 272, 281, 283, 284, 290, 292, 296, 297, 312, 315, 342
malaria, 210, 282, 312, 315, 316, 327, 334, 350
Malvinas (Falkland) Current, 25, 27
Malvinas (Falkland) Islands, 12
Malvinas Plateau, 10, 12, 17, 19, 25, 252–253
mammal, 24, 112, 116–118, 120–123, 124, 167
Manaus, 79, 81, 86, 93, 94, 95, 108, 137, 148, 275, 347
mangrove, 95, 145, 164, 168, 254, 256, 259, 260, 348
Manu Biosphere Reserve, 116
Maracaibo, 260
  Basin, 17, 109
  Gulf of, 77
  Lake, 32, 297
  oil-field, 19
Marajó, 138, 274
Marañon River, 19, 78
marine terrace, 15, 193, 250–252, 256, 312
marine transgression, 7, 251, 347
marsupial, 118, 120, 167
  Eocene, 39, 120
Martius, Carl F. P. von, 266
Mata Atlântica, 135, 147–150
Mata Decidua, 135, 150–154
matorral, 96, 160, 184, 190–191, 194, 195, 313
Maulino forest, 190, 196
measles, 282
Mediterranean-type, 96
  and biogeography, 349
  climate, 96, 104, 158, 160, 187, 188, 190, 219, 290, 349
  conservation, 188, 196–197
  diversity, 126
  endemism, 196
  environment, 184–196
  vegetation, 96, 188–194, 217, 349
mega-city, 324, 326, 350
mega-Niño, 317
mercury, 285
Mesozoic, 4, 7, 9, 10, 12, 13, 15, 16, 17, 18, 19, 38, 77, 112, 118, 223
  biogeography, 38–40, 120
  climate, 24–27
  tectonism, 27
metallic mineral, 19
methane, 334, 344
Mid-Atlantic Ridge, 3, 4, 12, 20, 120, 250
migration,
  biological, 23, 39, 41, 218, 222, 223, 310–311
  human, 39, 228, 282, 291, 302, 342, 343, 348, 350
Miocene, 13, 18, 24, 26, 32, 36, 39, 77, 113, 118, 120, 160, 162, 163, 187, 234, 236
  biogeography, 121, 123
  climate, 24, 26, 166, 187
  fauna, 122
  flora, 40, 187
MJO (Madden-Julian Oscillation), 52
mollisol, 106, 109, 110, 236, 292
Monte, 26, 27, 36, 169–177, 194, 232, 240, 347
  climate, 36, 169
  desert, 126, 158, 169–177, 178
  protected area, 177
  vegetation, 232, 245
moraine-dammed lake, 219
moraine (glacial), 60–72, 203, 209
muscovy duck, 271
MWP (Medieval Warm Period), 223, 317

Napo River, 34, 270
national park, 150, 196
natural selection, 92
Nazca plate, 3, 4, 13, 14, 15, 16, 17, 19–20, 27, 29, 41, 77, 161, 162, 201, 219, 249, 250, 256
Nazca ridge, 13, 252
NDVI (Normalized Difference Vegetation Index), 241, 243
Neogene, 19, 40, 87, 139, 163, 187, 188
neoglaciation, 72–73
Neotropical, 41, 118, 124, 172, 173, 243, 266
  fauna, 41, 112, 116, 117, 118, 146
  flora, 97, 223, 225
  vegetation, 94, 141–142, 151, 154, 229, 230
Neuquén Basin, 19, 171, 174
Nevado Sajama, 14, 29, 68, 70, 71, 202, 317
nitrogen, 101, 104, 140, 206
  deposition, 243, 341, 343, 349
  nitrogen oxide, 147, 343
Nolana, 167, 168
Nordeste, 55
Norte Chico, 158, 166, 167, 316
Nothofagus, 39, 40, 96–97, 190, 193–194
  forest, 217–230
NPP (Net Primary Productivity), 241, 243

Oligocene, 13, 24, 32, 42, 77, 119, 120, 121, 123
Olivillo forest, 188, 193
orchard, 267, 272
orchid, 94, 95, 167, 193, 209
Orellana, Francisco de, 281
Orinoco, 142, 266, 268, 273, 274
  delta, 38, 79, 259
  faunistic region, 113
Orinoco Basin, 24, 32, 37, 83, 98, 107,109, 110, 135, 141–142, 145, 146, 318, 346
Orinoco River, 32, 35, 42, 76, 78, 80, 82, 86, 108, 109, 259, 275, 308, 348
overfishing, 310
overgrazing, 34, 177, 207, 208, 243, 284, 297, 345
oxisol, 32, 98, 107, 108, 109, 110, 140, 236, 292

Pacaraima Plateau, 18
Pacific Ocean, 3, 17, 26, 27, 39, 47, 193, 219, 249
  climatic influence, 50, 80, 97, 185, 234, 236, 254–255
  coast, 348
  and ENSO, 191, 203, 305–307
Paleocene, 10, 24, 32, 120
Paleozoic, 4, 25, 60, 76
  fauna, 118
  tectonism, 7, 9, 10, 12, 13, 16, 17, 19, 25, 118, 219, 250
  wetland, 19
palm, 192, 267
  forest, 144, 192–193
  swamp, 94, 96
Pampa del Tamarugal, 158, 163, 168
Pampas, 36, 97, 109, 110, 116, 176, 234, 236, 244, 258, 280, 290, 296, 300, 349
Panama, 26, 95, 201, 266, 267, 268, 269, 271, 272, 275, 311
Pangea, 4, 9, 10, 12, 23, 24, 25, 38, 97, 112, 118, 119, 221
Pantanal, 36, 79, 82, 86, 96, 98, 117, 118, 268
papaya, 210, 271, 283, 290
Paraguay, 270, 346, 347
  and agriculture, 290, 292, 294, 295, 298, 342
  climate, 47, 54
  and climate change, 344
  and El Niño, 55
  population, 329, 341
  soil, 107, 108
  vegetation, 96, 148, 150
Paraguay Basin, 18, 86
Paraguay River, 35, 36, 76, 77, 78, 79, 80, 81, 82, 84, 86, 107, 254, 255
páramo, 118, 124, 125, 208
  diversity, 116
  fauna, 118, 124
  vegetation, 206, 207, 208
Paraná Basin, 7, 9, 12, 17, 18, 19, 24, 77, 83, 88, 109, 110, 254, 346
Paraná River, 35, 76, 77, 78, 80, 82, 84, 87, 107, 232, 237, 254, 255, 258, 347
passerine, 115, 116, 119, 120, 122, 124, 176
Patagonia, 36, 38, 46, 78, 218, 229, 230, 282, 297, 309
  and Andes, 16, 26, 36, 37, 60, 98, 219, 309
  carbon uptake, 241, 243
  climate, 26, 27, 46, 227, 229, 233–234

coast, 254, 255, 256, 257
desert, 169, 238
and disturbance, 228–229, 244, 245, 297
diversity, 114, 118, 124, 126, 173
fauna, 41, 113, 115, 116, 117, 118, 176
fire, 227
flora, 173
forest plantation, 230
glaciation, 37, 38, 258
and grazing, 243–244, 245, 297
ice field, 73, 79, 258
indigenous population, 228, 280
population, 228, 245, 329, 341
soil, 171, 219
tectonism, 4, 9, 12, 16, 17, 251
vegetation, 97, 158, 169, 173, 217, 220–221, 223, 225, 226, 227, 228, 229, 232–246, 254
volcanism, 31
Patagonian Shield, 107
peach palm, 267, 270
peanut, 270, 284, 290, 315
peat, 69
penguin, 39, 119, 169
periglacial, 203, 219, 254
Pleistocene, 203
peri-urban environment, 333, 350
permafrost, 27
Permian, 7, 9, 118
Perú, 9, 36, 83, 200, 281, 293, 297, 300, 302, 306, 312, 330
and agriculture, 110, 281, 283, 290, 294, 295, 297
Amazonia, 135, 292, 296, 298, 300
Andes, 13, 18, 36, 66, 69, 110, 124, 202, 211, 285, 307, 308
biogeography, 168
climate, 48, 71, 160, 161, 187, 202, 203, 307, 344
and climate change, 344
coast, 54, 96, 110, 118, 254, 256, 281, 282, 284, 313, 332
and coca, 293
conservation, 168
dam, 88
desert, 36, 47, 158–169
diversity, 123, 125, 166–167, 168, 206
and El Niño, 34, 54, 71, 160–161, 203, 305, 307–312, 315, 318
fauna, 116, 118, 123, 167–168, 271
fishery, 310–311
flora, 166–167, 204–205, 267
geomorphology, 161–162
glaciation, 60, 66, 69, 72
Holocene, 70
indigenous population, 282
and La Niña, 318
paleoclimate, 66–69
population, 72, 161, 329, 332, 341
tectonism, 18–19, 29, 250, 251–252
vegetation, 95–96, 158–160, 163–165, 167, 204–205, 208, 209, 312–313
volcanism, 204, 250
Perú-Chile Trench, 4, 13, 16, 18, 20, 26, 163
Peruvian Desert, 96, 159, 160, 161, 163, 164–169, 281, 313
Phanerozoic, 4, 7, 18
phosphorus, 104, 105, 109, 110, 141, 271, 292
phytoplankton, 146, 310
pineapple, 290
Pinus radiata, 195, 230, 345
plantation, 244
and forestry, 111, 230
timber, 229
workers, 282
plate tectonics, 3–12, 24, 95, 249
and biogeography, 38–41, 97–98, 221–223
and zoogeography, 118–120
Pleistocene, 4, 15, 26, 36, 139, 163, 193, 195, 202, 250, 251, 252, 256, 257, 259
climate, 34, 40, 161, 190
extinction, 125, 244
fauna, 39, 40, 41, 121, 122, 124
glaciation, 26, 27, 37, 60, 61, 63, 66, 69, 79, 94, 98, 141, 149, 203, 219, 223, 236, 254, 257, 258
overkill, 121
refugia, 124, 125
vegetation, 190
volcanism, 185
Pliocene, 13, 39, 120, 223, 259
biogeography, 123, 168
fauna, 39, 121, 122, 125
glaciation, 37, 60, 124, 236
tectonism, 38, 77, 112, 161, 202
Podocarpus, 97, 149, 221, 222
pollution, 150, 168, 169, 260, 326, 327, 332 , 335
air, 19, 327, 330–331, 334
industrial, 87, 88, 335, 350
urban, 87, 88
water, 19, 311, 326, 331–332, 348
Polylepis, 202, 205, 209
population (human), 105, 168, 185, 195, 226, 230, 245, 271, 286, 289, 314, 322, 324, 334
decline, 281, 282, 284, 285–286, 302, 341
doubling time, 341
indigenous, 146, 200, 212, 228, 266, 267, 275, 280–281, 282, 284, 302, 340, 341, 342
rural, 230, 266, 295, 324, 341, 342
South America, 72, 98, 126, 149, 265, 293, 294, 295–296, 324, 326–328, 329, 331, 341, 342, 347, 350
Portugal, 69, 279
potassium, 104, 109, 110, 140, 271
potato, 95, 103, 210, 281, 283, 290, 291, 296, 342
Precambrian, 4, 18, 19, 105, 138, 219, 236
craton, 4, 10, 13, 18, 76, 250
tectonics, 4–7
Precambrian Shield, 17
pre-European, 351
human impact, 265–276, 340
people, 41, 266, 270, 272
Pre-puna, 169
primary productivity, 39, 61, 243, 310, 313, 314
primate, 116, 117, 118, 120, 121, 124, 125, 146, 149, 271
Prosopis, 158, 168, 169, 172, 173,174–175, 191, 194
puna, 26, 172, 175, 206, 208
ecosystem, 160, 169
fauna, 116, 118, 124, 125
plateau, 14
volcanism, 31
Puya, 190, 194, 209, 210

Quaternary, 136, 167, 188, 234, 236, 249
climate, 24, 124, 142, 188, 318
and ENSO, 188
environmental change, 223–224
glaciation, 60–73
tectonism, 161
volcanism, 31
quinine, 210
quinoa, 10, 283, 284, 290
Quito, 280, 281, 322, 329, 334

rain shadow
related to the Andes, 26, 27, 36, 95, 97, 123, 187, 188, 192, 202, 206, 219, 221, 234
raised field, 272–273, 274, 275, 281, 283
refugia, 40, 121, 125
glacial, 124, 223
reptile, 7, 112, 114–115, 120, 121, 124, 146, 167, 175, 176, 312
restinga, 144
rías, 256, 257, 258, 259
Río Caquetá, 272
Río Colorado, 78, 170, 258
Rio de la Plata, 12, 18, 25, 35, 84, 98, 109, 228, 252, 253, 254, 258, 307, 315, 348
craton, 7, 258
estuary, 77, 80, 88, 228, 254, 255, 256, 258, 259, 260, 348
grasslands (RPG), 232–234, 241, 243, 244, 245
Rio de la Plata Plains, 232–246
Rio Guaporé, 84, 145
Rio Negro (Argentina), 81, 170, 171, 174
Rio Negro (Brazil), 37, 79, 84, 86, 108, 135, 137, 145, 266, 269, 275
riparian,
habitat, 221, 232, 236, 347
vegetation, 163, 164
river, 76–88
basin, 76–80
flow, 80–83
sediment transport, 83–88
rodent, 118, 123, 226, 316, 317, 334
fossil, 39, 318
Rodinia, 7, 9
Roraima, 18, 139, 146
RPG (Rio de la Plata Grasslands). *See* Rio de la Plata grasslands
rubber, 145, 286, 290
boom, 93, 147, 275

runoff, 79–84, 89, 96, 107, 163, 165, 196, 197, 270, 296–302, 307–309, 324
rural population, 230, 266, 295, 324, 341, 342
rural-to-urban migration, 291, 342, 343, 348, 350

SACZ (South Atlantic Convergence Zone), 46, 47
Salado River, 83, 238
Salar de Uyuni, 70, 71
salares, 26, 163, 170, 171
salinity, 26, 108, 163, 174, 238, 244
salinization, 168
salt marshes, 256, 259
Salvador de Bahía, 322
sandstone, 9, 10, 18, 86, 94, 98, 107, 139, 144, 146, 236, 259
Sandwich plate, 17
sanitation, 326, 327, 328, 331, 336, 341
Santiago, 185, 280, 328, 330, 333, 334, 350
Santillán, Hernando de, 281
São Francisco River, 35, 80, 259, 346
São Paulo, 324, 326, 327, 330, 335, 347
São Paulo Plateau, 12
savanna, 40, 41, 84, 88, 94, 96, 121, 125, 126, 136, 142, 144–147, 150–151, 158, 191–192, 194, 232, 268–270, 273, 274, 281, 285, 313–314, 343, 344, 348
  as dispersal corridor, 39, 123–124
Scotia plate, 4, 12, 16, 17, 219, 250
Scrub. *See* shrubland
sea level,
  change, 79, 249, 250–253, 347
  Holocene, 79
  Pleistocene, 193
  pressure, 46, 47
  rise, 334, 347, 348, 350
sea-surface temperature, 45, 47, 50, 65, 67, 68, 69, 91, 305, 306, 307, 308, 318, 344
Sechura Desert, 26, 161, 164, 168, 313
secondary forest, 146, 271, 272, 292, 314
sedimentation, 7, 13, 18, 23, 38, 65, 66, 259, 275, 326
  anthropogenic, 298, 300, 348
  estuarine, 254
  fluvial, 79, 80, 86, 88, 105, 172, 308, 315
  Neogene, 163
sheep, 41, 95, 97, 165, 228, 244, 245, 284, 285, 291, 297
shrubland, 34, 93–95, 96–97, 116, 118, 154, 152, 160, 173–174, 185, 190–193, 206–207, 223, 232, 240, 268–269, 272, 295, 314
  encroachment, 207, 245
Sierras Pampeanas, 9
Silurian, 7, 9
silver, 19, 285, 286
slash and burn cultivation, 107, 109
small farmer, 291, 300
smallpox, 281, 282, 341
snake, 114–115, 119, 120, 122, 124, 176
snow, 54, 61, 62, 64, 70, 76, 97, 208, 219
  avalanche, 225, 227
soil, 101–111
  and agriculture, 109–110, 346
  anthropogenic, 271–272, 285
  development, 24, 61, 101–109, 204
  erosion, 285, 296–302
  management, 283
  mountain, 106–108
  pH, 230, 343
  properties, 104–106
solar radiation, 23, 24, 46, 47, 55, 62, 95, 107, 175, 206, 313
solid waste, 326–328, 331, 332
Solimões. *See* Amazon River
SO (Southern Oscillation), 53, 305–307, 318
South America,
  biogeography, 38–41, 146–147, 166–168, 172–174, 221–224
  geomorphic process, 32–38
  southern, 39, 118, 217–230, 232–246
  temperate, 96–97, 114, 115, 118, 217–230, 232–246, 345, 348, 349
  volcanism, 27–32
South American plate, 3, 4, 12, 13, 15, 16, 17, 19, 20, 41, 161, 162, 201, 219, 249, 250, 256
South American summer monsoon, 25, 27, 47, 48
South Atlantic Convergence Zone. *See* SACZ
South Atlantic Ocean, 10, 25, 27, 52, 233
South Pacific anticyclone, 309, 313
South Pacific Convergence Zone, 46
soybean, 93, 107, 237, 246, 286, 292, 295, 296, 297, 298, 315, 342
Spain, 279, 282
Spanish conquest, 267, 281, 284
speciation, 40, 95, 98, 133–135, 210
squash, 266, 270, 284, 290
squatter settlement, 327, 333
steppe, 238–241
Storm Wave Environment, 254
STR (Soil Temperature Regime), 102–103, 107–110
subsidence (atmospheric), 47, 50, 51, 53, 160
subsidence (geological), 9, 17, 18, 19, 36, 38, 260
subsurface piping, 298
succession (plant), 105, 146, 195, 245, 283, 345
sugar cane, 107, 108, 109, 283, 284, 290, 291, 294, 315, 342
sunken field, 281, 284
swamp forest, 94, 193
sweet potato, 266, 270

tamarin, 149, 150
tapir, 39, 121, 270, 271
technology, 111, 266, 281, 283, 285, 342
tectonism, 3–20, 23–41
teleconnection, 52, 55, 161, 309
temperate forest, 97, 114, 115, 188, 190, 217–230, 345, 348, 349
  dynamics of, 224–229
tepuis, 18, 94, 139, 146
terrace (agricultural), 272, 283, 297, 302, 341
terrace (fluvial), 80–87, 236
terra firme, 139, 142, 144, 145, 266
terra preta, 271–272
Tethys Ocean, 10, 13, 17, 25, 26
thermocline, 305–306
Tocantis River, 80–81, 84, 88, 145, 259, 347
tomato, 210, 283, 290, 291
trade wind, 47, 91, 95, 136, 202,
treeline, 125, 194, 205, 227
tree plantation, 97, 108, 110, 111, 196, 197, 220, 229, 230, 243–245, 345
Triassic, 7, 9, 10, 118–119, 250
tropical rain forest, 135–147
Tropic of Capricorn, 4, 27, 102, 217, 290
Trujillo, 161, 164, 282
tsunami, 19
tuberculosis, 328
turbidity, 298, 311
turtle, 39, 114, 115, 176

Ucayali River, 78
ultisol, 98, 106–110, 140
ungulate, 38, 117, 120, 123, 271
upwelling, 27, 39, 47, 160, 186, 306, 310, 311
urban environment, 322–336
urbanization, 322–336
Uruguay, 12, 46, 88, 108, 115, 244, 254, 255, 290, 291, 292, 293, 300, 341, 345
  climate, 46, 47, 54, 234
  and climate change, 349
  and humans, 282
  vegetation, 148, 232, 243
Uruguay River, 35, 232, 236, 237, 258

Valdivian forest, 221
Vargas Llosa, Mario, 322
várzea, 139, 140, 145
vegetation, 91–98
  change, 41, 195, 218, 266, 342
  index, 241
  succession, 105, 146, 195, 196, 245, 269, 283, 345
  type, 91, 93, 96, 153, 185, 188, 192, 194, 196, 207, 208, 210, 211, 217, 223, 228, 229, 238–241
Venezuela, 10, 12, 18, 19, 34, 39, 61, 88, 94, 95, 97, 107, 109, 110, 139, 200, 203, 250, 259, 266, 267, 270, 273, 285, 290, 300, 307, 341
  and disease, 316
  and global change, 348
  and humans, 282, 318, 324, 329, 334
  vegetation, 125, 146, 150, 158, 204, 205
vicariance, 95, 119, 120, 125
vicariant event, 119, 125
volcanism, 3, 12, 13, 17, 27–32, 202, 219, 225, 226, 236

Waorani. *See* Huaorani
water resource, 302, 333, 346, 347, 350
  groundwater, 168, 331, 334
waterway management, 274–275

weathering, 18, 32, 34, 83, 91, 98, 101, 105, 139, 144, 185, 253
West Coast Swell Environment, 254–256
wetland, 19, 24, 36, 61, 69, 76, 82, 83, 88, 98, 116, 139, 141, 145, 168, 193, 200, 204, 283, 285, 316, 318, 346
  flora, 94, 193, 210, 211
  vegetation, 209–210
wheat, 105, 107, 238, 284, 286, 292, 294, 296, 297, 312, 315
white water, 137, 142, 145
woodland, 39, 40, 94, 95, 96, 97, 121, 135, 136, 145, 148, 150, 151, 152, 158, 169, 173, 174, 175, 190, 191, 192, 193–194, 205, 209, 217, 220, 221, 223, 227, 271, 283, 284, 285, 286, 291, 349

xeric
  adaptations, 121
  flora, 194, 206, 223
  habitat, 206, 207, 209, 221, 236
  soils, 104, 107, 109
  species, 167
  vegetation, 97, 104, 153, 172, 188, 190, 191–192, 217, 221, 227, 228, 230
xerophyte, 135, 152
Xingu River, 37, 84, 145, 272, 274

yam, 266
yardang, 36
Younger Dryas, 64, 69–71

Zonda event, 51
zoogeography, 112–126

Printed in the USA/Agawam, MA
December 10, 2010
555474.001